Joint Conference on Green Engineering Technology and Applied Computing 2019

IOP Conference Series: Materials Science and Engineering
Volume 551

Bangkok, Thailand
4 - 5 February 2019

Editors:

Shayfull Zamree Abd Rahim
Mohd Mustafa Al-Bakri Abdullah
Mohd Nasir Mat Saad
Muhamad Farizuan Rosli
Azlan Mohd Zain
Mohamad Shukor Talib

ISBN: 978-1-5108-9385-6
ISSN: 1757-8981

Printed from e-media with permission by:

Curran Associates, Inc.
57 Morehouse Lane
Red Hook, NY 12571

Some format issues inherent in the e-media version may also appear in this print version.

Printed by Curran Associates, Inc. (2019)

For permission requests, please contact the Institute of Physics at the address below.

Institute of Physics
Dirac House, Temple Back
Bristol BS1 6BE UK

Phone: 44 1 17 929 7481
Fax: 44 1 17 920 0979

techtracking@iop.org

Additional copies of this publication are available from:

Curran Associates, Inc.
57 Morehouse Lane
Red Hook, NY 12571 USA
Phone: 845-758-0400
Fax: 845-758-2633
Email: curran@proceedings.com
Web: www.proceedings.com

TABLE OF CONTENTS

IOP Conf. Series: Materials Science and Engineering **551** (2019) 011001 doi:10.1088/1757-899X/551/1/011001

Preface: Proceeding of Joint Conference On Green Engineering Technology & Applied Computing 2019

This joint conference is organized simultaneously of International Conference On Green Engineering & Technology 2019 (IConGETech2019) and International Conference On Applied Computing 2019 (ICAC2019). This joint conference invites papers from various fields in both green engineering & technology and Applied Computing field. The primary goal of the conference is to provide the opportunities for academicians, professionals, practitioers and policy makers in the engineering and computing fields to share their thoughts and empirical works both to those involved in their field or those interested in the subject being researched.

The objectives of this Joint conference are:

1. To provide a forum for researchers, educators, students and industries to share and exchange ideas and research findings in both fields of researches.
2. To give an opportunity to both academia and industries to communicate on problems faced in current research and the industries.
3. To create networks and stimulate potential collaborations between researchers in the same field of research.

International Conference On Green Engineering & Technology 2019 (IConGETech2019) will foster the progress in green engineering and related technology by contributing your expertise to what promises to be a very comprehensive and exciting meeting. The aim of the IConGETech2019 is exploring the advances and challenges in Green Engineering & Technology between academia, scientists and engineers by bringing together again the world class researchers and Industrial heads to discuss the latest developments and innovations in the fields of related field of research works.

International Conference On Applied Computing 2019 (ICAC2019) aims to address the main issues of concern within the applied computing area and related fields. This conference offers a great opportunity to bring together professors, researchers, and scholars around the globe, having a great platform to deliver the latest innovative research result and the most recent development and trends in applied computing. We cordially invite all prospective authors to submit their original manuscripts to ICAC2019.

IOP Conf. Series: Materials Science and Engineering **551** (2019) 011002 doi:10.1088/1757-899X/551/1/011002

ATTACHMENT – A

4. Names and affiliations of all editors (in the order as they should appear on the title page)

Ir. Dr. Shayfull Zamree Abd Rahim
Green Design and Manufacture Research Group,
Center of Excellence Geopolymer & Green Technology (CEGeoGTech),
School of Manufacturing Engineering,
Universiti Malaysia Perlis, MALAYSIA.
Email: shayfull@unimap.edu.my

Assoc. Prof. Dr. Mohd Mustafa Al-Bakri Abdullah
Center of Excellence Geopolymer & Green Technology (CEGeoGTech),
School Of Materials Engineering
Universiti Malaysia Perlis, MALAYSIA.
Email: mustafa_albakri@unimap.edu.my

Ir. Mohd Nasir Mat Saad
Green Design and Manufacture Research Group,
Center of Excellence Geopolymer & Green Technology (CEGeoGTech),
School of Manufacturing Engineering,
Universiti Malaysia Perlis, MALAYSIA.
Email: nasirsaad@unimap.edu.my

Muhamad Farizuan Rosli
Green Design and Manufacture Research Group,
Center of Excellence Geopolymer & Green Technology (CEGeoGTech),
School of Manufacturing Engineering,
Universiti Malaysia Perlis, MALAYSIA.
Email: farizuan@unimap.edu.my

Assoc. Prof. Dr Azlan Mohd Zain
School of Computing, Faculty of Engineering,
Universiti Teknologi Malaysia, MALAYSIA.
Email: azlanmz@utm.my

Dr. Mohamad Shukor Talib
School of Computing, Faculty of Engineering,
Universiti Teknologi Malaysia, MALAYSIA.
Email: shukor@utm.my

IOP Conf. Series: Materials Science and Engineering **551** (2019) 011003 doi:10.1088/1757-899X/551/1/011003

Peer review statement

All papers published in this volume of *IOP Conference Series: Materials Science and Engineering* have been peer reviewed through processes administered by the proceedings Editors. Reviews were conducted by expert referees to the professional and scientific standards expected of a proceedings journal published by IOP Publishing.

Epoxidation Reaction Parameter of Palm Olein for Synthesis of Dihydrostearic Acid (DHSA) Via Hydrolysis Reaction

Mohd Jumain Jalil[1], Noor Hidayu Abdul Rani[2], Aliff Farhan Mohd Yamin[3], Nur Radzi Adhary Anuar[1], Intan Suhada Azmi[1] and Abdul Hadi[1]

[1]Chemical Engineering, Universiti Teknologi Mara, Cawangan Pulau Pinang, Malaysia.
[2]Chemical Engineering, Universiti Teknologi Mara, Cawangan Johor, Kampus Pasir Gudang, Malaysia
[3]Mechanical Engineering, Universiti Teknologi Mara, Cawangan Pulau Pinang, Malaysia.

E-mail: mjumain0686@ppinang.uitm.edu.my

Abstract. Nowadays, most of the industries in the world such as plastic industries manufacture products based on petrochemicals resources. Epoxidation is a method where fatty acids are converted to a three membered epoxide ring that is called oxirane ring when the double bond of carbons reacts with active oxygen. Thereafter epixidized vegetable oil was hydrolyzed to form DHSA. Based on the experimental findings, the optimum conditions for epoxidations palm olein temperature at 55oC with the presence of sulphuric acid as catalyst with producing the highest yield of 80.33%. FTIR results shows that the disappearance of double bond carbon C=C group peak at 3,000 cm-1 and formation of oxirane ring in the spectrum of fingerprint region is indicated by the peaks between 815 to 890cm^{-1}. For DHSA synthesis, the conversion of oxirane is decreasing to 16.10% after 5 hours of heating at 60oC. The FTIR analysis shows the the epoxy group at 837 cm^{-1} had disappeared and the hydroxyl group which confirmed the formation of DHSA was observed at the region of 3200-3600 cm^{-1}.

1. Introduction

For past decades, chemical production was mainly focused on petroleum but it tends to depletes sooner or later [1]. Therefore, exploitation of renewable must be done by researcher around the world in order to maintain the synthesis of products based on chemicals [2]. Vegetable oil is one of the renewable resources that can be chemically or enzymatically modified as a replacement of petrochemical based product in industries [3]. Vegetable oil is only referred to plant oils that are liquid at room temperature and the oil is extracted by pressing or by solvent extraction [4] . There are many types of vegetable oil produced around the world and palm oil is one of them. Palm oil consists of unsaturated fatty acids such as oleic acid, linoleic acid and linolenic acid with one, two or three carbon-carbon bond present [5]. Oleic acid has the highest percentage in the palm oil, hence oleic acid is applicable for any further research for reaction and chemical modification [6]. Among the chemical modifications of oleic acid, epoxidation is an easy and effective method to produce intermidiate product that can be used in a variety of applications [7]. Epoxidation process is one of the newest method where fatty acids are converted to a three member epoxide ring that is oxirane ring when the double bond of carbons reacts with active oxygen [8].

IOP Conf. Series: Materials Science and Engineering **551** (2019) 012001 doi:10.1088/1757-899X/551/1/012001

Epoxidized oil can be used as a raw material for a various types of chemicals such as alcohols, alkanolamines, glycols, carbonyl compounds, olefinic compounds, and polymers [9]. Epoxidized oil has been widely used as polymer stabilizer and plasticizer in polyvinylchloride (PVC) [10]. These epoxidized oils are used as intermediate products to manufacture varies of end products that can be useful for industrial purposes. Hence, the economic value of the palm oil is increasing because of the conversion of palm oil into epoxidized palm oil [11]. Epoxidized oils are very important as intermediates in organic synthesis because of the high reactivity of oxirane rings can make them involve in some reactions [12]. By hydrolysis reaction of epoxidized palm olein, a high-value hydroxyl fatty acid called dihydroxystearic acid (DHSA) is produced. DHSA (C18H36O4) is also known as glycolic stearic acid that contains two alcohol groups a carboxylic group in a long hydrocarbon chain [13]. DHSA is derived from in-situ epoxidation of the unsaturated fat in the palm oil with peracetic acid in the presence of catalyst followed with hydrolysis of the epoxide with hydrogen donor such as water. DHSA has been used widely in industry nowadays, as primary and secondary emulsifiers, pigments coating agents and oil gelling agent in cosmetic formulation [14]. The first paragraph after a heading is not indented (Bodytext style).

2. Experimental

2.1. Material

The material used for the epoxidation are fatty acid palm oils as main solvent, acetic acid and hydrogen peroxide (30%) for peracetic formation , sulphuric acid as catalyst, and hydrogen bromide and crystal violet droplet as indicator for titration. For hydrolysis reaction to produce DHPA, distilled water and alumina as catalyst is used.

2.2. In-Situ Epoxidation

In-situ epoxidation is the most common process in epoxidation process. The epoxidation is using peroxy-acids as oxygen donor and carboxylic acid as the oxygen carrier has received concern on its commercial reputation [5]. The epoxidation reaction needs to be accelerated by using strong mineral acid as catalyst to avoid low yield of the epoxidation [15].

Basically, the peroxy-acid is generated in-situ by the reaction of concentrated hydrogen peroxide with carboxylic acid such as acetic or formic acid with the presence of a mineral acid as a catalyst to accelerate the process, according to the following reaction [11].

Peracetic Formation

$$H_2O_2 + HCOOH \leftrightarrow HCO_2OH + H_2O \quad (1)$$

Epoxidation Process

$$R\text{-}CH{=}CH\text{-}R' + HCO_2OH \leftrightarrow R\text{-}\underset{\diagdown\, O\, \diagup}{CH\text{-}CH}\text{-}R + HCOOH \quad (2)$$

Based on the equation (1) and (2), epoxidation process occurs when palm olein reacted with the peroxy-acid generated in-situ to produce epoxidized palm olein. Both formic and acetic acids are carboxylic acids and therefore it is possible to use either one of these acids for epoxidation [16].

2.3. In-Situ Epoxidation

DHSA is produces by reacting the epoxidized palm olein with water in the presence of catalyst [13]. This reaction is called hydrolysis reaction where oxirane ring opening of the epoxidized palm olein is done to produce crystal-like solid DHSA. The reaction equation for synthesis of DHSA is as shown.

IOP Conf. Series: Materials Science and Engineering **551** (2019) 012001 doi:10.1088/1757-899X/551/1/012001

R-CH-CH-R + H_2O ⟶ R-CH-CH-R (3)
\ / | |
O OH OH

Alumina is used as catalyst for the synthesis of DHSA. Alumina is proven to be an effective catalyst for oxirane ring opening by oxygen-containing and nitrogen-containing nucleophiles [17].

3. Results and Discussions

3.1. Epoxidation palm olein without catalyst

RCO% is calculated from the oxirane oxygen content (OOC) that can be obtained from the direct titration of hydrogen bromide. The RCO% is calculated for the reaction process with different temperature and presence of catalyst. The percentage RCO calculated for epoxidation without presence of catalyst with different reaction temperature shown in table 1.

Table 1. The percentage of RCO without presence of catalyst.

Temperature, (°C)	35°C	55°C	75°C
Time, (min)		RCO %	
30	12.8756	23.1760	19.3133
60	21.8884	25.1073	21.2447
90	26.3949	30.2576	29.6138
120	32.8327	38.6267	36.6953
150	50.2147	53.4335	51.5022
180	59.2275	68.8842	63.7340
210	52.1460	53.4335	54.7211
240	33.4764	39.2704	43.1331
270	25.1073	30.2576	28.9700
300	19.9571	27.6824	27.6824
330	18.0258	23.1760	20.6009
360	16.7382	20.6009	18.0258

From table 1, the percentage relative conversion to oxirane (RCO) is the highest at 30 minutes of the reaction for all three different temperatures that are 59.23% for 35°C, 68.88% for 55°C and 63.73% at 75°C. Percentage of RCO indicated the yield of epoxide from palm. Hence, at the temperature of 55°C, the reaction produces the highest yield of epoxide. Even though there are no presence of catalyst, it still shows high yield of epoxidation. This is due to the presence of acetic acid that acts as a catalyst during the peracetic formation in the epoxidation reaction [7].

3.2. Epoxidation palm olein with catalyst

The percentage RCO calculated for epoxidation with H_2SO_4 as catalyst with different reaction temperature is shown in table 2.

IOP Conf. Series: Materials Science and Engineering **551** (2019) 012001 doi:10.1088/1757-899X/551/1/012001

Table 2. The percentage of RCO with H_2SO_4 as catalyst.

Temperature, (°C)	35°C	55°C	75°C
Time, (min)		RCO %	
30	15.4507	30.2576	23.1760
60	30.2576	46.3520	48.2833
90	53.4335	61.1589	55.3649
120	61.1589	68.8842	65.6653
150	68.8842	77.2533	71.4593
180	70.8155	84.3349	76.6095
210	53.4335	67.5967	54.7211
240	38.6267	46.3520	36.6953
270	30.2576	38.6267	30.9013
300	23.1760	36.6953	27.6824
330	20.6009	30.9013	23.1760
360	15.4507	23.1760	19.9571

From table 2, the percentage relative conversion to oxirane (RCO) is the highest at 30 minutes of the reaction for all three different temperatures that are 70.82% at 35°C, 84.33% at 55°C and 76.61% at 75°C. Percentage of RCO shows the yield of epoxidation process. Hence, at the temperature of 55°C, the reaction produces the highest yield of epoxidation.

3.3. Effect of Reaction Conditions on the Reaction Rate of Epoxidation Palm Olein

3.3.1. Effect of Reaction Temperature.
From figure 1, it shows that reaction at temperature of 55oC produces the highest RCO of 84.34%. The finding shows that at moderate temperature will produces highest yield of epoxidation.

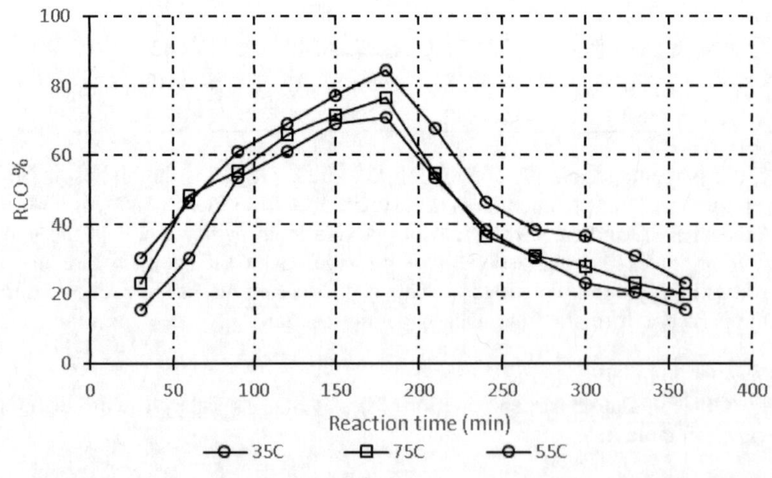

Figure 1. Effect of different temperature of reaction to RCO % of epoxidation.

The epoxidation rate is increasing with temperature, but it will gradually decrease if the temperature increased higher. It is because RCO will achieve maximum yield faster at high temperature and then will decrease gradually [18]. Other reason is due to the reversible reaction of peracetic acid causes by the high temperature. When the temperature is high, the rate of reaction is also high but the conversion to oxirane is lower because of the reversible reaction of peracetic acid that satisfy the Equilibrium Law. The Equilibrium Law (Le Chatelier's Principle) stated that when an equilibrium system is disturbed by stress such as temperature, the system will change necessarily to avoid the disturbance.

3.3.2. Effect of Presence of Catalyst.
The RCO is the highest at 55oC whether there is catalyst or no catalyst presence, therefore temperature 55oC is selected for comparison of catalyst for the RCO % of epoxidation reaction. Figure 2 shows the effect of presence of catalyst to the percentage of RCO.

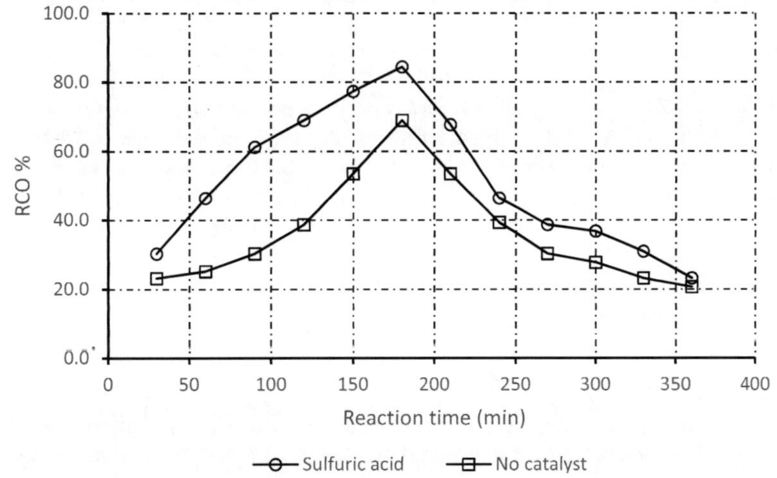

Figure 2. Effect of presence of catalyst to RCO % of epoxidation.

Various type of catalysts has been utilised in the epoxidation of vegetable oils such as mineral acid catalysts and solid catalysts. It is known that the presence of catalyst in the reaction will decrease the activation energy, which in turn increases the rate of reaction. By referring at Figure 2, it can be seen that the highest rate of reaction was observed when sulphuric acid was used as the catalyst that are with maximum 80.33% RCO. The epoxidation reaction was accelerated when mineral acids such as sulphuric acids used as catalyst because it can avoid low yield of the epoxidation calculated [19]. The result is same with previous research, which that sulfuric acid is determined as the best inorganic catalyst for this system that can produce a high conversion of double bonds to oxirane groups in epoxidation reaction [5].

3.4. Synthesis of DHSA
From the epoxidation process done earlier, the epoxidized oil with H_2SO_4 as catalyst and reaction temperature of 55°C is suitable to be chosen to produce DHSA. Epoxidized palm olein is reacted with distilled water with presence of alumina as catalyst and heated at 60°C. Figure 3 shows the RCO% of the hydrolysis reaction for 5 hours.

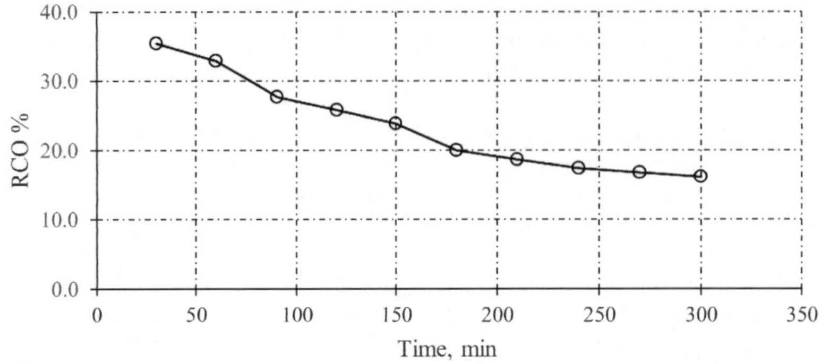

Figure 3. RCO% of the Hydrolysis Reaction for the Synthesis of DHSA.

Based on figure 3, the percentage of RCO is decreasing with time. After 5 hours, the RCO is decreasing to 16.10%. The progress of the hydrolysis reaction was followed by measuring the RCO and considered as complete when the RCO approaches zero but after 5 hours the RCO is not approaching zero yet [8]. For chemical characteristic, conversion of palm oil to DHSA resulted in lower iodine value (IV) because of conversion of unsaturated fat to saturated fat and higher hydroxyl value (OHV) because of formation of hydroxyl group. For physical characteristic, DHSA crystal size is affected by the cooling modes where higher cooling rate causing high super-saturation through higher temperature, producing smaller crystal and vice versa [20]. It can be concluded that hydrolysis need to takes longer time to decreasing the RCO to zero. The final DHSA product obtained is visually appeared as a white and waxy materials.

3.5. FTIR Analysis

The chemical properties examination of the crude palm oil and the epoxidized palm olein were accomplished by using Fourier Transform Infrared (FTIR) spectroscopy. Thermo Scientific Nicolet 6700 FTIR spectrometer were used to detect the presence of epoxide group that are available at wavenumber 750 - 880 cm^{-1} and 815 - 950 cm^{-1} based on theoretical FTIR analysis of palm oil. The disappearance of double bond group peak will be observed at 3000 cm^{-1} because of the conversion of double bond in linoleic acid to the epoxide group. To optimize the epoxidation reaction, the hydroxyl and double bond peak must be avoided to appear in FTIR spectrum [21]. The FTIR spectrum of the oleic acid and epoxidized palm olein are shown in figure 4 and 5 respectively.

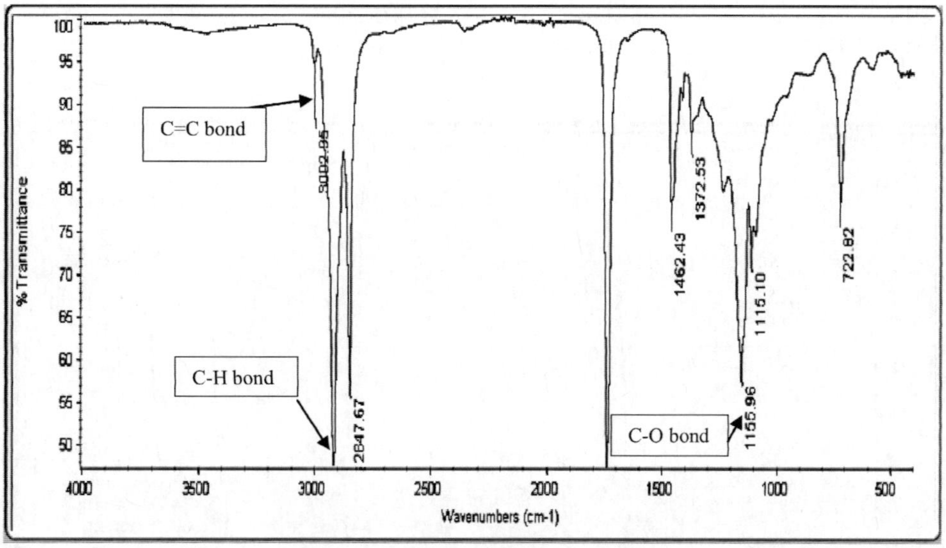

Figure 4. FTIR Spectrum for Palm Olein.

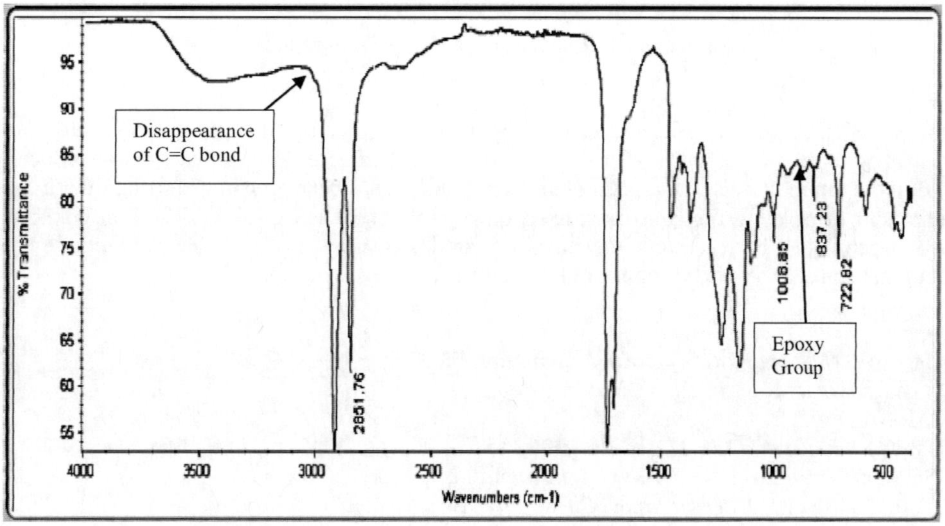

Figure 5. FTIR Spectrum for Epoxidized Palm Olein.

The disappearance of double bond group peak at 3,000 cm^{-1} demonstrate the conversion of double bond in linoleic acid to the epoxy group. Formation of oxirane ring in the spectrum of fingerprint region is indicated by the peaks between 815 to 890cm^{-1}[3]. Previous research has stated that to optimize the epoxidation reaction, the hydroxyl and double bond peak must be avoided to appear in FTIR spectrum.

Figure 6 shows the FTIR spectrum of the DHSA produced from epoxidized palm olein. For hydrolysis, the absorption band region of 1210-1320 cm^{-1} (C-O-C stretching) will disappear and hydroxyl group observed at 3200-3600 cm^{-1} [11]. After hydrolysis, the epoxy group at 837 cm^{-1} had disappeared. This shows that the epoxy ring from epoxidized palm olein has reacted by water to produce secondary hydroxyl compound to form DHSA [13]. The hydroxyl group which confirmed the formation of DHSA was observed at the region of 3200-3600 cm^{-1} as reported in previous research [11].

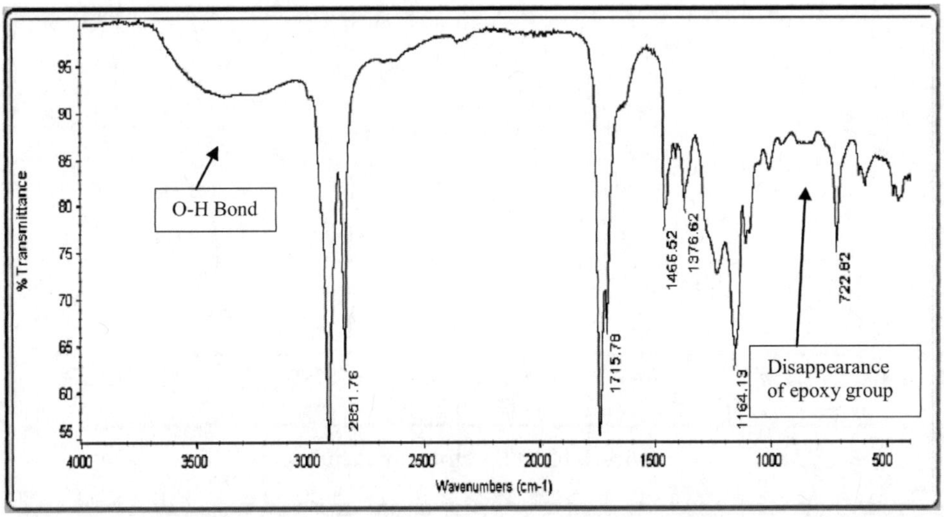

Figure 6. FTIR spectrum of the DHSA produced.

4. Conclusions

Epoxidized vegetable oils have been researched by many researchers in this past years because of they are derived from renewable and sustainable resources that can be used to replace petrochemical raw material in the future. In this study, palm olein from palm oil has been epoxidized using in-situ formed peracetic acid in order to produce epoxidized oleic acid. The DHSA can be synthesised from the epoxidized palm olein by reacting with water that can decreasing the RCO% and producing hydroxyl fatty acid that can also be used as ramaterial for industries.

References

[1] Jankovic M R *et al* 2014 *Ind. Eng. Chem. Res.* **53** 9357.
[2] Vanags E *et al* 2018 *J. Renew. Mater.* **6** 764.
[3] Derawi D *et al* 2014 *J. Anal. Sci.* **18** 584.
[4] Jalil M J *et al* 2014 *Adv. Mater. Res.* **906** 125.
[5] Saurabh T *et al* 2011 *Int. J. Adv. Eng. Technol.* **2** 491.
[6] Jalil M J *et al* 2017 *Recent Innov. Chem. Eng.* **10** 1.
[7] Derawi D *et al* 2010 *E-Journal Chem.* **7** 1440.
[8] Gupta S S *et al* 2016 *Int. Res. J. Chem.* **1** 17.
[9] Salzano E *et al* 2012 *Chem. Eng. Trans.* **26** 39.
[10] Sun S *et al* 2011 *Ind. Crops Prod.* **33** 676.
[11] Jalil M J *et al* 2014 *Adv. Mater. Res.* **906** 125.
[12] Jalil M J *et al* 2018 *Chem.Chem.Technol.* **12** 296.
[13] Rafizan A *et al* 2014 *J. Appl. Sci. Agric.* **9** 86.
[14] Koay G F L *et al* 2006 International Journal of Engineering and Technology **3** 115.
[15] Milchert E *et al* 2008 Czasopismo Techniczne. Chemia **105** 283.
[16] Goud V V *et al* 2007 Chem. Eng. Sci. **62** 4065.
[17] Piazza G J Nuñez *et al* 2003 Journal of the American Oil Chemists' Society **80** 901.
[18] Benaniba M T *et al* 2007 Eur. J. Lipid Sci. Technol. **109** 1186.

[19] Milchert E *et al* 2015 *Molecules* **20** 21481.
[20] Koay G F *et al* 2011 *J. Oleo Sci.* **60** 237.
[21] Performik A *et al* 2015 *J. Anal. Sci.* **19** 144.

IOP Conf. Series: Materials Science and Engineering **551** (2019) 012002 doi:10.1088/1757-899X/551/1/012002

Adipic Acid Crosslinked Nanocrystalline Cellulose/Chitosan Composite: Tensile Properties and Crystallinity

P.G. Gan[1], S.T. Sam[1], M.F. Abullah[1] and C.C. Chow[1]

[1]School of Bioprocess Engineering, Universiti Malaysia Perlis (UniMAP), Kompleks Pusat Pengajian Jejawi 3, 02600 Arau, Perlis, Malaysia.

E-mail: peigie023@gmail.com

Abstract. In the present study, empty fruit bunch derived nanocrystalline cellulose (NCC) was prepared and employed as a reinforcing agent to improve the tensile properties of the chitosan composites. Adipic acid was added as a crosslinker. The incorporation of NCC has greatly improved the tensile properties of composites. The crosslinking reaction has further improved the tensile properties. 3 wt% NCC/chitosan composites displayed the highest tensile strength which is 48.7MPa and 65.3 MPa without and with the addition of crosslinker respectively. The X-ray diffraction (XRD) results showed that the crystalline peak of chitosan composites was significantly improved by the addition of NCC however the crystallinity peak was decreased by the incorporation of adipic acid.

1. Introduction

In recent decades, renewable polymeric materials from natural resources have been widely explored as the substitutes for the non-renewable polymers. Chitosan is such an alternative due to its excellent film-forming capability, non-toxic, biodegradability and anti-microbial activity make it an ideal alternative for reducing the environmental problems caused by synthetic polymers. Chitosan is the second most abundant polysaccharide which obtained from the deacetylation of chitin [1]. The attractive properties of chitosan have inspired the researchers in the development of chitosan-based film composites where chitosan is utilized as a polymeric matrix and reinforced with other nanoparticle fillers.

Among the biopolymers, cellulose is of particular interest to be utilized as reinforcing agent due to their wide variety of source, low cost and biodegradability [2]. Thus, cellulose with diameters in the nanometre range known as nanocrystalline cellulose (NCC) is developed. NCC is a rod-like crystalline cellulose prepared from acid hydrolysis with a diameter between 5 – 70 nm [3]. Among the acids, sulphuric acid is the most common acid used. The use of sulphuric acid could lead to the formation of negatively-charged sulphate group on the surface of NCC which produced a highly stable NCC suspensions by electrostatic repulsion [4]. NCC as a reinforcing agent has been extensively applied in various polymer matrices for different applications due to its remarkable properties such as remarkable strength and stiffness, high aspect ratio and high biodegradability.

Crosslinking had been an important procedure in the formation of NCC/chitosan film in improving the properties of chitosan composites film. Yeng et al. [5] used epichlorohydrin as a crosslinker for chitosan composites and found out that the mechanical properties were improved significantly compared to non-crosslinked set. Therefore, in order to improve the mechanical properties of chitosan film composites for certain application, chemical modification by using crosslinker is introduced. In this study, adipic acid was employed as crosslinker for the production of NCC/chitosan film. The main

IOP Conf. Series: Materials Science and Engineering **551** (2019) 012002 doi:10.1088/1757-899X/551/1/012002

objectives of the study are to compare the tensile properties, morphology and crystallinity of both non-crosslinked and adipic acid crosslinked chitosan film composites with and without the addition of NCC.

2. Experimental

2.1. Material
Empty fruit bunch (EFB) was supplied by Taclico Company Sdn. Bhd. was purchased from Cleo Shanghai Pharmaceutical, China. Acetic acid, sodium hydroxide, adipic acid and sodium chlorite were purchased from Sigma-Aldrich whereas sulphuric acid was purchased from Merck.

2.2. Preparation of NCC from EFB fibres
EFB fibres were treated with 4wt% sodium hydroxide solution at 80 μ for 2 hours. After that, the fibres were bleached at 80 μ for 2 hours. Both treatments were repeated thrice. Acid hydrolysis was performed by using 60 wt% sulphuric acid at 45 μ for 1 hour. The suspension obtained was washed with distilled water and centrifuged repeatedly till the supernatant become turbid. The suspension was then sonicated and stored in refrigerator at 4 μ .

2.3. Preparation of NCC/Chitosan Composites
Chitosan powder was dissolved in acetic acid solution and heated at 60 °C for 4 hours. Glycerol (10 wt%) and NCC were mixed with the chitosan solution and heated for another 1 hour. The solution was poured into a plastic petri dish and dried in a non-air-circulating oven at 50 μ . For crosslinked composites, 5wt% of adipic acid was added.

3. Characterization

3.1. Tensile properties
The tensile test was performed using Instron universal tensile testing machine (UTM). A crosshead speed of 10mm/min was used in the test. Specimens of size 10 x 1 cm were cut from the films. The thickness of specimens was measured randomly using a micrometer and the average thickness was recorded. The data were averaged over at least 5 samples.

3.2. Field emission scanning electron microscopy (FESEM)
The fractured surface of non-crosslinked and adipic acid crosslinked NCC/chitosan film composites were observed by using FESEM. Gold-coated samples were analyzed using NOVA NANOSEM 450 with 5000x magnification and 10 kV accelerating voltage.

3.3. X-Ray Diffraction (XRD)
The crystalline structure of non-crosslinked and adipic acid crosslinked neat chitosan and NCC/chitosan film composites was examined by a desktop D2 Phaser X-ray diffractometer (Bruker Corporation, USA). The samples were scanned within the range of 5μ to 40μ with a speed of 0.1 s/step.

4. Results and Discussions

4.1. Tensile properties

Tensile properties of non-crosslinked and adipic acid crosslinked neat chitosan and NCC/chitosan film composites are presented in Figure 1. In general, the tensile strength of chitosan composites showed a significant improvement with the addition of NCC up to 3 wt%. The increment could be due to the good interaction between NCC and chitosan matrix. Besides, the reinforcing effect occurring through effective stress transfer between fillers and polymeric matrix also enhance the tensile strength significantly [6]. Nevertheless, the addition of higher concentration of NCC declined the tensile strength of the composites. It could be resulted by the agglomeration and uneven dispersion of NCC within the chitosan matrix. The addition of crosslinker has even improved the tensile strength by developing a more stable network between adipic acid and chitosan [7]. The non-crosslinked and adipic acid crosslinked composites with the addition of 3 wt% NCC displayed the highest tensile strength which is 48.7 MPa and 65.3 MPa, respectively.

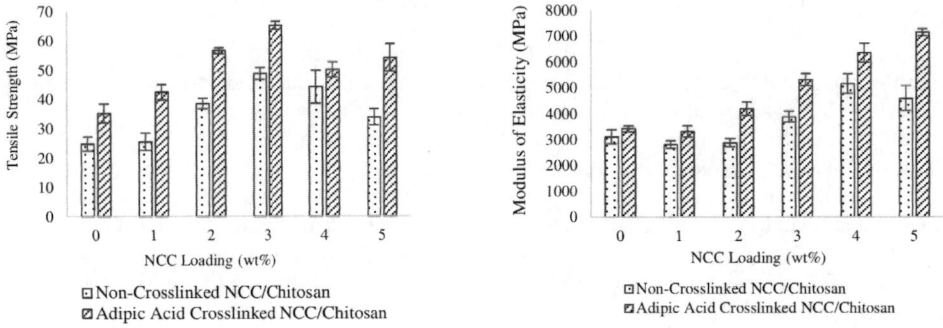

Figure 1. The (a) tensile strength and (b) modulus of elasticity of non-crosslinked and adipic acid crosslinked neat chitosan and NCC/chitosan film composites

The modulus of elasticity also displayed a similar trend as tensile strength in which the incorporation of NCC improved the modulus of elasticity significantly. The addition of highly crystalline NCC formed a strong intramolecular or intermolecular hydrogen bonding with matrix and improved the modulus of elasticity remarkably [8]. Moreover, the crosslinking process has further restrained the flexibility and chain mobility of polymer composites and increased the modulus of elasticity.

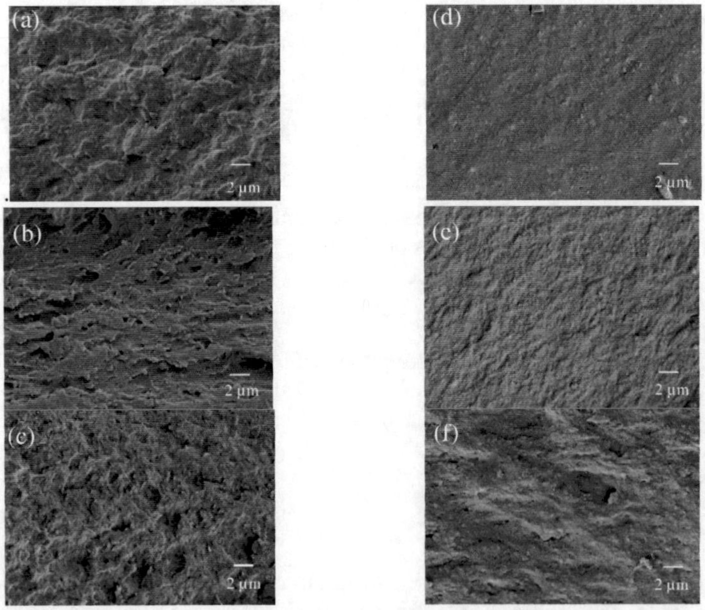

Figure 2. Fractured surface of the non-crosslinked (a) 0% NCC, (b) 1% NCC & (c) 5% NCC & adipic acid crosslinked (d) 0%, (e) 1% NCC & (f) 5% NCC/chitosan composites.

The fractured surface of non-crosslinked and adipic acid crosslinked neat chitosan composites in Figure 2 (a) and (e) showed a smooth and homogeneous surface. The incorporation of NCC resulted in a rougher and uneven surface with respect to the neat chitosan film. Adipic acid crosslinked NCC/chitosan composites presented smoother and more even surface compared to non-crosslinked set. The addition of adipic acid had improved the interaction between NCC and matrix. FeSEM analysis has proven that the enhanced tensile properties of composites were resulted by the NCC and crosslinking agent addition.

4.2. X-Ray Diffraction (XRD)

The XRD analysis (Figure 3) of non-crosslinked and adipic acid crosslinked neat chitosan and NCC/chitosan film composites showed intense peaks at $2\theta = 12.4\mu$, $22.5\ \mu$ and $34.7\ \mu$, which attributed to the hydrated "tendon" conformation of chitosan [9]. The addition of NCC has enhanced the intensity of the peak at $2\theta = 22.5\ \mu$. The adipic acid crosslinked composites showed a slightly lower peak at $2\theta = 22.5\ \mu$ compared to the non-crosslinked chitosan composites, which showed the effective crosslinking reaction between adipic acid and chitosan matrix [10]. Crosslinking reaction interrupted the intra and inters molecular hydrogen bonds of chitosan network which resulted a lower peak of composites [11].

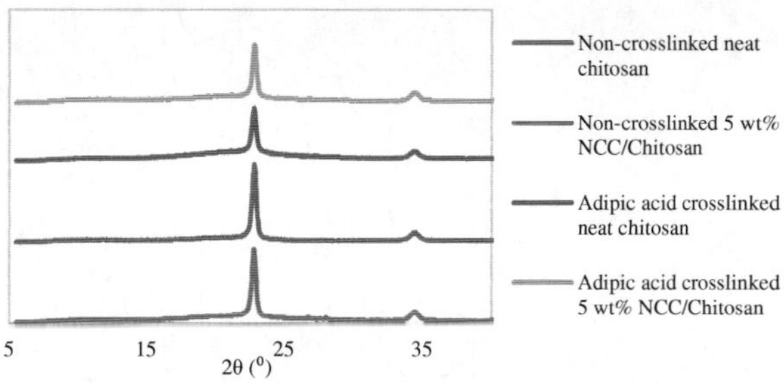

Figure 3. XRD pattern of non-crosslinked and adipic acid crosslinked neat chitosan and NCC/chitosan film composites

5. Conclusions

NCC/chitosan film composites were synthesized and chemical crosslinked using adipic acid. Tensile properties of chitosan film composites were greatly improved by adding NCC and adipic acid as crosslinker.

Acknowledgements

The authors would like to thank the Ministry of Higher Education for the Fundamental Research Grant Scheme (FRGS/1/2017/TK05/UNIMAP/02/10) for financial support.

References

[1] Kouchak M, *et al* 2014 *Iran. J. Basic Med. Sci.* **17** 14.
[2] Niazi M B K *et al* 2017 *Carbohydr. Polym.* **177** 258.
[3] Islam M T *et al* 2014 *AATCC J. Res.* **1** 17.
[4] Lu P *et al* 2012 *Carbohydr. Polym.* **87** 564.
[5] Falamarzpour P *et al* 2017 *Int. J. Mol. Sci.* **18** 1.
[6] Rubentheren V *et al* 2015 *Cellulose* **22** 2529.
[7] Rivero S G *et al* "Physical and Chemical Treatments on Chitosan Matrix to Modify Film Properties and Kinetics of Biodegradation" 2013.
[8] Ching Y C *et al* 2015 *BioResources* **10** 3364.

[9] Rivero S *et al* 2011 *J. Agric. Food Chem.* **60** 492.
[10] Fideles T B *et al* 2018 Open Acces Library Journal **05** 1.
[11] Julkapli N M et al 2010 *AIP Conference Proceedings* vol **1202** 106.

A Review on Process Parameters and their effects on Dissimilar Friction Stir Welding of Aluminium and Steel Alloys

Veerendra Chitturi[1], Srinivasa Rao Pedapati[1], and Mokhtar Awang[1]

[1]Department of Mechanical Engineering, Universiti Teknologi PETRONAS, Malaysia.

E-mail: srinivasa.pedapati@utp.edu.my

Abstract. Friction Stir Welding (FSW) presents numerous advantages over conventional fusion welding techniques such as eliminating the need for a shielding gas, requiring less energy per weld, and the lack of a flame or arc making it safer in the work place. Another advantage of FSW is its ability to join materials that are extremely difficult, or impossible to weld with conventional fusion techniques. The growing need for dissimilar welding relates to joining aluminium and steel base metals. Individually, both metals are utilized extensively in the automotive and aerospace industries due to various advantages respectively, including lightweight, high specific strength, and recyclability. Like conventional welding methods, the intermetallic compound formation represents the main problem issue in FSW. Heat cycle has a significant effect on Intermetallic Compounds thickening phase transformation and the elimination of the amorphous phase. The hybrid techniques utilized in FSW for Al-to-steel joints reveal high joint strength in conjunction with high costs of setup and equipment. These techniques modify the steel flow around the rotated pin by acting as an assisted heat source. Process parameters play the most important role in defining the quality of the weld to withstand. In this paper we are going to discuss the effect of different parameters on Dissimilar FSW.

1. Introduction

The advantages of Friction Stir welding are several in comparison with general welding techniques like, GMAW (Gas Metal Arc Welding), TIG (Tungsten Inert Gas Welding) and MIG (Metal Inert Gas Welding) etc. As it is a solid-state joining process, there is no need for an electrode or shielding gas and the energy required per weld is less than that of fusion welding. The welding of the materials is done with the heat developed between tool rotation and the traverse speed which means that there is no flame and guarantees the safety of the worker. The main advantage of Friction Stir Welding as shown in figure 1, is that it can join different materials whose melting points differ which are difficult to weld with the general welding techniques [1,2]. There is no melting of the parent material which can lead to the formation of unwanted and detrimental intermetallic compounds often present during the welding of two different materials such as steel and aluminium. National Aeronautics and Space Administration has adopted FSW for use in the circumferential welds of the spacecraft's fuel tanks. Other applications of Friction Stir Welding include the Boeing Delta IV heavy rockets used to launch the Mars Phoenix Lander in July 2007. General Motors has begun funding research in FSW and has even started using FSW for some spot-welding applications in production.

Recently, there has been much development in the FSW of dissimilar materials. Different materials like copper, titanium, magnesium, aluminium, steels are successfully welded by using FSW [3, 4]. The welding process for Al-Fe alloys are quite arduous and challenging as like welding process for other Aluminium and steel alloys, because of its low melting temperature and high thermal conductivity. Several applications, as such for Al-Fe alloys, necessitate the alloys to be welded with the help of FSW. In this regard, the FSW is one approach that prospectively demonstrates high potential, uncomplicated and comprehensive proficiency to weld these alloys. [5, 6, 7]. In addition, the mathematical modelling has substantially increased widespread applications and appositeness in materials science and engineering fields [8].

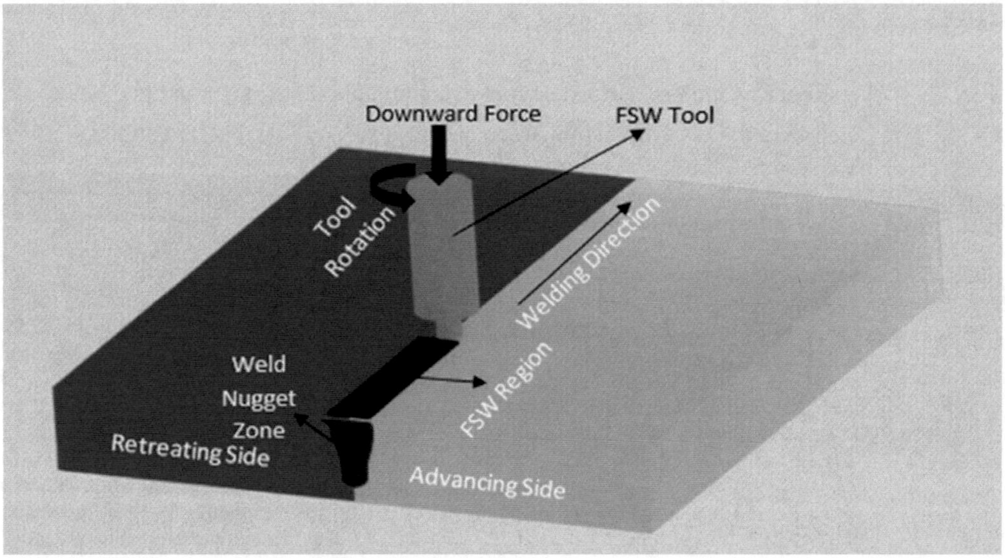

Figure. 1. Schematic diagram of Friction Stir Welding

2. Dissimilar Friction Stir Welding of Aluminium and Steel

2.1 Process Parameters of FSW

Several researchers have successfully welded different kind of materials. The main things that attract a researcher in Friction Stir Welding are different parameters and their ability to define the joint strength. In this paper the following parameters are discussed:

- Tool pin profile – Different tool pin profiles are used for different materials depending upon the need for strength and to obtain a perfect weld. Majorly used tool pin profiles are cylindrical, threaded cylindrical, tapered, tapered cylindrical. Now a days the new pin profile called the “tri-flute” is being used to weld dissimilar materials.
- Tool rotational speed – The main technique of FSW is material flow, which can be controlled using the tool rotational speed. The churning of the material starts on the advancing side and is thrown away to the retreating side to form the joint.
- Tool traverse speed – While the stirring is being done, the movement of the tool in the horizontal direction makes the stirred material movement to the back which results in the formation of the weld.
- Tool shoulder – The different shapes of the tool shoulder make an impact on the weld. The three different type of shoulders used are flat, concave and convex. The shape of the face of the tool shoulder also plays an important role in the stirring of material.

IOP Conf. Series: Materials Science and Engineering **551** (2019) 012003 doi:10.1088/1757-899X/551/1/012003

2.2 Effect of Tool Pin Profile

A cylindrical threaded tool was used to weld AISI 1018 steel and Aluminium 6061 plates of thickness 6 mm with the combined effects of FSW and fusion welding. The existence of the intermetallic phases Al_5Fe_2 and $Al_{13}Fe_4$ were found in the weld nugget. A perfectly stable heterogeneous joint was obtained using this combined effect process and the damage of the tool was identified in the microstructural analysis [9]. A cylindrical pin was used to attain 86% of the aluminium base metal joint strength with a small amount of IMC's (Intermetallic Compounds) formation at the starting part of the weld. Along the weld joint, it has been observed that there were no intermetallic compounds at the end of the weld. The stirring effect made many parts of the steel to scatter in the aluminium material and because of the rubbing motion on the faying surface the oxide film has been removed. The rotation pin was in contact with the face of the steel plate while welding.

A straight cylindrical pin was used to weld SS 400 and Aluminium 6061 plates to find the better conditions to operate and to find the suitable parameters. Charpy Impact test was used to determine the quality of the weld, depending on the achievable impact value. The lower the welding speed and the rotational speed the higher are the Charpy impact values. The acceptable tensile strength and good impact values are obtained with low traverse speed of 1 mm/s with a rotational speed of 560 rpm [10].

2.3 Effect of Tool Rotational Speed

Different tool rotational speeds are used to check the hardness of the stainless steel and aluminium 6014-T4 joints in different zones of the joint like nugget zone, Thermo Mechanically Affected Zone (TMAZ) and Heat Affected Zone (HAZ). It has been observed that the hardness value is less on the retreating side when compared to the thermo mechanically affected advancing side. Hardness in the weld nugget is different at various points because of the uneven scattering of the steel particles in the nugget zone. The researchers found the minimum hardness in the heat affected zone which is 5 to 12 mm from the nugget centre on the retreating side. They discovered that the fatigue properties of the joint were almost 30% lower than that of aluminium 6013-T4 base metal [11].

Along with different rotational speeds the tool offset also plays a major role in dissimilar friction stir welding of aluminium and steel alloys. Because of the offset position the churning and the flow of material mostly done on the aluminium alloy. Microstructural analysis have been done on different weld zones including the base metal. It has been found that the grain size and shape were mainly affected during the welding process. This has resulted in achieving the joint strength as close as 80% of the aluminium alloy base metal. The High-Speed Steel side had shown no plastic deformation during the tensile test. the whole deformation was on the aluminium alloy [12]. A new technique called friction stir spot welding was found where the rotational speed plays the important role. It helped to weld mild steel and 6061 aluminium alloy plates of thickness 1 mm. Different pins of length 1, 1.3 and 1.5 mm with different rotational speeds were used to produce sound joints between the plates. The FSSW was conducted on a round dent which was already made on the back of the plates. The authors found that the maximum shear load can be 3607 N [13].

2.4 Effect of Traverse Speed

The traverse speed is important when the joints are made with tool offset towards the aluminium side. 304 stainless steel and aluminium 5050 plates were welded with an offset of 1.5 mm towards the aluminum which helped in decreasing the voids caused in the nugget zone which has increased the tensile strength without offset. The authors achieved sound joints by changing the traverse speed. They have achieved a perfectly sound joint with a traverse speed of 80 mm/min. The annealed joint has increased the elongation and UTS by 10% [14].

A lap joint was made between 6061 aluminum and 316 stainless steel plates by using zinc as a filler material. Most of the zinc was moved to aluminium side during the stirring process, making retreating side as the zinc occupied side. EDS analysis found that there were tiny parts of $FeZn_{10}$ but there was no sign formation of intermetallic compounds when the pin is plunged into the zinc foil. The maximum strength of the joint was achieved with the influence of zinc layer mixing with aluminium and steel. The XRD patterns show that different solid solutions were found in the fracture parts of the joint [15].

6061 aluminium alloy and S235J steel are welded to improve the joint by changing the traverse speed with tool offset towards the aluminium side. There were tunnel defects with high traverse speed and if the offset was 0.2mm towards steel and the weld was not completed. 41.62 MPa tensile strength was achieved with a no offset and medium traverse speed. The best joints were achieved if the offset was 0.2 mm towards the aluminium side with low traverse speeds. The average tensile strength was 95.73 MPa. From all the experiments conducted it was found that the tunnel and void defects influenced in producing the joints which are low in tensile strength [16].

2.5 Effect of Tool Shoulder

Different tool shoulders have effects on the stirring process. A concave tool shoulder was used to weld steel DX54 and aluminium alloy AA5754-H22. The lap joint was innovative, and it has improved the quality of the welds. A wave shaped probe was used during the plunge, which created enough heat to stir the material with perfect interlocking. Several experiments are conducted with this new overlap weld technique and it was found that the welds are perfect with two passes. They achieved 50% of the tensile strength of aluminum base metal. Microstructural analysis shows that the intermetallic compound formation was uniform with a thickness of 1 μm throughout the weld joint [17].

A flat shoulder pin with a tilt angle of 1^0 was used to weld commercial pure aluminum and Interstitial free steel. They (Sorger, et al.) used different rotational speeds ranging from 600 to 1200 rpm with a fixed traverse speed of 100mm/min. In the weld nugget zone, it was found that Al_3Fe IMC's were observed, and the thickness of the IMC increased when the rotational speed was increased. The authors achieved a maximum tensile strength of 124 MPa with an elongation of 4.6% at the nugget zone of the joint [18, 19].

2.6 Defects in Dissimilar FSW with respect to Process Parameters

Weld quality in friction stir welding is affected because of various defects produced during the welding process. The friction stir welding defects includes Tunnel, Flash, Kissing bond, Void/Wormhole, Cavity/Groove, Cracks defect.

In figure 2(a), we can observe the formation of the tunnel and cavity which are formed because of insufficient heat and stirring between the plates. Flow in vertical direction is not possible if insufficient heat input is given which leads to tunnel defect. In figure 2(b), the formation of groove or and wormhole can be observed which is due to very low rotational speeds and high welding speed. Mechanical properties are mainly affected by the groove defect. Tensile strength is decreased because of the formation of grooves in the weld. In figure 2(c), 2(d) we can observe the crack line defect which are formed during FSW welding of dissimilar materials beneath the tool shoulder. Selection of perfect parameters play an important role in producing defect free weld during friction stir welding.

Lack of heat input during the stirring process and incomplete movement of the material leads to the development of root defects. While using the flat shoulder if the pin depth is inadequate then different kinds of grooves appear in the nugget zone. Crack-line defects are formed if the plunge depth is not perfectly aligned with a tilt angle. The downward force must be enough to plasticize the metal with tilt angle or else it will lead to crack-line defect due to lack of plunge depth. So, in general with smaller or higher tilt angles there is a chance for the formation of crack-line defect.

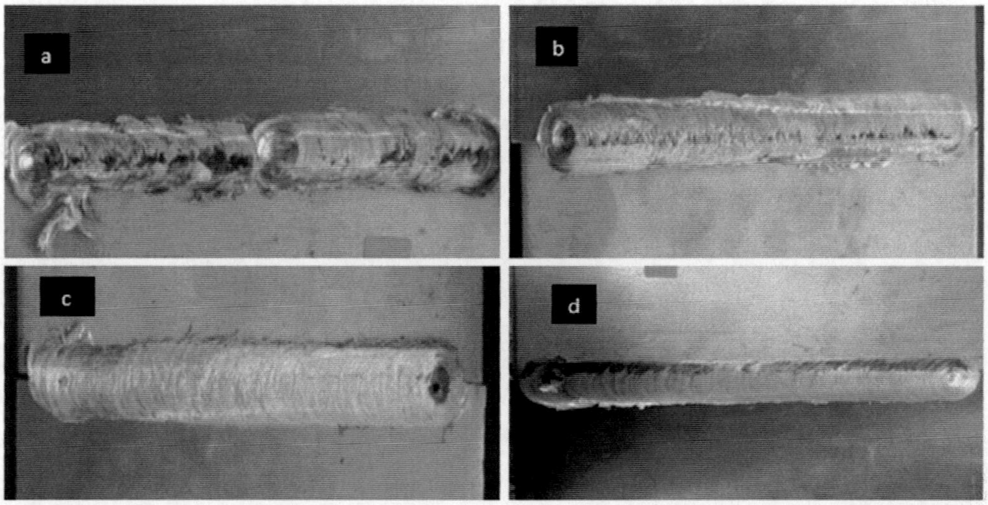

Figure. 2 Defects in FSW (a) Tunnel defect (b) Groove defect (c) and (d) Crack line defect.

3. Conclusions

The main challenge to weld dissimilar materials is selection of parameters because of the difference mechanical, chemical and physical properties. There is not enough research done on dissimilar materials such that the industries can rely and manufacture the products with this technology. The research of dissimilar materials should be extensively done so that it can move from a lab to industrial application. To achieve that goal the researchers must concentrate the combination of different materials and their performance in the long term. The focus of the researchers is on trying different combinations of the materials and the selection of different parameters to achieve good mechanical properties for a certain combination, instead of focusing of repeating the weld for a long-term reliability. The researchers who are working on Friction stir welding of dissimilar materials will face a lot of problems due to different welding conditions, formation of intermetallic compounds, welding defects which mainly occur in the nugget zone of the weld. If the focus is shifted towards selecting the perfect parameters and perfect material combination the FSW of dissimilar materials will be used in industries.

Acknowledgements

This work was financially supported by Universiti Teknologi PETRONAS, Perak, Malaysia under URIF.

References

[1] Mishra R *et al* 2005 *In Mater. Sci. Eng. Rep.* **50** 1.
[2] Ugender S *et al* 2014 *Proc Mat Sci* **6** 1600.
[3] Thomas *et al* 1991 *Patent Application No.* **9125978.8.**
[4] Sadiq Aziz Hussein *et al* 2015 *A review Materials Today Communications* **5** 1.
[5] Dawes C J *et al* 1995 *TWI Bulletin* **6** 124.
[6] Watanabe T *et al* 2006 *Quart. J. Japan Weld. Soc.* **24** 108.
[7] Nandan R *et al* 2008 *Int J Mater Res* **99** 434.
[8] Zhang H *et al* 2013 *Mater Des* **45** 206.
[9] Watanabe T *et al* 2006 *J Mater Process Tech* **178** 342.
[10] Scialpi A *et al* 2008 *Mater Des* **29** 928.
[11] Chen C *et al* 2004 *Int J Mach Tool Manu* **44** 1205.
[12] Thaiping Chen *et al* 2009 *J Mater Sci* **44** 2573.

[13] Uzun H *et al* 2005 *Mater Des* **26** 41.
[14] Coelho R S *et al* 2012 *Mater Sci Eng* **556** 175.
[15] Sun Y F *et al* 2013 *Mater Des* **47** 350.
[16] Habibnia M *et al* 2015 *Int J Adv Manuf Tech* **76** 819.
[17] Qixian Zheng *et al* 2016 *J ALLOY COMPD* **686** 693.
[18] Syafiq W *et al* 2017 *Intl. Conference on Mech., Manufacturing and Process Plant Engineering*
[19] Sorger G *et al* 2017 *Weld World* **7** 542.

IOP Conf. Series: Materials Science and Engineering **551** (2019) 012004 doi:10.1088/1757-899X/551/1/012004

A Comparison Study of Carbonized Kapok Fibres Treated by Sodium Hydroxide Solution and Hydrochloric Acid Solution as an Absorbent in Removing Oil Waste

Nik Ahmad Faisal Nik Mohd Azlin Shah[1], Mohammad Abdullah[1], Muhamad Azrul Aizad Mohamed Saadun[1], Siti Nurul 'Ain Zaiton[2], Hairul Amiza Azman[2] ,Diana Che Lat[3], Ahmad Khudzairi Khalid[4] , Nurhaslina Hambari1[1]

[1]Faculty of Chemical Engineering, Universiti Teknologi MARA Johor Branch, Pasir Gudang Campus, Bandar Seri Alam, 81750 Masai, Johor..

[2]2Faculty of Applied Sciences, Universiti Teknologi MARA Johor Branch, Pasir Gudang Campus, Bandar Seri Alam, 81750 Masai, Johor.

[3] Faculty of Civil Engineering, Universiti Teknologi MARA Johor Branch, Pasir Gudang Campus, Bandar Seri Alam, 81750 Masai, Johor.

[4]Faculty of Computer & Matematical Sciences, Universiti Teknologi MARA Johor Branch, Pasir Gudang Campus, Bandar Seri Alam, 81750 Masai, Johor.

E-mail: moham3767@uitm.edu.my

Abstract. Hydrophobic-oleophilic properties of Kapok (Ceiba pentandra) fibre enables it to show good oil absorption capacity. In this study, Kapok fibre was treated with two types of solvent which is Hydrochloric acid solution (HCl) and Sodium Hydroxide solution (NaOH). Both solutions function to remove lignin, pectin and wax that surrounds the outer surface of Kapok fibre. To achieve its carbonized properties, both samples of Kapok fibre were burned in a furnace at 450oC for approximately an hour. By doing this, it was found that activation time between carbonized Kapok fibre porosity shows a positive result, thus enhancing the capability of Kapok fibre as an absorbent. Absorption rate of two samples of Kapok fibre which is NaOH-treated Kapok fibre and HCl-treated Kapok fibre were evaluated on different types of oils which are diesel, lubricant oil, petrol and used vegetable oil. Mass used for each absorbent samples differ by using 0.2 g, 0.4 g, 0.6 g, 0.8 g and 1.0 g respectively. Dosage reusability and percentage of oil removal were examined for this two samples of absorbent. From the dosage reusability experiment it showed that both carbonized Kapok fibre can absorbed the oil after six cycle of using the same absorbent dosage. Meanwhile, the absorption oil experiment showed that HCl-treated Kapok fibre higher than NaOH- treated Kapok fibre. It is proven that HCL-treated Kapok fibre can perform better compared to NaOH-treated Kapok fibre.

1. Introduction

Oil pollution from food production either restaurant, hotel or cafeteria, household effluent, factories and accidental spills from ships have made a huge negative impact on society and severely affected the ecological environment [1]. Researchers have come out with numbers of ideas to clean up oil from polluted areas including mechanical extraction, in situ combustion and chemical degradation. It was proven these ideas were not the most convenient method. To suit a more economical way, the use of fibers from plants as an absorbent has been considered to concentrate, transfer and absorb spilled oil [2]. Studies on the characterization and the mechanism of adsorbents from natural organic materials has

IOP Conf. Series: Materials Science and Engineering **551** (2019) 012004 doi:10.1088/1757-899X/551/1/012004

been done in recent years [3-4]. The examples are rice husk, coconut husk, cotton, milkwood, silk-floss fiber, cattail, Kapok fiber, populus seed fibre, coconut hulls and sugar cane pulp [5]. Kapok trees are well known plants that are planted at most countries in Southeast Asia [4]. Kapok fibre has a fluffy texture, yellowish coloured, non-toxic and odourless [5]. The Kapok fibre is mainly composed of lignin, polysaccharide and has a wax covering the fibre surface which causes it to have a hydrophobic properties [2]. Moreover, Kapok fibers are in high demand to be used as oil adsorbent due to its hollow structure and hydrophobic characteristics [6-7]. In fact, Kapok has the advantages over traditional oil-absorbing materials: low cost, biodegradability and high sorption capacity [6-8].Therefore, the main purpose of this study is to compare between sodium hydroxide and hydrochloric acid as treated agent in enhancing the ability of carbonized Kapok in treating oil spill. Several experiments have been conduct to show the differences between this two types of carbonized Kapok in terms of its rate of sorption, and dosage reusability.

Experimental

1.1. Pre-treatment of Kapok using Sodium Hydroxide and Hydrochloric acid

Kapok that is used in this experiment is treated beforehand by using alkali and acid treatment. The purpose for these treatments are to ensure lignin, wax, and pectin that surround the outer surface of Kapok are bleached [3][9]. The treatments also increase the size of lumen in Kapok fiber so that it can trap more oil and increase sorption rate. About 0.5M of sodium hydroxide solution used to treat the Kapok. Sample raw of Kapok are soaked in the solution for one day. The sample of Kapok that treated with sodium hydroxide solution (NaOH-treated Kapok fibre) is neutralized with 0.5M of hydrochloric acid solution. The samples then were left for 20 minutes before dried in an oven for 100°C at an hour. The procedure was same for HCL-treated Kapok fibre but it started with soaking in 0.5M of hydrochloric acid solution and neutralized with sodium hydroxide solution.

1.2. Types of Oil

Various types of oil are used to test absorption rate of Kapok with different viscosity from different type of oil. Four different oil are used which is vegetable oil, diesel, petrol and lubricant oil.

1.3. Synthesis of Carbonized Kapok

After treating the raw Kapok fibre with sodium hydroxide and hydrochloric acid both samples were burned in a furnace at 450°C for approximately an hour. This purpose to achieve its carbonized properties.

1.4. Dosage Reusability

The main purpose for this testing is to know the retention of Kapok on oil absorbency. Four different types of oil are used which is diesel, petrol, used vegetable oil and lubricant oil. The sample was placed in the teabag which has 107.1 cm3 volume. About 0.5 g of adsorbent dosage used for both sample and soaked in four different types of oil for 20 seconds before dripped for 30 seconds. Afterwards, sample of Kapok is weighed and data is recorded.

1.5. Percentage of oil removal

The NaOH- treated Kapok and HCl-treated Kapok were soaked in different types of oil to measure a percentage of oil removal. It differs depends on viscosity of oil. Duration taken for soaking the samples into the oil are at least 15 min [7]. The percentage of oil removal was calculated using equation (1).

$$\text{Percentage of oil removal (\%)} = (\text{Final weight} - \text{initial weight}) / \text{initial weight} \quad (1)$$

IOP Conf. Series: Materials Science and Engineering **551** (2019) 012004 doi:10.1088/1757-899X/551/1/012004

2. Results and Discussions

2.1. Adsorbent Reusability capability

The adsorbent reusability of NaOH-treated Kapok and HCL-treated Kapok in a volume of a used vegetable oil (UVO), petrol, diesel and lubricant oil were conducted for a few cycles till the constant weight obtained. The result showed that for both adsorbent it can be reused as absorbent in removing all types of oil stated before for six cycle. The results have shown that without severe disruption of the fibre hollow lumen, the reusability of the Kapok can be increased. The carbonization of Kapok using NaOH and HCL as chemical treatment also shown its potential to become a good modification through the results obtained [4]. This is because the same adsorbent can be reused six time for oil removing. All the result showed in figure 1, figure 2, figure 3 and figure 4.

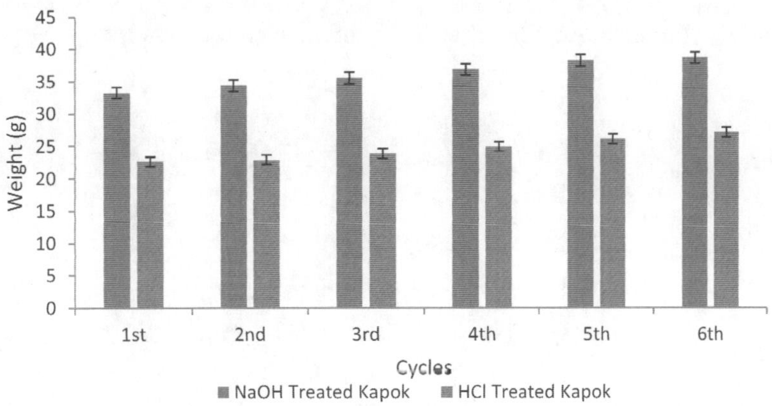

Figure 1. Adsorbent Reusability of Kapok on used vegetable oil waste

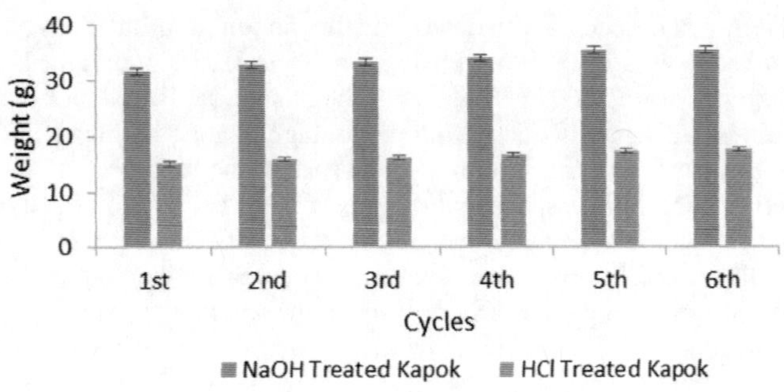

Figure 2. Adsorbent Reusability of Kapok on lubricant oil waste

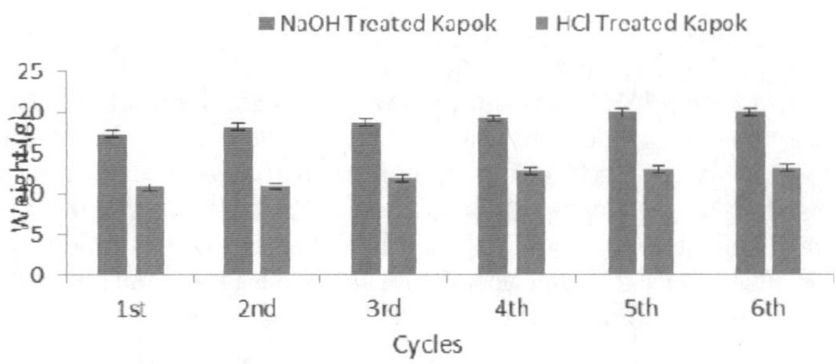

Figure 3. Adsorbent Reusability of Kapok on petrol waste

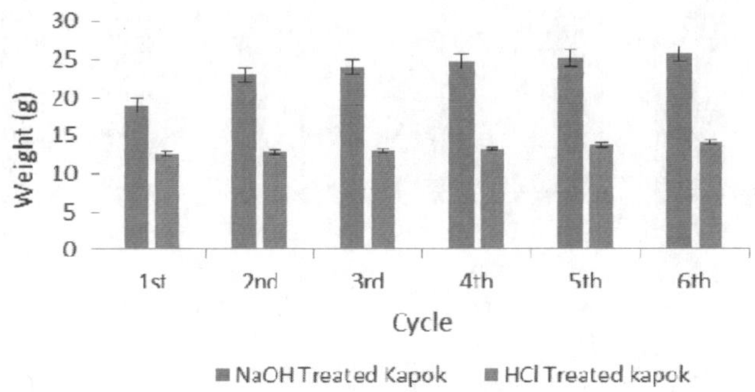

Figure 4. Reusability of Kapok on diesel waste

2.2. Percentage of Oil Removal

The percentage of oil removal is necessary to determine the amount of oil that has been dripped out from the adsorbent. The oil removal was calculated using NaOH treated Kapok and HCl treated Kapok at different weight of samples varying at 0.2g, 0.4g, 0.6g, 0.8g and 1.0g. Based on Figure 5, the percentage of oil removal, has shown a constantly decreasing percentage against an increasing weight of sample. This has resulted due the change in the amount of adsorbent. The reduction of oil removal may have caused by the contraction of fibre interspace and the presence of residual oil in fibre bundles [10]. The NaOH treated Kapok has recorded a removal percentage of 94.66% using 0.2g and 91.3% using 1g of Kapok. Generally, NaOH treated Kapok stated a constantly decreasing percentage of oil removal mainly due to the activation of porous through the carbonization of Kapok that allowed the fibres to store a huge amount of oil [11]. However, the characteristics of the Kapok did not imply on lubricant oil due to its very high viscosity. The removal percentage of oil by NaOH treated Kapok towards petrol and diesel are merely the same. Based on Figure 7, the percentage of oil removal using 0.2g of Kapok was 87.25% while 84.4% of removal percentage was recorded when using 1g of sample. The percentage is low due to the physical characteristics of petrol which is high volatility.

Meanwhile, the percentage of oil removal of untreated Kapok on lubricant oil showed a positive result mainly due to the high viscosity of the lubricant. The oil was observed to have gummy on the surface of the treated Kapok causing a heavier weight of absorbed Kapok. Based on Figure 6, 97.7% of oil was removed when using 0.2g of Kapok sample while 96% of removal were recorded when 1.0g of untreated Kapok was used. Furthermore, 98.4% of oil removal of diesel were tabulated upon using 0.2g and 94.19% of diesel (Figure 8) removal was recorded when using 1.0g. Undoubtedly, the result indicated a slight decrements of oil removal mainly due to viscosity properties of diesel.The result showed that the treated Kapok can hold much oil without releasing it in a long period of time. So, this gives an advantage to natural sorbent which has a lower oil removal percentage compared to other chemical sorbents. Economically, they are cheap and are easy to be obtained in Malaysia.

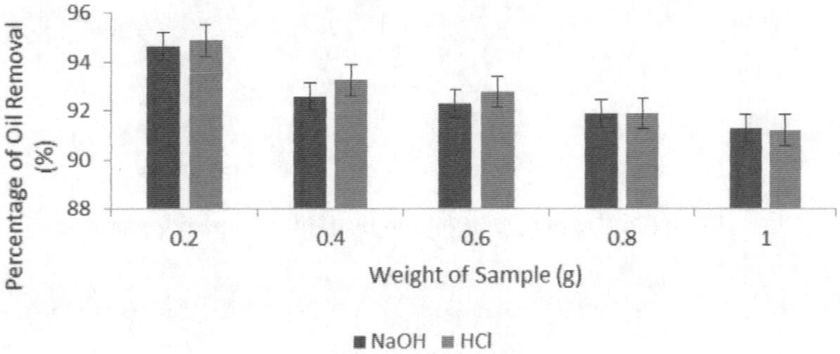

Figure 5. Percentage of oil removal towards used vegetable oil waste

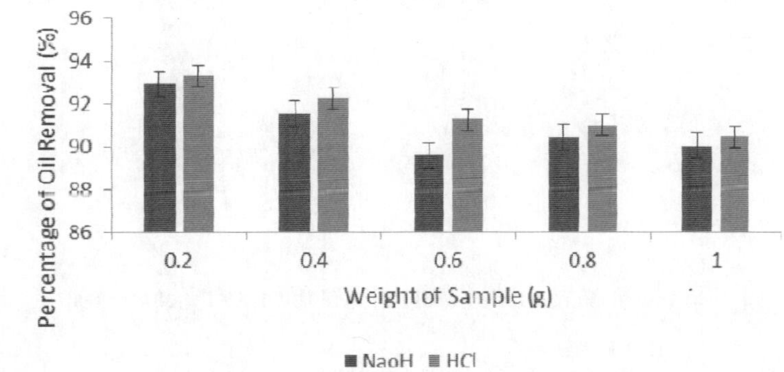

Figure 6. Percentage of oil removal towards lubricant oil waste

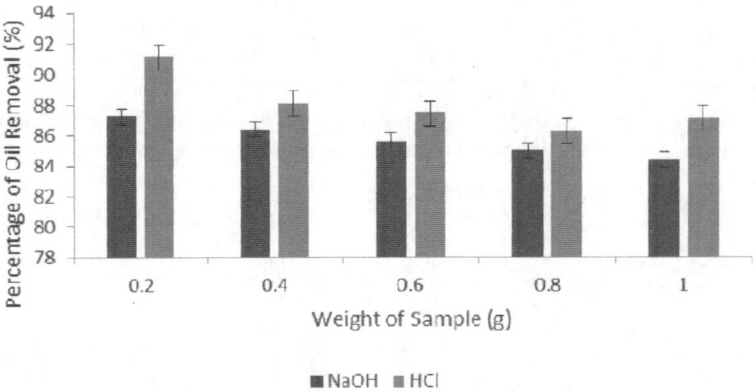

Figure 7. Percentage of oil removal towards petrol oil waste

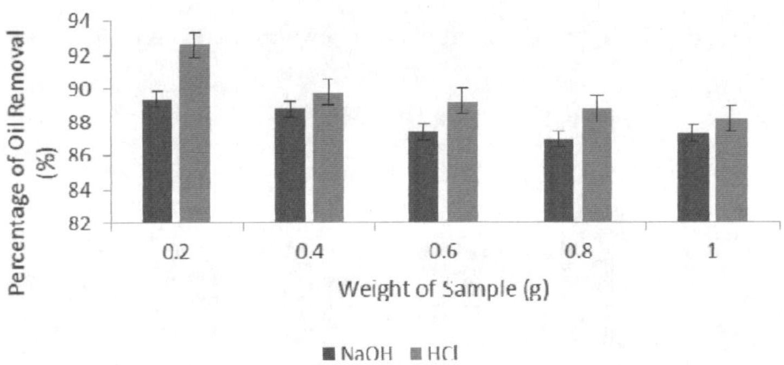

Figure 8. Percentage of oil removal towards diesel oil waste

3. Conclusions

In this study, chemically modified carbonized Kapok fibre has been successfully prepared by using sodium hydroxide and hydrochloride acid. It showed positive result for oil removal and an adsorbent reusability at six cycle. Besides, its properties contribute largely to oil absorption and retention. On top of that, testing of Kapok fibre with several types of oil shows that carbonized Kapok fibre can absorb high viscosity in longer period of time compared to oil with less viscosity. To conclude, chemically modified carbonized Kapok fibre using hydrochloric acid act as a better sorbent compared with the treated using sodium hydroxide.

References

[1] Wang, J *et al* 2013 *Mar Pollut Bull* **69** 91.
[2] Wang, J *et al* 2012 *J. Ind Crops Prod* **40** 178.
[3] Abdullah M *et al* 2016 *J. Appl. Environ. Biol. Sci* **6** 59.
[4] Chung J *et al* 2016 *Mater Lett* **93** 401.
[5] Dong T *et al* 2015 *Mar Pollut Bull* **91** 230.
[6] Lim T *et al* 2007 *Chemosphere* **66** 955.

[7] Abdullah M *et al* 2010 *J Hazard Mater* **177** 683.
[8] Annunciado T *et al* 2005 J . Environ Sci Technol **50** 1340.
[9] Wahi, *et al* 2013 *J. Sep Purif Technol* **113** 51.
[10] Deschamps G *et al* 2003 *J . Environ Sci Technol* **37** 5034.
[11] Ali N *et al* 2012 *J. Environ Technol* **33** 481.

IOP Conf. Series: Materials Science and Engineering **551** (2019) 012005 doi:10.1088/1757-899X/551/1/012005

Biological performance of integrated fixed film activated sludge (IFAS) process

N. M. Zahari[1], P. N. Khairuddin[2], H. A. Mohiyaden[3], L. M. Sidek[4], D. Mohamad[5]

[1,2,3,4,5]Civil Engineering Department, College of Engineering, Universiti Tenaga Nasional, Jln IKRAM-UNITEN, 43000 Kajang, Selangor.
[1,4]Institute Energy Infrastructure, Universiti Tenaga Nasional, Jln IKRAM-UNITEN, 43000 Kajang, Selangor.

E-mail: mubinzahari@gmail.com

Abstract. The aim of this paper is to explore the information and analyze the performance of bio ball and spawning brush media in order to compare the findings of the performance of both biomedia by using Monod Model Equation. The integrated fixed film activated sludge (IFAS) process uses aeration and biomedia of spawning brush and bio ball which are placed fixed in the reactors 1 and 2 respectively. The duration of the study was conducted in 10 days which consist of two-phase which the first phase is monitoring water quality parameter and apply the Monod equation to obtain the performance of the biomedia as the second phase. Chemical Oxygen Demand (COD) removal of reactor 2 has a slightly higher average of 61% as compared to reactor 1 with an average of 57.50%. Thus, the COD removal is more consistent in reactor 1 compared to reactor 2. From this results concluded that IFAS system has a good performance in I terms of low loading wastewater sources.

1. Introduction

Rivers is one of the main sources of ecosystem services for both living and non-living things. The river was utilized for both economic and social development in Malaysia which also leads to loss, degradation, and pollution of river ecosystem [1]. Water quality study of Malacca river showed that the overall water quality of Malacca river has exceeded IIB class which it has caused the death of thousand aquatic life due to the decrease of dissolved oxygen in the water [2]. Nowadays, there is much research on the method to improve the water quality for better environment [3-7]. Integrated Fixed Film Activated Sludge (IFAS) process is a relative one the technology which incorporates the use of a growth media either to be fixed or not within the suspended growth reactor. IFAS is a bioreactor setup by using a different type of biomedia and can be placed in the trickling filter of the setup or to be placed directly in the bioreactor itself [8]. Biomedia is highly effective as they are light and is easier to clean and maintain [9]. Attached biomass in the biological tank can grow on any type of surfaces such as rocks, sand, or plastic [10]. In this study, a laboratory scale of IFAS system was conducted in order to analyze the performance of the biomedia and the data obtained were based on standard river parameter and compared by using Monod equation model.

2 Methodology

2.1 Bioreactor setup

The first reactor consists of the spawning brush and the second reactor consists of the bio balls. Both reactors are filled up with 50% of the surface area covered with the biomedia. The water feed uses synthetic water and it will flow through the reactors by using a small standard size pump. The flow of the pump will be controlled and is set at a constant of 50% and 70%.

2.2 Material

Two types of media that will be used when conducting the lab testing. The biomedia are bio balls and spawning brush. Both biomedia takes up a volume of the concentration of 50% of the reactor.

2.3 Removal Efficiency of Chemical Oxygen Demand (COD)

The amount of Chemical Oxygen Demand (COD) removed in percent and the end value of effluent leaving a typical biogas digester into the environment. The influent and effluent concentrations are the two values that are entered the formula to obtain the answer of the removal efficiency.

2.4 Monod Model Equation

In the Monod equation, μ_{max} represents the specific growth rate (expressed in reciprocal hours) that is reached when all substrates are present in excess (i.e. when S becomes infinite). In practice, μ_{max} is approximated in batch cultures when the initial substrate concentration is much larger than the Monod constant Ks (Initial>Ks). Monod's Model equation is used to measured microorganism growth. Where μmax and Ks are empirical coefficients to the Monod equation.

2.5 Modelling Kinetic of the process

The use of mathematical and kinetic modeling in biological wastewater treatment is to estimate the performance of the reactor when is placed in similar operational conditions. For this study, the kinetic process was done in batch operation mode and varying the initial concentration of COD and monitoring its final concentration at different time intervals. The obtained data were analyzed using first order, Grau (second order), Monod, and modified Stover–Kincannon models in spreadsheet.

2.6 Biomedia removal performance analysis

The substrate removal performance of all biomedia was evaluated nine hours in a span of 4 weeks. Volumetric total substrate nitrogen conversion rate (VTR) was used as the principal indicator for evaluation of the filter media performance.

3 Results and Discussions

3.1 Dissolved Oxygen (DO)

Interestingly, there only is a slight difference of DO for both reactors. It can be seen in figure 1 and value are similar to one another. Simple calculations also showed that the average DO for both reactors are 8.1 mg/l, the maximum DO to be 8.4 mg/l and minimum DO is 7.9 mg/l for both reactors as well. For this study, DO level should be more than 4 mg/l to sustain the bacteria growth in the reactor. With that being said, the results and testing are valid as it is sufficient enough to sustain the growth

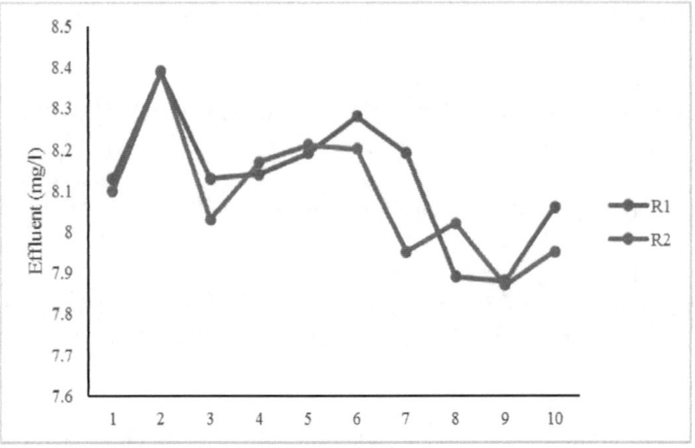

Figure 1. Dissolved Oxygen monitoring.

3.2 Temperature and pH monitoring

The DO obtained during the lab testing has an average of 8.1 mg/l. The difference is not that much and it is also important to take into consideration that the temperature for both researches is also different. The temperature of the whole process is standard room temperature with an average of about 26°C as shown in figure 2. Based on figure 3, the average pH for tank reactor 1 is 7.7 while the average for reactor 2 is slightly higher with a pH of 8.6. This indicates that the water in reactor 1 is more neutral and the water in reactor 2 is more towards alkaline. The minimum pH obtained is 6.6 for both reactors and the highest would be 12.3 for reactor 2 and 10.2 for reactor 1. This occurrence could be due to the sludge that resides at the below of the tank and affects the pH.

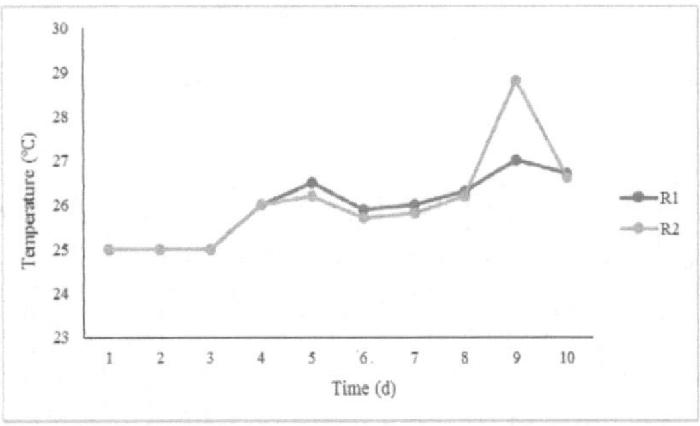

Figure 2. Temperature monitoring.

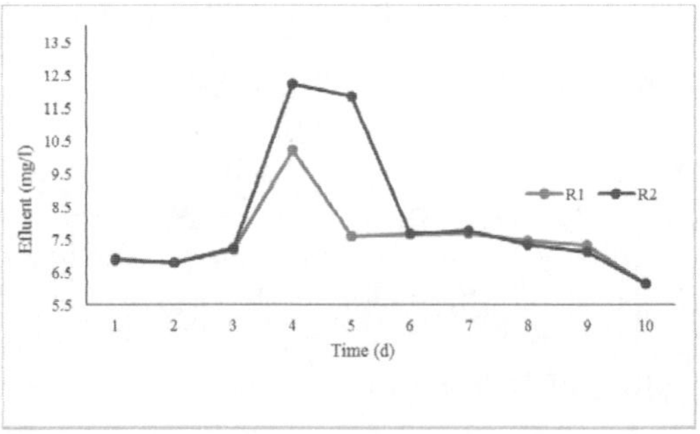

Figure 3. pH monitoring.

3.3 Carbonaceous Biochemical Oxygen Demand (cBOD)

Figure 4 shows that the initial DO and final DO for this lab testing dropped at least more than 1.0 mg/l. Though there was one day in particular that the DO did not drop more than 1.0 mg/l.

IOP Conf. Series: Materials Science and Engineering **551** (2019) 012005 doi:10.1088/1757-899X/551/1/012005

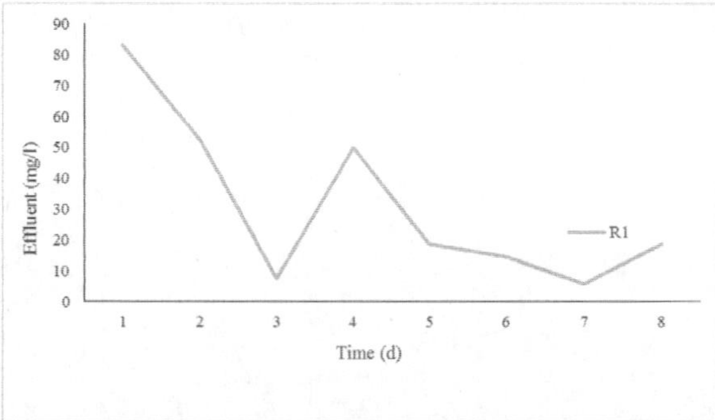

Figure 4. cBOD monitoring.

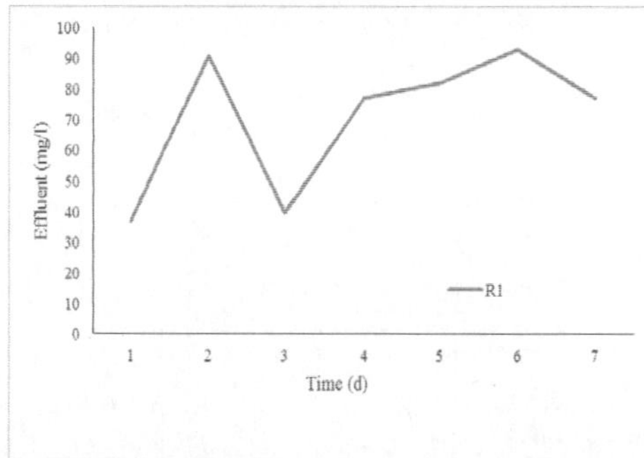

Figure 5. cBOD removal efficiency.

With the decrease in cBOD, it means that the oxidations of carbons are slowly decreasing. Considering that the initial cBOD is on the first day, the total number of days for the removal efficiency of cBOD is 7 days instead of 8. The cBOD removal efficiency as shown in Figure 5 is the amount of waste removed in percent.

3.4 Mixed Liquor Suspended Solids (MLSS) and Mixed Liquor Volatile Suspended Solid (MLVSS) monitoring

Figure 5, shows the results of the MLVSS during the testing. The results showed that the MLVSS is somewhat similar to the MLSS as during the 4th day the value of the MLVSS for both reactors shows a sudden increase as well followed by the next day. But then starts to drop on the 6th day and decreases again throughout the testing. Justifications that could be made for this occurrence would be similar to the previous justification made before for the MLSS. The increase in MLVSS would be the increase in COD removal as more microorganism is present in the aeration tank [11].

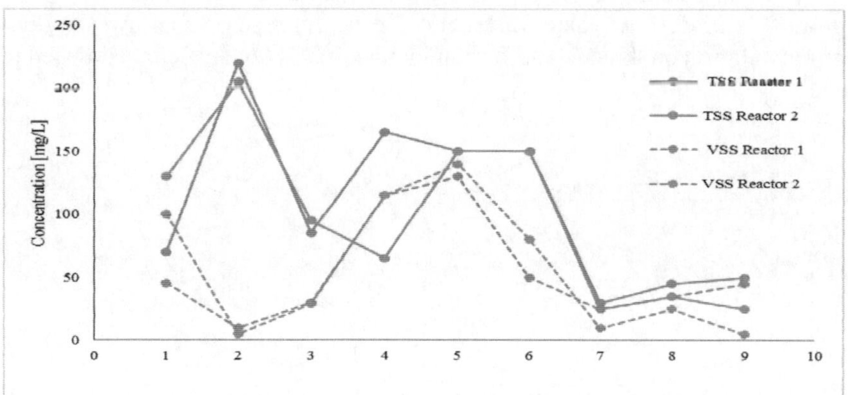

Fig. 6. Total Suspended Solid /Volatile Suspended Solid.

3.5 Chemical Oxygen Demand (COS) reduction rate

The graph results of the COD for both reactor 1 and 2 is shown in figure 7 and 8. It can be seen from the graph that it shows the COD decreases gradually in the overall testing.

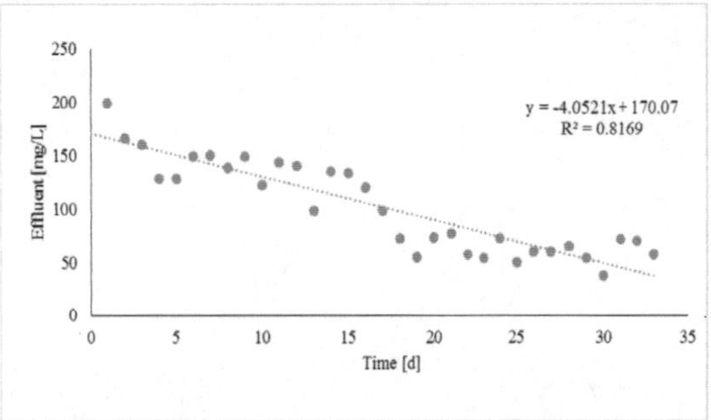

Figure 7. COD reduction rate (reactor 1).

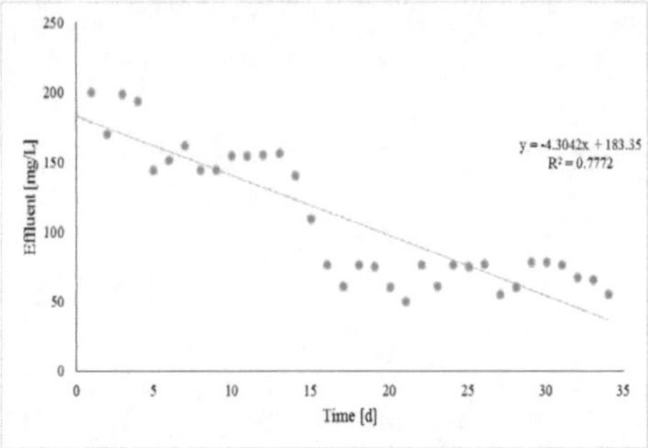

Figure 8. COD reduction rate (reactor 2).

Table 1. Removal Efficiency rate

Type	Average	Range
R1	57.50%	32%-70%
R2	61.00%	24%-64%

The removal efficiency indicates the performance of the ability to remove the contaminants present. The higher the removal efficiency the better. The reactor reaches it most optimum efficiency on the 7th day with a 92.99 mg/l. In terms of performance, tank reactor 2 would be more efficient as compared to tank reactor 1 as shown in table 1. COD removal efficiency reached between 83 and 92.5% and that the extended aeration process was 88 and 93.8%, where else the contact stabilization was 77 and 92% [12].

3.6 Monod Modelling Equation

The results for of R^2 for kinetic coefficient reactor 1 in figure 9 is 0.7376 and kinetic coefficient reactor 2 in figure 10 is 0.0589. R^2 value for kinetic coefficient reactor 2 is lowest value and not near to 1 [13]. Therefore, relation on kinetic coeffiction in reactor 2 is not suffienct and need to add another small scale value.

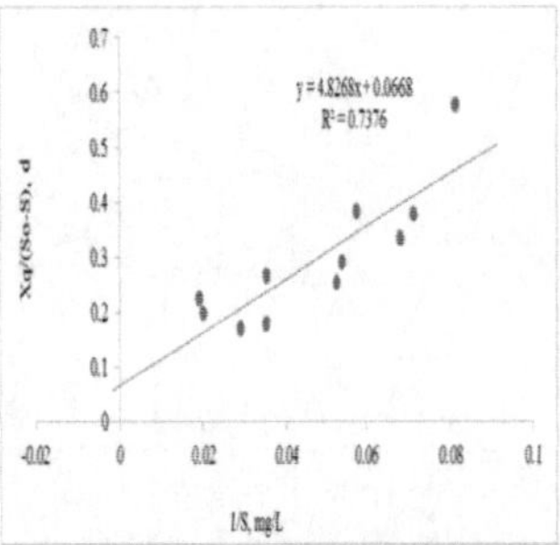

Figure 9. Kinetic coefficients reactor 1.

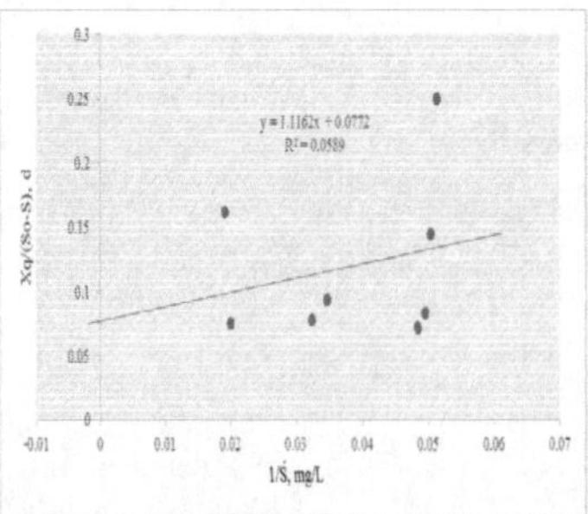

Figure 10. Kinetic coefficients reactor 2.

4 Conclusions

As conclusion, two different biomedia was used ands has done with some investigation in terms of removal efficiency and Monod coeffiecient.

Acknowledgments

This research project was supported by Universiti Tenaga Nasional (UNITEN) under the internal grant (UNIIG 2018).

IOP Conf. Series: Materials Science and Engineering **551** (2019) 012005 doi:10.1088/1757-899X/551/1/012005

References

[1] Khalid R M et al 2018 Ecosystem Serv, **29** 251-259
[2] SN Rosli et al 2018 Malaysian App Bio **44**
[3] Z Wu 2018 Sci. of the total Env. 612
[4] P Whitehead et al 2018 Sci. of the total Env. 631
[5] L Ngwira et al 2018 Phy. and Chem. of the Earth
[6] Q Wang et al 2018 Phy. and Chem. of the Earth
[7] G Martinsen et al Sci. of the Total Env.
[8] R L Pehrson 2002 Proc. of the Water Env. Federation **12**
[9] K M Sherman 2006 Proc. of the Water Env. Federation **8**
[10] J H Andrews 2000 Annual review of phytopathology **38**
[11] K Kumar 2014 Wat. Res. and Industry **5**
[12] Mardani S 2011 Iran. J. Environ. Health. Sci. Eng **8** 1 25-34
[13] N.M. Zahari 2015 Adv. Mat. Res **1113** 764-769

Prediction of Future Ozone Concentration for Next Three Days Using Linear Regression and Nonlinear Regression Models

Nazirul Mubin Zahari[1], Raja Ezzah Shamimi[2], Mohd Hafiz Zawawi[3], Ahmad Zia Ul-Saufie[4] and Daud Mohamad[5]

[1,2,3,5]Department of Civil Engineering, College of Engineering, Universiti Tenaga Nasional, Jln IKRAM-UNITEN, 43000 Kajang, Selangor, Malaysia
[4]Faculty of Computer and Mathematical Science, Universiti Teknologi MARA, Pulau Pinang, Malaysia.

E-mail: mubinzahari@gmail.com

Abstract. The aim of this research is to predict the ozone concentration level for the next three days. Linear regression model and a nonlinear regression model are used to measure the air pollution data and the result was compared. The performance indicator used to evaluate the accuracy of the methods is Index of Agreement (IA), Prediction Accuracy (PA) and Coefficient of Determination (R2). While Normalized Absolute Error (NAE) and Root Mean Square Error (RMSE) are for error measured. The results show that the prediction for the next three days. The highest ozone concentration of the linear regression model is 0.085ppm at Petaling Jaya, Selangor. While the lowest concentration for the linear regression model is 0.015 ppm at Klang, Selangor. Besides, the highest ozone concentration for the nonlinear regression model is 0.1 ppm at Petaling Jaya, Selangor for the second-day prediction. Comparison between the linear regression model and a nonlinear regression model indicates that nonlinear regression model can as an alternative method to the linear regression model.

1. Introduction

Malaysia is a newly developed country that experienced rapid urbanization and underwent an economic boom in the last ten years. Many industries have been growing and created new demands in housing, transportation, and urbanization as the population increases and cause environmental catastrophes especially pollution [1].

In term of air quality, Malaysia is in ranks 117th the worst country among 180 countries worldwide [2]. Air pollution structure contains six mains characteristics that always monitored by Department of Environment, which are ozone (O3), lead (Pb), carbon monoxide (CO), sulfur dioxide (SO2), nitrogen dioxide (NO2) and particulate matter (PM10). Air pollution is rick to human health, animal, damage crops and nature environment [3]. Studies have been proved that the dangerous impact of air pollution on human health specifically on patients suffering from cardiovascular and respiratory diseases [4]. Air pollution is the world's main environment risk for health and mostly this disease has suffered the population of developing countries [5]. In the past study, various types of statistical analysis have been applied to fit the air pollutant concentrations [3]. Based on the past study, researchers have applied regression models for predicting ozone concentration in different areas include in Malaysia [6-8]. This study focus on prédiction of linear regression and nonlinear regression model for future ozone

IOP Conf. Series: Materials Science and Engineering **551** (2019) 012006 doi:10.1088/1757-899X/551/1/012006

concentration level for the next three days. The final results will be compared to determine the best model.

2. Methodology

2.1 Study area

There is five air monitoring station in Malaysia that was selected for this research. It is to predict the future ozone concentration for the next day (D+1), next two days (D+2) and next three days (D+3). All the location selected based on the urban area, industrial area and the area that will influence the reading of air quality. This air monitoring station is monitored by the Department of Environment (DOE) Malaysia. The area of this study is Petaling Jaya, Perai, Seberang Jaya, Klang, and Nilai.

2.2 Linear Regression

The linear regression model is a statistical method to predict the relationship between of two variable which is dependent and independent variable. Linear regression attempts to model the relationship by fitting the linear equation to experimental data. This is the basic equation for the linear regression model:

$$\text{Response} = \text{constant} + \text{parameter} * \text{predictor} +.. + \text{parameter} * \text{predictor} \tag{1}$$

$$Y = bo + b1X1 + b2X2 + ... + bkXk \tag{2}$$

2.3 Nonlinear Regression

Nonlinear is extra complicated compared to linear to create. It is because the function is produced by a sequence of approximation and trial-and-error. To determine the equation is nonlinear or linear is if the equation does not meet the criteria of the linear equation, it is the nonlinear equation.

$$Y = E\,(b\,o + b1X1 + b2X2 + ... + bkXk) \tag{3}$$

2.4 Performance Indicator

The performance indicator for accuracy measures is the root mean square error (RMSE), index of agreement (IA), prediction accuracy (PA), and coefficient of determination (R2). While indicator for error measures are a normalized absolute error (NAE), root means square error (RMSE). This performance indicator is used to determine the best method in predicting ozone concentration for the next three days.

Table 1. Performance Indicator Parameter.

Performance Indicator	Equation	Description
Normalized Absolute Error (NAE)	$NAE = \frac{\sum_{t=1}^{n}(P_t - O_t)}{\sum_{t=1}^{n} O_t}$	NAE values close to 0 Method is correct
Root Mean Square Error (RMSE)	$RMSE = \sqrt{\frac{\sum_{t=1}^{n}(O_t - P_t)^2}{n}}$	RMSE values close to 0 Method is correct
Index of Agreement (IA)	$IA = 1 - \frac{\sum_{t=1}^{n}(P_t - O_t)^2}{\sum_{t=1}^{n}(\lvert P_t - \overline{O}\rvert - \lvert O_t - \overline{O}\rvert)^2}$	IA values close to 1 Method is correct
Prediction Accuracy (PA)	$PA = \frac{\sum_{t=1}^{n}(P_t - \overline{O})^2}{\sum_{t=1}^{n}(O_t - \overline{O})^2}$	PA values close to 1 Method is correct
Coefficient of Determination (R2)	$R^2 = 1 - \frac{\sum_{t=1}^{n}(P_t - \overline{O})^2}{\sum_{t=1}^{n}(O_t - \overline{O})^2}$	R2 values close to 1 Method is correct

IOP Conf. Series: Materials Science and Engineering **551** (2019) 012006 doi:10.1088/1757-899X/551/1/012006

3. Results and Discussion

3.1 Descriptive Analysis

The ozone concentration data for 2009 until 2015 at five air monitoring station which are Petaling Jaya, Perai, Seberang Jaya, Klang, and Nilai is summarised in table 2. The highest average ozone concentration is Petaling Jaya compared to others study area. This five study area shows the ozone concentration data is skewed to the right that shows it has an extreme event. Based on the kurtosis value, Petaling Jaya and Nilai have positive kurtosis value. While Perai Klang and Seberang Jaya have negative kurtosis value.

Table 2. Descriptive Statistic.

Descriptive Analysis	Study Area				
	Petaling Jaya	Perai	Seberang Jaya	Klang	Nilai
Air Monitoring Station	Sek. Keb. Bandar Utama, Petaling Jaya	Sek. Keb. Cederawasih, Taman Inderawasih, Perai	Sek. Keb. Seberang Jaya II, Seberang Jaya	Sek. Men. (P) Raja Zarina, Klang	Tmn. Semarak (Phase II), Nilai
Coordinate	(3.14,101.61)	(5.39,100.39)	(5.38,100.40)	(3.01,101.41)	(2.82,101.82)
N	2109	1677	2288	2102	2121
Mean	0.047	0.042	0.039	0.045	0.045
Variance	0.000	0.000	0.000	0.000	0.000
Std. deviation	0.021	0.0182	0.018	0.020	0.018
Minimum	0.0009	0.0010	0.0020	0.0000	0.0020
Maximum	0.125	0.105	0.105	0.124	0.119
Skewness	0.419	0.435	0.410	0.464	0.424
Kurtosis	0.119	-0.123	-0.336	-0.025	0.322

3.2 Model Development

Linear regression model and a nonlinear regression model were developed using IBM SPSS Statistic 23 as shown in table 3 and table 4. The model was developed for the next day, next two days and the next three days at five research area which are Petaling Jaya, Perai, Seberang Jaya, Klang, and Nilai. These models were used to predict the ozone concentration.

Table 3. Model Development for Linear Regression Model.

Day of Prediction	Model
The next day (D+1)	$O3,_{D+1} = b_o + b_1T + b_2H + b_3SO_2 + b_4NO_2 + b_5O_3 + b_6CO + b_7PM_{10}$
The next two days (D+2)	$O3,_{D+2} = b_o + b_1T + b_2H + b_3SO_2 + b_4NO_2 + b_5O_3 + b_6CO + b_7PM_{10}$
The next three days (D+3)	$O3,_{D+3} = b_o + b_1T + b_2H + b_3SO_2 + b_4NO_2 + b_5O_3 + b_6CO + b_7PM_{10}$

Table 4. Model Development for Nonlinear Regression.

Day of Prediction	Model
The next day (D+1)	$O3,_{D+1} = E\ (b_o + b_1T + b_2H + b_3SO_2 + b_4NO_2 + b_5O_3 + b_6CO + b_7PM_{10})$
The next two days (D+2)	$O3,_{D+2} = E\ (b_o + b_1T + b_2H + b_3SO_2 + b_4NO_2 + b_5O_3 + b_6CO + b_7PM_{10}$
The next three days (D+3)	$O3,_{D+3} = E\ (b_o + b_1T + b_2H + b_3SO_2 + b_4NO_2 + b_5O_3 + b_6CO + b_7PM_{10}$

IOP Conf. Series: Materials Science and Engineering **551** (2019) 012006 doi:10.1088/1757-899X/551/1/012006

3.3 Prediction Model

As shown in figure 1-5, the model developed by the IBM SPSS Statistic 23 were used to predict ozone concentration in the future. Linear regression model and a nonlinear regression model were applied to get the best fit value to predict the ozone concentration level for the next three days. Based on the graph below, show the prediction value for the next three days. This research is based on the air pollution data from five monitoring station in Malaysia. After that, the results of the linear regression model and a nonlinear regression model were comparing to get the best method.

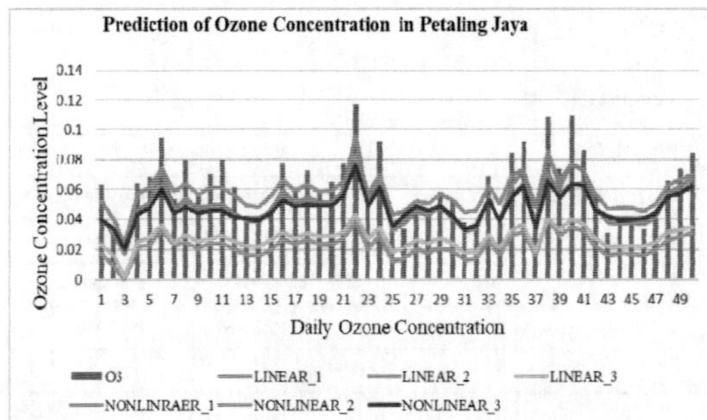

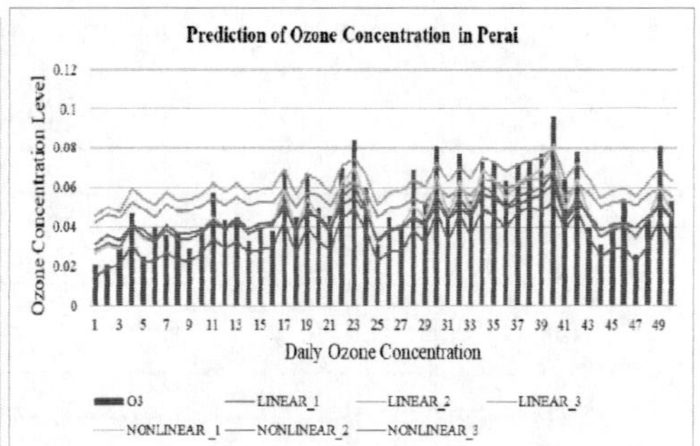

Figure 1. Prediction of Ozone Conc. in Petaling Jaya. **Figure 2.** Prediction of Ozone Conc. In Perai.

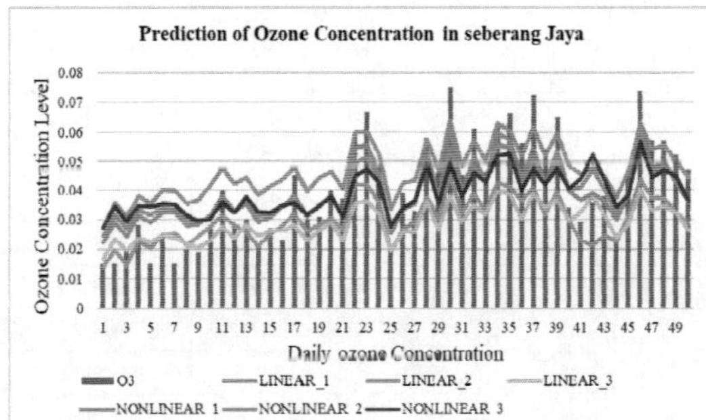

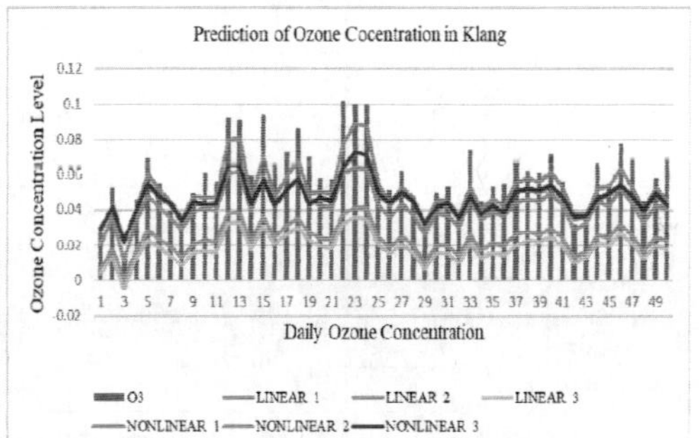

Figure 3. Prediction of Ozone Conc. in Seberang Jaya. **Figure 4.** Prediction of Ozone Conc. in Klang.

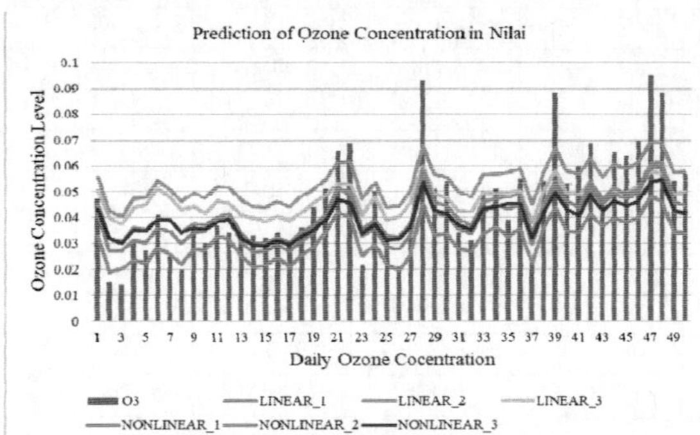

Figure 5. Prediction of Ozone Concentration in Nilai.

3.4 Identification Model

Table 5 shows a descriptive analysis on the performance indicator. The comparison between the two methods which is linear and non-linear. From the comparison of the performance indicators in Petaling Jaya, Perai, Seberang Jaya, Klang and Nilai, almost all the value of the NAE, RMSE, PA, R^2 and IA values between validation model and verification model are very close. The gap for each performance indicator is very small. The final value after compares to 0 and 1 between performances indicators also balance. So, we can conclude the nonlinear regression model can be the best alternative to a linear regression model to predict the ozone concentration in Malaysia. It is because the nonlinear regression model has slightly the same results with a linear regression model.

Table 5. Descriptive Statistic.

Sites	Prediction	Method	NAE	RMSE	PA	R^2	IA
Petaling Jaya, Selangor	D + 1	Linear	0.330202	0.018501	0.516835	0.266815	0.633424
		Nonlinear	0.311657	0.018625	0.478207	0.228422	0.652168
	D + 2	Linear	0.607149	0.033141	0.402198	0.161579	0.489003
		Nonlinear	0.332674	0.019401	0.382433	0.146088	0.571791
	D + 3	Linear	0.512534	0.029348	0.357862	0.127919	0.500480
		Nonlinear	0.340345	0.019767	0.343716	0.118006	0.534684
Perai, Pulau Pinang	D + 1	Linear	0.381158	0.018949	0.551498	0.303728	0.613635
		Nonlinear	0.285191	0.014568	0.536769	0.287721	0.692372
	D + 2	Linear	0.440168	0.020489	0.484774	0.234680	0.570974
		Nonlinear	0.304539	0.015398	0.468581	0.219263	0.611537
	D + 3	Linear	0.512534	0.029348	0.357862	0.127919	0.500480
		Nonlinear	0.340345	0.019767	0.343716	0.118006	0.534684
Seberang Jaya, Pulau Pinang	D + 1	Linear	0.307691	0.014158	0.633161	0.400485	0.753295
		Nonlinear	0.296841	0.014253	0.609894	0.371593	0.750966
	D + 2	Linear	0.390888	0.018531	0.545251	0.296996	0.619722
		Nonlinear	0.319681	0.015115	0.536364	0.287393	0.691487
	D + 3	Linear	0.426308	0.019958	0.498615	0.248364	0.580040
		Nonlinear	0.332746	0.015489	0.494118	0.243905	0.653095
Klang, Selangor	D + 1	Linear	0.342321	0.020378	0.606437	0.367364	0.663020
		Nonlinear	0.288836	0.017134	0.574277	0.329433	0.730889
	D + 2	Linear	0.651349	0.033240	0.523946	0.274219	0.514041
		Nonlinear	0.310164	0.018259	0.508485	0.258275	0.659017
	D + 3	Linear	0.734617	0.036782	0.485142	0.235106	0.482622
		Nonlinear	0.317200	0.018403	0.476569	0.226870	0.642043
Nilai, Negeri Sembilan	D + 1	Linear	0.356841	0.020533	0.446514	0.199150	0.555302
		Nonlinear	0.288206	0.016633	0.425909	0.181194	0.594215
	D + 2	Linear	0.363522	0.019390	0.386465	0.149187	0.533569
		Nonlinear	0.299020	0.016849	0.365729	0.133607	0.551138
	D + 3	Linear	0.308922	0.017003	0.336818	0.113319	0.484296
		Nonlinear	0.304960	0.017220	0.335922	0.112716	0.502955

4. Conclusions

In conclusion, nonlinear regression can be the alternative method to the linear regression. These two regressions are slightly the same in predicting ozone concentration level for the next three days in Malaysia. These prove that the linear regression model and the nonlinear regression model are approximately the same methods to predict the ozone concentration for the next three days in

Malaysia. Linear regression model and a nonlinear regression model can be used in prediction air quality data.

Acknowledgments
This study was supported by Universiti Tenaga Nasional internal grant (UNIIG) and special thanks to those who contributed to this project directly or indirectly.

References

[1] Afroz R *et al.* 2003 *Review of air pollution and health impacts in Malaysia* **92**
[2] EPI 2017 *Malaysia's Performance in Environmental Performance Index*
[3] Zawawi M H *et al. Journal of Scientific Research and Development* **3(4)** 106.
[4] Huebnerov Z *et al.* 2014 *Atmos. Pollut. Res* **5(3)** 471.
[5] Azam A G 2016 *J. Res. Med. Sci.* **21** 65.
[6] Ul-Saufie A Z *et al.* 2011 *International Journal of Applied Science and Technology* **1** 4
[7] Zahari N M 2018 *AIP Conference Proceedings* **2030** 020236.
[8] Isa M *et al.* 2018 *AIP Conference Proceedings* **765** 232.

IOP Conf. Series: Materials Science and Engineering **551** (2019) 012007 doi:10.1088/1757-899X/551/1/012007

Microstructure and Mechanical Properties of Cold Rolled AISI 1018 Low Carbon Steel

Khansaa D. Salman

Department of Electro-mechanics Engineering, Universiti of Teknology, Baghdad, Iraq.

E-mail: drkhansaa_ds@yahoo.com

Abstract. In this investigation, mechanical properties and microstructure of AISI 1018 low carbon steel in the form of strips with the dimensions about 50 ×25×4 mm subjected to cold rolling gradually by 20, 30, 40, and 50% were studied. The effect of the percentage of cold rolling on the mechanical properties of AISI 1018 low carbon steel was estimated by using tensile test, Brinell hardness test. The tensile test was done by using Universal Tensile Machine (UTM) to define yield strength and tensile strength as the percentage of increasing in the deformation, while the Brinell hardness number was obtained for each percentage of rolled deformation. Also, the microstructure of the specimens was analyzed by optical microscope. The results of this investigation show that the yield strength decreased with increasing the percentage of cold rolled as 300,233, 200, and 171 MPa, while the tensile strength decreased as 368, 297, 274 and 238 MPa respectively. While Brinell hardness number increased with increasing the percentage of rolling deformation as 145, 192.1,221 and 240.3 (HB) respectively. This is, in turn, show that increasing brittleness as a result of increasing rolling deformation. The photomicrographs of the specimens show that increasing in percentage of cold rolling leads to stretch the grains, which consist of fine ferrite grains with the pearlite grains.

1. Introduction

Since low carbon steel possess low mechanical properties such as hardness, tensile strength, and ability to cold working and don't respond to the heat treatment in which the properties may be increased by cold working. The main applications of low carbon steel are automotive components, structural sections and sheets using in bridges, pipe lines, and buildings. Cold working is an important process in steel production as well as to improve the mechanical properties or to obtain a desirable shape [1-3]. Cold rolling directly influences on the structure and properties material because of it is considered as no recrystallization can be occurred. The grains gradually extend in the direction at which the deformation occurs and then arrangement the crystallographic lattice can be happened in directional character. Due to the effect of deformation, bonding character for the other phases can be developed, such as pearlite structure or another inclusion. Hence, due to the deformation, crystallographic texture and structural arises which in turn causes the directional effect of the mechanical properties [4]. An important part for the strips and hot rolled is subjected to cold rolling; in cold rolling, the recrystallization cannot be occurred because of low working temperature [5]. The aim of cold rolling process is to produce strips or sheets with high surface finishing and the cold rolling products were then softened by annealing heat treatment before manufacturing [6]. A specific characteristic of the deformation for the structure is anisotropy of the mechanical properties and at the same time, the direction of grains arrangement is extremely depended on the deformation due to the cold rolling of strips and sheets [7]. There are many researches were published in this field, Nurudeen A. Raji and Olucke O. Oluwole [8] studied the effect of cold drawing on mechanical properties of 0.12 wt.% C steel using to manufacture the nails. Different degrees of cold drawing were used gradually by 20%, 25%, 40%, and 50%. The results of this work show that increasing the degree of cold working leads to increase Brinell hardness number and decreasing the yield strength, tensile strength, ductility, and toughness. Frederike et al. [9]. Were studied the effect of additional shear strain induced during the symmetrical rolling (SR) and asymmetrical rolling (ASR) on the mechanical properties and microstructure. ASR was carried out by using a rolling diameter ratio about 2:1. The results of this work show that a simple addition of shear strain induced by ASR affected extremely the microstructure and improving the mechanical properties. Whilst Y. Liu et al. [10] investigated the effect of cold –rolled on the mechanical properties and microstructure of SAE 1078 pearlitic steel. The cold rolling used in this study up to 90%, the percentage of cold rolling of the mechanical properties of AISI 1018 low carbon steel.

2. Methodology

2.1. Experimental Material

Material used in this work is AISI 1018 low carbon steel in the form of strips with 50×25×4 dimensions. Table.1 shows the chemical composition of AISI 1018 low carbon steel. Table.2 shows the mechanical properties of AISI 1018 low carbon steel.

Table 1. Show chemical composition of low carbon steel AISI 1018 wt%

C	Si	Mn	P	S
0.122	Less than 0.0005	0.442	Less than 0.0005	0.0269
Cr	Mo	Ni	Al	Co
0.0097	Less than 0.001	0.0119	0.0052	Less than 0.001
Cu	Nb	Ti	V	W
0.0135	Less than 0.001	Less than 0.0005	0.0009	0.0056
Ta	Sn	Zr	Zn	B
0.0082	Less than 0.001	0.0054	0.0036	Less than 0.0015
As	Pb	Se	Sb	Fe
0.0155	Less than 0.001	Less than 0.001	0.0071	Remain

Table 2. Mechanical properties of low carbon steel AISI 1018

Tensile Strength, MPa	Yield Strength, MPa	Vickers Hardness	Modulus of Elasticity, Gpa	Poisson's Ratio	Elongation, %
440	370	131	205	0.29	15

2.2. Cold Rolling Process

The specimens used in this work in strips form, the specimens were cold rolled by using rolling machine with many stages at different cold rolling percentages about 20, 30, 40 and 50%. Cold rolling was calculated by using the following equation [11].

$$C.W = (A_O - A_F)/A_O * 100\% \quad (1)$$

IOP Conf. Series: Materials Science and Engineering **551** (2019) 012007 doi:10.1088/1757-899X/551/1/012007

2.3. Mechanical Properties

Mechanical properties of AISI 1018 low carbon steel used in this work under cold rolling were defined by tensile test using Universal Tensile Machine (UTM) to evaluate yield strength, tensile strength, and ductility depending on the percentages of cold rolling. While the hardness number was obtained by using Brinell hardness test for each percentage of cold rolling using 750 Kg_f for 15 sec. Table. 3 show the mechanical properties of the specimens before and after cold rolling with different percentages.

Table 3. Show mechanical properties of the specimens with different % C.W.

C.W, %	BHN	σy (MPa)	σU.T.S (MPa)
0	126	325	448
20	145	300	368
30	192.1	233	297
40	221	200	274
50	240.3	171	238

3. Microstructural Examination

The microstructure reflected the mechanical properties of the material. The microstructure of the specimens was evaluated by optical microscope (OM) and scanning electron microscopy for each percentage of cold rolling. Microstructure examination was done for the specimens before and after cold working. After the cold rolling, the microstructure of the specimens was evaluated in the perpendicular section and in the parallel direction of cold rolling.

4. Results and Discussions

The mechanical properties essentially affected by the percentage of cold rolling as applicable for strip manufacturing.

4.1. Effect of cold rolling on tensile properties

Figure 1 show stress-strain curves for the different percentages f cold rolling. Different percentages of the cold rolling causes different deformation, from the Figure.1, it is noted that the increasing in stress leads to increase the strain, especially after proportional limit. After limit proportionality, the strain hardening will occurred as a result of increasing in stress due to the increasing the percentage of cold rolling. The yield strength is defined as the point at which the elastic deformation transforms to plastic deformation. The yield strength of the present alloy reduces with the increasing of percentage of cold rolling, which in turn reduces the ductility of material and causes the brittleness of material because of the strain hardening as a result of cold rolling.

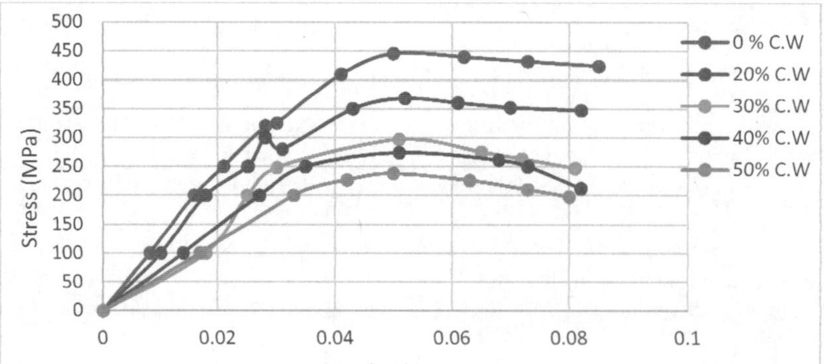

Figure 1. Stress-strain curves for the specimens with different C.W,%

IOP Conf. Series: Materials Science and Engineering **551** (2019) 012007 doi:10.1088/1757-899X/551/1/012007

Also, the increasing of percentage cold rolling reduces the tensile strength due to strain hardening and makes the material brittle as well as for the same reasons deal with yield strength. Figure.2 show the increasing in percentage of cold rolling leads to decrease tensile strength and yield strength.

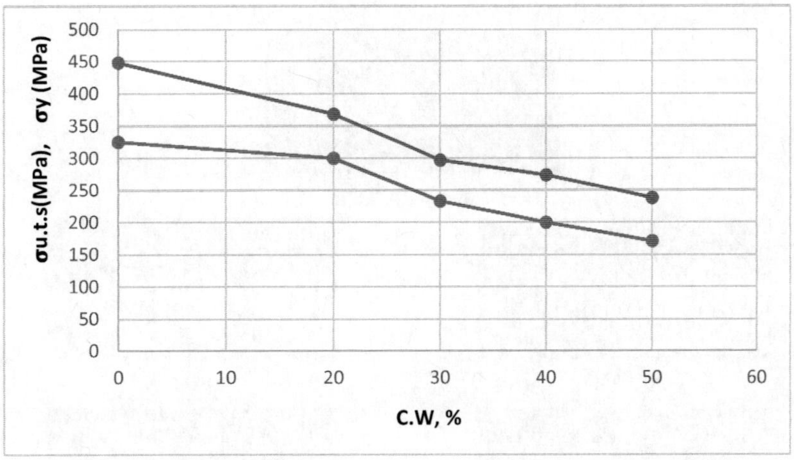

Figure 2. Ultimate tensile strength yield strength for different C.W,%

4.2. Effect of cold rolling on the hardness
Figure 3 shows that the increasing in percentage of cold rolling increases the hardness of low carbon steel due to the increasing of deformation. This is in turn corresponded to the strain hardening as a result of cold rolling which lead to reduce the thickness of steel strip.

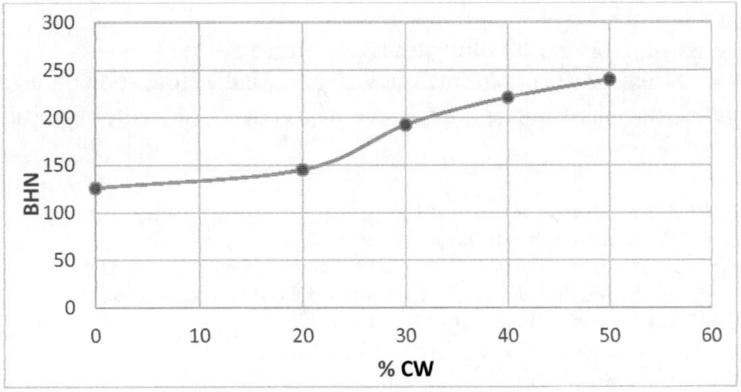

Figure 3. Brinell hardness number vs. C.W,%

4.3. Microstructure Analysis
During cold rolling, initially, elastic deformation occurred as a gradually elongation of the grains in the parallel direction of cold rolling. Increasing in percentage of cold rolling leads to the transition of the deformation from elastic deformation to the plastic deformation as a yield point is separated between

them. The specimens to evaluate the structure by optical microscope after cold rolling were consisted of the fine grains of the ferrite with minor grains of pearlite. Increasing the percentage of cold rolling leads to increase the elongation of ferrite and pearlite grains as shown in Fig. 4, which in turn extremely affected the mechanical properties; this is agreed with [12].

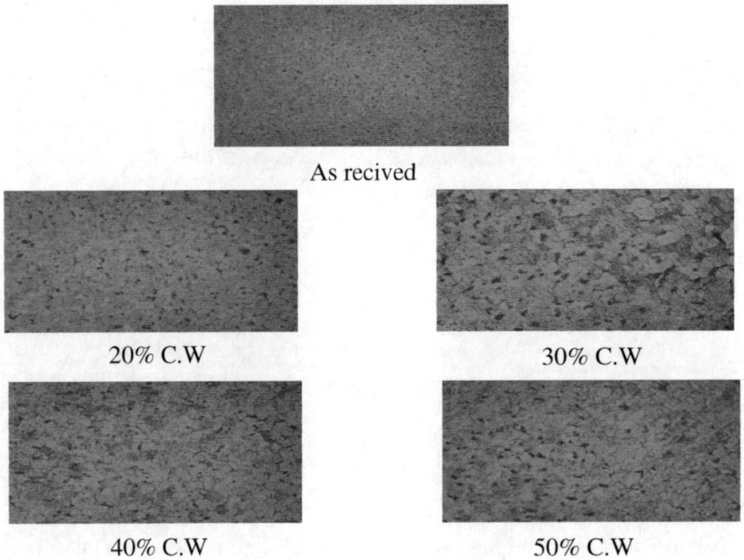

Figure 4. Photomicrographs of the specimen before and cold rolling.

5. Conclusions

From the research, we can conclude several conclusions which are;

i. Increasing the percentage of cold rolling leads to decrease yield strength, tensile strength, and ductility.
ii. Increasing the percentage of cold rolling leads to increase the hardness.
iii. Increasing the percentage of cold rolling leads to increase the elongation of the grains.
iv. After cold rolling the microstructure of the specimens consist of ferrite and pearlite.

References

[1] Ohashi T *et al* 2008 *Arch. Mater. Sci. Eng.* **30** 51.
[2] Schindler I *et al* 2006 *J. Achiev. Mater. Manuf. Eng.* **18** 231
[3] Kamwal T *et al* 2009 Journal of Pakistan Institute of Chemical Engineering **37** 51.
[4] Grachev S V *et al* 2005 Metallovedenic i Temiczeskaja Obrabotka Mtalow **1** 19.
[5] Schindler I *et al* 2004 *Arch. Civ. Mech. Eng.* **4** 104.
[6] Pawlak S J and Krzton H J 2006 *J. Achiev. Mater. Manuf. Eng.* **36** 18.
[7] Zhang X *et al* 2016 *Act. Mater.* **114** 176.
[8] Raji N A *et al* 2011 *Mater. Sci. App.* **2** 1556.
[9] Berrenberg F *et al* 2018 *Mater. Sci. Eng.* A **709** 1722.
[10] Liu Y *et al* 2018 *Mater. Sci. Eng.* A **709** 115.
[11] Huda Z 2009 *Eur. J. Sci. Res.* **26** 549.
[12] Janosec M *et al* 2007 *Arch. Civ. Mech. Eng.* **8** 29.

Investigating the effectiveness of Water Hyacinth Fiber Mat for Soil Erosion Control

M.F. Chow[1], H. Hashrim[2] , S.T. Chong[3] and Y.J. Ng[4]

[1]Sustainable Technology and Environment Group (STEG), Institute of Energy Infrastructure (IEI), Universiti Tenaga Nasional, Jalan IKRAM-UNITEN, 43000, Kajang, Selangor.
[2]Department of Civil Engineering, Universiti Tenaga Nasional, Jalan IKRAM-UNITEN, 43000, Kajang, Selangor.
[3]College of Engineering, Universiti Tenaga Nasional, Jalan IKRAM-UNITEN, 43000, Kajang, Selangor.
[4]College of Computing & Informatics (CCI), Universiti Tenaga Nasional, Jalan IKRAM-UNITEN, 43000, Kajang, Selangor.

Abstract. The usage of natural fiber mat has created much attention especially in soil erosion control sector due to low cost and sustainable materials. This study investigated the effectiveness of natural fiber mat made from water hyacinth stems for soil erosion control. The water hyacinth fiber mat (WHFM) was fabricated with mesh opening size of 30 mm × 30mm and tested at 30 degree slope inclinations under simulated rainfall with constant intensity. Results revealed that soil with treatment (covered by WHFM) significantly decreased sediment yields and volumes, with average 78.74% reduction efficiency compared to untreated soil (bare soil).

1. Introduction

Soil erosion is a major environmental problem in many regions of the world. Typically, the process of erosion begins with wind blowing or water falling on unprotected slopes, causing detachment of soil particles from small amount to a larger displacement of soil which moves down slope. Vegetation provides good protection for soil cover as it can reduce the impact of rain drops, as well as its roots are able to hold soil in place [1]. However not every places are covered by vegetation due to climate properties, causing vegetation establishment gets slow or difficult. Due to this matter, the applications of geotextiles are being introduced as alternative to vegetation. They are made from both natural and synthetic materials in different forms, sizes, and physical characteristics. In recent years, the usage of natural fibers as reinforcement in polymer composites has created much attention especially in soil erosion control sector due to low cost materials of engineering.

Erosion control mats can be effective in minimizing the erosive effect of rainfall when used to cover bare or newly planted soil. There are biodegradable materials that can be used to protect new plantings and reduces soil erosion such as coir and jute. Natural vegetation for sustainable erosion control and slope protection is a proven choice of soil conservation along hilly terrain. Vegetation growth in a troubled slope is occasionally met with problems such as absence of initial binding material in the soil, washing away by runoff, etc. In such conditions, natural fiber mat serve the purpose of protecting the soil and the seeds in the initial stage of vegetative growth [2]. They absorb moisture and are suitable in areas of low rainfall and situations where the establishment of vegetation takes a long time, leaving slopes susceptible to erosion. Natural fiber mats degrade and form organic mulch and this in turn helps in hastening the establishment of vegetation. Different natural fiber mat have different rates of degradation and this rate depends on site conditions.

Natural fibers are very abundant, especially those derived from water plant as they are very easy to grow up [3]. One of thus is water hyacinth or also known as Eichhornia Crassipes that grow very rapidly and floating on the water surface. Its growth is very difficult to control, which often provides some problems such as covering the surface of water thus affecting on decreasing the production in the fisheries sector, evaporating water rapidly and developing sector for growing mosquitoes. In most countries, water hyacinth is recognized as a big threat to agriculture, water mitigation, and aquatic ecosystem because of its rapid growth and widely spread forming dense floating mats that cover large areas of water surface [4].

Rapid urbanization process in developing country has increased the transports of numerous nutrients and pollutants into the freshwater bodies [5,6,7]. As a consequence, river or lake water pollutions tend to encourage the growth of weeds in the water bodies [8,9]. In Malaysia, it was found that water hyacinth is among four problematic weeds in the country. The favorable tropical climate and environmental factors had helped to trigger the massive growth of these weeds. Since water hyacinth is an invasive plant with rapid growth, control measures always needed to be taken to eliminate these plants from the water bodies. The common practice of authority bodies is to remove these plants routinely and end up burning it which leads to another pollution. The better way of handling this water hyacinth plant, is by recycling them as a new source of natural fibers product. Indeed, as natural fibers, water hyacinth fibers are very useful to be applied as reinforcement material in polymer composites due to their interesting characteristics; relative high celluloses contents and small cellulose diameter. From laboratory test results, woven water hyacinth can be utilized for erosion control in field applications because the moisture retained in woven water hyacinth was high thus it can help the growth of vegetation [10].

This paper studied the effectiveness of water hyacinth fiber mat as soil erosion control which evaluated in terms of soil volumes transported, sediment yield and soil loss reduction efficiency compared with the bare soil slope.

2. Methodology

2.1 Fiber mat development

The erosion mat used in the study was made from the stem of water hyacinth harvested from Sg. Ramal Lake in Kajang, Malaysia. The leaves and roots of water hyacinth were cut and removed from the stems. Meanwhile the stems were cut into half and dried under the sun for 10 hours. The fibers were not treated with any treatment as proposed by few researchers, as to maintain the characteristic of self-degradable of the natural fiber mat. Production of single spline rope was carried out through the process of twisting two even section of the stems in the same direction. As the twisting continued, the two sections of stems will begin to wrap around one another, forming a rope. As the near end of the first bundle reached, another two more section of stems were attached together by overlap the tails of the original stem sections with the heads of the new sections, so that the new stems are anchored into place. This process continues until desired length of rope achieved. The water hyacinth fiber mat (WHFM) weaving process follows the traditional open-weave pattern by interlacing the single splined fiber ropes at right angle with each other. The mesh opening size used for WHFM was 30 mm × 30mm. The coverage area of each WHFM is set as 1m (length) x 0.5 m (width).

2.2 Laboratory experiment

Two soil testbeds with slope of 30° and dimension of 1 m × 0.5 m were prepared for simulated rainfall test (as shown in Figure 1. The soil testbeds were setup for control plot (bare soil) and treated plot (soil covered by WHFM). Type of soil used was clay with bulk density measured was 1.36g/m³. The erosion control test was performed by using artificial rainfall which consists of two sprinklers installed on top of soil testbeds. The rainfall intensity was maintained constant for 30 minutes. The sediments yields were collected every 5 minutes and oven-dried before their masses are measured.

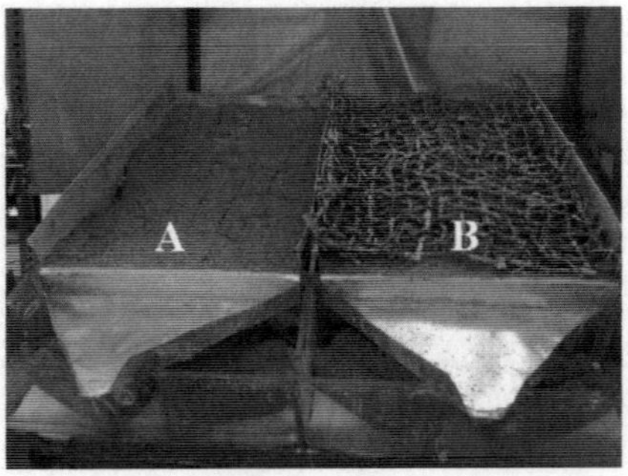

Figure 1. Experimental soil testbeds with slope of 30 degree setup for A) bare soil; and B) WHFM

3. Results and Discussion

3.1 Sediment yield

Sediment yield is defined as the amount of sediment per unit area transported by flowing water during specific rainfall event. The mass of sediments were measured after sediments were oven-dried. It was estimated by using equation (1) [11].

$$SY=\frac{S_m}{tAb} \quad (1)$$

Where SY is sediment yield (g/m².hr), S_m is mass of oven-dried sediment collected, A_b is area of soil test plot and t is duration of rainfall simulation (hr). The sediment yield results for both bare soil and WHFM testbeds are summarized in Table 1. It is observed that soil treated with WFHM shows significant reducing trends of sediments yield compared to the bare soil testbed.

Table 1. Sediment yields for bare soil and WHFM testbeds

Rainfall duration (minute)	5	10	15	20	25	30
Bare soil (g/m².hr)	4352.16	2070.72	1115.52	677.40	551.09	200.32
Treated with fiber mat (g/m².hr)	931.20	464.40	214.72	156.54	99.74	46.60

3.2 Wash-off sediment volumes

Sediment volumes wash-off during the simulated rainfall were calculated for every five minutes interval up to 30 minutes by using equation (2) and the results are presented in Table 2. It was observed that bare soil testbed showed significantly faster response to rainfall as higher soil volumes were transported compared to the soil testbed treated by WHFM.

$$\text{Sediment volume} = S_m \times \text{soil bulk density} \quad (2)$$

The sediment volumes transported during the simulated rainfall for both soil testbeds are plotted in Figure 2. It shows that the sediment volume were decreasing over time for both soil testbeds, specifically greater soil volume was washed off at the bare soil testbed. However, the soil testbed covered with WHFM showed much lower sediment quantity as the fibre mat act as a protective layer for the soil slope. This proved that the WHFM feature of remarkable amount of hollow cavity does contribute to its water absorption and reducing the raindrop impact onto the soil surface layer.

Table 2: Sediment volume wash-off during simulated rainfall test for bare & treated soil testbeds

Rainfall duration (minute)	5	10	15	20	25	30
Bare soil (cm³)	134.33	127.82	103.29	83.63	85.04	37.10
Treated with fiber mat (cm³)	28.74	28.67	19.88	19.33	15.39	8.63

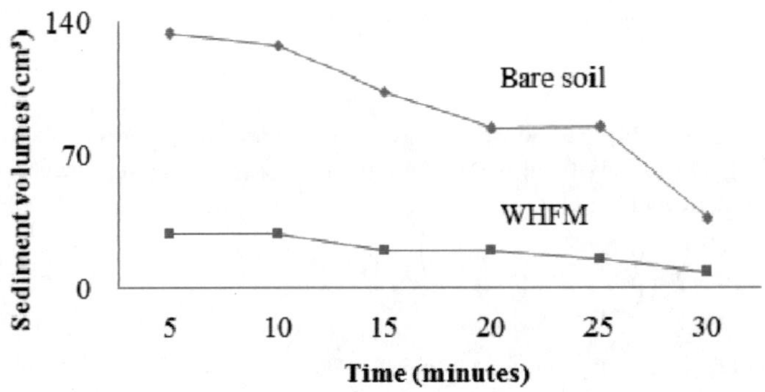

Fig. 2: Sediment volumes wash off during the simulated rainfall test for both soil testbeds

3.3 Soil loss reduction efficiency

The effectiveness of fiber mat in reducing the soil loss is determined by calculating the soil loss reduction efficiency (SLRE) using equation (3) as recommended by [12,13].

$$SLRE = \frac{SY_b - SY_g}{SY_b} \times 100 \quad (3)$$

SLRE is soil loss reduction efficiency (%), SY_t is sediment yield from bare soil (g/m².hr), SY_g is sediment yield from soil protected by fiber mat (g/m².hr).

The results of SLRE for soil testbed treated with WHFM at every 5 minutes during the simulated rainfall are tabulated in Table 3. The result revealed that WHFM has an average of 78.74% reduction efficiency for 30° slope.

Table 3: Soil loss reduction efficiency calculated for WHFM at every 5 minutes interval

Time (minute)	5	10	15	20	25	30
SLRE (%)	78.6	77.6	80.8	76.9	81.9	76.7

4. Conclusions

The performance of water hyacinth fibre mat (WHFM) as soil erosion control blanket has been investigated in this study using artificial rainfall. From the laboratory results, the sediment yields were significantly reduced when the soil testbed was treated with WHFM. The WHFM has reduced the impact of raindrops, acted as a shield between raindrops and soils by absorbing some amount of water. The soil loss had reduced by average of 78.74% when the 30 degree soil slope covered by WHFM compared to bare soil. This study has proved that WHFM is an effective and sustainable natural geotextile that can be implemented as soil erosion control for hilly slope condition.

Acknowledgements

The authors would like to thank the YCU grant (Project No. 201801001YCU/16) provided by Yayasan Canselor UNITEN (YCU) and Seed Fund provided by Tenaga Nasional Berhad (Project No. U-TG-RD-17-06).

References

[1] Chow M F *et al* 2015 *ARPN J. of Eng. and Appl. Sci.***10** 6419.
[2] Lekha K R 2004 *Geotext. Geomembr.* **22** 399.
[3] Zhou W *et al* 2009 *Bioresour. Technol.* **100** 5366.
[4] Tan L *et al* 2008 *Bioresour Technol.* **99** 4460.
[5] Chow M F and Yusop Z 2014 *Water Sci. Technol.* **69** 244.
[6] Chow M F *et al* 2013 *Water Sci. Technol.* **67** 1822.
[7] Haris H *et al* 2016 *IOP Conf. Ser.: Earth Environ. Sci.* **32** 012022.
[8] Chow M F *et al* 2016 *Environ. Earth Sci.* **75** 6.
[9] Chow M F *et al* 2017 *Water.* **9** 537.
[10] Artidteang S *et al* "Ground Improvement Case Histories: Compaction, Grouting and Geosynthetics" **25** 717.
[11] Javellonar R 2012 PhD Thesis, Central Luzon State University, Science City of Munoz, Nueva Ecija.
[12] Bhattacharyya R *et al* 2008 *Agric. Ecosyst. Environ.* **130** 50.
[13] Bhattacharyya R 2009 PhD Thesis, University of Wolverhampton, United Kingdom

IOP Conf. Series: Materials Science and Engineering **551** (2019) 012009 doi:10.1088/1757-899X/551/1/012009

Numerical Simulation on the Flame Propagation in Duct-Vented Gas Explosion: The Effect of Duct Length on the Secondary Explosion

Siti Zubaidah Sulaiman[1], Nurul Syazwana Noor Azmi[1], Azizan Ramli[1], Rafiziana Md Kasmani[2], Siti Kholijah Abdul Mudalip[1], Rohaida Che Man[1], Shalyda Md Shaarani[1], Zatul Iffah Mohd Arshad[1]

[1]Faculty of Chemical and Natural Resources Engineering, Universiti Malaysia Pahang, 26300 Kuantan, Pahang, Malaysia
[2]Faculty of Chemical Engineering & Energy Engineering, Universiti Teknologi Malaysia, 81310 UTM Johor Bahru, Malaysia

E-mail: szubaidah@ump.edu.my

Abstract. Duct venting is the most recognized explosion protection technique used in the industry. However, the key problem with this technique is the appropriate design of the duct size for an effective release material to prevent the secondary explosion that can cause further damage to the plant. Factors such as duct length and diameter have a significant effect on the duct venting mechanism that contributes to the second explosion. Thus in this study, the duct-vented gas explosion was simulated with the aim to investigate the influence of duct length focusing on the flame and pressure characteristics. From the simulation analysis, it shows that vessel interconnected with the duct with 0.05 m diameter and 0.5 m length experience a flame severe problem behavior to produce the secondary explosion effect.

1. Introduction

In theory, when an explosion event occurs in the vessel/pipe, the pressure will rapidly increase until it reaches the static pressure (vent panel pressure). This condition may cause the vent panel to rupture and allow the hot materials to be released from the vessel/pipes through vent duct. During the venting process, the unburnt gases (hot materials) flow at high velocity, causing it to choke. It then increases the amount of trapped unburnt gases inside the vessel and the duct [1]. The secondary explosions are the result of unburnt gases accumulation inside a duct re-ignited by the flame front. The secondary explosion mechanism inside the duct causes a backflow towards the main vessel and the interaction with the flame front will generate turbulent pressure wave and thus, enhancing the pressure inside the main vessel that eventually leads to detonation [1].

One of the major factors affecting flame and pressure characteristic that contributes to the strong explosion is a duct size [2]. Iida et al., 1985 has observed the flame and pressure characteristic in a smaller duct size (diameter: 5cm and length: 10 cm). They reported that fast flame or self-flame accelerations by the complex interaction between quenching effect and re-ignition process will lead to the violent secondary explosion [3]. Other studies supported the above hypotheses by using relatively narrow ducts (diameter: 5 cm and length: 10-20 cm) with a sharp vessel-duct area[4-6]. It was shown that the turbulent flame enhance the secondary explosion (i.e. with higher pressure amplitudes) during re-ignition [1]. Moreover, the effect of the duct size towards the secondary explosions was analyzed by the Baraldi et al., 2010 [7]. They reported that the secondary explosion occurred due to the reversal flow where the flame will travel backwards from duct to the main vessel, enhancing the burning rate by means of rapid pressure rise in the vessel and leads to the extensive damage[7].

From the above research on the vented gas explosion inside the duct that has been studied, most of the works are limited to smaller duct size with the length ranging from, 10-80 cm respectively, which does not represent the actual duct size in the industry. It should be noted that, beyond the limit, secondary explosion mechanism, as well as flame and pressure characteristics in the duct, will be different. Changing the duct length may change the flame behavior and hence, increasing the mass burning rate of spherical flames[7]. This would directly influence the flame and pressure characteristics in the duct venting. Therefore, in this study, the influence of duct length was studied thoroughly, as it has been recognized as factors contributing to the secondary explosion in a duct.

2. Experimental

The biogas flame propagation was simulated using ANSYS Fluent R16.2. The effect of the duct length towards the explosion pressure and flame speed for both vessels with and without duct were analyzed. The duct diameter was fixed at 0.05 m and three lengths;0 m, 0.5 m and 1.0 m were varied in this work. In order to simulate the venting process, the geometry is divided into two part (Part A and Part B) as shown in figure 1. Part A is a vessel, a place where the explosion occurred. Part B is a duct interconnected with the vessel. In this research, the biogas-air mixture was used at stoichiometric concentration for the explosion scenario.

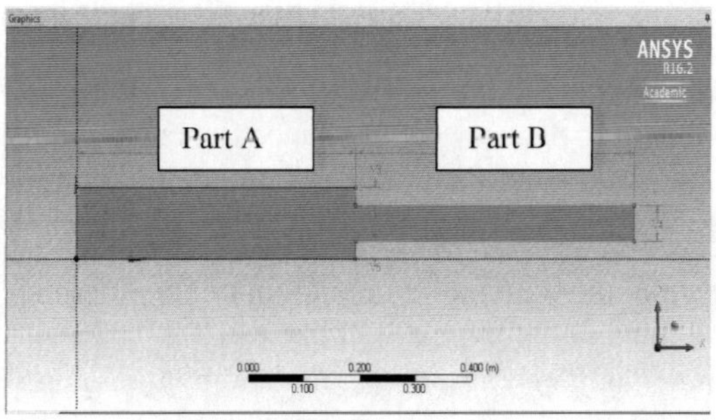

Figure 1. Geometry of the computational domain with surface body in Design Modeler.

The numerical model of premixed combustion was adopted in this study. Two-dimensional (2D) model was used and it was assumed that the turbulent flame is propagating under transient conditions. In this study, the premixed combustion model was used in which the fuel and oxidizer were mixed at the molecular level prior to ignition. The spark energy was applied to ignite the mixture. Carbon steel was used as the type of vessel wall. The operating condition was set up at the atmospheric condition and room temperature.

3. Results and Discussions

3.1. Effect of duct length on the pressure

Figure 2 shows the effect of three different duct lengths; 0m (ductless), 0.5m and 1.0m on the pressure. From the figure, it is clearly shown that the pressure trend was different for all ducts length. For instance, there are two pressure peaks were observed at duct length 0.5 m instead of 1 pressure peak at duct length 1 m, while the pressure in a ductless vessel just increases gradually without any distinct peak. To be noted, the 1st pressure peak was occurred due to the flame propagation into the duct. The second peak was suspected due to the secondary explosion. The overall pressure trend can be described based on the vented gas explosion mechanism (refer to figure 3). During the venting process, the unburnt gases (hot materials) flow from the vessel to the duct at high velocity, causing it to choke. It then increases the amount of trapped unburnt gases inside the vessel and the duct. When the flame propagated into the duct, it will ignite the unburnt gases leading to the external explosion [1]. This phenomenon caused to the increase of the pressure at all duct length as indicates in figure 2 (1st pressure peak).

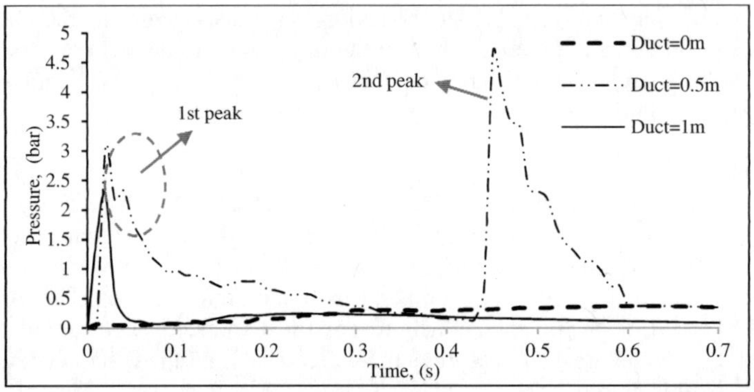

Figure 2. Pressure versus time for biogas-air mixtures at three different duct lengths.

However, the secondary pressure peak which was observed at duct length 0.5 m, was suspected due to the complex interaction between trapped unburnt gas-flame. During the external explosion, the pressure at the duct will be increased as compared to the vessel. At this instance, the flame will propagate back towards the main vessel and ignite with the trapped unburnt gas. This ignition will generate turbulent pressure wave and thus, enhancing the pressure inside the main vessel that eventually leads to secondary explosions. However, the second mechanism is not featured at duct length 1 m since there is only one pressure peak was recorded (see figure 2). The peculiar trend was suspected due to the heat loss to the duct wall. Increase the duct length from 0.5 m to 1.0 m leading to the increase of the heat loss to duct wall giving the intensity of the external explosion to be lesser which resulted to the weak reversal flow and no significant effect to the secondary explosion[1].

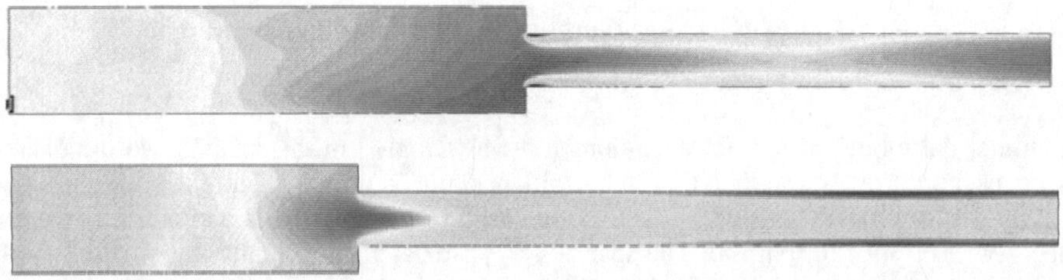

Figure 3. Flame propagation in duct venting: duct length 0.5 m (top) and 1 m (bottom).

3.2. Effect of duct length on flame speed

Figure 4 shows the flame speed trend in three different duct length ranged within 0 (ductless) to 1 m. It is clearly shown that the flame speed trend is comparable with all duct length. The flame speed increase before decreasing and later seed-up until reach to the highest flame speed at the third peak except for ductless. According to Zheng (2018), the flame acceleration is due to the flame stretch caused by the thermal expansion of the combustion product. Once the flame is stretch it will increase the flame surface area and cause to the increase of mass burning rate before. At this instance, flame starts to accelerate. However, the acceleration stops as the flame front touches to the sidewalls due to the quenching effect.

At this situation, the flame starts to stabilize, reduce the flame surface area and start to decelerate as represented by the first peak in figure 4. However, since the flame is kept propagating, it makes the flame surface area increase and start to re-accelerate until it reaches the highest speed before it starts to decrease as indicated in the second peak. The decreasing of the flame speed shows that the unburnt gas is completely burning.

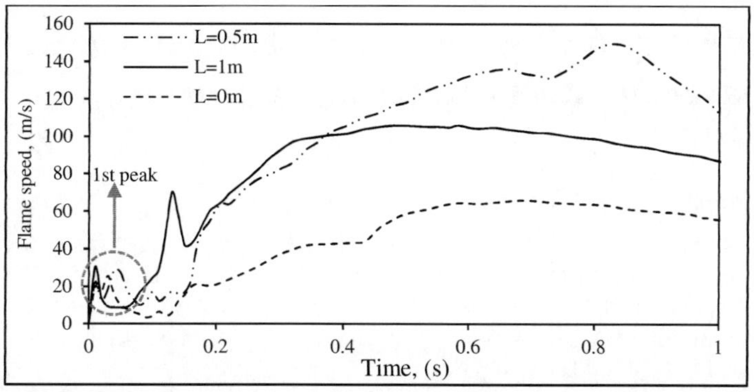

Figure 4. Flame speed versus time for biogas-air mixtures at three different duct lengths.

Furthermore, the highest flame speed was attained at the vessel connected with the 0.5 m length duct with the flame speed; 146.2 m/s followed by duct length 1 m (105.8 m/s) and 0 m ducts with the flame speed 37.43m/s respectively. It is clearly shown that, the shorter the duct length, the higher the flame speed. As mention earlier the quenching effect or heat loss to the wall play a major role on the flame propagation. At 1 m duct length, more heat losses to the wall cause the flame to quench. The quenching effect makes the turbulence flame to become more stable and then decelerates to about 28% as compared to the duct with 0.5 m. It is clearly shown that venting interconnected with the duct length 0.5 m giving worst case scenario which can result to the secondary explosion. and start to re-accelerate until it reaches the highest speed before it starts to decrease as indicated in the second peak. The decreasing of the flame speed shows that the unburnt gas is completely burning.

4. Conclusions

From the research, we can conclude several conclusions which are;

i. The complex interaction between trapped unburnt gas and flame is the main mechanism triggering to the secondary explosion incident.
ii. The quenching effect or heat loss to the wall play a major role on the turbulent flame which can lead to the secondary explosion effect.
iii. Vessel interconnected with the 0.05 m diameter and 0.5 m length duct experience the severe flame behavior that triggering to the secondary explosion.

References

[1] Zhang Q et al 2013 *Journal of Loss Prevention in the Process Industries* **26** 1230.
[2] Ponizy B et al 1999 *Combust Flame* **16** 259.
[3] Zhen Y et al 2018 *Journal of Fire Sciences* **36** 164.
[4] Kim N I et al 2006 *Combustion and Flame* **146** 283.
[5] Li J et al 2006 *Shock Waves* **16** 109.
[6] Salzano E et al 2002 *Journal of Hazardous Materials* **95** 233.
[7] Baraldi D et al 2010 *International Journal of Hydrogen Energy* **35** 12381.

IOP Conf. Series: Materials Science and Engineering **551** (2019) 012010 doi:10.1088/1757-899X/551/1/012010

Study of effective laying pattern of permeable interlocking concrete paver for storm-water management

M.F. Chow[1], J.W. Ng[2] and S.T. Chong[3]

[1]Sustainable Technology and Environment Group (STEG), Institute of Energy Infrastructure (IEI), Universiti Tenaga Nasional, Jalan IKRAM-UNITEN, 43000, Kajang, Selangor.
[2]Department of Civil Engineering, Universiti Tenaga Nasional, Jalan IKRAM-UNITEN, 43000, Kajang, Selangor.
[3]College of Engineering, Universiti Tenaga Nasional, Jalan IKRAM-UNITEN, 43000, Kajang, Selangor.

E-mail: mfchow@ikram.edu.my

Abstract. This study determined the skid resistance and runoff infiltration rate of different laying patterns of PICP, namely stretcher bond and herringbone patterns. The dimensions of PICP block are 10.3 cm (height) × 12.5 cm (width) × 25 cm (length) and arranged for a total coverage area of 1.8 m2. Pendulum skid resistance tester and ring method were used to determine the skid resistant and runoff infiltration rate, respectively. The herringbone pattern showed higher skid resistant value in either dry or wet condition compared to stretcher bone pattern. Herringbone pattern exhibited infiltration rate of 21450.7 mm/hr while stretcher bond pattern only has an infiltration rate of 18516 mm/hr. It is observed that herringbone pattern performed better than the stretcher pattern in term of storm-water management.

1. Introduction

Urban development is increasing around the world due to the growing human population. When the population of a city is increasing, more residential area and infrastructure is needed to cater the population, hence empty spaces are used for the construction of new buildings. Expansion of the existing structures to accommodate the growing amount of consumers can further causing the pervious surface to be reduced [1-4] for example the expanding of a single lane road into a double lane road.Over the decades, Malaysia have been struck by numerous floods with an increasing trend lately. [5, 6] proved that floods in Malaysia were mainly caused by inappropriate and insufficient drainage system and the researchers suggested that some of the precautions for flood mitigation was the use of flood preventive structures. Hence, permeable pavement plays an important role in stormwater management by infiltrating the surface runoff, especially in the urban area where lots of area are covered in impermeable surface. In addition, it act as a natural water filter by removing the suspended solids and pollutants [7, 8].

The common permeable pavements found around the world are porous asphalt (PA), permeable concrete (PC), permeable interlocking concrete pavers (PICP), plastic reinforcing grid pavers and concrete grid pavers (CGP) [9]. These permeable pavements are widely used in car park, access roads and driveway. Numerous of study have been done to compare the infiltration rate of PA, PC and PICP [9-11]. One of the studies showed that PICP reduced the most rainwater runoff with 46 % of

IOP Conf. Series: Materials Science and Engineering **551** (2019) 012010 doi:10.1088/1757-899X/551/1/012010

reduction, followed by PC with 35.6 % and PA with 12.7 % of runoff reduction [10]. However, some studies revealed the initial infiltration rate of PC was the highest among others [12]. Another study showed that PC was able to allow the rainwater to infiltrate the most, while PA was the second highest and last was PICP, yet the sequence shifted after two years of monitoring, PA provide the highest rate of infiltration and PICP provide the lowest rate [11]. By comparing the studies, it turned out that PC is not the best in terms of infiltration rate, the ranking is depending on many criteria for example slope, ratio of catchment area to the permeable area and maintenance [10]. Despite the different rate of infiltration, all of them performed much better than the impervious asphalt, and all three types of pavements were performed similarly during heavy rainfall (>50 mm) [12].

Most of the research studies proved that permeable pavements were perform excellent in surface water reduction, but very few studies investigated the skid resistance performance of permeable pavement. If the permeable pavement is used to construct a parking lot or driveway, the local authorities need sufficient information during the planning stage to allow them decide whether the use of porous pavement is foreseeable. But studies related to the effects of different laying patterns on PICP are limited, hence this study will determine the skid resistance and surface water infiltration rate of different laying pattern of PICP.

2. Methodology

2.1. Layout Patterns of PICP

Different types of layout pattern were reviewed based on previous studies in order to identify the best layout pattern in the study [13,14]. It was identified that the two common laying pattern of PICP are stretcher bond and herringbone patterns. The dimensions of PICP block are 10.3 cm (height) × 12.5 cm (width) × 25 cm (length). The PICP blocks were arranged in stretcher bond and herringbone patterns and filled with the fine aggregates.

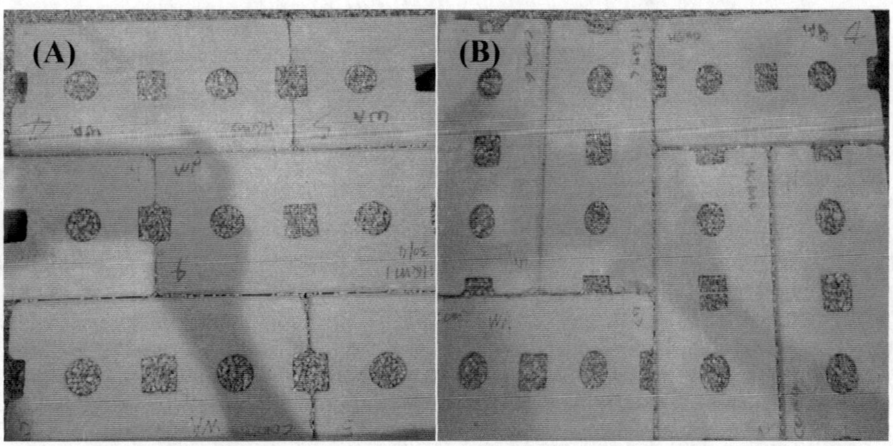

Figure 1. (A) Stretcher and (B) Herringbone bond patterns

2.2. Skid Resistance Test

Skid resistance test was conducted on the herringbone and the stretcher bond patterns as discussed in section 2.1. Both patterns were tested using the portable pendulum skid resistance tester in dry and wet conditions. The test was repeated three times to obtain the average skid resistance value of each pattern in both dry and wet conditions. Eq. 1 was used to calculate the average skid resistance value for each pattern. The average value of both herringbone and stretcher bond pattern for both dry and wet conditions were compared.

IOP Conf. Series: Materials Science and Engineering **551** (2019) 012010 doi:10.1088/1757-899X/551/1/012010

$$Average\ skid\ resistance = \frac{Total\ skid\ resistance\ value}{Number\ of\ samples} \tag{1}$$

2.3. Surface Infiltration Test

The ring method was used to study the runoff infiltration rate of the PICP. Small scale PICP pavement was constructed and tested for both herringbone and the stretcher bond patterns. The pavements were constructed together with sub base layers to simulate a small scale porous pavement. A rectangular aquarium was used to contain the sub base layer, base layer and concrete blocks during the experiment. The sub base layer was constructed using aggregates with sizes ranging from 25 mm to 4.75 mm. The aggregates were referred as #57 by the American Association of State Highway and Transportation Officials (AASHTO). Next, the base layer which was lay on top of the sub base was constructed using aggregates with sizes ranging from 9.5 mm to 2.36 mm (AASHTO categorized aggregates with this sizes as #8). Besides, the #8 aggregates were used to fill up the hole on the concrete blocks.

$$Infiltration\ rate = \frac{KM}{D^2T} \tag{2}$$

where

K = 4583666000 (constant, used to convert into mm/hr);
M = mass of water (kg);
D = inner diameter of ring (mm);
t = infiltration time (s)

$$Average\ Infiltration\ rate = \frac{Total\ infiltration\ rate}{No.of\ samples} \tag{3}$$

Eq. 2 was used to calculate the infiltration rate of each sample. An average infiltration rate was calculated using Eq 3 to represent the performance of the particular pattern and compared with each other afterward.

3. Results and Discussions

3.1. Skid Resistance

The skid resistant results for different laying pattern of PICP in both dry and wet conditions are presented in table 1. The average skid resistance values in dry and wet conditions were apparently similar. From the results showed in Table 1, the herringbone pattern showed higher average value in either dry or wet condition compared to stretcher bone pattern. Although the difference was not significant, but it showed that herringbone pattern was a better choice in term safety.

Table 1. Average skid resistance for Stretcher and Herringbone patterns in dry and wet conditions.

Pattern	Condition	
	Dry	Wet
Stretcher bond	15.7	15
Herringbone	20.2	22.2

3.2. Runoff Infiltration Rate

The results collected from the runoff infiltration test are summarized in Table 2. It is observed that the herringbone pattern performed better than stretcher-bond pattern in term of infiltration rate. Herringbone pattern exhibited infiltration rate of 21450.7 mm/hr while stretcher bond pattern only has an infiltration rate of 18516 mm/hr. This indicated that herringbone pattern was capable to infiltrate more runoff during rainfall event.

Table 2. Infiltration rate between stretcher bond and herringbone patterns.

Sample	**Infiltration rate (mm/hr)**
Stretcher bond	18516
Herringbone	21450.7

Previous study in Sweden showed that interlocking concrete paver (ICP) yield a infiltration rate of 18780 mm/hr [15]. The results reported by [15] was lower than the herringbone pattern's infiltration rate in this study. However, this might due to the different sizes of aggregates used and different thickness of bedding layer. The base course layer in the study by [15] is 50 mm thick, whereas the base course of herringbone is 30 mm thick. The sizes of aggregates used to construct the ICP are 4 mm to 8 mm for base course and 0 mm to 120 mm for sub base course, respectively. Meanwhile, the herringbone sample used 2.36 mm to 9.5 mm for base course and 4.75mm to 25 mm for sub base course.

Furthermore, a study by [16] indicated PICP provided a higher infiltration rate compared to CGP though the CGP have a higher percentage of open area. The PICP have a mean infiltration rate of 2000 cm/hr while the maintained CGP have a mean infiltration rate of 11.8 cm/hr. By comparing both findings by [16] and this study, PICP showed an infiltration rate of above 20000 mm/hr. This proved that PICP is capable of reducing peak runoff during rainfall. Belgard [17] in North America has tested the infiltration rate for PICP. According to their results, the highest infiltration rate was 2544 in/hr or 64617.6 mm/hr while the lowest infiltration rate was 785 in/hr or 19939 mm/hr. The lowest infiltration rate of 19939 mm/hr is comparable with the herringbone's infiltration rate results in this study.

4. Conclusions

The results from both experiments indicated that herringbone pattern performed better than the stretcher pattern. It provides a slightly higher skid resistance value together with a greater infiltration rate, the latter is important as heavy rainfall happened often in Malaysia, so adequate infiltration rate is one of the main consideration if the permeable pavements are to be constructed. On the other hand, the slightly higher skid resistance value of herringbone pattern might not be a significant finding if it is used for roadways construction, but it is useful for parking lots where the vehicles are travelling in low speed.

5. Acknowledgements

The authors would like to thank the YCU grant (Project No. 201801001YCU/16) provided by Yayasan Canselor UNITEN (YCU) and internal research grant (UNIIG 2017) (Project No. J510050716) provided by Universiti Tenaga Nasional.

References

[1] Haris H *et al* 2016 IOP Conference Series: Earth and Environmental Science **32** 12022.
[2] Chow M F *et al* 2014 *Wat Sci Tech* **69** 244.
[3] Chow M F *et al* 2013 *Wat Sci Tech* **67** 1822.
[4] Chow M F *et al* 2017 *Journal of Engineering and Applied Sciences* **12** 5379.
[5] Sani G *et al* 2014 *International Journal of Interdisciplinary Research and Innovations* **2** 59.

IOP Conf. Series: Materials Science and Engineering **551** (2019) 012010 doi:10.1088/1757-899X/551/1/012010

[6] Syed Muzzamil H S *et al* 2017 *Indian Journal of Science and Technology* **10** 2.
[7] Chow M F *et al* 2015 *Urban Water* **12** 240.
[8] Chow M F *et al* 2014 *Environ. Earth Sci.* **10** 4047.
[9] Huang J E *et al* 2016 *J. Environ. Eng.* **6** 142.
[10] Alyaseri I *et al* 2015 International Low Impact Development Conference **1** 1.
[11] Jalil M J *et al* 2014 *Adv. Mater. Res.* **906** 125.
[12] Jalil M J *et al* 2018 *Chem.Chem.Technol.* **12** 296.
[13] Rafizan A *et al* 2014 *J. Appl. Sci. Agric.* **9** 86.
[14] Koay G F L *et al* 2006 International Journal of Engineering and Technology **3** 115.
[15] Milchert E *et al* 2008 *Czasopismo Techniczne. Chemia* **105** 283.
[16] Goud V V *et al* 2007 *Chem. Eng. Sci.* **62** 4065.
[17] Piazza G J Nuñez *et al* 2003 Journal of the American Oil Chemists' Society **80** 901.

Evaluation of particle removal efficiency in sedimentation tank incorporating bottom grid structure

M.F. Chow[1], A. S. Ruzaiman[2], F. A. Rashid[2] and S.T. Chong[3]

[1]Sustainable Technology and Environment Group (STEG), Institute of Energy Infrastructure (IEI), Universiti Tenaga Nasional, Jalan IKRAM-UNITEN, 43000, Kajang, Selangor.
[2]Department of Civil Engineering, Universiti Tenaga Nasional, Jalan IKRAM-UNITEN, 43000, Kajang, Selangor.
[3]College of Engineering, Universiti Tenaga Nasional, Jalan IKRAM-UNITEN, 43000, Kajang, Selangor.

Abstract. A series of comparative experiments, with and without BGS, were carried out in a 2mx1mx1m tank. Samples used were taken from highway runoff and the BGS tested was in 1.3cmx1.5 cm size. The experimental results indicated that the particle removal efficiencies with the application of BGS were about 6% higher than when tested without the BGS for flow rates ranges from 0.2 l/s to 0.5 l/s. SPLIT was developed in the laboratory to test the effectiveness of the system to retain and treat stormwater particles. A series of experiments were conducted with different inflow rates and particle concentrations. The results showed that SPLIT system was able to remove >80% of the particles (measured as TSS concentration) from the laboratory-simulated stormwater runoff. The particle size analysis also demonstrated that reductions in the d50 of the samples in the laboratory testing from 300 μm (influent) to approximately 50 μm (effluent). Reduction in other pollutants that associate with particles is depending on their concentration association by particle size. The laboratory testing results had proved that SPLIT system was able to retain and remove the particles and pollutants effectively under various inflow rates.

1. Introduction

Urban stormwater runoff carries large loads of particles that are considered a major pollutant with deleterious effects on receiving water bodies[1-3]. Particles washed-off from paved areas contain various sorbed pollutants, and much of the pollutant load associated with stormwater runoff is carried by sediment. These sediment and sediment-associated contaminats are a substantial cause of degradation to receiving waters with associated toxic effects to aquatic organisms [4, 5]. Various treatment methods have been introduced to remove the suspended and colloidal particles, such as detention basins, sedimentation tanks, ponds, wetlands, biofiltration, such as grassy swales and strips, vortex or swirl concentrators, slow sand filters, multi chambered treatment trains and silt traps [6]. Removal of these sediments can therefore result in the removal of a large portion of associated pollutants. The sediments and associated pollutants captured in the pretreatment devices can be removed for disposal during maintenance operations. Sediment removal efficiency of tanks is one of the most informative parameters for measuring the treatment efficiency performance of such facilities, and it is affected by many factors such as influent pollutant concentrations, runoff magnitude, particle size, settling velocity and facility size. Removal efficiency is usually greater for higher influent pollutant concentrations. Because pollutant concentrations tend to decrease as runoff progresses, enhancing initial runoff (i.e., first flush) treatment can improve overall performance of a treatment facility [7-9].

The lack and/or cost of the land dictate that retention ponds are generally small. With short retention times and high inflow rates, it is challenging to obtain high removal efficiencies by gravity settling only, because of large undesirable disturbances by inflow generated turbulence. Turbulence, eddies and diffusion will re-suspend and move the particles back up into the water column. The efficiency of discrete particle settling can be improved by increasing the contact area between the particles and the settling basin surface by inserting lamella plates or settler tubes of different shapes [10,11]. Potential challenges in applying lamellar plate technology to enhancing particle settling in open-water bodies, such as stormwater ponds, are the needs to prevent sediment resuspension and clean out the settled particles. Thus, it is necessary to develop a new technology enhancing the removal efficiency of solids under high inflow rates and reducing scouring of settled particles. In gravity settling processes, turbulence is generally considered as an adverse factor disturbing particle settling and causing sediment resuspension. Past studies have shown that near the boundary vertical vortices would increase suspended particle flux toward the boundary, but a critical issue remains: how to practically generate controlled vortices serving to enhance sedimentation and keep the approaching particles stay on the boundary (bottom bed) without resuspension.

In order to improve the efficiencies of the sedimentation tank, the issues are explored by introducing a bottom grid structure (BGS) which is designed to convey the trapped particles into a protected location and a number of baffles are generated to scatter the kinetic energy of incoming flow. This paper is aimed to determine the efficiency of particle removal in sedimentation tank with and without the presence of proposed BGS.

2. Methodology

2.1 Sediment Particle & Litter Interceptor Tank (SPLIT)

The sedimentation tank (SPLIT) proposed in this study has a dimension of 2m (length) x 1m (width) x 1m (depth) as shown in the Figure 1. The SPLIT system comprises an inlet pipe, a circular sedimentation tank with a bottom grid structure, two settling chambers, two separators and an outlet pipe, wherein the circular tank (2) with bottom grid structure (3) at the bottom layer above the collecting cone (4), in which the circular tank is connected with an inlet pipe (1) and having an overflow opening (5), the two settling chambers (6) with bottom grid structures are separated by two baffle walls (7) in between, and the last settling chamber having an outlet pipe (8) to discharge the water.

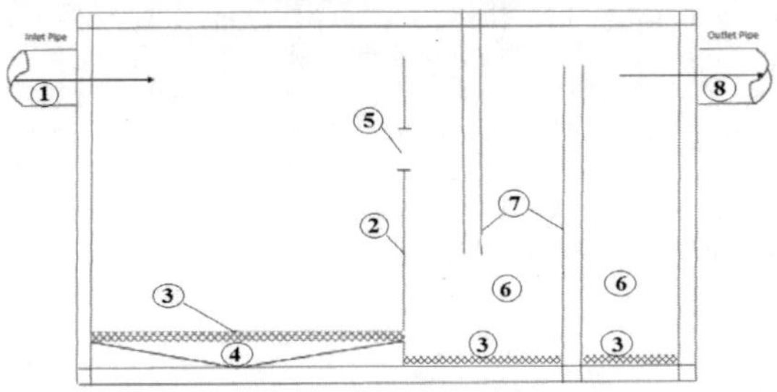

Figure 1. Side view of SPLIT system.

2.2 Laboratory experiment

A schematic diagram of the experimental setup is shown in Figure 2. During experiments, particles will be released from top of the inlet pipe, about 40 cm upstream of the settling tank inlet mouth. In such an arrangement, particles should be well mixed by the incoming fast turbulent flows. After passing through the settling tank, the flow exited through a 0.5 (W) × 0.1-m (D) opening located at the far end of the tank at 0.7 m above the tank bed. Before returning the flow into a large water supply tank, it was filtered through a 63-μm fine screen to prevent particles from returning into the circulation system. The flow was then pumped back to the settling tank via a 51-mm pipe. Between the pump and the inlet, a flow meter and a control valve will be installed, the latter serving to regulate flow rates. The particle samples will be sorted using sieves with mesh sizes of 300, 212, 150, 75, and 63 μm. Any particles smaller than 50 μm will be discarded to prevent the risk of their passing through the capture mesh. The sieved particles will then mixed together again with the weight ratio of 0.2 each. In general, the total 200g sample was used in each experiment, which would meet the experimental design for injecting particles continually for 10 min. The duration of each individual experiment was 15 min, which is much longer than the particle residence time in the test tank, and should contribute to obtaining consistent particle removal rates in the settling tank. As shown in Figure 2, a mixer is inserted near the bottom of the solids feeding pipe to make sure that the injected particles are well blended with the incoming flow in the inlet pipe. After the experiment is finished, the mixture of the retained particles and the captured flow will be drained through a bottom drain and the particles will be captured by a screen located below the exit of the drain pipe. The captured particles will be oven dried overnight at a temperature of 40°C (to prevent burning) and weighed to determine their total weight. The tests were carried out for many experiments under different flow rates of 0.2, L/S, 0.3 L/s, 0.35 L/s, 0.4 L/s and 0.5 L/s.

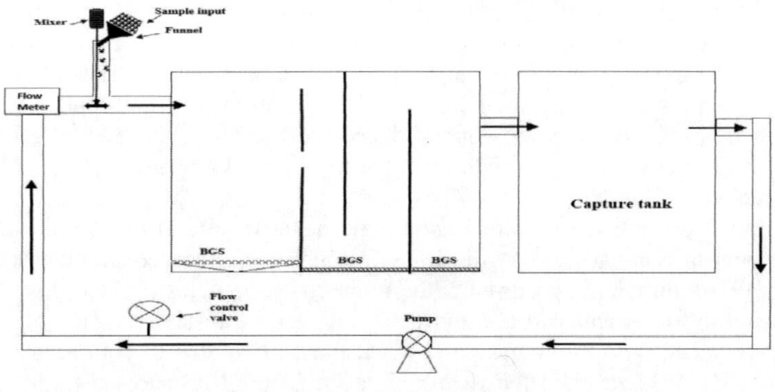

Figure 2. Experimental setup of SPLIT system in the laboratory.

3. Results and Discussion

The results of sediment removal percentage for sedimentation tank with and without BGS are presented in table 1. The efficiencies of removal rate range was between 98.21% to 99.52 % and 96.0% to 99.35% for sedimentation tank with and without BGS, respectively. The sediment removal percentages are also plotted for different flow rates as shown in Figure 3. It shows clearly that when the inflow rate was low, the tested BGS seems not having a large affects on the settling activities. This was due to particle settling behaviour itself was related to gravity, in which the particles could settle down naturally when the inflow rate was low. However, as the inflow rate increases, the BGS started to give better performance in term of trapping and preventing the re-suspension of sediment. The improvement in the sediment removal percentage indicates the efficiencies of the BGS to retain the settled particles in the tank.

Table 1. Sediment removal percentages under different flow rates with and without BGS.

Flow rate (L/s)	0.2	0.3	0.35	0.4	0.5
BGS	99.52	99.57	99.01	98.72	98.21
Without BGS	99.35	98.15	98.01	97.24	96

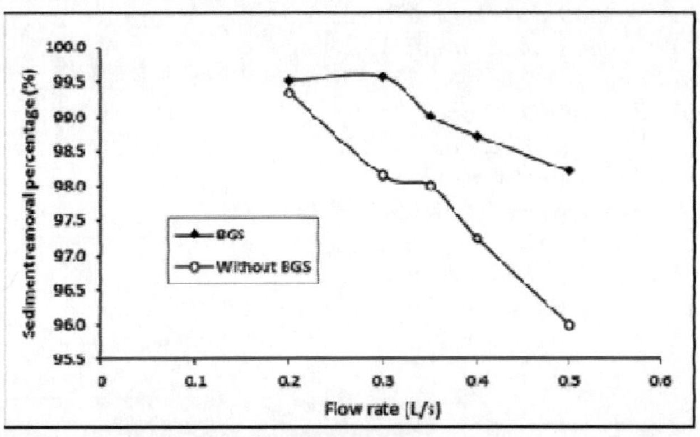

Figure 3. Sediment removal percentages against different flow rates for tank with BGS and without BGS system.

The efficiencies of the BGS started to decline at inflow rate 0.35 l/s. It was possibly due to the vigorous flows restrained the particles from being shifted into and retained in the BGS cells. For very strong turbulent flow in experiment without BGS, the by-passed particles were found in a large amount compared to other experimental with BGS cells. This shows that these particles not able to settle on the settling tank. It was expected because the increased turbulent would generate a negative influence on the settling behaviour. The BGS was designed to capture sediment and prevent re-suspension when dealing with rapid flows. This was likely occur in stormwater ponds at its sediment fore bays, where the flow cross-section area was rather small and the flow velocities were high in most cases. Going deep into the pond, where the flow area increase to a great extent, the flow velocities would reduced, thus resulting the BGS effectiveness to deplete. However, the fast flowing velocities at the sediment fore bays induced positive effects of the vortices that would shift more particles into the cells.

The cell walls provide protection that acted against the negative effects of bottom turbulent disruption which increases the bottom roughness when the BGS was presence, thus consequences in greater rate of removal. Even though extra turbulence occurred at the top surface, particles could still settle on the tank bed in a much larger area due to vertical vortices and cell protection. Particles retained easier when the flow speed changed from very fast flow to a much calm zone, most likely due to optimised combination of the vortices and cell protection. However, if the flow was too high until it causes disturbances at the bottom flows, all particles would be flushed away. As clarified before, where almost all particles could settle down and stay on the bed naturally at slowest velocity discharge, it was because of the same positive and negative effects were exhibited on the particle settling and its negative turbulence could possibly cause a larger influences on the effectiveness of the BGS.

4. Conclusions

The results indicated that the particle removal efficiencies with the application of BGS were about 6% higher than that tested without the BGS for flow rates ranges from 0.2 l/s to 0.5 l/s. The combination of the induced vortices and the preventative BGS cells resulting in better performance of the cells especially when inflow discharge was high. However, the BGS did not appear to have obvious effects in the weak flows since the turbulence produced were not strong that allows the particles to settle to the bottom naturally. The uses of BGS under high flow rates were proved to efficiently enhance the sedimentation and reduce resuspension possibilities of the earlier settled sediment. This would indicate a significant implications for a storm water management operation.

Acknowledgements

The authors would like to thank YCU grant (Project No. 201801001YCU/16) provided by Yayasan Canselor UNITEN (YCU) and Seed Fund provided by Tenaga Nasional Berhad (Project No. U-TG-RD-17-06).

References

[1] M F Chow *et al* 2014 *Water Sci. Technol.* **69** 244.
[2] M F Chow *et al* 2013 *Water Sci. Technol.* **67** 1822.
[3] Chow M F *et al* 2015 *Urban Water Journal.* **12** 240.
[4] Chow M F *et al* 2017 *Eng. & Applied Sci.* **12** 5379.
[5] M F Chow *et al* 2016 *Environ. Earth Sci.* **75** 6.
[6] Aldheimer *et al* 2003 *Water Sci. Technol.* **48** 113.
[7] Bertrand Krajewski J L *et al* 1998 *Water Res.* **32** 2341.
[8] Chow M *et al* 2014 *Environmental Earth Sciences* **72** 4047.
[9] Chow M F *et al* 2017 *Journal of Civil Engineering* **21** 1087.
[10] A M Razmi, *et al* 2008 *J.Appl.Fluid Mech.* **2** 1.
[11] Mahdi Shahrokhi *et al* 2012 *Applied Mathematical Modelling* **36** 3725.

Investigation of tribological properties of lithium grease with SiO2 nanoparticles

Adam **Rylski**[1] **and** *Krzysztof* **Siczek**[2]

[1]Lodz University of Technology, Institute of Materials Science and Engineering, 90-357 Lodz, Stefanowskiego Str. 1/15, Poland

[2]Lodz University of Technology, Department of Vehicles and Fundamentals of Machine Design, 90-357 Lodz, Stefanowskiego Str. 1/15, Poland

Abstract. The aim of the work was to investigate the effect of the SiO_2 nanoparticles concentration in the lithium grease and of contact pressure on the spinning friction coefficient in the contact zone between a steel plate and a ball. The invvestigations were carried out using the a tribotester for the spinning friction torque in the contact between the ball and the plate. Ball rotated relative to the plate axis at a constant speed and was loaded by the a force increasing in a stepwise manner for the consecutive series. The contact zone was lubricated with a pure lithium grease - pure or with the grease with suspended SiO_2 nanoparticles. The friction coefficient decreased with the increase of contact pressure for the pure lithium grease. For the concentration μ = 6wt.% of SiO_2 nanoparticles in the grease friction coefficient decreased and for the other concentrations it increased with the increase of the contact pressure. If the values of average contact pressure were below 550 MPa, the friction coefficient values for pure lithium grease were higher than those for lithium grease with suspended SiO_2 nanoparticles. For higher values of the contact pressure, the values of the friction coefficient for the pure lithium grease could be lower, equal or higher than those for lithium grease containing SiO_2 nanoparticles depending on the value of SiO_2 the concentration.

1. Introduction

The use of nanoparticles as a lubricant additive for various devices is more and more popular [1]. Nanoparticle additives can reduce friction and wear due to different mechanisms. Such mechanisms can be linked to direct and indirect effects of nanoparticle additives on the friction and wear processes [2]. Studies described in [3–5] show that nanoparticle additives can decrease friction and wear in ball bearings. Further studies [6–8] show that nanoparticle additives can also form a protective film over friction surfaces. Liu et al. [9] describe the formation of nanoparticle additive depositions on mating surfaces which compensate the loss of material. Nanoparticle additives can also decrease the roughness of mating surfaces due to their abrasive action [10]. Nair et al. [11] report a study on the effect of nanoparticle additives on journal bearings using the finite element method.

Shenoy et al. [12] study the effects of nanoparticle lubricant additives on the performance of externally adjustable fluid film bearings. The studies described in [11, 12] point out an increase in the load-bearing capacity and stability of journal bearings using a nanoparticle-dispersed lubricant. Binu et al. [13] describe a novel method for evaluating the load-bearing capacity of journal bearings operating with lubricants containing nanoparticle additives. It was found that variations in lubricant shear viscosity due to the nanoparticle additives can be simulated accurately using a modified Krieger-Dougherty viscosity model. According to Nam et al. [14] the most popular method for preparation of nano-grease is the 'Direct mixing method'. In this method, the nano-particles are directly mixed with grease under heavy mechanical stirring. Rashed & Nabhan [15] investigated tribological behaviour of mineral 20W-50 and semi-synthetic 15W-50 oil dispersing by different amounts (0.5 and 1.0%wt. oil) of SiO_2 nanoparticles. Experiments were carried out using tribometer test-rig under normal load at different temperature values from range 40 °C to 100 °C. The addition of SiO_2 to engine oils had not reduced the friction coefficient to a great extent. Li et al. [16] dispersed ultrasonically 0.3 wt.% of SiO_2 nanoparticles in ST5W/30 mobile oil. Such nanolubricant was studied using a reciprocating tribotester and a four-ball tribotester. As a result the frictional coefficient

IOP Conf. Series: Materials Science and Engineering **551** (2019) 012012 doi:10.1088/1757-899X/551/1/012012

decreased in relative to pure ST5W/30 gas mobile oil. Peng et al. [17] found that the SiO_2 nanoparticles oleic dispersed in acid exhibited the same excellent tribological behaviour as diamond nanoparticles in liquid paraffin. During friction the micro-grooves are formed and then more spherical nanoparticles roll into the contact area like tiny bearing, balls which reduce the sliding friction. Rawat et al. [18] studied frictional behaviour of a paraffin grease with dispersed SiO_2 nanoparticles. The paraffin grease was developed using paraffin oil as a base oil and 12-lithium hydroxy stearate as a thickener. The concentration of the thickener was of 14 wt.%. The SiO_2 nanoparticles were synthesized by a modified sol–gel method and dispersed in paraffin grease by the in-situ method. The frictional characteristics of the paraffin grease with SiO_2 admixture were studied with use of a four-ball tester. The addition of SiO_2 nanoparticles to the paraffin grease enhanced its tribological performance as compared to the pure paraffin grease. The maximum reduction of the friction coefficient was ~ 20% at concentration of 0.03% wt.%. He et al. [19] prepared SiO_2 nanoparticles encapsulated in lithium grease. Friction tests were conducted with use of a a four-ball friction tester for the pure lithium grease and for the one containing SiO_2 nanoparticles. The addition of the SiO_2 nanoparticles to the grease reduced the friction coefficient of the base grease. When the content of the SiO_2 nanoparticles in the grease was of 0.3 wt.%, the value of the friction coefficient decreased by 26% compared with that for the base grease. For the contentof the SiO_2 nanoparticles 0.3 wt.% the grease exhibited the lowest average friction coefficient at the load of 342 N, and the value of the friction coefficient decreased by 39% as compared with that for the base grease. The aim of work was to investigate the effect of the SiO_2 nanoparticles concentration in the lithium grease and of the contact pressure on the spinning friction coefficient in the contact zone between steel plate and ball.

2. Materials and methods

2.1 Samples preparation

Steel balls rotating against fixed steel plate counterparts were used. Two steel balls with diameters of 14 and 8 mm with polished surface were used. Each ball rotated with a constant speed and was loaded by the tribotester head with a constant force. The mating counterpart contained the steel plane plate immersed in an epoxy resin and chemically cured in the steel ring. The contact zone was lubricated with lithium grease with or without the addition of 2, 4, 6, 8 and 10 wt.% of SiO_2 nanoparticles of a size 10–20 nm and delivered by SIGMA-ALDRICH [20]. The grease was mixed with the SiO_2 commercial product using the High-Speed Planetary Mixer THINKY SR-5000. The grease was then treated with ultrasonic waves in the GT Sonic VGT-800 Ultrasonic Cleaner for only 2 min., to avoid agglomeration or the structural break of the nanoparticles.

2.2 Investigation of the friction coefficient

The friction coefficient in the contact zone between the rotating steel ball and the steel plate lubricated with the lithium grease containing different content of SiO_2 nanoparticles was estimated using the available tribotester [20] (Figure 1). This tester allowed the measurement of the spinning friction torque in the contact zone between ball (4) and its counterpart (5), on the base of the twisting angle of the string (7) supported by the aerostatic bearing in the tabble 9. The twisting angle was read on theshield (8). Measurements were carried out at a constant rotating speed of the shaft 1 equal to 36 rpm and at a constant load of the set of weights 3 increasing in a constant step manner for the consecutive series from 7.2 N to 27.2 N. The friction coefficient was estimated using the modelled contact zone between steel ball (4) and steel plane plate (5). The courses of the friction torque values against an average contact pressure were obtained for the 0, 2, 4, 6, 8 and 10 wt.% of SiO_2 nanoparticles in the lithium grease. The averaged contact pressure p_{aver} and radius a of the contact zone were estimated using equations (1) and (2) [21].

$$p_{aver} = F/(\pi a^2) \tag{1}$$

$$a = [3Fr(1-\nu^2)/E]^{1/3} \tag{2}$$

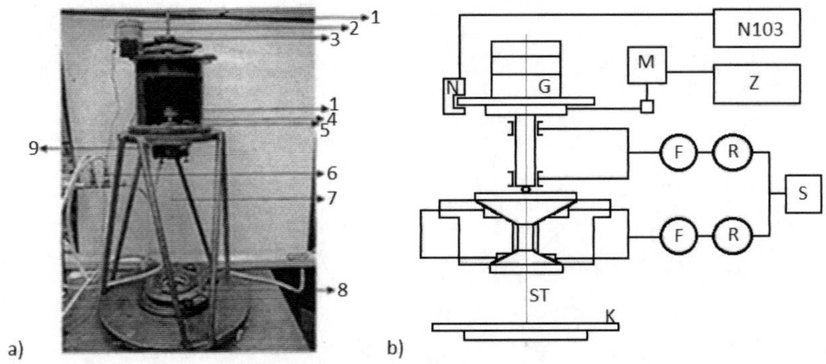

Figure 1. The ball-on-plate tribotester for measurement of the spinning friction torque; a) photo of the tester. 1 - shaft, 2 - electric motor, 3 - weights, 4 - ball, 5 - plane steel counterpart, 6 – outlet of compressed air, 7 - string, 8 - shield for measurement of twisting angle, 9 - table with aerostatic bearing [22]; b) Scheme of the tester. F - filter, G - weight, M - electric motor, N - tachometer, K - shield for measurement of twisting angle, S - compressor, ST – string, R - reducer, Z - power supply.

3. Results

Results of measurements of the friction coefficient an against average contact pressure for the ball radius r = 7 mm and pure lithium grease are given in Figure 2. The value of the friction coefficient decreased from 0.25 to 0.07 with the increase of contact pressure. The courses of the friction coefficient against average contact pressure for the ball radius r = 7 mm at different concentrations of SiO_2 nanoparticles in the lithium grease are shown in Figure 3. The value of the friction coefficient decreased from 0.16 to 0.116 with the increase of the contact pressure for the concentration of SiO_2 nanoparticles ε = 6wt.%. For the other SiO_2 concentrations the friction coefficient was increasing with increasing the contact pressure, from 0,073-0,08 for ε = 2wt.%, from 0.072-0.094 for ε = 4wt.%, from 0.086-0.1 for ε = 8wt.%, and from 0.081-0.096 for ε =10wt.%, respectively. The courses of the values of the friction coefficient against an average contact pressure for the ball radius r = 4 mm and different concentrations of SiO_2 nanoparticles in the lithium grease are shown in Fig. 4. The value of the friction coefficient decreased from 0.16 to 0.116 with the increase of the value of the contact pressure for the concentration ε = 6wt.%. For the other concentrations the friction coefficient increased with contact pressure, from 0.067-0.117 for ε = 2wt.%, from 0.091-0.1 for ε = 4wt.%, from 0.073-0.093 for ε = 8wt.%, and from 0.1-0.11 for ε = 10wt.%, respectively.

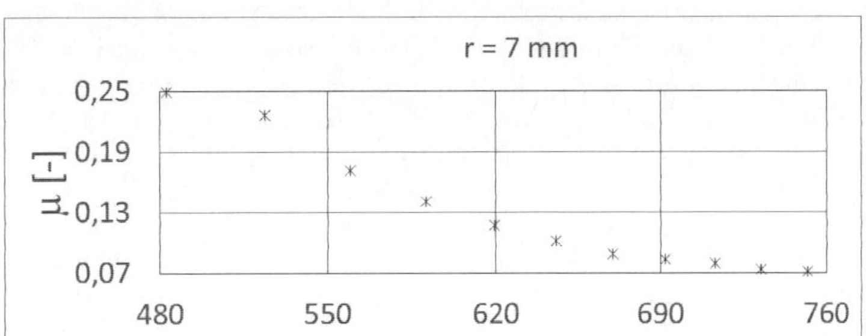

Figure 2. Plot of the the values of the friction coefficient against average contact pressure for the ball radius r = 7 mm and concentrations of SiO_2 nanoparticles in the lithium grease of 2, 4, 6, 8 and 10wt.%.

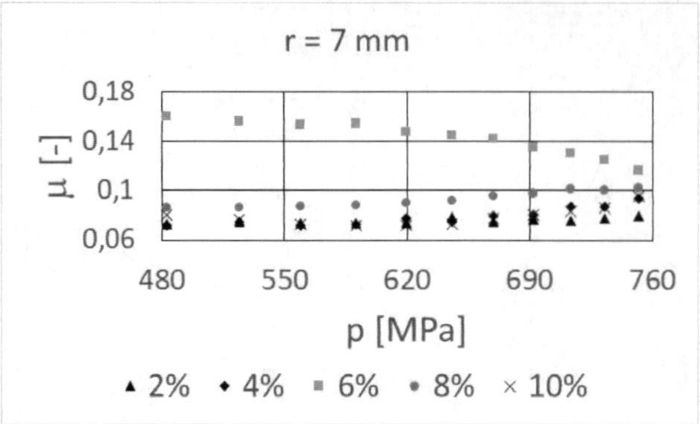

Figure 3. Plot of the values of the friction coefficient against an average contact pressure for the ball radius r = 7 mm and concentrations of SiO_2 nanoparticles in the lithium grease of 2, 4, 6, 8 and 10wt.%.

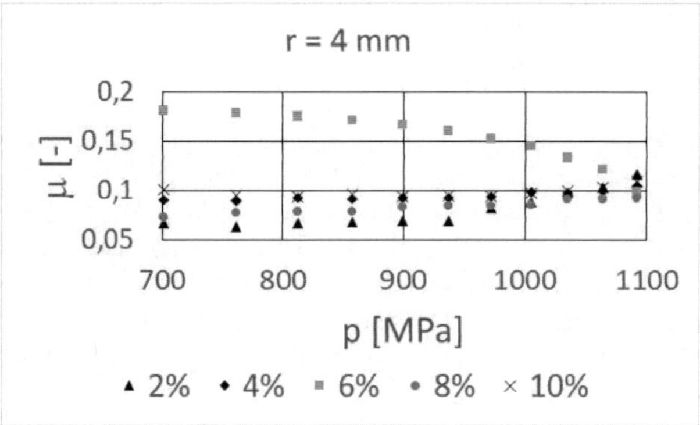

Figure 4. Plot of the values of the friction coefficient against an average contact pressure for the ball radius r = 4 mm and concentrations of the SiO_2 nanoparticles in the lithium grease of 2, 4, 6, 8 and 10wt.%.

4. Summary

The value of the friction coefficient value decreased with the increase of an average contact pressure for the pure lithium grease used as lubricant. The addition of SiO_2 nanoparticles had a different effect depending on the SiO_2 nanoparticle concentration ε in the lithium grease and on the range of the value of the contact pressure. For the case of ε = 6wt.%, the value of the friction coefficient decreased and for the other SiO_2 concentrations it increased with the increase of the value of the contact pressure. If the values of the contact pressure p_{aver} were below 550 MPa, the values of the friction coefficient for pure lithium grease were higher than those for lithium grease with admixture of SiO_2 nanoparticles. For higher values of the contact pressure p_{aver}, the values of the friction coefficient values for pure lithium grease could be lower or higher than those for lithium grease with admixture of SiO_2 nanoparticles depending on the concentration of the latter. This result is in agreement with the results reported in [22].

References

[1] Z J Zhang *et al* 2014 *Lubricants* **2** 44.
[2] K Lee *et al* 2009 *Tribol. Lett.* **35** 127.
[3] L Rapoport *et al* 2002 *Ind. Lubr. Tribol.* **54** 171.
[4] F Chinas Castillo *et al* 2003 *J. Tribol. T ASME* **125** 552.
[5] Y Y Wu *et al* 2007 *Wear* **262** 819.
[6] Z S Hu *et al* 2002 *Wear* **252** 370.
[7] B M Ginzburg *et al* 2002 *Russ. J. Appl. Chem.* **75** 1330.
[8] Z Xiaodong *et al* 2007 *Lubr. Sci.* **19** 71.

[9] G Liu *et al* 2009 *Tribol. Lett.* **17** 961.
[10] X Tao *et al* 2009 *J. Phys. D Appl. Phys.* **29** 2932.
[11] K P Nair *et al* 2009 *IJNP* **2** 251.
[12] B S Shenoy *et al* 2012 *Tribol. Int.* **45** 38.
[13] K G Binu *et al* 2014 *Procedia Mater. Sci.* **6** 1051.
[14] S R Nam *et al* 2013 *Energy* **60** 195.
[15] A Rashed *et al* 2018 *Rome Italy Oct* **20** 30.
[16] X Li, *et al* 2006 *Appl. Surf. Sci.* **252** 7856.
[17] D X Peng *et al* 2009 *Tribol. Int.* **42** 911.
[18] S S Rawat *et al* 2018 *Appl. Nanosci.* **10** 018.
[19] Q He *et al* 2017 *Nanomater Nanotechnol.* **7** 1.
[20] J Burcan *2010 Tribologia* **4** 47.
[21] D A H Hanaor *et al* 2015 *Int. J. Solids Struct.* **59** 121.
[22] A Eman *et al* 2018 EGTRIB. J. **15** 61.

IOP Conf. Series: Materials Science and Engineering **551** (2019) 012013 doi:10.1088/1757-899X/551/1/012013

Effect of Mixing Parameter on Epoxy-Organo Montmorillonite Composites: Flexural and Microscopy Analysis

Yusrina Mat Daud[1,2], Azlin Fazlina Osman[1,2], Farah Farhana Zainal[1,2], Mohammad Firdaus Abu Hashim[1,2], Hasri M.Si[3], Hartati S.Si[3], Ng Soo Li[2], and Sinar Arzuria Adnan[2]

[1]Chemical Engineering, Universiti Teknologi Mara, Cawangan Pulau Pinang, Malaysia.
[2]Chemical Engineering, Universiti Teknologi Mara, Cawangan Johor, Kampus Pasir Gudang, Malaysia
[3]Mechanical Engineering, Universiti Teknologi Mara, Cawangan Pulau Pinang, Malaysia.

E-mail: yusrina@unimap.edu.my

Abstract. The effect of two mixing parameters during preparation of epoxy composites on speed 250 rpm and 500 rpm, mixing time 60, 90, 120, 150 and 180 minutes were observed in this study by flexural and microscopy analysis. Preparation epoxy composites were based on epoxy diglycidyl ether of bisphenol A (DGEBA) and montmorillonite (OMMT) filler. It was found that mixing speed at 500 rpm performed higher flexural strength as compared to 250 rpm at mixing time 150 minutes. This was due to an improved mixing speed by appropriate time as observed through Scanning Electron Microscopy (SEM).

1. Introduction

Epoxy is well known matrix materials used commonly in composites development due to their ease processing. Method of preparation nanocomposites is very important in ways to produce a good nanocomposites. The processing of epoxy nanocomposites have varies techniques depends on types of applications. Some literature referred as synthesis nanocomposites.

There are four general approaches to synthesized layered silicates/polymer nanocomposites i) Solution approach, ii) In-situ polymerization, iii) Melt intercalation and iv) Sol gel technology. However, numerous processing technique highlighted in-situ polymerization and it have been practiced by many authors in an attempt to obtain a homogeneous dispersion of nanofiller in the nanocomposite. In-situ polymerization by mechanical mixing are the most common technique that used to synthesized epoxy –clay nanocomposite [1]. It has been reported widely for advanced epoxy nanocomposites filled nanoclay such organo-montmorillonite. H.Alamri in his research epoxy nanocomposites used mechanical mixer for 10 minutes with rotation speed 1200 rpm [2]. A.Jumahat and W.Liu prepared their sample at same duration time 120 minutes at 400 rpm and 1000 rpm respectively [3-4]. While, I.Isik in his study epoxy montmorillonite composites used 120 minutes to achieved desired properties at no specific rotation speed [5]. However, less of researcher reported on the effect of parameter of mixing and time of mixing towards mechanical properties.

In this study, the effect of time and rotation speed of stirrer was considerably investigate at optimum montmorillonite content on epoxy diglycidyl ether of bisphenol A (DGEBA) [6].

Experimental

1.1. Material

Epoxy matrix resin, diglycidyl ether bisphenol-A (DGEBA) was sourced from Euro Pharma (M) Sdn Bhd. Organo modified montmorillonite (OMMT) used was surface modified contains 25-30 wt% octadecyldiamine is from Sigma Aldrich, and curing agent Isophoronediamine (IPDA) were sourced from Grechem Material Technology.

1.2. Sampel Preparation

The mix ratio of epoxy and hardener amount was determined by stoichiometry. OMMT was dried in vacuum oven at temperature 60̊ C for 24 hours before mixing onto epoxy resin. Then epoxy and OMMT were mixed together for 60, 90, 120, 150, 180 minutes at mixing speed of 250 rpm and 500 rpm (Refer Table 1). Followed by adding curing agent IPDA and degassed for about fifteen minutes using the two stage vacuum pump before poured onto the silicon mold and cured at room temperature.

Table 1. Formulation Design Epoxy Organo-montmorillonite Composites

Sample	Epoxy: IPDA	Filler	Speed (rpm)	Time (minutes)
A	100:23	3 phr	250 rpm	60, 90, 120, 150, 180
B	100:23		500 m	

1.3. Testing and Characterization

Three point flexural test were performed at Instron machine according to ASTM D790 with specimen dimensions 130×20×5 mm at crosshead speed of 1 mm/min and support span of 50 mm to study the flexural strength and modulus. Scanning electron microscope (JEOL Ltd., Japan) were tested at an accelerating voltage 10-20 kV to study the fracture surface of specimens. The specimen were coated with platinum coating.

2. Results and Discussions

2.1 Flexural Strength and Modulus

Effect of mixing times to mechanical stirring speed at 250 rpm and 500 rpm was investigated by three point bending testing. Figure 1 shows an ultimate flexural strength at different mixing time for 250 rpm and 500 rpm in bar graph and the data attained was displayed in table 1. It was observed at minimal mixing time 60 minutes the flexural strength for 250 rpm and 500 rpm does not show very significant differences where the value is about 7.0 MPa and 7.8 MPa respectively. However as mixing time increase to 90 minutes and 120 minutes the strength start to increase from 7.2 MPa to 8.2 MPa for rotation speed 250 rpm. The same situation applies for the rotation speed 500 rpm until the mixing time 150 minutes and reaching the highest bending strength at 13.7 MPa. It was observed that after 150 minutes the flexural strength was decreased to 9.0 MPa differs by 250 rpm that decrease after 120 minutes.

The bending response through modulus elasticity also shows the similar trend in which the maximum bending strength of 250 rpm and 500 rpm at 120 minutes and 150 minutes respectively has achieved highest modulus. It was observed a longer mixing time more than 150 minutes decrease the modulus elasticity.

Table 2. Values of flexural strength and modulus elasticity of different mixing time and speed for epoxy organo-montmorillonite.

Rotation Speed	250 rpm	500 rpm	250 rpm	500 rpm
Mixing Time (minutes)	Flexural Strength (MPa)		Modulus Elasticity (MPa)	
60	7.0	7.8	1001	1020
90	7.2	8.2	1032	1065
120	8.2	10.0	1053	1060
150	7.0	13.7	1144	1549
180	5.8	9.0	600	895

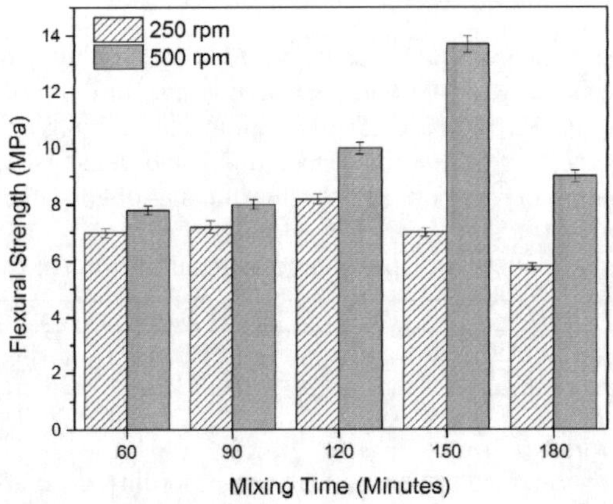

Figure 1. Effect of mixing time on different speed content on flexural strength of epoxy organo-montmorillonite composites.

Overall, the highest bending strength and modulus achieved at mixing time of 150 minutes with mechanical stirring speed 500 rpm may be explained by the fact that dispersion of organo-montmorillonite in the epoxy matrix is better at high velocity within the range time. It is because the at 500 rpm it promotes a good radial flow for drawing the material to be mixed from the top and the bottom with high turbulence at high shearing forces to lead pre-intercalation process efficiently [7]. The enhancement in bending strength also may be ascribed by effectiveness mixing time and speed for better dispersion of organo-montmorillonite filler throughout the matrix that helps increasing of interaction matrix/filler interface. Meanwhile, the lower flexural strength of epoxy organo-montmorillonite in mechanical stirring 250 rpm were expected due to non-homogeneity within the matrix by low shearing forces in short time mixing and tend to agglomerate such degrade the interfacial adhesion between the matrix and filler as can be seen in fractographic Figure 2(a).

2.2 Fractograph Analysis
Image of the fracture surface effected by three point bending testing for the optimum flexural strength achieved by mechanical mixing speed 250 rpm and 500 rpm have been analysed using scanning electron microscope (SEM) and shown in Figure 2. Generally a rougher surface is seen upon adding organo-MMT into epoxy matrix [8–11]. The fracture surface of 250 rpm show considerably different fractographic features as shown in Figure 2 (a) and magnify image at in (b) compared to figure 2 (c) for 500 rpm at 150 minutes and magnify image in (d). As can been seen from figure 2 (d) lower speed 500 rpm, it promote much rougher than 250 rpm. According to Lin et al. (2006) more energy was required for a rougher surface to break which support the high mechanical properties obtained for epoxy organo-montmorillonite at mechanical mixing speed 500 rpm [12].

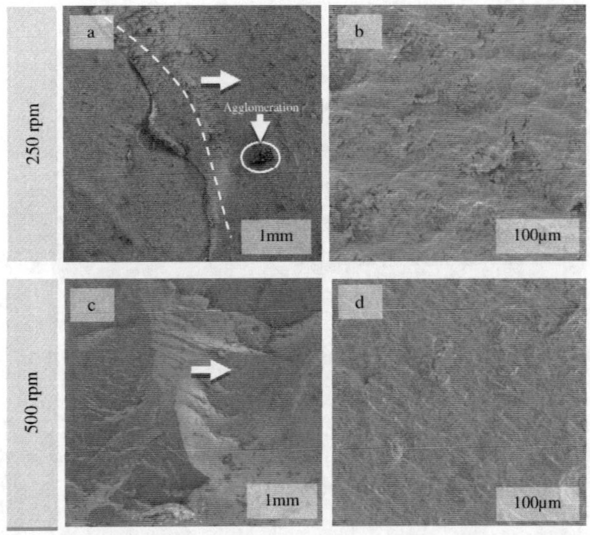

*** The dotted lines mark crack propagation along the fracture surface*

Figure 2. SEM micrograph of the fractured surface epoxy organo-montmorillonite at optimum flexural strength for 250 rpm and 500 rpm.

3. Conclusions

In this study, the influence of mixing speed and varies time from 60 to 180 minutes was studied by means of flexural properties and fractographic analysis. The samples was prepared using mechanical stirrer at speed 250 rpm and 500 rpm. It was observed that mixing speed at 500 rpm performed higher flexural strength as compared to 250 rpm at mixing time 150 minutes.

Acknowledgements

The authors would like to thank the staffs of Center of Excellence Geopolymer & Green Technology (CEGeoGTech), School of Materials Engineering, Universiti Malaysia Perlis (UniMAP) for their involvement in the research. This work was supported and funded by the Universitas Negeri Makassar Kampus FMIPA UNM, Makassar, Indonesia.

References

[1] Zunjarrao S C *et al* 2006 *J. Mater. Sci.* **41** 2219.
[2] Alamri H *et al* 2013 *Comp. Part A Appl. Sci. Manufact.* **44** 23
[3] Jumahat A *et al* 2012 *Pro. Eng.* **41** 1607.
[4] Liu W *et al* 2005 *Comp. Sci. Tech.* **65** 307.
[5] Isik I *et al* 2003 *Polymer* **44** 6371.
[6] Daud Y M *et al* 2016 *AIP Conf. Proc.* **1885** 1.
[7] Daud Y M *et al* 2016 *Key Eng. Mater.* **673** 55
[8] Castrillon P D *et al* 2015 *Comp.Struc.* **133** 70.
[9] Subramanyan A K *et al* 2007 *Comp. Part A Appl. Sci. Manufact.* **38** 34
[10] Zainuddin *et al* 2010 *Mater. Sci. Eng. A* **527** 7920.
[11] Lin L Y *et al* 2006 *Comp. Sci. Tech.* **66** 2116.
[12] Lin J *et al* 2006 *Comp.Struc.* **7** 30.

The transplanter tools for small paddy fields: System of Rice Intensification (SRI) compliance

M.S. Imran[1], M.S. Abdul Manan[2], A.N.M. Khalil[3], S. Hamat[4] ,R.N. Ahmad[5] and M.K. MdNaim[6]

[1,2,3,4,5]School of Manufacturing Engineering, Universiti Malaysia Perlis (UniMAP), Kampus Tetap Ulu Pauh, Jalan Arau-Changlun, 02600 Arau, Perlis, Malaysia
[6]Centre for Diploma Studies of the University Malaysia Perlis, Malaysia

E-mail: syediimran@unimap.edu.my

Abstract. The objectives of the research are to develop manual planting tools compliance with System of Rice Intensification cultivation guideline for small fields and improve SRI farmers working posture. Several problems were recognised to assist in developing design requirement. A current product called Planting Tools is identified as a base product to reverse engineering so it can plant a single young rice seedling in each planting point. Findings: a simple planting tool with smaller size compare to the current product is developed. It consists of several components that can be divided by four main functions which are the hopper, vertical tube, penetrator and handle. The hopper is where the rice seedling is placed. The seedling will fall by gravity through the vertical tube and positioned inside the penetrator. The operator needs to position the penetrator to a specific point. The seedlings will be planted less than 10mm from tops soil upon pressing the lever which connected to linkages. Therefore, a study of the penetrator strength in the static analysis was performed by using finite-element modelling technique.

1. Introduction

Since System of Rice Intensification (SRI) planting method is introduced in Malaysia, local farmers have been applied the technique and gain the good result of increasing in paddy yield. This method originated in Madagascar by French agronomist, Father Henri de Laulanié and spread to other countries worldwide including Malaysia. A case study conducted in privately owned paddy field known as SRI Lovely at Sik Kedah shows that local farmers have been practice the SRI method since 2010 and it proves to increase the paddy yield with minimum water usage and at minimum cost since no chemical fertiliser and pesticide being used. SRI transplanting method plays a significant role. Start from growing young seedling from the nursery, transplanting techniques which minimise possibilities of root injuries, wider spacing between plants, weeding process which also help to aerate the paddy fields soil and organic rice cultivation management. SRI technique will ensure good development for the overall growth of plants if all of the SRI technique is followed.[1] However, after more than 20 years since the SRI method been introduced, there is no fully automated paddy transplanter for SRI. There are some effort to develop those machine and some success in developing semi-auto machine [2]. It still not widely applied by SRI farmers in Malaysia due to few additions in SRI technique by farmers. During the interview session with local farmers, they tried to minimise heavy machinery allowed to enter their paddy field to maintain soil condition. This will add in design requirement to create simple, lightweight tool or machinery. The development of SRI transplanter tool is expected to fill the gap and hopefully drive others researcher to develop automated transplanter machine that fit with SRI technique. It also one of the mechanisms that studying for further development of fully automated SRI transplanter machine. The primary objective of this paper is to develop a manual planting tool that SRI compliance and also to improve the working posture of local farmers during the transplanting process.

2. Problem Identification

Base on observation in SRI Lovely paddy field located at Kedah. There are 6 techniques in practicing SRI method which is (i) the use of young seedling, (ii) planting it near the topsoil, (iii) plant young seedlings with 25 cm spacing, (iv) plough the field to create aerated soil, (v) minimum water level, (vi) organic fertilizer and organic pest control. However, there is no concrete method where some farmers will advance different technique to suit their field condition [1]. The innovation of SRI planting stick is to support the first three methods. Currently, farmers grow young seedling using nursery bed and transplant it afterwards by hand to the field. The problem with this method is the separation young seedling from it nursery soil may damage the root. It also will take a few minutes before the farmers plant it on topsoil. This will increase the risk of transplanting shock which will delay roots growth subsequently overall plant development [3]. Since there is no transplanter machine or tools available in the area, the transplanting process is done by the conventional way using a hand. Besides that, the farmers also need to plant the seedling close to the topsoil so the root can benefit the fertility of topsoil. It also observed that farmers need to bend their body repeatedly to complete the task as shown in figure 1. The farmers suffered discomfort in some of the body parts due to repetitive bending during the working process [4].

Figure 1. Transplanting process done by local farmers of SRI Lovely paddy field.

3. Product Requirement

Two significant requirements are developed base on problem identification which is: (i) A planting tool needs to be able to transplant a young seedling [5] (below 14 days old) near the topsoil (less than 10 mm of depth) without removing the nursery soil. There is an explanation for planting the young seedling along with the nursery soil which is to reduce the disturbance of seedlings root growth to a minimum so that the plants can adapt themselves to their new environment. Figure 2 shows the x value is target less than 10 mm. Just enough to make sure the seedling did not tilt and fixed in position after planted. (ii) The working posture for the user is standing to eliminate awkward working condition as shown previously in figure 2.

IOP Conf. Series: Materials Science and Engineering **551** (2019) 012014 doi:10.1088/1757-899X/551/1/012014

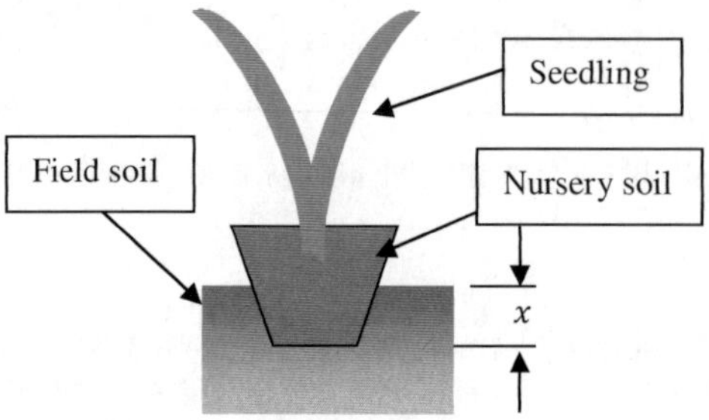

Figure 2. Target result.

A current Planting Tool is studied as shown in figure 3. The existing product is a tubular tool which inserts cylinder shapes crops seedlings into the ground when the lever is pressed. This Planting Tool is manufactured in China and made to plant crops such as vegetable or small plant. It can be bought at RM 200. The specifications of the tools are compared and analysed. Three areas are identified and measured so the redesign process will occur to meet SRI planting techniques.

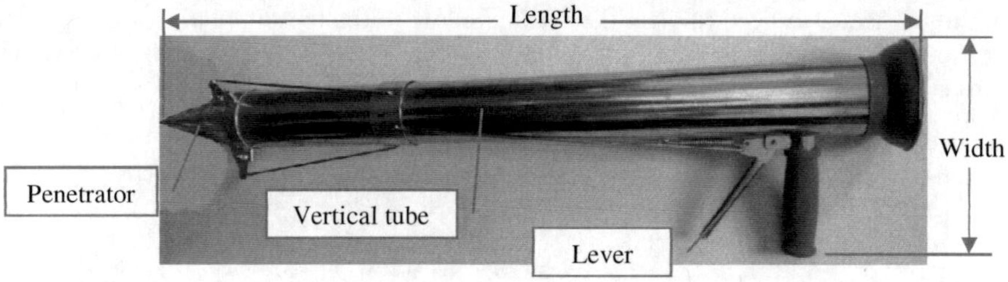

Figure 3. Current product

Table 1 shows the dimension and its relation to how the current product will function. The overall dimension will determine the working posture of the farmer. The shape and size of the vertical tube will determine the shape of seedling nursery soil and overall size allowable into the tube. While penetrator (refer Figure 3) depth will determine the maximum planting depth. Base on the measured dimension, it can be noticed that the vertical tube and penetrator is oversize to meet product requirement. The product weight is 1.62 kg.

Table 1. Current product dimension.

No	Item	Dimension (mm)	Design relation
1	Overall Dimension	930 (Length) × 240(Width)	The working posture of the operator
2	Vertical tube (Cylinder)	75 (Diameter)	Maximum seedlings size
3	Penetrator	140 (Depth)	Maximum planting depth

The current product is further analyzed, and the material is made from 90% stainless steel and others are made from plastic and soft foam (can be seen as blue colour in figure 3). It is noticed that almost all components are permanently joint by welding method (as shown in figure 4). This makes the product very sturdy but with disadvantage which is the component is difficult to maintain. For example, if any part needs to be replacing, another part that weld together will be broken or unusable in the disassembly process. As a result, the current product is not easy and costly to maintain.

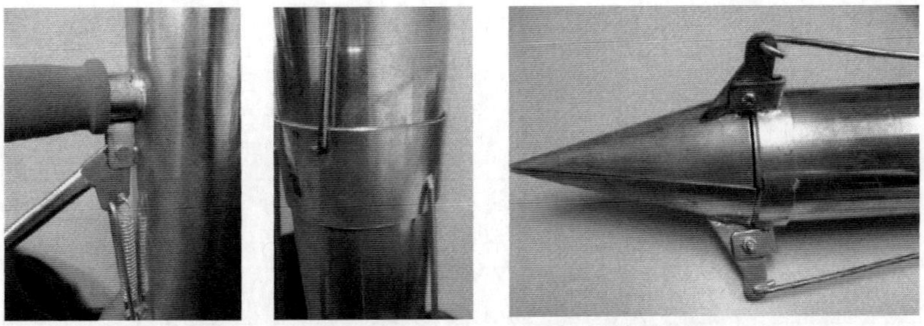

Figure 4. Close up view of the welded part.

4. New Product Development

In general, the SRI Planting Tools (as shown in Figure 5) consists of an aluminium square tube which is to insert a seedling along with nursery soil into the ground. The handle design is upgraded to improve the comfort level when holding the tool. The different shape of the vertical tube is to suit with nursery tray cavity shape which is square as shown in Figure 6. The process sequence is shown in a series of simulation picture shown in Figure 7.

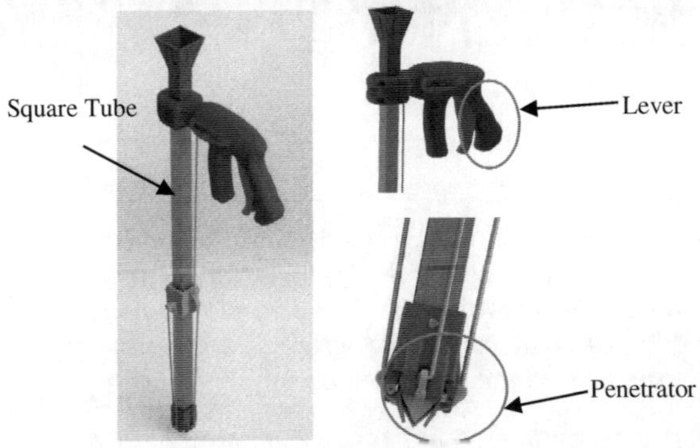

Figure 5. Assembly drawing.

Figure 6. Nursery tray.

Table 2 shows a dimension of newly designed planting tools. The overall dimension is similar to the previous dimension since the tool is used in a standing position. Since the shape and size of the vertical tube is change, almost all parts need to redesign to meet the new specification. The size of penetrator is significantly reduced, and the purpose is merely to scratch the topsoil so that the seedlings will securely be positioned. Possibility the scratched soil to concealed back is predicted since the field is moist. Thus, the penetrator dimension is designed above 10 mm of depth. There are few advantages come with the newly designed SRI Planting Tool which is:

i. All components are easily dismantled for maintenance or part replacement.
ii. The aluminium square tube is a standard part which is readily available in the local aluminium workshop. It can be cut or replaced to alter the length, so it ergonomically fit according to user preference.
iii. The total weight is 810 gram, a reduction by 50% compared to existing Planting Tool.
iv. All parts are from a non-corrosive material such as plastic, aluminium and stainless steel.

A full-scale working model is developed for future analysis and product testing. Plastic component is 3D printed (fused deposition modelling) using ABS (Acrylonitrile butadiene styrene) as a material.

Table 2. Improved product.

No	Item	Dimension (mm)	Design relation
1	Overall Dimension	900 (Length) × 278(Width)	The working posture of the operator
2	Vertical tube (Cylinder)	25.6 × 25.6 (Outer dimension)	Maximum seedlings size
3	Penetrator	15 (Depth)	Maximum planting depth

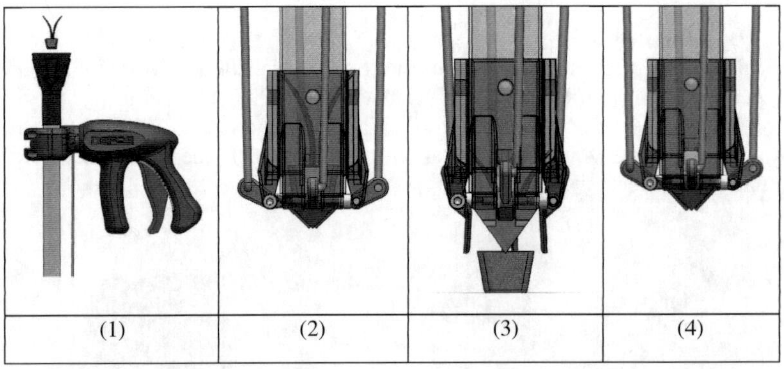

Figure 7. Assembly drawing. (1) The seedling is inserted inside the hopper 2) the seedling immediately felled by gravity into the bottom of the square tube. (3) The operator will press the lever which is a linkage that will open up the penetrator. 4) Seedling is planted to the ground and spring mechanism will retract and close the penetrator for next cycle.

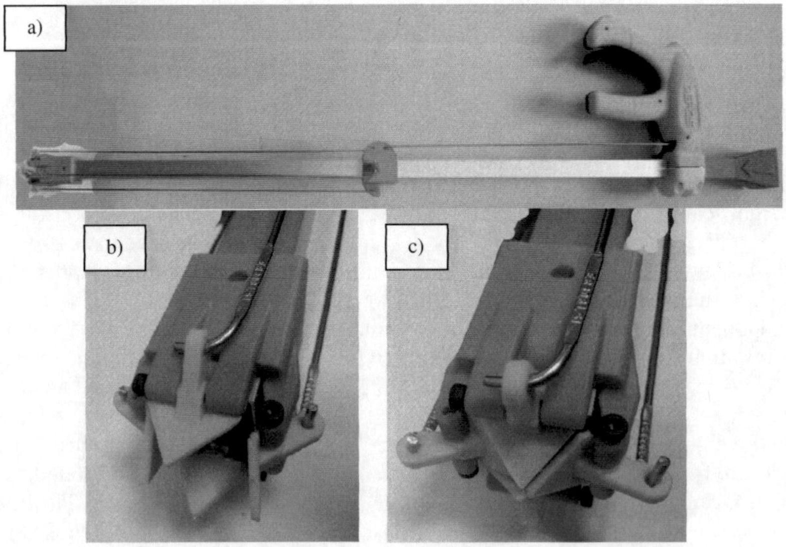

Figure 8. Working model of SRI Planting Tool. a) Assemble product. b) Open penetrator. c) close penetrator.

5. Finite Element Modelling

In this work, a study of the single penetrator strength using finite element modelling in the static analysis was performed in the commercial software Abaqus. An elastoplastic model was applied during analysis to simulate the behaviour and damage of the polymer type ABS in penetrator. The performance of the penetrator was analyzed in term of stress distribution, plastic strain, and failure of material where the failures are modelled with two damage initiation criteria: the ductile damage criterion and the shear damage criterion. To combine and visualize a damaging behaviour for the elastic-plastic model (isotropic elasticity and Von Mises plasticity), it requires the damage evolution law in displacement and energy type for both ductile damage criterion and shear damage. Thus, the damage evolution law is specified as the function strain failure and function fracture energy for ductile damage and shear damage [6] respectively in ABS property of penetrator besides isotropic elasticity and Von Mises plasticity.

In order to develop a material model for this polymer, a tensile test was performed with three specimens at 50 mm/min (equivalent strain rate of 0.0083 1/s) to obtain the engineering stress-strain data and then converted to a true stress-strain curve during the calibration process. In the material calibration process the elastic and plastic of ABS material data is auto calculated by Abaqus and recorded in material property library. After converting to a true stress-strain curve it shows that the average maximum yield stress is 43 MPa while the modulus of elasticity is 1350 MPa with 0.35 of Poisson's ratio. A second order tetrahedral elements (C3D10M) with an average element size of 1 mm were assigned to the penetrator region as the optimal parameters for an element type and mesh size in term of accuracy and simulation time. The penetrator was assigned a force of 50 N that respect to the perpendicular of the soil contact surface in penetrator during SRI planting process. The mass/inertia of the penetrator is neglected as the simulation is performed in static analysis. The simulation was performed in non-linear step mode with the starting incremental size of 0.1 is assigned and had a time length of 1 s. The boundary conditions were defined where all the movement and rotation towards any axis was fixed by using symmetry encastre at the two inner surface pin hole in penetrator. In post-analysis results visualisations presented on Figure 9(a) Von Mises, it shows that the maximum stress produced at critical region (joint area) of penetrator is 32.6 MPa lower than the yield stress of the ABS material. Hence, the equivalent plastic strain, PEEQ can be approximated as 0. The condition of material failure can be observed in damage initiation criteria which the ductile damage criterion defined as stress triaxiality, TRIAX and shear damage criterion as shear stress ratio, SHRRATIO that can be recalled as a field output in Abaqus. Figure 9(b) shows that stress triaxiality, TRIAX is 0.911 and for shear stress ratio, SHRRATIO in Figure 9(c) is 0.885 at the critical area of the penetrator. Thus, no failure occurred for both ductile and shear damage in penetrator as both value are below than 1 [7].

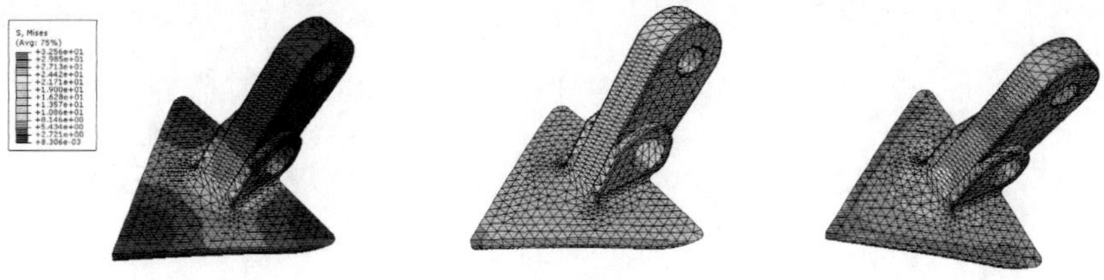
a) b) c)

Figure 9. Post-analysis of penetrator in FE-modelling. (a) Von Mises. (b) Ductile damage. (c) Shear damage

6. Conclusions

The designed tool is in the experimental stage, and the devices shown are more a 3D printed working model. Although few mechanical refinements may need for further development, the planting tool promises many advantages. It is hoped that further testing will facilitate with a new invention for large scale SRI compliance planting technique.

Acknowledgements
This work was financially supported by FRGS from Ministry of Education (MOHE), Malaysia and Faculty of Applied Sciences UiTM, Shah Alam, Selangor.

References

[1] Tejendra Chapagain and Eiji Yamaji 2010 *Paddy Water Environ* **8** 81.
[2] Redmond Ramin Shamshiri *et al.* 2018 *Indian J Sci Technol.* **11** 1.
[3] Akihiko Kotera et al. 2004 *Plant Prod Sci* **7** 62.
[4] Pragya Ojha and Seema Kwatra Mater 2017 *International Journal of Current Microbiology and Applied Sciences* **6** 3439.
[5] Pasuquin E, Lafarge T. and Tubana B. 2007 *Field Crop Res* **105** 141.
[6] Dan-Andrei Serban *et al.* 2015 *Procedia Eng* **109** 97.
[7] Dassault Systémes *Abaqus Analysis User's Manual 6.11* 2011.
[8] Hadi H *et al.* 2012 *International Review of Mechanical Engineering* **6(3)** 462.
[9] Zailani Z A *et al.* 2012 *International Review of Mechanical Engineering* **6(6)** 1262.

Evaluation of Kerf as a Crack Arrestor in a Plate under Tension Loading using FEA

M. Firdaus[1], M.S. Abdul Manan[2], A.F. Annuar[3], M.T. Mustaffa[4] and M.E. Azamudin[5]

[1,2,3,4,5]School of Manufacturing Engineering, University Malaysia Perlis, Pauh Putra Main Campus, 02600, Arau, Perlis, Malaysia.

E-mail: m.saifuldin@unimap.edu.my

Abstract. Crack arrestor means to prolong the life cycle of the crack. There are many methods has been studied to arrest crack but the use of a kerf as a crack arrestor has not been widely studied. Initial study has stated that the value of SIF at the crack tip can be reduced with the presence of the kerf in the same body, but detail dimension of the kerf that will produce best arresting effect is unknown. The finite element analysis was used to model the finite plate which contains different kerf geometries and a crack to obtain crack's tip stress intensity factor. Four types of kerf's configuration have been considered which are; single central kerf, double central kerf, single edge kerf and double edge kerf. Each type used different kerf's geometries selection that is predicted will produce higher kerf-crack interaction. The results of SIF were calculated as a percentage reduction when compared to SIF of a single edge crack without any kerf. Results show double edge kerf type is the most suitable for crack arrestor. Maximum percentage SIF reduction of 34.27% can be achieved when the crack is the shortest and separation to width ratio equal to 0.1. If the longer crack is detected on a structure, highest percentage SIF reduction can be achieved by fabricating maximum kerf length with separation to width ratio in the range of 0.15 and 0.25.

1. Introduction

One of the crack arresting methods in fracture mechanics is by using kerf fabrication parallel to an edge crack. The introduction of the kerf reduces the fatigue crack growth rate at the early stage and increase its fatigue life cycles. This is based on fatigue test study [1] that has been conducted using specimens with an edge crack and a kerf under tension loading. The presence of the kerf reduces the stress intensity factor (SIF) of the crack tip and thus retards the crack growth rate. This is due to the kerf-crack interaction which produces similar shielding effect near the crack tip as the crack-crack interaction [2]. The advantage of using kerf is it can be easily machined near the crack region such as by using wire electrical discharge machine (WEDM) [3]. The early experimental study of crack arrestor using kerf [1] has been completed but detail dimension of the kerf that will produce best arresting effect is not clear. It is important to study the sizes and configuration of kerf that will produce best solution for crack arresting method. Therefore, the main objective of this study is to find the best dimension and configuration of a kerf as a crack arrestor using finite element analysis.

Kerf is a slit or a gap that produced when the amount of material is removed by machining tool including its tool cutting tolerance [1]. Figure 1 shows how the kerf fabricated by using Wire cut Electrical Discharge Machine (WEDM). Crack arrestor is a mechanism that used to prolong the life cycle of crack and total fatigue in the crack body [4]. There are many methods to arrest the crack growth in metal, such as drilling (stop-hole), welding, routing and sealing, and stitching [5].

Most of the analysis which relate to the cracks in a body will apply the theories of Linear Elastic Fracture Mechanics (LEFM) [6]. LEFM assumes that linear elastic theory can be applied to bodies containing cracks even though a plastic zone exists at the crack tip. Crack grows in the body due to the increase of the stress at the crack tip called Stress Intensity Factor (SIF) and the energy release at the crack tip [7]. The higher the value of SIF means the higher chances of crack will grow depending on its life cycle.

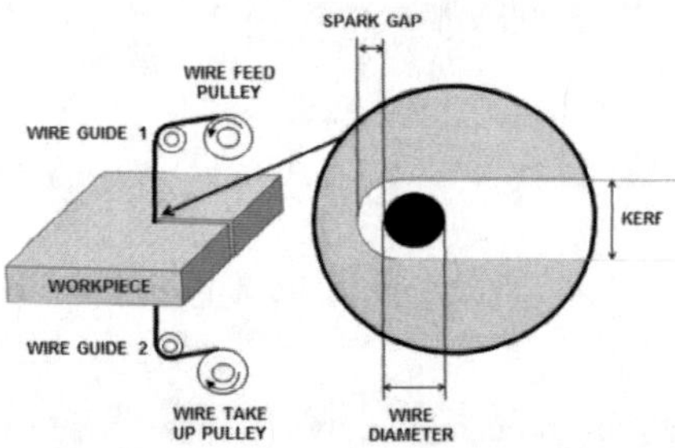

Figure 1. Kerf fabrication by using WEDM [1].

The study of kerf-crack interaction [8] demonstrated that the value of SIF is decreased when kerf is added to the crack body. This study shows that the kerf is also can be considered as one of the crack arresting method. Meanwhile, this finding is also supported by the study of crack-crack interaction [9]. The study demonstrated the crack interaction produces shielding effect which reduces the stress normal to the crack surface.

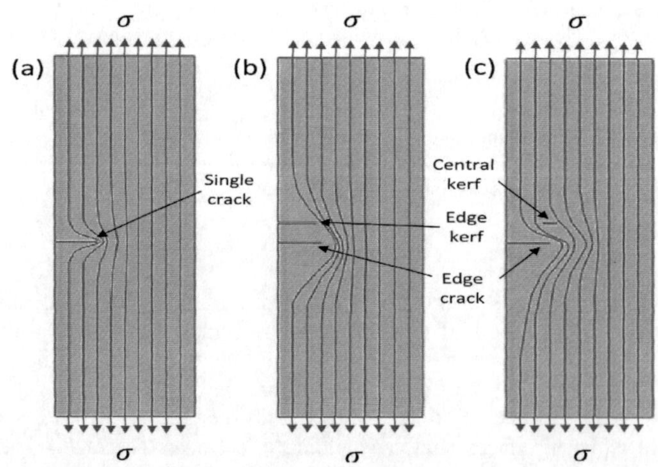

Figure 2. Stress flow around (a) single crack, (b) an edge crack and edge kerf, (c) edge crack and a central kerf.

It is predicted that the crack interaction disturbs the stress flow when passes near the crack tip. The presence of another crack or kerf parallel to the crack will divert this stress flow away from the crack tip. Two types of kerf's configurations were considered in order to divert this stress flow which is edge kerf and central kerf as shown in Figure 2. Figure 2 (a) shows the stress flow around the edge crack. The stress contour near the crack tip is closely distributed. Figures 2 (b) & (c) show the stress flow near the crack tip is slightly diverted. This phenomenon will reduce the SIF value as it depends on the

IOP Conf. Series: Materials Science and Engineering **551** (2019) 012015 doi:10.1088/1757-899X/551/1/012015

stress value at the crack tip. Therefore, the work focuses on these two major types of kerf geometries as a basis for its configuration. Another modification for each type is by introducing another parallel kerf so that it will have two similar kerfs. The purpose is to ensure smooth and stable stress flow diversion.

2. Methodology

This research is mainly used finite element analysis software to model different types of specimens to obtain their SIF values. The package software used in this study is Abaqus [10]. Mild steel with yield strength of 210 MPa was selected as a material. An experiment was conducted to obtain actual size of the kerf such as its gap and diameter at the kerf tip. Ten samples of thickness 4 mm were cut using WEDM using wire diameter of 0.25 mm and measured with microscope.

Every FE model has an edge crack followed by different types of kerfs. There are four types of kerfs that were used in the FE model: single edge, double edge, single central, and double central. Figure 3 shows the FE model for double edge kerf and double central kerf. Single edge kerf and single central kerf have only one kerf as highlighted in Figure 2. The design of the specimen for FE model follows the ASTM standard [11]. Initially the FE models without any kerf were run using FEA to obtain their SIF values. SIF values for the 5 FE models with different crack to width ratio a/W, are shown in Table 1. The SIF values were used as a reference for comparison with FE models that contain kerf. The parameter of the FE models used for the edge and central kerfs are shown in Table 2.

Figure 4 shows the flowchart to complete all types of FE models for FEA. To get the accurate result in package software, the domain integral method [12] was used to get the value of SIF at the crack tip. This is because domain integral using area instead of line as for contour integral method [13]. The pressure value of 150 MPa was applied as a tensile stress. The model was partitioned and meshed using quad normal type but at the crack tip region sweep mesh was used to ensure accuracy of the FEA results. Each model was submitted and run using the FEA software. The FE model will be modified if an error occurs during the process. The SIF values of the successful model were recorded for analysis.

Table 1. SIF value for single edge crack.

Crack length, a/W	Stress intensity Factor, $MN/m^{3/2}$
0.1	15.75
0.2	23.57
0.3	37.92
0.4	55.59
0.5	83.14

Table 2. Parameter for FE models for all types of kerfs.

Crack length, a/W	Kerf length, b/a		Distance between crack and kerf, d/W
	Central	Edge	
0.1	0.20	0.5	0.05
0.2	0.25	0.6	0.10
0.3	0.30	0.7	0.15
0.4	0.35	0.8	0.20
0.5	0.40	0.9	0.25

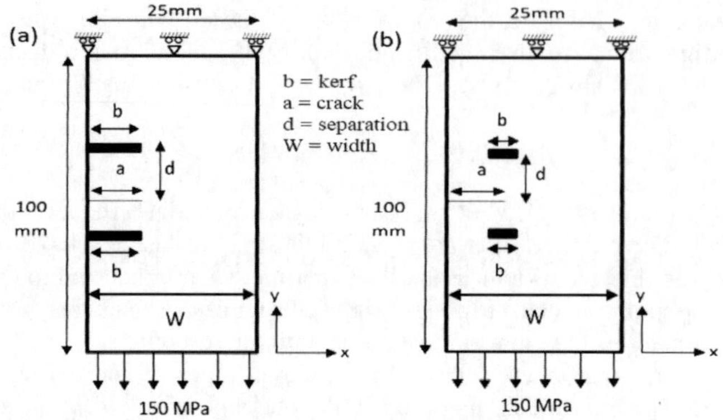

Figure 3. Finite element model for (a) double edge kerf, and (b) double central kerf.

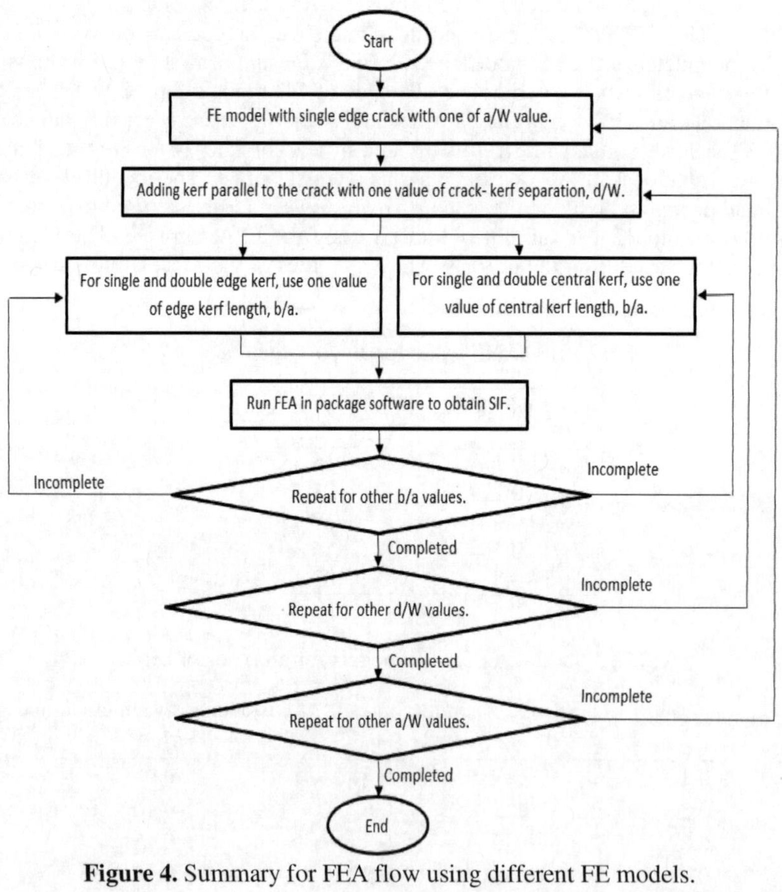

Figure 4. Summary for FEA flow using different FE models.

IOP Conf. Series: Materials Science and Engineering **551** (2019) 012015 doi:10.1088/1757-899X/551/1/012015

3. Results and Discussion

The main result in this study is the value of SIF at the crack tip with the presence of the kerf. The results of SIF were calculated as a percentage reduction when compared to SIF of a single edge crack without any kerfs. Table 3 summarizes maximum percentage SIF reduction obtained from each type of kerf from all parameters in Table 2. Obviously, the use of double edge kerf produces the highest SIF reduction. Figure 5 shows deformed and underformed of the FE model for the double edge kerf model. Figure 6 shows results of percentage SIF reduction versus kerf to crack ratio, b/a for various separation to width ratio, d/W and crack to width ratio, a/W values. All figures show similar trend where the percentage SIF reduction increases as the kerf length increased. All results shows highest SIF reduction can only be achieved at maximum kerf to crack length ratio (b/a = 0.9). This is the maximum possible b/a value that can be modelled as beyond 0.9, the new crack will initiate and propagate from the kerf's tip instead of at crack's tip [1].

The most significant result is for the case of shorter crack length (a/W = 0.1) as shown in Figure 6(a). The maximum result is when d/W equal to 0.1 with 34.27% reduction. SIF reduction becomes less significant when the crack length is the longest (a/W = 0.5) as shown in Figure 6(e). This is because at longer crack the SIF will increase rapidly as it is closer to fracture phase. Figures 6(b)-(d) show that highest SIF reduction for other a/W values can be achieved using d/W values in the range of 0.15 and 0.25.

Table 3. Maximum SIF reduction values for all types of kerf.

Type of Kerf	Maximum SIF reduction, %.
Single central	22.07
Double central	29.9
Single edge	16.94
Double edge	34.27

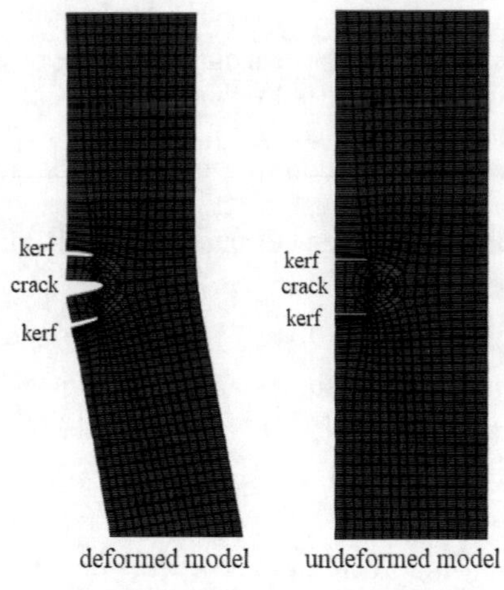

Figure 5. Double edge kerf model.

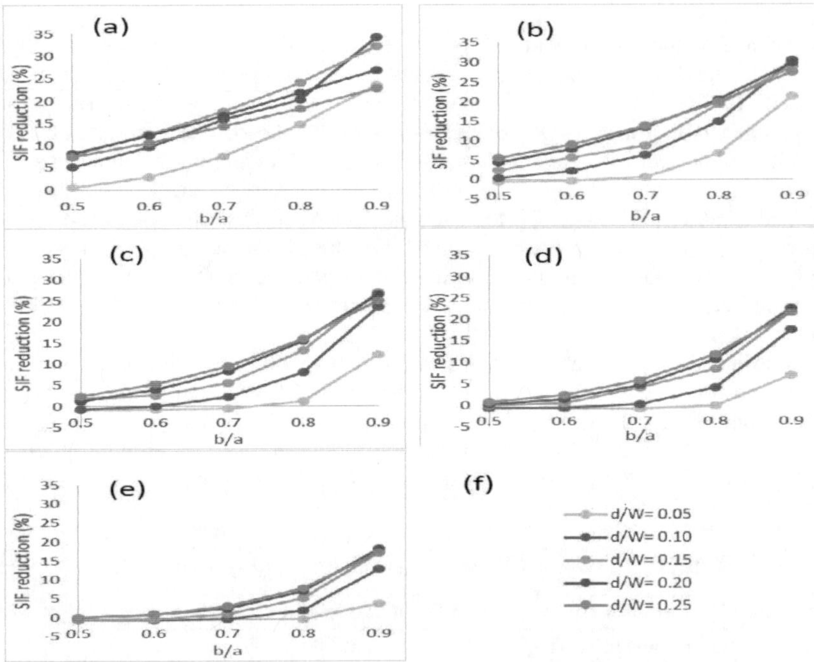

Figure 6. Double edge kerf SIF reduction results with various d/W and b/a values; (a) a/W= 0.1, (b) a/W= 0.2, (c) a/W=0.3, (d) a/W= 0.4, (e) a/W= 0.5 and (f) graphs legend.

4. Conclusions

A presence of kerf in the crack body would be reducing the value of SIF value at the crack tip. The best method of using kerf as a crack arrestor is by fabricating double edge kerf parallel to the edge crack. Maximum kerf length (b/a =0.9) needs to be used as it will produces good kerf-crack interaction. Maximum percentage SIF reduction of 34.27% can be achieved when the crack is the shortest and separation to width ratio, d/W equal to 0.1. If the longer crack is detected (a/W = 0.2, 0.3 & 0.4) on a structure, highest SIF reduction can be achieved by fabricating maximum kerf length with d/W in the range of 0.15 and 0.25.

Acknowledgements

The author would like to acknowledge the support from the Fundamental Research Grant Scheme (FRGS) under a grant number of (FRGS/1/2017/TK03/UNIMAP/03/2) from the Ministry of Higher Education Malaysia.

References

[1] Azamudin M E *et al.* 2018 *International Review of Mechanical Engineering* **12** 183.
[2] Annuar A F *et al.* 2014 *International Review of Mechanical Engineering* **8** 766.
[3] Tosun *et al.* 2004 *J Mater Process Tech* **152** 316.
[4] Song P S *et al.* 2004 *Int J Fatigue* **26** 1333.
[5] Fanni M N *et al.* 2015 *Ain Shams Eng J* **6** 987.
[6] Anderson T L 2017 *Fracture mechanics: fundamentals and applications* CRC press.
[7] Chudnovsky A 2014 *Int J Eng Sci* **83** 6.
[8] Annuar A F 2015 *Master Thesis* Universiti Malaysia Perlis.

[9] Daud R *et al.* 2012 *International Conference on Applications and Design in Mechanical Engineering*.
[10] ABAQUS 2012 *User Manual* Dassault Systemes Simulia Corp, Rhode Island, USA.
[11] ASTM E647-15e1 2016 *American Society for Testing and Materials.*
[12] Castle C 2012 University of Wisconsin Milwaukee.
[13] Mahanthesa M I *et al* 2016 *Int J Res Eng Technol* **5** 78.

Investigation on Antioxidants Compounds Composition Contains in *Leucaena Leucocephala* (*Petai Belalang*)

Z Zarina[1], C M Ruzaidi[2], S T Sam[3] and A M Mustafa Al Bakri[4]

[1]Faculty of Engineering Technology, Universiti Malaysia Perlis (UniMAP), Kampus UniCITI Alam
Sungai Chuchuh, Padang Besar, 02100 Perlis Indera Kayangan
[2]School of Fundamental Science, Universiti Malaysia Terengganu, 21030 Kuala Nerus, Terengganu.
[3]School of Bioproses Engineering, Universiti Malaysia Perlis (UniMAP), Kompleks Pusat Pengajian Jejawi , 02600 Arau, Perlis, Malaysia
[4]Center of Excellence Geopolymer & Green Technology (CEGeoTech), School of Materials

Email: zawawi_zarina@yahoo.com

Abstract. The main objective of the current work was to investigate the unknown compound composition contains in *Leucaena Leucocephala (Petai Belalang)* to prove that there were antioxidants properties presents in the leaves and seeds of the plant. High-performance liquid chromatography (HPLC) and Fourier-transform infrared spectroscopy (FTIR) were used to characterize the samples. It was found that for both leaves and seeds sample of *Leucaena Leucocephala,* the spectrum had shown the main existence of alcohol (O-H bond. As the O-H bond is the main element of antioxidant and phenolic compound, the FTIR analysis shows for the high possibilities for the present of antioxidant and phenolic compound in the leaves and seeds of *Leucaena Leucocephala.* Further analysis using HPLC, both seed and leaf sample shows for the presence of the same compounds, identified as *apigenin, caffeic acid, formic acid, isorhamnetin, keampferol, luteolin* and *quercetin,* but by comparing the peak, can be concluded that the compounds in seed are higher than the compounds in leaf.

1. Introduction

Antioxidant is one of the most widely used groups of elements for maintaining the physiological functions of human body such as the liver, kidney, digestive system, and prevention of cardiovascular diseases and cancer [1]. Natural antioxidants such as polyphenols and carotenoids are usually derived from plants. Most plants contain compounds that possess antioxidant activity which occur in all parts of the plant, for examples wood, bark, stems, leaves, fruit, roots, flowers, pollen and seeds [2].

The advantages of Leucaena Leucocephala such as it can be found easily in large amount and available throughout the year. Leucaena Leucocephala is a tropical tree which has been categorized into the group of tree legumes. Since long time ago, many tree legumes have been recognized for their capacity to enhance the productivity and sustainability of tropical agricultural systems, both for developed and less developed countries of the world. Studies show that tree legumes can provide fuel wood, nutrient-rich, erosion control and land stabilization. Other than that, it also was taken as food and as fencing materials for farmers [3].

FTIR has proven to be a valuable tool for the characterization and identification of compounds or functional groups (chemical bonds) present in an unknown mixture of plants extract. In addition, FTIR spectra of pure compounds are usually so unique that they are like a molecular "fingerprint".

For most common plant compounds, the spectrum of an unknown compound can be identified by comparison to a library of known compounds [4].

High performance liquid chromatography (HPLC) is a versatile, robust, and widely used technique for the isolation of natural products, HPLC is a chromatographic technique that can separate a mixture of compounds and is used in phytochemical and analytical chemistry to identify, quantify and purify the individual components of the mixture. Currently, this technique is gaining popularity among various analytical techniques as the main choice for fingerprinting study for the quality control of herbal plants [5]. The present of bioactive compounds in plants was discovered to act as protection against bacterial, fungal and pesticides pathogens.

The common application of antioxidant from plant sources is usually being focused only on cosmetic and medicine industry. Indeed, at present, general research in looking for others used of antioxidant from plant's sources is still in its infancy.

In this work, the main types of equipment used for characterized and identified the antioxidants are fourier-transform infrared spectroscopy (FTIR) and the high performance liquid chromatography (HPLC), both of which are technically classified as efficient method for separation techniques related to plants materials [6,7].

2. Methodology

2.1 Sample preparation

Samples of mature seeds and leaves were collected from different locations around Perlis by neglecting the effect of different location and morphological variation of the plants. The seeds were separating from its pod and was washed using water before undergoes for drying process. While for the leaf, it were separate from its twig before undergoes drying process at 45°C for 72 hours (3 days). Special precaution with proper actions and accurate awareness should be strictly considered to assure that potential bio active compound are not lost, distorted or destroyed during the preparation of the extract from plant samples.

2.2 Fourier-transform infrared spectroscopy (FTIR)

The preparation of samples for FTIR analysis was done by collecting the fresh matured seeds and leaves from plant before undergoes drying proceed in the oven at 45°C for 72hours (3 days). The dried samples were blended using laboratory blender and was sieved (63µm) to get the uniform powder.

By using microspatula, about 1/8" of the solid sample and about 0.25 0.50 teaspoons of Potassium Bromide (KBr) were taken carefully and thoroughly mixed in a mortar while grinding with the pestle. The mixture powder was placed just enough amount to cover bottom in pellet die and pressing the die at the pressure of 5000-10000 psi. The pressed sample was carefully removed from die and placed in the FTIR sample holder [6].

FTIR analysis was carried out using a PerkinElmer FT-IR/FIR spectrometer frontier. The peaks obtained were analysed by using a standard IR spectra table.

2.3 High performance liquid chromatography (HPLC)

Reverse-phase high performance liquid chromatography (RP-HPLC) analysis of samples was performed using PerkinElmer A-10 solvent and sample module, Altus with a PDA detector, and chromatographic separations were performed on a LiChrosorb Rp-18 column (5µm). The composition of solvents and the gradient elution conditions used were as described by Uzelac et al. (2005) and Butsat et al. (2009) with some modifications [7][8].

The mobile phase consisted of purified water with acetic acid (pH 2.74) (solvent A) and acetonitrile (solvent B) at a flow rate of 0.8 ml/min. Gradient elution was performed as follows: from 0 to 5 min, linear gradient from 5 to 9% solvent B; from 5 to 15 min, 9% solvent B; from 15 to 22 min, linear gradient from 9% to 11% solvent B; from 22 to 38 min, linear gradient from 11% to 18% solvent B; from 38 to 43 min, from 18% to 23% solvent B; from 43 to 44 min, from 23% to 90% solvent B; from 44 to 45 min, linear gradient from 90% to 80% solvent B; from 45 to 55min, isocratic at 80% solvent B; from 55 to 60min, linear gradient from 80% to 5% solvent B and a re-equilibration period of 5 min with 5% solvent B used between individual runs.

Spectra were recorded from 200 to 600 nm. Phenolic compounds in the samples were identified by comparing their relative retention times and UV spectra with those of standard compounds and were detected using an external standard method [9].

3. Result And Discussion

3.1 FTIR Analysis

Fourier Transform Infrared Spectroscopy (FTIR) identifies chemical bonds in a molecule by producing an infrared absorption spectrum. The spectra produce a profile of the sample, a distinctive molecular fingerprint that can be used to screen and scan samples for many different components. FTIR is an effective analytical instrument for detecting functional groups and characterizing covalent bonding information.

In this part of analysis, both leaves and seeds sample from *Leucaena leucocephala* tree was undergoing for FTIR analysis. The result was shown in figure 1.

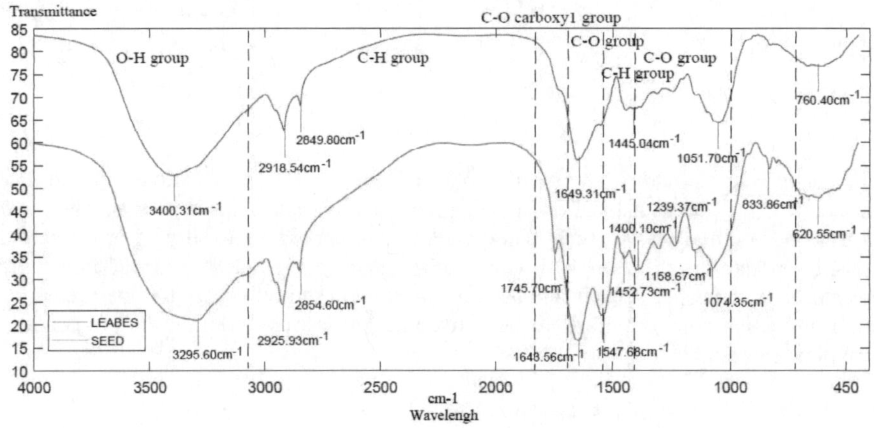

Figure 1. FTIR analysis.

Table 1. FTIR peaks value for extract of *L.Leucaena* extract.

Functional Group	Wavelength range (cm^{-1})
O-H	3600-3200
C-H	3100-3000
C-H	3000-2850
C-O	1820-1670
C=O carbonyl	1700-1819
C=C	1620-1669
C-C	1600-1400
C-H	1480-1350
C-O	1300-1000
1,2-disubstituted	1000-900
1,3-disubstituted	1100-600

Fourier Transform Infrared Spectroscopy (FTIR) identifies chemical bonds in a molecule by producing an infrared absorption spectrum. The spectra produce a profile of the sample, a distinctive molecular fingerprint that can be used to screen and scan samples for many different components. FTIR is an effective analytical instrument for detecting functional groups and characterizing covalent bonding information.

Figure 1 shown for the IR spectrum appeared for leaves and seeds of *L.Leucaena* when analysed using FTIR. Table 1 recorded the range wavelength to distinct functional groups of organic compounds that present in leaves and seeds of *L.Leucaena*.

Peak shapes are sometimes very useful in recognizing what kind of bond is present. The rounded shape of most O-H stretching modes occurs because of hydrogen bonding between different hydroxyl groups [10].

Because protons are shared to varying extent with neighbouring oxygen, the covalent O-H bonds in a sample of alcohol all vibrate at slightly different frequencies and show up at slightly different positions in the IR spectrum. Instead of occur as single sharp peak, a whole lot of them usually all smeared out into one broad blob. Since C-H bonds (single bond) are actually a week bond, the peak appear differently for an ether, and an O-H peak is very easy to distinguish in the IR spectrum [11].

For the leaves sample, the peak observed were 3400.31, 2918.54, 2849.80, 1649.31, 1445.04, 1051.70, and 670.50 cm^{-1}. While for seeds sample, the peak observed were 3295.36, 2925.93, 2854.60, 1745.70, 1643.56, 1547.68, 1452.73, 1400.10, 1305.04, 1158.67, 1074.35, 833.86 and 620.55 cm^{-1}. The peaks proven for the existence of organic compounds (O-H bond), alkane (C-H bond), aldehyde (C-H), alkene (C-H bond), alkene (C=C bond), Cyclopentanone (C=O bond), Amine (C-N bond), vinyl ether (C-O bond), and tertiary alcohol (C-O bond) which all represent of hydrocarbon group.

As O-H bond is the main element of antioxidant and phenolic compound, the FTIR analysis shows for the high possibilities for the present of antioxidant and phenolic compound in the leaves and seeds of *Leucaena leucocephala.* The various functional groups observed in both leaves and seeds of *L.Leucaena* probably indicate the presence of carbohydrates ((CH2O)n), glycogen ($C_{24}H_{42}O_{21}$), starch ((C6H10O5)n), lipids (–COOH)), glycogen ($C_{24}H_{42}O_{21}$) and cellulose ((C6H10O5)n) [11].

3.2 HPLC Analysis

This part of research aims to prove and identified the presence of antioxidant compounds in samples with reversed phase High Performance Liquid Chromatography (HPLC). There are seven (7) type of compounds that was used as standard. The chosen for the standard was done by referred to previous study by Tradit et al where the same compounds are assuming to contains in tested sample [12]. The peak for standards was represented in figure 3 and table 1.

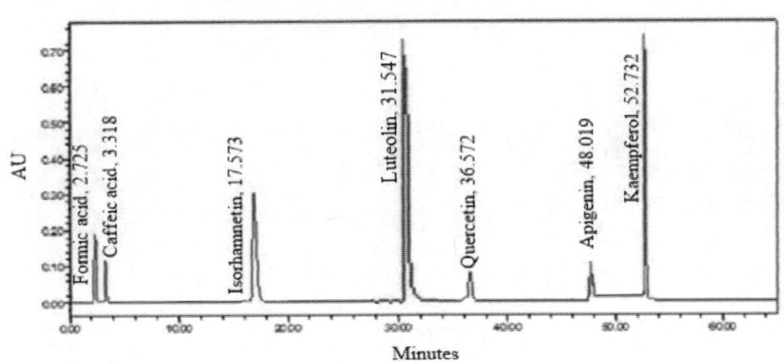

Figure 2. Run standard for HPLC analysis

IOP Conf. Series: Materials Science and Engineering **551** (2019) 012016 doi:10.1088/1757-899X/551/1/012016

Table 2. Peak versus retention time (RT).

Compound	Retention time (RT)
Apigenin	48.019
Caffeic acid	3.318
Formic acid	2.725
Isorhamnetin	17.573
Kaempferol	52.732
Luteolin	31.547
Quercetin	36.572

The samples, which is seed and leaf extract was run by referred to the previous standard. The peak which appeared on time same as retention time of the standard representation for the specific compounds that is contained in the sample. For both seed and leaf sample, there were proven for the presence of antioxidant (bioactive compounds). The results are shown in figure 3 and figure 4. Evidence for the presents of bioactive compounds (antioxidant) was provided by the result of HPLC analysis.

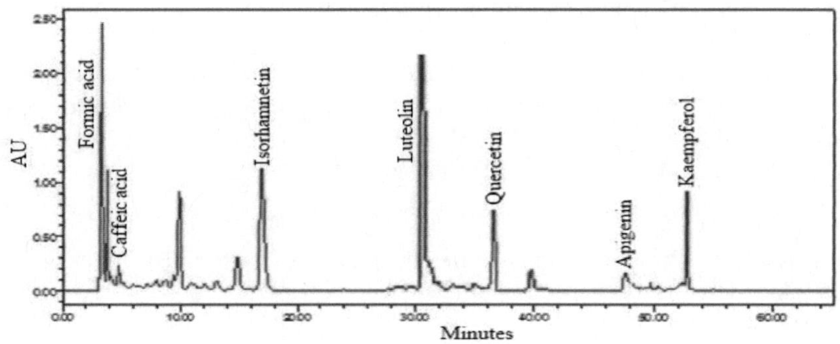

Figure 3. HPLC analysis of seed sample

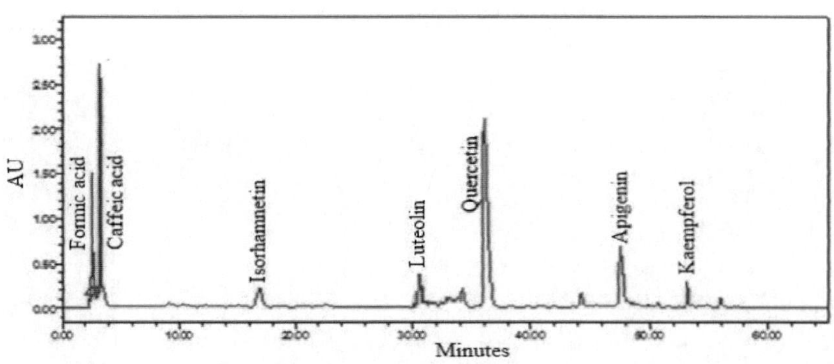

Figure 4. HPLC analysis of leaf sample.

The compounds were identified by comparing with standards of each identified compound using the retention time, the absorbance spectrum profile and also by running the samples after the addition of pure standards.

Despite both seed and leaf sample shows for the present of same compounds, identified as *apigenin*, *caffeic acid*, *formic acid*, *isorhamnetin*, *keampferol*, *luteolin* and *quercetin*, but by comparing the peak, can be concluded that the compounds in seed are higher than the compounds in leaf. The results obtained agreed with the other work carried out by Al-Anbaril and Hassan, 2015 where was mentioned that the phenols in seeds were higher more than leaves [11].

The result has proved for the present of antioxidants in seeds and leaves of *Leucaena Leucocephala* that will be proposed as bio additives in high density polyethylene (HDPE). This compounds are expected to perform as UV protection for HDPE when exposed to the UV light as the general function of antioxidants is significantly to delay or prevent oxidation of oxidizable substrates when present at lower concentrations than the substrate [13].

4. Conclusions

The result obtained from this study has laid an important platform from which to investigate the antioxidant compound composition in *Leucaena leucocephala.* As the results has revealed in the present of antioxidant in samples, identified as *apigenin, caffeic acid, formic acid, isorhamnetin, keampferol, luteolin* and *quercetin*, it will become a platform to propose the plant for further used.

References

[1] D Wilson *et al* 2017 *Antioxidants* **6** 81.
[2] D P Xu *et al* 2017 *Int. J. Mol. Sci.* **18** 20.
[3] R C Gutteridge *et al* 2007 *Forage Tree Legum. Trop. Agric.* 7.
[4] S Sasidharan *et al* 2011 *Afr J Tradit Complement AlternMed.* **8** 1.
[5] N M Abdelhady *et al* 2017 *European J. Med. Plants.* **17** 1.
[6] S Sangeetha *et al* 2014 *Res. J. Biotechnol.* **9** 65.
[7] O Kaisoon *et al* 2011 *J. Funct. Foods* **3** 88.
[8] S Butsat *et al Int.* 2016 *Food Res. J.* **23** 180.
[9] O Kaisoon *et al* 2011 *J. Funct. Foods* **3** 88.
[10] D Cozzolino 2015 *Antioxidants* **4** 482.
[11] L S Pillai *et al* 2014 *J. Pharmacogn. Phytochem.* **2** 120.
[12] R A Hassan *et al* 2014 *Afr. J. Tradit. Complement. Altern. Med.* **11** 67.
[13] D M Kasote *et al* 2015 *Int. J. Biol. Sci.***11** 982.

IOP Conf. Series: Materials Science and Engineering **551** (2019) 012017 doi:10.1088/1757-899X/551/1/012017

A Review of hybridization of carbon nanotube into graphene for gas sensor application

Nur Hanim Khairul Azman[1], Muhammad Mahyiddin Ramli[1,2], Siti Salwa Mat Isa[1, 2]

[1]School of Microelectronic Engineering, Universiti Malaysia Perlis, Pauh Putra Campus, 02600 Arau, Perlis, Malaysia
[2]Center of Excellence Geopolymer & Green Technology (CeGeoGTech), School of Material Engineering, Universiti Malaysia Perlis (UniMAP), P. O. Box 77, d/a Pejabat Pos Besar, 01000 Kangar, Perlis Malaysia.

Abstract Hybridization of two mechanically strong materials which is carbon nanotube and graphene can provide specific advantages over other materials in gas sensor research and can relate to the chemical properties. Consequently, graphene and carbon nanotube show extreme sensitivity towards changes in the local chemical environment that originates from the susceptibility of the electronic structure to interacting molecules. Furthermore, a sensitivity and chemical selectivity to a diverse array has made them perfect candidates for consolidation into the design of gas sensor. The development of hybridization of carbon nanotube and graphene based gas sensor still in its infancy, but attempts to make the gas sensor more quantitative are important for further advancement of gas sensor innovation.

1. Introduction
The atmospheric air, there are various kinds of chemical species that are common and manufactured. Some of which are crucial to our daily life while many others are harmful. To response the changes in the local chemical environment, the researchers have been looking for affordable alternatives to obtain a better gas sensor.

Basically, there are various kinds of gas sensors so far developed. Gas sensors, which have been in incredible interest to make sure safety, health, environmental reservation, energy saving and others. In order to be utilized in practice, gas sensors should fulfill many requirements which depend on the purposes and condition of sensor operation. Among the requirement, the important would be sensing performance such as sensitivity, selectivity, rate of response and recovery rate in the chemical environment. Furthermore, there is also a strong demand for the ability to monitor and control the ambient environment, especially with the increasing concern of the global warming.

The key importance in research and development of gas sensors is the sensing materials. Advance progress in the synthesis and fundamental understanding of surface phenomena has produce extraordinary favor about formation of nanomaterials into sensor design [1]. Nanomaterial sensor is needed to serve as a cheap, small, users-friendly and reliable gas sensing device [2].

More recently, carbon based materials called Carbon Nanotubes (CNTs) and Graphene have been identified to be promising sensing material due to their unique and excellent electrical properties and mechanical properties. Besides, there have a low operation temperatures compared with other gas sensors. The gas sensing mechanism generally attributed to the adsorption and desorption of gaseous molecules on the surface, leading to changes in the conducting properties. The unique structures and outstanding electronic properties of CNTs and Graphene which are small size, large specific surface area, high electron mobility and high sensitivity to electrical perturbations from gas molecules [3].

2. Graphene

Carbon is one of the essential building blocks of life on earth and is the 15th most abundant element on earth and the 4th most abundant in the universe. Since Iijima discovered CNTs in 1991 [4], the research and development by utilizing carbon based material has got a superb attention. The world has more valued this carbon material when the duo from the Manchester University successfully created the wonder material called graphene in 2004 using the scotch-tape mechanical exfoliation technique.

Graphene consists of two-dimensional array of carbon atoms covalently attached via sp^2 bonds to produce a honeycomb sheet. Crystal structure of graphene consists of hexagonal rings forming thin parallel plates. The outermost electron shell of a carbon atom has four valence electrons, three of it are used by the covalent bonds. The graphene are tied together by weak van der Waals forces. The graphene layer structure allows sliding movement parallel graphene plates. Weak bonding between the plates determines softness and self-lubricating properties of graphite [5]. Figure 1 shows the graphene structure (Figure (a)) and the SEM image (Figure (b)) seen randomly aggregated wrinkled sheets closely associated with each other.

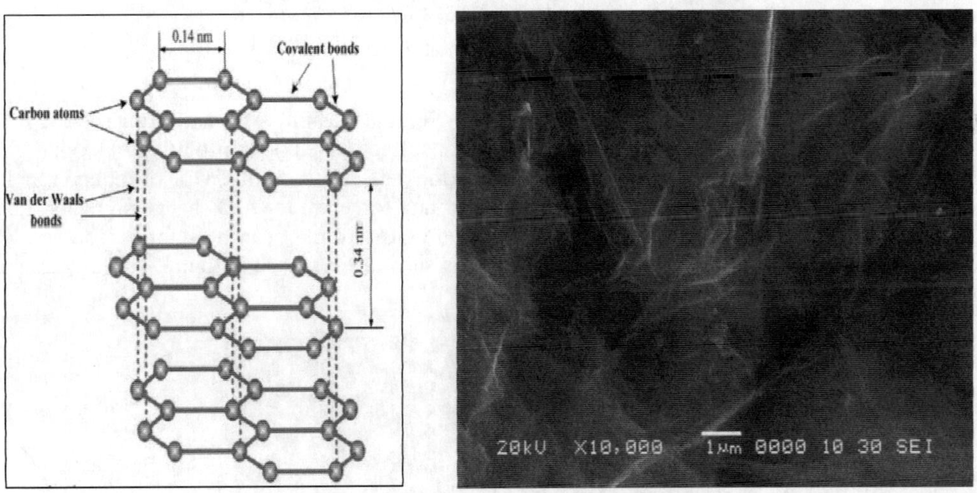

Figure 1. (a) Graphene structure [5]. (b) SEM micrograph of Graphene Oxide (GO) [17]

According t(**(a)** ı Choudhuri, et al. [18] graphene sheet 'Bulk Synthesis of Graphene Nanosheets', GO was synthesizeu by using the Modified Hummers method. This method can produces cheap and large quantities of graphene as well as easy to manipulate for device fabrication. Modified Hummers method yielded a graphene oxide. The oxide group oxygen and hydroxyl come from a chemical reaction between mixtures of concentration acid with an oxidation agent by a heat process. A freestanding graphene presented the possibility of high sensitivity towards detection of chemical analytes. Graphene is shown to exhibit inherently low electrical noise at room temperature and high electron mobility [12].

3. Reduced Graphene Oxide (rGO)

The oxidation process is to decrease or losing the electron, while the reduction process is increase or gaining the electron. Figure 2 shows acronyms for oxidation and reduction process [7]. According to V. H Pham et al. (2012) [6] graphene sheet 'Chemical reduction of an aqueous suspension of graphene oxide of nascent hydrogen' concentration between of aluminium foil (Al foil) and acid chloric (HCl), Al foil, sodium hydroxide (NaOH) and Zink (Zn) powder and NaOH are used to reduce GO. This method is simple and effective that known as nascent hydrogen. Figure 3 shows SEM image of rGO deposited on silicon substrate.

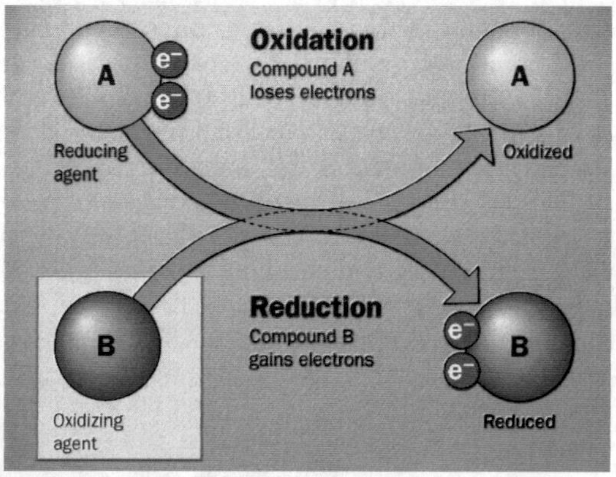

Figure 2. Basic oxidation and reduction [7]

Others metal can be used to reduce GO such as iron (Fe), magnesium (Mg) and manganese (Mn) for this reason of nascent hydrogen that can produce a metal surface based on the solution of oxidation [6]. The higher is the pH of GO suspension, the larger is the zeta potential of GO as a function and the electrostatic force among GO sheets [6]. If the electrostatic force among GO sheets is greater than or equal to the total of gravity and van der waals force, the GO hydrosol can be stabilize forever. SEM images of the rGO deposited by spin coating on a silicon substrate reported in figure. Some overlapping of flakes is noticeable [16].

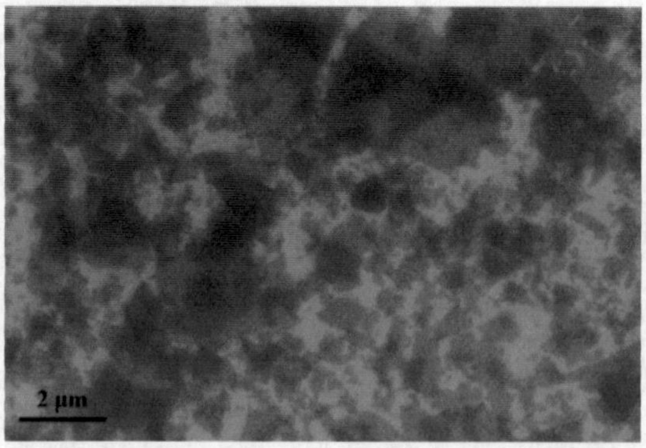

Figure 3. SEM image of rGO deposited on silicon substrate [16]

IOP Conf. Series: Materials Science and Engineering **551** (2019) 012017 doi:10.1088/1757-899X/551/1/012017

The electronic properties of graphene have been found to be extremely sensitive to the local chemical environment. Reduced graphene oxide has also been reported as a useful material for developing gas sensor. The first study of graphene use in gas sensing was reported in 2007 by Novoselov's group. The adsorption of ppm of gases caused the gas sensor to show concentration-dependent changes in resistivity.

4. Carbon Nanotubes (CNTs)

Carbon nanotubes (CNTs) are another nanoscale allotrope of carbon. They can be thought of as ribbons of graphene that have been rolled into a tube. While only nanometers in diameter, CNTs can grow onto millimeters in length. Due to the strength of the bonds in a hexagonal carbon lattice, nanotubes are one of the strongest fibers ever discovered. Primarily, CNTs are divided in two different types, determined by the number of the walls of the nanotube either Single Wall Nanotube (SWNT) or Multiwall Nanotube (MWNT) [15].

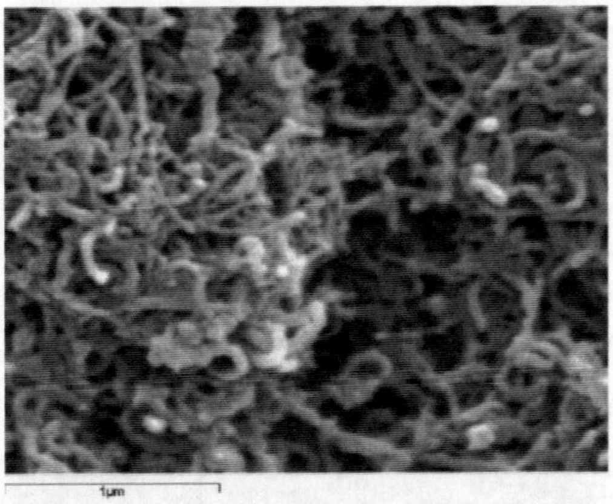

Figure 4. SEM image of CVD grown multiwall CNT bundles [19]

Figure 4 shows SEM image of CVD grown multiwall CNT bundles. The image shows CNTs appear randomly oriented in the form of a porous mat [19]. CNTs possesses exceptionally high surface to volume ratio. These characteristics lead the CNTs a strong and promising candidate for high sensitivity gas sensor [15]. Dong *et al* [13] used carboxylic acid functionalized CNTs for detection of 10 ppm carbon monoxide (CO) gas through weak hydrogen bonding with the attached carboxylic group. Recently, CNTs based electronic gas sensor either in the form of chemiresistors or back-gated field-effect transistors have been created to detect various gases [12]. In the sensing process, basically CNTs accept electrons in the sensing. The response times of CNT is mainly governed by the generation of the charge carriers in the defect sites and depends on how fast these carriers can be transported from the defect sites to the electrodes [12]. Sensor conductance was investigated in the presence of nitrogen dioxide and ammonia vapors [19]. Figure 5 shows the electrical response of SWCNT device to gas molecules.

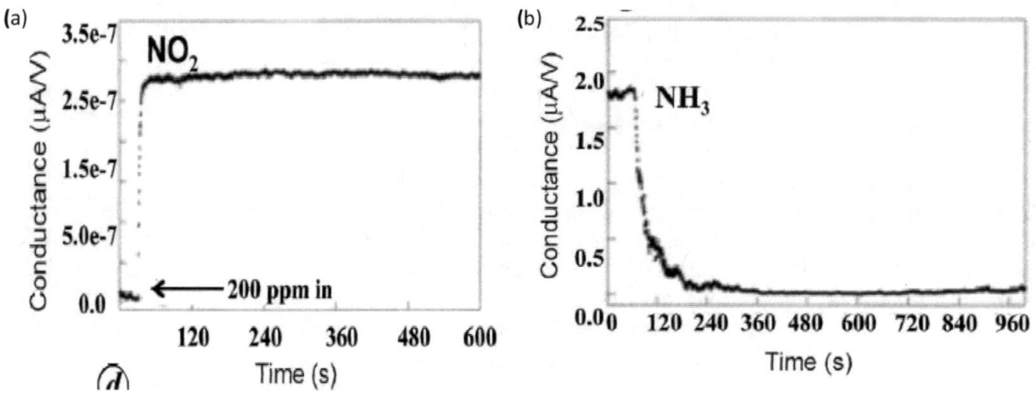

Figure 5. Conductance versus time response for 200 ppm NO_2 and 1% NH_3 exposures respectively [19]

5. Hybridization carbon nanotube (CNTs) and graphene gas sensor.

An advanced carbon based gas sensing material consisted of CNTs and graphene hybrid film were introduced. These intelligent and unique graphene-carbon nanotube consist of CNTs connected perpendicularly to graphene layers, could overcome some of the shortcomings and limitations of individual components [8-10]. The hybrid CNTs and graphene structure potentially shows outstanding properties that could not be attained either material alone.

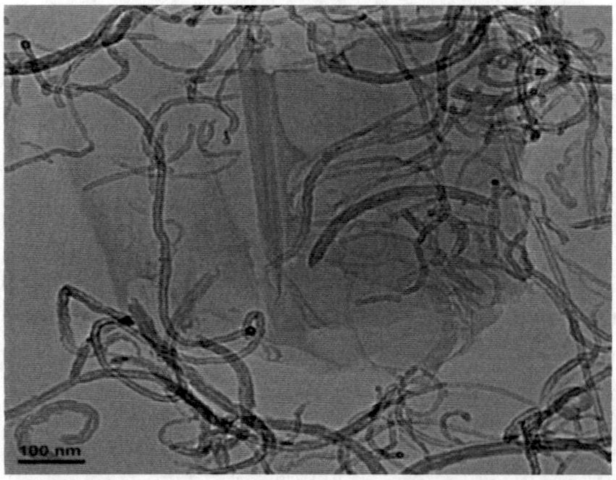

Figure 6. TEM micrographs of G-MWCNTs [17]

Report indicated that the chemically bonded graphene leaves on CNTs show good optoelectronic and gas sensing properties with reduced sheet resistance [14]. Gas molecules concentration dependence of electrical property was analyzed and the sensor will proposed based on carrier transportation and adsorption isotherm respectively for different gas molecules. Variety of gas sensors have been proposed to detect different types of chemical gas molecules such as Nitrogen Dioxide (NO_2), Ammonia (NH_3), Methane (CH_4), Hydrogen (H_2), Hydrogen Sulphite (H_2S) and Oxygen (O_2). TEM images of G-MWCNTs hybrids are shown in figure 6. As observed from the micrograph, the MWCNTs are seen to be incorporated into graphene layers and the surface morphology exhibits a mixed structure with MWCNTs dispersed between different GO sheets [17].

The gas sensitivity (S) was determined by $S = (R-R_0) / R_0$, where R_0 is the initial resistance value measured in the pure gas and R is the resistance value measured in the presence of gas mixture. Figure 7 shows NO_2 sensing response signal from CNTs/rGO hybrid film devices operated at room temperature. The results shows, CNTs/rGO hybrid films take advantage of the high sensing performance on NO_2 gas. CNTs/rGO gas sensor responded to 10 ppm NO_2 at room temperature with a sensitivity of 20% after 60 minutes of exposure [20].

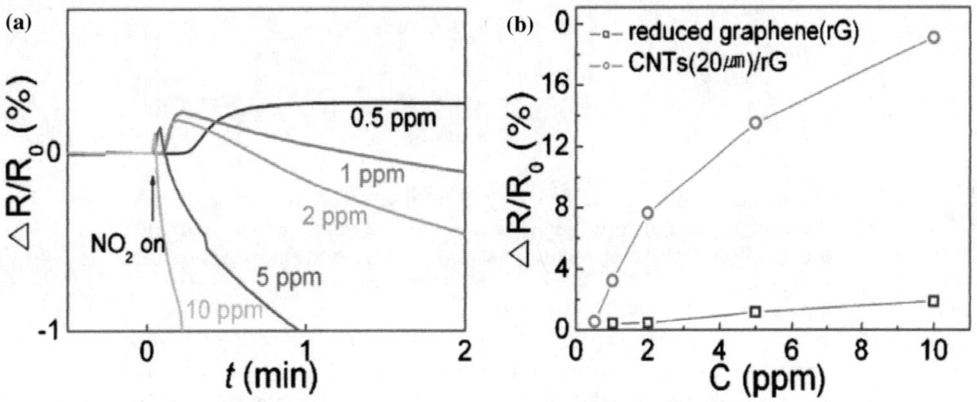

Figure 7. (a) NO_2 sensing behaviours of CNTs/rGO hybrid film devices at NO_2 concentrations from 0.5 – 10ppm. (b) Resistance response of rGO/CNTs [20]

6. Gas sensor array

An array module containing different types of gas sensors to sense the gases. If the target gas was assorted with other gases, the gas sensor array containing different sensors was more efficient when measuring a specific gas than a single element [11]. Another capability of the gas sensors is the sensors can be manufactured to be very small. In addition, portability and low power consumption are important characteristics needed for this application. GO was used to spin coat a Si substrate and standard photolithography was used to create interdigitated arrays of Ti/Au electrodes [19]. Liu and co-workers [21] to decorate RGO with Pd nanoparticles via a solution chemistry method. An array of sensors using oxygen plasma treated multiwalled CNTs decorated with Pd nanoparticles [22]. Figure 8 shows typical room temperature responses of the sensors to H_2S.

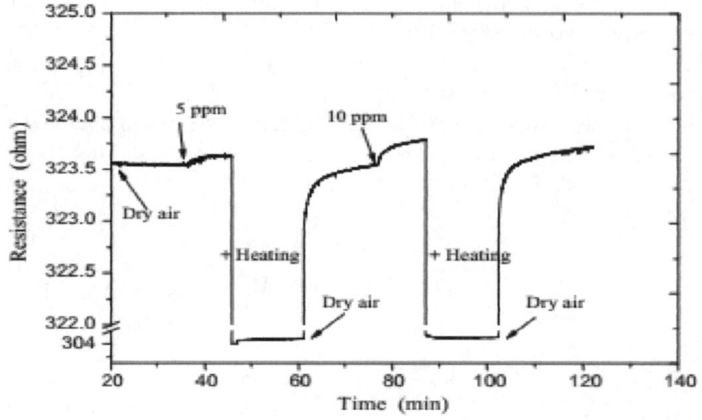

Figure 8. Response and recovery of a Pd-CNT sensors towards H_2S. Sensor respond at room temperature and heating at 150℃ is applied during initial phase of recovery to promote desorption of gas species [22]

7. Conclusion

The unique properties of graphene, reduced graphene oxide and CNTs offer various advantages and threatening the dominance of established nanomaterials in potential application of gas sensors. In this review, selectivity and sensitivity remains an important issue as many studies show the difficulty of making carbon nanomaterial absolutely selective and sensitive for different gas. CNTs and graphene sensors provide high sensitivity up to sub ppb level for a range of gases a room temperature. Hybridization of graphene and CNTs is considerably extending the possibilities of gas sensing. They presented possibility of high sensitivity towards detection of chemical analytes. Portable sensor arrays technology lies in the development that allow detecting and screening multiple analytes gas. Improved modification methods are being investigated and developed for hybridization of graphene CNTs sensor.

References

[1] Franke M E *et al* 2006 Small **2** 36.
[2] Patolsky F *et al* 2006 Anal. Chem. A **78** 4260.
[3] Jang Y T *et al* 2004 Sensors and Actuators B **49** 118.
[4] Kauffman D R *et al* 2008 Angew. Chem. Int. Ed. **47** 6550.
[5] S.Iijima 1991 Nature **354** 56.
[6] D.Kopeliovich "Graphite Solid as Lubricant" 2012.
[7] Pham V H *et al* 2012 J. Mater. Chem. **680** 749.
[8] Weber State University "Graphene Oxide" 2015.
[9] Kim Y S *et al* 2012 Nanotechnology **23** 015301.
[10] Woo S *et al* 2012 Electrochim. Acta. **59** 509.
[11] Mani V *et al* 2013 Biosens. Bioelectron **41** 309.
[12] Kim E *et al* 2012 Sensor **16262** 16273.
[13] Misra A 2014 Current Sciense **107** 419.
[14] Dong K Y *et al* 2013 Nanoscale Res. Lett. **8** 12.
[15] Wan N *et al* 2013 Carbon **53** 260.

[16] Ling Y "carbon nanotube gas sensor conductance model, sensing mechanism analysis and applications in flexible sensors and wireless sensors." 2013.

[17] Cittadini M *et al* 2014 Carbon **54** 453.

[18] Shahriary L *et al* 2014 J. Nanotechnol. **2014**.

[19] Choudhuri S "Bulk Synthesis of Graphene Nanosheets" 2012.

[20] Llobet E 2013 Sens. Actuators, B **179** 32.

[21] Jeong H Y *et al* 2010 Appl. Phys Lett **96** 213105.

[22] Li W *et al* 2011 ACS Nano **5** 6955.

IOP Conf. Series: Materials Science and Engineering **551** (2019) 012018 doi:10.1088/1757-899X/551/1/012018

Comparative Studies of Rib Waveguide Material for Quantum Communication Application

Roshidah Yusof [1], Norshamsuri Ali[1,3], Piotr Kolenderski[2], Karolina Slowik[2] and N. A. M. Ahmad Hambali[1]

[1]School of Microelectronic Engineering, Universiti Malaysia Perlis, 02000 Arau, Perlis, Malaysia.
[2]Faculty of Physics, Astonomy and Informatics, Nicolaus Copernicus University, Grudziadzka 5, 87-100 Torun, Poland.
[3]Green Materials for Electronic Application Group, CEGeoGTech, Universiti Malaysia Perlis, 01000 Kangar, Perlis, Malaysia.

Abstract. Recently, the needs to achieve a very high traffic capacity and superfast quantum computing led to advancement in optical waveguide technology. The selection of material with a high refractive index and transparent within the telecommunication wavelength range are crucially needed to achieve these. In this paper, the performance of two promising quantum materials, Silicon on Insulator (SOI) and Lithium Niobate on Insulator (LNOI) rib waveguide were studied. The mode analysis was conducted by using a finite element method to observe the confinement electromagnetic wave across rib waveguide. Apart from the variation of a material index, the width and height of core were optimized to achieve single mode propagation at the wavelength of 1550 nm. Based on the simulation work, it is shown that both material structures were able to produce single mode propagation with SOI showing higher confinement compared to the LNOI rib waveguide structure. LNOI structure was able to provide wide range of propagation signal wavelength.

1. Introduction

Currently, the rapid advancement in integrated optics has led to the acknowledgment of optical waveguides as one of the key element in optical integrated circuits. The optical waveguide has been used extensively in a telecommunication system to carry the information in a form of photon efficiently compared to copper-based interconnecting system. With an ability to produce more broadband, low loss and minimal power consumption these promising optical devices offer a bright future for quantum communication. For the past decades, research on optical waveguide has been conducted comprehensively for various applications such as switches, filter, and detectors[1-3]. These waveguide functions are determined by their design structure and material. Rib optical waveguide is one of the practical and extensively used solutions for the past few years compared to other optical waveguides such as a ridge, buried and planar waveguide [4]. It consists of three elements which are a substrate, a thin buffer layer on top of the substrate and a rectangular confinement layer as shown in Figure 1.

The refractive index of the rectangular layer has to be larger than the surrounding area (buffer layer) in order to produce light confinement based on total internal reflection phenomena. Relative refractive index contrast, Δn is given by the ratio of the index between the rectangle confinement layer and the buffer layer. The bending structure is one interest since there where a critical point of total internal reflection and centripetal force of electromagnetic wave. As the refractive index increase, the

IOP Conf. Series: Materials Science and Engineering **551** (2019) 012018 doi:10.1088/1757-899X/551/1/012018

minimum bending radius is reduced. Selection of high index material such as Silicon and Lithium Niobate is an alternative to improve the performance of quantum communication. Additionally, the rib waveguides structure must be designed to operate within fundamental mode either TE00 or TM00. On the contrary, the higher order model led to higher propagation loss[4]. This work aims to compare and analyze single mode propagation of rib waveguide for two different material which are Silicon on Insulator and Lithium Niobate on Insulator.

1.1 Rib waveguide structure
Rib waveguide structure consists of 3 important elements: substrate (A), buffer layer (B) which act as a cladding and confinement layer (C) with a micro-thick slab of height (h) which acts as a core. In a quantum communication system, it is crucially required to obtain a single mode propagation across the waveguide. The selection of material and geometrical optimization are the key element to achieve this. Based on Figure 1, width (W) and height (H) of the core, height of slab (h) and an etched depth are among the important parameters to be optimized in order to achieve single mode fiber geometries.

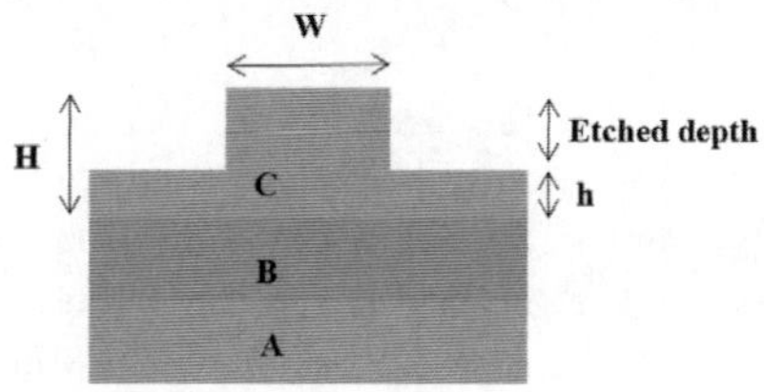

Figure 1. Rib waveguide cross section.

1.2 Silicon on insulator
Silicon is known as an ultimate semiconductor material with a broad transmission region within Infrared wavelengths. Due to its low absorption loss at 1310nm and 1550nm, silicon becomes a very promising material for quantum communication. Due to its high refractive index contrast, silicon is able to provide strong light confinement. Silicon is highly compatible with integrated circuit manufacturing processes known as CMOS fabrication[4,6]. In order to operate within fundamental mode properties at a minimal propagation loss, the thickness of the silicon core has to be sufficiently small.

1.3 Lithium Niobate on Insulator
Recently, research on Lithium Niobate (LN) based photonic devices has been conducted due to its wide transmission window spanning from UV to mid-IR wavelength region and strong electro-optical coefficient. With an ability to produce high second-order optical nonlinearity, it leads to the tremendous progress in nonlinear nanophotonics technology. Unfortunately, the small index contrast between waveguide core and cladding results in large device dimensions and large bending radii which limits the ability for dense integration. Recently, the commercialization of LN on insulator (LNOI) substrate offers a brighter alternative to enhance the performance of LN [7-9] mainly for microscale LN waveguides and its application in nonlinear wavelength conversions.

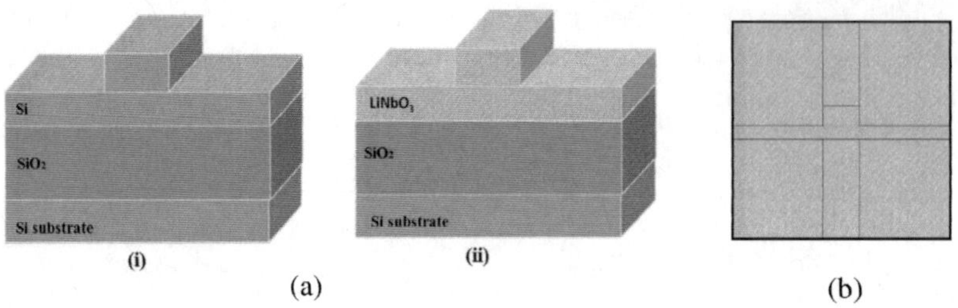

Figure. 2 (a) (i) SOIand(ii) LNOI rib waveguide structure consist of Silicon Substrate, SiO_2 layer and Si (a) and LiNbO3 (b) which act as a confinement layer. **Figure 2(b)** 2D rib waveguide structure consist of SiO_2 structure and confinement layer (Si or LiNbO3). The condition is set under the standard air condition ($n_{air} = 1$) on top of confinement layer

1.4 Mode Analysis

In order to identify the modes of the waveguide, we have employed the eigen-mode solutions to the Maxwell equations with an appropriate boundary condition of the waveguide geometry. Currently, a number of numerical techniques can been used to analyze the waveguide modes such as the beam propagation method, effective index method and finite element method [6] . In this work the finite element method is used to conduct mode analysis as this method is used to obtain the polarization independence single mode waveguide with a smallcross section as shown in Equation 1.1. This method has been introduced by Chan et al[5].

$$\frac{W}{H} \leq 0.05 + \frac{(0.94+0.25H)r}{\sqrt{1-r^2}} \tag{1.1}$$

where $r = h/H$ (h/H < 0.5 for small cross section) , 1.0 um ≤ H ≤ 1.5 um

The thickness of the core, H must be greater than the thickness of the slab, h ($H>h$) to allow light confinement at the ridge. Apart from that, the variation of thickness and width of core affect the effective indices of refraction, n_{eff} of the structure.

2. Geometrical and material optimization

The simulation of 2D rib waveguide has been performed to observe and analyze the mode propagation across the structure by using finite element analysis software. In this simulation, the dimension of the core is given by 0.5 μm x 0.3 μm (WxH) and the thickness of the slab, h was set at 0.2 μm. The value of h is set smaller than H to optimize the mode confinement at the core. Based on Equation 1.1, width, W of the core was varied from 0.5 μm to 1 μm to obtain a small cross section polarization independent single mode properties. Based on Figure 3, the refractive index of Silicon, n_{Si} and Silicon Oxide, n_{SiO2} are set at 3.5 and 1.5, respectively. Whereas for LNOI based waveguide structure, the refractive index of LN, n_{LN} and Silicon Oxide, n_{SiO2} are set at 2.2 and 1.5, respectively. Hence, the value of refractive index differences, Δn for both structures are given by2.0 and 0.7, respectively.

The operating free space wavelength, λ is set as 1550 nm for both SOI and LNOI structure. Alternatively, it has been optimized to 755 nm for LNOI, respectively as these material are optimally operated within this region. The mode analysis study was conducted to compute the propagation constant, wave number and mode confinement within the core area for a given frequency, f_o. These fundamental mode values are obtained based on the largest propagation constant and electric field distribution.

3. Results and Discussion

From the simulation of 2D SOI and LNOI rib waveguide, the geometrical structure has been optimized to obtain fundamental mode. The optical operation wavelength of SOI was set at 1550 nm whereas for LNOI was set at a pump wavelength of 1550 nm and second harmonic wavelenght of 755 nm, respectively.

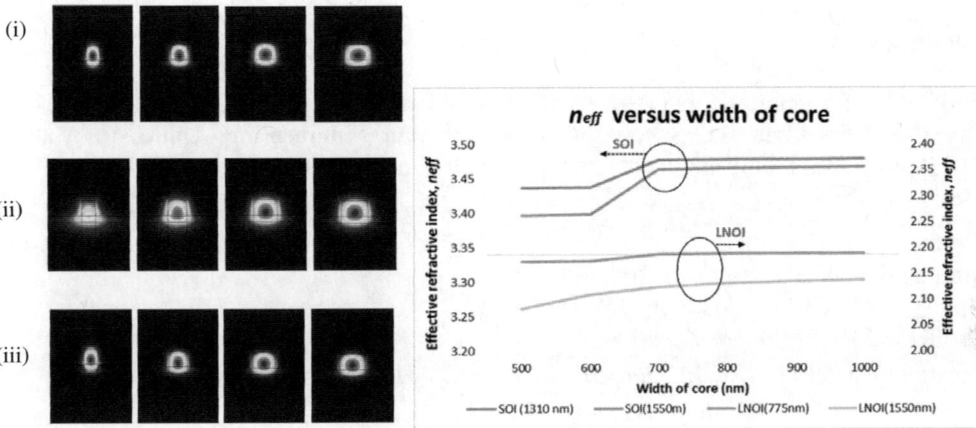

(a) (b)

Figure 3 (a) Mode distribution across the (i) LNOI waveguide at 755 nm , (ii) LNOI waveguide at 1550 nm and (iii) SOI waveguide at 1550 nm as the width is varied to 500 nm, 700 nm, 900 nm and 1000 nm (from left to right). **Figure 4(b)** Graph of effective refractive indiex versus width of core for the confined mode of SOI and LNOI rib waveguide.

Figure 3 shows the cross section view of optical mode field distribution across the waveguide structure for (a) LNOI at 755 nm (b) LNOI at 1550 nm and (c) SOI at 1550 nm with a variation of cores' width, *W* from 500 nm to 1000 nm. Consequently, the correlation between effective refractive indices and a variation of the width of the core are shown in Figure 4 (b). Based on Figure 4 a(i) and a(iii), the optimum single mode distributions were found consistently across the SOI waveguide structure. However, in LNOI waveguide at 1550 nm, the single mode distributions were only found as the minimum width, W of 900 nm was applied. These optical modes were presented as a solution of eigenvalue problem derived from Maxwell's equation. From the simulation, the fundamental mode is selected based on the highest propagation constant, β which reflected the maximum confinement of an electric field across the structure. The normalized propagation constant is intepreted in a form of a modified effective refractive index, $n_{eff.}$ As shown in Figure 4(b), as the width of the core was optimized, the value of n_{eff} varied. The graph shows that the value of n_{eff} increase significantly as the width of core increased and it seems to be less sensitive to the wavelength variation especially for SOI waveguide. A as the wavelength is varied at 1310 nm and 1550 nm for SOI and 755 nm and 1550 nm for LNOI, there are only a slight variation. The maximum value of n_{eff} for SOI waveguide at 1310 nm and 1550 nm are given by 3.4786 and 3.4469, respectively. Whereas for LNOI waveguide at 775 nm and 1550 nm are given by 2.1890 and 2.1384, respectively. Equation 1.1, it indicates that the single mode condition is dependent on the size of waveguide but independent to the propagation length.

4 Conclusions

From the studies, it can be conclude that both material structures were able to produce single mode propagation with SOI showing higher confinement compared to the LNOI rib waveguide structure. Secondly, geometrical structure optimization is crucially required in order to obtain single mode propagation. The LNOI rib waveguide gives the flexibility of wavelength range for single mode waveguide structure design. In this design structure, usually Second Harmonic Generation (SHG) or Spontaneous Parametric Down Conversion have modal phase matching so if the fundamental is single mode then the SHG should also be a single mode eventhough the waveguide supports a multimodes at this wavelength[13].

Acknowledgements

This work was financially supported by FRGS/1/2015/SG02/UNIMAP/02/4 from Ministry of Education (MOHE), Malaysia. School of Microelectronic Engineering, UniMAP, Malaysia and Faculty of Physics, Astonomy and Informatics, Nicolaus Copernicus University, Poland.

References

[1] Marpaung *et al* 2015 *Optica* **2** 76.
[2] Han *et al* 2016 *Appl. Phys. Lett.* **108** 141106.
[3] Chollet F 2016 *Micromachines* **7** 18.
[4] Fonstad N D C 1985 *IEEE J. Quantum Electron* **21** 315.
[5] Pogossian *et al* 1998 *J. Light. Technol.* **16** 1851.
[6] Chan *et al* 2005 *J. Light. Technol.* **23** 2103.
[7] Aalto *et al* 2018 *Optical Society of America.*
[8] Wang *et al* 2017 *Opt. Express* **25** 6963.
[9] Wang *et al* 2018 *Optica*, **5** 1438.
[10] Boes *et al* 2018 *Laser Photonics Rev.* **12** 1700256.
[11] Bazzan *et al* 2015 *Phys. Rev.* **2** 040603.
[12] Amnon Yariv *et al* *Oxford Publication* 130-139.
[13] Horn *et al* 2013 *Sci. Rep.* **3** 2314.

Electric Field and Voltage Distribution (EFVD) Analysis for 132kV Bushing Transformer Using Finite Element Analysis

N. Abd. Rahman[1], M. Isa[2] , MNKH. Rohani[2], HA.Hamid[2] and MMAB Abdullah[3]

[1]Pusat Pengajian Kejuruteraan Mikroelektronik, Universiti Malaysia Perlis, 02600 Arau, Perlis, Malaysia.
[2]Pusat Pengajian Kejuruteraan Sistem Elektrik, Universiti Malaysia Perlis, 02600 Arau, Perlis, Malaysia.
[3]Centre of Excellence Geopolymer and Green Technology, University Malaysia Perlis, 01000 Kangar, Malaysia.

E-mail: nurulazmi@unimap.edu.my

Abstract. The electric field and voltage distribution (EFVD) are an important parameter for assessing high voltage bushing transformer performance. In order to understand how EFVD being distributed along the bushing transformer, Finite Element Method (FEM) has been employed for simulation purposed. 2 dimensional (2D) object designed using Computer Aided Design (CAD) using actual dimension of 145kV bushing transformer from manufacturer data sheet. The model is then exported to Finite Element Method (FEM) software for further analysis. Using FEM, the EFVD can be obtained. From this study, an understanding of electric field and voltage distribution along bushing transformer and between aluminium foils are attained.

1 Introduction

Electrical energy is the major requirement for all over the world. Nowadays, it can't be neglected the importance of electrical energy in various sectors such as industrial production, agriculture, health facilities, machineries and transportation [1, 2]. The modern day of life is so much dependent on electrical energy so that it is not affordable to have an outage of electricity even for a few minutes. There are three main component for power system. It begins with generation, then transmit the energy through transmission line and finally distribute it to consumers.

Transformer is considered as an among the highest investment in power system. Transformer plays and important role in generation, transmission and distribution. It's main function is to convert high voltage to low voltage or vise versa depending on whether to transmit or to distribute the electrical energy. There are two category of transformer ie. power transformer and distribution transformer. Usually the rating for power tranformers are 500kV, 275kV, 132 kV and distribution transformer rated at 33kV and 11kV.

Bushing transformer are the fundamental component of high voltage power. Figure 1 shows the position of bushing transformer. As shown there, HV bushing are the connection between an incoming line and LV bushing are the connection for an outgoing to the lower voltage. There are three types of main insulation in bushing transformer [2]; polymer, resin impregnated paper (RIP) dan oil impregnated paper (OIP). In this paper, only OIP types of bushing transformer will be discussed. High voltage bushing are an integral part of electrical equipment for carrying one or more high-voltage conductors through a grounded barrier, e.g. a transformer tank or wall [2]. Therefore, bushing must be designed to cater electrical, mechanical, thermal dan environment stress [2]. The bushing position at transformer is shown in figure 1.

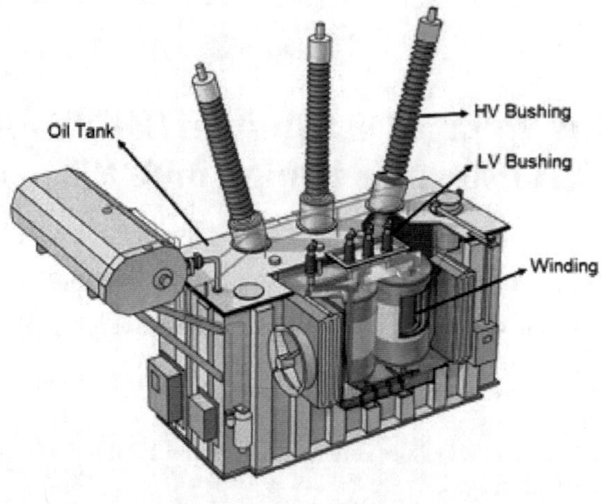

Figure 1. Power transformer

Electrical stress is the most important types of stress to be considered compared to other stress as mention above [2] . Based on [5], it is reportted that, 25% to 30% failure of bushing transformer contribute to transformer failure. Hence, it is important to study and analyze the effect of Electric Field and Voltage Distribution (EFVD). There are several research but found very little focusing in bushing transformer [1-6]. Most of the research shared their EFVD results to be discussed. In this paper, finite element methord (FEM) software is employed to evaluate the performance of EFVD for 132kV OIP bushing transformer. The results from simulation shows good agreement with several result from [3-6].

2. Methodology

2.1 Modelling of Bushing

Computer Aided Design (CAD) software was used to design the bushing transformer model. The dimension for bushing transformer is taken from [1]. It is a model GOB 145kV OIP type. The structure of bushing transformer is depicted in Figure 3. The construction of bushing transformer consist of this six main items and it is shown in Figure 2.

Figure 2, shows the 2D and 3D bushing transformer design using CAD software. In order to reduce the computer processing time, only half of the bushing or 2 dimensional analysis will be consider. It is shown in Figure 2. The dimension for the modeled bushing is taken from [1, 2].

IOP Conf. Series: Materials Science and Engineering **551** (2019) 012019 doi:10.1088/1757-899X/551/1/012019

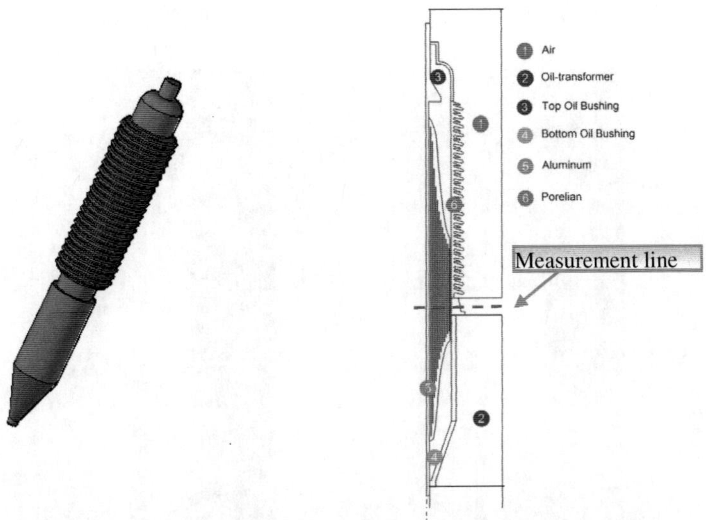

Figyre 2. Bushing transformer modeled in 3D and 2D model.

2.2 Finite Element Analysis

Finite Element Method (FEM) software using a numerical method used to solve problem in engineering and mathemathics. It is useful to solve problems with complicated geometries, loadings, and material properties where analytiacal solutions can be obtained. The module of FEM used to simulate was AC/DC electrostatic model. The equation related to electrostatic are given below :

$$\boldsymbol{E} = -\boldsymbol{\nabla V} \tag{1}$$

$$\boldsymbol{\nabla . D} = \boldsymbol{\rho_f} \tag{2}$$

$$\boldsymbol{D} = \boldsymbol{\varepsilon . E} \tag{3}$$

$$\boldsymbol{\nabla . \varepsilon \nabla V} = \boldsymbol{0} \tag{4}$$

Where **D** is the electric displacement field, **E** is electric field, $\boldsymbol{\rho_f}$ is the charge density and **V** is electric potential. After the model drawn and imported to FEM software, the permittivity and electrical conductivity were assigned to the model with the value shown in Table 1 [tesis]. Then, the boundary setting is done. The bushing boundaries along the zero axis which is in red colour are taken as electric potential. Aluminium foils, OIP boundaries and porcelain sheds are set as continuity. Air and oil outer boundaries are set as electric insulation. Metal flange and last foil are grounded. The Relative Permittivity, ε_r for air is 1, oil is 2.2, porcelain is 5, oil impregnared paper (OIP) is 4 dan aluminium foils is $\boldsymbol{10^8}$.

3. Results and Discussion

3.1 Electric Field and Voltage Distribution

Figure 3 shows the simulated results for electric field and voltage distribution in the OIP bushing transformer.

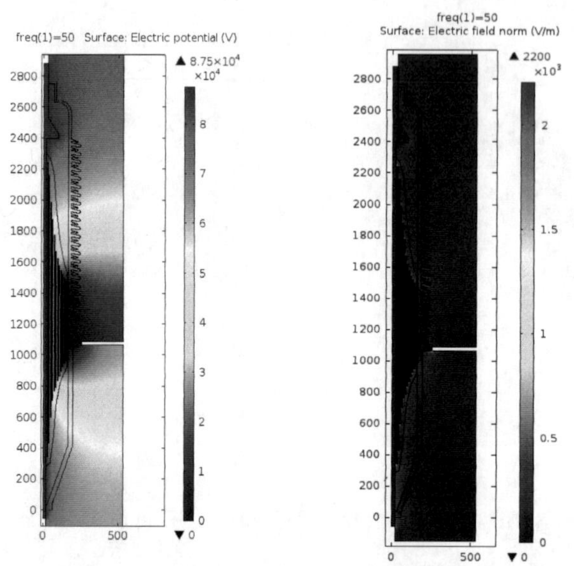

Figure 3. (a) Voltage distribution for OIP bushing (b) Electric field for OIP bushing.

Figure 3 (a) shows the voltage distribution in terms of surface plot for voltage distribution for bushing transformer. On the other hand, Figure 3 (b) shows a surface plot for electric field distribution for bushing transformer. For voltage distribution, the stress is low at the bushing transformer flange. This is because of the flange is connected to ground. The voltage distribution also distritubed uniformly at each aluminium foils. However for electric field, the intensity become higher especially at the sharp edge of bushing transformer.

In Figure 4(a) , the voltage distribution is measure at flange bushing transfore as represented by dash line Figure 2. The value of voltage at last foil is zero because the last foil is connected to the bushing flange and both of it are grounded. It is also clear that the distribution of voltage is uniform between each 12 aluminium foils.

On the other hand, in Figure 4(b), the electric field is uniformly distributed between foils as voltage distribution. The value of electric field for the last foil also zero as it is connected to ground. It is proven that the foils is actually used to distribute electric field so that there is no cencetrated electric field at certain point in insulation.

IOP Conf. Series: Materials Science and Engineering **551** (2019) 012019 doi:10.1088/1757-899X/551/1/012019

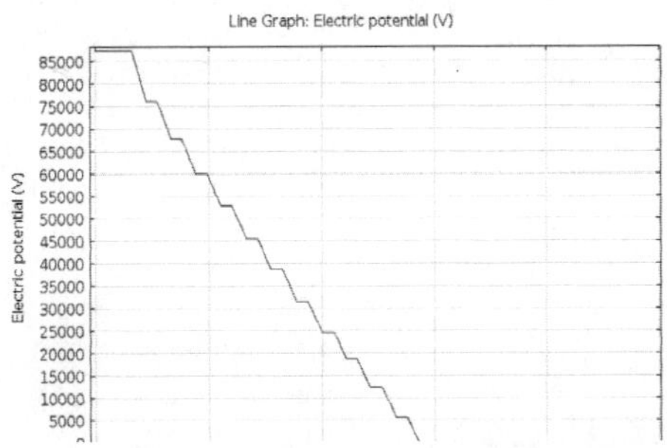

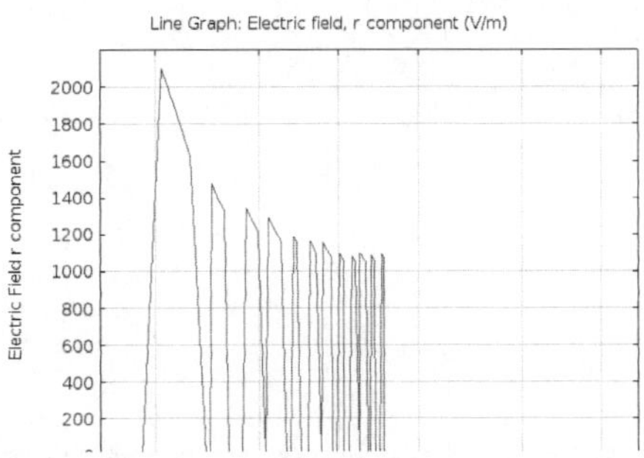

(a) (b)

Figure 4. (a) Voltage distribution for OIP bushing measured along flange (b) Electric field for OIP bushing measured along flange

4. Conclusions

From the research, we can conclude several conclusions which are;

i. In this paper, the simulation of EFVD have been successfully carried out to 2D axial-symmetry geometry decribing OIP bushing transformer.
ii. The results of EFVD shows an agreement with other researchers.
iii. The aluminium foils plays an important role in reducing the electric stress for bushing transformer.

References

[1] Z. Ahmed 2011 *Analysis of Partial Discharge in OIP Bushing Models, Thesis*
[2] R. Anguraja *et al.* 2017 *Third International Conference on Current Trends in Engineering Science and Technology*
[3] Jyothi *et al.* 2010 *IEEE T Dielect El In* **17** 931.
[4] G.J. Pukel *et al.* 2006 *International Conference on Condition Monitoring and Diagnosis, CMD*
[5] A. K. Lokhanin *et al.* 2002 *CIGRE Report* **15** 1.
[6] Illias *et al.* 2012 *IEEE International Conference on Power and Energy (PECon).*
[7] Rohani *et al.* 2015 *IEEE Student Conference on Research and Development (SCOReD)*

Graphene Multi-layer on Plastic Substrate

Muhammad M. Ramli[1,2], Siti S. Mat Isa[1,2], Simon S. Henley[3]

[1]School of Microelectronic Engineering, Pauh Putra Campus, University Malaysia Perlis, 02600 Perlis, Malaysia
[2]Center of Excellence Geopolymer & Green Technology (CeGeoGTech), School of Material Engineering, Universiti Malaysia Perlis (UniMAP), P. O. Box 77, d/a Pejabat Pos Besar, 01000 Kangar, Perlis Malaysia.
[3]Advanced Technology Institute University of Surrey Surrey, United Kingdom

E-mail: mmahyiddin@unimap.edu.my

Abstract. Vacuum filtration method was used in order to fabricate a homogeneous and uniform thin film of multi-layer graphene on plastic substrate. This self-regulating technique allows the number of graphene layer to be controlled thus controlling the film thickness by simply varying either the concentration of the graphene in the suspension or the filtration volume. Measurement of the sheet resistance as a function of graphene concentration in solution shows the percolation behavior of multi-layer graphene films.

1. Introduction

Graphene is a single layer of carbon atom arranged in honeycomb lattice like structure. It has been proved that graphene exhibits fast response and high sensitivity for detection of small concentration of toxic gases in room temperature due to very high volume-exposure ratio [1]. It has improved the sensitivity of the device by increasing charge carrier concentration [1].

Recently, the demand on low cost and flexible device has improved greatly. The use of plastic substrate become preferable as reduce the production cost and make it easy to wear this device as an 'easy-to-wear' item that can be used in any locations and activities for safety purposes. In this paper, the fabrication of graphene on plastic substrate is discussed in details.

2. Experimental Details

2.1. Graphene oxide production

First, graphene oxide (GO) was produced from graphite flake (NGS Naturgraphit GmbH, carbon contents >85%) using the modified Hummers method. The graphite powder (2.0 g) was added into a solution of concentrated hydrogen sulphide (H_2SO_4) (20 mL), potassium persulfate ($K_2 S_2O_8$) (1.0 g) and phosphorus pentoxide (P_2O_5) (1.0 g) mixture that were completely dissolved at 80 °C. Then the mixture was kept at 80°C for 4 and half hours using an oil bath. After that, the mixture was cooled to room–temperature and diluted with 1 L de–ionised (DI) water and left overnight. Subsequently, the resulting product was filtered (0.2 μm Milipore membrane) and washed with DI water until the pH of filtrate water become neutral in order to remove any residual acid. Then the oxidation process of Hummers method was performed. The pre – oxidised graphite was then dispersed by stirring into cold H_2SO_4 (80 mL) in an ice bath. Then, potassium permanganate ($KMnO_4$) (10 g) was added by stirring it slowly to keep the temperature of reaction mixture below 20 °C. After that, the mixture was stirred at 35 °C for 2 hours and diluted with DI water (160 mL) to give a dark brown solution. The addition of water in concentrated H_2SO_4 released a large amount of heat. Hence water was added into an ice bath in order to keep the temperature below 50 °C. Then, the mixture was further diluted by DI water (500 mL) and stirred for 2 hours. After that, the hydrogen peroxide (H_2O_2) (30%, 8 mL) was added slowly into the mixture and the colour of the mixture turned into yellowish from dark brownish previously. The resulting product was then washed with 10% hydrochloric acid (HCl) (1L) by stirring and the supernatant was decanted after overnight. The remaining product was then centrifuged and washed with DI water until the pH of the solution became neutral. This yellow – brown homogeneous suspension was stable for several months and used for reduction. This GO suspension was then further diluted with DI water to produce a set of different concentration that varied from 0.00004 mg/ml up until 0.01 mg/ml. The variation of concentration was critical in order to investigate the film conductivity as a function of concentration.

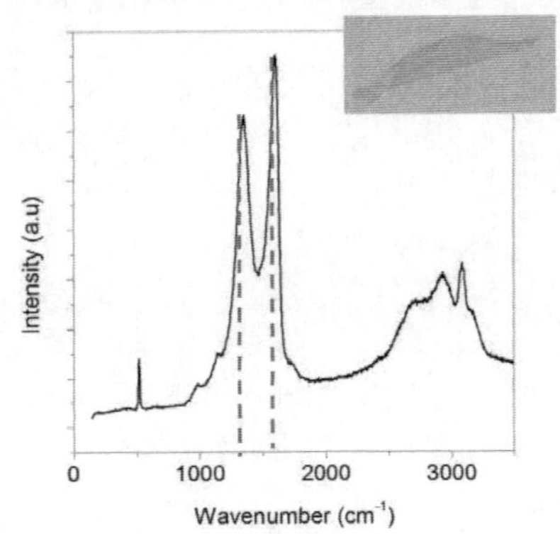

Figure 1. Raman spectrum shows the D and G peaks. (Inset) Graphene sheet on SiO_2.

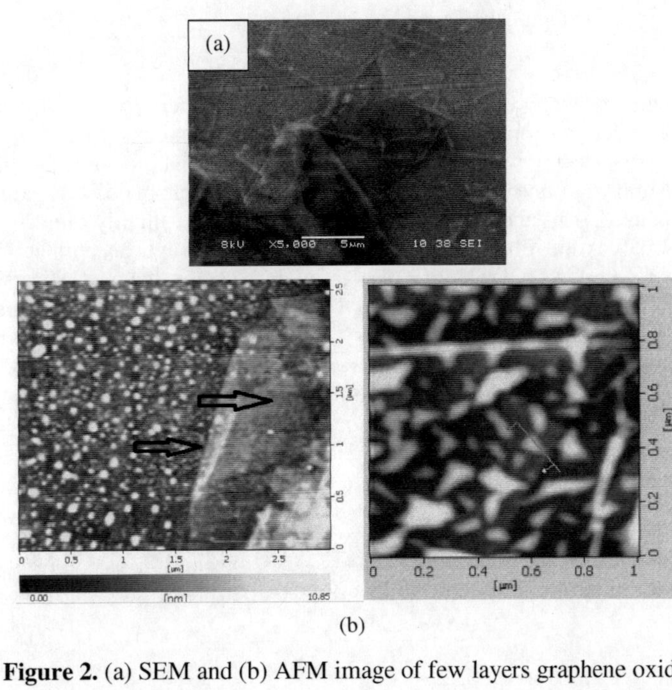

(b)

Figure 2. (a) SEM and (b) AFM image of few layers graphene oxide.

IOP Conf. Series: Materials Science and Engineering **551** (2019) 012020 doi:10.1088/1757-899X/551/1/012020

2.2. Thermal treatment

Peracetic Formation GO is electrically insulating because of the disruption of the sp_2 bonded graphitic structure by the oxygen atoms. A reduction process is needed to make GO conducting by removing oxygen functional groups. By exposing GO to reduction agent such as hydrazine [7] or through thermal treatment [8], GO can become conducting. Hydrazine reduction is believed to remove the oxygen functional groups [9]. However in doing so, hydrazine also introduces a significant amount of nitrogen-containing functional groups to the GO. The presence of these groups may decelerate the sensing response of the device when expose to various adsorbates [9]. Hence, the reduction process using thermal treatment method was preferred in order to increase GO electrical conductivity. Thermal reduction of GO was carried out in vacuum oven (Technico Ltd., India). The GO solutions were then placed in vacuum oven and treated at 100 °C, 200 °C and 300 °C for 1, 2 and 3 hours. After heating, samples were cooled to room temperature.

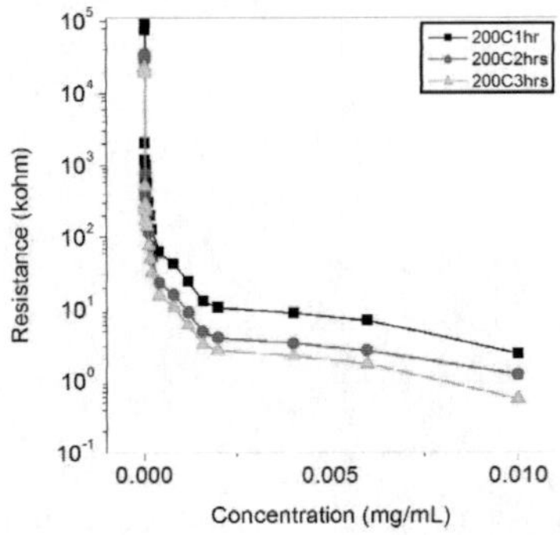

Figure 3. Electrical resistance of different concentration of graphene oxide solution for different annealing time.

2.3. Sample preparation

The production of homogeneous and thin film of r-GO was crucial. There are several methods to produce uniform film. Methods such as drop casting [12], rapid freezing by spraying [13] and dip coating [14] from r-GO colloidal suspension were used to obtain a thin and homogeneous film. In order to reproducibly achieve uniform thin films with a controllable layer (thickness) over large areas, vacuum filtration method was used. This method was widely used to deposit highly uniform r-GO thin films [15]. This method involves the filtration of r-GO suspension through a commercial polycarbonate membrane (0.1 μm pore size). As the solvent passes through the pores, the r-GO sheets were trapped on the membrane surface thus forming an interconnected network (Figure 2(a)). The permeation rate of the solution was controlled by the accumulation of r-GO sheets on the pores so as the number of tubes increases at a given location on the porous membrane, the rate of filtration decreases, but does so to a lesser degree at thinner or uncovered regions. The process is therefore self-regulating, which allows reasonably good nanoscale control over the r-GO sheets by simply varying either the concentration of the r-GO in the suspension or the filtration volume. Here, 0.5 ml of each solution was sucked through a membrane, coating a circular area with a diameter of 13 mm. Hence, the density of the r-GO sheets or thickness can be controlled by varying the concentration of the suspension filtered through the membrane.

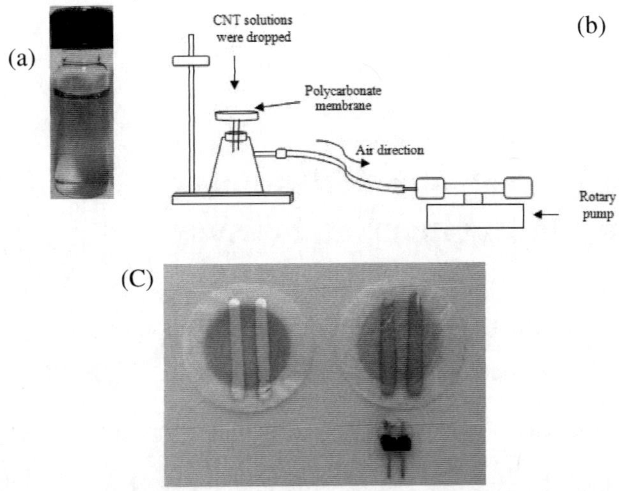

Figure 4. Graphene oxide suspension in 100% concentration. b) Illustration image of vacuum filtration set-up. c) Plastic membrane that was used as a substrate with graphene layer and gold electrodes for further potential applications.

3. Conclusions

A colloidal suspension of reduced-graphene oxide was produced by modified Hummer's method. The graphene sheets images was observed using SEM and AFM. Vacuum filtration method was successfully applied to produce conductive and uniform thin film of reduced-graphene oxide sheets. The optimum temperature for annealing treatment was observed at 200 °C for 3 hours.

Acknowledgment

This work was supported by Fundamental Research Grant (FRGS) funded by Ministry of Higher Education (MoHE) of Malaysia Government.

References

[1] Eason G *et al* 1955 *Phil. Trans. Roy. Soc. London* **A247** 529.
[2] Clerk Maxwell J A 1892 *Oxford: Clarendon* 3rd ed. **2** 68.
[3] Jacobs I S *et al* 1963 *New York: Academic* 271.
[4] Yorozu Y *et al* 1982 *IEEE Transl. J. Magn. Japan* **2** 740.
[5] Young M 1989 *Mill Valley CA: University Science.*

IOP Conf. Series: Materials Science and Engineering **551** (2019) 012021 doi:10.1088/1757-899X/551/1/012021

Development And Evaluation Of Immersive Card Games As Booster Memory In Engineering Science

Nur Ismalina Haris[1], Amarul Talip[2,3], Mohd Zulfabli Hasan[1]

[1]Mechanical Engineering Department, Politeknik Tuanku Syed Sirajuddin, Perlis, Malaysia.
[2]Green Design and Manufacture Research Group, Centerof Excellence Geopolymer and Green Technology (CEGeoGTech), Universiti Malaysia Perlis, Perlis, Malaysia.
[3]PP Diploma, Universiti Malaysia Perlis, Perlis, Malaysia.

E-mail: nurlyena@gmail.com

Abstract. Engineering science is one of the core courses that must be taken by students in engineering. There is no doubt that too many laws and concepts to be learned and understood will cause students to feel difficult and less interested in pursuing engineering science. Many studies have shown that learning motivation and understanding can be achieved through educational games. Therefore, the main purpose of this study was to help engineering students learn and understand concepts by using educational card games through immersive technology. Card development and evaluation methods have been done while pre and post test have been conducted to see the effectiveness of the card against respondents. As a result, it shows that card designs play a role in enhancing student understanding while providing positive attitudes towards the use of the educational card games with immersive technology in engineering science learning. Notwithstanding, the outcome of the survey of the respondents involved in the test shows the enthusiasm and determination to use the card games. As a result, it can be highlighted and applied to all students.

1. Introduction

Engineering science is one of the many courses that emphasizes the concept and law contained in Physics such as Zero-Thermodynamic Law, First Newton Law, Bernoulli's Law and Ohm's Law. There are many laws and concepts that need to be understood and memorized traditionally will cause students less interest in this engineering science. Infact, most of the law is interconnected with others such as as the First, Second and Third Newton Laws. Misunderstanding one of Newton's Law will be resulted in no continuity between the three Newton Laws.

A study conducted on 1,400 students and a group of Physics teachers has agreed that Physics is cumulative which is if do not understand a concept or theory it is difficult to understand other concepts and theories e.g Newton's theories of law. In fact, they also point out that there are too many things to learn and formulas to remember in Physics [1].

Thus, one of the activities that can be highlighted to encourage collaborative learning among engineering students is through games in education. Games in education are seen to have some positive features that can bring interest in learning in science. It is one way that can be used to remember and understand the basic facts and principles of science. It offers advantages over traditional learning approaches [2]. The spirit of learning also can be enhanced through games as well as exposing students to more creative and fun learning forms and reducing boredom in teaching and learning [3].

Game cards are widely used in science and engineering education. Referring to [4], card games are more effective than traditional methods and very helpful in improving memory and understanding. The study uses card games in biological education where the field is seen to have many complex concepts and vocabularies that need to be understood and remembered.

Meanwhile previous literature [5], also developed a card games named Voyager: Satellites since 2001 for school children aged between 9 and 13 years and also for general public. The card is used to convey information about Earth, space science and satellites in fun and entertaining way. The card games has received encouraging response from teachers and the community and a large number of Voyager card games: Satellites have been printed and circulated throughout the United Kingdom.

Along with technological advances, card games have also undergone a transformation where it has been created with the added use of technology. The advantages offered by the card games with this technology have gained attention in education such as web-based learning, android application, creative learning through Virtual Reality technology and the latest Augmented Reality (AR) technology along with the Industrial Revolution 4.0. The term Augmented Reality is defined in multiple ways and the most popular definition by Azuma proposed in 1997 that

"Augmented reality is a field in which 3D virtual objects are integrated into a 3-D real environment in real time." [6].

The study conducted by [7] shows that the use of interactive multimedia technology with AR technology in the subject of fitting skills in vocational colleges successfully enhances the learning process, helping students to perform practical tests and can be used as a teaching tool to generate active learning environment between teachers and students.

2. Methods

This study was divided into two phases. Phase 1 was the development of immersive card games and phase 2 was the evaluation of of effectiveness immersive card games as booster memory in Science Engineering. Figure 1 below shows the flow of methodology framework of this study. This framework was adapted from similar study [4,8].

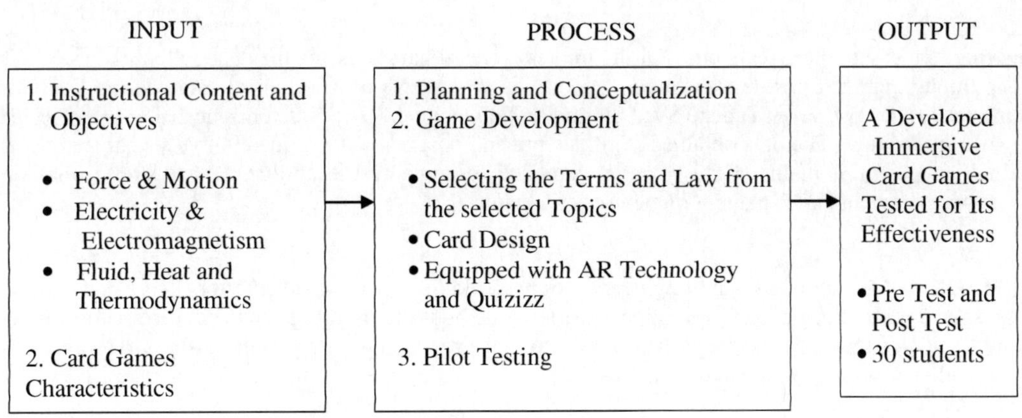

Figure 1. Flow of methodology framework.

Referring to Figure 1, there are four steps involved to develop and evaluate the effectiveness of card games: instructional content and objectives, planning and conceptualization, game development and pilot testing and evaluation of the effectiveness of the card games.

2.1 Step 1: Input
Instructional content and objectives: Three main topics in Engineering Science were selected that are Force and Motion, Electricity and Electromagnet, Fluid, Heat and Thermodynamics. The card games must have features such as pocket size, easy to carry and play.

2.2 Step 2: Process
Planning and conceptualization: Research and studies, preparation of materials also other pertinent aspects were established in this step. Four important basic elements were considered to develop this immersive card games [9]:

- Mechanics: Action, Rules & Chance
- Story/Picture: Force & Motion, Electricity & Electromagnetism, Heat & Thermodynamics
- Aesthetics: Looks, Audio, Video & Feels
- Technology: Augmented Reality and Quizizz

Card design: In this project three important basic concept in Engineering Science were selected; Force and Motion, Electricity and Electromagnetism and Fluid, Heat and Thermodynamics, so there are 48 cards and each topic consists 16 cards as shown in Figure 2. The stable and suitable AR techniques to enhance the visual learning that is HP Reveal were choosen to equip the card games. This card games also have been linked to Quizizz to test students understanding and also can be use as booster memory.

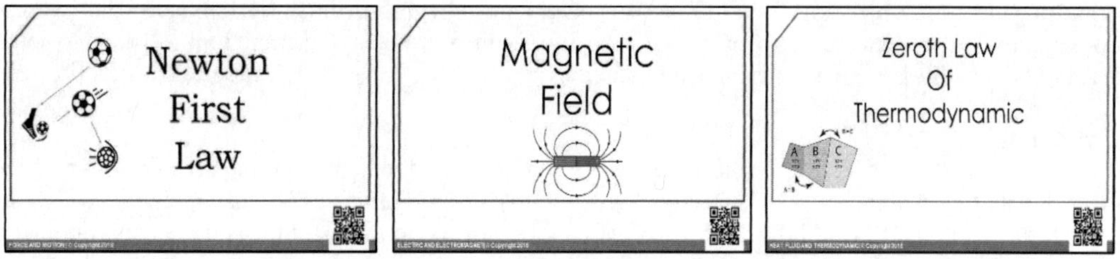

Figure 2. Sample of three cards from three topics

2.3: Step 3: Output
In order to measure the effectiveness of the immersive card games usage, the data collection process is done through pre and post test execution. A total of 30 students of Mechanical Engineering Diploma semester 1 (DKM1A) from Tuanku Syed Sirajuddin Polytechnic, PTSS Session June 2018 were selected to undergo this test. The data obtained from this pre and post test will be quantitatively analyzed using paired sample t-test method. This t-test is intended to compare the abilities of students to answer questions before and after playing immersive card game.

2.3.1 Pre-Test
These DKM1A students take the Engineering Science course in the first semester. Force and Motion topics have been selected and conventional learning methods are used together with discussions. After completing this topic, students are given 7 days to make revision before pre-test. After 7 days, pre test was conducted among 30 students. The researchers found that marks obtained from pre-test were low and some were zero.

2.3.2 Post Test
After pre-test, researchers continue teaching and learning assisted by this card game. Students are given the opportunity to play this card game in several groups. Students are required to send scores obtained from the game together with pictures and videos as evidence. Post-test is done after completing the teaching and learning sessions.

IOP Conf. Series: Materials Science and Engineering **551** (2019) 012021 doi:10.1088/1757-899X/551/1/012021

2.3.3 t-Test
Reliable sample t-tests have been performed to see the effectiveness of card games usage on teaching and learning. Table 1 shows the experiment information carried out.

Table 1. Experimental Information

Number	Item	Value
1	Sample, n	30
2	Degree of Freedom, *df*	29

3. Results and Discussions
The data has been analyzed has been shown in Table 2 and Table 3 below.

Table 2. Descriptive statistics for pre test scores, post test and differences between them

	Pre Test	Post Test
Mean	5.8667	7.2000
N	30	30
Standard Deviation	1.38298	1.24291
Std. Error Mean	0.25250	0.22692

Table 2 above shows the mean score of student performance before using the card games is 5.8667 and is smaller than the mean of performance after using card games of 7.200. This means that student performance before and after using cards games has a significant different.

Table 3. T-test on pre-test and post test scores

	Perbezaan
Mean	-1.3333
N	30
Standard Deviation	2.12267
Std. Error Mean	0.38755
t	-3.440
df	29
Sig. (2-tail)	0.002

Likewise, based on Table 3, the t-values for student performance comparison before using card games and after using card games are $t = -3.440$ and significant level $p = 0.002$. This significant level is smaller than 0.05 ($p > 0.05$). Therefore, the findings show that there is a significant difference between student performance before applying card games and after using card games.

4. Conclusions

The increment of scores in post-test after using this immersive card games shows that this card games can enhance students' understanding and memory in Science Engineering courses. This game card has been designed with the addition of Augmented Reality Technology elements where the advantages of this technology can enhance the player's visual memory in addition to the fun of play. The findings support the study by [10] as these students are Gen-A that to be among the most intelligent generation, highly materialistic and obsessed with technology. This card games also have been linked to Quizizz to test students understanding and also can be use as booster memory. In fact, this card games can be played with fellow students to help their independent learning time.

References

[1] Ornek F *et al.* 2008, *International Journal of Environmental and Science Education* **18** 165.

[2] Ellington H *et al.* "Games and Simulations in Science Education," in *Games and Simulations in Science Education*" 1981.

[3] Subramaniam R *et al.* 1999, *J Biol Educ* **33** 103.

[4] Gutierrez A F 2014 *CBE—Life Sciences Education* **13** 76.

[5] D. R. Smith *et al.* 2009, *Physics Education* **44** 479.

[6] Azuma R T 1997, *Presence: Teleoperators and Virtual Environments* **6** 355.

[7] Ismail M E *et al.* 2017 *Online Journal For Tvet Practitioners, Advanced Center for Technical and Vocational Education* **2** 2.

[8] Bulagang A F *et al.* 2017 *Indian Journal of Science and Technology,* **10** 39.

[9] S. Jesse 2008 ELSERVIER **41** 46.

[10] McCrindle M *et al.* 2014, *Understanding the Global Generations* **1** 12.

IOP Conf. Series: Materials Science and Engineering **551** (2019) 012022 doi:10.1088/1757-899X/551/1/012022

Handover Delay Analysis on Cellular Communication for Connected Car Application in Urban Area

Azarul Fahmin Ab Hamid[1], Md Tasyrif Abdul Rahman[1], Anas Rahman[1] and Zairul Ameer Tan[1]

[1]School Mechatronic Engineering, Universiti Malaysia Perlis (UniMAP), Kampus Pauh Putra, 02600 Arau, Perlis, Malaysia

E-mail: tasyrif@unimap.edu.my

Abstract. The study on delay of cellular communication is conducted by considering handover session in urban area. The experiment was performed on three distinct devices to study the handover delay based on 2G and 3G connectivity. Handover delay experienced by a moving vehicle was measured based on the data transmission latency during communication transfer session from previous cell network area that is passed by the vehicle to a new cell network area. From this experiment, the best selection of wireless communication module by considering the handover delay in urban area is determined by suited module to be use on connected car system.

1. Introduction

Connected car these days play an important role for the end user or buyer[1]. Research have shown studies on technologies towards car information system and connected car system is suitable choice [2-11]. According to Kleberger et al[12]., connected car consists of four domains: the vehicle itself, the in-vehicle network and the ECUs, the portal of the automotive company and the communication link between them. This paper focuses on the communication link between the vehicle and the database to ensure the reliability of cell network information sharing. As highlighted by Junyeong et al[13]., it is important to improve the quality of communication links due to the increase in vehicle speed and data packet size. To overcome the poor communication link during vehicle moving in urban area road, experiments on the handover delay of cellular communication were performed and its methodology is discussed in this paper.

2. Experimental Setup

To test the reliability of the cellular network in the communication link, communication node hardware was developed using three devices; Arduino Uno, GSM modem and GPS to communicate between the vehicle and database. Three different types of modem models (with specific band and coverage) have been selected for GSM to test their communication performance; SIM900, SIM800L and SIM5360E. Note that for their communication coverage, SIM900 and SIM800L use the 2G signal while SIM900 uses 1800 MHz as the system band in this experiment. Additionally, they were set as network band for SIM800L and 900MHz using the AT+CBAND AT command. Due to the 3G coverage capability of the modem, SIM5360E was set to use the UMTS 2100MHz as a network band. Each communication node performs the AT+CSQ AT command to obtain the communication modem that determines the received power in the GSM antenna and conducts the ping request for the communication latency test. GPS GY-NEO6MV2 was used in this communication node to determine the latitude and longitude of the vehicle

IOP Conf. Series: Materials Science and Engineering **551** (2019) 012022 doi:10.1088/1757-899X/551/1/012022

between its location and the base station. Eventually, all three GSM modems was set to send all data to the webserver at www.000webhost.com.

3. Experimental Environment

The communication link was tested in urban area specifically at the Sekolah Derma road, Perlis as shown in figure 1(A). This area was selected based on the building density and their height (3 floors or more). Additionally, from open signal apps (figure 1(B)), the area also has a large amount of base station. This is desirable in minimizing propagation problem based on the distance between vehicle and base station.

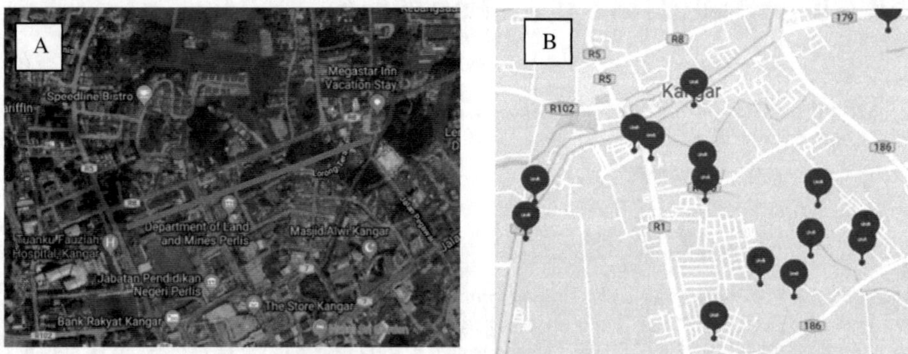

Figure 1. Sekolah Derma road in satellite and open signal view

4. Handover Delay

In regards of spreading problems in mobile communications, handover delays occur when the communication node moves between two or more cell networks covering the base station[14]. Communication handover is a process when the base station cell network transfers the current cell network area into a new cell network area that passes through by the vehicle as illustrated figure 2[15]. The ongoing communication link will have a delay in the data transfer sequence during this transfer process. The experiment on three communication modules aims to determine the differences in the average delay that affect the latency of communication between them.The experiment was conducted by examinated the ping on the server (www.google.com) with 32-bit data. Latency is defined as a period to successfully request the Hypertext Transfer Protocol(HTTP). Delay of handover session is measured when the handover process occurs between two base stations.

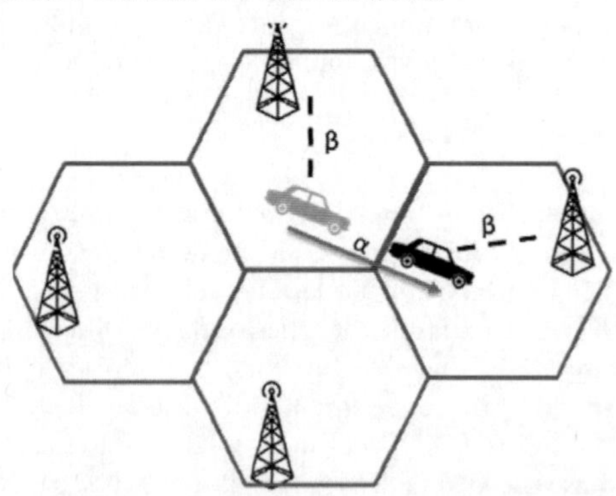

Figure 2. The vehicle moving from one cell network area to another cell network area where the α is vehicle movement direction and β is the communication link between the vehicle and base station[15].

IOP Conf. Series: Materials Science and Engineering **551** (2019) 012022 doi:10.1088/1757-899X/551/1/012022

5. Experimental Result and Discussion

The latency results from the three communication modules experiment is illustrated by separating three different vehicle velocity at 40km/h, 60km/h and 80km/h as shown in figure 3. The results show handover process that occur when $\log(L_p)$become zero and delay of the communication link in this situation is 80ms>delay>20ms. However, the SIM5360E receive lowest affect of handover delay in this experiment.

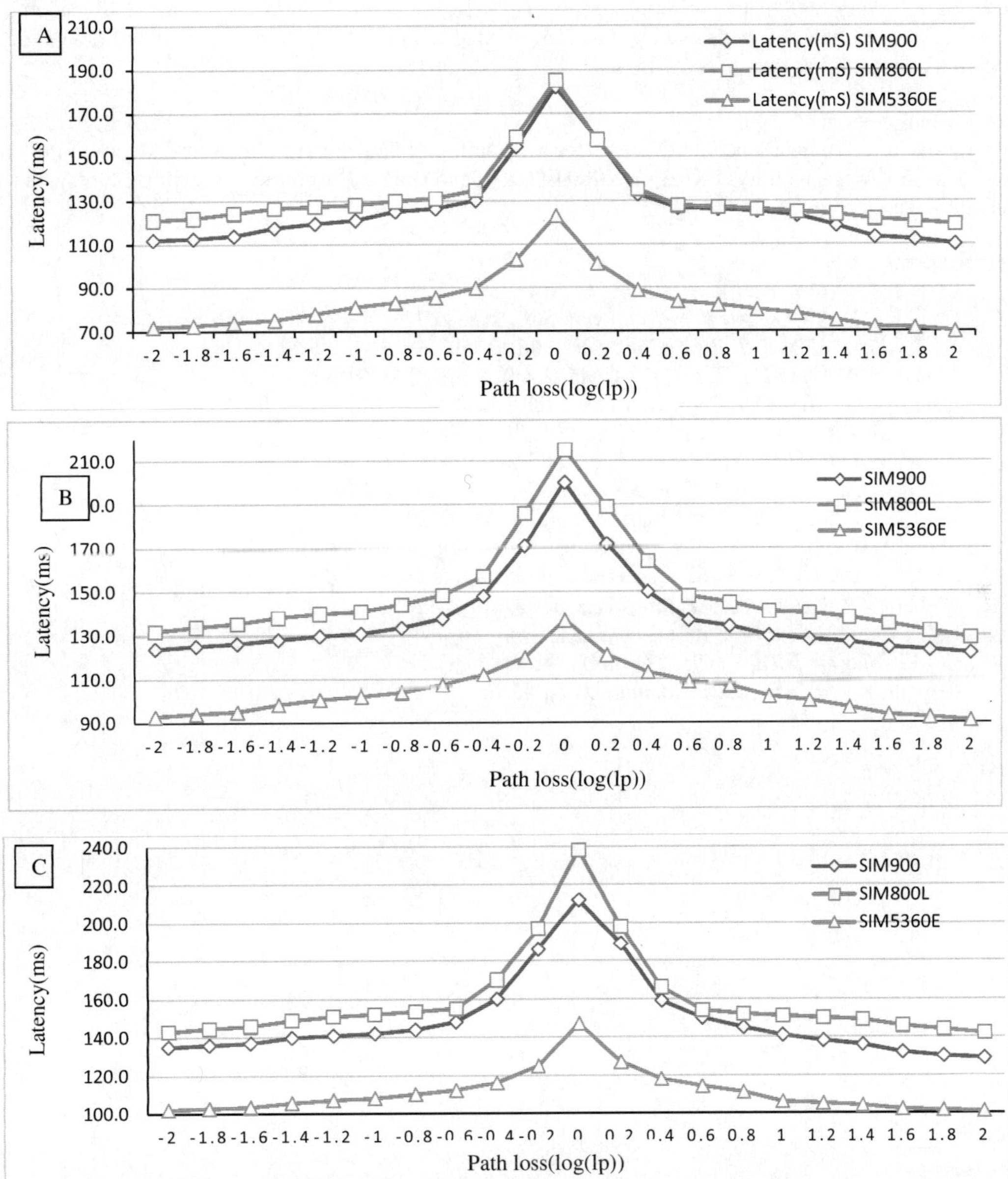

Figure 3. Path Loss vs Latency data for three different communication modules with vehicle speed at A=40km/h, B=60km/h and C=80km/h.

6. Conclusions

This study explored full potential of three communication modules. Three module SIM800, SIM900 and SIM5350E were analysed based on the handover delay that took place in urban environment. The studies show that SIM5360E had the lowest latency of communication link and better handover delay process. In the meanwhile, 5G technology will be introduced in next few years. With this implementation of communication technology, the amount of base station is expected to increase too. With huge amount cell network base station, the performance of handover process will become significantly important. From this study, it can be concluded that the best communication module to be placed in connected car system is SIM5360E.

Acknowledgements

I would like to extend our sincere gratitudes for the funding provided by the Universiti Malaysia Perlis (UniMAP) (Geran Penyelidikan Khas - No.9004-00064) and UniMAP Automotive Racing Team, which enable this research to be carried out.

References

[1] Lu N *et al* 2014 *IEEE Internet Things J.* **1(4)** 289.
[2] Hamid A. F. A. *et al* 2017 *J. Phys. Conf. Ser.*, **908**, 12079.
[3] Norizan A *et al* 2017 *J. Phys. Conf. Ser.*, **908** 12069.
[4] Mohamad M L *et al* 2017 *J. Phys. Conf. Ser.* **908** 12042.
[5] Azmeer M *et al* 2017 *J. Phys. Conf. Ser.* **908** 12051.
[6] Zaidie M N A *et al* 2017 *J. Phys. Conf. Ser.* **908** 12058.
[7] Zaman Z. I *et al* 2017 *AIP Conf. Proc.* **1885**, 1.
[8] Rahman A *et al* 2018 *IOP Conf Ser Mater Sci Eng.* **428** 012049.
[9] Aziz S *et al* 2018 *IOP Conf Ser Mater Sci Eng* **429** 012075.
[10] Manaf E H A *et al* 2018 . *IOP Conf Ser Mater Sci Eng* **429** 012048.
[11] Manaf E H A *et al* 2018 AIP Conference Proceedings **2030** 020115.
[12] Kleberger P *et al* 2011 *IEEE Intell Veh Symp Proc* **4** 528.
[13] Bok J *et al* 2014 *Int Conf Adv Commun Technol* **16** 1.
[14] Li M *et al* 2004 *IEEE VTCfall, Orlando*, **44** 51.
[15] Doppler K *et al* 2009 *IEEE Commun Mag.* **47** 42.

Path Loss Analysis Considering Doppler Shift Effect on Cellular Communication for Connected Car Application at Rural Area

Azarul Fahmin Ab Hamid, Md Tasyrif Abdul Rahman, Anas Rahman and Zairul Ameer Tan

School Mechatronic Engineering, Universiti Malaysia Perlis (UniMAP), Kampus Pauh Putra, 02600 Arau, Perlis, Malaysia

E-mail: tasyrif@unimap.edu.my

Abstract. In this paper, the study on the loss of pathways in cellular communication is carried out by experimentation and simulation on three different modules in rural areas. These three-communication module has different carrier frequencies to determine the effect of the doppler shift to the loss of the moving vehicle communication path. The simulation and experimental output are compared to improve the accuracy of these studies. Simulation studies are based on the proposed model for the modified path loss. For the experimental portion, the path loss was determined based on two things: the measured received power on communication modules antenna and transmit power from base station. From this study, the best selection of the communication module is selected for use in connected cars by considering the fading effect of doppler shift.

1. Introduction

Nowdays vehicle technology has grown rapidly, this is including information systems that can improve the capability of analysis on vehicle condition [1-10]. Basic information like the location of the car and the performance of the motor vehicle itself can be monitored. Vehicles nowadays have employed several Engine Control Unit(ECU), from just one in the last few decade, to enable them to share information with one anaother [11]. In the 1990s, with the introduction of 802.11 wireless communication standard, ECU is no longer limited to internal communication within the vehicle system, but can also share information between different vehicles [12]. This technology has enabled the connected car system to support ITS(Intelligence Transport System).

According to Kleberger et al, the connected car consists of three domains: the vehicle itself, which consists of the in vehicle network and the ECUs, the automotive company's portal and the communication link between them [13]. This paper focuses on the communication link between the vehicle and the database to ensure the reliability of information sharing on cellular networks. As Junyeong et al. suggested, improving the quality of communication link is important as vehicle speed and data packet size increases [14].

2. Existing Studies

There are few solutions proposed to improve the quality of communication. Jyothi et al has recommended the use Global Communication System Mobile (GSM) SIM900A for advancing mobile communication, instead of Blueetooth and Zigbee system. Two main parameters must be analyzed before the communication module is implemented on the connected car: path loss and doppler shift effect [15]. Although there are many models available to analyze the path loss model, not all of them can be used in certain unique environments[16]. Studies have also been carried out on the loss of track between transmitter and receiver in urban and suburban environments [17]. Significantly, there are only a few studies that consider the influence of the doppler effect in the loss of the pathway.

3. Communication Node Development

Communication node hardware has been developed with three devices that are Arduino Uno, GSM modem and GPS. For GSM, three different models (with separate band and coverage) have been selected to test their communication performance; SIM900, SIM800L and SIM5360E. Note that SIM900 and SIM800L are using 2G signal for their communication coverage. In this experiment, SIM900 was using 1800MHz as the system band. Alternatively, for SIM800L and 900MHz, they were set as network band by using the AT+CBAND AT command. Lastly, SIM5360E was set to using UMTS 2100MHz as network band due to the modem 3G coverage capability. The experiment setup for these three GSM modems can be referred in Figure 1. Each communication node will perform the AT+CSQ AT command to obtain the communication modem Received Signal Strength Indicator (RSSI) which determines the received power in the GSM Antenna and perform ping request to test the latency of communication link. In this communication node GPS GY-NEO6MV2 has been used to get a latitude and longitude of the vehicle in deciding the distance between vehicle location and base station. At the end, all the data from these three modules was sent to webserver at www.000webhost.com.

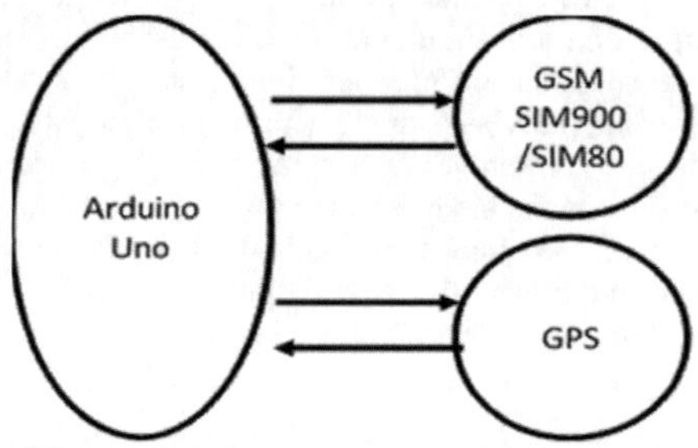

Figure 1. SIM900 Communication node

4. Analysis on measure and simulation path loss of communication that correlate to distance and vehicle speed

For communication link performance analysis, the experiment used only one method; measure track loss based on RSSI data in each communication module type. This communication module has been tested by moving the vehicle around the countryside as shown in Figure 2. During this test, three different speeds were used: 60 km/h, 90 km/h and 120 km/h.

Figure 2. Jalan Pauh-Jitra rural area beside paddy field road

The distance between vehicle and base station along the road was determined by referring to the Open Signal App.

Next, this doppler shift parameter was employed in calculating path loss in Tuned Free-Space path model using the following formula[19]:

$$L = K_1 \times log_{10}(f_d) + K_2 \times log_{10}(d) + K_3 + 32.45 + 20 \times log_{10}(d) + 20log_{10}(f) \quad (1)$$

Where L is a path loss result that obtain from Tuned Free-Space model by using K_1, $K_2$and K_3 that is median of optimum values minimum error between tuned path loss model and measured data obtained from Hong et al. analysis result on signal measurement at rural area. The value of K_1=5.0, K_2=19.2 and K_3=9.0.corrosion.

5. Results

Based on the 3rd Generation Partnership Project (3GPP) in Technical Specification Group Radio Access Network, Evolved Universal Terrestrial Radio Access (E-UTRA) and Base Station (BS) radio transmission and reception (Release 15) that state, the standard transmitting power for macro cellular communication base station is 46dBm. The result in Figure 3, where track loss data are generated based on RSSI measurement from the communication module and power transmission from the base station. The difference in the loss of the path in the communication link are shown to increase with the increase of frequency.

Based on the 3rd Generation Partnership Project (3GPP) in Technical Specification Group Radio Access Network, Evolved Universal Terrestrial Radio Access (E-UTRA) and Base Station (BS) radio transmission and reception (Release 15) that state, the standard transmitting power for macro cellular communication base station is 46dBm. The result in Figure 3, where track loss data are generated based on RSSI measurement from the communication module and power transmission from the base station. The difference in the loss of the path in the communication link are shown to increase with the increase of frequency.

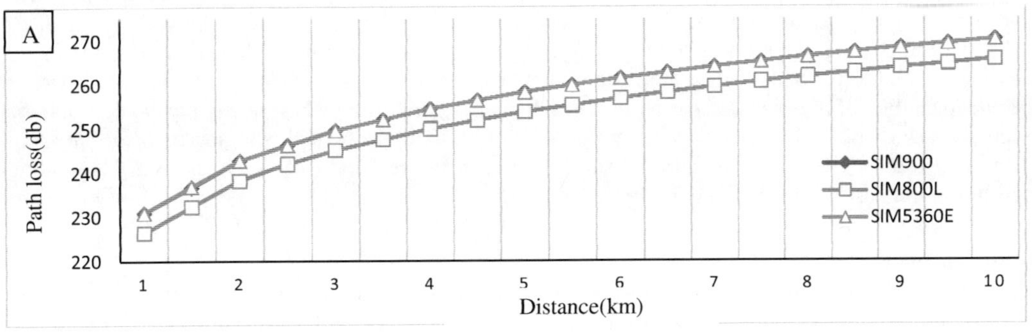

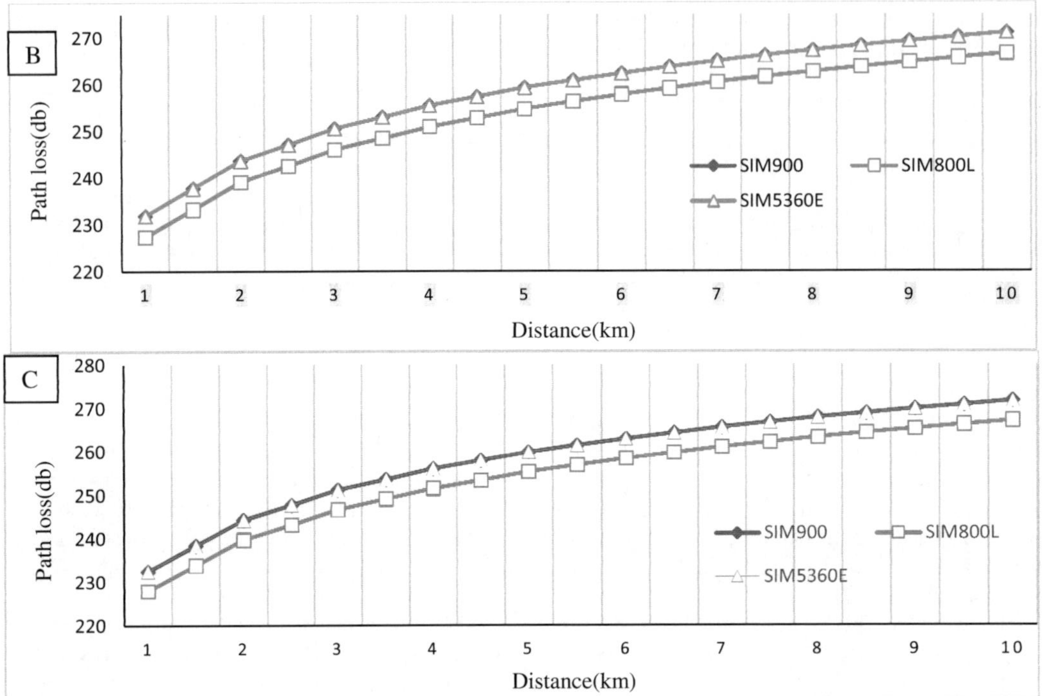

Figure 3. Path Loss vs Distance Tuned Free-Space path loss model data for three different communication modules (A=60km/h, B=90km/h and C=120km/h)

6. Conclusions

In this paper, the effect of the propagation path loss on the connected car communication link was investigated taking into account the doppler effect. The results of the experiment showed that SIM800 is the most suitable due to the low frequency of the communication module. This shows that the lower the frequency band, the better signal propagation link. However, further research is necessary to improve the understanding of the communication link in urban or suburban areas before any conclusions can be reached. Data from this study can therefore be applied to network technologies that specialize in connected automotive communication systems with another component, such as signal encoding algorithms and network technologies.

Acknowledgements

We would like to extend our sincere gratitudes for the funding provided by the Universiti Malaysia Perlis (UniMAP) (Geran Penyelidikan Khas - No.9004-00064) and UniMAP Automotive Racing Team, which enable this research to be carried out.

References

[1] Hamid A F A *et al* 2017 *J. Phys. Conf. Ser.* **908** 12079.
[2] Norizan A *et al* 2017 *J. Phys. Conf. Ser.* **1** 12069.
[3] Mohamad M L A *et al* 2017 *J. Phys. Conf. Ser.* **908** 12042.
[4] Azmeer M *et al* 2017 *J. Phys. Conf. Ser.* **908** 12051.
[5] Zaidie M N A *et al* 2017 *J. Phys. Conf. Ser.* **908** 12058.
[6] Zaman Z I *et al* 2017 *AIP Conf. Proc.* **1885** 1.
[7] Rahman A *et al* 2018 *IOP Conf. Ser. Mater. Sci. Eng.* **429** 012049.

[8] Aziz S *et al* 2018 *IOP Conf. Ser. Mater. Sci. Eng.* **429** 012075
[9] Manaf E H A *et al* 2018 *IOP Conf. Ser. Mater. Sci. Eng.* **429** 012048.
[10] Manaf E H A *et al* 2018 *AIP Conference Proceedings* **2030** 020115.
[11] Prashant A *et al* 2015 *International Conference on Circuit, Power and Computing Technologies* 7.
[12] Coppola R *et al* 2016 *ACM Comput. Surv.* **49** 46.
[13] Kleberger P *et al* 2011 *IEEE Intell. Veh. Symp. (IV)* 528.
[14] Bok J *et al* 2014 *Int. Conf. Adv. Commun. Technol.* 1.
[15] Phillips C *et al* 2013 *IEEE Commun Surv Tutorials. V* **15** 255.
[16] Alam N *et al* 2010 *IEEE Veh. Technol. Conf.* 1.
[17] Amornthipparat A *et al* 2008 *Second International Conf. ICCE.* 336.
[18] Wei H *et al* 2010 *5th Int. Commun. Netw. China ICST Conf.* 1.

IOP Conf. Series: Materials Science and Engineering **551** (2019) 012024 doi:10.1088/1757-899X/551/1/012024

Warpage optimisation using virgin and recycled polycarbonates: a case study of front panel housing

Nur Aisyah Miza Ahmad Tamizi[1], Shayfull Zamree Abd Rahim[2], Nurul Hidayah Mohamad Huzaim[3], Norshahira Roslan[4]

[1,2,3,4,]School of Manufacturing Engineering, Universiti Malaysia Perlis, KampusTetapPauh Putra, 02600 Arau, Perlis, Malaysia.
[2]Design and Manufacture Research Group, Centerof Excellence Geopolymer and Green Technology (CEGeoGTech), Universiti Malaysia Perlis, 01000 Kangar, Perlis, Malaysia

E-mail: aisyahmiza94@gmail.com

Abstract: Plastic product are commonly produced using an injection molding process. A large amount of product can produce within a very short time. One of the defect that often being encountered in the production is warpage. By determine the appropriate parameter setting by using optimisation method, the quality of the moulded part can be improve. The warpage can be minimise but difficult to eliminate. The part can be smaller than the actual size cause by the warpage. Due to this defect will cause the plastic part difficult to be assembled. Considering the increasing application of these thermoplastic materials, the use of recycled polymers has increased too. Environmental issues, the subsequent tendency to save resources and the high costs of polymers have made the recycling an inevitable option. This study provides a scientific knowledge that will be useful to plastic manufacturing industries in injection moulding process especially by using virgin and recycle material. The parameter that being significant from the studied that been review were mostly melt temperature, packing time, cooling time and packing pressure.

1. Introduction

Plastic product produce by using an injection moulding process which is categorise as one of the manufacturing process in the industry. This study presents about the injection moulding defect of warpage as their output response. In addition, the warpage defect was highlighted because this defect can reduce the quality of the part and affect the assembly process of the product [1, 2]. Plastic housing components are widely used in many assembly manufacturing processes to produce a final product [3]. The major quality characteristics of concern such as weight, dimension and thickness for the plastic housing components are need to be taken care as to pass the quality control [4, 13]. In the other hand, mould quality significantly affects the injection moulding procedure by affecting the material quality of the product.

Many of the children's toys and automobile parts made from this injection moulding process. Most of thermoplastic product has specified such as complex dimension, tight tolerance and high production volume. Therefore, the materials used as to make the part also examine from the reviewed journal. The

IOP Conf. Series: Materials Science and Engineering **551** (2019) 012024 doi:10.1088/1757-899X/551/1/012024

simulation analysis or experimental were also examined by review the method of the researcher used in their studies. The simulation were studied by observing the method that had been used as the optimisation method. In industry, there are many types of process other than injection moulding such as blow moulding, compression moulding and rotational moulding. Each type of them brings their own advantages but the most preferable in industry is injection moulding.

Plastic products commonly produce by using injection moulding. Within a very short time, it can produce a large amount of product with low production cost. By comparing than previously thought, plastic injection moulding is more complex. A significant economies of mass production gain as a benefit of this process [5]. The plastic parts produce defects cause by the parameter settings, inappropriate mould design and material used. Warpage, shrinkage, sink marks, short shot and so on are the defects found by many researchers [3-4, 6-8]. The quality of plastic parts produced by injection moulding can be affected by the process parameters, material of the part and part design [10]. Geometric reduction on size should be controlled as to define shrinkage, because warpage are the problem effect of the variation in shrinkage [11]. Warpage is considered as one of the most difficult defect to control and it caused by a non-uniform change of internal stress [12].

2. Injection Moulding Process

A plastic part produces by injecting the raw molten plastic material into a mould and it is called an injection moulding process. In addition, the hydraulic system is used as to run the machine. The main unit that was used in injection moulding processes were driven unit, plasticising unit and clamping unit. The clamping unit is used as to open and close the mould. This part also will move and the plate will be fixed as the mechanism has been clamped.

The plasticising unit will heat the raw plastic material until it melts. Then the molten plastic will inject into the mould by the injection unit. The drive unit provides the power for this unit. Within a very short time, it can produce a large amount of product with low production cost. Moreover, Short cycle time production, the light weight of products, and high surface quality are their several advantages and solution for plastic injection moulding industries (PIM). By comparing than previously thought, plastic injection moulding is more complex. The plastic parts produce defects cause by the parameter settings, inappropriate mould design and material used. Sink marks, warpage, short shot, shrinkage and so on are the defects found by many researchers [3-5, 6-9, 13].

The process will become more difficult in plastic injection moulding difficult when the plastic part was very thin. Besides that, a thin plastic part was a demand in the industry features [14-16]. Plastic injection, packing, cooling, ejection and process part quality control applications describe the procedure of injection moulding. The plastic product was ejected after stable of the interior cavity had achieved. This case of a defect will affect the accuracy of plastic part and quality [18]. A minimum warpage of a thin shell plastic part resulted can be accepted by an important factor which is dimensional stability. As to improve the injection moulding's quality, warpage need to be reduced [19-24].

At the first stage of plastic injection moulding, a closed mould is injected with granular plastic resin which are melted in the barrel machine. The mould needs to be opened first so that the melted material that has been solidified in the mould can be ejected out. In order to produce higher quality of the moulded parts, an appropriate setting of injection moulding machine required and also the old design need to be reconsidered. So that the part defect can be minimised as examples shrinkage, warpage and so on. Geometric reduction on size should be controlled as to define shrinkage, because warpage are the problem effect of the variation in shrinkage [11].

3. Important Parameter in Injection Moulding Process

A thin shell is commonly used in industrial production of injection moulding. The parameters condition takes place in determining the level of warpage. As to minimise the warpage until reach the acceptable stage, the warpage need to be reconsidered in more proper way [17]. Many studied had been done over this year as to handle the warpage of a thin shell plastic products [21-22, 25-27]. There are a few parameters need to be considered in injection moulding.

Plastic parts that produce by using an injection moulding process was one of the important methods in polymer processing [10]. Another method as to reduce defects can be done by controlling the parameter settings. Previously, the trial and error approach is used onto the process parameter in PIM process relied but eventually for the complex manufacturing processes is not suitable and effective to use this approach [28-29].

Parameters such as injection pressure, mould temperature, melt temperature and injection time are taken into consideration that effecting minimum warpage. In terms of critical process parameters, Response surface methodology software can create a predictive model for warpage. In order to find the optimum process parameter values, response surface model coupled with a more efficient of genetic algorithm [10].

4. Material Used In Injection Moulding Process

Considering the increasing application of these thermoplastic materials, the use of recycled polymers has increased too. Environmental issues, the subsequent tendency to save resources and the high costs of polymers have made the recycling an inevitable option [25-27,29-30]]. In order to achieve 'go green environment', many waste plastic part need to be recycle but the defect was hardly to be diminish. Many researcher have study about virgin-recycle blend ratio material but focus into mechanical properties as the output [31-33] and one of them use Polycarbonate as the material of the study but focus on tensile stress as the output response [34]. There are two recycled polymer that can be differentiate. First recycled polymer are come from the materials of a plastic waste that lead biological and environmental problem and second is the plastic waste that comes from the manufacturing process in a factory known as the scrap materials. Examples of the scrap materials generated by the factory were gates and runners which are discarded from a part and have a noticeable weight regarding the weight of the main part [30].

5. Optimisation in Injection Moulding Process

The most important factor contribute in warpage was shown that packing pressure factor was the result. The result also shown that filling time and gate location gave a small impact toward warpage [35]. GA also was a technique as to obtain optimal conditions. Taguchi methods [36, 37], hybrid optimisation [31], and DOE [18] show a numerical simulation on shrinkage by several of researches and the result shown that packing time as a critical factor. Packing pressure also shown as more significance other than packing time that contributes toward shrinkage of the moulded [38, 39]. For **Error! Reference source not found.** shows the optimisation made by using RSM, GA, PSO and Taguchi methods.

IOP Conf. Series: Materials Science and Engineering **551** (2019) 012024 doi:10.1088/1757-899X/551/1/012024

Table 1. The optimisation made by using RSM, GA, PSO and Taguchi method.

Title	Method	Part	Result
Huang *et al.* [20]	Taguchi, response surface methodology, neural network, and genetic algorithm	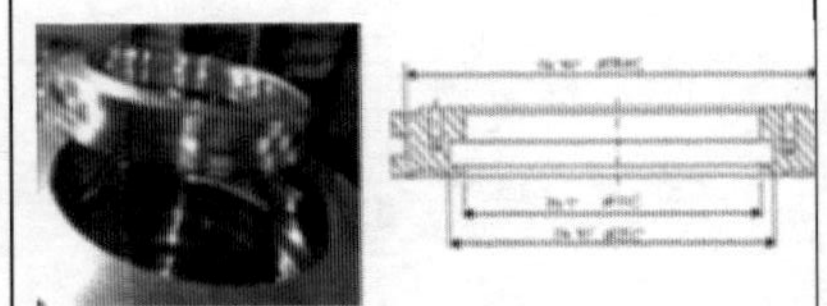	Compared to the traditional methods, there exists advantages and shortness of the three models, but all of them can effectively find out the settings of manufacturing parameters through experimental designs, which dramatically reduce the number of experiments
Kurtaran and Erzurumlu [10]	Response surface methodology and genetic algorithm	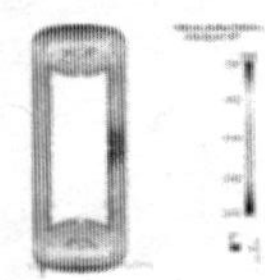	GA has reduced the warpage of the initial model significantly. Warpage is improved by about 46%.
Ozelik and Erzurumlu [40]	Response surface method and genetic algorithm	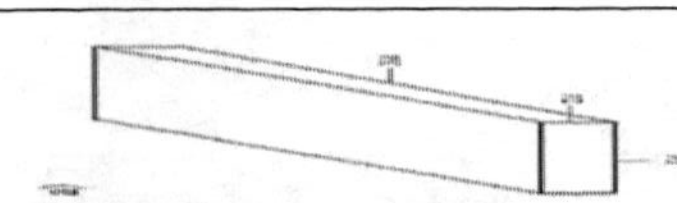	Seen that maximum warpage on the thin shell plastic parts model, 0.0977 mm before the optimisation, is reduced to 0.0582 mm by 40.4% after optimisation
Sun and Wu [39]	Response Surface Methodology and Genetic Algorithm	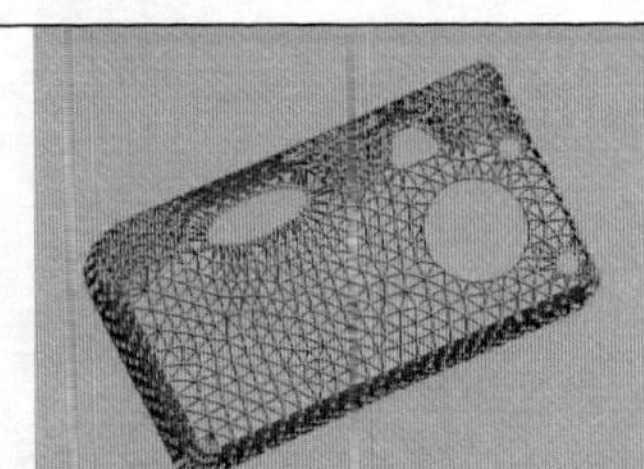	Warpage has been improved by about 30%. The optimisation process reduced the warpage of the initial model significantly.

IOP Conf. Series: Materials Science and Engineering **551** (2019) 012024 doi:10.1088/1757-899X/551/1/012024

Table 1. The optimisation made by using RSM, GA, PSO and Taguchi method.

Title	Method	Part	Result
Türköz *et al.* [41]	Response Surface methodology and Genetic Algorithm	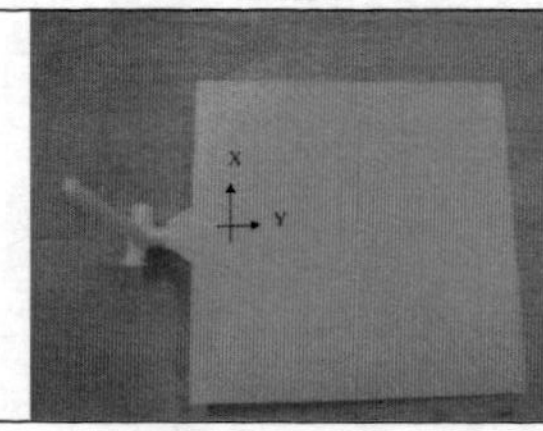	At the optimal conditions, the dimension shrinkage variation of thin-wall moulded plate has been improved about 85% and 63% in the transversal direction and the longitudinal direction, respectively.
Nasir *et al.* [42]	Response Surface methodology		This study has applied the RSM method to estimate the optimal solutions of injection moulding process that lead to minimise shrinkage. The mathematical model of the shrinkage has been carried out to correlate the dominant the injection moulding parameters of the plastic injection moulding process thick plate using ABS material.
Yin *et al.* [43]	Neural Network Model and Genetic Algorithm		For its efficiency and flexibility, the proposed BP/GA optimisation method can be used generally to optimise defects of plastic as well as other considerations such as production cycle, cost and so on during PIM.
Chen *et al.*[44]	Response Surface Methodology	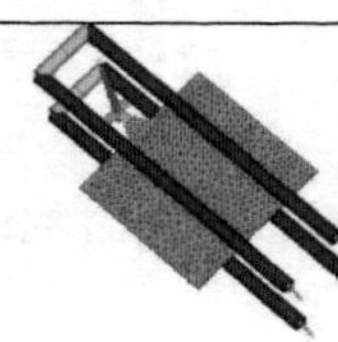	The minimum values of dimension shrinkage variation obtained at the optimal condition in both transverse and longitudinal direction were 0.081 and 0.286 mm,

6. Summary and future works

Plastic have been widely used in industrial as to produce many parts in especially in automotive, spacecraft and other industrial products. Other than increasing of oil price, the industrial also demand in the usage of recycled plastic as to reduce cost of the production. When examining the mechanical and rheological properties of the recycle plastic, it tends to have less desirable properties compare to virgin plastic and lead to defects. Warpage hold the highest rank of factor that form warpage in plastic geometry part. Therefore, to overcome this a further study need to be done.

Other than that, an injection moulded thermoplastic part with a thin-shell feature need important procedures in injection moulding processes in industry. Then, the major factors that affect quality of thermoplastic found that machining parameters in addition to moulding material, part and mould designs produce by injection moulding. The numerous mathematical model has been proposed due to the complexity of injection moulding and many analysis studies of different stages of the injection moulding process were done [45].

Most of the researchers used RSM and GA as the optimisation method in order to minimise the warpage nor shrinkage for the plastic part. Therefore, these two optimisation methods which were RSM and GA were choose as the optimisation method in this study and it also had been used by many researchers in their studies. Less time will be spent as to find the most appropriate and optimal solution to more complex engineering problems when this GA method was applied to the neural network optimisation. This method also will help the engineering designer in reducing the cost and time spent in trial and error in the experiment and analysis. By reducing the defects occurrence to the plastic parts, the quality of the product can be increase. Therefore, optimisation needs to be determined precisely in injection moulding processes as to give a better result in producing plastic part.

In conclusion, by introduced the RSM and genetic algorithm can resulted in generate the efficient optimisation methodology in minimised a recycled plastic parts in warpage aspect. The plastic part made by using injection moulding. The successful optimisation parameters resulted in plastic part will reduce warpage defect and minimise the cost by using recycled plastic materials.

References

[1] Chen C C *et al* 2014, *Int. J. Adv. Manuf. Tech. Polym.* **45** 14-15
[2] Gu Y *et al* 2001 *Adv. Poly. Process. Tech.* **20(1)** 14-21
[3] Ehsan M *et al* 2011 *Ipcsit. Com* ***5*** 570-576
[4] Öktem H *et al* 2012 *Adv. Manuf. Tech.* **61(5-8)** 519-528
[5] Isafiq M *et al* 2016
[6] Shi H *et al* 2013 *Int. J. Adv. Manuf. Tech. Polym. Eng. Sci.* ***65*** 343-353
[7] Tzeng C J *et al* 2012 *Int. J. Adv. Manuf. Tech. Polym. Eng. Sci.* **63** 691-704.
[8] Kusić D*et al* 2013 J. *Grum, Poly. Test* ***32*****(3)** 583
[9] Wang X *et al* 2013 *Mater. Des.* **47** 779–792
[10] Kurtaran H *et al* 2005 *Mater. Process. Tech.* ***169*****(2)** 314-319
[11] Shoemaker J *et al* 2006a Moldflow Design Guide : A Resource For Plastics Engineers, Hanser
[12] Guo W *et al* 2012 *Mech. Sci. Tech.* ***26*****(4)** 1133-1139
[13] Oktem H *et al* 2018 *Mater. Des.***28(4)** 1271-1278
[14] Chiang K T *et al* 2014 *Adv. Manuf. Tech.* **35(5-6)** 468-479 (2007)
[15] Liao S J *et al* 2004 *Polym. Eng. Sci.* **44(5)** 917-928 (2004)
[16] Fathullah M *et al* 2011 *I.RE.M.E* ***5*****(7)** 1189-1195
[17] Nasir S M *et al* 2013 *I.RE.M.E* **7(5)** 977-990
[18] Kramschuster A *et al* 2005 *Polym. Eng. Sci.* ***45*****(10)** 1408-1418.
[19] Lee B H *et al* 2006 *Polym Plas. Eng. Tech.* **34** 793–811
[20] Huang M C *et al* 2001 *Mater. Process. Tech.* **110(1)** 1-9
[21] Bushko W C *et al* 1995 *Stokes, Polym. Eng. Sci.* **35(4)** 351-364
[22] Jacques M S *et al* 2000 P*olym Eng Sci.* ***22*****(4)** 241-247

[23] Fan Z 2000 *Soc. Plas Eng.* **46**
[24] Shoemaker J 2006b Moldflow Design Guide : A Resource For Plastics Engineers, Hanser
[25] Gu F S *et al* 2014 *Mater. Des.* **62** 189-198
[26] Liu S J 1996 *Polym. Eng. Sci.* ***36*(6)** 807-818.
[27] Matsuoka T *et al* 1991 *Polym. Eng. Sci.* ***31*(14)** 1043-1050
[28] Nasir S M *et al* 2012 I.RE.M.E *7***(5)** 977-990
[29] Chen W C *et al* 2014 *Eng. Manuf.* **15(8)** 1583-1593
[30] Rahimi M *et al* 2014 *J. Mater. Process. Tech.* **214(11)** 2359–2365
[31] Fei N C *et al* 2013 Isrn Industrial Engineering
[32] Scaffaro L. *et al* 2012 *Eur. Polym. J.* **48(3)** 637-648
[33] Gu F *et al* 2014 *Mater. Des.* **62** 189-198
[34] Mohamad N F 2013 *Doctoral Dissertation*
[35] Ozcelik B 2009 *Mater. Des.* **30(2)** 367-375
[36] Alkaabneh F A *et al* 2013 *Int. J.Adv.Manuf. Tech. Polym. Eng. Sci.* **66** 679-694
[37] Fathullah M *et al* 2011 I.RE.M.E **5(7)** 1278-1286
[38] Alam M M *et al* 2013, *Sci And Res*, **2(2)** 107-110
[39] Sun B *et al* 2010 *Comput. Eng. Tech.*
[40] Ozcelik B. *et al* 2005 *Int.Commun. Heat. Mass.* **32(8)** 1085-1094
[41] Türköz N *et al* 2014 **115** 826-832
[42] Nasir S M *et al* 2016 *Eng. Mater* **700** 12-21
[43] Yin F *et al* 2011 *Mater. Des.* **2(6)** 3457–3464
[44] Chen C C *Int.J.Adv.Manuf. Tech. Polym. Eng. Sci.* **45** 11-12
[45] Mehat, N M *et al* 2011 *Mater. Process. Tech.* ***211*(12)** 1989-1994

IOP Conf. Series: Materials Science and Engineering **551** (2019) 012025 doi:10.1088/1757-899X/551/1/012025

Warpage Optimisation On The Front Panel Housing Moulded By Acrylonitrile Butadiene Styrene (ABS) (Virgin and Recycle) Material: A Review

Nurul Hidayah Mohamad Huzaim[1], Shayfull Zamree Abd Rahim[2], Norshahira Roslan[3], Nur Aisyah Miza Ahmad Tamizi[4]

[1,2,3,4]School of Manufacturing Engineering, Universiti Malaysia Perlis, KampusTetapPauh Putra, 02600 Arau, Perlis, MALAYSIA.

[2]Green Design and Manufacture Research Group, Center of Excellence Geopolymer and Green Technology (CEGeoGTech), Universiti Malaysia Perlis, 01000 Kangar, Perlis, MALAYSIA.

E-mail: hidayah_huzaim@yahoo.com

Abstract. Injection molding process is well known in plastic molding industries. The development of this process had causes the materials used to mold plastic product to increase. However, this downside of this large amount of plastic consumption also leads to the fact that plastic waste fills the landfill. This cause researches on recycling plastic to increase. Even though injection molding process has many advantages, but the occurrence of defects on plastic part molded is one of the most important issues to be resolved. Warpage defect is one of the defect that normally occurred on molded part. This paper presents a review of the current research on the outlook of injection molding process and the effect of parameter processing on plastic molded part. The effect of parameter processing on mechanical properties of plastic mold parts with a mixture of virgin and recycled material has been numerously studied. However, studies on the physical appearance and quality of plastic mold parts by a mixture of virgin and recycled material have yet to be explored.

1. Introduction

Plastics industries were one of the fastest growing major industries in the world [1]. The amounts of the plastic products used are increasing every year. Meaning that, the widespread use of plastic materials has increased the quantities of plastic waste sent to landfills and raw material extracted. Thus, recycling of plastics is important to be implemented in order to reduce the waste of the plastic. However, the performance or quality of products made from recycled polymer will be differ from products made of virgin polymers, due to several reactions occur. An alternative to maintain the properties of product is to blend virgin material with recycled material [2]. There are many researcher studies on the impacts of percentage of recycled material by focusing on strength and mechanical properties of the part produced during injection molding process [3–5].

Material selection, part and mould design as well as processing parameters in the injection moulding process influence the quality of the plastic product and improper combination of it can lead to numerous production issues, reduce the competitive price advantage and also reduce the company's profitability [6–7]. One of the common thermoplastic used in industries and highly demand in the market in recent year is Acrylonitrile Butadiene Styrene (ABS). This is due to its mechanical properties, dimensional stability and chemical resistance performance. In order to produce plastic product of low cost with high value, Recycled Acrylonitrile Butadiene Styrene (r-ABS) can be used as a mixture with virgin ABS.

The quality of the product is the concern of the manufacturers and customers, the high quality of the product and the great rate of production are the key to the success of the industry [7]. Thus, the identification of the root cause of the defects directly contributes to the elimination of part defects as well as the improvement of the quality of the moulded parts. Many researchers found that injection moulding parameters have an important impact on the quality of plastic parts. The researchers address injection moulding issues such as shrinkage, warpage, weld lines, sink marks, and residual stress caused by process parameters. Their study also shows that processing parameters that influenced the quality of the plastic parts are packing pressure, melt temperature and mould temperature [8–10].

In fact, it is noted that product defects, with process instability and so forth, are due to a combination of processing parameters that are not appropriate. To solve these problem, many researchers conducted injection moulding simulation experiments by using Computer-Aided Engineering (CAE) simulation software such as Autodesk Moldflow Insight (AMI), C-MOLD and Moldex3D and there are also used experimental design for the optimisations process and optimisation method such as Taguchi method, Response Surface methodology (RSM) and Genetic Algorithm (GA) [11–18].

In recent years, the performance of recycled ABS in injection moulding process has triggered many researchers to investigate the effects of recycling process on physical and mechanical properties of r-ABS and virgin ABS when mixed according to the loading ratio. Performing the injection moulding process is quite challenging as we have to control the processing parameter involves during the manufacturing process efficiently in order to obtain a high quality product.

2. Injection Molding Process

Injection moulding process is segmented into several levels, which are the operating cycles, including plasticisation, filling, packing, cooling and ejection. Initially, screw shearing and feed pipe heating the plastic pellets before plasticised it in to melt. Then, to fill the cavities of the mould, the melt will be injected into the runner system. Lastly, after cooling process, the product will be ejected out. The quality of final moulded parts is highly unpredictable due to it will be affected by several factors in the injection moulding process. Some of the significant impacts were the process parameter, design of the mould and the properties of plastic material [19–21].

Consequently, several process parameters which include the melt temperature, mould temperature, injection pressure, injection velocity, injection time, packing pressure, packing time, cooling temperature, and cooling time all potential effect the quality of injection-moulded plastic products [22].

3. Parameters in Injection Molding Process

Numerous researcher found that the injection moulding process parameters have crucial effects on the calibre of the plastic parts, most of researcher looked into the problems of the injection moulding part including the shrinkage warpage, weld lines, sink marks, and residual stress made by process parameters. From their studies, the main parameters regarding the quality of the plastic parts are packing pressure, packing time, cooling time, melt temperature, and mould temperature [8–10], [23–24]. When packing time was fixed, the warpage and shrinkage will be more affected by the cooling time compared to packing pressure. Thus, the quality of the warpage and shrinkage will be increased due to the cooling time increased [25]

Liao *et al.*[26] determining the most proper condition for a thin walled part according to shrinkage and warpage problems. A cell phone cover was applied to the analysis. Melt temperature, mould temperature, packing pressure and injection speed were required as the injection parameters. The experimental tests were built in Taguchi experimental method. Cyclone scanner, polycad and polyworks were used to determine the shrinkage and warpage values. As a result of experiments packing pressure was found as the most important parameter.

Ozcelik and Sonat [27] investigated on the warpage and structural analysis of thin shell phone cover made of PC/ABS. Determination on the effect of warpage in different thickness values was used by the Taguchi method. From the analysis, the warpage values were found by Moldflow Plastic Insight (MPI) 4.0 software. Mould temperature, packing pressure, packing time and melt temperature were the parameters that involved in this analysis. Packing pressure was found as the main effect parameter on the warpage of PC/ABS material. Same goes to Chen *et al.*[28] they have performed similar research and they have found that packing pressure and melt temperature were used to be the most important parameters for thin shell parts.

Increasing in melt temperature of the material will not give improvement to the flow of molten material in the diaphragm gate of the thin section towards the mould cavity. Nevertheless, a concentric sprue will be the fed to the diaphragm to allow a uniform flow of molten material to all parts of the gate, so that it will make the maintenance of it much easier. This will improve the injection time in the cavity of the molten material.

Besides that, mechanical properties of product produced from recycled plastic also can be improved by optimizing the processing parameters in the injection molding process. Mehat and Kamaruddin [29] found that melt temperature with 43.34% of contribution is the most significant parameter in yield at stress of the recycled tool box tray that made of 100% recycled Polypropylene.

Wavare and Ubale [30] found that cooling time is the most influencing parameter in plastic injection moulding process of PP Homopolymer (1110MAS) in order to control the warpage and improve the quality of the manufactured specimen. Other than that, according to ANOVA analysis in the case study by Chen and Kurniawan [24], cooling time and packing time is the most significant processing parameter to obtain an optimal value for shrinkage and warpage on the moulded parts of optimisation on the processing parameters in an injection moulding process. Chen *et al.*[11], concluded that the packing time should be controlled earlier than the cooling time or increase at the same time in order to produce a better and lower warpage.

Other than that, Wang *et al.*[31] analyze on the optimization of processing parameter in plastic injection molding process with the used of Taguchi optimization method to obtain the strength compression of the product to be manufactured which is a brake booster valve body. The compression strength and curing percentage of the brake booster valve get a high impact from the mold temperature with 12% of improvement was obtain from the test. From the previous studies show that mold temperature affects the surface and strength of the plastic part. Mold temperature is the temperature required for the mold to maintain. When the mold temperature is increased, it will allow a slower injection rate without the plastic getting cold. Mold temperature will affect the surface quality, shrinkage and deformation. Higher elongation at break values during tensile test will be obtained when the mold temperature is low, as the temperature gradient during cooling is high and fast, thus prevent the crystallization processes [32]. These show that each of parameter processing influence the part produce by injection molding process. However, most researchers are investigating optimization of processing parameters to minimize the problems related to dimensional stability or mechanical properties of the part, which is made only from virgin materials that are scarcely found in recycled materials.

4. Recycled Material in Injection Molding Process

The use of recycled material in the market of plastic industry in Germany has been wandering by the designers as the product from recycled material can be rejected due to their surface appearance other than their physical properties. Nevertheless, in that respect are some countries in the US such as Michigan have been reached nearly to 100% of recycling rates where it turns out that there are others still have a concerned regarding the recycling plastic waste which changes in the market attitude [33]. There are various methods of polymers recycling such as mechanical recycling and chemical recycling has been used by most of the studies. Still, mechanical recycling is the most desirable from the

IOP Conf. Series: Materials Science and Engineering **551** (2019) 012025 doi:10.1088/1757-899X/551/1/012025

perspective of industrial point as it is low in price and reliability [34]. Thus, in order to reduce the cost of a production, the recycle material will blend in with virgin material to produce a new plastic product with minimum occurrence of defects. Besides processing parameter and percentage of mixture of virgin and recycled blends, the number of recycling process also give influence to the mechanical properties of plastic product. Plastic recycling and recovery are divided into four categories. First is primary recycling where involve in mechanical reprocessing into a product with equivalent properties. Next is secondary recycling which is mechanical processing into products requiring lower properties. Tertiary recycling is the recovery of chemical constituents and lastly is quaternary which is the recovery of energy [30].

Scaffaro *et al.* [4] studied on the effect of postconsumer content on ABS (pc-ABS) and of reprocessing cycles on the physical properties of virgin – recycled ABS blends and also with investigation on rheological, thermal and mechanical properties of injection-moulded samples after up to three reprocessing cycles. In this study, the material of virgin ABS was supplied by Starex that been used in the frame of this work and from ground phones the pc-ABS has been obtained by injection moulding of the same material. Evaluation on the effect of pc-ABS amount and number of the reprocessing steps on physical properties in this study has been done by recycled the ABS/pc-ABS blends up to three times. In investigation of rheological measurement, it showed that the viscosity of all the blends were slightly decreases has been influenced by the third reprocessing cycle as shown in Figure 1. While in Heat Distortion Temperature (HDT) test, the values for HDT of the materials does not influenced significantly by the presence of pc-ABS and the reprocessing steps up to two. Besides that, the effect on the reduction of the relaxation times has been proved on the orientation which with consequent more relevant decrease of HDT, especially at high pc-ABS amount in R3 blends. Addition of the smallest amount of pc-ABS in the first reprocessing does not give effect towards the decreasing of the tensile properties of ABS/pc-ABS blends as it did not cause significant changes. This can be said that crosslink of the PB phase and the opposite degradation effect of chain scission were give a balance results. For flexural properties, pc-ABS were almost not give any effect but slightly influenced by the recycling operations. This can be explained that the injection moulding operation was inciting the degree of orientation of ABS and this would make to the flexural stiffening of the structure since the bending stiffness is more affected by the degree of orientation in the surface layers of the specimens. Other than that, addition of pc-ABS and increasing the number of reprocessing steps were give effect to the impact resistance of the blends as it significantly decreased. Thus, correlated with the degradation of the PB phase will cause to the worsening of this property as in thermo-mechanical stresses undergone by the material during processing and to the photo-oxidation undergone by pc-ABS.

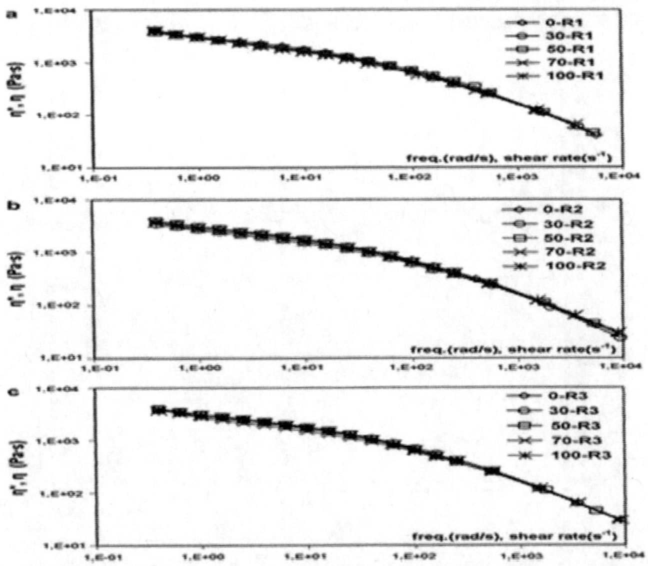

Figure 1. Viscosity curves of ABS/pc-ABS blends processed up to three times: (a) 1 reprocessing cycle; (b) 2 reprocessing cycles; 3 reprocessing cycles [4].

Kuram *et al.*[5] studied on the effect of recycling process on rheological, thermal chemical and mechanical properties on Polycarbonate/Acrylonitrile–Butadiene–Styrene (PC/ABS) and Polyamide6/PC/ABS (PA6/PC/ABS) blends. In this work, the composition of 75wt%/10wt%/10wt% was for PA6, PC and ABS, respectively, and 5wt% of compatibilier in the ternary blends of PA6/PC/ABS. The influence of the reprocessing number of the properties will be obtained by reprocessed the blends five times. After the virgin blends were moulded, only five specimens will be used for each test of the virgin samples. While the remainder of the specimen was remanufactured again as the first recycled specimen and the process were repeated to bring on other recycled specimens. From the solution obtained by the Differential Scanning Calorimetry (DSC) which been used to investigate the miscibility and melting behaviour of blends, the melting temperature and chemical structure of blends did not significantly shift. In the degradation and thermal behaviour of blends, the effects shown by Thermal Gravimetric Analysis (TGA) curves were similar to each other for virgin and recycled specimens as shown in Figure 2. From this study, it has been observed that impact strength and Melt Flow Index (MFI) of PC/ABS were decreased, while the elastic modulus increased with the recycling process. But, tensile strength and flexural properties of PC/ABS did not significantly vary by the recycling operation. Other than that, the elastic modulus and tensile strength of PA6/PC/ABS will continuous decreases as the recycling process increasing. But as the recycling process increasing, the MFI, impact strength and flexural properties of PA6/PC/ABS were increased gradually. Therefore, it can be said that recycled PC/ABS and PA6/PC/ABS can be applied as the replacement of virgin ones without a substantial alteration in the mechanical properties and it can be applied as the production in commercial functions and roles that does not need hygienic and aesthetic attributes. Lastly, it can also be concluded that the mechanical properties of PC/ABS will be improved with the addition of PA6 polymer.

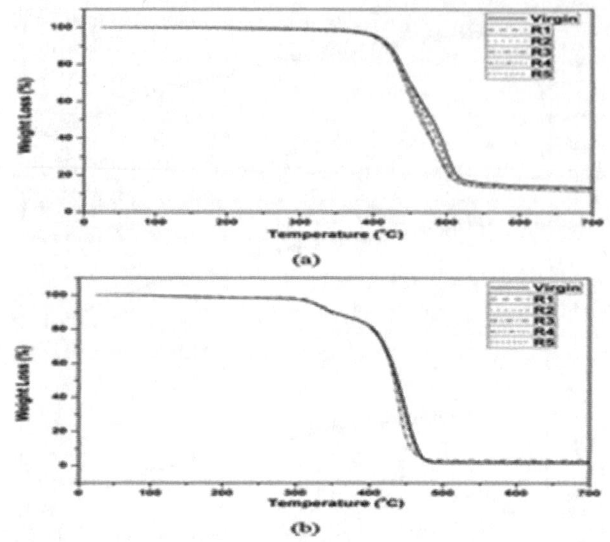

Figure 2. TGA curves for virgin and recycled specimens of (a) PC/ABS binary and (b) PA6/PC/ABS ternary blends. TGA: thermogravimetric analysis; PC: polycarbonate; ABS: acrylonitrile–butadiene–styrene; PA6: polyamide 6 [5].

From the literature review above, it is generally dependent on the purity of the recycled plastic material and the property requirements of the plastic product being produced whether or not the virgin polymer can be replaced with recycled plastic. The result obtained from most of the studies show that the impact strength and Melt Flow Index (MFI) will be decreases as the recycling process of the material increases.

5. Application of Optimization Method in Injection Molding Process

There are some researchers that have been studied on improving warpage by using different optimisation method such as Taguchi method, Design of Experiment, Artificial Neural Network, artificial intelligence, Response Surface Methodology (RSM) and mould flow analysis in order to reduce the values of warpage and the summary of the previous studies

There are numerous researchers that are using statistical method which is Taguchi method in order to improve the quality of product that been manufactured by optimising the processing parameter.

Wavare and Ubale [30] analysed on parametric of PP Homopolymer (1110MAS) in plastic injection moulding process by using Taguchi method to improve the quality of manufactured specimen which were optimising the shrinkage and warpage. The material used for the specimen was PP Homopolymer (1110MAS) with dimensions of 169×16×3.2 mm and it has been modelled in PRO/Engineer. In this study, determination of effect of injection moulding process parameters on shrinkage and warpage of the specimen will be obtained by calculating the S/N with the categories of smaller is better. The volumetric shrinkage has been minimise by 3.75% which least than experimental values. While for warpage it has been reduce to 2.54% which least than experimental value. Besides that, validation of the analysis was confirmed by the confirmatory experiment with the use of the optimal values parameter and the result obtain were close enough to the experimental values. The most influencing parameter to controlling the volumetric shrinkage was melt temperature followed by injection time, injection pressure, cooling time and holding pressure. While for controlling the warpage, the most influencing parameter was cooling time followed by melt temperature and injection time. Therefore, from the optimisation of parameters in injection moulding process by using Taguchi OA, S/N ratio and ANOVA method were very powerful tools to minimise warpage and volumetric shrinkage and can be successfully employed to improve injection moulding of other plastic parts with complex geometry.

Nasir *et al.*[13] analysed optimisation on the processing parameters in thermoplastic processing by using RSM to obtain an optimal value of shrinkage on the moulded parts. The measurement of the shrinkage of the moulded part will be simulated by Autodesk Moldflow Insight (AMI) software and specification of injection moulding machine of Nessei NEX100 have been set in this software to get a more accurate results. The main effects of shrinkage of the moulded part will be estimated by using Design Expert 7.0 software in the process of DOE. RSM were used to define the relationship between various responses and various processing parameters. A quadratic model of the RSM will be adequate by the CCD and produced empirical model to obtain minimise shrinkage. The plastic part was used the material of ABS or called as Polylac PA-777B and the variable parameters were melt temperature, coolant inlet temperature, cooling time and packing pressure. From DOE, the main factor that contributed to the shrinkage was packing pressure, with the determination for dual and single gates as shown in

Figure 3 and Figure 4, are 0.9816 and 0.9786 respectively. While the determination coefficient, R^2 were the dual gates and single gate of 0.9854 and 0.9849, respectively. P-value were obtained from ANOVA and ensure that it was less than 0.0001 for both types of gates. Minimum value of shrinkage will be obtained from the combination software. Hence, from the investigation it can be concluded that the main factor of parameters was packing pressure and result of shrinkage of single gate were better than the dual gates.

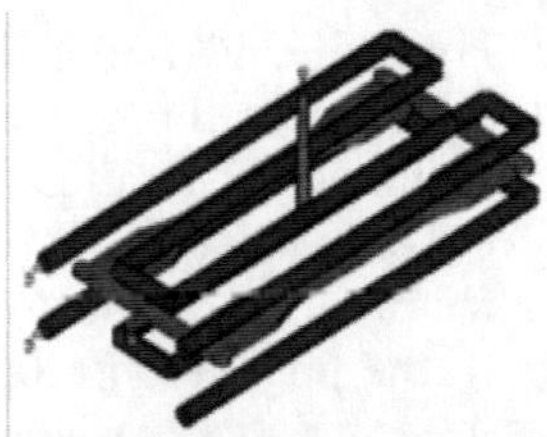

Figure 3. Mold design for dual gate [13].

Figure 4. Mold design for single gate [13].

Chen *et al.* [11] investigated optimisation on the processing of Plastic Injection Moulding (PIM) using Taguchi method, RSM and GA to obtain an optimal value of shrinkage and warpage on the moulded parts. Computer Aided Engineering (CAE) simulation software and DOE were used to obtain the data that can be conducted in ANOVA as the result of optimal parameter will be used in RSM. Then, by combining the GA with the regression model, the best combination of optimal process parameter will be gained. The injection moulding machine Victor Taichung VS-80 were used in this case study. The plastic part as shown in Figure 5 moulded by using PBT-2100 has been used as a case study. Length and warpage parameter combination of highest S/N ratio result for the shrinkage was 0.02% nearly closest to the target value which is 170.5. While for the Cpk, the value increased by 69.93% from 1.53 and the standard deviation decreased by 9.09% from 0.022. The average value for warpage and standard deviation were decreased by 64.3% and 38.095% respectively. Thus, this will have reduced the cost for the injection and time consuming of the mould design to the manufacturing process.

Besides that, Chen *et al.* [36] also investigated optimisation on the processing parameters in an injection moulding process using Taguchi method, RSM and hybrid GA–PSO to obtain an optimal value for shrinkage and warpage on the similar moulded parts as shown in Figure 5. Data were processed using the Backpropagation Neural Network (BPNN) combined with GA. ANOVA and DOE were used to screen the experiment setting by simulation. Then, the RSM and GA were used to determine an appropriate parameter for injection moulding process. Then, the RSM and GA were used to determine an appropriate parameter for injection moulding process. Melt temperature, injection velocity, packing pressure, packing time and cooling time were selected as variable parameters. The closest to the target value (170.5) was the length average value of 170.483. The Cpk length value increased 67.37%, standard deviation decreased 40.91%. While for the warpage, the average value decreased 72.826% and for the standard deviation decreased 66.67%. From this result, the warpage values decrease as the length value approached the target value. Thus, the time required and injection cost to be moulded and the process of manufacturing will be reduce.

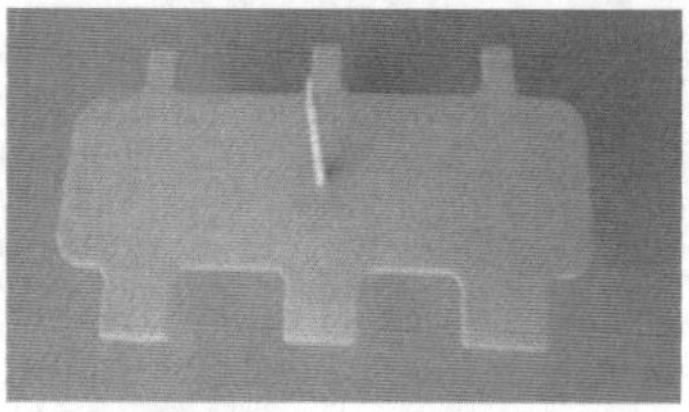

Figure 5. Plastic part [11].

Xu *et al.* [37] investigate on multi-objective optimisation of Multiple-Input Multiple-Output (MIMO) to obtain solutions for different objectives of multiobjectives of system and to find the best Pareto in injection moulding by using PSO. MIMO in plastic injection moulding process was used in optimisation process parameter to assisting engineers. The thin-walled part as in Figure 6 is use in the experiment was high density Polyethylene with dimension of 80 m length and 3 mm thickness. The parameter used in this experiment were mould temperature, injection time, injection pressure, melt temperature, cooling time, holding time and holding pressure. Hydraulic vibrating dynamic plastic injection moulding machine were used for the injection moulding process. From the results of the experiments, the optimal parameter for PSONN were maximum allowable iteration number (10K), particle population (100), parameter dimension (122), optimal value of inertia weight (0.9), error limit (0.01), ultimate value of inertia weight (0.3) and acceleration factor of 2. While for MOPSO, the acceleration factors (1.49), ultimate value of inertia weights (0.9), the population (200) and ultimate value of inertia weights 0.3. Improvement to PSONN give 32%, 43% and 14% for volume shrinkage, flash response and product weight severally. While for MOPSO, the improvement gives 42%, 30% and 21% for volume shrinkage, flash response and product weight respectively. Therefore, from this study know that PSONN model and MOPSO will provides an accurate estimations and predictions. Then PSONN will provide solution in form of Pareto-optimal with different combinations of optimal solutions. Thus, the approach that have been proposed was effective and feasible for optimisation process parameter in MMO plastic injection moulding and give advantages on costs and quality.

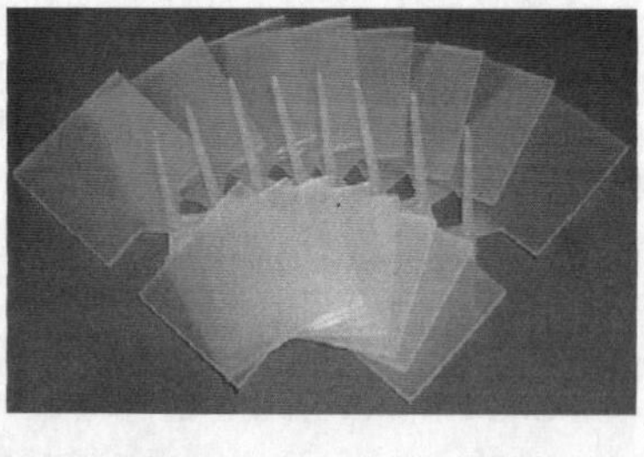

Figure 6. Thin walled plastic parts [37] .

Liu *et al.* [38] investigated injection conditions for a thin- walled wax pattern by using RSM to obtain an optimal value of shrinkage variation of thin walled wax in the injection process. ANOVA and DOE were used to get the mathematical model and to obtain optimal range of design space. Multiple response method was used to gain the best setting for the minimum shrink. A MPI55 injection moulding machine were used in this experimental works and to obtain the mean of the shrinkage a CMM with 0.1um resolution was used. The material that have been used as case study was crystalline wax (KC4017B). From the experiment, the optimal value of the shrinkage was 0.98935 as melt temperature become the most significant factors with major contribution percentages 43.67% followed by the mould temperature (23.77%), holding time (16.99%) and packing pressure (11.85%). The optimal setting for mould temperature, melt temperature, packing pressure, and holding time is 74°C, 30°C 25 bar, and 5 s, respectively were indicates from RSM analysis. Thus, from this case study it can be conclude that to reduce the shrinkage of the wax part, the relationship between the process and the shrinkage of thin walled will be describe by the mathematical regression model with the used of RSM.

Many researchers have studied on the optimization of parameter processing in plastic injection molding process, in order to reduce defects that will occur on the plastic part produce, such as warpage and shrinkage. From the studies by some researcher, it shows that by using optimization method such as RSM, PSO and GA were proven improve the quality of plastic part produced. However, most of the researchers only focused on the quality of plastic part produced by virgin material. Research on quality of plastic part produced by recycled material are still lack of supports and verifications from optimization method and real experiment data.

6. Summary and Future Works

The need of the plastic has been continuously increase together with the increases of population in worldwide which will affect the waste problem. The amount of polymer material wasted which is feed system of the part and scrap generated along the process due to quality problems is very high [39]. Thus, recycling process should be implemented for both economic and environmental reasons. To overcome these issues, there are many researcher focuses on the effect of virgin-recycled blends material and number of reprocessing cycles on mechanical properties of part produced by performing experiment [3], [29], [40–43]. While, there are only several researcher that are studied on the effect of processing parameters on the plastic part produced by virgin-recycled blends material [29], [44]. However, the effect of processing parameters and virgin-recycled blends on the quality of part (warpage) is hardly to find.

One of the most successful engineering thermoplastics is ABS [4]. ABS is widely employed in automotive industry, mobile phones, electronic equipment and other commodities[4]. Products that are made from ABS are fast moving and high demand, therefore by recycling and reuse of these blends are very important. The effect of virgin-recycled ABS blends on reprocessing cycles and mechanical properties have been extensively studied [4], [5], [43]. However, it is difficult to find a study on the effect of virgin-recycled ABS blends on warpage that affects product quality. The comparison should be done between the plastic part produced by virgin material and various combination of virgin and recycled material in term of the length, width and depth. In addition, most of them are using real experimental work without the aided of simulation software and artificial intelligence optimisation method.

The moulded plastic parts may not be functional or visually acceptable if there is warped occurred on the plastic parts [45]. It will affect the assembly process of the plastic part. The formation of warpage in plastic moulded part is influenced by the most influential processing parameter in the injection moulding process [13], [27], [45–47]. Thus, formation of warpage will affect the productivity, quality and cost of production in plastic injection moulding industry [31], [37], [47–48]. Traditionally, trial and error approach are often used in plastic injection moulding process in order to obtain an optimal combination of processing parameter. To overcome these issues, the use of artificial intelligent methods in simulation studies and experimental works will reduce both the time consuming and cost, while producing the best quality of product. Most studies based on optimization method have been concerned with optimizing processing parameters to minimise the warpage of the part produced using virgin materials only. However, there are lack of research on optimising processing parameters in order to minimising warpage compared to the mechanical properties of plastic parts produced by combination of virgin and recycled material, especially on ABS.

References

[1] A. Mohd *et al* 2016 *Indian J. Sci. Technol.* **9** 9.
[2] S. Chen *et al* 2011 *Polym. Plast. Technol. Eng.* **50** 306–311.
[3] C. Meran *et al* 2008 *Mater. Des.* **29** 701–705.
[4] R. Scaffaro *et al* 2012 *Eur. Polym. J.* **48** 637–648.
[5] F. Kuram *et al* 2015 J. *Elastomers Plast.* **48** 164–181.
[6] T. C. Chang *et al* 2001 *Polym. Eng. Sci.* **41** 703–710.
[7] N. C. Fei *et al* 2013 *ISRN Ind. Eng.*
[8] W. C. Bushko *et al* 1996 *Polym. Eng. Sci.* **36** 658–675.
[9] K. M. B. Jansen *et al* 1998 *Polym. Eng. Sci.* **38** 5.
[10] M. C. Huang *et al* 2001 *J. Mater. Process. Technol.* **110** 1–9.
[11] W. C. Chen *et al* 2015 Taylor & Francis Group, London, 341-346.
[12] W.-C. Chen *et al* 2016 *Int. J. Adv. Manuf. Technol.* **83** 1873–1886.
[13] S. M. Nasir *et al* 2016 *Key Eng. Mater.* **700** 12–21.
[14] Fathullah M *et al.* 2011 *International Review of Mechanical Engineering* **5(7)** 1278
[15] Fathullah M *et al.* 2011 *International Review of Mechanical Engineering* **5(7)** 1189
[16] Nasir S M *et al.* 2012 *International Review of Mechanical Engineering* **6(3)** 372
[17] Nasir S M *et al.* 2013 *International Review of Mechanical Engineering* **7(5)** 977.
[18] Isafiq M *et al.* 2016 *MATEC Web of Conferences* **78** 01084.
[19] K. Park 2005 *Polym. Plast. Technol. Eng.* **43** 1569–1585.
[20] H. Li *et al* 2007 *Int. J. Adv. Manuf. Technol.* **32** 927–931.
[21] M. Kurt *et al* 2010 *Int. J. Adv. Manuf. Technol.* **46** 571–578.
[22] H. Kurtaran *et al* 2006 *Int. J. Adv. Manuf. Technol.* **27** 468–472.
[23] M. H. Othman *et al* 2012 *Appl. Mech. Mater.* **229–231** 2536–2540.
[24] W. C. Chen *et al* 2014 *Int. J. Precis. Eng. Manuf.* **15** 1583–1593.
[25] L. Balasz, I. Sors, 1989 Design of. Hungary: Elsevier Science Publishers.
[26] S. J. Liao *et al* 2004 *Polym. Eng. Sci.* **44** 917–928.
[27] B. Ozcelik *et al Mater. Des.* **30** 367–375.
[28] C. P. Chen *et al* 2009 *Expert Syst. Appl.* **36** 10752–10759.
[29] N. M. Mehat *et al* 2011 *J. Mater. Process. Technol.* **211** 1989–1994.
[30] A. Waware *et al* 2016 *Intro to Stats.* **4** 323–331.
[31] Y. Q. Wang *et al* 2014 *Mater. Des.* **56** 313–317.
[32] R. Navarro *et al* 2008 *J. Mater. Process. Technol.* **195** 110–116.
[33] A. F. Ávila *et al* 2003 *Polym. Degrad. Stab.* **80** 373–382.
[34] K. Hamad *et al* 2013 *Polym. Degrad. Stab.* **98** 2801–2812.
[35] J. Hopewell *et al* 2009 *Trans. R. Soc. B Biol. Sci.* **364** 2115–2126.
[36] W. Chen *et al* 2016 *Int J Adv Manuf. Technol.* **83** 1873–1886.
[37] G. Xu *et al* 2012 *Int. J. Adv. Manuf. Technol.* **58** 521–531.
[38] C. H. Liu *et al* 2014 *Appl. Mech. Mater.* **607** 185–192.
[39] C. Javierre *et al* 2007 *Waste Manag.* **27** 656–663.
[40] D. Bhattacharya *et al* 2014 *Procedia Eng.* **97** 186–196.
[41] D. Marulanda *et al* 2014 *J. Elastomers Plast.* **46** 355–367.
[42] J.Abdullaha *et al* 2016 *Int. J. Mech. Prod. Eng.* **4** 76–81.
[43] S. D. Mancini *et al* 1999 *Mater. Res.* **2** 33–38.
[44] F. Gu *et al* 2014 *Mater. Des.* **62** 189–198.
[45] S. Taghizadeh *et al* 2013 *IJST, Trans. Mech. Eng.* **37** 149–160.
[46] N. M. Mehat *et al* 2013 *ISRN Ind. Eng.*
[47] Q. H. Sun *et al* 2013 *Appl. Mech. Mater.* **377** 133–137.
[48] C. Chen *et al* 2009 *Int. J. Adv. Manuf. Technol.* **45** 1087–1095.

IOP Conf. Series: Materials Science and Engineering **551** (2019) 012026 doi:10.1088/1757-899X/551/1/012026

Optimization of Shrinkage and Strength of the Molded Part Produced using LDPE (Virgin and Recycled) material: A Review

Norshahira Roslan[1], Shayfull Zamree Abd Rahim[2], Nurul Hidayah Mohamad[3] Huzaim and Nur Aisyah Miza Ahmad Tamizi[4]

[1,2,3,4]School of Manufacturing Engineering, Universiti Malaysia Perlis, Kampus Tetap Pauh Putra, 02600 Arau, Perlis, MALAYSIA.
[2]Green Design and manufacture Research Group, Center of Excellence Geopolymer and Green Technology (CEGeoTech), Universiti Malaysia Perlis, 01000 Kangar, Perlis, MALAYSIA.

E-Mail: shahiraroslan@yahoo.com

Abstract. The injection molding process is seen to be one of a versatile and important manufacturing process. This is due to its ability in producing product with complicated design and producing part in large quantity. The advantages of this process had caused an active research on its quality improvement and process. This paper reviews the part quality issues happened in the processes. Along with the reviews of the parameters which affects the defects occurrences. Besides that, recycled plastic application in injection molding process was also reviewed. At the end, it is seen that as compared to other thermoplastic materials, LDPE is least popular and discussed as a subject of research. This paper organizes past research studies in few categories firstly the defects occurrences in plastic injection molding, followed by application of recycled material in molding plastic part, application of optimization method in injection molding, next the application of LDPE material in molding plastic part. Lastly, this paper ends with a brief summary and future work.

1. Introduction

Plastic materials have a wide range of ability in producing part with various shapes and complex design Things such as water bottles, food container which comes in various shapes, color and sizes are made possible with the plastic injection process. Hence causing the products made of plastic to be high in demands. Since the plastic injection molding involves manufacturing part in mass quantity, thus the amount of part rejected and go to waste is abundance in a day. The discard of part can cause waste in money, time and also effort in the improvement of the part. Hence, a simple definition to the quality improvement is the ability to reduce the amount of waste.

Besides quality improvement, the recycling of plastic may also help in reducing the waste. As mentioned previously, the discard of part can cause loss in money, hence it is seen that the ability in reusing this unwanted part may become a big advantage to the manufactured. However plastic recycling was restricted by restrictions of standards, specifications and trust issues on the use of recycled plastic. Most of this happened because of perception that recycled plastic produce a really bad quality of part as compared to the use of virgin material.

Although the use of recycles raises a lot of doubt to the consumer and manufacturer, researchers on the other hand show interest in studying its potential. Besides that, it is seen that many researches focus on the improvement of the recycles plastic properties in material perspective when applied the additives. The cost of applying additives to recycled materials however become a main considerations as the manufacturer had to paid more for the manufacturing cost even though the recycles improves due to the additives help [1]. Thus, another alternative arises from the manufacturing perspectives, where process parameters settings control involves.

There are numbers of factor that can affect a quality performance of a molded part as shown in Figure1 below. This includes factors such as the mold design, material selection, machine parameters and operator and working parameters. These variables in the process can easily results in waste due to not fulfilling the quality specification [2-5].

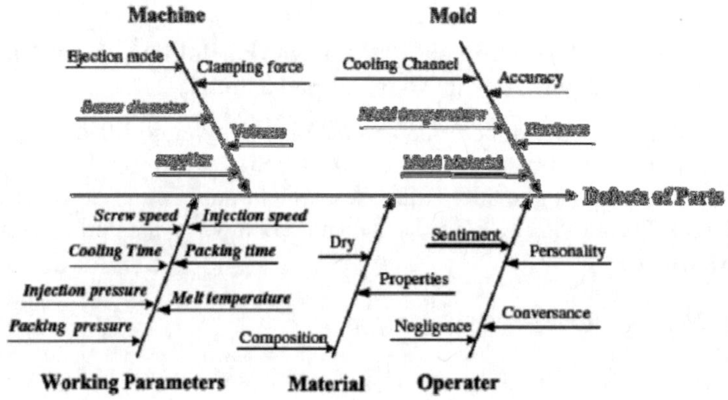

Figure 1: Fishbone diagram on parameter affecting the defects of part [6].

In practice, when the part mold is secure to the machine, the molding engineer has to set the parameters of the machine as refer to the material specifications provided by the material's supplier, sometimes lead to the extent based on the mold designer and also his experience. Next, the engineer needs to performed mold trial procedure where the tested parts were produced and the machine parameters were adjusted. This trial and error continue until the design specifications of the molded part fulfilled. Such procedure does not only require long time but also has to be repeated for different materials, machine and mold specifications. However, these steps are necessary in order to achieve a quality part. Hence it is seen that due to its complex manufacturing process, the trial and error process that is costly and yet time consuming was not suitable to be applied. Besides that, trial and error is seen to be impossible in verifying the real optimal processing settings [7].

2. Defect Occurrences in Plastic Injection Molding Process

As the mold was fully filled by the injected plastic, the screw stays stationary in a while in order to keep the plastic inside the mold under pressure (holding time). In this stage additional plastic material was injected into the mold, thus compensate for shrinkages happened which caused by the cooling effects. Hence it is seen that parameters involved in injection molding process can potentially effect in shrinkage occurrences. All of these factors are worth to pay attention to in identifying its influence on shrinkage [8].

Once most of the plastic has been injected by applying the injection pressure, next the machine will shift to inject the material by packing pressure. This pressure applied allows a final filling into the cavity by packing the molecule together. Packing pressure continue to apply until the gate freezes. A study on a casing of a cellular phone was performed by S.J. Liao et al. [9] recorded that the most influential process parameter is the packing pressure. It is discussed in his study that as the packing pressure high, melted material can squeeze into the cavity and compensate the shrink volume as the thermoplastic resin cool down. Besides that, mold temperature also seen as one of the influential factors. The study seen that a high mold temperature results in the plastic in the cavity to cool down from a high temperature condition. Thus, the finished product has higher density and low shrinkage. A similar results was achieved by [10], a high holding pressure does seen to cause shrinkage reduction in both parallel and normal flow direction.

Another study performed by A.Waware and S.Ubale [11] in the year of 2016, investigated the effect of applying the parameter optimization on the warpage and shrinkage of plastic part. The optimization techniques applied in the study is Taguchi. In the study it was found that the melt temperature influenced the shrinkage occurrences the most. Meanwhile for warpage, the holding pressure was the most influential parameter. Similar results was achieved by few other researchers [12-13], where the melt temperature was one of the most important factor affecting the shrinkage and warpage of part. Shrinkage of plastic part was also studied by G.J. Kang [14]. Besides that, Taguchi optimization was also applied in this study. Four parameters were selected for the study. They are the melt temperature, mold temperature, cooling time, packing time, and lastly packing pressure. In the study, it was found that the most affecting parameter for shrinkage is the injection time. While for warpage is the cooling time. By applying the optimal parameter settings, it is seen that the quality characteristic of shrinkage and warpage improved.

Not only the shrinkage, but a weld line can also affect the appearance of the molded part. Beside that the weld line also may affect the strength of the produced part. One of the advantages of the injection molding method is a multi-gated mold and insert can be designed to fit the design of the part needs and shapes. By this means that a formation of weld lines during the process of injection molding is unavoidable. When the design of the products causes the option of changing the weld lined location by changing the gates to be impossible, it can lessen the weld line visibility by adjusting the processing condition.

A study on molded plastic strength was studied B.Ozcelik [15] in year 2011. 3 parameters were studied, they are packing pressure, melt temperature and injection pressure. Properties such as the extension at break, charpy impact strength and maximum tensile load were investigated. Taguchi orthogonal array was employed and optimal conditions of each response were recorded. Based on the regression analysis, shown a linear relationship between molding parameter with the mechanical properties. It is also seen that injection pressure and melt temperature were highly affected the maximum tensile load and extension at break. While, the charpy impact strength is the melt temperature and packing pressure.

Melt temperature and packing pressure was also seen to be the an influential factors in a study conducted by Dar et al [16] on Polycarbonate. The study seeks to measure yield and post yield effect by differentiating the process parameters of injection molding. The variable studied were mold temperature, injection velocity, cooling time, melt temperature and packing pressure. It was seen that an increase in melt temperature from 533 K to 593 K caused an increased in yield stress. Besides that, this study shows that when the flow velocity increases the material viscosity reduces. In every change on the melt temperature value, the mold temperature was held constant at 303K. These differences in temperature reduce the material's residual stress. In each test conducted, the yield stress lied in the range of 0.21 to 0.35 MPa. Upon reaches the value of 343K, the reading was constant but was increased from 343K until 413K. It was observed that, an increased in packing pressure of 30MPa, the yield stress was increased at 1.1MPa.

Based on the few mentioned researchers, it can be concluded that a different process parameter has different effect on the quality of plastic part. Therefore, achieving the suitable process parameter settings is seen to be an important task in the molding industries. Trial and error method had been a popular applied to find a suitable molding condition. Even though this method has been applied for many years, however this method depends on the operator experience and time consuming. Achieving an optimal process condition for a part is a major advantage since it can improve not only the quality but also reducing the amount of waste. In the industries, lesser waste means the manufacturer can save money on producing new quality part. It is seen that, not only by maintaining good quality, ability to fully utilizing the material can also be an advantage to the manufacturer.

3. Application of Recycled Material in Molding Plastic Part

Thermomechanical degradation affects the chain breakage of the material's polymeric chains. This happens due to the radical chain reaction which leads to a series of changes in the structural properties of the polymers. Changes on the material's structural properties worry the manufacturer to apply the recycled plastic in molding new part. This due to the reason that, quality of the molded part might reduce.

A research conducted by Aurrekoetxea *et al* [17] investigate Recycled Polypropylene (PP) fracture behavior on injection molding induced morphology. The material used was Polypropylene (PP). The steps of recycling were conducted up to six times in simulating the recycling process. The specimens were tested on its microhardness, and hardness. Besides that, fracture test was also conducted and fracture surface was also been analyzed. The test revealed a higher elongation caused the core breaks, while at small elongation level the skin layer of the specimen broke. Besides that, it was also seen that first recycled PP to six times recycled PP specimens fractured brittle. Charpy specimens broke immediately as the maximum load was achieved and at small deflections. As compared to recycled, virgin shown higher plastically deformed zone at crack tip and micro-hardness values. Otherwise, in both materials, skin layer has shown lower micro-hardness values and smaller plastic zone extension than the core region.

Another study on reprocessing cycle of recycled material was performed by R. Scaffaro et al. [18]. In rheological test conducted, it was determined that viscosity of ABS compared to the virgin material until two processing cycle was not substantially different. However, in the third cycle, it was seen that the viscosity was reduced slightly. In the first reprocessing, the material tensile decreased. Although the amount of pc-ABS added increased it did not have any significant changes. In contrast, the materials flexural properties were slightly affected by the recycling operations, and were almost did not influenced by the amount pc-ABS added. Meanwhile, there was decreased in the impact resistance when the reprocessing cycles increased and by increasing the amount of pc-ABS. The worsening of the property was related with the degradation of the PB phase, likely due to thermo-mechanical stresses.

Based on the study, it is seen that degradation of plastic does highly affected by the number of reprocessing operations. Thus, further affect the recycle plastic's rheological and also mechanical properties of the molded part. Although it is proven by few researches that reprocessing the plastic few times affect the part quality, the research of plastic recycle ability doesn't stop there. Other research found that applying an ideal recycle to virgin ratio is acceptable and safe depending on its application. In some company they completely ban the use of recycled material or placing a maximum limit of recycled material added to the raw material.

In 2008, C. Meran et al. [19] conducted a study on the recycling possibility and using recycled polypropylene (PP), high-density polyethylene (HDPE) and low-density polyethylene (LDPE) in relation to the tensile strength exchange ratio. The test specimen in this study is shown in the Figure 2.17 below. The recycled material used was mixed with the virgin materials in different proportions and were examined by using the tensile test. In the experiment, the test specimens of each mixture were broken and the average values were the recorded. Besides that, the material usability was observed and its optimal points were searched. The tensile strength decreases to 15% as the mixture proportions of LDPE decreases. While for HDPE, as the mixture proportion decrease, the tensile strength decreases to 5%. Lastly, high and low density polyethylene mixture decreases, the tensile strength decreases to 3%.

C. Javierre et al. [20] conducted a study on the effect of percentages of the recycled material on mechanical behavior and the safety factor of the part. The HDPE recycled material was sorted into different percentages of 20%, 40%, 60%, 80% and 100%. The characteristic of the formed specimen from raw and recycled were both tested using tensile machine. CAE tool was utilized in studying the effect of the recycled percentage on mechanical tests. Simulation analysis was performed with the FEM model of the container's part, with fully recycled material and also with raw material. The addition of different recycled material percentage during injection molding processes implies an important change in the stiffness and safety factor. 100% recycled HDPE blend decreases strength up to 40% in regard to raw HDPE. An important factor to be remembered was there must be an increase in 20% of material strength for it to perform its functionality well with fully made of recycled material. Lastly, with 40% of recycled material, minimum effect was achieved on safety factor in regard to stress.

In 2014, D.Bhattacharya and B.Bepari [21] conducted a study on the feasibility and recyclability of polypropylene (PP) in injection molding based on grey relational analysis. The parameters studied were the ratio of virgin to recycled, injection temperature, injection pressure and injection speed. To obtain the elongation at break and tensile strength value, universal Testing Machine was used. VSP (Vicat Softening Point) was determined by using microprocessor controlled HDT/VSP apparatus. While, the density of PP sample was measured by using Dhona. In this research, the number of experiments conducted was reduced by applying the box-behnken model. While, in order to obtain the response's weight, a grey entropy method was used. It is seen that, the most contributing factor is the injection speed, while the least contributing factor is the injection temperature. Up to 5% of adulteration the product maintains the properties tantamount to the virgin one. Lower level of injection temperature and lower level for injection pressure and higher level of injection speed must be selected with a virgin to recycled ratio of 95:05.

The addition of recycled material percentage to raw causes change in the mechanical [20] and also the material rheological behavior [22]. It is seen that the utilization of a certain recycled material gave a minimum influenced to the safety factor or sometimes gave a less significant influence on the material's behavior. However, it is also seen that when a percentage of recycled was applied along with a specific parameter setting can also produce a good property as the raw material.

4. Application of Optimization Method in Plastic Injection Molding

Optimization of injection molding parameters has proven to be effective in many studies conducted in improving the quality appearance of product. Warpage and shrinkage had seen to be reducing when applying the optimization method. The optimization method was not only limited to the use of Taguchi even though it is popularly applied. However, the introduction of other optimization method was seen to achieve even a better part quality.

A study was conducted by Chen et al. [23] on the warpage analysis of plastic spur gear. Five processing parameters, the holding pressure, injection pressure, holding time, plastic temperature and mold temperature were studied. In this study the Taguchi and RSM optimization method were implemented. Taguchi concluded the holding time as the most influenced factor. By comparing both method results, it is seen that RSM produced the best optimal combination. They were packing time of 19s, packing pressure of 70MPa, injection pressure of 105s, plastic temperature of 190°C and lastly mold temperature of 60 °C. In measuring the accuracy between the experimental and simulation, the error count was only 2.7% difference.

GA optimization was performed by S.Yoewono and A.Kaswadi [12] in producing the optimum plastic injection process cycle time. Full factorial design was used with 4 factors and 2 levels. The factors were melting temperature, holding pressure, injection pressure, and holding. While the responses studied were cycle time, wall shear stress and volume shrinkage. Moldflow simulation was used in running all 16 run orders and the response was produced. In order to produce the relation between factors with cycle time, a linear regression analysis was conducted of the cycle time responses. The optimum parameters were determined by optimizing linear regression equation in GA. The result for both simulated and experiment were compared. The best parameter settings determined by GA were holding pressure of 16MPa, injection pressure of 20MPa, melt temperature of 180°C and holding time of 8s, with 14.11 of the cycle time. By simulation both wall shear stress and shrinkage did not change significantly and stay below allowed maximum limit.

S.Sudsawat and W.Sriseubsai [24] performed an optimization method in minimizing the shrinkage of wrench specimen and also the warpage. In the study, Response Surface methodology (RSM), Genetic Algorithm (GA) and Firefly Algorithm (FFA) was applied. 7 parameters were studied, they are mold

IOP Conf. Series: Materials Science and Engineering **551** (2019) 012026 doi:10.1088/1757-899X/551/1/012026

temperature, injection pressure, melt temperature, packing pressure, packing time, flow rate profile and cooling time. RSM was applied in stabling the experimental run. Meanwhile, the GA and FFA were applied in minimizing the volume shrinkage as well as the warpage. It is seen that cooling time and melt temperature is the most influenced factor for both shrinkage and warpage. At the end of the study, it is concluded that FFA optimization produced better results than the GA method.

Referring to the previous section, it is seen that most research on the plastic recycling focused on improvement of the recycled plastic properties from material perspectives. Hence another alternative from manufacturing perspective arises. In few, process optimization was seen to have a decisive influence on quality improvement of molded part, but its influence is underestimated. Few studies were performed in investigating the influence of the injection parameters on mechanical properties and on the defect occurrence [25-26].

In 2011, N.M. Mehat and S.Kamaruddin [13] employed the Taguchi method in enhancing the mechanical properties of recycled plastic. By applying L9 Taguchi Orthogonal Array, the trials were reduced to nine trials with five repetitions. Four parameters for this study were, packing pressure, packing time, injection time and melt temperature. The material used in this research was the Polypropylene (PP), injected by a Battenfeld TM750/210 molding machine. In obtaining the influential processing parameter on the quality of product and the optimal process parameters, the S/N ratio and ANOVA was conducted. The outcome of this study shows that, a 75% to 25% of recycled to virgin polypropylene product has better flexural modulus as compared to the virgin polypropylene. While the same product decreases 3.4% in flexural strength. Thus, shows that by optimizing the processing parameter can elevate the degradation in the mechanical properties of recycled plastic.

Another study on PP was performed by F.Gu *et al* [27]. The influence of recycled plastic proportion and processing parameters on the mechanical characteristic was investigated. Four factors were studied in this research, they were mold temperature, packing pressure, melt temperature and injection speed. The L9(34) orthogonal array was applied which further reduce the amount of experiment conducted from 81 to 9. It was observed that, lower melt temperature achieved a good flexural result and achieved a good outcome in most mechanical performance and a good result obtained with higher packing pressure. On the other hand, better result was obtained for tensile properties when a lower packing pressure was used. From the experiment indicate that, the best blending composition for adapting to recycled plastic content was PP90. Which, shown to be an improvement on 100% virgin PP. Besides that, tensile and flexural properties improved from 0.36% to 2.66% as 10% recycled PP was added, without significantly deteriorated the impact properties (loss about 2.06%). This study shows that, if the controlled variable changed, the mechanical properties or improvement on the specimen may also be changed.

J. Abdullaha et al. [28], studied the effect of process setting on the quality characteristics of plastic tray. This study employed the Taguchi orthogonal array. The quality characteristic tested was the tensile strength, flexural strength and shrinkage. Results optimal process settings in optimizing shrinkage were packing pressure of 70%, melting temperature of 513.15 K, injection pressure of 15 MPa and packing time of 30 s. Next in optimizing tensile strength the optimal settings were, packing pressure of 85%, injection pressure of 15 MPa, melting temperature of 518.15 K and packing time of 30 s. lastly, the optimal settings for flexural strength were, packing pressure of 80%, injection pressure of 14 MPa, melting temperature of 518.15 K and packing time of 10 s. It was determined that, the most significant parameter that effect the mechanical properties and shrinkage were the packing time. Lastly, for the tensile strength, the top contributor determined was the melting temperature. Through confirmation seen that the quality characteristic of the material has almost similar behavior with expected performance and the percentages of errors are less than 10%.

Taguchi optimization had been used as one of the statistical tools that helps to find the suitable parameters condition as seen in few mentioned researches. The application of this method had attracted the attention of many researcher in the past 20 years, and in recent years it has been vastly applied in various field such as in mechanical component design [29], process optimization [30-31] and also in manufacturing system [32]. This popularity was gained due to its practicality in developing a

IOP Conf. Series: Materials Science and Engineering **551** (2019) 012026 doi:10.1088/1757-899X/551/1/012026

high-quality system which provides less experimental run with an optimal process setting of the control parameters. Thus, prove the points that statistical tools are suitable and efficient to be applied in improving the molded part quality. Not only even though Taguchi able to provide optimal settings, few researches show a better optimization results by other statistical tools. Such as the artificial intelligence method of Particle swarm optimization and Firefly Algorithm (FFA). These tools which adopt the natural behavior have slowly gaining its reputation in parameter optimization study. However, the application of the artificial intelligence tool was hardly found in optimizing the parameter for a recycled material.

5. Application of the LDPE Material in Molding Plastic Part

Polyolefins (PP, HDPE, LDPE) are massively used thermoplastic material through the world. Due to its various applications in producing many types of product. Data from 2003 recorded approximately 21.37 million tones consumed the three polymers in Western Europe alone. Which constitute about 56% of total thermoplastic[33]. Since the three thermoplastics are applied in various amount of application, thus clearly show how important and valuable study on these materials.

Study on Low Density Polyethylene (LDPE) was performed by S.K. Lal and H. Vasudevan [34] which optimized the injection molding process parameters of low density polyethylene(LDPE).Taguchi method is used to investigate the effects of melting temperature, injection pressure , refilling pressure and cooling time on the shrinkage of LDPE. Signal to noise ratios were used for determining the optimum combinations of the process conditions for shrinkage. The result showed that melting temperature of 190 °C, injection pressure of 55 MPa, refilling pressure of 85 MPa and cooling time of 11 s. gave minimum shrinkage for LDPE. Cooling time was found to be most effective factor for LDPE followed by refilling pressure. Injection pressure was found to be the least effective factor. From the findings, it can be stated that Taguchi method is a powerful tool for evaluating the defect of shrinkage in the plastic injection molding.

The study on LDPE material doesn't stop there, the potential of recycled plastic applied in manufacturing part was studied by many researchers. Their shows the ability of the recycled and most of study was conducted on evaluating the recycle material's mechanical ability.

In 2008, C. Meran et al. [19] conducted a study on the recycling possibility and using recycled polypropylene (PP), high-density polyethylene (HDPE) and low-density polyethylene (LDPE) in relation to the tensile strength exchange ratio. The recycled material used was mixed with the virgin materials in different proportions and were examined by using the tensile test. In the experiment, the test specimens of each mixture were broken and the average values were the recorded. Besides that, the material usability was observed and its optimal points were searched. Based on the experimental results, it is seen that, the recycled material of all three material are 100% usable. PP was seen to be the most succeed to be fully recycled. This because the tensile strength for PP decreases to only 15% as compared to the virgin. While for HDPE, tensile strength decreases to 24%. Lastly for LDPE, the tensile strength decreases to 36%. The result also recorded that all three material doesn't have problems during pressing and it even achieved to 100% of the pressing level.

Similar study was performed by Bajracharya and Manalo [35] in 2014 where the mechanical properties of recycled material was studied. The recycles plastics used in this study were recycled low density polyethylene (rLDPE), recycled high density polyethylene (rHDPE), recycled polypropylene (rPP) and other proprietary additives. The variables for this study were the temperature across the barrel and also melt temperature. The plastic pallets were injected with barrel temperature of 220°C, 210°C and 205°C for front, mixing and feeding temperature. The temperature of melt was set for 23°C, 30°C, 40°C, 60°C, 80°C and 100°C. It is seen that the thermal stability of mixed recycled plastic is similar to virgin plastic. It is also seen that there was reduction of tensile strength for PP as compared to MRP when the temperature as around 60°C. It was seen that, an increased in temperature from 23°C to 100°C caused the modulus and tensile strength reduced at a significant amount of 75% and 80%. Besides that, there is a higher reduction of the tensile modulus than tensile strength at elevated temperature.

Even though the addition of additives able to improves a certain wanted properties of the recycled, and few researches had been performed on the addition of fillers, however there is an increase in cost [22]. In addition, a complex relationship between processing parameters needed numerous repeated experiments, which is seen to be time consuming and costly. Thus, it seen that the statistical methods are efficient to be apply in finding an optimal processing condition and shorten the time consumed. Most of the research on recycled which applied the optimization method used the Taguchi [13-28].Indeed the Taguchi able to lessen the trials numbers in experiments. However, in many studies on processing condition optimization of the virgin material, artificial intelligence had started to be applied. Its ability had been proved by many optimization study [36-37]. However, for LDPE material, the study performed by using artificial intelligence is hardly found for both virgin and recycled material.

6 Summary and Future Works

Plastic materials have wide range of usability, hence causing its application to increase in demands year by year. This past decade had witnessed the injection molding fields continue to grow and more developed. Thus, it is seen that a review on the field would help researchers to have ideas on identifying the path to further improve the injection molding.

The review highlights the effects of injection molding parameters on the part quality. Although it can be solved by finding a suitable parameter setting, however the trial and error method which based on the molder's experience is costly and time consuming. In order to overcome the effect on cost, few researchers investigate the potential of using the recycle material. The usage of recycled material derived from rejected parts and gating system is seen to be cost efficient. However, the reprocess of this recycled plastic is seen to have an issue to its mechanical properties. Few companies set a specific limit on recycled percentage. Few of the reviewed study also highlight a safe recycle percentage can be achieve depending on its application. Where it seen that for different material, shape of design and application the recycle percentage has to be adjusted. Hence finding the efficient way of achieving the right percentage is beneficial. Similar to virgin, process parameter on recycled material such as melt temperature also affect the manufactured part quality.

In order to overcome the appearance as well the recyclability issues, the optimization tools is seem to be one of the best solutions. Not only it able to improve the shrinkage happen on the molded part, but in certain cases also help on achieving a safe molding setting by using the recycled material. However, most optimization method applied on recycled was the Taguchi. Which in most studies it delivers a really good results, unlike the virgin material, the application of Artificial intelligence (AI) is hardly found on recycled material. Although Polyolefins (PP, HDPE, LDPE) are studied in many researches. However, among the three materials, study on LDPE is hardly found. Hence it is seen that. LDPE material can also be studied in future works. Also, the potential of AI on recycled material should also be studied, so it can be applied in the manufacturing field.

References

[1] Ramirez E *et al* 2006 *J. Appl. Polym. Sci.* 100 3696–3706.
[2] Nasir S M et al. 2013 International Review of Mechanical Engineering 7(5) 977.
[3] Fathullah M et al. 2011 International Review of Mechanical Engineering 5(7) 1189.
[4] Nasir S M et al. 2012 International Review of Mechanical Engineering 6(3) 372.
[5] Fathullah M et al. 2011 International Review of Mechanical Engineering 5(7) 1278.
[6] Tidke Y P *et al* 2006 *Int. J. Res. Appl. Sci. Eng. Technol.* 2 58–65.
[7] Kale H P *et al* 2015 *Int. J. Sci. Res.* 4 722–725.
[8] Isafiq M et al. 2016 MATEC Web of Conferences 78 01084.
[9] Liao S J *et al* 2004 *Polym. Eng. Sci.* 44 917–928.
[10] Pomerleau J *et al* 2006 *Polym. Eng. Sci.* 1275–1283.
[11] Wavare A *et al* 2016 *Int. J. Eng. Res. Gen. Sci.* 4 323–331.
[12] Yoewono S *et al* 2017 *Chinese J. Mech. Eng.* 30 398–406.
[13] Mehat N M *et al* 2011 *J. Mater. Process. Technol.* 211 1989–1994.

[14] Kang G J *et al* 2014 *Appl. Mech. Mater.* 541–542 359–362.
[15] Ozcelik B 2011 *Int. Commun. Heat Mass Transf.* 38 1067–1072.
[16] Dar U A *et al* 2016 *J. Appl. Polym. Sci.* 1–9.
[17] Aurrekoetxea J *et al* 2003 *Polym.* 44 6959–6964.
[18] Scaffaro R *et al* 2012 *Eur. Polym. J.* 48 637–648.
[19] Meran C *et al* 2008 *Mater. Des.* 29 701–705.
[20] Javierre C *et al* 2012 *Adv. Mater. Res.* 445 935–940.
[21] Bhattacharya D *et al* 2014 *Procedia Eng.* 97 186–196.
[22] Javierre C *et al* 2007 *Waste Manag.* 27 656–663.
[23] Chen D C *et al* 2017 *Smart Sci.* 5 206–213.
[24] Sudsawat S *et al* 2017 *Indian J. Eng. Mater. Sci.* 24 228–238.
[25] Mehat N M *et al* 2011 *Mater. Manuf. Process.* 26 202–209.
[26] Shen Y K *et al* 2002 *Int. Commun. Heat Mass Transf.* 29 97–105.
[27] Gu F *et al* 2014 *Mater. Des.* 62 189–198.
[28] Abdullaha J *et al* 2016 *Int. J. Mech. Prod. Eng.* 4 76–81.
[29] Shim H et al 2009 *Eng. Fail. Anal.* 16 1955–1963.
[30] Mohan N S 2005 *Compos. Struct.* 71 407–413.
[31] Ariffin M K A M *et al* 2009 *Sci. Res. Essay* 4 1109–1116.
[32] Mahfouz A *et al* 2010 *Simul. Model. Pract. Theory* 18 471–482.
[33] Achilias D S *et al* 2007 *J. Hazard. Mater.* 149 536–542.
[34] Lal S K *et al* 2013, *Int. J. Eng. Res. Dev.* 7 35–39.
[35] Bajracharya R M *et al*, "Effect of elevated temperature on the tensile properties of recycled mixed plastic waste" (*23rd Australasian Conference on the mechanics of Structures and Materials (ACMSM)*, 2014), pp. 281–286.
[36] Fei N C *et al* 2013 *ISRN Ind. Eng.* 2013 1–11.
[37] Kurtaran H *et al* 2006 *Int. J. Adv. Manuf. Technol.* 27 468–472.

IOP Conf. Series: Materials Science and Engineering **551** (2019) 012027 doi:10.1088/1757-899X/551/1/012027

Optimization Parameters to Reduce the Warpage Defect of Plastic Injection Molding Process for A Thin-Shell Part Using Design of Experiment

H Radhwan[1], S M Nasir[1,2], M M Rashidi[3], K Kamarudin[4] and Abdellah el-hadj Abdellah[5]

[1]School of Manufacturing Engineering, Universiti Malaysia Perlis, Kampus Tetap Pauh Putra, 02600 Arau, Perlis, Malaysia.
[2]Green Design and Manufacture Research Group, Center of Excellence Geopolymer and Green Technology (CEGeoGTech), Universiti Malaysia Perlis, 01000 Kangar, Perlis, Malaysia.
[3]Faculty of Mechanical Engineering, Universiti Malaysia Pahang, 26600 Pekan, Pahang, Malaysia.
[4]University Tun Hussien Onn Malaysia, Parit Raja, 86400 Batu Pahat, Johor.
[5]Laboratory of Mechanics, Physics and Mathematical Modelling (LMP2M), University of Medea, Medea 26000, Algeria.

Email: radhwan@unimap.edu.my

Abstract. Development of light, small and thin plastic products that possess high strength characteristic such as electronic devices have become one of the tremendous demands in the plastic injection molding industry nowadays. However, smaller and thinner wall part design has increased the possibility for the parts to warp. The aim of this study is therefore to determine the best set combination of molding parameters that could reduce the warpage defect. There are six parameters that have been selected in this study which are mold temperature, melt temperature, packing time, cooling time, injection time and packing pressure. Taguchi orthogonal array is used to simplify the experimental runs. The analysis is done by applying S/N ratio approach and ANOVA method. Based on the results obtained from the analysis, it is found that the best set combination parameters give out the smallest warpage value.

1. Introduction

Nowadays, the development of plastic products as consumer products such as communication and electronic devices like mobile telephones, portable computers, and etc. has made these products to be small, thin and light. Smallest and thinnest plastic parts mean the possibility of parts to warp will definitely increase. In plastic injection molding, the production of thin walled parts is exceptionally troublesome due to the fact that melted plastic cannot fill the mold cavity easily [1-3]. In order to optimize the performance of the plastic injection molding process, the best parameter setting is very important. Therefore, this study will be using parameter design in Taguchi method to find the optimal injection molding processing conditions for improving the quality characteristics of the plastic part produced or to be more specicific; this method will be employed to facilitate the experimental design for warpage reduction analysis [2, 4]. There are various tools and technique of optimization to produce an optimized setting for plastic production such as Response Surface Methodology (RSM), Taguchi and so on [5-9]. In this study, Design of experiment is used to conduct the optimization process. DOE is the most powerful quality improvement technique to reduce process variation, enhances process effectiveness and process capability [4]. DOE of Taguchi orthogonal array, S/N ratios, and ANOVA was used to exploit the effects of process parameters the selected thin shell part, and they have been used throughout this study in both simulation and analysis [2]. Six parameters which comprise of mold temperature, melt temperature, packing time, cooling time, injection time and packing pressure were used as model variables. Taguchi has been used by researchers [2, 4, 10-14] respectively as an optimization method to improve defects problem specifically warpage defect, and this technique gives a proficient way to optimize the quality characteristics and cost in the manufacturing process. Tang. S. H et al. [12] in his study found out that the melt temperature as the most significant factor affecting warpage. Meanwhile, B. Ozcelik and T. Erzurumlu [15] in their study on the effects of injection molding parameter in warpage identified that the most influential parameter on warpage on thin shell PC/ABS material was packing pressure. The same parameter, which is the packing pressure had also become the most significant factor contributing to warpage based on Taguchi optimization technique applied by Oktem.H et al.[14], Huang and Tai [13] and Liao et al. [16].

2. Methodology

The steps involved in the project begin with the 3D Modelling until confirmation run. It was shown in table 1.

Table 1. Steps involved in the methodology

3D Modelling	• Drawing the selected part by using CATIA
Gating System Design	• Defining gate location/size, runner size and cooling channel
Parameters and respective levels selection	• Applying Minitab software
Experimental Design	• Applying Taguchi Orthogonal Array
Experimental Implementation	• Simulation analysis using Moldflow software
Interpreted data and Optimization	• Applying Taguchi method and ANOVA
Confirmation run	• Compare predicted value with confirmation value

2.1 3D Modelling

For the analysis, a cellular phone cover was chosen to be the model of study. It was designed with CATIA V5R16. The part's dimensions which consider its length, width, height, and thickness are 130, 70, 11 and 1 mm respectively. The model that has been drawn by using CATIA V5R16.

2.2 Gating System Design

The number and location of the gates need to be determined. The best gate location should be placed where the user cannot see the gate marks after assembling the products. Figure 1(a) shows the gating system design for the part study, cellular phone housing. In this study, the gate type chosen was the side gate with a two-plate mold. The dimension for the runner's system is the circular size with 6 mm, 6.5 mm and 8 mm. Figure 1(b) shows the part meshing with dual domain and cooling channel for a thin shallow plate. Based on the meshing, the thin shallow parts are divided into 9576 surface triangle elements. The proper cooling system was designed to assist the whole mold system, and the diameter for each channel is 6 mm, and the space in between channel is 45 mm.

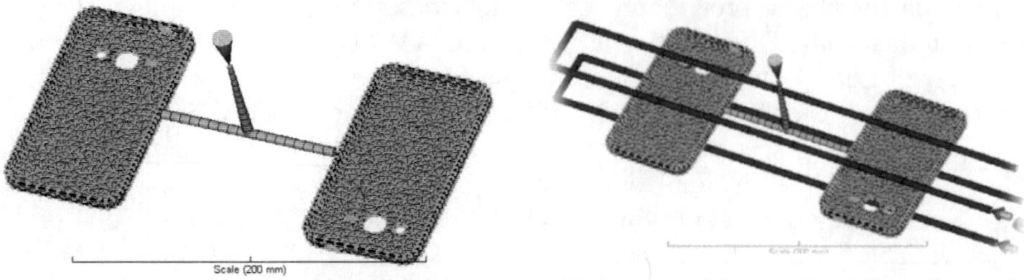

a) Gating system design. b) Part meshing and cooling channel

Figure 1. Gating system design.

2.3 Selection of Parameters

Based on previous researchers, there are several parameters that are significant to be optimized for better performance of plastic model upon warpage defect. In this study, the parameters selected to be optimized are mold temperature, melt temperature, packing time, cooling time, injection time and packing pressure. Table 2 shows the selected parameters used in this study and their respective level. The value for each level is determined based on the recommended process setting upon the selection of material in Moldflow software.

Table 2. The process parameters and their levels.

Experimental Factors	Experimental Level		
	1	2	3
A: Mold Temperature (°C)	70	90	120
B: Melt Temperature (°C)	300	320	340
C: Packing time (Sec)	8	10	12
D: Cooling time (Sec)	8	10	12
E: Injection time (sec)	0.5	0.8	1
F: Packing pressure (MPa)	120	140	170

2.4 Experimental Design

The experimental design is done by using the Taguchi method. There are six parameters selected to control the injection process: mold temperature (A), melt temperature (B), packing time (C), cooling time (D), injection time (E) and packing pressure (F). Every factor consists of three levels each where a Taguchi orthogonal array L27 (3^6) was used to suit the inputs. Taguchi Orthogonal Array Variance for L27 (3^6) generated by Minitab.

3. Results and Discussions

3.1 Taguchi Analysis Result

S/N (signal-to-noise) ratio is one of the measurement indexes to find the optimal process values for improving quality characteristics. Data points were analyzed using the "smaller-the-better" approach since this research is focused on minimizing the warpage in the injection molding process within optimal process parameters. The S/N ratio was calculated using equations (1) and MSD using equation (2). MSD is the mean square deviation, where y represents the value of result and n is the number of tests in one trial [17].

$$S/N = -10 \log (MSD) \quad (1)$$

$$\text{where MSD} = \frac{1}{2}\sum_{i=1}^{n} yi^2 \qquad (2)$$

Minitab software has been used in this study to analyze the results obtained from the Moldflow simulation, in order to identify the best parameter to be optimized. The warpage results obtained in the simulation process were then used to be analyzed in Minitab to get the S/N ratio, mean and standard deviation values. The result is summarized and tabulated in table 3. Table 4 shows the response table for the signal to noise ratios which presented the best set of combination parameters and it is determined by selecting the highest level value for each factor. It shows here that the optimal process parameters are A(3), B(1), C(3), D(3), E(2) and F(3) for PC/ ABS material. Besides, this table also ranks every factor used based on the highest delta. From this S/N ratios response table, melt temperature (B) has become the first rank, followed by mold temperature (A), cooling time (D), packing time (C), packing pressure (F) and injection time (E) seems to fall on the last rank. Hence, it can be concluded that melt temperature is the most effective factor to reduce the warpage while injection time has become the insignificant one in this study. Based on the table, the recommended setting result of warpage was produced by a combination of A3, B1, C3, D3, E2, F3 and each one represents mold temperature 120 □C, melt temperature 300 □C, packing time 12 seconds, cooling time 12 seconds, injection time 0.8 seconds, and packing pressure 170 MPa.

Table 3. Summary of result

Run	Result Of Analysis		Mean	Msd	S/N Ratio
	27	30			
1	0.7945	0.7730	0.78375	0.01520	2.11563
2	0.7916	0.7727	0.78215	0.01336	2.13357
3	0.7947	0.7759	0.78530	0.01329	2.09867
4	0.8391	0.8244	0.83175	0.01039	1.59980
5	0.8355	0.8216	0.82855	0.00983	1.63332
6	0.8392	0.8149	0.82705	0.01718	1.64843
7	0.8692	0.8520	0.86060	0.01216	1.30354
8	0.8707	0.8482	0.85945	0.01591	1.31484
9	0.8605	0.8407	0.85060	0.01400	1.40490
10	0.7903	0.6926	0.70095	0.01181	3.08564
11	0.7121	0.6937	0.70290	0.01301	3.06139
12	0.7131	0.6946	0.70385	0.01308	3.04965
13	0.7649	0.7471	0.75600	0.01259	2.42896
14	0.7689	0.7511	0.76000	0.01259	2.38313
15	0.7682	0.7483	0.75825	0.01407	2.40300
16	0.8472	0.8270	0.83710	0.01428	1.54382
17	0.8352	0.8151	0.82515	0.01421	1.66870
18	0.8608	0.8412	0.85100	0.01386	1.40083
19	0.7111	0.6928	0.70195	0.01294	3.07314
20	0.712	0.6937	0.70285	0.01294	3.06201
21	0.7129	0.6944	0.70365	0.01308	3.05212
22	0.7279	0.7115	0.71970	0.01160	2.85641
23	0.7182	0.7017	0.70995	0.01167	2.97486
24	0.7207	0.7043	0.71250	0.01160	2.94373
25	0.7857	0.7682	0.77695	0.01237	2.19159
26	0.8045	0.7831	0.79380	0.01513	2.00499
27	0.7932	0.7746	0.78390	0.01315	2.11418
		TOTAL	20.9097	0.34274	60.5509
		MEAN	0.77443	0.01269	2.24262

Table 4. Response table for signal to noise ratio.

Level	A	B	C	D	E	F
1	1.695	2.748	2.193	2.208	2.244	2.213
2	2.336	2.319	2.265	2.076	2.249	2.250
3	2.697	1.661	2.270	2.444	2.235	2.265
Delta	1.002	1.087	0.077	0.368	0.013	0.052
Rank	2	1	4	3	6	5

3.2 Analysis of Variance (ANOVA)

From ANOVA result in table 5 below, the F0.05,2,26 =3.37 for a level of significant parameter with 95% confident interval that is equal to 0.05. Therefore, Injection time (E) [Fstatistics = 0.10 < F 0.05,2,26 = 3.37], and Packing pressure (F) [Fstatistics = 1.61 < F 0.05,2,26 = 3.37] does not give a significant effect to the warpage as their F-statistic values are lower than 3.37. While mold temperature (A) [Fstatistics = 569.13 > F 0.05,2,26 = 3.37], melt temperature (B) [Fstatistics = 662.28 > F 0.05,2,26 = 3.37], packing time (C) [Fstatistics = 4.10 > F 0.05,2,26 = 3.37], and cooling time (D) [Fstatistics =76.77 > F 0.05,2,26 = 3.37], give a significant effect to the warpage, with mold temperature (B) giving the highest significant level. The melt temperature (B) contributes the highest percentage value which is 50.4% followed by mold temperature (A) 43.31%, and cooling time (D) 5.84%, as the most influence factor for warpage defect. Meanwhile packing time (C) contributed 0.31%, packing pressure (F) contributed 0.12% and lastly injection time (E) only contributed 0.14% and thus, it make the packing time, packing pressure and injection time to be the least significant factors for the warpage defect in this study.

Table 5. ANOVA table

Source	f	SS	MS	F-Statistic	Percentage
A	2	4.63835	2.31918	569.13	43.31%
B	2	5.39749	2.69875	662.28	50.4%
C	2	0.03344	0.01672	4.10	0.31%
D	2	0.62570	0.31285	76.77	5.84%
E	2	0.00085	0.00043	0.10	0.0076%
F	2	0.01312	0.00656	1.61	0.12%
Error	14	0.05705	0.00656		
Total	26				

3.3 Confirmation Test

The confirmation runs were conducted to measure the reliability of optimization solutions obtained from the software analysis. The comparison of test results between the theoretical prediction and the confirmation test results was the final consideration that will evaluate whether the optimum parameters predicted were in the allowable range as shown in table 6. The margin of error from the prediction and simulation results was set below than 10%. Margin error was calculated using the equation (3) below:

$$\text{Margin Error} = \frac{(\text{Confirmation test} - \text{Predicted})}{\text{Predicted}} \times 100 \quad (3)$$

Table 6. Comparison Test Result.

Response	Prediction (Minitab)	Simulation (Confirmation Test)	Error Margin (%)
Warpage	0.667937	0.69205	3%

From the result, it can be concluded that for both responses, the margin error is below 10%. This means that the confirmation test is accepted since it has minimized the defects for shrinkage.

4. Conclusions

Based on the results obtained from the analysis, it is found that the best set combination parameters to be used are: mold temperature (A) of 120 °C, melt temperature (B) of 300 °C, packing time (C) of 12 s, cooling time (D) of 12 s, injection time (E) of 0.8 s and packing pressure (F) of 170 MPa. These optimum combination parameters successfully reduce the warpage value to 0.6922 mm for ambient temperature 27°C and 0.6919 mm for ambient temperature 30 °C. By calculation, these results give the margin error as low as 3% when compared to the predicted value, and hence it can be said that the predicted result is accepted and reliable. According to ANOVA result, with a 95% confidence interval, it is concluded that the significant parameters that affect warpage are melt temperature, mold temperature and packing time. Meanwhile, injection time and cooling time give the least percentage that contributes to warpage defect. Among all factors, melt temperature has become the most significant factor and injection time is the least factor that affecting the warpage of the thin shell part in this study.

Acknowledgments

We specially thank the reviewer(s) for their useful advice and comments. The authors acknowledge the School of Manufacturing Engineering, Universiti Malaysia Perlis for the lab facilities. Special thanks to those who contributed to this project directly or indirectly.

References

[1] N. A. Shuaib *et al.* 2012 *International Review of Mechanical Engineering* **6**, 865.
[2] H. Radhwan *et al.* 2014 *International Review of Mechanical Engineering* **8**, 1057.
[3] Shuaib, Norshah Afizi, *et al.* 2014 *Trans Tech Publications* **594**, 676.
[4] M. Fathullah *et al.* 2011 *International Review of Mechanical Engineering* **5** 1295.
[5] M Isafiq *et al.* 2016 *MATEC Web of Conferences* **78** 01084.
[6] S M Nasir *et al.* 2013 *International Review of Mechanical Engineering* **7(5)** 977.
[7] M. Fathullah *et al.* 2011 *International Review of Mechanical Engineering* **5(7)** 1278.
[8] S M Nasir *et al.* 2012 *International Review of Mechanical Engineering* **6(3)** 372.
[9] M. Fathullah *et al.* 2011 *International Review of Mechanical Engineering* **5(7)** 1189.
[10] R. Hussin *et al.* 2013 *International Review of Mechanical Engineering* **7**, 453.
[11] B. Ozcelik and I. Sonat 2009 *Mater Des* **30**, 367.
[12] S. H. Tang *et al.* 2007 *J Mater Process Tech* **182**, 418.
[13] M. C. Huang and C. C. Tai 2001 *J Mater Process Tech* **110**, 1.
[14] H. Oktem, T. Erzurumlu and I. Uzman 2007 *Mater Des* **28**, 1271.
[15] B. Ozcelik and T. Erzuruml 2006, *J Mater Process Tech* **171**, 437.
[16] S. J. Liao *et al.* 2004 *Polym Eng Sci* **44**, 917.
[17] R. K. Roy, *A primer on the Taguchi Method* (New York, USA: Van Nostrand Reinhold,1990).

Design and Analysis of Jigs and Fixtures for Manufacturing Process

H Radhwan[1], M S M Effendi[1,2], Muhamad Farizuan Rosli[1,2], Z Shayfull[1,2], K. N. Nadia[1]

[1]School of Manufacturing Engineering, Universiti Malaysia Perlis, Kampus Tetap Pauh Putra, 02600 Arau, Perlis, Malaysia.
[2]Green Design and Manufacture Research Group, Centerof Excellence Geopolymer and Green Technology (CEGeoGTech), Universiti Malaysia Perlis, 01000 Kangar, Perlis, Malaysia.

E-mail: radhwan@unimap.edu.my

Abstract. This final year paper report is about the design and analysis of jigs and fixtures for manufacturing process. For this project, it is focus on the perspective of the job handling to produce the part. Currently, they implement the manual handling to assembly each component. Improvement that is trying to do is by replace the manual handling for the process by designing the semi auto jig for this process. The objectives of this project are to investigate the collected data and analyse the design of jig and fixture to ease work handling. The methodology of research procedures covered on this project consists of collect data, brainstorming and interprets data, design concept, and documentation. The design of jig and fixtures is using Unigraphics NX7.5 software. The data that can be studied include the analysis of Finite Element Analysis (FEA) using CATIA software, the analysis of design ergonomic, the time study analysis and cost analysis. At the end of this project, this research will be able to give understanding about jig and fixture, design by using Unigraphics NX7.5 software and finally able to do the analysis for the selected part of product by using FEA and other relevant analysis

1. Introduction

Over the past century, manufacturing sectors play an important role to the economic development of a country. Jigs and Fixtures are devices that used to facilitate production work in industry especially that involve in machine. The perfect jigs and fixtures can work repeatability and interchangeability to produce the same parts in production. In manufacturing industry, jigs and fixtures are most important device that can assist the workers in their production process become easier. Jig and fixture is important tool using in industry. Tool that are carries the main forces will form the final shape of the workpiece. [1]. The difference between jig and fixture is in the way of the tool is guided with the workpart [2]. . Fixture are essential element in production operation as it always important in industry such as automated manufacturing, inspection, and assembly operation [3]. One of the principles that can be follow for designing the jig and fixture for this project is as states by Nee et al. [4] The general factors to be considered when designing jig and fixture are shape, material and state of workpart, pre-machined surface tolerance, type of operations and the machine tools used, workpiece handling, ergonomics and safety considerations. However, the other studies are different from Taufik et. al. [5] . The design parameter such as maximum deformation, maximum shear stress, number of contact faces, and maximum holding force were presented. It is found that the gripping ability is the important factor that affected to the clamping and holding the workpart perfectly during machining operation. Besides that as explained by Sawita et. al.[6] which has done a design and finite element analysis of jig and fixtures for the design are very useful for this project methods. In addition, the dimension of design is very important. This statement can be proves by the research from Peshatwar et. al. [7] which has presented a fixture design system of eccentric shaft for ginning machine. Designer should design a fixture according to dimension required by industry to fulfil production target. The important consideration that taken when designing this jig and fixture before making any decision also gathered from the research study by Saari et. el. , product design and development requires that engineers consider trade-offs between product attributes for the cost, weight manufacturability, quality and performance. The optimum design is the usually one that compromise are acceptable, but understanding the impact of design decision on all relevant attributes is difficult. [8]. According to Yugeskumat et. al. [9] on the journal titles an advance method of jig and fixtures planning by using CAD methods, it review methods and techniques for the geometry analysis of fixture feasibility in product development which the synthesis methods including geometrical analysis and fixture assembly planning are surveyed. The decision making for selection of material used is based on the reviewed research from Kharmiza Japar [10], states that a jig and fixture are made from a variety of materials. Some of which can be hardened to resist wear. It is sometimes necessary to use nonferrous metals like phosper bronze to reduce wear of the mating parts or nylon or fibre to prevent damage to the workpiece.

2. Methodology

The overview on the overall research methodology and operation steps as shown in Fig.1 are required in achieving the objectives of this project.

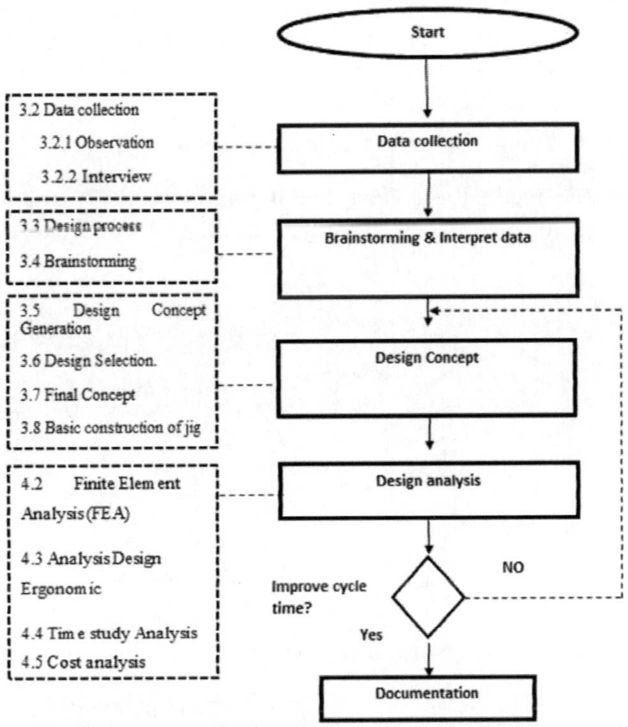

FIGURE 1. The overall research methodology.

3. Data Collection

The product planning is based on the data collected on observation from the current process. The process that needs to do the improvement for the jig and fixture are at insertion of the e-ring as shown in Fig.2.

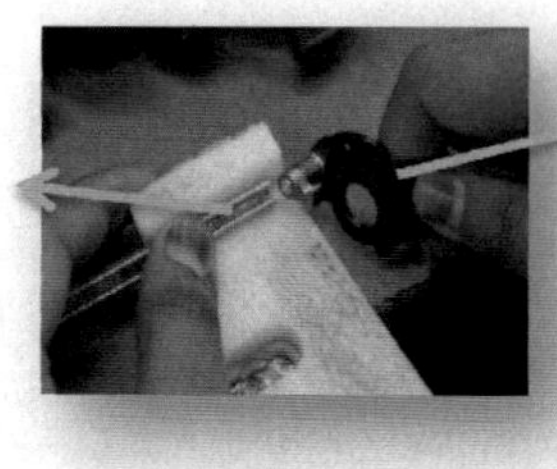

FIGURE 2. Manual E-Ring insertion

From the data taken for this operation time, this process took 12.15 seconds to complete the job. So, from the collected data, the target cycle time that needs to improve is 8 seconds,

From observation, this process is highly recommended for improvement because compared to other process in this operation line, this process are difficult with the uses of manual handling tool and operators always take a time for the job done. To do this process, operators need to make sure that the e-ring must insert in their position and grip to the part very well. Currently, operator always faces a problem to insert the e-ring. Sometime the e-ring is loose and missing at the end of the process. This will cause the reject part and need to rework. So, it shows that this process was selected to make an improvement because it has high possibilities to improve the overall time of the process with the new design and more advance of the jig and fixture. With the use of jig and fixture, it not only can improve the cycle time but it also can improve productivity and help to make the operator job become much easier.

The author used qualitative method to collect data through interviews. Based on that method, the researcher was able to analyze the gap of design study. The data collected can be shown in Fig.3.

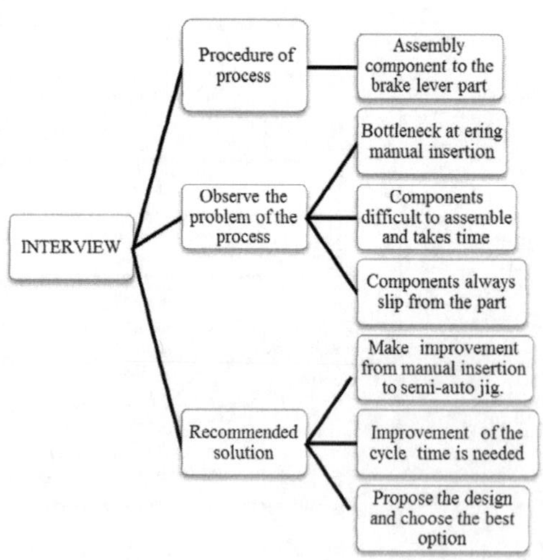

FIGURE 3. Interview data

IOP Conf. Series: Materials Science and Engineering **551** (2019) 012028 doi:10.1088/1757-899X/551/1/012028

4. Design Process

The engineering design process is one of the steps required for the development of product. Step in designing a jig and fixture is by evaluation of its functional requirements in order to find the balanced combination of design characteristics at a reasonable cost. At this step, tool design process should be clearly state the problem to be solved and achieve as needed. The design process also included the analysis that involve to the product such as the clamping position, the load distribution, the material selection and more analysis and information from the design.

This process also will explain on how the semi-auto e-ring insertion jig will designed and how the analysis will be performed. CATIA CAD software is used to designing and analyse for FEA. The targeted result for the FEA analysis is to find the value of deformation when the load applied and to estimate the value of maximum allowable strength based on the material yield strength. These values are very important in order to determine if the design of the part is suitable and meet requirement.

5. Brainstorming

Brainstorming is how the rough idea can be drafted to form the design. One of the methods for brainstorming the idea is through the benchmarking of related product for designing. Benchmarking is the study of existing products with functionality similar to that of the product under development or to the sub problems on which need to be focused. Benchmarking can reveal existing concept that have been implemented to solve a particular problem. The designer obtained the related products in order to discover the general concepts on which that can be absorbed for the new product.

6. Design Selection

Design selection is the process to choose the concept that has been evaluated according to the requirement and the criteria needed.[11,12] When comparing the relative strengths and weakness of the concept, select the best concept that can be used for further investigation. Generally, concept selection is the process of narrowing the alternative concept that has been considered.[13,14] In order to choose the final concept for the jig, the screening method was choosing.[15] This method can help designer to make a decision for the best design. For the beginning, the design of jig comprises of three designs which it have the different concept design.

Table 1. Concept screening matrix

Selection criteria	Concepts		
	First Concept (Datum)	Second Concept	Third Concept
Ease of handling	0	-	+
Ease of use	0	+	+
Durability	0	+	+
Ease of manufacture	0	-	-
Portability	0	+	+
Cleaning ease	0	0	0
Maintenance ease	0	+	+
Production time	0	-	+
Sum +'s	0	5	6
Sum 0's	8	1	1
Sum –'s	0	3	1
Net score	0	2	5
Rank	3	2	1
Continue?	No	Yes	Yes

From the Table I shown above, the result shows the best option can be selected by choosing the high rank. The third concept was the first rank which is the most acceptable for this requirement.

IOP Conf. Series: Materials Science and Engineering **551** (2019) 012028 doi:10.1088/1757-899X/551/1/012028

Concept scoring always uses to select the final concept selection which use the rating score to rate the design. For this design, the concept scoring need to be skipped because the best design was selected from the design screening which it is shows that the Third Design as the dominant among others. So that, the Third Design will be proceed for further research.

7. Final Concept

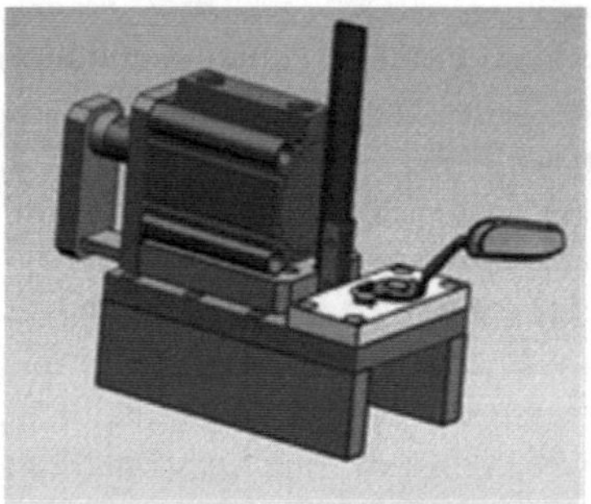

FIGURE 4. Final concept

The final concept of design as shown in Fig.4 are selected based on the decision that has been made during concept selection which is here the third concept is the best option.This final concept design which is the semi-auto insertion jig are much more better than second design which is the assembly part for this design also are more simple.

8. Basic Design Principle for Jig and Fixture

Basic principle while designing of jig and fixture is supporting, locating and clamping and workholding. From the final design of semi-auto insertion this principle can be shown in Fig.5.

FIGURE 5. Classification for Basic Principle.

The supporting of the part should be rigid. From this semi-auto insertion jig, make sure the design for supporting is sufficiently stiff to secure the accuracy while inserting the e-ring to main lever part. Locating of the part must be "fool proofing" which means that main lever part must lead into jig and fixture in a proper position at the same time not interfere with loading and unloading the workpiece. The foolproofing device that is used for this parts is pins which clears can correctly position parts and prevent incorrectly parts from entering this jig. This semi-auto insertion jig using screw clamps. Clamping screw is suitable for this jig because it can be used for light clamping. The basic screw clamp uses the torque to hold a part into places by direct pressure on another part. The holding power for this jig is using the bolts and nuts. The bolts and nuts are easy to use and every part can assemble to each other's.

9. Economical

When designing the jig and fixture it is important to make sure the product is simple in construction, give high accuracy and rigid. For this project, the economic in terms of the cost analysis of the product will calculated to satisfy this condition.

10. Design for safety

When designing jig and fixture, make sure that the product is safe and convenient in use. Some factor of safety that must be followed is:

1. Sharp corner on the body of jig or fixture should be avoided. In this project, while designing the semi-auto insertion jig, every design part are using chamfers in every single corners to make sure that no sharp corner that can harm operator while using the jig.
2. Bolt and nut should be inside the body and not protrude the surface. The surface of this semi-auto insertion are designed well as all the bold is put inside of the jig which is make the work surface clear and can avoid any injury and obstacle while doing the job.

11. Results and Discussions

11.1. Finite Element Analysis (FEA)

CATIA uses the displacement formulation of the finite element method to calculate component Von Mises stress, principal stress, tension and compression, and displacement under internal and external loads. The main parameter on this analysis is the material and the force on the jig. The material selections for each part are steel because the actual materials to produce the part are made from a composition of steel. The static force represents the force applied by the operator to part while insert the e-ring. All these parameters had been analyzed by searching for the limit force of the part and the relevant size and further it has been designed in CATIA to simulate the load on the jig. Table 2 shows the force reacted on the jig.

TABLE 2. Comparison of Several Studies with Task similar to Those Being Investigated (in Newtons) [16]

		Mean	Std Dev	5th Percentile	20% of 10th Percentile (Kodak)	Normalized Peak Force at ACGIH Action Limit
Schoorlemmer and Kanis (1992) [1]						
Thumb Push (free posture)	Male	86.9	43.0	16.0	6.3	39.1
	Female	68.5	36.1	8.9	4.4	30.8
Thumb Push (90° elbow flexion)	Male	81.3	26.7	37.2	9.4	36.6
	Female	64.1	32.4	10.6	4.5	28.8
Astin (1999) [1]						
Index-Finger Push (forward)	Male	52.6	18.0	22.9	5.9	23.7
	Female	39.3	14.9	14.7	4.0	17.7
Index-Finger Press (down)	Male	50.9	18.4	20.6	5.4	22.9
	Female	35.2	14.9	10.6	3.2	15.8
Hertzberg (1973)						
Thumb Push	Male	77.4	16.7	49.9	11.2	34.8
Index Push	Male	56.8	12.7	35.8	8.1	25.6
Dickson (1972) [2]						
Index Push (Dominant)	Male	45.1			9.0 [3]	20.3
Index Push (Non-dominant)	Male	43.1			8.6 [3]	19.4
Army (MIL-STD-1472D) [4]						
Grip Push (up)	Male			98.0	19.6 [5]	n/a [6]

[1] Participants were seated
[2] Only 50th percentile male were reported
[3] Values represent 20% of the mean
[4] Only 5th percentile male were reported
[5] Values represent 20% of the 5th percentile
[6] Mean was not provided, cannot be calculated

To decide the force distribution values are getting from Table II which is the limit force for the thumbs is 36.6 N. So that, the distributed load for part on the jig assumed as 40 Newton. The factor of

IOP Conf. Series: Materials Science and Engineering **551** (2019) 012028 doi:10.1088/1757-899X/551/1/012028

safety that has been choose is 2 which is it is used for well-known material under reasonably constant environmental condition and also subjected to loads and stresses that can be determined readily.

For OCTREE tetrahedron mesh, the meshing size is 14 mm and for the absolute sag is 1.4 mm. This parameter is used to all jig and fixture design when doing the nanalysis because different meshing parameter will influence the result when computing.

11.2. Analysis result

Static case analysis are deformation analysis, Von Mises stress analysis, translation displacement analysis, and principle stress. In this result that we gain from the analysis, several factors we should understand to make sure our product will be okay without failure. Other than that we also identify the modulus young, yield strength and density in this result that effect to product. This data are very important to identify the failure factor at the product before we could send to factory to produce it. The condition of each product is 40N acted downward. This analysis will show the areas that have high stress when the structure deformed. The material that is selected for all the part is steel. The steel material properties are shown in Table 3:

TABLE 3. Material Properties of Steel

Material properties	Value
Young modulus	$2 \text{ X } 10^{11}$ (N/m^2)
Poisson ratio	0.266
Density	7860 kg/m^2
Thermal expansion	$1.17 \text{ X } 10^{-5}$Kdeg
Yield strength	$1.25 \text{ X } 10^8$ (N/m^2)

This value $1.25 \text{ X } 10^8$ N/ of allowable stress is useful to identified either the part will deform if the load is applied. If the value of stress is higher than the value of allowable stress, the part considered as plastically deforms. So, from the data analysis using the FEA, if the von misses stress value is not exceed the allowable stress, yielding will not occur. Von misses stresses are used to predict the yielding of material under any loading condition which is if the maximum value of distortion energy per unit volume in material is smaller than distortion energy per unit volume required causes yield strength, the design part is safe.

11.3. Von Misses Stress

Base plate. The von misses maximum value for this part is is $2.73 \text{ X } 10^6$ N/m². This value is lower than the allowable stress value $1.25 \text{ X } 10^8$N/ . Thus, the base plate will undergo plastic deformation because the base plate can handle the maximum stress given by the load given by the push of main lever using finger force. It is elastic deformation which is after the load applied, the structure will a bit deformed but after the load is removed the base plate will back to its original shape.

E-Bit .The von misses maximum value for this part is is $8.7 \text{ X } 10^4$ N/m². This value is lower than the allowable stress value $1.25 \text{ X } 10^8$N/ . Thus, the e-bit will undergo elastic deformation at the end of the tool. It can handle the stress given by the pneumatic cylinder but it is elastic deformation which after the load is applied, the structure will slightly deformed. But after remove the load, the e-bit will back to its original shape structure.

Connector .The von misses maximum value for this part is is $1.5 \text{ X } 10^4$ N/m². This value is lower than the allowable stress value $1.25 \text{ X } 10^8$N/ . Thus, the connector will undergo elastic deformation. Same as the e-bit, the force applied also given by the pneumatic cylinder. The connector also can handle the stress given by the pneumatic cylinder but it is elastic deformation which after the load is

IOP Conf. Series: Materials Science and Engineering **551** (2019) 012028 doi:10.1088/1757-899X/551/1/012028

applied, the structure will slightly deformed. But after remove the load, the e-bit will back to its original shape structure.

Base plate support (B-P Support).The von misses maximum value for this part is is 1.0 X 10^5 N/m². This value is lower than the allowable stress value 1.25 X 10^8N/ . Thus, the base plate support will undergo elastic deformation. The base plate support can handle the stress given by the load from upward because the this plate is act as the support to all part. But it is elastic deformation which after the load is applied, the structure will slightly deform. But after remove the load, it will back to its original shape structure.

LR jig .The von misses maximum value for this part is 2.26 X 10^5 N/m². This value is lower than the allowable stress value 1.25 X 10^8N/ . Thus, the LR jig will undergo elastic deformation. The applied load to this LR jig is given by the force from the finger to put the main lever on this plate which is this LR jig will support the main lever while the e-ring inserted to the part. It can handle the stress given by the load from upward but it is elastic deformation which after the load is applied, the structure will slightly deform. But after remove the load, it will back to its original shape structure.

11.4. Displacement Analysis

Displacement analysis can be describe as the better approximation of displacement function is the one that yields a lower potential energy. In structural analysis, the approximation are represented by consideration that displacement are small and can be neglected in equilibrium equation, the load are conservative, independent on displacement and the support of the structure remain unchanged during loading. The materials behave nonlinear or linear material model cannot be used if stress exceeded some value because the load may change their orientations according to displacements and support may change during loading.

Base Plate. From the analysis the translational displacement is 7.97 X 10^{-5}. As the value is very small, the displacement of the structure is considered as negligible because the maximum displacement of this part is 0.05mm in order to maintain the insertion of e-ring to the main lever. This can avoid from dislocation of e-ring to the main lever.

E-Bit .The translational displacement is 1.61 X 10^{-5}. As the value is very small, the displacement of the structure is considered as negligible. E-Bit are very important part which is it is used to carry the e-ring and insert to the main lever. So, the maximum displacement that is allowable for this part is about 0.01mm because it is important to make sure the insertion of e-ring is accurately to the main lever. If the displacement value is higher after the load is applied, this will cause failure for the insertion of e-ring due to the high deflection of the ebit from its original size.

Connector. From the analysis, the translational displacement is 2.85 X 10^{-7}. As the value is very small, the displacement of the structure is considered as negligible. This part is also important which is it is useful to carry the E-Bit that act as a tool to insert the e-ring. The small displacement value can avoid the dislocation of e-ring to be inserted to the main lever.

B-P Support.The translational displacement is 2.85 X 10^{-5}. As the value is very small, the displacement of the structure is considered as negligible. B-P support is the part that used to support all the part that assembled under it. So the value of displacement is very important to make sure that this part can support all the applied load that exerted . so, this small value of displacement shows that the part are excellently support the other part without a lot of changes.

LR Jig. The translational displacement is 8.62 X 10^{-6}. As the value is very small, the displacement of the structure is considered as negligible. This part is also important as the place to support the main lever during the e-ring insertion. Because of the load applied is not too much, so the displacement value is not too high. This can help to maintain the main lever in its position which is the insertion of e-ring are easier and successfully inserted.

The finite element analysis is very important in manufacturing engineering. With some help by technology, it could determine the product result by using CATIA software. This software can help to try an error the product until biggest improve can do at the product. Furthermore, this could reduce the percentage to get failure.

12. Conclusions

This semi-auto insertion jig has added a lot of changes to the job handling in process operation. It help to ease the handling for the job by eliminate the manual insertion to semi-auto. The conclusion from the result data are as follows:

i. The FEA analysis shows that the von misses stresses and translational displacement for each part of material structure is stable which the results are not exceeds the required maximum value for each condition when 40N loads was applied.

ii. For ergonomic analysis, the risk factors identified before implement this jig can be eliminate. It shows that the risk of injury can be avoided.

iii. From the time study, when eliminate the manual handling, the cycle time can be improved from 12.15 seconds to 8 seconds which is this will result the time to produce the required quantity of part are shorted to 34.16%.

iv. Lastly, the cost analysis shows that the cost of production can reduce and productivity is increase which is the parts per hour can be increased to 51.88%, the cost of labor expenses reduced to 34.16%, and the cost per part can be reduced to 14.29%.

The objective of the study are achieves at the end of this project which is the data analysis are completely collected, the design the jig and fixture to ease work handling is successfully developed and finally all the analysis result is successfully done.

References

[1] Krsulja M *et al* 2009 *J. Adv. Eng.* **6** 6.
[2] Hoffman E G, "Jig and Fixture Design." 2004.
[3] Soo W H, "Development of Modular Jig and Fixture for Catia." 2007.
[4] Nee A Y C *et al* 1987 *Robot Comput. Integr. Manuf.* **3** 195.
[5] Taufik R S *et al* 2012 *Malikussaleh Industrial Engineering Journal* **1** 19.
[6] Sawita D D *et al* 2014 International Journal of Research in Advent Technology **2** 1.
[7] Peshatwar S V *et al* 2013 Int. J. Eng. Res. Appl. **3** 1591.
[8] Saari S M *et al* 2008 International Journal of Research in Advance Technology **6** 10.
[9] Hong S W, "Development of Modular Jigs and Fixture for Catia."2007
[10] Kharmizan J, "Design Jigs and Fixture for Surface Grinding Long Part." 2007
[11] Fatima S B A *et al* 2018 AIP Conference Proceedings **2030** 020070.
[12] Azri N M *et al* 2018 AIP Conference Proceedings **2030** 020069.
[13] Zaidi N A et al 2017 AIP Conference Proceedings **1885** 020005.
[14] Zain F *et al* 2017 AIP Conference Proceedings **1885** 020023.
[15] Zhafri Z M *et al* 2018 AIP Conference Proceedings **2030** 020071.
[16] Nussbaum M A *et al*, "Strength Capabilities and Subjective Limits for Repetitive Manual Insertion." 2001.

IOP Conf. Series: Materials Science and Engineering **551** (2019) 012029 doi:10.1088/1757-899X/551/1/012029

A Modified Artificial Bee Colony Algorithm for Function Optimization Strategy

Chang-feng Chen[1] , Nasarudin bin Daud[2] , Kai-qing Zhou[3] and Tadiwa Elisha Nyamasvisva[4]

[1,2,4]Faculty of Engineering, Science and Technology, Infrastructure University Kuala Lumpur, Kuala Lumpur 43000, Malaysia
[3]College of Information Science and Engineering, Jishou University, Jishou, 416000, Hunan, China

E-mail: changfeng1993@gmail.com

Abstract. Artificial Bee Colony (ABC) algorithm has received wide attention due to the outstanding performance in solving complex optimization problems. However, the further application of ABC is still hindered due to some bottlenecks, such as slow convergence velocity, poor global search ability and low accuracy of solution. To overcome these problems, an improved ABC (CT-ABC) is proposed in this paper by using both a chaotic map and a tolerance based search equation (TSE). Some standard test functions are chosen to verify the performance by implementing the proposed modified ABC algorithm, basic ABC and another modified ABC (WABC). Simulation results reveal the proposed CT-ABC algorithm is able to obtain the faster convergence speeds and better accurate approximate solution.

1. Introduction

Swarm Intelligence (SI) [1] is briefly defined as the self-organizing, adaptive, and social division of labor as well as intelligent behavior of social life in group living creatures. Being a classical SI algorithm, ABC was proposed by Karaboga in 2005, which based on the bees foraging behavior [1].Compared to other SI algorithms, ABC has shown good features and signs. However, some clear disadvantages such as "slow convergence and low accuracy of solutions" [2], hinder the further researches of ABC. In order to address these shortages, an improved ABC (CT-ABC) is proposed in this manuscript to improve the convergence speed and accuracy of solutions as well as balance exploration and exploitation ability by using chaotic map to adjust the inertial weight ω and tolerance based search equation (TSE) to help the colony get grid of the local optimum.

To verify the feasibility and properties of CT-ABC, four famous standard benchmark functions are applied to test CT-ABC's performance and results are compared with basic ABC and one ABC variant (WABC). The rest of this paper is arranged as follows: in Section 2, the original ABC algorithm is presented. Section 3 described and analyzed the proposed CT-ABC in detail, which include three parts. The simulation experiments are conducted and results are discussed in Section 4. The conclusion of research and future study is summarized in Section 5.

2. Basic artificial bee colony algorithm

In ABC algorithm, there are three kinds of bees [1-6]: employed bees, onlooker bees, and scout bees. Employed bees take charge of moving nectar to the hive and dancing in waiting area to share information of nectar source with onlooker bees. Onlooker bees wait in the hive to get information from employed bees and decide which food source to select. Scout bees searched the new food source in the neighbourhood of the hive [2-5].

The fundamental steps of basic ABC algorithm are described as follows:

1. Initialization of parameters (N , D, $Limit$, $iter$, $iter_{\max}$);

2. Generate the initial solutions $X_m(m=1,2\cdots,N)$

$$X_m^n = X_{\min}^n + rand(0,1)(X_{\max}^n - X_{\min}^n) \quad (1)$$

3. Search new solutions in the neighbourhood;

$$new\_X_m^n = X_m^n + rand()(X_m^n - X_k^n) \quad (2)$$

4. Update the solutions according to Greedy Selection;

$$new\_X_m = \begin{cases} 1, f(new\_X_m) \geq f(X_m) \\ 0, f(new\_X_m) \leq f(X_m) \end{cases} \quad (3)$$

5. Onlooker bees choose the employed bees to follow;

$$P_n = fit_n / \sum_{k=1}^{N} fit_k \quad (4)$$

6. After limit times search, the solutions still no update. The employed bee will transform into scout bees, and generate a new solution randomly.

$$X_n(k) = X_{\min} + rand(0,1)(X_{\max} - X_{\min}), Bas_n \geq Limit \quad (5)$$

3. The proposed modified artificial bee colony

In this section, three improved strategies are utilized to modified the original ABC. First, a inertial weight based adjustment strategy is applied to enhance the diversity of the nectar resource. Second, a chaotic map is selected to tune inertial weight. Thirdly, a tolerance based search equation (TSE) is proposed to avoid the colony from sinking into local optimum and change the search direction a when it needs.

3.1. Inertial weight based ABC

In order to improve the diversity of the nectar source, Lei et al, add an inertial weight ω to the basic ABC, which was inspired by PSO evolution equation and its improving strategy. This process can be defined by equations (6) and (7) as follow:

$$new\_X_m^n = \omega * X_m^n + rand()(X_m^n - X_k^n) \quad (6)$$

$$\omega = (\omega_{start} - \omega_{end}) * (iter_{\max} - iter) / iter_{\max} + \omega_{end} \quad (7)$$

Where ω_{start} is the initialized value of inertial weight, and ω_{end} is the final value. $iter$ stands for the number of iteration, and $iter_{\max}$ is the max iteration number of the algorithm [6].

In conclusion, this strategy can transfer the local searching process of colony into global searching and mean while balances the two sides, which can make the converge process close the optimal solution more effectively .

IOP Conf. Series: Materials Science and Engineering **551** (2019) 012029 doi:10.1088/1757-899X/551/1/012029

3.2. Chaotic map and tolerance search equation based adjustment mechanism

The chaotic map has the characteristics of sensitivity, non-repetition and ergodicity; It can help the algorithm to increase the population diversity and achieve high-quality solutions. Here, we choose sine map to tune inertial weight [5]. The logistic equation is given as follows:

$$\omega_{k+1} = A * \sin(\pi\omega_k),\ \omega_k \in (0,1),\ 0 < A \leq 1,\ \ k = 1,2,\cdots,M_{\max} \tag{8}$$

Where k is the current iteration number.

In the basic ABC, the employed bees will abandon the original food source when it can't be updated with limit times search, which will decrease the accuracy of solution. So, we propose a tolerance based search equation to decide when we need adjust the search direction. The tolerance search equation as follows:

$$P_a = \frac{\exp(T) - 1}{\exp(10) - 1} \tag{9}$$

Where P_a is the probability of making adjustments. If $P_a > rand()$, the colony will abandon the original food source and adjust it's search direction.

With the help of chaotic map and TSE strategies, the algorithm can make full use of the potential leading ability to help the colony step out of the local optimum.

4. Experiments

4.1 Test problems

In this section, in order to verify the validity of the proposed CT-ABC algorithm, we apply four classical benchmark functions as test functions. Descriptions for the aforementioned test problems are listed in Table 1, where D is the solution space dimension, Opt denotes the global optimum[2,4-7].

Table 1. Five classic benchmark functions in experiments

Function	Formulation	Search range	Opt	D
Sphere	$f_1(x) = \Sigma_{i=1}^{D}(x_i^2)$	[-100,100]	0	30
rosenbrock	$f_2(x) = \sum_{i=1}^{D-1}(100(x_{i+1} - x_i^2)^2 + (x_i - 1)^2)$	[-30,30]	0	30
rastrigin	$f_3(x) = \sum_{i=1}^{D}(x_i^2 - 10\cos(2\pi x_i) + 10)$	[-5.12,5.12]	0	30
griewank	$f_4(x) = \frac{1}{4000}\sum_{i=1}^{D} x_i^2 - \prod_{i=1}^{D}\cos(\frac{x_i}{\sqrt{i}}) + 1$	[-600,600]	0	30

4.2. Comparison of CT-ABC with ABC and WABC

The comparison results on the solution accuracy and convergence speed including mean fitness value and value's standard deviation (std) which are listed in Table 2 [7]. We mark the best mean of the objective values on each test functions with bold font, and mark the best standard deviations of them with underlined.

Table 2. Results of the basic ABC, WABC, CT-ABC over classic benchmark functions

F	ABC		WABC		CT-ABC	
	Mean	Std.	Mean	Std.	Mean	Std.
f_1	71.4394	67.4375	59.3259	57.1724	**55.1843**	52.1769
f_2	3.23684e-07	2.01436e-13	2.41975e-06	1.69174e-17	**1.03776e-05**	1.01436e-09
f_3	13.569	12.134	12.651	10.087	**11.363**	9.542
f_4	15.3759e-11	14.2517e-13	5.1963e-09	3.2751e-11	**4.7407e-06**	2.6524e-19

As the results shown in Table 2, we can see that no matter in the mean of fitness value or the standard deviation, CT-ABC can achieve the best value compared to other two algorithms. So we can get the initial conclusion that proposed strategies can help the original ABC to improve the accuracy of solution, as well as solve existed poor global search ability problem.

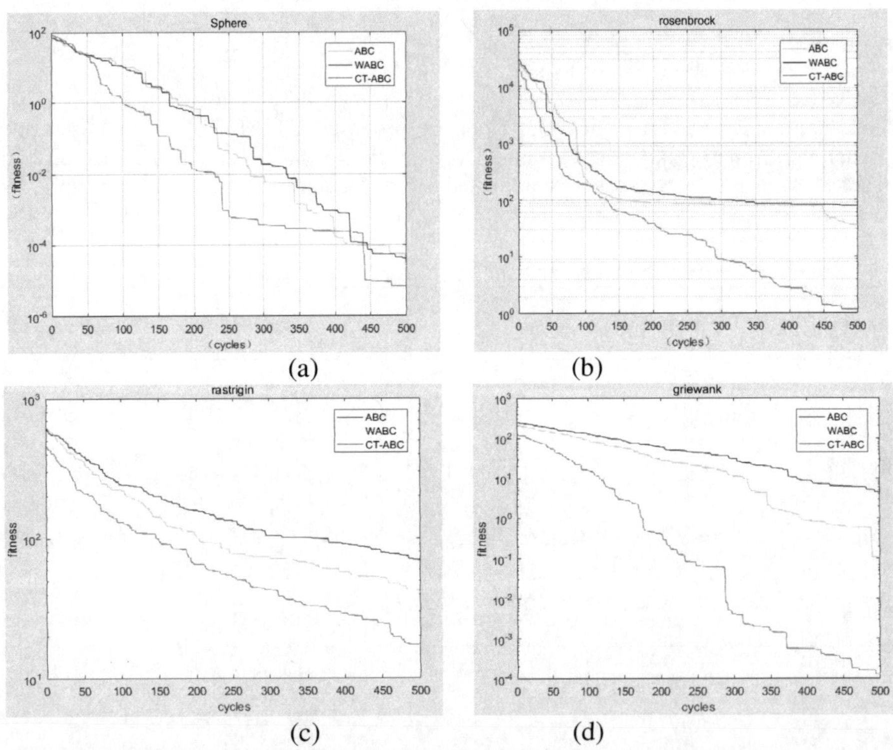

Fig. 1 (a-d) Convergence curves of ABC, WABC, CT-ABC on first four (f1-f4) test functions.

Fig. 1 graphically present the convergence characteristics of ABC, WABC, as well as CT-ABC on four selected test functions. Through analyzing the experiment result of Sphere function in Fig.1 (a), we can find in the first 50 iterations three algorithms almost achieved the same solution, however CT-ABC performed much better than other two algorithms in the following process except an abnormal change between the 400 - 440 iterations. From Fig.1 (b) we can see in the rosenbrock function, the proposed CT-ABC can achieve the best solution in each iteration of the whole process. Both ABC and WABC fall into local optimum in the early time of the optimization, respectively, in 100th and 350th iteration. After comparing the performance of CT-ABC with ABC and WABC in Fig.1 (c) and (d), we can see CT-ABC has a better performance in both griewank and rastrigin functions. The experimental results confirm that proposed strategies can significantly improve the accuracy of solution and increase the convergence speed on all test functions.

5. Conclusion and future study

In the present work, in order to enhance the search performance of ABC and acquire good global solution, we proposed a modified ABC algorithm which is based on chaotic map inertia weight adjustment strategy and tolerance-based search equation (TSE). To verify the performance of CT-ABC, we compared it with basic ABC and inertial weight-based ABC on four representative test functions. Experimental results show that the CT-ABC combined chaotic map strategy and TSE strategy can get better convergence accuracy as well as faster convergence speed.

As future work, the study of CT-ABC algorithm will be expanded to include modifying the way of getting initial solution. In addition, the CT-ABC algorithm will be applied to various optimization fields such as pattern recognition and artificial intelligence, etc.

References

[1] Karaboga D *et al* 2014 *Artif. Intell. Rev.* **42** 21.

[2] Gao W. F *et al* 2012 *Comput. Oper. Res.* **39**. 687.

[3] Ma L *et al* 2016 *Neural Comput. Appl.* **27** 387.

[4] Ranjan T *et al* 2018 *Int. J. Pavement Res. Technol.* **11** 509.

[5] Gao W *et al* 2012 *J. Comput. Appl. Math.* **236** 2741.

[6] Lei X *et al* 2010 *IEEE Fifth Int. Conf. Bio-Inspired Comput. Theor. Appl.* 514.

[7] Wang F *et al* 2018 *Inf. Sci. (Ny)* **436–437** 162.

IOP Conf. Series: Materials Science and Engineering **551** (2019) 012030 doi:10.1088/1757-899X/551/1/012030

A review on applied statistical and artificial intelligence techniques in crime forecasting

Alif Ridzuan Khairuddin1[1], Razana Alwee[1] and Habibollah Haron[1]

[1]Applied Industrial Analytics Research Group (ALIAS), School of Computing, Faculty of Engineering, Universiti Teknologi Malaysia, 81310, Johor Bahru, Johor, Malaysia

E-mail: alifjam1991@gmail.com

Abstract. Crime forecasting is an important component of crime analysis towards providing early information about possible crime occurrences in the future. Different models have been proposed to assess different crime data structures and representations. From the literature study conducted, there are several types of crime forecasting models that have been introduced such as statistical model and artificial intelligence (AI) model. Recent trends indicate that researchers have shifted their interest towards AI model due to its flexibility in handling variations in crime data structures. The study found that AI model is capable of capturing nonlinearity pattern of crime data in which statistical model fails to achieve. Moreover, the structure of crime data is mostly nonlinear. Thus, an AI model is favoured among researchers towards developing a robust crime forecasting model. This paper provides a review on the background, trends, and challenges on applied statistical and AI model in crime forecasting.

1. Introduction

Forecasting is widely applied in estimating the degree of risks or uncertainties in many fields including but not limited to geographical, finance, economic, engineering, security, health, and various interdisciplinary fields [1]. In the real world, crime is a part of society in which the occurrences are unpredictable by law enforcement agencies [2]. Forecasting provides useful information to government and law enforcement agencies in planning for crime prevention measures. The term crime analysis refers to a discipline practiced in the policing community [3]. The analysis of crime data may help relevant stakeholders to understand the behavior of crime trends and forecast future values based on past observations [4]. This is crucial as crime forecasting assists the authorities by providing useful information for enforcing early crime prevention measures [5].

Crime forecasting is a promising solution to tackle crimes which affect society's livelihood and properties [6]. The main objective of crime forecasting is to efficiently and realistically predict and forecast crime patterns towards providing significant insights about possible future crime trends based on past crime data. Crime forecasting is beneficial to the police and authority as the method is capable of providing early information regarding possible crime occurrences in the future. Advantages of crime forecasting include preventing the recurrence of crimes in specific areas or regions by analyzing the pattern of past crimes occurrences, providing insights for allocating appropriate resources within community for better policing coverage, and providing useful information to authority for planning efficient solutions in crime prevention measures. This paper discussed the background, development, trends, and challenges of applied statistical and AI models in crime forecasting. The rest of the paper is organized into the following sections. Section 2 provides literature study in crime forecasting. Section 3 provides recent trends in crime forecasting and its challenges. Finally, Section 4 concludes the paper and highlights future works.

2. Criminology literatures on crime forecasting

In criminology study, researchers attempt to forecast crimes based on two objectives encompassing identifying potential crime hotspots in specific areas or regions and identifying future crime patterns. Different researchers employ different methods in crime forecasting, depending on the nature and types of data. In crime forecasting, the most widely utilized crime data representation is time series data representation.

Time series data is a set of observations with a discrete-time stochastic nature [7]. Time series data is advantageous as the representation is able to extrapolate time series well ahead of time, into the future [8]. In addition, time series based models are useful when there exists limited knowledge on the underlying data-generating process and in the absence of satisfactory explanatory models that relate the estimation variables to explanatory variables [9]. In time series based model, there are two types of analysis including univariate and multivariate analyses. Univariate analysis involves one variable of time series data while multivariate analysis employs multiple time series data in model development. In most works related to crime forecasting, multivariate analysis is favored in developing a crime forecasting model [8, 10-12]. This is attributed to the ability of multivariate analysis in finding cross-correlations between multiple time series data and discovering distinctive patterns of crime data that never occurred in the past [8, 13-14].

In literature, several types of crime forecasting models have been introduced such as statistical model, AI model, and a mixture of both models, known as hybrid model. Examples of statistical model variations include linear regression, exponential smoothing, and autoregressive integrated moving average (ARIMA). On the other hand, examples of AI model variations cover mostly machine learning-based techniques such as artificial neural network (ANN) and support vector regression (SVR). Table 1 lists existing statistical- and artificial intelligence-based crime forecasting models proposed by different researchers.

3. Recent trends in crime forecasting and its challenges

From the literature study conducted, research on crime forecasting is quite limited. However, crime forecasting is not new because the field has been studied in the last 30 years. Even though crime forecasting shows promising results in estimating future crime rates, the method is rarely applied by most countries [5]. This is due to challenges faced by researchers, associated with crime data and model accuracy. In the real world, historical crime data are limited and rather difficult to obtain [8]. Meanwhile, a majority of available crime data contain 'noise' and are sometimes incomplete. This makes it harder for researchers to develop effective crime forecasting models due to the insufficiency of data. Another reason for the lack of interest by most countries in implementing crime forecasting is related to accuracy, which is plague by the nature of the available crime data. This may eventually to inconsistency of accuracy performance and varying results when the method is employed on different crime datasets.

In the last decade, researchers have gradually shifted their research interest from statistical model into AI based model in crime forecasting. This is associated to the incapability of statistical model in handling abrupt changes in varied environments or systems [26]. This incapability negatively affects the accuracy performance of the forecasting model. Further, statistical model is a parametric technique, which assumes the obtained data are stationary and linear [8]. In other words, the crime data must be first transformed into stationary and linear data prior to feeding into the forecasting model. Another reason for the shift of interest into AI is linked to the inability of statistical model to capture the nonlinearity pattern of crime data [27]. In most cases, a high volume of crime data contains a mixture of both linear and nonlinear patterns. Such mixtures consequently contribute to worsening the performance of statistical crime forecasting models.

Table 1. Existing crime forecasting models proposed by researchers

Reference	Crime model	User data description
[2]	• Artificial Neural Network (ANN) • Nonlinear Autoregressive with Exogenous Input	Crime record data from Selangor and Kuala Lumpur, Malaysia in 2013.
[8]	• Particle swarm optimization (PSO) • Support Vector Regression (SVR) • ARIMA	Historical USA crime rate dataset with external economic data from 1960 to 2009.
[10]	• Exponential Smoothing • ARIMA	Weekly burglary counts over a nearly five year period for the City of Pittsburgh.
[11]	• Interpolation method • Kernel density estimation (KDE)	Crime dataset in Arlington, Texas from 2007 to 2008.
[12]	• Artificial Neural Network (ANN) • Support vector machine (SVM)	Crime dataset comprising aggregated counts of crime and crime related events categorized by police department in USA.
[15]	• Autoregressive (AR)	Crime dataset in Chicago, USA from 2001 to 2014.
[16]	• ARIMA	50 weeks of property crime data in China.
[17]	• Linear regression • Adaptive regression • Decision stump algorithms	Crime and communities dataset in UCI machine learning repository.
[18]	• Random walk • Brown's simple exponential smoothing • Holt's two-parameter linear exponential smoothing	Crime data collected from computer aided system and offense report at Pittsburgh from 1991 to 1998.
[19]	• Space-time Autoregressive	USA regional and state rates of violent and property crimes.
[20]	• Eigenvector Spatial Filtering • Generalized Linear Mixed Model	Vehicle burglary crime data in Plano, Texas from 2004 to 2009.
[21]	• Linear Regression	Complaint dataset from Internet Crime Complaint Centre (IC3) from 2000 to 2005.
[22]	• Artificial Neural Network (ANN) • Probabilistic model	Crime related and spatial dataset in Pakistan.
[23]	• Fuzzy Logic • Self-organizing Map (SOM)	Monthly crime data of 20 county police bureaus in Taiwan from 2003 to 2004.
[24]	• Genetic programming	Crime and communities dataset in UCI machine learning repository.
[25]	• Random forest • Support vector machine (SVM) • Correlation analysis	Crime dataset in India from 2001 to 2012.

To overcome such drawbacks, several nonlinear models have been introduced by researchers to capture the nonlinearity of crime time series data through employing AI techniques. This is because AI techniques contain some nonlinear functions which are able to detect nonlinear patterns in data, thus improving forecasting performance [27]. In addition, AI techniques have the capability to execute global searching, rendering higher efficiency in handling complex environment [27]. Due to that, AI models are able to handle complex patterns in real world crime data effectively. Although AI models look promising, several limitations exist. Firstly, the performance of AI models heavily depends on the structure of data. Secondly, certain AI models such as ANN only work well when dealing with huge datasets but observe varied performances when dealing with smaller datasets. Lastly, it also suffers overfitting issue. Overfitting occurs when AI models take into account the noise of data during data training. Noisy crime data prevent the development of robust AI models.

To overcome the problems in statistical and AI models, researchers have integrated both statistical and AI techniques in their proposed crime forecasting models. From the study conducted, by applying both statistical and AI techniques into a crime forecasting model, the model is capable of capturing both linear and nonlinear patterns in crime data [8]. Findings from existing studies indicate that the hybrid model is effective as compared to individual statistical and AI models, as the hybrid model is capable of overcoming individual limitations resulting in better forecast outcomes. In overall, the literature study conducted found that the structure of real world crime data itself poses a major challenge in producing a good forecast. Further, the study observed that there is no universal model that is able to handle all types of crime data well, instead, researchers tend to develop crime forecasting models that fit their crime datasets. In most cases, the structure of crime data in specific case studies may vary. Thus, different approaches have been proposed to forecast different datasets.

4. Conclusion and future works

Crime forecasting is a huge milestone in crime domain as a mean of preventing offenses from being committed. Forecasting crimes offer the authority with insights of upcoming trends and community with an awareness of occurrences of crimes. An early development of crime forecasting began with an implementation of statistical techniques to forecast crime. Over time, with the introduction of AI technique, researchers have begun to shift away from statistical techniques. Attributed to exponential growth of technology over the last decades, it is now possible to utilize the full potential of AI technique in forecasting crimes. Study found that different researchers have introduced different models to handle different patterns of available crime datasets efficiently. Statistical model is good when handling linear distribution with small datasets. On the other hand, AI model is suitable for handling complex data distributions.

When selecting suitable models to be used, researchers should consider the crime dataset distribution, size, structure, and its representation. In addition, researchers are encouraged to implement multivariate analysis to discover new crime patterns, which could yield greater accuracy for the selected forecasting model. Based on these findings, a good crime forecasting model should possess a low generalization error, low bias trade-off, high variance trade-off, and should be insensitive to outlier. Such features should be sought by researchers in developing a robust crime forecasting model. For future works, the authors will implement several AI techniques to forecast crime rates. The performance of each selected AI technique is then compared and benchmarked. The analysis will then be conducted using a multivariate analysis approach. Through the approach, several external factors such as unemployment or gross domestic product that significantly affects crime rates pattern will be considered towards discovering a new crime pattern with improved model accuracy.

Acknowledgements

This work was supported by Fundamental Research Grant Scheme (FRGS) by Malaysian Ministry of Education (MOE) under grant number R.J130000.7828.4F825 for School of Computing, Faculty of Engineering, Universiti Teknologi Malaysia (UTM).

References

[1] Jain A and Kumar A M 2007 *App. Soft Comp.* **7** 585-592
[2] Ghazvini A, Nazri M Z B A, Abdullah S N H S, M N Junoh, and Kasim Z A 2015 *Proc.7th Int. Conf. of SoCPaR* p 334-339
[3] Ferreira J, João P and Martins J 2012 *The Elec. J. Info. Sys. Eval.* **15**
[4] Shrivastav A K and Ekata D 2012 *Int. J. of Comp. Sci. and Eng. Tech.* 415-421
[5] Ismail S and Ramli N 2013 *Procedia-Soc. and Behavioral Sci.* **91** 654-660
[6] Huang Y-L, Lin C-T, Yu Y-S, Hsieh W-H and Pai S-M 2015 *Proc. of NCITPA*
[7] Oliveira J F L and Ludermir T B 2015 *Proc. of IEEESMC* p 3867-3872
[8] Alwee R 2014 *Swarm Optimized Support Vector Regression with Autoregressive Integrated Moving Average for Modeling of Crime Rate* PhD, Faculty of Computing, UTM
[9] Khashei M and Bijari M 2011 *App. Soft Comp.* **11** 2664-2675
[10] Huddleston S H, Porter J H and Brown D E 2015 *J. of Business Research* **68** 1810-1818
[11] Hart T and Zandbergen P 2014 *Policing: An Int. J. of Police Strategies and Management* **37** 305-323
[12] Yu C-H, Ward M W, Morabito M and Ding W 2011 *Proc. of 11th IEEE Int. Conf. of ICDMW* p 779-786
[13] Preez J and Witt S F 2003 *Int. J. of Forecasting* **19** 435-451
[14] Han M and Wang Y 2009 *Expert Sys. with App.* **36** 1280-1290
[15] Cesario E, Catlett C and Talia D 2016 *Proc. of IEEE 14th ICPIC, 2nd ICBDICCSTC* p 795-802
[16] Chen P, Yuan H and Shu X 2008 *Proc. of Int. Conf. of FSKD'08* p 627-630
[17] McClendon L and Meghanathan N 2015 *Machine. Learn. and App. : An Int. J. (MLAIJ)* **2**
[18] Gorr W, Olligschlaeger A, and Thompson Y 2003 *Int. J. of Forecasting* **19** 579-594
[19] Santitissadeekorn N, Lloyd D J and Short M B 2015 *Euro. J. of App. Math.* **27** 451-478
[20] Shoesmith G L 2013 *Int. J. of Forecasting* **29** 191-201
[21] Chun Y 2014 *Geo. Analysis* **46** 165-184
[22] Omar A, Kwanbunbumpen A and Alijani D 2007 *Info. Sys. Edu. J.* **15**
[23] Mahmud N, Zinnah K I, Rahman Y A and Ahmed N 2016 *Proc. of 19th ICCIT* p 414-418
[24] Su L-Y 2010 *Comp. and Math. with App.* **59** 737-744
[25] Castelli M, Sormani R, Trujillo L and Popovič A 2017 *J. of Ambient Intel. And Humanized Comp.* **8** 29-36
[26] Baliyan A, Gaurav K and Mishra S K 2015 *Procedia Comp. Sci.* **48** 121-125
[27] Rather A M, Sastry V and Agarwal A 2017 *J. of OPSEARCH* 1-22

IOP Conf. Series: Materials Science and Engineering **551** (2019) 012031 doi:10.1088/1757-899X/551/1/012031

Multi-Perspective and Low-Cost Coverage Network Design for Implementing in Campus Network : A Case Study

A. Tongkaw[1,*]

[1] Faculty of Science and Technology, Songkhla Rajabhat University, Thailand

E-mail: aumnat.to@skru.ac.th

Abstract. This research proposed multi-perspective and coverage network design for implementing in two campuses of a university include physical and the logical design. This research shows the TOP model to concern three elements aspects of network design. The results of the research feature the technical perspective (T), the network designer considering the VLAN design and network management software. For the organizational perspective (O), the coverage network is necessary; designing a network depending on a practical value appropriately under the investment budget and quality of services (QoS). For the personal perspective (P), the network designer needs to recognize the applications that meet the requirement of the users, whether it is WIFI, VoIP, or CCTV— the authentication software to protect the network and still maintain system security.

1 Introduction

The Thai government has the policy to develop Thailand 4.0, focusing on innovative new ways of adding value, reducing labor, and bringing technology to all levels of education from elementary to higher education. In higher education or university, The government has pushed students to use classroom technology and management. Basic installation in design is essential to save as much money as possible. Universities have limited budgets to invest in; they must be designed to be cost effective and efficient.

Songkhla Rajabhat University is located in Amphoe Mueang, Songkhla Province. There are seven facades, and the building is scattered. Besides, agricultural buildings, gymnasiums, swimming pools, fitness, and sports facilities are also located in separate buildings. Network design requires to support three applications, including: VoIPs, access points, and a CCTV camera. If we use the same link, the data will be broadcasted.

There are much research talking about how to implement VLAN in the existing network. [1] The efficient VLAN design will reduce the risk of the network by using Simple Network Management Protocol (SNMP), decrease number of the routers, and also confine broadcast domains for reducing the traffics. However, there is a limit deployed in the real world situation especially in the campus network. [2] A research described a simulation-based study on network architecture using inter-VLAN routing and security. [3] The research showed the ability of the control and handled the reliable operation, however, it was a simulation-based study. [4]

This document is a presentation of network design and application by under considering of multi-perspective and low-cost coverage network design for Songkhla Rajabhat University Campus network. The quality of QoS services will be considered in this application.

2 Research Method

The research used mix methods to conduct this research. Both quantitative and qualitative data collected to support the framework and developed the new knowledge. The quantitative data included: a number of VLANs and software that implement in two campuses. The qualitative data included a model of multi-perspective represented in the TOP view point. A used a multi-perspective model to study the problem of deploying IT in an organization. [5] The results of the research indicate that the Organization (O) and Personal Systems (P) factors overlap with the Technical System (T). Also, the factors related to the organization system (O) and the people involved in the personal system (P) can help solve complex IT problems. Research used multidimensional models to study the problems of IT system installation in educational institutes. [6] As a result, the researcher has shown the factors that come from outside the school. The researchers analyzed the factors at various levels and used multidimensional models to study IT installation problems in both government and private sectors. [7] The problem is that the level of IT planning at various levels is not enough to consider multi-perspective views. The use of multi-perspective models is used to plan the installation of the network because the model is a clear pattern in determining the cause of the problem. Models can also explain the fundamental technical and organizational relevance of the problem. Multi-perspective models can also explain the depth of the problem and the culture of the organization. This paper demonstrated the design of VLAN in the Songkhla Rajabhat University in two campuses, main campus and Satun campus by considering the multi-perspective model.

3 Results and Discussions

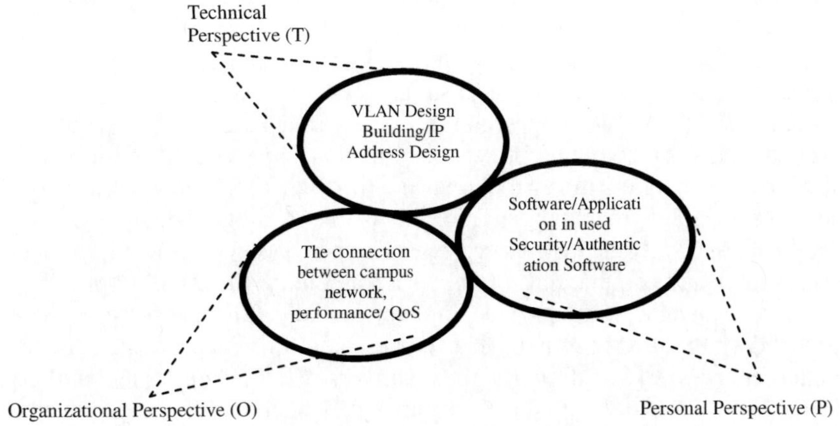

Figure 1. Adapted TOP model

The multiple perspective approach proposed by Mitroff and Linstone presenting the three parts. [8] The result can be expand those three parts of the multi-perspective TOP model: Technical perspective (T), Organizational perspective (O), and Personal perspective (P). Fig. 1. Shows adapted TOP model and details.

3.1 Technical perspective (T)

The technical perspective (T) included the coverage network design by using VLAN architecture and configuration. The main campus of Songkhla Rajabhat University consists of 77 buildings. All three switch models are S5750C, S2910PoE, and S2928. There is a set Virtual Switch Unit (VSU) switch to increase performance and to prevent failover. VLANs are designed to be 52 VLANs. Some buildings have more than one VLAN, depending on the number of devices. The VLAN design has a different IP address, such as / 22, / 24; the IP class B is designed for end users based on building numbers. For example, IP is 10.59.8.8., the IP address of the building VLAN number 59. This design has the advantage of being able to determine which user is accessing the building. If there is a problem, it is convenient to track the cause of the problem immediately, and easier to track down the IP address if sorted by building, it will be more comfortable to find and simple to control devices and policies.

3.2 Organizational perspective (O)

The construct of the organization perspective (O) views emerged from the uses of a digital system to manage the whole system known as E-University. However, the university had network design problems such as the obsolete of network infrastructure. The network infrastructure activated for over twenty years. Since then, the network was designed as a centralized system. In the long run, the design of the network is not up to date. Today, it cannot be compatible with new devices because of modern equipment and new technology. Today's technology is more efficient to manage data; The traditional system did not have a network monitoring system when problems such as broadcasting, loop, phishing of the device, occurred. Sometimes the problems emerged from the wrong settings. The problems made the campus network disrupted; Later, the university invested 18 million baht for installing new networks, infrastructures and implementing new systems to change the university. The implementation of this is to encourage the University to use both hardware and software cost-effectively. The new system has been selected for both open source and commercial software such as Linux, Elastix IP Phone.

The University planned is to provide internet service, full access across the university including the service to another campus, Satun Province. Satun Campus will have Internet access to Songkhla Rajabhat University main campus and will be able to use the same functions as the main campus had, such as MIS, e-document, e-meeting, and other ERP. Songkhla Campus has 77 buildings. It needs to be designed for the Internet expertly by covering all the buildings. Design the internet on a limited budget is very difficult. It needs to consider the environment and installation techniques and carefully selects the device, wiring cable selection, Length of the cable, choice of core switch, choice of distributed switch, selection of software authentication and configuration, user management, VLAN design, license, and MA use. Moreover, the access point devices need to be installed to the point where they are the most cost-effective without any additional cost.

3.3 Personal Perspective (P)

Satun campus is a new campus from Songkhla Rajabhat University. Previously, the branch did not have internet access and Internet infrastructure. The computer center at the main campus needs to design to link between the main campus and Satun campus. For personal perspective (P) view, it is necessary to design the Internet in Satun campus to fit the requirement of the users in that campus network. The demands of the users are as same as the main campus, such as the user of the Access point, VoIP, CCTV, MIS, and ERP.

Satun campus network consists of a firewall connected to another firewall at Songkhla Rajabhat University. Two firewalls will connect through VPNs site-to-site between the Songkhla and Satun campus. The two campuses used different networks and located on a different campus. The network was connected with a 300 Mbps lease line; The users at Satun can use VoIP calling from Stun to Songkhla campus by using four-digit numbers and use other applications such as MIS, database, journals under the requirement of teachers and students.

Regular internet packets used for standard applications will have authentication by ID to Songkhla, but if another application, it will have to run over another leased line, TOT leased line, with a speed of 1000 Mbps; The Telephone Organization of Thailand (TOT) line was used for particular applications such as multimedia, i.e., youtube or streaming at high speeds. The packet allocation will make the operation more efficient by using the routing function of the firewall on

IOP Conf. Series: Materials Science and Engineering **551** (2019) 012031 doi:10.1088/1757-899X/551/1/012031

both sides of the server. It is a VM or Virtual Machine to host the following: VoIP uses opensource Elastix V4.4 and runs on Ubuntu 16, authentication software, SKYRADIUS, run on Ubuntu 16, DHCP server runs on Ubuntu 16, respectively.

The VLAN class division classified by location and application for example in each building by physical and if that building contains a specific application, they also divide by application.

Inside the Satun campus building; During the building use fiber optic single mode connected to the hotel. Student dormitory has a switch between the building. VLANs are divided into 16 VLANs, namely VoIP, AP, Library, CCTV. This division of VLANs makes the performance more efficient and cost-effective and supports network expansion in the future.

For authentication software, Songkhla Rajabhat University, the main campus used RG-SAM+ and Satun campus used SKYRADIAS. We used RG-SAM+ is a proven comprehensive authentication and accounting system based on RADIUS systems. The RG-SAM+ has integrated with most of the available authentication modes including 802.1X and web. With the single unified authentication platform, the RG-SAM+ fully supports access control, accounting/billing, log management, and unified external interfaces for all the authentication methods. The user is only required to login once then able to enjoy automatic login in the future without all the tedious repeated login procedure again.

4 Conclusion

This paper discusses the design and implementation of the campus network, taking into consideration the multi-perspective model of coverage network that based on the design principles of both hardware and software. It takes into account several factors, including security, accessibility, ease of use, the network expansion, the increase in the number of users in the future, and more importantly, must support the technology in the future to speed up the network, reduce broadcast. For the technical perspective (T) the network designer needs to consider the VLAN design using VSU technology reduced the risk including using Simple Network Management Protocol (SNMP). Network management software can view the traffic of each network device, such as a switch and access point. For the organizational perspective (O), the coverage network is necessary to design a network depending on a practical value appropriately under the investment budget and quality of services (QoS). It needs appropriately designed and implemented to be the standard network design. However, the management and updating of the firmware in all network devices always need to be done. For the personal perspective (P), the network designer needs to consider on applications that meet the requirement of the users, whether WIFI, VoIP, or CCTV, the authentication software to protect the network and still maintain system security. Form the research results, the multi-perspective model can be applied with low-cost coverage for campus network and works effectively. Suggestions for future research and development should be included performance testing in each applications and the review of the users.

Acknowledgments
The authors wish to thank the Computer Centre of Songkhla Rajabhat University that provides the information and experiences in testing and implementing.

References
[1] D. A. J. AL-Khaffaf, "Improving LAN Performance Based on IEEE802.1Q VLAN Switching Techniques," *J. Univ. Babylon*, vol. 26, no. 1, pp. 286–297, 2018.
[2] M. Yu, J. Rexford, X. Sun, S. Rao, and N. Feamster, "A survey of virtual LAN usage in campus networks," *IEEE Commun. Mag.*, vol. 49, no. 7, pp. 98–103, Jul. 2011.
[3] S. Somasundaram and M. Chandran, "A Simulation based study on Network Architecture Using Inter-VLAN Routing and Secure Campus Area Network (CAN)," vol. 6, no. 3, pp. 111–121, Mar. 2018.
[4] Giovanni and N. Surantha, "Design and Evaluation of Enterprise Network with Converged Services," *Procedia Comput. Sci.*, vol. 135, pp. 526–533, Jan. 2018.
[5] H. A. Linstone, "Multiple perspectives: concept, applications, and user guidelines," *Syst. Pract. Action Res.*, vol. 2, p. 307, 1989.
[6] U. Fredriksson, "Innovative use of ICT in schools based on the findings in ELFE project," *Educ. Inf. Technol.*, vol. 13, p. 83, 2008.
[7] M. Metcalfe and E. Hobson, "Concern solving not problem solving," presented at the 12th Australasian Conference on Information Systems, 2001, pp. 455–461.
[8] Ian I. Mitroff and Harold A. Linstone, *The Unbounded Mind Breaking the Chains of Traditional Business Thinking*. Oxford University Press, 1993.

RAD Design and Data Management Systems of Natural Resources and Local Wisdom

S. Tongkaw[1], W. Inkaew[2], and A. Tongkaw[1]

[1]Faculty of Science and Technology, Songkhla Rajabhat University, Thailand
[2]Faculty of Management Science, Songkhla Rajabhat University, Thailand

E-mail: sasalak.to@skru.ac.th

Abstract. This research proposed the Rapid Application Development (RAD) design and data management systems of natural resources and local wisdom from the real data. The objectives were to: 1) collect local resources such as biological of local plant, animal, and the wisdom in the community; 2) study, organize and create an information system collection; 3) create a web-based application to present information of the data. The research used Rapid Application Development to plan and combining process from existed data to the web-based application for managing natural resources and local wisdom for data management. Natural resources and local wisdom data management at Tambon Ramdang is a case study of the successful implementation from the RAD design. The conclusion showed the lesson learn from this research and extended future work in this area.

1. Introduction

Technology and society can help each other to increase economic growth. It is highly complex data and interrelated with a massive value with social responsibility. There are many processes drive technological advancement; however, this paper picks up two theories for drawing the model from both scientific and social sources. The new technology is the result of rational thought independent of social influences. If the technology fit closely with the social and organizational factors, it will be effective work from a clear socio-technical view point [1]. However, social constructivism suggested that technological advance is the result of social influences. This research believed that social constructivism comes first and second, rational-objective view [2]. There is a research proposed SDLC database design collecting medicinal plants from Saudi Arabia [3]. It is few studies related with other database design. Therefore, this research tried to design an appropriate model and implement in the real situation by involved social influences.

From local resource operations to conservation of plant and natural resources and the treatment of Thai culture and wisdom and biodiversity. For example, a research created a local digital resources by using GIS and data management of champedak and local wisdom of Champedak gardener, satun province [4]. This research surveyed local plants in Tumbon Ramdang. Identify the source code of the original plant species and rare, endangered species, keep records not to lose, easy to find, also used for monitoring to grow plants. Alternatively, to propagate the plant. Community personnel have learned from the study of physical resources. Participants observed the survey information and record of their flora, fauna, and cultural heritage in the area. It is part of Thailand to raise awareness and establish a sustainable academic knowledge based. That will lead to conservation and preserve sufficiency economy as well. Plant Genetic Resources Center, as well as other resources including many organisations and universities involve this project by linking information between the units that participate in the royal initiative.

2. Research Method

2.1. Method

Rapid Application Development (RAD) was developed to address the weaknesses of structured design [5]. With the use of tools and techniques to help develop the system quickly under the real situation and high quality systems, fast development and delivery, and low costs [6]. The villages have the considerable amount of data, and the user can also try the prototype. Creating the prototype module also generates the system quickly and provide the key to validate data [7]. The RAD methodology has introduced various techniques and tools to support the development of systems that can be implemented at various stages in the Systems Development Life Cycle (SDLC) less than Structured Systems Analysis and Design Method (SSADM). Figure 1 shows the RAD process which include the data collection process, analysis and quick design, prototype cycle (the cycle of demonstrate, refind and develop), testing, and development.

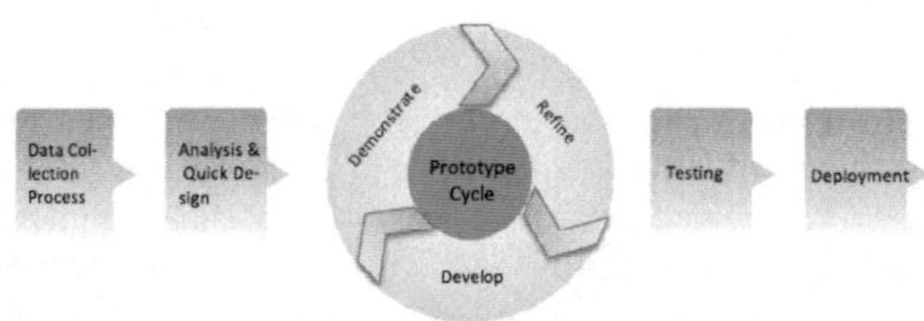

Figure 1. RAD process.

The development methodology for RAD has been broken down into several different approaches, including phased development-based methodology, prototyping-based methodology, and throw-away prototyping-based methodology. This research explains in phased development-based methodology approach. A research also suggested that RAD prototyping become important when the new technology such as software or hardware development was discovered [8].

This research used RAD development instead of SDLC. The process of the RAD development cycle consists of five parts: (1) first, the data source was collected by the people, students, SAO administration, (2) second, design by technical computer system using existed spreadsheet which contained the data source, (3) third, develop web-based system by using PHP and MySQL, (4) testing program, by implement in the real situation and complete the system test before the training, then make changes to the rapid development tools, (5) deployment, is the process of implementation and maintenance. This research combined the SDLC process with a RAD development cycle, and it can be seen as follows (Figure 2).

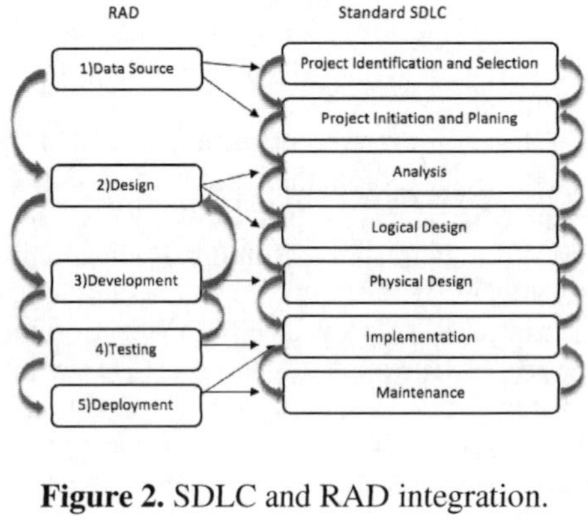

Figure 2. SDLC and RAD integration.

Figure 2 shows the comparison RAD model and SDLC, a software development process. It is iterative development by using prototypes instead of a step-by-step design like the SDLC of waterfall framework. The first stage, data from natural resources, i.e. varieties of the plant, animals, local wisdom of gardeners, and village information in spreadsheet form were used to analysis and quick design. The real data from Ramdang area made the software to be written faster and easier to change requirements. The second stage, the rapid demonstrate, refine data, created a prototype from preliminary existed data, and build the prototype. This stage, the researchers integrate the interview and activities to collect the data from the real situation in Tambon Ramdang. The third stage, the researcher developed the web-based applications followed the designed model combined with local wisdom data. The final stage, this technique merge testing and implementation into the same stage, various structure techniques, especially data-driven information engineering, with the prototyping design to decrease time-consuming of software developing process.

2.2. Data Collection

The government tried to collect natural resources systematically, and the objective of the systems is to collect data by using information systems for easy access and fast search. The vision of the chief executive of the Ramdang Subdistrict Administrative Organization (SAO) is to use modern information technology to store information and display the data. The local resources, both physical and biological, are collected. All nine worksheets used to collects manually from field surveys by community registra, biological registra or villager.

The existed data is complex and various forms. For collecting the original data Tambon Ramdang, SAO, provided the appointment of the Local committee for local resources. There are four resources: plant data, animal data, biological data, and cultural data and local wisdom data. The researchers had many meetings with personnel interview and community members in the village, such as community leaders, village heads, teachers and school administrators. They were involved in providing feedback and implementation. They also cooperated with Songkhla Rajabhat University to carry out their activities. Next step, the members of SAO selected the outstanding of biological resources, the cultural and intellectual resources of the area. There is an activity to complete resource level Kor.1. Then, they surveyed the preserved area of local resources, local animal, the biological resource at "Ta-Kae cemetery area" and identified the code of animal and plant species by providing the label.

This research used many programs for designing tables suitable for the variety of databases and most of them are useful for this cases, for example, MySQL Workbench. However, because we have existed spreadsheet files as a starter guideline. Each column of the spreadsheet was transformed into variables in MySQL table. Then, we defined the properties of each variable and defined the size of the primary key, secondary key of each table. After finished designing each table neatly, we established all relationships between tables. There are 31 tables and their relationships with Primary Key (PK) and Foreign Key (FK) in each table, showing in figure 3.

The researchers implemented the web-based system a few weeks later with PHP language and MySQL database. We created a system for local resources management as well as other functionality, and it eventually delivered at the end of a three-month period. A further three months was spent providing a software module in the SAO and offered a course administration. The third phase was spent in the maintenance program as well as making some further adjustments and training module. The project also had the fourth phase preparing for other developers in terms of re-engineering and upsizing the system onto an industrial-strength database platform.

IOP Conf. Series: Materials Science and Engineering **551** (2019) 012032 doi:10.1088/1757-899X/551/1/012032

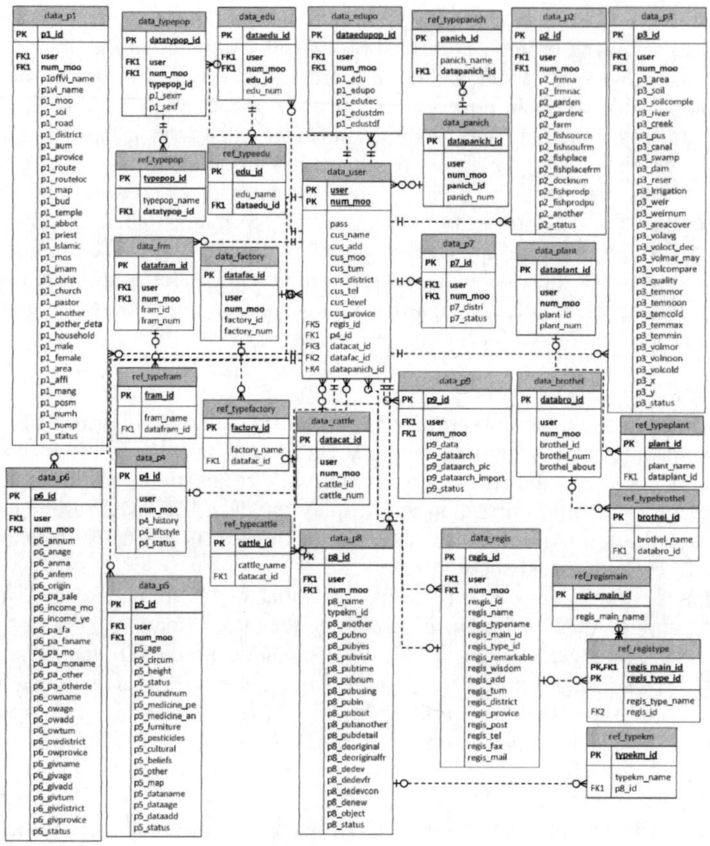

Figure 3. ER Diagram.

3. Conclusion

This research proposed the Rapid Application Development (RAD) design and data management systems of natural resources and local wisdom from Tumbon Ramdang data. It showed that RAD could reduce the steps and time of design and implementation of the systems. The tables from the spreadsheet were handy for RAD design and MySQL developments and implementations. Moreover, it worked efficiently regarding the real situation, and accurate existed data and social participates. The benefits of this research are the SAO members can participate in the community personnel knowledge and understanding of plant necessary genetic information. They knew how to conserve genetics and collect the local resources database. Regarding system development, local resources were carefully stored in a database storage system which could be easily searched and it could be published to the people under their priority. Villagers or other participants can use information of local resources database such as sort, search, and save in digital form with high performance. For social impact, the local resource information will be useful for Ramdang students concerning learning with data storage. Understanding the recording of the local plant, animal information, local wisdom resources and create their opportunities to take advantage of the data stored.

Acknowledgments

The authors wish to thank the Chief Executive of the Ramdang Subdistrict Administrative Organization, Mr. Udom Takkara, for advicing this study and provide the structure of the data sources.

In addition, the authors wish to thank the Computer Centre of Songkhla Rajabhat University that provide the information and experiences in testing and implementing.

References

[1] Tongkaw A 2013 *Procedia Soc. Behav. Sci.* **93** 1467.
[2] Thamhain H J 2005 *Management of technology : managing effectively in technology-intensive organizations* (New Jersey: John Wiley & Sons Inc.).
[3] Hassan A *et al* 2017 *Int. J. Adv. Comput. Sci. Appl.* **8** 5.
[4] Tongkaw A *et al* 2018 *3rd Int. Symp. Expertise of Engineering Design* Johor Bahru Malaysia.
[5] Beynon-Davies P *et al* 1999 *Eur. J. Info. Syst.* **8** 3.
[6] Beynon-Davies P *et al* 2002 *Inf. Sw. Tech.* **44**, 10.
[7] Pop D P *et al* 2014 *Procedia Eng.* **69** 1172–1179.
[8] Kordon F *et al* 2003 *Des. Autom. Embed. Syst.* **8** 4.

E-learning prospect on improving academic performance in Omani Universities

Ragad M. Tawafak, Awanis Romli, Ruzaini Abdullah Arshah

Faculty of Computer Systems and Software Engineering, Universiti Malaysia Pahang, 26300 Kuantan, Pahang, Malaysia

E-mail: raghad@buc.edu.om

Abstract. This paper will discuss the impact of E-Learning prospect on the improvement of learner academic performance of Oman universities. The study explains the needs to expand working on E-Learning models and continue use of face to face traditional educational process. The real problem faced education is a need for several applications and technologies to improve their outputs and improve on teacher-student prospect. This paper aims to present an expanded review of different E-learning models to evaluate the prospect of their academic performance. The method of this paper divided into two phases. In the first phase, including the analysis of different technologies and most used tools affected in the educational process to measure the acceptance of learning outcomes. The Second phase, show the testing and validation of learner prospect of E-Learning to enhance the learning process and improve the academic performance.

1. Introduction

Within terms of learning and teaching, the word "E-Learning" has many other names of use like "online learning", "blended learning", or "distance of learning" [1]. In real term the E-Learning meaning in this paper highly related to the technology enhancement learning design [2, 3]. Within the recent huge number of internet users and the large need for using technologies in education in Oman, the indeed needs to use E-Learning models in educational process [4]. This way of learning carried the teachers to use conventional education strategies. One of these strategies, the learning process connected with maximum numbers of students who like to use mobile in learning for the facility of using and speeding of getting results [1, 10]. Another strategy is to use the official model of learning as Moodle, MOOC, or UCOM for student learning and assessment [4]. During the evolution of technology, and the variety of tools offered, there are different requirements for IT development because of the need to develop software, for better assessment way and prospect on academic performance [4, 5]. E-Learning is the type of learning mixed between face-to-face contacting and online learning to reduce classroom time problem, increase the availability of electronic materials with evidence and improve education processes [6]. The adoption of E-learning, to multi-activities of loading course requirements of material, assignments, guidelines, peer moderation and assessment results, etc. all these facilities increased in huge expanding with few last years[6, 7].

The overall use of existing E-learning models could illustrate the level of student's grades and skills, even though, the way of course content material handled by expert teachers in E-Learning with student interactivity that provides effective learning skills freely without being committed to a specific space and time [4, 8, 9]. These requirements need to be well observed and respected within the academy regulations [2].

There are many qualifications offered in Oman higher educations including (diploma, higher diploma, bachelor, master, and Ph.D.). The generalized use of E-Learning in academic programs with academic majors could improve the online evaluation of their improvements in academic and knowledge skills. All these academic programs and management issues are required to develop the relationship between face to face in education and the technology development in the learning process. These called type of blended learning that work for managing requirements and results using (LMS) in addition to an E-Learning model like UCOM, MOOC, or Moodle to increase the faculty performance through facilities of declaration attached, announcements to sections, evaluate and assessment techniques, open access of electronic material etc. [4, 6, 11]. The assistance applications that help in E-Learning process could be linked with Google drive forms and Google classroom, with another application like video programs, office assignments. The continuous use of learning needs a lot of skill acquisition and enhancement of using perfect technology and the one that usually interests of users. This need become important for professional work and tangible outcomes. What is the best solution for traditional slow systems with a lot of hard timing, far away location, and careers to recurring their need to develop new skills? E-Learning or use blended learning technology to prospect improve the academic performance in significant values [12].

2. Methodology

2.1. E-Learning model design

In recent years, the need for E-learning techniques in the academic programs of Universities was proposed. The educational process in higher levels has faced different types of problems in the recent, leading to learners missing the ability to understand and collect huge material from different resources that need basic knowledge of technologies and searching steps. Therefore, the use of E-Learning model can reduce these problems and improve their experience in technology use and enhance their skills and performances in significant results. The E-Learning model of UCOM in this study [6], use waterfall sequence in analysis steps, as shown in Figure 1. Which starting with an analysis of all previous models and their main factors effects on using technologies in teaching. Then develop the mechanism of mix use and the link between multi-type and use of applications in the same course material discussion and assignments between the factors of teachers, students, and course contents moderator. The last step is the cycling during the whole course is worked with evaluation report to assess and re-design the changing in material based on evaluation feedback and understanding level. These three steps used to ensure that all the major factors (members) are accessed in the right way to prevent the error of misunderstanding or course dropping by the students.

2.2. Participant Information

The full description of the E-Learning model for each step between the three-member cycles describes the sequence of learning procedure. The active model applied in Buraimi university college-Oman. BUC. This study based on the University Communication Model (UCOM) model published by Tawafak, et al, (2018) as adapted model [6]. The participants of this study were the Universities that activated (UCOM) as a means of E-Learning in addition to using Google classroom as the complementary stage. The total number of participant (n=53), survey samples collected from Information Technolgy department at BUC. See table 1.

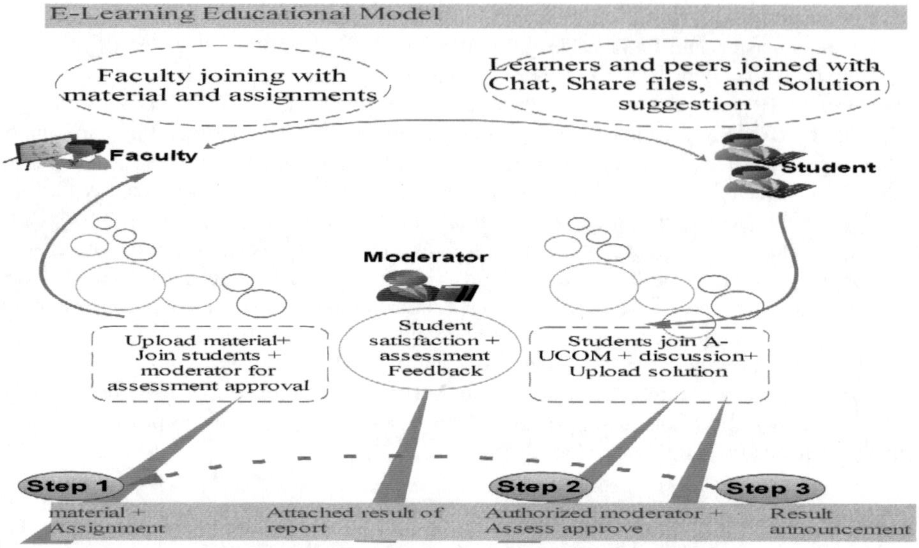

Figure 1. E-Learning model methodology progress

Table 1. Demographic information of BUC participants

Characteristic	Type	BUC	Percent
Gender	Female	34	64%
	Male	19	36%
Degree	Bachelor	17	32%
	Diploma	36	68%
Age	18-22	15	62%
	Above 23	21	68%
Knowledge skills	High	19	36%
	Normal	24	45%
	Low	10	19%

2.3. Methodology acceptance

The data were collected from two types of sections, morning and evening sections. The questionnaire includes geographic information about the participants and College in the first stage. In the second stage, there are 11 question items affected on improving academic performance in Oman universities using E-Learning process. See table 1 and table 2.

In this model, a survey used to evaluate the E-Learning prospect in improving the academic performance of students in universities. Therefore, the results of question analysis tested by SPSS program. This statistical program used Likert scale method, 1 for strongly agree to 5 for strongly disagree to measure many points like mean, std deviation and variance for each item tested as shown in table 3. In addition, the program identifies the reliability and validity of the questionnaire. Therefore, the reliability factor defined in Cronbach's alpha result that measured in this survey (0.961), which is highly accepted whenever it's greater than 0.76% [13].

Table 2. Survey items of improving academic performance [6]

No	Construct items
Q1	I anticipate good grades in courses with E-Learning is mixed used in a course.
Q2	I anticipate better grades in classes using E-Learning heavily compared to where they are not used at all.
Q3	E-Learning model efficiently allows faculty-student collaboration.
Q4	E-Learning model is trusted by faculty to improve learning outcomes.
Q5	It is worth to recommend E-Learning for other learners.
Q6	I'm interested to use the E-Learning in the future.
Q7	I have learned a lot with E-Learning model.
Q8	I intend to utilize E-Learning for various purposes such as self-development.
Q9	I intend to use E-Learning frequently even after graduation.
Q10	I had sufficient chance to interact with other learners using E-Learning model
Q11	E-Learning model is a very enjoying experience.

Table 3. Mean, Std, and Variance statistical analysis for BUC

No.	Mean	Std. Deviation	Variance	No.	Mean	Std. Deviation	Variance
Q1	1.8125	0.6927	0.48	Q7	1.6563	0.54532	0.297
Q2	1.75	0.508	0.258	Q8	1.7813	0.6592	0.434
Q3	2	0.67202	0.452	Q9	1.875	1.186	0.242
Q4	1.75	0.62217	0.387	Q10	1.7093	0.6578	0.414
Q5	1.9688	0.73985	0.547	Q11	1.3613	0.7292	0.534
Q6	2	0.71842	0.516				

3. Results and Discussion

The results of table 3, approve the usability of E-Learning to increase active communication between factors and support teachers with guidelines of soft copies and student feedback intention after use to measure the improvement of academic performance [2, 6, 10]. In the study of [12], they used smartboard as techniques to enhance learning but not to improve the academic performance. In addition, the researchers of [9], they compared between traditional and E-Learning only in programming course as limited study, while in this paper it's for different courses and many items tested. The E-Learning prospect better in this paper than the factors examined in [8], with specific goal of enhancement. At the end this paper comes out with E-Learning models compaired deeply with improving academic performance factor as the main objective of learving achievements.

4. Conclusions

This paper presents the prospect of improving the academic performance in Oman universities. This kind of E-Learning model support the university outcomes with significant improvement of academic performance and to be matched with the benchmarked requirement. This model measured the academic performance of the students from the type learning process used in the university and the way of assessment in a cycle of three factors (teacher, moderator, and student) for professional purposes, explaining the suitable learning process based on the understanding and satisfactory feedbacks. The SPSS program showed high reliability and validity of the examined survey.

Acknowledgments

This work is supported by University Malaysia Pahang Term No. (PGRS180301). These supports are appreciated.

References

[1] Aldiab A *et al* 2017 *Energy Procedia* **110** 574.
[2] Tawafak R M *et al* 2018 *Adv. Sci. Letters* **24** 10.
[3] Tawafak R M *et al* 2018 *7th Int. Conf. Software and Computer Applications* Kuantan Malaysia.
[4] Tawafak R M *et al* 2018 *J. Math. Sci. and Tech. Educ.,* **14** 6.
[5] Wilby K J *et al* 2017 *Currents in Phar. T&L.* **9** 1.
[6] Tawafak R M *et al* 2018 *IEEE Access* **6** 66481.
[7] Hone K S *et al* 2016 *Comp. & Educ.*, **98** 157.
[8] Tawafak R M *et al* 2018 *Educ. and Info. Tech.* **1** 25.
[9] Malik S *et al* 2016 *J. Info. Tech. Educ. Res.* **16** 1.
[10] Abdalla E *et al* 2018 *J. Eng. and App. Sci.* **13** 15.
[11] Abdalla E *et al* 2017 *Int. J. Inform. Vis.* **1** 4.
[12] Gurunathan P *et al* 2014 *Proc. ICAICT*, Elsevier.
[13] Hair J F *et al* 2013 *Long Range Planning* **46** 1.

Environmental Risks in Supply Chain: Recommendations and Directions for Future Research

Arwa Mukhtar[1], Awanis Romli[1], Mansoor Abdullateef[2] and Hael Al-bashiri[1]

[1] Faculty of Computer Systems and Software Engineering, Universiti Malaysia Pahang, Kuantan 26300, Malaysia
[2] Faculty of Computer and Cyber Science, Software Engineering, University Prince Mugrin, Al-Madinah 41499, Saudi Arabia

E-mail: arwakanani@gmail.com

Abstract. This paper explores the environmental risks in supply chains and provides recommendations for subsequent studies. Typically, the environmental responsibility is a core part of sustainability which focuses attention directly to protecting and restoring the environment. Furthermore, environmental awareness is essential to design globally distributed supply chain networks. Therefore, the aim of this research is to explore the environmental risks and to identify the available solutions along with a projection of future trends. The basic research steps have been followed to complete this work. The Findings of this research describes the environmental risks that arises during supply chains operations. Furthermore, recommended solutions have been highlighted to address the identified risks. The Blockchain technology is regarded as a suitable solution for the environmental risks issue. The motivation for recommending this technology is its capabilities to contribute effectively towards greening the supply chain operations.

1. Introduction

Firms are continuously seeking to manage supply chain activities, in a bid to improve production processes and to gain competitive advantage [1]. Supply chain management (SCM) refers to activities related to production management, and delivery of products [2]. The process starts from raw materials suppliers through to the end consumers. However, over the years, this traditional flow has expanded thereby exposing the flow of chains to risks. Notably, the expansion factors contribute to the depletion of resources, and to the environmental pollution [3]. Specifically, environmental risk may be defined as the environmental damages which ensue from everyday supply chain operations [4]. Hence, managing this risk aims at protecting and restoring the environment's integrity. This is mainly because they have consequences on the shared ecosystem and it may also impact the natural ecosystem and corporate reputation [3]. The environmental damage is considered as one of the drivers and emerging challenges in supply chain management [4]. Furthermore, industries consider the environmental risks as the most important risk factors [3].

Nowadays, there is a growing need to deal with the challenges made by supply chain operations on the global environment and control their negative impact. Especially the context that related to complex supply chains. It is well known that both developed and developing countries are challenged with the increasing amount of CO_2 emissions. Some of the notable countries include USA, Japan, China, the European Union (EU) countries and Malaysia [5]. Till this moment, deducing appropriate means to mitigate this environmental impact have become a major issue for supply chains, especially those seeking competitive advantage for their operational performance in the marketplace [6]. Firms such as Hewlett Packard, Walmart, Patagonia, and Tesco, have benefited from considering green practices across their supply chain processes. Specifically, their approaches involve carbon control of assets and infrastructure, waste reduction through process optimization, the use of energy-efficient vehicles and recycling [7]. Therefore, this study aims to explore the environmental risks issue across supply chains and provides recommendations and useful insights to overcome these issues in subsequent supply chain processes. In addition, possible directions for future research are also identified and presented herein.

This study is presented as follows: the next section, section 2 illustrates the research methodology while Section 3 discusses the environmental risk factors in details. Section 4 investigates the available environmental risk solutions whereas Section 5 offers recommendations and future research agenda. On the other hand, Section 6 concludes the research findings.

2. Research Methodology

This paper goal is reviewing the environmental risk factors in supply chain and provides recommendations for future researches in the area. In order to structure the literature review, the mind map has been designed to set the paper structure. To achieve the defined goal of this research, a review protocol was designed based on [8] as shown in figure 1. It is comprised of three main processes namely: determination of the target resources of search, specification of the keywords, and selection of the relevant articles. Searching the literature was achieved by using popular database sources such as Web of science, GoogleScholar and Scienedirect. Defining the search terms for data collection purposes, the following keywords were used: "supply chain"," Environmental Risks "and "supply chain management" with combinations of these words including: Environmental AND Green Supply Chain AND Environmental Supply Chain. The initial search results in both relevant and irrelevant studies in SCM and environmental risks literature. Hence, the selection and inclusion criterion are based on the scholar's title, keyword and abstract relevancy. In conducting the defined process, certain technologies have been used such as Endnote, which is used to search and manage the process of collecting the related studies and filtering them.

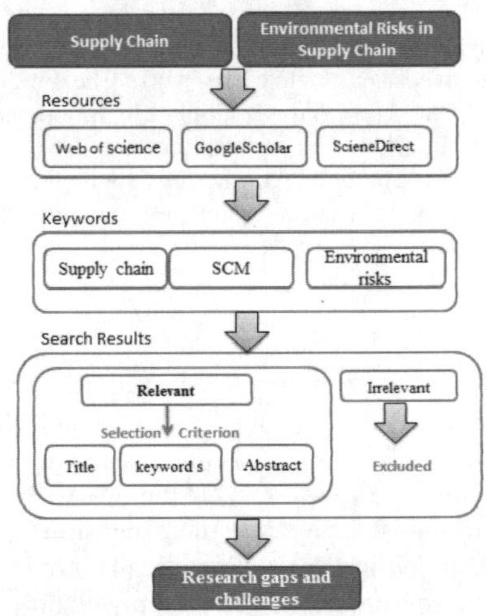

Figure 1. Review process adopted from [8].

3. The Environmental Risk Factors

Minimizing the negative effect that the supply chain has on the environment is called a green supply chain [7]. This negative effect includes the hazardous materials used during supply chain operations [4], environmental pollution [10], energy consumption [3], and carbon emissions [6]. Greening the supply chain mainly aims at integrating the environmental practices into the ordinary supply chain flow. A review of literature on green supply chain risk management reveals a list of the environmental risk factors that may be caused by supply chain operations.

Zhao [4] defined carbon emissions as the atmospheric gases spread which results in the effect of the greenhouses. However, the measurement of the carbon footprint within supply chain networks is a challenging issue for many organizations [6]. Environmental pollution is the contamination that ensues from production operations on water, soil or air due to the different production operations [11]. The risk associated with energy consumption refers to inefficient use of energy in the production processes, or during goods and services delivery [3]. On the other hand, natural disasters are considered as one of the environmental risk sources when such natural disasters cause disruptions. Examples of these are fire, earthquakes, hurricanes, flood, and storms [12]. Another potential source of risk is the environmental accident, which are accidents cause by production operations, and which affect the environment. These included fires and explosions [3]. Similarly, hazardous waste generation describes the production of unwanted substances during operations which may harm the human health plus the environment [13]. Furthermore, water scarcity is also considered as a potential source of environmental risk and it refers to the risk caused by lack of sufficient available water resources to meet the demands of water usage (for energy production, manufacturing, and transportation) [3]. Finally, other potential risks involve the risk of heat waves and droughts which is the vulnerability caused by increase in temperature due to climatic changes [3].

4. Investigation of the Available Environmental Risk Solutions

This section reviews the current state of the art in the environmental sustainability. The authors will discuss several previous studies found in the literature that examined the environmental issues [3, 10,14-17]. Giannakis and Papadopoulos [3] calls for integrated approaches in sustainability and risk management to enhance the supply chain sustainability. The study differentiates between the environmental risks and the other supply chain risks. The study findings revealed that the internal environmental risks may be considered as the most essential factors for many industries. Zhao [4] proposed a tool to help in identification and analysis of manufacturers strategies that reduce the materials and carbon missions. On the other hand, Urata, Yamada [5] developed a global supply chain network model for production and logistics in the Asian countries which aims at balancing of CO2 volumes and cost. In their research, they defined a mixed integer programming (MIP) problem that determines the supplier and factory location and aims at reducing the environmental impact ratio.

In a particular study, Song, Ming [10] used a trial evaluation laboratory (DEMATEL) and decision-making method to identify the critical risk factors. Furthermore, the method assists managers of supply chains to concentrate on the most influential risk factors for SSCM performance. The results showed that the most significant risk factor for SSCM is failure in the right supplier selection. Sharfman, Shaft [14] examined the management of cooperative supply-chain environmental (CSCEM) concept using an online exploratory survey and theoretical sampling. Interviews with 14 leading-edge firm were conducted to examine a model of CSCEM antecedents. They reported that the consideration of the adoption CSCEM is an important condition for firms to transit towards adopting more sustainable practices.

In a similar study, Yu and Solvang [15] proposed a stochastic programming model with multi-product multi-echelon for reverse logistics design. The results obtained from the sensitivity analysis showed the implications of the profit and carbon emission requirement relationship. The study also suggested that more environmental sustainability factors should be included such as land and water pollution. Mavi, Kazemi [16] studied supplier selection and evaluation in green supply chains. In their study, the fuzzy DEMATEL method was used to evaluate the logistic factors of green implementation. Observations from the study showed that environmentally friendly packaging is more influential than other factors. In another vein, the work of Rahimi and Ghezavati [17] designed a recycling construction and demolition (C&D waste) reverse logistics sustainable network to reduce the negative environmental impacts of waste. Results showed the model flexibility based on risks severity.

The subsequent section presents the risk factors as indicated by F1 to F6 to represent Environmental pollution, Natural disasters, Energy consumption, Environmental accidents, Carbon emissions, and Hazardous waste generation, respectively. Table 1 presents some studies that discussed the environmental risks issue and the proposed solution in each study along with their limitations.

5. Recommendations and future research agenda

Reviewing the literature of environmental sustainability, the following information has been derived and certain conclusions has been highlighted:

Despite the growing awareness on the need to incorporate sustainable principles into the supply chains [18], there is still the major challenge of missing implementation mechanisms for environmental integration in supply chain operations from literature [19-21]. Although Giannakis and Papadopoulos [3] pointed to the limited researches that study the topic of supply chain sustainability risks and risk management strategies, Calatayud [21] had refuted the assertion by using new technologies such as Blockchain technology for better supply chain risk management. In fact, the findings of the study conducted by Sharfman [14] revealed that Blockchain technology appears as a good solution for digital supply chain integration. Notably, due to its ledger and smart contracts, it offers tools for building more flexible and cost-effective supply chains network. In addition, Blockchain traceability offers an enhanced visibility throughout the product life cycle [22, 23]. Likewise, Price [24] stated that Blockchain is the right solution to solve core problems in supply chain transparency.

Specifically, unlike traditional centralized systems, Blockchains create decentralized networks with trust-less parties [25]. Blockchains facilitate the possibility for tracking the materials and to determine the provenance of a product from its origin to the end customers [26].

Based on the highlighted points, it is believed that Blockchain can effectively contribute to greening the supply chains. This is mainly because the practical implementation of sustainable environmental factors in new evolutionary technologies such as Blockchain presents an open research area filled with opportunities.

Table 1. The existing approaches and their limitations.

	Environmental Risks						Solution	limitation
Ref.	F1	F2	F3	F4	F5	F6		
[14]	√	√	√	√	√	√	Cooperative supply-chain environmental management (CSCEM)	Only successful CSCEM has been examined.
[4]					√		Game Theory	Consumer preference factor was not considered in the study. The manufacturers are considered as one entity.
[16]	√	√	√			√	Supplier Assessment	The shortage of study respondents. Lack of validation method such as case study
[5]					√		A mixed integer programming (MIP)	Limited number of products, countries, fixed cost (factory location), foreground data (cost and CO2 volumes) , risk types (lead time and scale).
[3]	√	√	√	√	√	√	Risk Management Framework	A small sample size. Lack of in-depth investigation in risk management related to sustainability topics.
[10]	√	√	√	√	√	√	A Novel Method Of Rough Weighted DEMATEL.	The evaluations are based on expert's judgments. No differentiate between positive and negative influences in the methodology.
[15]					√		Stochastic Programming Model	The Social consideration was not considered.
[17]	√					√	A mathematical (mixed integer linear programming) model	The social impacts are excluded. The model is not a multi-product model

6. Conclusion

The environmental damage which ensue from supply chain everyday operations is a major issue for supply chain management. Due to its impact in the natural ecosystem, managing this risk will add competitive advantage for corporate reputation and operational performance. Blockchain as a fast growing technology is able to contribute in greening the supply chains. The technology is able to track sub-standard products and identify their origin. Furthermore, the data availability removes redundancies and bottlenecks which thus decreases resource consumption. The technology is a promising approach that defines energy sources and energy storage systems, proffer solutions to environmental challenges and contribute to greenhouse gas reductions thereby boosting energy efficiency. In addition it helps to ensure that products are sold in an environmental friendly manner. Nowadays, leading-edge firms such as Oracle, SAP, Walmart and IBM are already developing Blockchain based cloud management solutions for the supply chain. Despite those successful trials, there is still the need for future supply chain applications of Blockchain technology to ensure the integration of certain factors. These include environmental considerations, enhancement in best business, and promotion of global market transformations.

Acknowledgment
The research reported in this study is conducted by the researchers at University Malaysia Pahang (UMP), it is funded by RDU160361 grant. The researchers would like to thank UMP for supporting this research.

References
[1] Cohen S *et al* 2013 *Strategic Supply Chain Management: The Five Core Disciplines for Top Performance* (McGraw-Hill).
[2] MacCarthy B L *et al* 2016 *Int. J Ops. & Prod. Mgmt.* **36** 12.
[3] Giannakis M *et al* 2016 *Int. J Prod. Econ.* **171** 455.
[4] Zhao R *et al* 2012 *J Loss Prev. Proc. Ind.* **25** 6.
[5] Urata T *et al* 2015 *Procedia CIRP* **26** 664.
[6] Sundarakani B *et al* 2010 *Int. J Prod. Econ.* **128** 1.
[7] Andiç E *et al* 2012 *Res. Cons. Recycl.* **58** 50.
[8] Hujainah F *et al* 2018 *Info. Sw. Tech.* **102** 85.
[9] Sisco C *et al* 2011 *Supply chain sustainability: A practical guide for continuous improvement* (United Nations Global Compact).
[10] Song W Y *et al* 2017 *J Clean Prod.* 143 100.
[11] Blackburn W R 2012 *The sustainability handbook: The complete management guide to achieving social, economic and environmental responsibility* (Routledge).
[12] Waters D 2011 *Supply chain risk management: vulnerability and resilience in logistics* (Kogan Page Publishers).
[13] Dües C M *et al* 2013 *J Clean Prod.* **40** 93.
[14] Sharfman M P et al 2009 *Bus. Strat. Env.* **18** 1.
[15] Yu H et al 2017 *J Clean. Prod.* **164** 1248.
[16] Mavi R K et al 2013 *Polish J Envir. Sts.* **22** 2.
[17] Rahimi M *et al* 2018 *J Clean Prod.* **172** 1567.
[18] Su C M *et al* 2016 *J Clean Prod.* **134** 469.
[19] Reefke H *et al* 2017 *Omega* **66** 195.
[20] Kudla N L *et al* 2012 J *Pur. Sy. Mgmt.* **18** 4.
[21] Calatayud A 2017 *The Connected Supply Chain: Enhancing Risk Management in a Changing World* (Inter-American Development Bank).
[22] Wu II ct al 2017 *Info.* **8** 4.
[23] Imbault F et al 2017 *1st Ieee Int. Conf. Environment and Electrical Engineering and 17th Ieee Industrial and Commercial Power Systems Europe* (Eeeic / I&Cps Europe).
[24] Price S L 2018 *Sustainable Supply Chains* (Better Global Outcomes with Blockchain).
[25] Christidis K *et al* 2016 *Ieee Access* **4** 2292.
[26] Omran Y *et al* 2017 *IPSERA* (Budapest Balatonfüred).

A Brief Conceptual View on Classification Using Support Vector Machine

***Nur Fatin Liyana* Mohd Rosely[1], *Azlan* Mohd Zain[1], *Yusliza* Yusoff[1] and *NorFadzlan* Yusup[2]**

[1]School of Computing, Faculty of Engineering, Universiti Teknologi Malaysia, Johor
[2]Faculty of Computer Science and Information Technology, Universiti Malaysia Sarawak, 94300 Kota Samarahan, Sarawak

Email: yanarosely@gmail.com

Abstract. Classifier is an important element for a classification process to classify features or samples of a dataset successfully. The existence of classifier is important, especially involved multiple class dataset with huge number of features or samples. Support Vector Machine (SVM) classifier catches more attention over other classifiers as it is widely used in various fields. Therefore, this paper aims to give a brief conceptual view of SVM in classification. The important elements and principle of SVM are also described. Along with that an appropriate evaluation and validation for SVM performance are also included. As a result, SVM concept in solving real world classification problem is identified for future reference.

1. Introduction

Classification is a supervised learning process that the input space is mapped into predefined classes and it can be done thorough supervised learning algorithm. In classification case, the learner act as function to map a vector that corresponding to its class using input-output examples. These examples consist of an input object and a desired output value. In order to accomplish this, a help from classifier is needed. A classifier is an abstract meta-class classification that describes common behavior or structural features for a set of instances. Figure 1 shows what classifier is.

Figure 1. Classification Conceptual View.

Based on the Figure 1, the input represents the data from an individual. It can be an expression profile of a tumor, meteorological measurement, digitized fingerprint scans or any type of data that technically consist several of the class. The output represents the class of the individual. In order to produce accurate classification output, a classifier is needed. There are many alternative classifiers such as k-nearest neighbor (KNN) algorithm, artificial neural network (ANN) and support vector machine (SVM). Among them, the SVM has a bias among the researchers [1, 2, 3, 4, 5, 6, 7, 8]. Further discussion about support vector machine is discussed in Section 2. Section 2 is organized into three subtopic of SVM. Performance validation offered to evaluate the SVM is viewed in Section 3. Section 4 briefly review a few related studies to SVM by previous researchers.

2. Support Vector Machine

Support vector machine is the state of the art classification techniques [9]. SVM is known as classification techniques with computational advantages over their contend showers [10]. SVM promise generalization performance in classification and it depends on the principle of structural risk minimization [4, 6, 11].

2.1. SVM Elements and Principle

In SVM there are two main elements; a) support vectors represent the points in datasets b) decision boundary represents hyperplane. The distance between all these points and the line is the maximum possible among the possible decision boundaries that lead to produce the most optimal one. The line that successfully segregates the points into classes better among the lines is defined as the optimal. Table 1 shows possible scenarios in SVM that lead to different solutions to obtain most optimal hyperplane.

Table 1. Scenario and Solution Using SVM.

Scenario	Situation and Solution
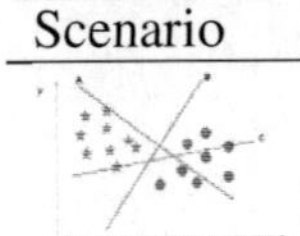	Situation: To identify the right hyper plane to classify star and circle. Solution: Follow thumb rule- Select the hyperplane which segregates the two classes better. Hyperplane B is selected.
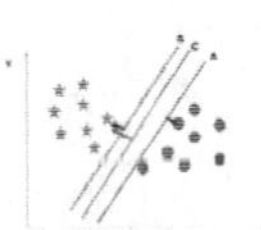	Situation: Three hyper-planes (A, B, C) to segregate the classes (Star and Circle) well. Solution: Maximize the distance between nearest data point (either class) and hyperplane. Based on figure, hyperplane C is high as compared to both A and B. Then, hyperplane C is selected.
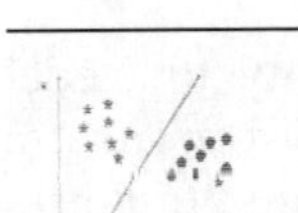	Situation: Unable to segregate the two classes using a straight line. One of the stars lies in the territory of circle as an outlier. Solution: Ignore outlier and find the hyperplane that has maximum margin. Hence, the hyperplane as shown.
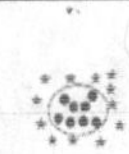	Situation: Cannot have linear hyperplane between two classes. Solution: Using kernel trick.

Based on Table 1, Scenario 1 till 3 involves linear separable data and considered as linear classification. Meanwhile, Scenario 4 is unable to be classified using linear decision boundary. In this situation, kernel plays an important role. Technically, the kernel takes two inputs, spits out them and grouped how alike they are. There are two types of kernel, single and multiple. Referred to [12] single kernel needs to choose proper parameters and multiple kernel is required to search either linear or non-linear combination of predefined base kernels by maximizing the margin maximization. Therefore, multiple kernel provides more flexibility in solving similarities of dataset compared to single kernel.

2.2. General Framework

Generally, SVM performance is depending on value of parameters in the training phase. Besides kernel function, regularization parameter (c) and gamma (g) are the important parameter in SVM. The c parameter determines the trade-off cost between the training error and the complexity of the model. Meanwhile, g parameter defines the non-linear mapping from the input space to some high dimensional feature space. Figure 2 reflects the general framework of SVM.

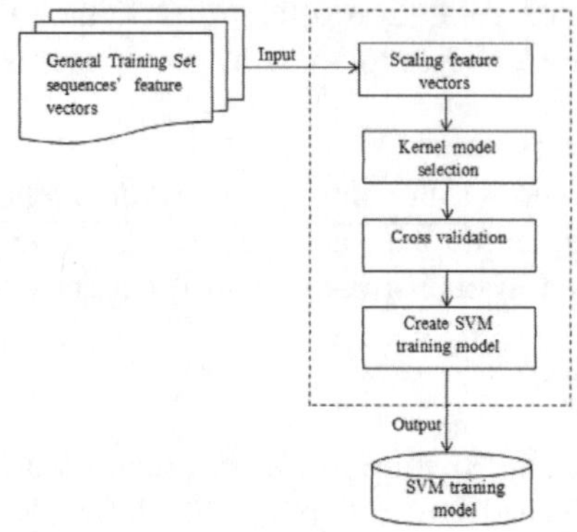

Figure 2. General Framework of SVM.

Relate in solving problem using the SVM, the general equation is defined as equation (1), equation (2) represents the hyperplane.

$$W(\alpha) = \sum_{i=1}^{m} \alpha_i - \frac{1}{2} \sum_{i,j=1}^{m} \alpha_i \alpha_j y_i y_j \langle x_i \,.\, x_j \rangle \quad (1)$$

where,

$\alpha_i \geq 0, i = 1, \dots, m \qquad \sum_{i=1}^{m} \alpha_i \, y_i = 0$

$$f(x) = < w.x > +b = \sum \alpha_i y_i \langle x_i \,.x \rangle \text{ with } \quad w = \sum \alpha_i y_i x_i \quad (2)$$

where,

weight (w) represented only by data points

kernel function $K(x_1, x_2) = \langle \emptyset(x_i) . \emptyset(x_j) \rangle$

2.3. SVM for Classification

The kernel implicitly consists of nonlinear transformation reflect SVM has the ability to model nonlinear relationship. SVM do not suffer from deficiency therefore SVM superior performance in classifying high dimension and sparse data [13]. Besides, SVM has advantages compared to others classifier because the value of c and g is chosen optimally [14]. As an example, neural network (NN) traps to local minima and reflects NN not robust over different samples.

Sahran et al. [15] proposed Support Vector Machine-Recursive Feature Elimination (Absolute Cosine) (SVM-RFE(AC)) to overcome overfitting issue due to the high dimensional texture of individual tissue components. SVM-RFE(AC) is integrated embedded procedure of SVM-RFE with filter method of AC. This integration helped prevent redundancy in selected features of SV-RFE and a un-optimised classier in the AC. The results show SVM-RFE(AC) superior to SVM and SVM-RFE. It has proven the proposed SVM-RFE(AC) able to identify of the most crucial texture of each tissue component.

Qi et al. [12] applied SVM as classifier where the process involves feature selection and multiple kernel boosting frameworks based on PSO. Then it was tested through experiment that using HyperSpectral Image (HSI) datasets as a benchmark. The results reflect the proposed approach can achieve better accuracy and efficiency than state of the art methods. Differ to Rajamohana and Umamaheswari [16] used SVM to classify the performance of proposed method which is improved binary particle swarm optimization and shuffled frog leaping algorithm. The main of objective of the study is to avoid spammers manipulate the product or service reviews in their favor either devalue or promote purposed.

3. Performance Validation

Generally, cross validation and performance measurements are used to evaluate and validate how good a classification model. As SVM is one of classification models, these also applied to SVM. Normally, cross validation involved training and testing data and performance measurement relate to evaluation of the final result of model performance.

3.1. Cross Validation

Cross validation is divided into two types, exhaustive cross validation and non-exhaustive. Exhaustive cross validation guarantee learn and test on all possible ways to divide the original sample into training and a validation set. Leave-one-out and leave-p-out are categorized into this type. Meanwhile, non-exhaustive cross validation does not compute all ways of spitting the dataset. Holdout method and k-fold cross validation are methods of non-exhaustive cross validation. Table 2 summarizes the cross validation methods.

Table 2. Summary of cross validation methods.

Method	Description
Leave-One-Out	1. Specific case of k-fold cross validation with K=N (number of data) 2. High computation time, however a good practice to validate data.
Leave-p-Out	1. In N data, N-P is taken in one iteration and the remaining are used as validation. 2. The high value of p produces more exhaustive.
Holdout Method	1. Simplest types of cross validation and using the training set only. 2. Short time to compute, but have high variance for evaluation.
k-Fold Cross Validation	1. A single subsample is reserved as validation data for testing part and remaining k-1 are treated as training data. 2. Less matter how the data gets divided and selection not bias.

3.2 Performance Measurement

Basically,the performance measurements are based on accuracy (Ac), sensitivity (R), precision (P), specificity (Sp) and F-Measure (F1).. The Ac is a metric proportion of identifying the class label of the new and invisible data. R (also known as recall) and Sp is used to evaluate the obtain feature subset. The R represents the positive and Sp represents the negative. They are divided into true positive (TP), true negative (TN), false negative (FN) and false positive (FP). True represent the correct classification and false is the incorrect classification. The P is used to evaluate the feature subset to all available features. The F1 is the combination of P and R and it known as harmonic mean. The formula of these measurements as stated as follows

$$\text{Accuracy (Ac)} = \frac{TP + TN}{TP + TN + FP + FN} \tag{3}$$

$$\text{Specificity (Sp)} = \frac{TN}{TN + FP} \tag{4}$$

$$\text{Sensitivity (R)} = \frac{TP}{TP + FN} \tag{5}$$

$$\text{Precision (P)} \quad \frac{TP}{TP + FP} \tag{6}$$

$$\text{F-Measure (F1)} = \frac{2TP}{2TP + FP + FN} \tag{7}$$

4. Related Studies

Based on the above description, this section will discuss several recent studies related to SVM. Arumugam an Jose [4] applied SVM with filter Decision tree to solve problem related to computational time of labelled training data. The modification taken by recuperate the data points that are near to the decision boundaries. This study uses 10 different dataset from https://www.csie.ntu.edu.tw/~cjlin/libsvmtools/datasets/binary.html to evaluate the proposed method. They are leukemia, Duke breast cancer, Wisconsin Prognostic Breast (WPBC), Breast Cancer Wisconsin (Diagnostic) Data Set (WDBC), Gisette,IJCNN1, epsilon, susy, KDDcup 2010 and Higgs. 10 fold cross validation is used as performance validation.

Bouazza et al [7] used SVM as a classifier to perform studies of feature selection that are applied to five different microarray datasets. The studies focused on the notion of maximum margin and kernel unction of SVM. The TP, TN, FP, FN and accuracy are used as the performance measurement. Basically, the accuracy of SVM is compared to others classifier such as K Nearest Neighbors (KNN), Linear Discriminant Analysis (LDA), Decision Tree for Classification (DTC) and Naïve Bayes (NB).

Dash [6] used SVM with Hybrid Harmony Search and Pareto optimization in order to overcome the difficulties in the selection of the most significant features. SVM proof not easily get stuck at local optimum and good approach for binary classification problem. The result through accuracy, specificity and sensitivity shows SVM performed better than other classifier (KNN, NB and ANN).

5. Conclusion

From the Section 2, the definition of the SVM concept for classification is described conceptually. The main concept and important element in SVM are explained. Through this, the readers able to notify the basic needed for a standard SVM. Meanwhile, Section 3 provides the available evaluation and validation methods for a SVM model. As conclusion, by having better understanding of the basic conceptual of SVM, the development to enhance and modify the SVM will be easier.

Acknowledgements

Special appreciation to reviewers for useful suggestions. The authors greatly acknowledge the Research Management Centre, UTM and Ministry of Higher Education Malaysia (MOHE) for financial support through the UTM Zamalah. Also the Research University Grant (RUG) for financial support through grant vot. No. Q.J130000.3551.05G41.

References

[1] Jiang H *et al* 2017 *Knowl Based Syst.* **138** 79.
[2] Jain I *et al* 2018 *App. Soft Comput.* **62** 203.
[3] Apolloni J *et al* 2016 *App. Soft Comput.* **38** 922.
[4] Arumugam P *et al* 2018 *Materials Today*: Proceedings **5** 1.
[5] Ahila R *et al* 2015 *App. Soft Comput.* **32** 23.
[6] Dash R 2018 *J King Saud Univ. Comp. & Info. Sci.* DOI.
[7] Bouazza S H 2018 *Procedia Comp. Sci.* **127** C.
[8] Tuo Y *et al* 2018 *Molecular Med. Rpt.* **17** 3.
[9] Vapnik V *et al* 1997 *Neural Info. Proc. Sys.* **9** 281.
[10] Cristianini N *et al* 1999 *Adv. Neural Info. Proc. Sys.* **11** 204.
[11] Mishra S *et al* 2015 *Int. J Mod. Sci.* **1** 2.
[12] Qi C *et al* 2017 *Neurocomputing* **220** 181.
[13] Kusairi R M *et al* 2017 *Int. J Adv. Sci. Eng. Infor. Tech.* **7** 4-2.
[14] Ren Y *et al* 2010 JCP **5** 8.
[15] Sahran S *et al* 2018 *AI in Med.* **87** 78.
[16] Rajamohana S P *et al* 2018 *Comp. & Elec. Eng.* **67** 497.

IOP Conf. Series: Materials Science and Engineering **551** (2019) 012036 doi:10.1088/1757-899X/551/1/012036

Teaching ICT Skills to Children and the Empowerment of Female College Students in STEM in Japan

Kaori Tsukazaki[1], Takeshi Shintoku[2], and Takaaki Fukuzoe[3]

[1]Department of Liberal Arts and Sciences, National Institute of Technology, Kagoshima College, 1460-1 Shinko Hayato Kirishima Kagoshima, Japan
[2]Department of Information Engineering, National Institute of Technology, Kagoshima College, 1460-1 Shinko Hayato Kirishima Kagoshima, Japan
[3]Department of Electronic Control Engineering, National Institute of Technology, Kagoshima College, 1460-1 Shinko Hayato Kirishima Kagoshima, Japan

E-mail: tukazaki@kagoshima-ct.ac.jp

Abstract. According to the Japanese government, "Society 5.0" is defined as "a future society that Japan should aspire to" and we will be able to solve our social problems with the use of Big Data, IoT, AI, and robots there. For this reason, the Ministry of Education, Culture, Sports, Science and Technology (MEXT) has focused its attention on teaching computing skills from early stages in formal education. It was decided to teach computing skills at elementary schools in 2020 and junior high schools in 2021. It will be particularly important to encourage female students to be involved in using computer skills in the future and this paper has its purpose to introduce a project of female students at a national institute of technology in Japan who teach programming a robot to younger generations. Their goal is to foster interest in STEM, particularly among female pupils and introduce them to role models through workshops because the number of female students is very low in this field in our country.

1. The Background

1.1. Society 5.0 and Teaching ICT Skills to Children in Japan

The Japanese government has started making efforts to build "Society 5.0" which was first proposed in "the 5th Science and Technology Basic Plan" in Japan. We will solve our social problems with the use of Big Data, IoT, AI, and robots in the future society as shown in figure 1[1]. For this reason, the Ministry of Education, Culture, Sports, Science and Technology (MEXT) has focused its attention on teaching computing skills from the earliest stages of compulsory education. It was decided to teach computing skills at elementary schools in 2020 and junior high schools in 2021. The curriculum at all elementary schools and junior high schools needs to be designed to suit the government guideline. It follows that it is compulsory to teach computing skills in their curriculum. It is worth noting that computing skills are regarded to be as important as language skills for children to acquire in the new guideline. The goals are to enable them to acquire basic skills in using the computer and make them aware of the ICT in their everyday life. They need to be able to think in a logical way through using the computer in the classroom [2].

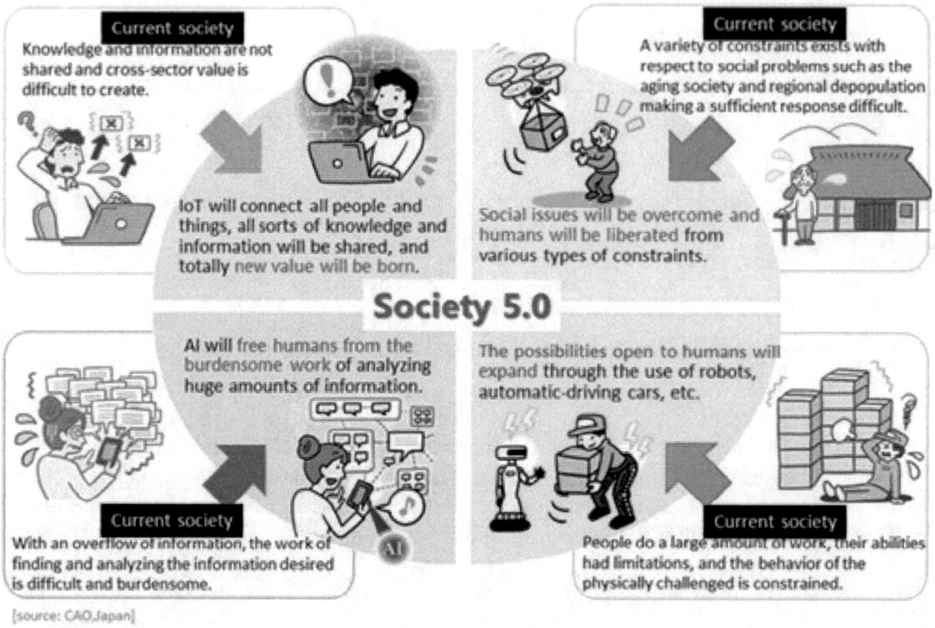

Figure 1. Society 5.0 [1].

1.2. Problems to Be Solved before the Launch of Teaching ICT at Elementary Schools

In reality, there are some problems to be solved before starting teaching computing skills formally at the earlier stage of school education. First, we lack professionals to teach computing skills to children. Teachers are not sufficiently skilled to programme the computer, and it is needless to say that they are too busy to start studying it. In addition, Kuno (2016) states that teaching programming is not systematic in formal education in Japan [3] and there is a lack of computers available in the classroom at present as Igarashi (2017) pointed out in "Software Japan 2017", a conference held by Information Processing Society of Japan [4]. For example, one of the authors was told by an elementary school teacher in a local city that elementary school students had little opportunity to use the computer. In this way, we had a request from the local boards of education for our students to become mentors for children in the computer classroom at the weekend next year. In the next section, let us look at a group of female student volunteers at a national institute of technology who teach children programming a robot with the computer.

2. The Project

2.1. Gender Issues in STEM

Some female students at National Institute of Technology, Kagoshima College, launched a project to increase the number of female students in their institution. "Gender equality" is one of the Sustainable Development Goals of the United Nations to be achieved by 2030 [5]. According to the UNESCO Science Report, Japan has the least number of female researchers among OECD countries, which is only 15%. In addition, we have the second smallest number of female researchers in engineering and technology out of 75 countries, which is only 5.3% [6]. As for undergraduate students in Japan, female students are only 15% in engineering [7]. This means that there are few role models in this field for girls in younger generations to meet, which creates a vicious cycle whereby female students do not enter such courses as STEM, especially engineering.

National Institute of Technology, Kagoshima College has the least number of female students among 51 national colleges of technology in Japan, which is currently only 10 %. A national college of technology is a five-year college, which is a combination of a high school and a junior college and a two-year advanced course follows it so that you can receive a bachelor's degree. The group of the female students at NIT, Kagoshima College would like to impact girls in younger generations and their goal is to increase this number up to 20% in the near future.

2.2. Female Student Volunteers at a National College of Technology

This is how the female college students have been trying to change the situation at NIT, Kagoshima College. They set up a chapter of a student-run international volunteer organisation which was established by Australian university students in 2008 for the purpose of increasing the number of female students in the STEM fields. Currently, it has the headquarters in Australia and more than 30 chapters all over the world in places such as Europe, Africa, North America and Asia Pacific. There are three chapters in Japan: the first one in Tokyo, the second one in Hokkaido and the third one in Kagoshima.

In their workshop, they normally use a line-trace robot to teach programming a computer. First, they show children how a robot traces a black line on a white sheet of paper. And then, participants do it themselves. To enable their robot to move properly without falling or going off the line, they are asked to check on figures of a program on the computer in order to adjust the robot's speed and turns. At the end of a two-hour workshop, their robot moves playing a melody they programmed. According to the size of a workshop, the number of volunteer students varies from 3 to 13 and participants are from less than 20 to 80.

They first had a workshop in March 2017. Since then, the female students have been regularly implementing computer programming workshops for younger students in local elementary schools and junior high schools even before 2020 when computer programming will be taught in formal education. Their purpose is to foster children's interests in STEM through teaching programming robots and to introduce them to role models through workshops. They will continue this project until the number of female student's increases up to 20% in their college.

2.3. The Result and Suggestions

This project can empower college students themselves as well as children. First, they can learn from the feedbacks of the participants. When they organise a workshop, they have meetings to tailor them to the age of the participants. They make a plan to adapt their workshop to be easy enough for children to understand, which also help them develop their ICT and communication skills. Afterwards, they try to improve their workshop based on the feedbacks. When they go to local elementary schools, they receive a warm welcome. Children look very excited during the workshop. Surprisingly, after a while, they also receive a bunch of thank-you compositions from the 80 ten-year-old elementary school students. To take some examples, their feedbacks are as follows: "How great you are to programme such a wonderful robot in a day!" "My dream is to be a cook, but I want to make a robot, too." "I am surprised that 5 girls solve a problem when their robot does not move." "How cool a robot is!" "It is a lot of fun to learn about robots. Please come to see us again." According to their teacher, it is a great experience for them to see something programmed by the computer because they have almost no opportunity to 'touch' the computer. It should also be added that one of the female students at the elementary school visit NIT, Kagoshima College with her mother to attend a programming workshop again. It seems that the female students at NIT, Kagoshima College manage to give a positive impact on her.

Second, it is possible to provide opportunities to learn to be a global citizen through this project. Since they belong to a global organisation, international members communicate with each other. When launching a chapter, they contacted with Australian members. Since then, they have kept in touch. Moreover, some of the students attended an international conference in Australia to make a presentation on this project and meet Australian university student volunteers. They shared their volunteering experiences, brushed up their teaching skills in ICT, and learned the leadership together. Others went to universities in the USA and the UK for a short term to learn communication skills and teaching ICT skills for this project.

Finally, their workshops have a good reputation in that children have fun learning IT skills in the workshops. They are asked to participate in a variety of local events in Kagoshima. It is very fortunate that they receive financial support from leaders in the industry who appreciate the importance of educating female students to be engineers. They are provided opportunities to visit their factories to have a look at the latest technologies which will surely be a part of "Society 5.0" mentioned in the previous section. Their support also enables them to go abroad such as Australia, the USA and the UK. In this way, working for the younger generations seems to lead to the female college students' own empowerment.

3. Conclusion

We may be able to conclude that this approach is innovative in Japan in that female college students are involved in the project for teaching ICT skills to children. Their purpose is to foster children's interests in STEM and show them role models in this field. We hope that they will be able to reach as many girls in younger generations as possible so that the number of female students in STEM will increase in the near future, although the number at present is much lower than that of other countries.

Acknowledgements
This work was supported by JSPS KAKENHI Grant Number JP17K18686.

References

[1] Cabinet Office n.d. *Society 5.0* Online:
https: //www8.cao.go.jp/cstp/english/society5_0/index.html

[2] Ministry of Education Culture Sports Science and Technology Japan (n.d.) *Shin gakushu shido yoryo* Online:
http://www.mext.go.jp/component/a_menu/education/micro_detail/__icsFiles/afieldfile/2018/03/30/1375607_01.pdf

[3] Kuno Y 2016,April *IPSJ Magazine* **57**(4) Online:
https://www.ipsj.or.jp/magazine/9faeag000000n6gt-att/programming-edu01.pdf [In Japanese]

[4] Igarashi T 2017, February,3 *Software Japan 2017* Online:
http://www.ipsj.or.jp/event/sj/sj2017/download/citp/Session5.pdf

[5] The United Nations (n.d.) *Sustainable Development Goals* Online:
http://sustainabledevelopment.un.org/sdgs

[6] UNESCO 2017 *Science Report Towards 2030*

[7] Ministry of Education Culture Sports Science and Technology Japan 2018 *Gakko kihon chosa* Online:
https://www.e-stat.go.jp/stat-search/files?page=1&layout=detalist&toukei=00400001&tstat=000001011528&cycle=0&tclass1=000001117475&tclass2=000001117501&tclass3=000001117502&tclass4=000001117504&stat_infid=000031738711&second2=1 [In Japanese]
https://www8.cao.go.jp/cstp/english/society5_0/index.html

IOP Conf. Series: Materials Science and Engineering **551** (2019) 012037 doi:10.1088/1757-899X/551/1/012037

A Review on Simulation and Modelling for Patient Flow in Emergency Department

Nidal Hamza [1],Mazlina Abdul Majid [1] , Khalid Adam[2] and Noor Akma Abu Bakar[1]

[1]Faculty of Computer Systems & Software Engineering, Universiti Malaysia Pahang, Malaysia
[2]Faculty of Electrical and Electronics Engineering, Universiti Malaysia Pahang, Malaysia

E-mail: nidalhamza@hotmail.com

Abstract. At the moment, most of the hospitals are more interested in patient satisfaction because this has been identified as a main issue of quality of service indexes. In most of the Asian countries' the type of registration system being operated in the healthcare systems are a sort of mixed-mode. Specifically, there are scheduled patients, as well as patients who just walk in and register. Unfortunately, this type of registration structure could increase the patient waiting period even in emergency cases. Interestingly, various methodologies have been explored in order to shorten this undesirable waiting time. This paper focuses on exploration of the characteristics and significance of existing models and simulation techniques from the literature. The combination of agent-based modelling (ABM), DES and integration of DES and ABS technique has been chosen as the solution to modelling patient emergency waiting time in the emergency department realistically. The proactive and independent characteristics of agents in both approaches will contribute to the good representation of patient emergency waiting time in the especially in the emergency department.

1. Introduction

Simulation is one of the most robust approaches that have been explored for studying the seemingly problematic nature of some complicated service and manufacturing systems [1]. In general, simulation models are constructed to evaluate the changes in a system. For example, healthcare is one of the most important service sectors. Hence the level of service provided by the healthcare sector is of great importance and it has been often evaluated through simulation studies. This is particularly because most of these important sectors could be sometimes complex. On the other hand, the cost involved with the health care sector is reasonably large. As such, different optimization processes may be carried required. These days, it is possible to obtain credible forecasting and predictions based on results generated from simulation processes [2].

In another vein, modelling and simulation and modelling make it easier to reproduce a prototype of actual systems. This may then be used for purposes including performance simulation for optimization of certain technological processes, education and safety engineering purposes, testing, as well as healthcare and video games. As such, simulation and modelling are becoming more common in the healthcare sector [3]. In fact, the need to improve the performance quality of services rendered by the healthcare sector has suggested over time. As such, healthcare providers make it a priority to provide quality services to patients especially in form of improved treatment among others [4]. Among the various factors that may influence a patient's satisfaction level, waiting time has been identified as very crucial, as reported in literature [5,6]. Many researchers have argued the effectiveness of modelling and simulation in healthcare departments, such as emergency departments (ED) [7]. Therefore, this paper aims to present a review on simulation with respect to patient's waiting time in the emergency department (ED), so as to reduce the waiting time.

The paper is organized as follows. Section 2 presents the emergency department simulation while section 3 summarises the existing literature about ED improvement by simulation modelling to shorten the patient's waiting time. Section III is followed by the conclusion of this review and the recommendations for subsequent studies.

2. Emergency Department Simulation

Healthcare is one of the notable sectors that have been extensively evaluated in recent times [7,8]. One of the major issues associated with healthcare systems is challenge from overcrowding of patients in the emergency department. This has been particularly investigated severally in many countries, especially the U.S. [8,9]. Overcrowding at the ED is such a serious problem that it holds a critical influence on patient safety. In fact, in some cases, it is capable of reducing the chances of survival for patients who are severely injured, because of time length for which they have to wait [9, 10].

3. Overview of Existing Literature

The attention of different nations has been drawn to the consequences and negative impacts of overcrowding and long waiting times in ED. As such different potential solutions have been put forward [11]. Generally, the usual causes of overcrowding in the ED include non-availability of sufficient resources [12], and increasing demand for services from the ED [13]. Notably, decision-making in EDs is usually not very easy and it is in fact a very delicate process. This is particularly based on its potential effect on the quality of services that may be obtained. It could also influence mortality levels as well as the number of unattended patients in the ED [14,15]. Generally, the challenges faced by the decision makers in the ED includes how to reduce congestion, how to shorten the patient's waiting time, and the less availability of the resources required in EDs for maintaining high standards of patient services. On the other hand, emergency care units may be somewhat complex due to the irregularities in patient arrival times, as well as lack of pre-knowledge about the type of treatment they would require. It should however be reminded that the major duty of an ED is to attend patient's emergency and life-threatening conditions rather that to manage less or mildly injured or sick patients. [14]. Figure 1 illustrates the patient flow in ED to three-parts input, processing, and output.

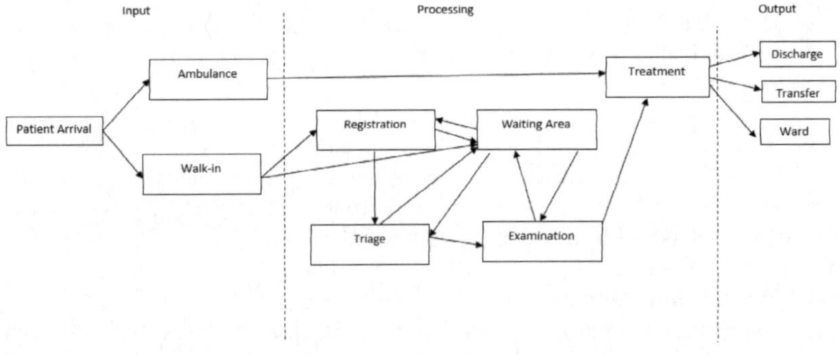

Figure 1. Overview of the patients flow in the emergency department.

Several methods have been explored for modelling the ED. Notably, three-modelling approaches have been explored such as agent-based modelling (ABM), Discrete Event Simulation (DES), and an integration of DES and ABS. In a particular study [15], the agent-based modelling (ABM) framework was proposed for use, to simulate the behaviour of patients leaving the ED of government hospitals unnoticed. This was termed leaving without being seen (LWBS). Observed that the number of LWBS patients was reduced by about 42.14%. Likewise, there was reportedly a decrease of around 6.05% in the patient's average length of stay (LOS) [15]. ABM technique is a facile tool that may be used to represent any particular environment. It is also a very important tool to assist decision-makers to evaluate the relative influence of their control systems.

As a solution to overcoming the challenges associated with patient waiting times and other related challenges, the introduction of fast track, or entirely different section was proposed for low-acuity patients in the ED. This was done through a queue-based Monte Carlo simulation in MATLAB [9]. The model combined the principles of queuing theory and broadened the discrete event simulation in order to cover the time-based arrival rates. In addition, the ED habitation and nursing resource demand were designed and evaluated using the Emergency Severity Index (ESI) levels of patients, instead of the number of bed spaces in the department. The result of the simulation revealed that the introduction of a separate fast track with an additional nurse helped to reduce the overall median waiting times.

Generally, from literature, it was observed that authors made use of a DES model in which patients just walk-in to clinics without prior schedule [16]. A case study was carried out on a walk-in clinic situated in central Texas. The research work by [8] presents the application of Business Process Model (BPM) in healthcare sector, using the Business Process Model and Notation (BPMN), coupled with a multidimensional Agent Based Model (ABM) of multidimensional organizational network of resources and geography.

The research work by [17], proposed a technique for incorporating discrete event simulation (DES) and agent-based simulation (ABS) in order to reduce patient waiting time. This was the first time such approach was attempted to solve this critical problem faced by orthopaedic departments. From the data generated, the behaviours of patients were modelled and integrated into an extensive agent-based simulation. The proposed technique helped the orthopaedic department in their analysis, and to carry out necessary modifications as required. It also helped them to be more detailed and to come out with credible results

Other than that, [18] are tried to extend the application domain of metamodels into decision-making in the EDs. This was done by fabricating a discrete event simulation (DES) model in combination with other suitable metamodels. This was used as an innovative support system for decision-making in order to improve the flow of patients, and to reduce congestion by introducing changes to the number of ED resources (i.e., the number of receptionists, nurses, residents, and beds).

4. Conclusions

Based on the review of the existing literature, it summarizes and concludes the findings from the articles. There are individual and combined techniques which has been used to model patient emergency waiting time in the emergency department. It can be seen that DES is the most popular and most frequently used by the researchers to model patient emergency waiting time. On the other hand, the integration between ABS and optimization method is the most recent and it is fast becoming a more powerful approach. However, there are few reports so far with respect to modelling patients waiting time in the hospital by using the decision-making techniques. Thus, this paper proposes that this approach should be used to model flow of patient in the ED of hospitals so as to improve the waiting time by using the simulation techniques discussed in this literature review. Furthermore, future studies should focus on investigating the simulation models for ED simulation technique which will show a good representation of the real-world patient flow in emergency department (ED) in term of waiting time. As future work, will be using the simulation technique to calibrated more EDs to validate the result adaptability. More health care sub systems, such as EDs, out-patient service units, and in-patient units should be simulated and connected together I subsequent studies. On the other hand, simulation model may be combined with optimization algorithms to optimize the performance of ED systems.

Acknowledgements
This research is financially supported by Fundamental Research Grant Scheme (FRGS) (FRGS/1/2018/ICT04/UMP/02/4) under Malaysian Government and internal research grant (RDU190167) under Universiti Malaysia Pahang. Appreciation to the reviewers and editors for considering this paper.

References

[1] Paul J A *et al* (2012) *The Journal of Emergency Medicine* **43**(6) 1119-26
[2] Bocciarelli P *et al* (2014) *Proceedings of the 2014 Winter Simulation Conference* 3012-23
[3] Freeman K *et al* (2008) *Journal of Peri Anesthesia Nursing* **23**(6) 387-393
[4] Eilers G M (2004) *Journal of American College Health* **53**(1) 41-48
[5] Rohleder T R *et al* (2011) *Health Care Management Science* **14**(2) 135-145
[6] Bernhart M H *et al* (1999) *Social Science & Medicine* **48**(8) 989-996
[7] Richardson L D *et al* (2001) *Academic Emergency Medicine* **8**(11) 1056-63
[8] Uriarte A G *et al* (2017) *Operations Research for Health Care* **15** 102-122
[9] Jun J *et al* (1999) *Journal of the Operational Research Society* **50**(2) 109-123
[10] Trzeciak S *et al* (2003) *Emergency Medicine Journal* **20**(5) 402-5
[11] Paul S A *et al* (2010) *Simulation* **86**(8-9) 559-571
[12] El-Masri S *et al* (2012) *Journal of Medical Systems* **36**(6) 3917-23
[13] Afilal M *et al* (2016) *IFAC-PapersOnLine* **49**(12) 721-6
[14] Nidal H *et al* (2018) *National Conference for Postgraduate Research 2018 (NCON-PGR 2018)*
[15] Bai J *et al* (2018) *Health Care Management Science* **21**(1) 1-24.
[16] Yousefi M *et al* (2018) *Brazilian Journal of Medical and Biological Research* **51**(3)
[17] Huang D *et al* (2018) *The American Journal Of Emergency Medicine* **36**(10) 1874-79
[18] Reese H D *et al* (2017) *Proceedings of the 2017 Winter Simulation Conference* 2764-73

IOP Conf. Series: Materials Science and Engineering **551** (2019) 012038 doi:10.1088/1757-899X/551/1/012038

Simulation and Modelling the Human Crowd Evacuation

Noor Akma Abu Bakar[1], Mazlina Abdul Majid[1] and Khalid Adam[2]

[1]Faculty of Computer Systems & Software Engineering, Universiti Malaysia Pahang
[2]Faculty of Electric and Electronic Engineering, Universiti Malaysia Pahang

E-mail: noorakma215@gmail.com

Abstract. The operational research (OR) become one of emerging areas and significance and its relevance to be used in the simulation and modelling. To simulating and modelling the crowd evacuation, the most important elements to have in a realistic model is the appropriate simulation technique. To simulate the evacuee's movement in crowd is still in research and challenge because of the emergence and become a complex task and dangerous for the real and actual case. The computational simulation technique is required in order to model the crowd evacuation as one part of OR and become a solution to represent the fire crowd evacuation in the closed space e.g. Building relates to human movement and its states. The theories and concept of computational method allows for creating, analysing and experimentation. The techniques; Agent Based Simulation (ABS), Social Force Model (SFM) and the hybrid SFM/ABS has been proposed for this research work. SFM is the well-known and popular technique for crowd evacuation while ABS is best-known, intelligent and appropriate to imitate the human movement. This paper provides a review of this research work from an OR perspective and the outcomes of a review of the computational simulation techniques literature are presented, using a proposed conceptual model will be valuable for future researchers, and modellers alike.

1. Introduction

For a several decades, the investigation of simulation and modelling has been extensively conducted in numerous areas such as in traffic flow, entertainment, crowd management, industrial operations, human actions and behaviour and other related applications [1-4]. Generally, the real system of the phenomenon or scenario will be imitated, and no real simulation needed for a computational simulation model[1]. The basic guidelines and general principles required for developing and designing the simulation model. The general idea of a simulation model is a tool to represent the real system or actual scenario. The prediction and decision making also one important element in the simulation model [1]. Throughout literature review, several organization such as the industries, academicians, and investigators researched about the human behaviour and movement as well [2].

Simulation model is used as a support tool for prediction and decision-making of the effects of any changes or optimizing the technique in terms of time, cost, effort, and identity safety. In order to produce a model and the results, the simulation models are being tested as the simulation modelling implements and experimental test [3-5]. The, computational simulation models able to obtain more understanding in a current system through the testing scenario (what-if scenario) using specific software tools. Hence, the simulations model processes have become a useful tool to overcome the issues especially in the modelling the human movement in the crowd evacuation such as building on fire [6]. The organization of this paper which is describes the significance for this study. The research work is organized as follows. Section 2 examines the related work of the research. Section 3 presents the methodology. Section 4 describes the discussion of the crowd evacuation model. Finally, Section 5 is about the discussion and conclusion for this paper.

2. Related Works

The crowd evacuation happens because of the scenarios like accident, tsunami, earth quake, congestion, clogging, egress behaviour, panic escaping, effects of obstacles, jamming, and fire [5, 7]. Thus, the need of simulation model is to define, comprehend and forecast the movement in group of crowd evacuation. The literature analysis describes that several methods have been used to simulating and modelling the human crowd evacuation in the building. This is related to the most popular OR techniques such as Discrete-event Simulation (DES) and Agent-based Simulation (ABS) [8] and the existing crowd evacuation simulation techniques such as Social Force Model (SFM), Cellular Automata (CA), Lattice Gas (LG) and etc. Furthermore, for the future work, we will come out with the significant attribute factors and performance measurement parameters which are important for this research work in order to have a closer representation simulation models on the fire crowd evacuation. The crowd can be defined as follow; when the group of objects such as people, vehicle and so on and the objects interact each other [9]. The movement state changed from normal to emergency with fast speed compared to normal [10]. Hence the crowd evacuation is the process of egress which happens in the crowd with one goal which is egress to the exit or to the safe places.

2.1. Simulation Techniques

The case study for this research work is the crowd evacuation for fire in the building and related to the human movement. Thus, this section discusses the appropriate techniques for the simulation models. This is the first phase for this research work which is to study and identify the simulation techniques. Notably, the simulations enable to predict pedestrian flows are important for the evaluation of architectural designs and operational plans [11]. The simulation techniques significant for this research work in order to modelling the crowd evacuation for building on fire of the pedestrian movement can be simplified as follows; The three simulation techniques have been proposed for modelling the crowd evacuation of fire emergency scenario which are SFM, ABS and hybrid SFM/ABS. Then, the paper will discuss the significance and motivation of these three techniques. The first one is, the three techniques are already chosen because of the other techniques such as Cellular Automata (CA) & Lattice Gas (LG) is not appropriate to use which make pedestrian walk at or within a fixed node or grid. This approach does not allow everyone to move around in an unrestricted manner (not realistic) [12, 13]. SFM and ABS have been chosen because they can deal with individual behaviour to cater the issues on crowd evacuation with the pedestrian movement. The simulation techniques are as follows; Agent-Based Simulation - A newer simulation approach, namely as Agent Based Simulation (ABS), emerged later in the 1990s [12,13] compared to other techniques. ABS able to model the human behaviour and movement without queue and it is claimed that with realistic and suitable for modelling the pedestrian traffic flow [4, 14]. Social Force Model - The most popular simulation technique in crowd evacuation is Social Force Model and simplified as SFM. SFM is a behaviour modelling method which developed with Molnar and Vicsek. The subject is the human pedestrian and the movement in the crowd. The social force motivates the pedestrian for motion planning in crowd. Human behaviour modelling refers to computer-based which imitate either the behaviour of a single person or the collective actions of a team. SFM is continuous in time, space and appropriate for pedestrian dynamic. Proposed Human Crowd Evacuation Model - The critical review from our research work on the literature, it can be summarized as follows; the selection of the simulation method relies on the experience on the simulation techniques and individual assessment of the modeller. Other than that, the question arises as when and what issues SFM can be chosen and how about the ABS and when it should be preferred? ABS has been chosen by and SFM has been chosen by in several simulation models of crowd evacuation in the building e.g. shopping mall, office building, and airport, then at the railway and subway station, road and tunnel evacuation, event disaster evacuation (e.g. flood, tsunami, earthquake), and the vehicle evacuation (e.g. flight, bus and ship).

3. Methodology

In this section, it describes the flow of methodology of this research work. Firstly, the case study description and secondly is data collection on the crowd evacuation, and the computation simulation techniques. The final one is the designing the model for the fire crowd evacuation. The details of the processes are as follows; Search strategy; throughout this process, the term 'crowd evacuation' 'crowd modelling' 'evacuation model' 'simulation model' denotes as the keywords in this paper. In any literature review, the choice of papers and the related inclusion/exclusion criteria with the goal to minimize this by requiring that every paper selected for full-text reading. It is mainly emphasizing on the conclusions of previous studies and the predominant thrust in narrative literature reviews. In this data collection phase also include the process of identifying the significant elements for the crowd evacuation simulation model together fit with the proposed computation simulation techniques. Then, designing the conceptual model for the crowd evacuation model. The last phase it the results of the simulations.

4. Simulation Model

It is important to realize that a crowd evacuation simulation model consists of the crowd evacuees' modules. The modules consist of general movement in the crowd evacuation of the pedestrian (human). Other than that, the significant element for the model required the movements of the people for egress and the evacuation route plan. The theory on the influence of the social force from SFM model is needed for the crowd evacuation model; humans are communicating and interacting to each other during the evacuation process. The models contain simulation technique and each model also has this movement state of evacuees. The basic state should have for each evacuee is from normal can be idled or evacuated if the fire alarm triggered. The idle state from evacuation will be happen while waiting for route searching. The conceptual model for the fire crowd evacuation in the building can be reviewed as in figure 1. The models with three different techniques which are SFM, ABS and hybrid model developed to simulate the case study. Each simulation model required the important elements which obtained from the literature review. The model should have attributes factor, the constant variables and the performance measurement parameters (PMP). These elements will be described further in the next publication as one of our future works. Important to realized that each simulation technique for a model contain the rules based on the specified technique such as SFM, ABS and hybrid model.

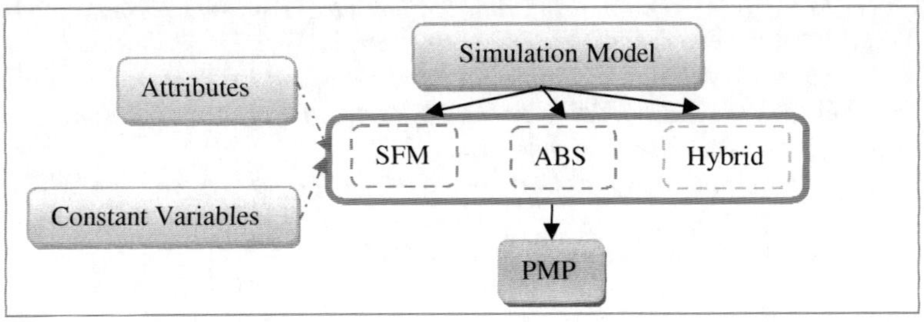

Figure1. The conceptual model for fire crowd evacuation.

5. Conclusions

The conceptual model relates to the OR element which is simulation and modelling for the described case study. In order to have the simulation model for replicating the real system, the simulation techniques required; SFM, ABS and hybrid model. These models needed the attributes as the factors before the simulation will be ran together with the constant variables to obtain the simulation results. The results will be analysed using statistical analysis based on the performance measurement parameters (PMP). The expected result shows that the hybrid model obtains the better result compared to the other two traditional and single simulation technique (SFM and ABS). As conclusion, this research work focuses on the simulation model for modelling the case study to evaluate the performance of the simulation techniques. The simulation techniques have been proposed as SFM, ABS and hybrid model. Throughout the literature, SFM and ABS are the most frequently chosen by the researchers. In addition, the literature indicated that hybrid model with the better ways, powerful, popular choice, and flexibility as expressed by [2] for the better result. The investigation on the performance of each simulation technique is based on the attribute factors and the output for the simulation model is the performance measurement parameters (PMP). The future works will reveal the specific attributes, constant variables and PMP with the simulation and statistical results. This will proof that hybrid model is the better way for modelling the real system which is will be collected through the simulation results.

Acknowledgements
This research is financially supported by Fundamental Research Grant Scheme (FRGS) (FRGS/1/2018/ICT04/UMP/02/4) under Malaysian Government and internal research grant (RDU190167) under Universiti Malaysia Pahang. Appreciation to the reviewers and editors for considering this paper.

References

[1] Majid M A *et al* (2013) *International Journal of Multimedia and Ubiquitous Engineering* **8** 329-338
[2] Ijaz K *et al* (2015) *International Conference on Modelling and Simulation* 111-116.
[3] Wang X *et al* (2014) *IEEE International Conference on Digital Home* 347-352
[4] Viswanathan V *et al* (2014) *The European Physical Journal B* **87** 1-11
[5] Bakar N A A *et al* (2018) *Advanced Science Letters* **24** 7611-7616
[6] Wageningen-Kessels F V *et al* (2016) *Physica A: Statistical Mechanics and its Applications* **443** 272-285
[7] Wu S *et al* (2016) *Proceedings of the 33rd International Conference of CIB W78* **12** 1-11
[8] Majid M A *et al* (2016) *International Information Institute (Tokyo). Information* **19** 4213
[9] Helbing D *et al* (2013) *arXiv preprint arXiv:1309.1609*
[10] Johansson F *et a*l (2015) *Physica A: Statistical Mechanics and its Applications* **419** 95-107
[11] Davidich M *et a*l (2013) *Transportation Research Part C: Emerging Technologies* **37** 210-222
[12] Radianti J *et a*l (2014) *Applied Intelligence* **42** 3-23
[13] Anh N T N *et a*l (2012) *International Conference on Autonomous Agents and Multiagent System* **7068** 371-383
[14] Xiong M *et al* (2013) *New Generation Computing* **31** 211-235

IOP Conf. Series: Materials Science and Engineering **551** (2019) 012039 doi:10.1088/1757-899X/551/1/012039

Deep Learning Layer Convolutional Neural Network (CNN) Scheme for Cancer Image

Z Zainudin[1], S M Shamsuddin[1] and S Hasan[1]

[1] School of Computing, Faculty of Engineering, Universiti Teknologi Malaysia

E-mail: zanariah86@gmail.com

Abstract. Recent years, in medical image especially cancer detection used whole slide digital scanners, called as histopathology image (images of tissues and cell) where it can now be keep in digital images. Consequently, using Deep Learning will help pathologist in cancer detection (cancer cell known as mitosis). In this paper, we are using Deep Learning Layer Convolutional Neural Network (CNN) for cancer classification using histopathology image and used AMIDA dataset which are related on female breast cancer dataset. Mitosis is an important parameter for the prognosis/diagnosis of breast cancer. However, using histopathology image for cancer detection is a challenging problem that needs a deeper investigations. This problem occurs when to classify mitosis because mitosis is small objects with a large variety of shapes, and they can thus be easily confused with some other objects or artefacts present in the image. In this paper, the objective to find the suitable layer for Deep Learning Convolutional Neural Network and reduce the loss rate.

1. Introduction

Cancer has become the most popular health problem that can be dangerous to human life especially for women. This is supported by the statistic from Europe and United States health reports in year 2018 and year 2017. In year 2017, table 1 show that 30% women's from 852,630 women's has been estimated getting breast cancer. Table 1 also shows that year 2018, it stated that breast cancer is the number one based on the population statistics in Europe and United States population. Based on this report, the study conducted by National Cancer Registry (NCR), US women are more exposed on breast cancer where the number of the new cases are slightly increases, where 878,980 new estimated cases involved in breast cancer are reported to NCR for year 2018 and the number keep increasing in last two years. The number of new cases of female breast cancer was increased to 852,630 in 2017[1,2].

Table 1. Estimating New Cancer Cases in the US in 2018 and 2017.

Female New Cases	**2018 (878,980 person)**	**2017 (852,630 person)**
Breast	30%	30%
Lung & Bronchus	13%	13%
Colon & Rectum	8%	7%
Uterine Corpus	7%	7%
Non-Hodgkin Lymphoma	4%	4%

*supported by Europe and United States health reports

Su et. al., 2015 [3] has done experimental using deep convolutional neural network and the experiments excution time is 2.3 seconds using the High Performance Computers (HPC). Meanwhile, Adriana Romero et. al, 2014 also stated that using trained model network are verfy effective and sugessted that using the spatial information to get better result, combine the output feature and max pooling steps and adding the new layer to deep learning to improve kappa agreement score. Chan et. al., 2014 [4] used Deep Learning called PCANet and the experiments results shows that for pattern recognition 99.58% for Extended Yale B Dataset, and 86.49% accuracy using 60% Extended Yale B Dataset for training. On the other side, Couprie et. al., 2013 [5] also tested using Deep Learning for Stanford Background Dataset and getting some over segmentation for some images. Sukre et. al., 2015 [6] also using Deep Learning Restricted Boltzmann Machine and has better performance measure. Wahab et. al, 2017 [7] also mentioned that using Deep Learning Convolutional Neural Network getting better training for the image dataset. Some researchers also stated that using complex image can effects the performance measure of classification.

Based on previous researchers, it can be concluded that using Deep Learning algorithm can improve performance detecting the cancer cell. It also support the evidence where the misclassification of the images are based on many criteria such as quality of images, lacking of physician eye, over or poor segmentations using histopathology image and many other cause where it can lead to incorrect misclassification

2. Research Methodology

In this subsection, the research methodology for these studies is presented below. The main goal of these studies to detect/classify the cancer cell correctly. In this paper, the research framework were carried out based on:

- **Phase One**: Preliminary Investigation
- **Phase Two**: Dataset Collection
- **Phase Three**: Cancer Image Detection using Deep Learning Layer Convolutional Neural Network (CNN) algorithm
- **Phase Four**: Performance measurement
- **Phase Five:** Result and discussion

The first phase in this research is the preliminary investigation which also called as literature review where it can give a clear understanding how to setup the deep learning algorithms and to know the main objective of this research. In phase one, main objective in this research work is been identified where the aims to detect/classify the cancer cell correctly. This is really important to get research gap and proper solution where it can solved the histopathology image problem. After that, phase two will be started as the dataset applied in this studies were been collected from public dataset available in previous studies. In this paper, dataset have been collected from freely database such as

AMIDA dataset. In phase three, experimental has been run to create the deep learning classifier for each class in each dataset, able to find good accuracy and able to reduce the loss rate using AMIDA dataset. Performance measurement is the fourth phase where all the experiment result will be evaluated when used the Deep Learning Layer Convolutional Neural Network. The proposed research methodology is been illustrated in figure 1 for detailed explanation.

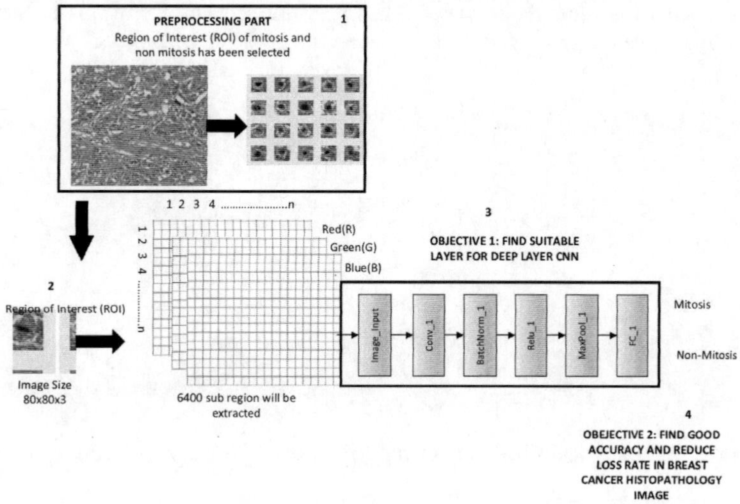

Figure 1. Accuracy and Loss Rate using 20-Layer CNN.

The CNN architecture for this research contain few layers and parameters which is described more detail below:

- *Input Layer:* Input layer where the size and the pixel of the input image is been defined. The height, weight, and the number of colour channels of that image is been sorted out whether is RGB (3) or grayscale (1). Input layer is where the network been specified to execute data augmentation. Data augmentation is used when image dataset are small, where the image data been transformed, flipping or been cropped in the training set. Input image are been set to 80x80x3.
- *Convolutional Layers*: Convolutional layer will convolves from previous layer where input image from the input layer to several feature map in the output image. For this deep learning model, we used three layers which is the first two layer will learns all the feature maps and initialized the Gaussian distribution with standard deviation of 0.001. Then, the third layer will learns all the filters from the Gaussian distributions with standard deviation of 0.001. The outcomes from the neurons of a convolutional layer more often will pass form of nonlinearity model.
- *ReLU Layers:* ReLU layer is an activation function for nonlinearity model. The ReLU layer will performed a threshold operation to each elements, where the input value less than zero will be set to zero based on the equation (1) below:

$$f(x) = \begin{cases} x, & x \geq 0 \\ 0, & x < 0 \end{cases} \tag{1}$$

- *Max Pooling Layer*: This layer will pool all maximum value from all the elements inside previous layer
- *Fully Connected Layer/Output Layer*: The last layer for CNN is fully connected Layer. It depend on the output of the classification, for example in this research we used two output mitosis and non-mitosis.

3. Experimental results and Discussions

The main objective of this studies to find the most suitable layer for Deep Learning Layer to detect the mitosis from histopathology image and also reduce Loss Rate. The AMIDA dataset is split into two type of dataset which include training dataset contain 70% and testing dataset containing 30%. The goal to split the data is to ensure the learning generalization has been achieved by feeding the classifier to new datasets rather that the same dataset which has been trained before. In this experiments, two different type layer of deep learning is been used.

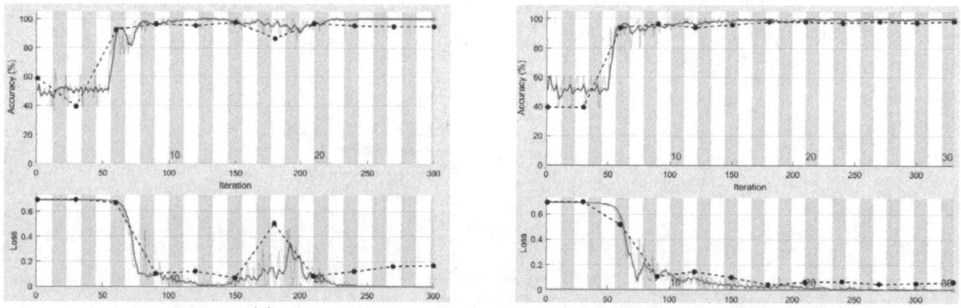

(a) (b)

Figure 2. Accuracy and Loss Rate using (a)13-Layer CNN and (b) 20-Layer CNN.

Figure 2(a) gives the accuracy of CNN using deep learning algorithm of Convolutional Neural Network (CNN) using 13- Layer CNN. Figure 2(b) gives the accuracy of using deep learning algorithm and show the loss using deep learning Convolutional Neural Network (CNN) using 20-Layer CNN. Based on the all the figure above, we can see that using CNN classifier has a better result by using 20-layers CNN give better result compared to 13-layers CNN. Table 2 shows the detail experiments where we can compare the accuracy from two type of layer for each iteration.

Table 2. Experimental Result of 13-layer CNN and 20-Layer CNN.

.Iteration	13-layer CNN	20-layer CNN
60	92.40%	94.12%
120	95.34%	94.12%
180	86.52%	98.04%
240	95.34%	97.30%
300	94.85%	97.55%

This is due to more training layer where all the images getting more training compared to13-layer CNN. This result from this experiments, shows some significant value where it have proven using 20-Layer CNN result also helps in reducing Loss Rate.

4. Conclusions

In this paper, Deep Learning Layer Convolutional Neural Network (CNN) been used for detecting the breast cancer mitosis and non-mitosis performance comparison between two types of CNN layer and more hidden layer are better as we can see that from the experimental results. In this paper, we used two different type layer of CNN which is 13-layer and 20-layer to see the different for each layer either give the best result or not. In conclusions, we are able to achieve the objective of this studies which we able to find suitable layer Deep Learning Convolutional Neural Network for histopathology image and also success in reducing the Loss Rate. This is due to CNN architecture allows the more training inside the hidden layer which is include the Convolutional Layer, ReLU layer, Softmax Layer, and Fully Connected Layer. Therefore, for future work we will improved some work such as: 1) use GPU machine for running deep learning of CNN algorithm for better computational time, and 2) tackle the imbalanced number of distribution of mitosis and non-mitosis.

Acknowledgements
This work is supported by Ministry of Education (MOE) and Universiti Teknologi Malaysia (UTM). This paper is financially supported by MYBRAIN, Grant No. 17H62, 03G91, 13H82. The authors would like to express their deepest gratitude to the researchers who providing the AMIDA datasets to ensure the success of this research.

References

[1] Howlader C K N *et al* 2015 *Cancer statistics review 1975-2012: introduction* 1–101
[2] Siegel R *et al* 2014 *CA. Cancer J. Clin.* **64**(1) 9–29
[3] Su H *et al* 2015 *IEEE 12th Int. Symp. Biomed. Imaging* 55–58
[4] Chan T *et al* 2014 *arXiv Prepr.* 1–15
[5] Couprie C *et al* 2013 . *Pattern Anal. Mach. Intell. IEEE Trans.* **35**(8) 1915–1929
[6] Sukre K M *et al* 2015 *Int. J. of Technical Research and Appl.* **31** 111–116
[7] Wahab N *et al* 2017 *Comput. Biol. Med.* **85** 86–97

IOP Conf. Series: Materials Science and Engineering **551** (2019) 012040 doi:10.1088/1757-899X/551/1/012040

The Performance Review of mRMR for Gene Selection and Classification of DNA Microarrays

Norfadzlan Yusup[1] and Azlan Mohd Zain[2]

[1]Faculty of Computer Science and Information Technology, Universiti Malaysia Sarawak (UNIMAS), 94300 Kota Samarahan, Sarawak, Malaysia
[2]Applied Industrial Analytics Research Group (ALIAS), School of Computing, Faculty of Engineering, Universiti Teknologi Malaysia (UTM), 81310 Skudai, Johor Darul Takzim, Malaysia

E-mail: fadzlanyusuf@gmail.com

Abstract. There are two main stages in the classification of DNA microarray data. The first stage is known as gene selection and the second stage is the classification of selected genes. The number of genes produced in high-dimensional microarrays is enormous, and only some of these genes help to identify a particular disease. The selection of relevant or informative genes that provide sufficient information about the condition is therefore essential. Gene selection is vital in reducing the data dimensionality which can ease the workload of the computer and increase the high classification performance. In this paper, we review the recent performance on one of the most popular filter based gene selection technique, maximum relevance minimum redundancy (mRMR). We also discuss several current improvements on the mRMR method.

1. Introduction

For Deoxyribonucleic Acid (DNA) microarrays are extensively used in the medical field and considered vital because they enable researchers to 1) identify what caused the diseases 2) how the diseases are classified and 3) how the diseases are treated. The classification is an essential problem in the DNA microarray [1] and currently has received the most attention in the context of cancer research. Gene selection is the process of choosing the most relevant genes in a microarray dataset, and it is an essential task for classification and dimensionality reduction. If the datasets contain irrelevant genes, it will affect not only the training of the classification process but also the accuracy of the model.

Functional classification accuracy is achieved when the model correctly predicted the class labels. Generally, there are three categories of gene selection methods which are Filter, Wrapper and Embedded. Currently, there are two new methods for gene selection; Hybrid and Ensemble method. Classification is a data mining technique that extracts models(classifier) describing significant data classes classifiers where it predicts categorical class labels [2]. There are many types of classifiers used for classification of DNA Microarrays such as Support Vector Machine (SVM), Naïve Bayesian (NB), Random Forest (RF) and Artificial Neural Network (ANN).

2. Filter-based Gene Selection

In the filter method, the gene selection process is independent of the learning process. Gene selection using filter method tend to select redundant features because it did not consider the interactions between features. Once the best features are selected, it will be ranked and evaluated by using either univariate (e.g. Relief F) or multivariate (e.g. mRMR) filter method. Filter method did not necessarily use with classifiers; therefore, it usually used as a pre-processing step [3]. Filter method is computationally less complicated and faster than the wrapper method.

Filter method such as mRMR is employed to generate a subset of relevant gene. mRMR was initially proposed by [4] to reduce the number of genes selected in microarray data for classification. mRMR was also introduced to solve the redundancy issues in ranking approach where the features could correlate among themselves. The fundamental problem with redundancy is that the feature set is not a comprehensive representation of the characteristics of the target phenotypes [4][5]. In mRMR, features that are different to each other are maximized (e.g. mutual Euclidean distances) or their pair-wise correlations are minimized to expand the symbolic power of the feature set. These minimum redundancy criteria are supplement by the usual maximum relevance criteria such as maximal mutual information with the target phenotypes[6].

3. Gene Selection using mRMR

In classifying DNA microarray data, [7] defined the mRMR process as follows. The mRMR method used two mutual information (MI) operations: one between cancer classes and each gene to measure the relevancy, while the second mutual information between every two genes to calculate the redundancy. Therefore, S denotes the selected genes and Rl measures the relevancy of a group of selected genes S that can be defined as follows:

$$\mathrm{Rl} - \frac{1}{|S|}\sum_{G_x \in S} I(G_x, C), \tag{1}$$

Where $I(Gx\ , C)$ represents the value of mutual information between an individual gene Gx that belongs to S and the cancer class $C = \{c1, c2\ \}$, where $c1$ and $c2$ denote the normal and tumor classes. When the selected genes have the maximum relevance Rl value, it is possible to have a high dependency (i.e., redundancy) between these genes. Hence, the redundancy Rd of a group of selected genes S is defined as,

$$\mathrm{Rd} = \frac{1}{|S|^2}\sum_{G_x, G_y \in S} I(G_x, G_y), \tag{2}$$

Where $I(Gx, Gy)$ is the mutual information between the xth and yth genes that measures the mutual dependency of these two genes.

4. Classification Performance and improvement on mRMR technique

In this section, we review some of the recent classification performance of DNA microarrays data using mRMR as a gene selection technique. In [7], mRMR was employed to select the most informative genes in six DNA microarray datasets. In this research, mRMR was used to predict top relevant genes that give 100% accuracy. In [8], mRMR with SVM and kNN classifiers were used to choose n top genes of the relevant genes in microarray data. From the result, the number of selected genes by mRMR-SVM is smaller than those of mRMR-kNN with more than 70% classification accuracy achieved in four datasets. In the research by [9] and [10] also achieved excellent results of mRMR classification using SVM with an average accuracy of 93.06% and 84.15% for all datasets with 50 top-ranked selected genes. In [11] the performance of mRMR experimented on six DNA microarray datasets. Based on their experiments, mRMR with NB classifier performs better than 1NN classifier concerning classification accuracy and the number of selected genes. The performance of mRMR is also better than Relief F and Fast Correlation-Based Filter (FCBF) concerning classification accuracy and the number of selected genes.

The authors in [12] review the performance of four feature ranking (IG, Relief-F, mRMR, SVM-RFE) based on top 10 and top 50 features in nine DNA microarray datasets. The authors also review another three standard feature selection technique (CFS, FCBS and INTERACT). Using mRMR as feature selection, the results of Distribution optimally balanced stratified cross-validation (DOB-SCV) with SVM classifier is better than C4.5 and NB particularly for Colon, Gli85 and Ovarian. The authors in [13] analyze the feature selection techniques of Random Forest Feature Selection (RFFS), Random Forest Feature Selection-Grid Search (RFFS-GS), and mRMR algorithm for comparative experiments in two gene expression datasets. Regarding classification accuracy and Area Under Curve (AUC), mRMR achieved the best results in breast cancer datasets. In [14] the authors compare the performance of four feature selection techniques of mRMR, Chi-Square, Relief F, Effective Range Based Feature Selection (ERGS) with their proposed inter feature effective range overlap technique with top 10 to 60 features selected using NB classifier. With top 10 selected number of features, the results of mRMR are better than other techniques mainly in 11_Tumors, Brain Tumor 1 and Prostate Tumor datasets. The mRMR was also employed in [15] where it identifies the top relevant gene that gives 100% classification accuracy using SVM classifier. mRMR is an advanced processing (filter) stage in which a new subset of data is subsequently transmitted to the wrapper algorithm (e.g. Genetic Bee Colony (GBC).

There are a few improvements that have been proposed by some researchers to improve the performance of mRMR further. Although mRMR is a fast and greedy heuristic, it does not guarantee to find a globally optimal solution. In [16] mRMR was extended by adding a greedy search for mRMR. The experiments used five DNA microarray datasets, and the performance was evaluated based on selection time for 200 features. The proposed approach outperforms the mRMR drastically concerning computation time across three different platforms (CPU, GPU and Parallel Computing). The mRMRe parallelized was applied in [17] with Random Forest classifier to classify Cancer Genome Project (CGP) and Leukaemia microarray data. From the research, the reported selected number of genes is 10, and the out of bag (OOB) error was 0.06 and 0.02 for CGO and Leukemia data respectively. From the literature, the number of the selected gene varies among the researchers. The researchers specified no standard number of selected top *n* genes. For *n* number of the top-ranked gene is usually chosen between 10-50, 50-100, and more than 100 genes. A high number of genes chosen usually gives high classification accuracy [7] [18] [15]. In the research [7] [9], it was reported the standard number of the top-ranked gene to be selected is 50. Researchers mostly prefer the classification of using SVM classifier due to its stability, and it shows good results with mRMR compared to another classifier such as NB and *k*N[8]. From our review, the mRMR method is usually employed as a filter in the hybrid filter-wrapper classification because using mRMR single-handedly does not guarantee to find a globally optimal solution. In hybrid filter-wrapper approach, mRMR was used to reduce the number of the feature set. Then, this reduced feature set will be fed to the next process using various metaheuristics algorithm [19][20].

5. Conclusions

In the literature, mRMR is one of the most popular multivariate filter technique as gene selection in classifying DNA microarrays. The advantage of the mRMR filter method is it can reduce the redundancy in the feature set. However, mRMR does not guarantee to find a globally optimal solution. Therefore mRMR is usually applied as a pre-processing (filter) process in the hybrid filter-wrapper feature selection model. mRMR was employed to reduce the number of irrelevant genes in microarray datasets, and then the selected genes are fed to the wrapper process. Using wrapper (e.g. metaheuristic algorithms) directly is inefficient due to the high dimensionality of microarray data. Several improved mRMR techniques have been implemented to extend mRMR performance. These improved techniques should be explored further in various type of microarray datasets to find the most informative genes in classifying DNA microarrays.

Acknowledgements
The authors would like to thank the editors and reviewers for their valuable comments. We also would like to thank Universiti Teknologi Malaysia (UTM) for providing Transdisciplinary Research Grant (TDR), Grant No: Q.J130000.3551.05G41

References

[1] Miller S *et al* 2015 *Methods in Microbiology.* **42** 395.
[2] Han J *et al* 2012 *Data Mining: Concepts and Techniques*. Morgan Kaufmann.
[3] Armina R *et al* 2017 *J. of Physics*. 892.
[4] Ding C *et al* 2003 *Comp Sys Bioinf. Proc of the IEEE Bioinfo Conf.* Stanford, CA, USA, **3** 523.
[5] Zhou K Q *et al* 2016 *Art Intel Rev*. **45** 405.
[6] Shayfull Z *et al* 2014 *Advances in Polymer Technology.* **33** 1.
[7] Alshamlan H *et al* 2015 *BioMed Res Int.*
[8] Mohamed N S *et al* 2017 *Exp Sys with App*. **90** 224.
[9] Alomari *O A et al* 2018 *Applied Intelligence*, June.
[10] Angulo A P 2018 *Inf (Switzerland)*, **9** 1.
[11] Wang A *et al* 2015 *J. of Comp Inf Sys.* **5** 1563.
[12] Bolón-Canedo V *et al* 2014 *Inf Sc*. **282** 111.
[13] Zhang Y *et al* 2018 *BioMed Res Int.* 7538204 .
[14] B. Chandra *et al* 2018, *IEEE Conf on Comp Intel in Bioinfo and Comp Bio*. 1–6.
[15] Alshamlan H M *et al* 2015 *Comp Bio and Chem*. **56** 49.
[16] Ramírez-Gallego S *et al* 2017 *Int J. of Intel Sys*. **32** 134.
[17] Kusairi R M *et al* 2017 *Int J. on Adv Sci, Eng and Info Tech*. **7** 1595.
[18] Du W *et al* 2013 *Int J. of Data Mining and Bioinfo*. 7 58.
[19] Lazim D *et al* 2017 *Art Intel Rev.* 10.1007/s10462-017-9580-4.
[20] A. Deris *et al* 2017 *J. of Physics: Conf Series*. **892** 1.

IOP Conf. Series: Materials Science and Engineering **551** (2019) 012041 doi:10.1088/1757-899X/551/1/012041

AI for Heart Rate Measurements for Sport Performance: A review

Sharifah Zarith Rahmah Syed Ahmad[1], Yusliza Yusoff[1], Azlan Mohd Zain[1], Ruhaidah Samsudin[1], Nurzal Effiyana Ghazali[2]

[1]School of Computing, Faculty of Engineering, Universiti Teknologi Malaysia, Johor
[2]School of Electrical, Faculty of Engineering, Universiti Teknologi Malaysia, Johor.

E-mail: alishazel@gmail.com

Abstract. Increasing sport performance in athletic has been a major task for coaches around the world. The ability to improve athletes' performance benefits not only the coaches but to the athletes themselves. Heart rate analysis for sport performance has been investigated for many years. Heart rate plays an important role to measure the athletes' ability to perform in sports of their choices. The heart rate information will not only aid in training the athletes according to personal abilities but it will also assist in monitoring athletes' health by reducing the chances of cardiovascular mortality risk. This paper briefly outlines the potential of Artificial Intelligence (AI) in learning information contain in human heart rate measurement for improving the athlete's sports performance.

1. Introduction

Heart rate monitoring plays important roles in as sports performance training aid. Stork [1] explains that heart rates always related to physiological limits, thus heart rate are suitable to measure athletes' performance. Wearable technologies such as smart sport watches, power and cadence meters, and heart rate monitors band are connected to training technologies has made heart rate measurement becomes easier [2]. The heart rate determines several information for athletes, it has been use to trace medical condition [3], achieve desire result in planned training [4], monitoring athlete's performance [5] and to prevent sudden cardiac death [6]. There are considerably several factors that affected the athlete's performance among those are physical, technical, tactical and psychological as shown in Figure 1. The heart measurement is apart of cardiovascular system [3].

2. Heart Rate Measurement

Heart rates (HR) is a measurement of heart beats; the state of HR depends on how many times a heart beats per minute (BPM). HR and parameters such as oxygen intake (VO2) and lactate concentration are normally used to predict individual fitness level [7].Typically, in order to improve a person aerobic fitness, it was suggested that the HR of a person during training should be consistent. The value should represent 70% of their maximum HR as the value is unique for each individual [3]. The equilibrium for HR for individual is as follows:

$$hr_{min} \leq hr\ (u,t) \leq hr_{max}\ [beats.min^{-1}] \quad (1)$$

Where;

$hr\,(u,t) = hr_{min}$ Resting Heart Rate
$hr\,(u,t) = hr_{max}$ Maximum Heart Rate
u Exercise intensity
t Time, Duration

The heart rate equilibrium has been used by few researchers to measure HR for athletes [1].

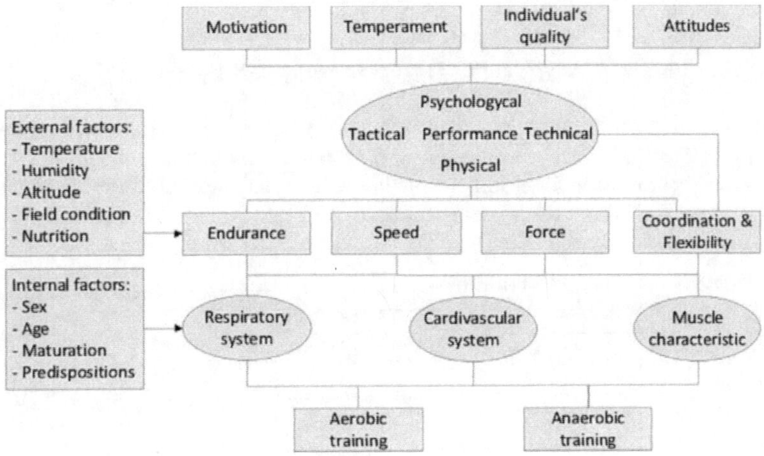

Figure 1. Key components of sports' training performance [3].

Adjunction to HR measurement, Heart rate variability (HRV) is variation in the time interval between sequential heartbeats or commonly known as R-R interval [8]. HRV allows predicting cardiovascular outcomes. Conclusively, HR, HRV and R-R interval plays important roles in reading of cardiovascular abilities during psychical activities. The conventional HR Measurement are by using index finger and middle finger on the pulse nodes such as wrist, lower neck or wind pipes. In clinical setting electrocardiograph or ECG is use as it generates a graph pattern which shows heart function where it is more reliable to calculate HRV and R-R interval.

In recent years, HR monitor is accessible commercially for normal consumer. Practical wearable device like chest electrodes band HR monitoring system, sport shoes, sports watches, HR band and head phones. All these devices provide the data is real time and viewable providing it is connected to mobile phones or personal computers. The more recent technology in wearable devices are smart clothing where it caches health information namely HR information to mobile, mobile cloud from time to time and alerted authorities if anything happens to wearer [23][24].

3. Heart Rate Variability (HRV)

Heart rate Variability (HRV) is use to monitor autonomic nervous system where it is affected by stress, cardiac diseases or pathological states [9]. ECG recording also allows researcher to study the R-R interval as it always associates with body position of athlete which affecting Blood Pressure [10]. There are 4 ways to conventionally measure HRV for analysis; The time domain methods, Frequency domain methods, Non-Linear and Rhythm patterns analysis. All these methods required analysing graph produce by electrocardiogram (ECG) [11]. Figure 2 displays R-R interval estimation from ECG graph.

IOP Conf. Series: Materials Science and Engineering **551** (2019) 012041 doi:10.1088/1757-899X/551/1/012041

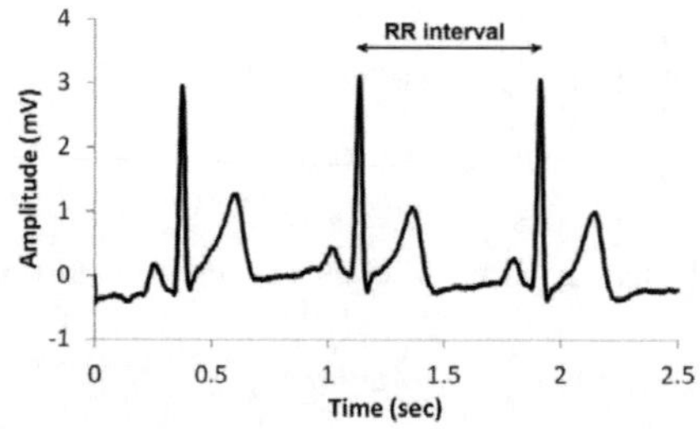

Figure 2. A typical ECG signal showing R-R Interval [25]

Figure 3 shows steps taken to measure HRV by 2 conventional methods; Time domain method and frequency methods. The ECG recorded requires 5 minutes under stable condition for frequency domain methods or 24 hours for time domain methods. The data was digitalised, edited and follow with reading of Normal to Normal (N-N) data sequence.

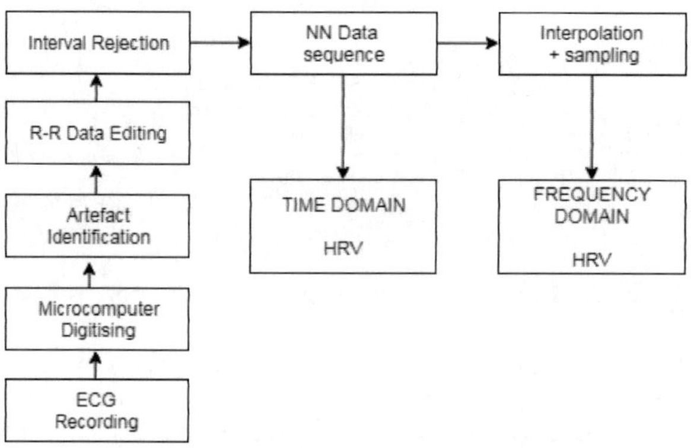

Figure 3. Steps to obtain individual ECG signal for HRV Analysis[11]

4. Artificial Intelligence and Heart Rate Measurement for Sports Performance

Artificial Intelligence (AI) is statistical application to through human machine collaboration [12]. This section briefly discusses on possibilities of information produce by learning Heart rate (HR) information by using AI. HR information plays important role in sport performance as physical activities influence the HR patterns. Thus, researchers start to predict HR rate by using Support Vector Regression [13] and Artificial Neural Network (ANN) to measure HR responses to exercises [14]. This is due to the nature of ANN where input and output data is easy to construct [15].

Xiao [22] uses Evolutionary Neural Network to model HR prediction. The HR prediction is based on physical activities data retrieved from alive HR monitor. García-García [16] uses Principle Component Analysis to calculate physical activities intensity using Multi-Axial Accelerometer and HR data. The study produces energy expenditure information and suggested that the class should be balance in order to produce maximum accuracy. The physical activities intensity is important in sport performance as it will contribute in learning the effectiveness of the training undergone.

Some researcher studies athletic aerobic endurance by controlling athlete's HR. It prevents overworking of athlete's heart. In order to control athlete's heart rate, the training was normally undertaken in controlled environment condition. Exercise machine such as treadmill, rowing, elliptical and bikes are normally use for aerobic endurance studies. For example, the automated real time HR control system during treadmill normally realised by continuously adjusting controllable variable such as treadmill speed and training programmes [17]. These methods are normally reprogrammed the exercise machinery by using algorithm such as Non-Linear Fuzzy control [4], Fuzzy PI [18] as fuzzy is known as a simple method to map an input to an output [19].

Besides physical activities and heart rate control, HR information combine with AI are used to form training model for athletes. Adaptive Particle Swamp Optimization (APSO) are used together with HR and training duration to build their training model; it is suggested that APSO is the best method for producing high performance training model [20]. Similarly, Particle Swamp Optimization is use to generates training impulses. Training impulses is outcomes from calculating training volume in minutes and training intensity by overlooking average HR beat. The studies conclude that by using PSO and training impulse it helps improving athlete's capability after long absence from training [26]. In addition, Bat algorithm helps to expand athlete's current formation by suggesting fitness training plan [2] and Artificial Immune System was suggested to rule out athlete's preparedness for training [21]. This shows that implementation of Artificial intelligence to learn new information from HR measurement is robust and should be consider for further venture.

5. Conclusion

Section 1, 2 and 3 discuss how to obtain Heart Rate(HR) and HR variability information. The section also discusses the potentials of the HR information to enclose vital information that can be converted into elaborate information. From section for of this paper concludes there are some advantages combining HR or HRV measurement with AI, among those is;

- Predicting physical activities such as resting, sitting, running, cycling and swimming.
- Learning energy expenditure and power output produce by human body while exercising.
- Program training cardio machinery such as treadmill, stair mill, bikes to be able to control user heart beat to prevent injuries and maximizing training abilities.
- Building training models that are suitable for the condition of athletes and predict athletes' readiness for competition.

As conclusion, it is clear that HR measurement is one of the main important information that should be used to elevate athlete's abilities. Furthermore, the usage of AI improve the understanding of hidden data contain in HR measurement.

Acknowledgements

Special appreciation to reviewers for useful guidance and comments. The authors greatly acknowledge the Research University Grant (RUG) for financial support through grant Vot. No. Q.J130000.2751.03k42.

References

[1] Stork M *et al* 2017 *Int Conf on Applied Elec (AE), IEEE* Pilsen. 1-4.
[2] Fister I *et al* 2015 *Neurocomputing*. **149** 993.
[3] Fister Jr I *et al* 2015 *Applied Mathematics and Computation*. **262** 178.
[4] Pătraşcu A *et al* 2014 *18th Int Conf on Sys Theory, Control and Comp*. Sinaia, Romania.
[5] Su S W *et al* 2010 *The Open Medical Informatics Journal*. **4** 81.
[6] Almeida M B *et al* 2003 *Revista Brasileira de Medicina do Esporte*. **9**(2) 113.
[7] Achten, J. *et al* 2003 *Sports medicine*. **33**(7) 517.
[8] Billman G E *et al* 2015 *Frontiers in physiology*. **6** 55.
[9] Tarvainen M P *et al* 2014 *Comp methods and prog in biomed*. 113(1) 210.
[10] Bringard A *et al* 2017 *European journal of applied physiology*. 117(4) 619.
[11] Malik M *et al* 1996 *European heart journal*. 17(3) 354.
[12] Yu B *et al* 2018 *Frontiers of Information Technology & Electronic Engineering*. 19(1) 6.
[13] Füller M *et al* 2015 *Int Conf on Info and Comm Tech for Ageing Well and e-Health*. Lisbon, Portugal. 106
[14] Ludwig M *et al* 2018 *Frontiers in Physiology*. **9** 778.

[15] Yusoff Y *et al* 2017 *Artificial Intelligence Review*. 1-36.
[16] García-García *et al* 2011 *Int Conf on Art Intel in Med in Europe*. Berlin, Heidelberg, 70-79.
[17] Hunt K J *et al* 2018 *Opt Control App and Methods*. 39(2) 503.
[18] Hunt K J *et al* 2016 *Biomedical Signal Processing and Control*. **30** 31.
[19] Adnan M M *et al* 2015 *Art Intel Review*. 43(3) 345.
[20] Kumyaito N *et al* 2018 *BMC research notes*. 11(1) 9.
[21] Me E *et al* 2011 *Comp in human behaviour*. 27(5) 1499.
[22] Xiao F *et al* 2010 *Fourth Int Conf on Genetic and Evol Comp. Shenzhen, China.* 198 .
[23] Patel S *et al* 2012 *J. of Neuro eng and Rehabilitation*. 9(1) 21.
[24] Ullah K *et al* 2016 Int Conf on Intel Syst Eng (ICISE). Islamabad.
[25] Cornforth D J *et al* 2014 *Frontiers in bioengineering and biotechnology*, 2 (34)
[26] Fister I *et al* 2019 *Adv in Intel Sys and Comp*. Springer. 837

IOP Conf. Series: Materials Science and Engineering **551** (2019) 012042 doi:10.1088/1757-899X/551/1/012042

The Comparison of Semantic Suffix Tree Clustering and Suffix Tree Clustering Algorithm Influence on the Accuracy Rate of an Indonesian Question Answering System

Dininta Isnurthina, Muhammad Ihsan Jambak, and Novi Yusliani

Faculty of Computer Science, Sriwijaya University, Palembang, Indonesia.

E-mail: dini.dininta@gmail.com

Abstract. This research analyses the comparison between Semantic Suffix Tree Clustering (SSTC) algorithm and Suffix Tree Clustering (STC) algorithm in clustering documents on Indonesian Question Answering System. The fundamental difference between the two algorithms is that SSTC considers the meaning of the words to generate clusters and readable labels rather than relying only on string matching like STC. As such, SSTC is able to return a more specific and accurate clusters, and has potential to increase the accuracy rate of a question answering system. However, contrary to the hypothesis, comparison results shows that the accuracy rate of Indonesian Question Answering System after the documents is clustered by SSTC is lower than by STC. The accuracy rate degradation occurred in almost every question category, except Definition category. In average, the accuracy rate obtained by Indonesian Question Answering System with SSTC is only 23.31%, while Indonesian Question Answering System with STC is able to obtain 83% accuracy rate. This significant difference indicates that Semantic Suffix Tree Clustering algorithm is not suitable in the context of document clustering on Indonesian Question Answering System.

1. Introduction

According to Zulen and Purwarianti [1], Question answering system is a system that automatically answers questions in natural language. A question answering system can utilize a database or documents collection (local or web) as a source for its answers, and typically, it consists of three main components; question analyser, document/passage retriever, and answer finder. There has been an extensive amount of research on question answering system, but only a few considered adding document clustering component, particularly on the Indonesian-based system. Purwarianti et al. [5] developed an Indonesian question answering system for factoid questions using inverse document frequency (idf) to retrieve the source documents. The experimental result shows that idf technique is more suitable than the term frequency-inverse document frequency (tf-idf) technique. On the other hand, Yusliani [6] did a research on Indonesian non-factoid questions and the document retriever component implemented tf-idf technique alongside cosine similarity. From 90 questions collected from 10 Indonesian people and 61 source documents, the system returned MRR value of 0.7689, 0.5925, and 0.5704 for each definition, reason, and method category respectively.

Using the same samples from the two previous researches, Rahmansyah [3] experimented on a new Indonesian question answering system by adding a document clustering component, implementing Suffix Tree Clustering (STC) algorithm from Zamir and Etzioni [2]. The STC algorithm is proven to be more feasible than standard clustering methods because it makes use of common phrases found in group of documents to cluster and label them. By adding document clustering component using STC, Rahmansyah succeeded in developing an Indonesian question answering system with 83% accuracy rate. However, it can be assumed that the question answering system's accuracy rate could be further increased by considering the semantic similarity between each words from the source documents. Janruang and Guha [4] developed Semantic Suffix Tree Clustering (SSTC) algorithm that combines string matching and semantic similarity equation to cluster documents. The result of performance comparison between each algorithm in clustering 26.800 datasets from Dmoz.com is SSTC gained a precision rate of 0.81 while STC gained 0.68.

Based on the conclusion from Janruang and Guha [4], this research focuses on the addition of document clustering component using Semantic Suffix Tree Clustering algorithm in an Indonesian question answering system, assuming that it will increase the accuracy rate, and compares it with the result from Rahmansyah [3].

2. Related Works

2.1. Suffix Tree Clustering

According to studies from Zamir and Etzioni [2], STC algorithm has a higher average precision rate than other clustering methods that rely on term frequency distribution such as Single-Pass, K-Means, Buckshot, Fractionation, and Group-Average Agglomerative Hierarichal Clustering (GAHC). This is likely due to the phrases usage by the algorithm itself, as phrases are usually more informative than an unorganized string collection. In STC, phrases can be interpreted as a sequence of one or more words. STC algorithm has two main phases, the first one is base cluster identification for all documents using suffix tree. The suffix tree model considers document $d = w_1 w_2 \ldots w_m$ as a words set w_i, not characters $(i = 1, 2, \ldots, m)$. The second phase is combining base clusters that overlap in one cluster.

Perera [9] implements the STC algorithm in his question answering system to cluster its information. The system is evaluated with MRR and has 0.73 average accuracy rate for 10 sets of questions where each set consists of 35 questions.

2.2. Semantic Suffix Tree Clustering

Janruang and Guha [4] compared SSTC algorithm and Semantic Lingo, a technique developed by Sameh and Kadray [7] which uses semantic similarity to extract phrases in the documents into cluster labels as well. The experiment result showed that SSTC algorithm is able to cluster the documents into a more specific and readable labels compared to Semantic Lingo. On the other hand, Shabbir et al. [8] compared SSTC to conventional Lingo algorithm by clustering web search results using 'scholarships' as its topic. After testing with 5 different queries, the average amount of cluster labels returned by SSTC is 4, due to its nature to return more specific clusters, while Lingo returned 2 clusters. In terms of precision rate, SSTC excelled with average precision of 79%, leaving Lingo with average precision of 75%.

3. Research Methodology

3.1. Data

The data that are used in this research are collected from following researches.

- 3116 pairs of Factoid questions and answers sample, 221 source documents, and research result from Purwarianti et al. [5], conducted in Department of Information and Computer Science, Toyohashi University of Technology.
- 350 pairs of Non-factoid questions and answers sample, 61 source documents and research result from Yusliani [6], conducted in School of Electrical and Informatics Engineering, Institut Teknologi Bandung.
- Research result from Rahmansyah [3], conducted in Faculty of Computer Science in Universitas Sriwijaya.

IOP Conf. Series: Materials Science and Engineering **551** (2019) 012042 doi:10.1088/1757-899X/551/1/012042

3.2. Pre-processing

Pre-processing is the first step to process data before the main process is conduced. Question inputs are pre-processed to help the Question Analyser process in obtaining the expected answer types (EAT) and keywords. Answer types are used in Answer Finder process, while keywords are needed by Document Retriever process. In this stage, there are four stages of pre-process, consist of case folding, tokenizing, stop words removal, and stemming.

3.3. Question Analyzer

The aim of Question Analyser is to analyse the questions entered by user to the system and obtain the keywords, and expected answer types (EAT).

3.3.1. Expected Answer Types (EAT).

To determine the EAT, the system will accept only eight question words in Indonesian, which are "*siapa*" (who), "*siapakah*" (who), "*apa*" (what), "*apakah*" (what), "*dimana*" (where), "*kemana*" (where), "*bagaimana*" (how), dan "*bagaimanakah*" (how). Clue words are also used to determine the EAT. This research implements the method of classifying EAT from previous research by Zulen and Purwarianti [1].

3.4. Document Retriever

The Document Retriever process will retrieve all documents that contain the keywords which has been obtained previously. The retrieving method only checks the keywords occurrences in each document. If a document has one of the keywords inside, then the document will be considered to have the required answer.

3.5. Document Clustering using SSTC

SSTC algorithm clusters semantically similar documents into the same cluster. SSTC consists of three phases; pre-processing and semantic suffix tree construction, tree pruning, and cluster identification.

3.5.1. Semantic Suffix Tree Construction.

Semantic suffix tree is a new data structure that extends the suffix tree and is constructed based on the meaning of word strings. To determine whether a pair of words is similar semantically, the synonym sets of each word are compared. Basically, if there is any intersection between the synonym sets, the pair of words is deemed similar. The synonym sets can be obtained from an online dictionary or database. The semantic similarity equation is shown as equation (1).

$$\begin{aligned} &SemSim(w_a, w_b) = 1 \; if \; |synset(w_a) \cap synset(w_b)| \geq 1 \\ &SemSim(w_a, w_b) = 0 \quad otherwise \end{aligned} \tag{1}$$

Where w_a is first word to be compared, w_b is second word to be compared, $synset(w_a)$ is synonym set of w_a, and $synset(w_b)$ is synonym set of w_b.

3.5.2. Tree Pruning.

The first semantic suffix tree constructed after the previous phase is too huge to be executed. Therefore, tree pruning process is conducted to reduce the number of nodes without omitting the concept or meaning of each node and sub-tree. The tree pruning process uses pre-order traversal that traverse from left to right to update and delete a branch using the depth-first traversal method. Additionally, after the pruning process is finished, the tree is compactified to further reduce the length of tree by merging a sub tree that has only one child or no documents and suffix link.

3.5.3. Cluster Identification.
In this algorithm, each sub-tree is assumed to be a concept cluster and each node is a member that has a set of documents. The post-order traversal technique is used to calculate cluster and label. To reduce the number of useless clusters, the similarity of each cluster is checked by comparing the documents set between two cluster and finding if there is any intersection, as shown by equation (2).

$$ClusSim(C_a, C_b) = 1 \ \text{if} \ |C_a \cap C_b| = |C_a|, \ C_a \ \text{is deleted}$$
$$\text{or} \ |C_a \cap C_b| = |C_b|, \ C_b \ \text{is deleted}$$
$$ClusSim(C_a, C_b) = 0 \quad \text{otherwise} \tag{2}$$

Where C_a is first cluster to be compared and C_b is second cluster to be compared.

3.6. Answer Finder
The clustered documents become the source to find the answer for entered question. This research uses the surface expression method based on research by Yusliani [6] to find the answer.

4. Results
The experimental result is achieved by comparing answers returned by the system to the actual answers from the questions and answers collection to determine whether or not the returned answer is correct. Afterward, the percentage result is calculated using equation (3).

$$Percentage = \frac{Amount\ of\ correct\ answers}{Amount\ of\ returned\ answers\ by\ the\ system} \ x\ 100\% \tag{3}$$

Subsequently the average of all percentage result from each question type is calculated to become the result of this research. The average equation is described as follows.

$$\bar{X} = \frac{\sum_{i=1}^{n} x_i}{n} \tag{4}$$

Where x_i is the value of x in i index, $\bar{X}$ is average, and n is total amount of samples.
The entire percentage result of all question types is shown in the Table 1 below.

Table 1. Experimental result of Indonesian Question Answering System usingSSTC algorithm

Question Type	Returned Answers by the System	Correct Answers	Percentage (%)
	Factoid		
People	546	82	15,01
Organization	527	120	22,77
Location	524	123	23,47
Name	508	86	16,92
Date/Time	506	138	27,2
Quantity	505	119	23,56
	Non-Factoid		
Definition	110	83	75,45
Reason	119	29	24,3
Method	121	28	23,14
Total	**3466**	**808**	**23,31**

IOP Conf. Series: Materials Science and Engineering **551** (2019) 012042 doi:10.1088/1757-899X/551/1/012042

Table 2. Experimental result of Indonesian Question Answering System using STC algorithm [a]

Question Type	Returned Answers by the System	Correct Answers	Percentage (%)
Factoid			
People	500	447	89
Organization	500	421	84
Location	500	405	81
Name	500	435	87
Date/Time	500	410	82
Quantity	500	396	79
Non-Factoid			
Definition	110	107	97
Reason	119	102	85
Method	121	79	65
Total	**3350**	**2795**	**83**

[a] Source: Rahmansyah [3]

5. Conclusions

The comparison between Indonesian Question Answering System with SSTC and with STC showed a huge difference in terms of the accuracy rate, particularly in answering the factoid questions. However, for the non-factoid questions of definition, the accuracy rate gap is not overly significant. The reason is the question sentences in definition category satisfy the surface expression rules, hence they are easier to find in the documents.

The poor accuracy rate of Indonesian Question Answering System with SSTC in answering questions from other categories is caused by several factors as follows.

1. The documents order affects the order of the returned answers by system. After the documents are clustered, their order follows the order of their cluster respectively. As a result, the correct answer could also be placed in the bottom of answers order if its source document is in the bottom of documents order.
2. The incomplete stop words dictionary could prompt the returning of irrelevant keywords and cause the retrieval of irrelevant documents. Problem arises when an incorrect answer sentence contains more keywords than the actual answer sentence.
3. The surface expression method from Rahmansyah [3] is not the proper method to be implemented in this research's Question Analyzer component. In some cases, the question analyzer result obtained a false expected answer type (EAT), therefore the system returned a false answer.

References

[1] Zulen A *et al* 2011 *25th Pacific Asia Conf on Lang, Info and Comp.* 622
[2] Zamir O *et al* 1998 *Proc. of the 21st Annual Inter ACM SIGIR Conf on Res and Develop in Info Retrieval.* Melbourne, Australia. 46
[3] Rahmansyah A 2015 *Implementasi Suffix Tree Clustering Pada Sistem Tanya Jawab Bahasa Indonesia Untuk Pertanyaan Factoid Dan Non-Factoid* (Indonesia: Universitas Sriwijaya)
[4] Janruang J *et al* 2011 *First IRAST Int. Conf. on Data Eng. and Int Tech(DEIT).* Bali, Indonesia.
[5] Purwarianti A *et al* 2007 *Int Conf on Art Intel and App. 25th Multi-Conf.* Innsbruck, Austria.
[6] Yusliani N 2010 *Sistem Tanya Jawab Bahasa Indonesia untuk Non Factoid Question* (Indonesia: Institut Teknologi Bandung)
[7] Sameh A *et al* 2010 *Int J. of Res and Rev in Comp Sc (IJRRCS).* **1** 2.
[8] Shabbir U *et al* 2015 *13th Int Conf on Frontiers of Inf Tech.* Islamabad, Pakistan.
[9] Perera, R 2012 *IEEE 4th Int Conf on Tech for Edu.* Hyderabad.

IOP Conf. Series: Materials Science and Engineering **551** (2019) 012043 doi:10.1088/1757-899X/551/1/012043

Conceptual Design of a Generic Nodal Abstraction (GNA) for a Human-Agent Collaboration Systems

Khudhair Abbas Mohammed[1], Mazlina Abdul Majid[1] and Mohd Sharifuddin Ahmad[2]

[1]Faculty of Computer Systems & Software Engineering, University Malaysia Pahang, Kuantan, Pahang, Malaysia.
[2]College of Computer Science and Information Technology, Universiti Tenaga Nasional, 43000 Kajang, Selangor, Malaysia.

E-mail: khudhair86@outlook.com

Abstract. The issues of modelling software agents in collaborative systems present some problems and difficulties caused by the diversity of tasks and the procedures attached to them. Consequently, such diversity poses a critical challenge to implement scheduled tasks. In this paper, we propose collaborative human-agent system in which a software agent collaborates and cooperates with its human counterpart in several job positions of an organization. The goal of the GNA approach is to improve collaboration between humans when performing tasks within a workflow process. The GNA is built based on the concept of a node which comprises of a human, agents, and their shared functions. This concept espouses the notion that agents could be deployed to assist humans in many common workflow processes and to handle mundane tasks. Furthermore, a multi-agent system is advocated as an application or platform for managing distributed data. Specifically, this paper describes a conceptual design of a generic approach based on the notion of nodes to construct a Generic Nodal Abstraction (GNA) approach that is deployable as a modeling and simulation platform for organizational processes.

1. Introduction

An agent is a software entity having autonomous behaviour that provides convenient outcomes in several domains such as transportations, e-commerce, and healthcare. The software agent technology could be applied in a collaborative process to resolve problems that humans face while performing their jobs (e.g. heavy workloads). Developing an agent system that possesses the ability to collaborate in a natural way with humans is considered as a fundamental challenge [1].

Nowadays, many organizations work with complex software systems which have become gradually incapable to deal with the huge amount of information. Therefore, there has been a demand to deploy software agents to assist humans in performing their tasks automatically. For instance, autonomous agents infer their own decisions based on stored resources in a wide range of areas. This ability could be exploited to transform "modular" software toward its model of perfection and enable a system to be mainly self-managing, as they can be provided with the knowledge to deal with special situations [2].

Collaborative agents work together to implement some work processes that leads to the achievements of some common goal. Subsequently, to design a collaborative agent, there are critical factors that need to be considered such as the cognitive model of the agent and the execution of reasoning tasks. For example, Huang et al. proposed a self-awareness agent model that contains two personality traits (super-ego and ego) and includes an external learning mechanism and internal cognitive capacity [3]. The model claims to provide better collaborative performance over classical agent models and improved agent reasoning capabilities.

Consequently, in this paper, we propose Generic Nodal Abstraction (GNA) approach of agents that assist humans in doing their jobs based on the concept of nodes. The term, "nodal abstraction" comprises a human and one or more software agents with their mediator agent, both work collaboratively in a virtual workspace, i.e. a node. This paper also shows how such node is deployed to develop an applicable human-agent model. This approach should be able to arrange several agents to perform tasks cooperatively and achieve the goal more efficiently.

The paper is organized as follows: In Section 2, we highlight and describe the related work in this area. In Section 3, we describe a preliminary concept of a node, with which a model of collaboration is built and present an example of interacting nodes. Finally, in Section 4, we conclude the paper with a brief discussion of future research extensions to improve our system.

2. Related Work

The literature offers a plethora of approaches that have been proposed to model a multi-agent system, but there is no unique method for agent modelling. Therefore, researchers have continuously work on the conceptualization of complex systems, which composed of several agents. However, some models only attempt to emulate the external process of a system and neglect the internal processes which involve how humans and agents collaboration within the environment.

2.1 Agent Modelling

In mid-1950's, the concept of the agent was started by John McCarthy [4] and established by Oliver G. Selfridge and several years later by [5]. Usually, an architecture of an agent defines its main structure so that every agent has a specific internal architecture which is different from other agents. Consequently, the internal architectures of such agents are presented by many modelling languages, for instance, the Unified Modelling Language (UML) [6]. Therefore, a new abstraction of distributed management system is required to fill the gap due to centralized approaches drawbacks. In the current work, a Generic Nodal Abstraction (GNA) approach for a human-agent collaboration system process is presented for the management work flow process.

Humans aspire to collaborate in order to achieve their commons goals, but they may face a less risky, less challenging level of relationship. As their relationships improved over time, these might enable them to collaborate on other issues. For these different levels of relationships, humans' willingness is needed before choosing the most realistic level for a given set of participants. When humans need to join a venture, they hold themselves accountable to a level of relationship that is beyond their current capabilities.

3. The Generic Nodal Abstraction (GNA) Approach

In this approach, the key points discussed are the importance of modelling a generic architecture for intelligent agents, which has the ability to develop any kind of agent-based systems. Additionally, we propose an innovative model that consists of intelligent nodes. Each node comprises a human and one or more agents to build flexible human-agent collaboration, utilizing multi-agent systems (MAS) benefits and promote human-agent communication.

Our approach operates based on collaborative nodes which consist of a model of a human's job position within an organization, software agents, workflow, and resources. The collaborative nodes enactment architecture realizes the human position and its functions with selected functions of agents. The functions are modelled as actions with which intelligent agents perform to assist its human counterpart in cooperative processes.

A high-level general flowchart of the proposed approach shown in figure 1 illustrates the significant steps to establish the nodes.

IOP Conf. Series: Materials Science and Engineering **551** (2019) 012043 doi:10.1088/1757-899X/551/1/012043

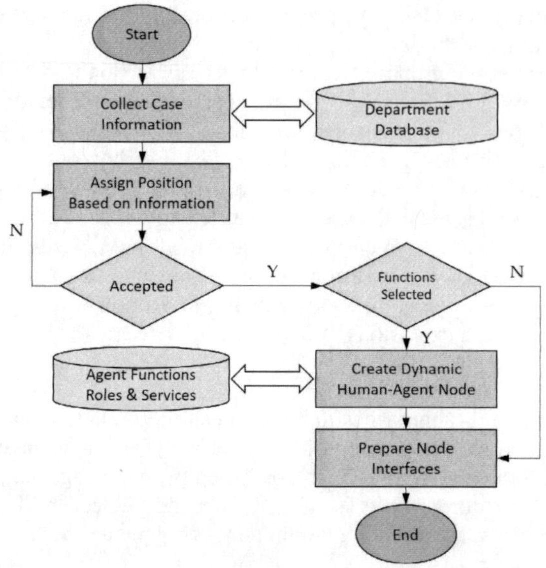

Figure. 1. Flowchart of GNA approach

The outcome of implementing the GNA approach is a human-agent collaborative model which espouses the notion that agents could be deployed to assist humans in many common workflow processes to handle mundane tasks. Progressively, agents with their mediator agent are tightly coupled to their human counterpart. The human-mediator agent together embodies a symbiotic relationship between a human and his/her agents. Note that, the Mediator agent in the node acts as consultant for its human counterpart and it is a link between the human and a number of agents created in the node. We model such symbiosis as a virtual node which consists of the human and one or more agents with services of the mediator agent. Figure 2 shows the concept of a virtual node which constitutes the human-mediator agent and his/her agents.

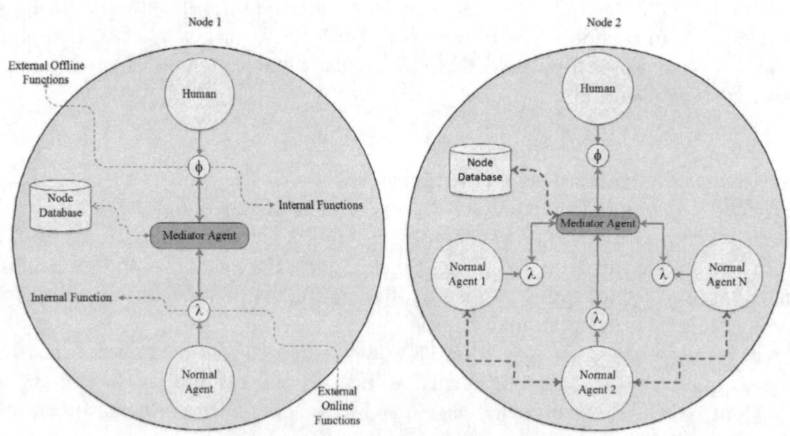

Figure.2. Generic Nodal Approach Concept

In Node 1 of the figure, ϕ represents a human's set of functions and λ represents an agent's set of functions. Node 1 shows a human with only one assistant agent via a mediator agent, presumably to handle all the mundane tasks of its human counterpart. In Node 2, the human is assisted by three agents, the number of agent is selected according the node mediator agent which handles one or more specific tasks for its human counterpart. The human's and agents' functions are selected during the create mode, in which relevant human and agent functions are selected to instantiate a node. Depending on the functions selected, the node represents a unique intelligent entity, which behaves and operates within a collection of other unique nodes to achieve some common goals. Consequently, in the create mode, a user could define the following nodes:

- Node 1 (NH): A human-only node, when no agent's functions are selected,
- Node 2 (MDA): A mediator agent with one agent node, when no human's functions are selected,
- Node 3 (HMDA): A human-mediator agent with one agent node, when relevant human's and an agent's functions are selected,
- Node 4 (HMDAs): A human-mediator agent with various agents node, when relevant human's and many agents' functions are selected.

The significant roles of mediator agents are to meet its design objectives, ability to receive data from the environment, choose suitable agent or many agents, and act to change such an environment. Accordingly, the objectives and tasks assigned to the mediator agents are performed when such agents follow the 'route' of the target step-by-step. Almost all the route is searched over the Internet, which can be a proactive search or by using the Semantic web.

4. Conclusion And Further Work

A software agent is considered as an indispensable application in most workflow processes management. The issues of modelling software agents face critical obstacles and difficulties. Initially, the design started from architectural step until it becomes a comprehensive model. Several challenges occur to determine agent tasks especially when the model implements scheduled tasks practically. However, the architectural design of the Generic Nodal Abstraction (GNA) approach is presented to be a fundamental solution. Moreover, our GNA approach significantly reduces the workload process by implementing a multi-agent system in order to assist its human counterpart. Additionally, this approach offers a solution for the traditional agent modelling which being applied in various organizations. In our future work, we shall develop a variety of nodes and implement several workflows to animate some business processes

Acknowledgements

This study is supported and financed under Postgraduate Research Scheme Grand (PGRS 160319) from University Malaysia Pahang (UMP).

References

[1] Chen W *et al* 2007 *Proc. of the AAAI, Distributed Intel Sys,* Virginia, USA, 38.
[2] Albus J S 2001 *Eng of Mind : An Intro to the Sci of Intel Syss.*
[3] Huang C Y *et al* 2011 *9th World Congress on Intel Control and Automation,* Taipei.759.
[4] Bradshaw J M 1997 *Software agents*: MIT press.
[5] Kim M *et al* 1999 *IEEE Int Fuzzy Sys. Conf Proc* Seoul, South Korea. **1** 321.
[6] Gonçalves E J T *et al* 2015 *J. of Sys and Soft.* **108** 77.

IOP Conf. Series: Materials Science and Engineering **551** (2019) 012044 doi:10.1088/1757-899X/551/1/012044

Weather Data Analysis Using Hadoop: Applications and Challenges

Mohammed Adam Ibrahim Fakherldin[3 *], Khalid Adam[2], Noor Akma Abu Bakar[1] and Mazlina Abdul Majid[1]

[1]Faculty of Computer System & Software Engineering, Universiti Malaysia Pahang
[2]Faculty of Electric and Electronic Engineering, Universiti Malaysia Pahang
[3]Faculty of Computer Science and Information Systems, Jazan University

Email: khalidwsn15@gmail.com

Abstract. Weather data is very crucial in every aspect of human daily life. It plays an important role in many sectors such as agriculture, tourism, government planning, industry and so on. Weather has a variety of parameters like temperature, pressure, humidity and wind speed. The meteorological department deployed sensors for each weather parameter at different geographical locations to collect data. This data is stored mostly in the unstructured format. Thus, a big amount of data has been collected and archived. Therefore, storage and processing of this big data for accurate weather prediction is a huge challenge. Hadoop an apache product it used to support big data sets in a distributed environment. Hadoop has greatest advantages over scalable and fault-tolerant distributed processing technologies. This paper explains a system that uses the historical weather data of a region and apply the MapReduce and Hadoop techniques to analysis these historical data.

1. Introduction

Big Data has become one of the buzzwords in IT, during the last couple of years. Originally it was created by companies which had to manage fast increase rates of data such as web data, data resulting from scientific or business simulations or other data sources. Some of those business companies' models are basically based on indexing and using this large amount of data. The challenges to handle the fast growing of data amount on the web e.g. lead Google to develop the Google File System [1] and MapReduce [2].

Furthermore, most of the cities have become smart. Thus, many sensors devices utilized in smart city can be used to measure weather parameters [3]. Which led the weather department collect and analysis huge amount of data like temperature [4]. These different sensors value such as temperature, humidity to predict the rain fall etc. Consequently, when the number of sensors/devices increases, the data becomes high volume and velocity [5]. Therefore, there is an essential of a scalable analytics tool to process massive amount of data.

However, the conventional method of process the data is very slow. Compare to process the sensor data with Hadoop framework which remove the scalability bottleneck. Hadoop is a framework used for handling huge amount of data. Mainly the processing engine is MapReduce, which is currently one of the most common big data processing frameworks available. MapReduce is a framework for performing highly parallelizable and distributable algorithms across big data sets using commodity computers cluster [6]. Thus, using Hadoop/MapReduce the temperature can be analyses without scalability problems. The speed of processing data can improve rapidly when across multi cluster distributed network.

2. Related Works

[7] describe the analysis of huge amounts of climatic data by using MapReduce with Hadoop. Huge amounts of climatic data collected, stored and processed for accurate prediction of weather. Climatic data collected by using different types of sensors to store the following parameters temperature, humidity etc. weather datasets collected from National Climatic Data Center (NCDC). Daily Global Weather Measurements 1929-2009 (NCDC, GSOD) dataset is one of the biggest datasets available for weather forecast. Its total size is around 20 GB. Results show that temperature analyzed effectively by Using MapReduce with Hadoop.

[6] gives a detailed description of build a platform that is extremely flexible and scalable to be able to analyse petabytes of data across an extremely wide increasing wealth of weather variables. Data processed by Apache Hadoop and Apache Spark. Experiments performed to select the best tools among Hadoop using Pig and Hive Queries.

[8] explains the meteorological data storage as well as analysis platform based on Hadoop framework with the help of online logistic regression algorithm for prediction. This platform is based on distributed file system HDFS which includes distributed database HBase, data warehouse management and useful query processing tool Hive, data migration tool Sqoop. The best data mining prediction algorithm regression also integrated into the system. This architecture has an ability of mass storage of meteorological data, efficient query, and analysis, climate change prediction.

3. Experimental Setup and Results

The experiments were carried out in a physical cluster environment, the researcher used three computers. Hadoop cluster on Linux Ubuntu 14.04 where one computer ran a NameNode and ResourceManager and the remaining ran Datanode and DataManager. Each of the computers has the following configuration: Core i7 processor, 4 GB main memory, and 1 TB disk space as shown in figure 1. The researcher used a Hadoop-2.7.1. The max replication factor "dfs.replication.max" is used to set the replication limit of blocks.

Figure 1. Hadoop cluster

The proposed System use dataset of NCDC contains the following parameters: station number, station name, date, country, Precipitation, Temperature, and Wind and so on as shown in figure 2. The data files are stored in HDFS as shown in figure 3. Then, weather files are split and goes to different mappers. The output of each mapper is a set of pairs (key, value) where key is consists of station name, date and value is contains the parameters: Precipitation, Temperature, and Wind. Then the output of mappers is merged and sort by key. Finally, all results sent to the reducers. For each reducer

IOP Conf. Series: Materials Science and Engineering **551** (2019) 012044 doi:10.1088/1757-899X/551/1/012044

calculate Average (monthly, yearly, and seasonal), Maximum (monthly, yearly, and seasonal), and Minimum (monthly, yearly, and seasonal), for each parameter precipitation, Temperature, and wind in different stations. Each reducer stores the final results in HDFS as show in figure 4.

This stage incorporates the goals of data stored in HDFS. This is finished utilizing MapReduce framework. The fundamental procedure of MapReduce is shown in figure 2. MapReduce structure takes a split-apply-reduce combine strategy. The last output is the value which has more key-value occurrences.

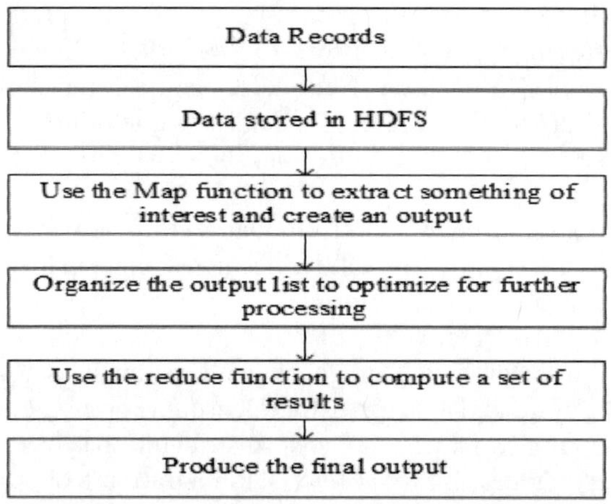

Figure 2. MapReduce flow chart process

Table 1. Weather Dataset into ASCII Format

```
Wban Number, YearMonthDay, Time, Station Type, Maintenance Indicator, Sky
Conditions, Visibility, Weather Type, Dry Bulb Temp, Dew Point Temp, Wet Bulb
Temp, % Relative Humidity, Wind Speed (kt), Wind Direction, Wind Char. Gusts (kt),
Val for Wind Char., Station Pressure, Pressure Tendency, Sea Level Pressure, Record
Type, Precip. Total
03013,19960701,0053,AO20,-,CLR                    ,10SM  ,-,64,60.1,35,
87 , 7  ,180,-,0 ,26.30,-,162,AA,-
03013,19960701,0153,AO20,-,CLR                    ,10SM  ,-,64.9,60.1,35,
84 , 10 ,190,-,0 ,26.30,6,153,AA,-
03013,19960701,0253,AO20,-,CLR                    ,10SM ,-
,62.1,60.1,34.9, 93 , 8  ,200,-,0 ,26.29,-,150,AA,-
03013,19960701,0353,AO20,-,CLR                    ,10SM  ,-,60.1,59,34.7,
96 , 3  ,310,-,0 ,26.29,-,151,AA,-
03013,19960701,0453,AO20,-,CLR                    ,10SM  ,-,59,57.9,34.6,
96 , 0  ,000,-,0 ,26.30,5,154,AA,-
03013,19960701,0553,AO20,-,CLR                    ,10SM  ,-,64,61,35, 90 ,
0  ,000,-,0 ,26.30,-,155,AA,-
03013,19960701,0653,AO20,-,CLR                    ,10SM ,-
,66.9,62.1,35.2, 84 , 6  ,310,-,0 ,26.31,-,162,AA,-
03013,19960701,0753,AO20,-,CLR                    ,10SM  ,-,72,63,35.4,
73 , 5  ,310,-,0 ,26.31,3,160,AA,-
03013,19960701,0853,AO20,-,CLR                    ,10SM  ,-,75.9,63,35.5,
64 , 6  ,270,-,0 ,26.31,-,156,AA,-
03013,19960701,0953,AO20,-,CLR                    ,10SM  ,-,80.1,64,35.7,
58 , 7  ,270,-,0 ,26.30,-,150,AA,-
```

IOP Conf. Series: Materials Science and Engineering **551** (2019) 012044 doi:10.1088/1757-899X/551/1/012044

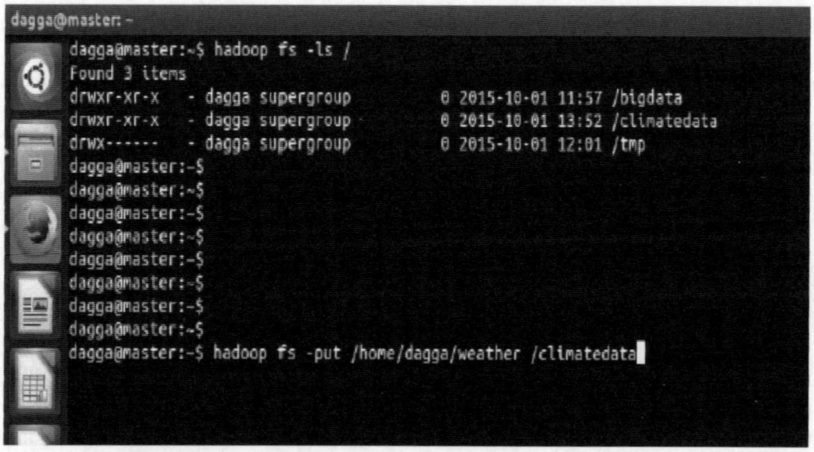

Figure 3. Push the dataset into climatedata folder

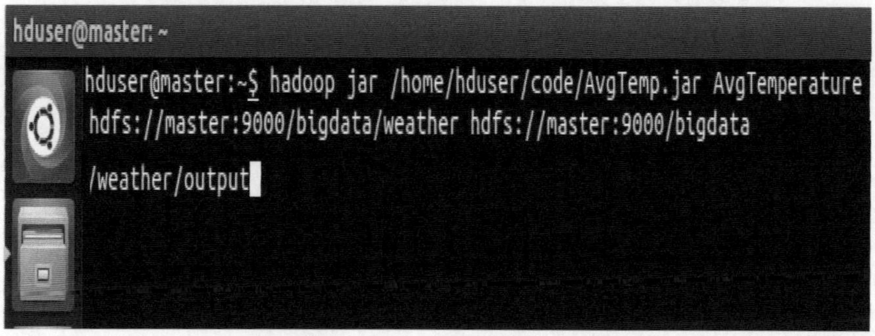

Figure 4. Run the weather dataset

From the above figures 3 and 4, the last output will be climate information as it has increasingly key events. MapReduce is an effective way of data resolution of large amount of data in a cluster. The MapReduce program is made of a map () procedure that performs percolating and sorting the required data into the DataNode. This module follows a split-apply combine strategy.

4. Conclusion

In case of using traditional systems, to process millions of sensors data it is time consuming. Today IoT and the meteorological department uses various of sensors devices to collect data (e.g. temperature, humidity etc). Hadoop/MapReduce is a framework for processing huge amount in distributable way across large number of computers cluster. Using this framework, the sensors data can be analyzed efficiently. The major benefit of Hadoop framework speeds up the processing of huge data. Where the volume of data is increasing every day.

References

[1] Ghemawat S, Gobioff H, and Leung S T 2003 *The Google file system* **37** 5
[2] Dean J and Ghemawat S 2008 MapReduce: simplified data processing on large clusters *Commun. ACM,* **51** 1 107–113.
[3] Ge M, Bangui H, and Buhnova B 2018 Big Data for Internet of Things: A Survey *Futur. Gener. Comput. Syst.*
[4] Ismail K A, Majid M A, Zain J M, and Bakar N A A 2016 Big Data prediction framework for weather temperature based on MapReduce algorithm *Open Systems (ICOS), 2016 IEEE Conference* 13–17.
[5] Hammad K, Fakharaldien M, Zain J, and Majid M 2015 Big data analysis and storage *International Conference on Operations Excellence and Service Engineering* 10–11.

[6] Dagade V, Lagali M, Avadhani S, and Kalekar P 2015 Big data weather analytics using Hadoop *Int. J. Emerg. Technol. Comput. Sci. Electron* **14** 2.
[7] Riyaz P and Varghese S M 2015 Leveraging map reduce with hadoop for weather data analytics *IOSR J. Comput. Eng* **17** 3.
[8] Ramya M, Balaji C, and Girish L 2016 Environment Change Prediction to Adapt Climate-Smart Agriculture Using Big Data Analytics *Int. J. Adv. Res. Comput. Eng. Technol.* **4** 201.

IOP Conf. Series: Materials Science and Engineering **551** (2019) 012045 doi:10.1088/1757-899X/551/1/012045

Consideration of Canny Edge Detection for Eye Redness Image Processing: A Review

M. R. H. Mohd Adnan[1], Azlan Mohd Zain[1], Habibollah Haron[1], Mohd Zulfaezal Che Azemin[2], Mahadi Bahari[3]

[1]Applied Industrial Analytics Research Group (ALIAS), School of Computing, Faculty of Engineering, UTM Johor Bahru, Johor, Malaysia
[2]Kulliyyah of Allied Health Sciences, International Islamic University Malaysia, IIUM Kuantan, Pahang, Malaysia
[3]Azman Hashim International Business School (AHIBS), UTM Johor Bahru, Johor, Malaysia

mohdridhwanadnan@gmail.com

Abstract. Eye redness can be taken as a sign of inflammation which may suggest severity and progression of a specific disease. In image processing, there is apportioning a digital image into relevant features in sets of pixels where is called image segmentation. The image that consists of numerous parts of different colors and textures need to be distinguished in this process. In each digital image, the transformation of images into edges was using edge detection techniques. It represents the contour of the image which could be helpful to recognize the image as an object with its detected edges. The Canny edge detector is a standard edge detection algorithm for many years among the present edge detection algorithms. This paper focuses on important canny edge detection for detecting a region of interest (ROI) in eye redness images.

1. Introduction

Eye redness is a symptom that can be associated with a broad group of ocular diseases [1]. The excess of blood vessels with increased diameter may cause an effect in ocular surface [2]. The clinician can diagnose ocular surface clues earlier and the development of validated ocular disease scales can be purified [3-4]. Bulbar redness is the occurrence of an inflammation in a blood vessel in the sclera (white area), which causes conjunctivitis, ocular xerosis, glaucoma, scleritis and keratitis [1]. Pterygium is a wing-shaped, vascular and fleshy growth which encroaches from bulbar conjunctiva and progresses toward central cornea. Pterygium has been established as one of the factors which cause induced-corneal astigmatism, alteration of the anterior corneal parameter, and changes in corneal topography findings. However, based on [5] findings, the exact etiologies of pterygium are still debated.

The manual process that optometrists have to face is tedious, time-consuming, highly subjective, and nonrepeatable [6]. The first step is to obtain an image acquisition of the patient eye. Then, the images must be analyzed in detail, searching for indicators of the symptom, such as the aforementioned red hue. Finally, the optometrist compares the patient's eye with a given grading scale, in order to obtain the final evaluation. Grading scales are collections of images that show the ocular disease grading scales [15-17].

Segmentation may also depend on various features that are contained in the image. It may be varying by color or texture [7]. Redness can be taken as a sign of inflammation which may suggest severity and progression of a specific disease. Several attempts have been made to quantify redness of the eye and to demonstrate the importance of automating the redness grading which will reduce individual variability due to low inter-grader repeatability [8-9]. A standard objective grading can ensure the reliabilities of the grading as individual variance are still needed to be considered. Subjective grading based on redness is somewhat extremely difficult to reproduce; hence the repeatability also can be questionable [5].

Image analysis techniques are crucial in measuring redness objectively [10]. The advantage of using image analysis method is that we can eliminate the individual variability and bias, which might happen in subjective grading [18-19].

Image processing algorithms involved in many various areas in order to detect the true edges of the image. Hence, edge detection is very crucial tasks in image processing because it might bring the major information contributors to an image. It represents the contour of the image which could be helpful to recognize the image as an object with its detected edges [11]. Among the present edge detection, the Canny edge detector was the best-selected edge detection algorithm. This paper focuses on important canny edge detection for detecting a region of interest (ROI) in eye redness images.

This paper outlines as follows: Image processing in eye redness briefly discussed in section 2. Section 3, shows the eye redness segmentation by applying the canny edge detection and region of interest in eye redness. Finally, section 4 presents the conclusion and potential improvement.

2. Image Processing in Eye Redness

The fundamental development of image processing in eye redness involves five phases which are image acquisition, image pre-processing, image segmentation, description and representation, and recognition and interpretation.

The first phase is image acquisition which needs the real experimental data to be analyzed. The second phase is image pre-processing that will be used to normalize image brightness, correction of non-uniform illumination, reduce noise and reduce image artifacts.

The image segmentation which defined in phase 3 is to partitioning digital image with relevant features where is easier for further process. This process will partition an image into regions or set of pixels. Edge detection and region of interest (ROI) are one of the ways in the segmentation process also known as image analysis. This phase is very tedious and crucial in image processing because it might affect the phase 4 to obtain significant features.

Description and Representation defined in phase 4 involved the most relevant features contributed to an image. In this phase, description is referred to feature extraction process where the feature (color or texture) has been applied to the image. Whilst, Feature selection process where the not relevance features will be eliminated is referred to as representation.

Finally, phase 5, involves the recognition and interpretation that employed the meaningful features created in phase 4. Recognition is assigning to Classification, which is a label to an image. Interpretation is assigning the meaning to an ensemble of the image. In this phase, the experimental image data will be classified according to the eye redness grading scales.

3. Eye Redness Segmentation

As the edge is one of the major information contributors to eye redness image, hence the edge detection is a very important step in many of the image processing algorithms [12]. For the edge detection, eye redness image is given as input and the edge map is obtained as output. It represents the contour of the image, which could be helpful to recognize the image as an object with its detected edges [13]. In the ideal case, applying the edge detector to an image gives the different edges that are connected to form the outline of the object [14].

From the literature by [11], there are three main criteria belongs to canny edge detection:

i) Low error rate: All edges in images should carefully be a response.

ii) Good Localization: Distance between the actual edges found should be minimum.

iii) Single Response: Response to a single edge. The algorithm mainly has five steps [11].
Step 1: A horizontal and vertical gradient can be computed for every pixel in an image.
Step 2: The calculation of each pixel in the images was based on image information in step 1.
Step 3: Non-Maximal Suppression is counted as zero in this step.
Step 4: The gradient magnitude either high or low threshold is depending on histogram calculations.
Step 5: Hysteresis thresholding is used to obtain the weak and strong edges. It is important to measure the proper edge map. Strong edge is referred to superior pixel with high threshold while those pixel whose lays between the weak and high threshold is referred to as a weak edge.

A region of interest (ROI) is a portion of an image defined for further analysis or processing. ROI of pterygium fibrovascular tissue was selected manually and was defined as the visible area of pterygium which is from the limbus to apex of pterygium [5]. Figure 1 shows examples of pterygium ROI in all range of redness grading. Blue lines indicate the pterygium ROI.

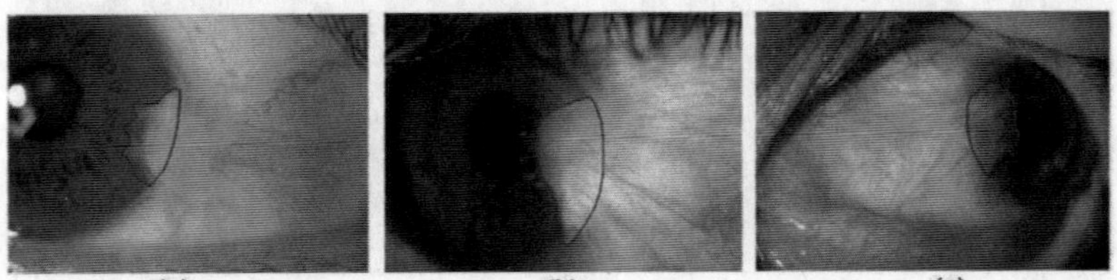

(a) (b) (c)

Figure 1. Examples of pterygium fibrovascular tissue. Blue lines indicate the pterygium ROI: (a) Grade 1, (b) Grade 2, and (c) Grade 3 [5].

Figure 2 shows the system to detect ROI. The development of the software is applying platform dot net framework with C# language. The extent of the software will be made for the next phase which is feature extraction and feature selection of the eye redness image.

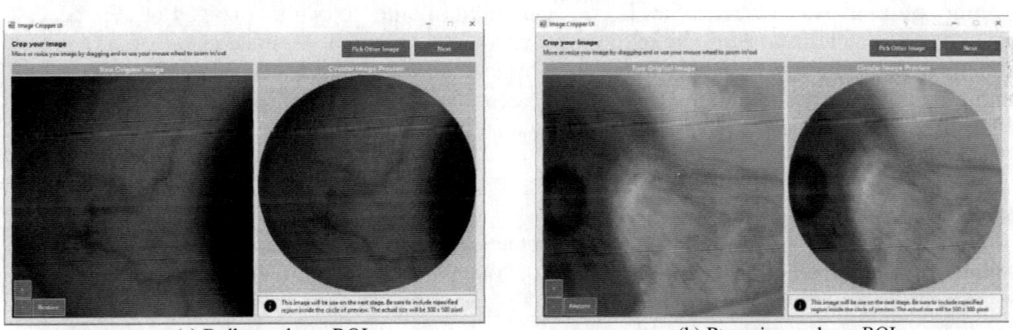

(a) Bulbar redness ROI (b) Pterygium redness ROI

Figure 2. The system developed to detect ROI

Figure 3 shows the image segmentation of bulbar redness. The first step depicts the original photograph was taken of bulbar redness. Image enhancement and remove noise have been implemented to remove eyelashes and pupil. Next, the canny edge detection has been applied. Strong edge is referred to superior pixel with high threshold while those pixel whose lays between the weak and high threshold is referred to as a weak edge (non-edge). In the final step, the reddish color (dashed rectangle) is the ROI of the eye redness image. The most relevant features will be selected based on the final step.

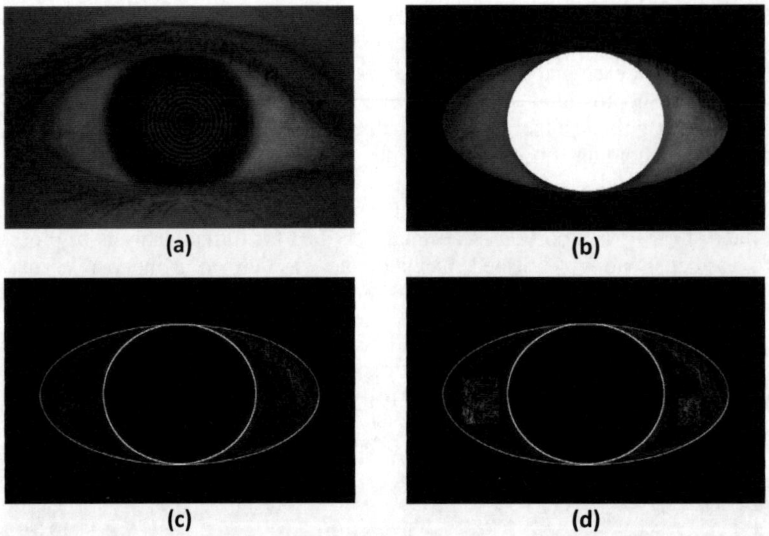

Figure 3. Image segmentation of bulbar redness. (a) Original image of bulbar redness. (b) Image enhancement and remove noise. (c) Illustrates of canny edge detection. (d) Redness color (dashed rectangle) are the ROI area was assessed.

4. Conclusions

The ability of canny edge detection has been widely used in order to extract the ROI of eye redness. Image segmentation leads to partitioning the eye redness images with better edge map. A future line of research will be on the development of improvement in edge detection, feature extraction, and feature selection in order to quantify the redness through most relevant features because it provides much more distinct marked edges and thus has better visual appearance than the standard existing.

Acknowledgment

This work was supported by the Research Management Centre, UTM under the Research University Grant (GUP) with identification number Q.J130000.2528.19H65.

References

[1] Sánchez Brea M L, Barreira Rodríguez N, Mosquera González A, Evans K and Pena-Verdeal H 2016 Defining the optimal region of interest for hyperemia grading in the bulbar conjunctiva *Comp. and Math. Methods in Med.* 1-9

[2] Brea M L S, Rodríguez N B, Maroño N S, González A M, García-Resúa C and Fernández M J G 2016 On the development of conjunctival hyperemia computer-assisted diagnosis tools: Influence of feature selection and class imbalance in automatic gradings *Artif.l Intel. in Med.* **71** 30-42

[3] Zhao W J, Duan F, Li Z T, Yang H J, Huang Q and Wu K L 2014 Evaluation of regional bulbar redness using an image-based objective method *Inter. J. of Ophthal.* **7**(1) 71-76

[4] Rosas-Romero R, Martínez-Carballido J, Hernández-Capistrán J and Uribe-Valencia L J 2015 A method to assist in the diagnosis of early diabetic retinopathy: Image processing applied to

detection of microaneurysms in fundus images *Comp. Med. Imaging and Graphics*, **44** 41-53
[5] Hilmi M R, Che Azemin M Z, Mohd Kamal K, Mohd Tamrin M I, Abdul Gaffur N and Tengku Sembok T M 2017 Prediction of changes in visual acuity and contrast sensitivity function by tissue redness after pterygium surgery. *Current Eye Research* **42**(6) 852-856
[6] Rodriguez J D, Johnston P R, Ousler III G W, Smith L M and Abelson M B 2013 Automated grading system for evaluation of ocular redness associated with dry eye *Clinical Ophthal.* **7** 1197-1204
[7] Yogamangalam R and Karthikeyan B 2013 Segmentation techniques comparison in image processing *Inter. Journal of Eng. and Tech. (IJET)* **5**(1) 307-313
[8] Azemin M Z C, Tamrin M I M, Hilmi M R and Kamal K M 2016 Inter-Grader Reliability of a Supervised Pterygium Redness Grading System *Adv. Sci. Letters* **22**(10) 2885-2888
[9] Derakhshani R, Tankasala S P, Crihalmeanu S, Ross A and Krishna R 2016 A comparative analysis of wavelets for vascular similarity measurement *Inter. Joint Conf. on Neural Networks* (IJCNN) 3870-3876
[10] Hajare P A and Tijare P A 2011 Edge detection techniques for image segmentation *Inter. Journal of Comp. Sci. and App.* **4**(1) 49-53
[11] Kabade A L and Sangam D V 2016 Canny edge detection algorithm *Inter. Journal of Adv. Research in Electronics and Comm. Eng.* **5**(5) 1292-1295
[12] Rong W, Li Z, Zhang W and Sun L 2014 An improved CANNY edge detection algorithm *Inter. Conf. on Mech. and Automation* (ICMA) 577-582
[13] Romano V, Steger B, Brunner M, Kaye A, Zheng Y, Willoughby C E and Kaye S B 2017 Detecting Change in Conjunctival Hyperemia Using a Pixel Densitometry Index *Ocular Immunology and Inflammation* 1-6
[14] Kaur E K, Mutenja V and Gill E I S 2010 Fuzzy logic based image edge detection algorithm in MATLAB *Inter. Journal of Comp. App.* **1**(22) 55-58
[15] Park I K, Chun Y S, Kim K G, Yang H K and Hwang J M 2013 New clinical grading scales and objective measurement for conjunctival injection *Investigative Ophthal. and Visual Sci.* **54**(8) 5249-5257
[16] Wolffsohn J S 2004 Incremental nature of anterior eye grading scales determined by objective image analysis *British Journal of Ophthal.* **88**(11) 1434-1438
[17] Winder R J, Morrow P J, McRitchie I N, Bailie J R, and Hart P M 2009 Algorithms for digital image processing in diabetic retinopathy *Comp. Medical Imaging and Graphics* **33**(8) 608-622
[18] Astvatsatourov A and Mösges R 2015 Image-Based Assessment of Allergic Inflammation under Conjunctival Provocation. *ICIMTH* 15-18
[19] Esa N M, Zain A M, Bahari M, and Yusuf S M 2018 Comparative Study of Segmentation and Feature Extraction Method on Finger Movement *Inter. Conf. of Reliable Inf. and Comm. Tech.* 117-127

Comparison of dimensional reduction using the Singular Value Decomposition Algorithm and the Self Organizing Map Algorithm in clustering result of text documents

Muhammad Ihsan Jambak*, **and Ahmad Ikrom Izzuddin Jambak**
Faculty of Computer Science, Sriwijaya University, Palembang, Indonesia.

E-mail: jambak@unsri.ac.id

Abstract. Dimension reduction has two methods namely feature selection and feature extraction. Dimension reduction using the feature selection method has a better influence than the feature extraction method on the cluster results. However, there is still a need for feature extraction methods to reduce dimensions. For this reason, an alternative algorithm is needed from the feature extraction method. Self Organizing Map (SOM) is one of the artificial neural network models that has a special nature that is effectively able to create spatial internal representations of input data, or in general to create smaller data dimensions. This research was examining the capability of SOM compared to Singular Value Decomposition (SVD) in reducing data dimension of text documents before there were clustered by k-Means. Results show that SVD still better than SOM in cluster quality index but SOM faster than SVD in computation times.

1. Introduction

High-dimensional data is data that has a large number of features. In high-dimensional data, it allows complex data to occur which is data that has noise, anomalies (outliers), missing meaning (missing value) and discontinuities between data [1, 2]. Some examples of data that have high dimensions are text document data, image data, and gene expression data. In the text document clustering process, each dimension or feature of the data has an important influence. The most frequently used algorithm for clustering is the k-Means algorithm that groups data into groups of k, with k being the desired number of groups [3]. The k-Means algorithm has a faster data processing time compared to other clustering algorithms.

Each data feature will affect the location of a data when grouped into a class. However, when a data has many features and is too diverse, clustering algorithms such as the k-Means algorithm are not able to find the closeness or similarity between data so that the data cannot be grouped properly. This is often called curse of dimensionality [4]. To get good results, high-dimensional data must go through the initial processing or pre-processing, namely dimension reduction. Dimension reduction is a process of reducing random variables or nonessential variables with certain considerations [5].

Dimension reduction has two methods namely feature selection and feature extraction. Dimension reduction using the feature selection method has a better influence than the feature extraction method on the cluster results using the k-Means algorithm [6]. The feature extraction algorithm that is often used is Principal Component Analysis (PCA). However, the PCA algorithm is often used as a dimensional reduction algorithm in several studies. This indicates that there is still a need for feature extraction methods to reduce dimensions. For this reason, an alternative algorithm is needed from the feature extraction method.

Artificial Neural Network (ANN) is a calculation model that is inspired by the workings of nerves in the human body. The ANN does not stand as a specific algorithm, but rather a framework that forms the basis of various other calculation models [7]. Self Organizing Map (SOM) is one of the ANN models that are trained to use unsupervised data. In contrast to other ANN algorithms, SOM has a special nature that is effectively able to create spatial internal representations of input data [8]. In general, the purpose of mapping the SOM algorithm is to create smaller data dimensions.

[9] merged SOM with the Density-Based Spatial Clustering of Applications with Noise image segmentation algorithm (DBSCAN), in this study SOM was used as a reduction in the dimensions of image data which is high dimensional data. Others study were conducted by [10] and [11] who implemented SOM to reduce high-dimensional data, the results of this study showed the ability of SOM to project non-linear data into lower dimensions. Based on the description above, it is necessary to do research on the SOM algorithm as an alternative to the PCA algorithm in reducing feature extraction dimensions.

2. Research Methodologies

2.1. Data and Pre-processing Data

The data test used English language journals that are downloaded through the several websites which is stored in the ASCII file (* .txt). With total 40 journals and text terms obtained from pre-processing results amounted to 35,000 words. The pre-processing stage is the conversion of data formats, a series of actions to convert input data from text to numerical so that the clustering process can be carried out. Stages carried out on are case folding, tokenizing, stop words removal, and stemming. Furthermore, the terms of the research test data are given weights using the tf-idf (term frequency-inverse document frequency).

2.2. Singular Value Decomposition

Singular Value Decomposition is a form of factor analysis on the matrix. In SVD matrix contains the frequency of occurrence of keywords decomposed into three matrix components [12]. The first matrix component (U) describes the line entity as an orthogonal matrix vector. The second matrix component (S) is a diagonal matrix containing the scalar value of the matrix. And the third component (V) is the column entity matrix as an orthogonal matrix vector.

The singular value decomposition of the weighted matrix will be used in the linear transformation approach. SVD is basically to estimate rank from the matrix. If matrix H is known with dimensions m x n , where $m \geq n$ *and* $rank(H) = r$, as follows SVD equation is :

$$H = USV^T \tag{1}$$

H	: matrix from weighting process	m	: term
U	: left eigenvector	n	: *document*
V	: right eigenvector	r	: rank
S	: singular value		

where, $$U^T U = V^T V = I_n \tag{2}$$

and fulfill the conditions, $$S = diag(\sigma_1, \ldots, \sigma_n) \tag{3}$$

where, $\sigma_1 > 0 \ for\ 1 \leq i \leq r$ and $\sigma_j = 0\ for\ j \geq r+1$

2.3. Self Organizing Map

Self-Organizing Map is one of the techniques in a neural network that uses unsupervised learning methods that use competitive learning methods. SOM aims to map high-dimensional data into the form of low-dimensional data by assuming a structured topology makes new class units based on data similarity [8, 13]. To project data into lower dimensions, SOM uses a weight vector, which is initially randomly determined and then becomes a reference in grouping input vectors. Euclidean Distance is used to measure the distance between the input vector and the weight vector, with the input vector closest to the value with the weight vector will be selected as the winning neuron. The weight vector will then be updated based on the results of the first iteration using the alpha / learning rate multiplier which will always decrease if there is no change.

According to Haykin [14] there are three important components in SOM, namely: Competition, for each input pattern neurons calculate the value of each discriminant function that gives the basis for competition. Certain neurons with the smallest value of the discriminant function are declared winners; Cooperation, winning neurons determine the spatial location of the excited neuron topology environment to provide a basis for cooperation in a neuron environment; Synaptic Adaption, excited neurons decrease the value of the discriminant function associated with the input pattern through the corresponding weight adjustment so that the response of the winning neurons to the next application with the same input pattern will increase.

Following are the steps of SOM: (1). Determine the initial weight value randomly (0 to 1) a number of k dimensions; (2). Determine the value of learning rate and the number of iterations, in this study an experiment will be conducted first to determine the learning rate value and the best number of iterations; (3). Calculate the distance between input data and weights using the Euclidean Distance formula; (4). Determine the winning neurons, namely neurons whose weight values are closest to the input data; (5). Update the weight value; (6). Repeat steps using data to i + 1 until the iteration reaches the specified amount.

2.4. k-Means Clustering

k-Means is one of the frequently used partitioning methods due to its simplicity, this technique group objects into k groups. To do this clustering, the value of k must be determined first. The process that k-means is to choose k centres as a centroid for use in clustering data. The centroid point will determine success in the k-means algorithm. If the point of the centroid selected at the beginning is wrong, then the results obtained will not be optimal. In k-means the centroid point is often a problem that must be overcome because of random selection. The k-means algorithm classifies data with the number of clusters (k values) that have been predetermined; it is possible that the k value is the number of clusters desired or the value of k entered randomly. Entering the number of clusters that are not ideal in the data can cause poor clustering quality, allowing data not to be in the actual cluster.

The steps in the k-Means algorithm are as follow [2, 5, 15, 16]: (1) Determine the number of clusters (k) as the number of clusters to be formed; (2) Generate k centroids (starting point) randomly; (3) For each data, locate the nearest cluster centre, so each cluster centre has a subset of datasets; (4) For each cluster k, find the centre of the cluster area, update the location of each cluster centre to the new value of the centre area; (5) Repeat steps 3 and 5 to the data in each cluster becomes centralized or completed.

2.5. Davies Bouldin Index

For the success of the clustering process then one of the important issues is cluster validation. Although in general clustering validation is categorized into external clustering validation and internal clustering validation; but we realize that the purpose of clustering is to group objects in the same or most similar cluster and different objects in different clusters. For this reason, internal evaluation is more suitable.

The internal clustering validation step is based on 2 criteria, the first is Compactness, the members of each cluster must be close to each other. The typical measure of compactness is the variant. The second is Separateness which measures how different or separate clusters are from other clusters [17, 18]. One calculation of internal clustering validation is the Davies Bouldin Index (DBI) with the smaller the index value, the better the clustering results. DBI is based on the similarity of the R_{ij} size of the cluster. R_{ij} is a cluster with S_i as dispersion size and d_{ij} as distance.

$$R_{ij} = \frac{S_i + S_j}{d_{ij}} \quad (4)$$

where, $$d_{ij} = d(v_i, v_j) \quad (5)$$

and, $$S_i = \frac{1}{|C_i|} \sum_{x \in C_i} d(x, v_i) \quad (6)$$

Therefore, the Davies Bouldin Index (DBI) is: $$DBI = \frac{1}{n_c} \sum_{i=1}^{n_c} R_i \quad (7)$$

with n_c is amount of cluster, then: $$R_i = max_{j=1,\dots,n_c; i \neq j}(R_{ij}) \quad (8)$$

As a comparison reference, the original data is first clustered with k-Means by conducting various experiments to get the most optimum k value. The measurement to determine the most optimum k value is to calculate the DBI value and find the lowest value. Next, the original data is reduced using either SVD or SOM. Each data that has been reduced is clustered with k-Means and each DBI value is measured.

In addition to DBI values that describe quality rather than clusters, computational time was also observed and evaluated for each test both in the original data clustering by k-Means and in clustering data reduced by SVD and SOM. Furthermore, all clustering performance both without dimension reduction and dimension reduction by SVD and SOM were compared with statistical tests, Analysis of Variance (ANOVA).

3. Results

Testing is done with the help of RapidMiner Studio Version 9.0 software. The first test is testing the original data to obtain the most optimum k value. There have been 12 parameter variations were used where for each variations was tested 10 times; and obtained the optimum k value at k = 3. The variation of the k-Means parameters were done by changing the values of determined or not good start value; use or not local random seed value; and determining random seed value. Furthermore, the value of k = 3 is used in subsequent tests for data that have been reduced by SVD and SOM, with variations in the same k-Means parameter settings.

Table 1. Clustering results before and after reduction

Parameter Variations - #	k-Means		SVD + k-Means		SOM + k-Means	
	Average DBI	Average Time (sec)	Average DBI	Average Time (sec)	Average DBI	Average Time (sec)
1	4.344	18	0.395	43	0.762	32
2	4.489	16	0.395	42	0.762	31
3	4.391	15	0.395	45	0.762	28
4	4.542	15	0.395	57	0.750	30
5	4.725	20	0.395	66	0.762	33
6	4.491	14	0.395	43	0.762	28
7	4.613	14	0.395	42	0.762	31
8	4.275	14	0.395	45	0.762	30
9	4.524	15	0.395	44	0.762	29
10	4.602	14	0.395	43	0.762	30
11	4.590	21	0.395	43	0.762	28
12	4.522	17	0.395	48	0.762	28

But the dimension reduction process adds the total computing time significantly as well, which originally was 16.08 seconds to 46.75 seconds by SVD and becomes 29.83 seconds by SOM. Or the dimensional reduction process by SVD has an impact on increasing computing time almost 3 times longer; while dimension reduction by SOM is faster than SVD because it only adds 1.8 times longer computing time.

4. Conclusion

SVD reduction techniques work well reduce the data by processing the entire data matrix (using input term weighting), by reducing the multiplication of the matrix solutions to a minimum, as a result SVD produces a good representation of cluster input data. Although the SOM reduction technique is also able to provide significant results but not as good as the SVD, but SOM has an advantage over SVD, which is significantly faster computation time.

Because in this study the focus was on high-dimensional data originating from text documents, then feature extraction reduction techniques could be considered less suitable to process text data dimension reduction. This is because the extraction technique only takes dataset with the highest variance. While the text dataset presented with weights word has a different variance values in each document, so that the input data used do not represent all data values.

References

[1] Chen T C *et al* 2009 *World Congress on Computer Science and Information Engineering* **pp. 670-673**.

[2] Jambak M I *et al* 2018 The Impacts of Singular Value Decomposition Algorithm Toward Indonesian Language Text Documents Clustering *International Conference of Reliable Information and Communication Technology* 173-183.

[3] Jin X and Han J 2017 K-means Clustering *Encyclopedia of Machine Learning and Data Mining* ed: Springer 695-697.

[4] Keogh E and Mueen A 2017 Curse of dimensionality *Encyclopedia of Machine Learning and Data Mining*, ed: Springer 314-315.

[5] Han J *et al* 2011*Data mining: concepts and techniques*: Elsevier.

[6] Hasanah SIR *et al* 2018 Comparison of dimensional reduction using singular value decomposition and principal Component analysis for clustering results of Indonesian language text dsocuments *The 2nd International Conference on Applied Sciences, Mathematics, and Informatics (ICASMI)*, Bandar Lampung

[7] van Gerven M and Bohte S 2017 Editorial: Artificial Neural Networks as models of neural information processing *Frontiers in Computational Neuroscience* 11.

[8] Kohonen T 1990 The self-organizing map *Proceedings of the IEEE* 78 1464-1480.

[9] Chen Q *et al* 2017 Towards a Hybrid Approach of Self-Organizing map and density-Based spatial clustering of applications with noise for image segmentation *Developments in eSystems Engineering (DeSE) 10th International Conference* 238-241.

[10] Campoy P 2009 Dimensionality reduction by self organizing maps that preserve distances in output space *Neural Networks IJCNN 2009. International Joint Conference* 432-438.

[11] Cong T *et al* 2013 Applications of ANNs in flow and heat transfer problems in nuclear engineering: a review work *Progress in Nuclear Energy* **62** 54-71.

[12] Leskovec J *et al* 2014 *Mining of massive datasets*: Cambridge university press.

[13] Kohonen T 2013 Essentials of the self-organizing map *Neural Networks* **37** 52-65.

[14] Haykin S 1999 Multilayer perceptrons *Neural networks: a comprehensive foundation* **2** 156-255.

[15] Madhulatha T S 2011 Comparison between k-means and k-medoids clustering algorithms *Advances in Computing and Information Technology*, ed: Springer 472-481.

[16] Nur'aini K *et al* 2015 Combination of singular value decomposition and K-means clustering methods for topic detection on Twitter *Advanced Computer Science and Information Systems (ICACSIS), 2015 International Conference* 123-128.

[17] Ristevski B *et al* 2008 A comparison of validation indices for evaluation of clustering results of DNA microarray data *Bioinformatics and Biomedical Engineering ICBBE 2008. The 2nd International Conference* 587-591.
[18] Wiwie C *et al* 2015 Comparing the performance of biomedical clustering methods *Nature methods* **12**, 1033.

Indoor Air Quality (IAQ) Sensing Device Prototype: Software-Hardware Architecture and Performance

A Nor Azhar[1*], M Rahmah[2], AH Roslina[3], and Z Fauziah[3]

[1]Department of Computer Science, [2]Department of Graphic & Multimedia Technology, [3]Department of Software Engineering, Faculty of Computer Systems & Software Engineering, Universiti Malaysia Pahang, Lebuhraya Tun Razak, Gambang, 26300 Kuantan, Pahang, Malaysia

Email: nazhar@ump.edu.my

Abstract. In a green environment, air quality is among the prominent aspects to be considered to prevent pollution and in turn maintain safe air. Sick Building Syndrome (SBS) and Building Related Illness (BRI) are the problems that must be encountered by building occupants. Existing devices are quite expensive, and more attention is needed on monitoring air quality. Therefore, this paper discusses a software-hardware architecture proposal to overcome the existing IAQ data retrieval performance problem. The objective of this work is to analyse the most recent IAQ sensing device developed by the researcher, to evaluate the performance attribute, and to implement the software-hardware architecture based on the Performance-Driven Software Architecture Refactoring (PDSAR) method. In order to assess the overall performance of the architecture proposed, the prototype of the Indoor Air Quality (IAQ) sensing device was developed. The testing was performed during a period of six months. The results demonstrate that the average performance for 540 readings is 76.67%. This shows that the architecture can be accepted as a generic IAQ sensing device development structure. The device act as an alternative, cheaper and safer device which can be employed by all building owners in the future.

1. Introduction

Air quality inside a closed room concerns many people. Many government bodies, especially science and technology, urge industries and building managers to increase the building health through Green Building Initiatives. However, this effort is far from being accomplished, primarily due to cost constraints. Indoor Air Quality (IAQ) detection is among the approaches used to overcome Sick Building Syndrome (SBS) and Building Related Illness (BRI). Recent research adopted educational [1], factory [2], vehicle [3] and residential premises [4] as domains. In this paper, we discuss the sensing device focusing on architecture and its performance. We also propose a novel architecture, and report on how it performs, while the IAQ device is running. IAQ was defined as an environmental issue in a confined space which comes from pollutants, and it harms occupant's health. These harmful pollutants (gasses, volatile organic compounds (VOCs), and pollen/dust) all contribute to SBS, resulting in an uncomforted feeling while in that space. Sensing devices are cheaper and more reliable through-out the year. However, the performance of this architecture has not been tested well.

Based on [5], Performance-Driven Software Architecture Refactoring (PDSAR) can be defined as a discipline that collects approaches, methodologies, and tools aimed at introducing performance

IOP Conf. Series: Materials Science and Engineering **551** (2019) 012047 doi:10.1088/1757-899X/551/1/012047

assessment in the context of software architectures. It focuses on the top end of the architecture of software, and on the performance itself. Most of the architecture in research on IAQ sensing devices lacks these attributes.

2. Related Works

Several researchers compare overall strengths, characteristics, and performance. Since we focus on performance, the data transfer is needed for performance analysis. We also put in mind the economic value of the device, since it would reduce the reliability factor. This ensures the effective usage of energy based on eco-design methodology [6].

We selected six most recent sensing devices to compare performance. Several criteria were chosen, related to performance analysis. Sensor main device is the "brain" of IoT devices, controlling all sensors connected to them. For every device, we compare the architecture, since it is important to make sure the structure of the developed solution can be implemented successfully in other domains. The performance results were also chosen as among the attributes to be compared in determining the best solution. The data transmission protocol and time frame were chosen to determine how good the devices managed to sense the environment and store the data.

Table 1. Comparison between six recent IAQ sensing devices

Criteria	**IEQ Toolbox** [7]	**Proactive IAQ** [8]	**SPIRI** [9]	**IAQ IOT** [10]	**Cloud IAQ** [11]	**WSN IAQ** [12]
Sensors main device factor	9 Arduino, XBee	1 PIC microcontroller PIC16f877a, ESP8266	10 Raspberry Pi Zero W	3 IDT smart gas sensor, Raspberry Pi	5 Grove, GSM Module	6 Arduino UNO
Top down approach architecture	Partial, Wireless Mesh Network	No	No, 6LoPAN	No, Based on IDT Architecture	No, Based on GSM	No
Performances	Good	No Discussion	Good	No Discussion	Bland-Altman statistical analysis model	Success
Data transmission protocol	Calibration with National Instrument DAQ card	Wi-Fi	Ripple (RPL), CoEP (Constrained Extensible Protocol)	Wi-Fi and Bluetooth	GPS and GSM	Wi-Fi. WSN, SD Card
Data transmission timeframe	1 minute	Not stated	1 minute	Not stated	3.6 seconds (Average)	1 second

Table 1 summarizes the findings of most recent works by prior work. It seems that only SPIRI fulfills the performance based architecture, and cloud IAQ separately does the after session performance analysis. Architectures in these works complement with software-hardware architecture, but none of them have a top-down approach to rely on device performance over time, due to problems in cloud computing devices, as stated by al-Sharafi [13].

3. Proposed Architecture and Implementation

Inspired by PDSAR architecture, we developed the architecture for the focused domain. In Figure 1, the flow of the data transfer is presented. The sensor reads the environment/contamination value, while the data performance rate is calculated. The lower the value of data performance, the better the IoT device would be. On the other hand, any error in the data leads to performance issues, which contributes to the reliability of the architecture. It is an indicator of how well the sensors communicate with the database. If the performance of data errors is high, the architecture should be readjusted. This approach is inspired by agent-based works in cloud computing [14]. The PDSAR has demonstrated a good sign in real time software architecture, which is able to perform top-down approach checking, while compromising with the performance of the devices. This generic architecture supports the dual/more architecture running simultaneously to create a high-reliability performance in software-device architecture. We propose a top-down architecture inspired by PDSAR, as shown in Figure 2. The performance is compromised based on the current architecture, and it then changes with time, based on the variables in Figure 1.

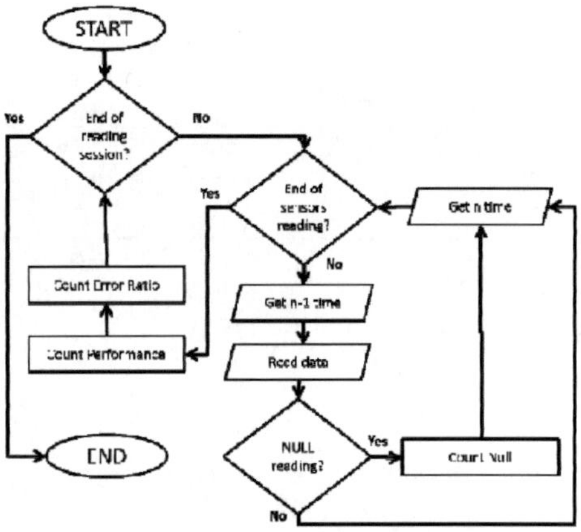

Figure 1. Flowchart for IAQ sensor device – Data Management

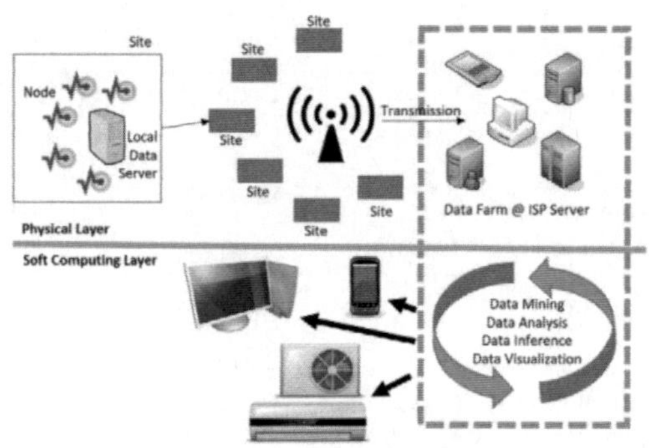

Figure 2. Proposed architecture and implementation

In this general architecture, we performed simultaneous sensor data processing through separate entities from the storage. This is due to a report from a recent IAQ device researcher, mentioning that the board or the main factor of the device could potentially reduce the performance. To reduce the burden on the sensing device main board, sensor data from nodes are transferred and accumulated to the local data server. During this stage, the burden of the device was reduced significantly. Figure 1 shows the flow diagram of the proposed architecture. The performance of data transfer was counted based on three factors: null, unacceptable and data transfer rates. Two sessions of the data performance log were set up, namely, from sensors node to local server, and from the local server to data farm.

This action was followed up with interchangeable architecture, which is based on the data performance. The device works to capture the data, and not to accumulate them. Local data server does the work, as shown in Figure 3. The data reading gap is optimized, especially in reading and analysing. We had five architectures working. We tested simultaneously, and performed modifications based on the performance of the readings. These five architectures were changed based on the performance of the data transfer throughout time. It is controlled by local (site), and data farm servers. The redefined flowchart is shown in Figure 4. The changes factor compromise on several occasions: (a) lack of power, or the ability to transfer data slowly because the device cannot accumulate the data in the timeframe; (b) failure of sensor nodes, which affects the other nodes due to incomplete data; and (c) Failure of the server, which affects the reliability of IAQ reading.

There are two types of switch conditions: (a) NULL wise, a signal for single or multiple sensor errors; and (b) performance wise, based on the transmission rate calculation. If the rates arrive at the threshold, switching will occur. We tested 20 nodes of sensors, which have five nodes on each server. The sensors are temperature, humidity, environmental, dust sensor PM2.5, carbon monoxide and sound/noise. The main board we chose is Lolin NodeMCU, which is based on the ESP 8266 chip. For the Bluetooth connection, HC05 was chosen due to the capability to read and write in real time. The time frame for sensor readings was set every 8 hours to the server farm. However, in a local site, the timeframe is five seconds.

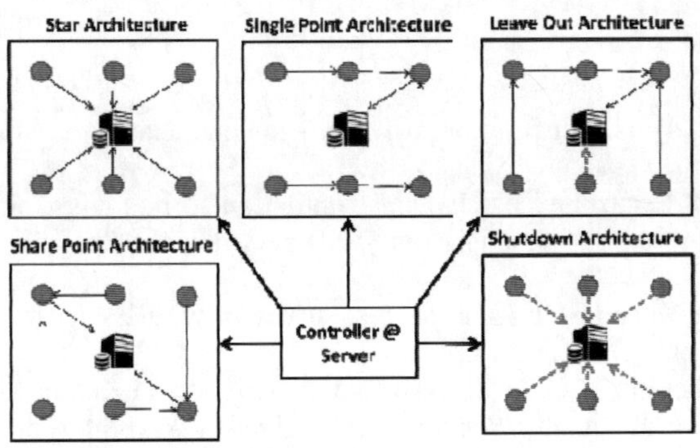

Figure 3. Transition architecture for data transmission

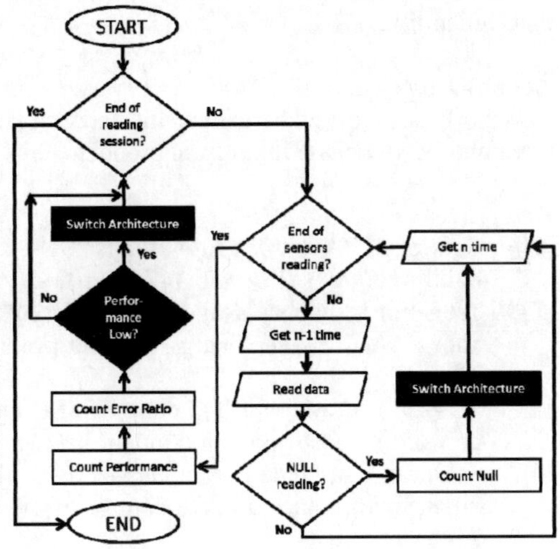

Figure 4. Redefine flowchart for IAQ sensor device – Performance Based

4. Result and Discussion

Simulations of the device were developed to evaluate the preliminary performance test. Later on, we tested the device on-site, to assess the architecture proposed. The testing was performed during a period of six months. An interruption on a sensor was performed to the sensor to check its reliability. Every time the architecture makes a transition, it accumulates the changes. The same goes to the null switch counter (when the sensors give over 5 null or zero values). Table 2 shows the summary of sensors log as the result of the IAQ device performance. The average performance count was on the percent of perfect readings versus the readings counted. With a result of 76.67% accuracy by the device, the proposed architecture can be accepted as one of the ways of measuring the performance, and ensures the quality of data sensing.

Table 2. Architecture performance log

A	1 – 3 month period				4 to 6 month period				Overall
	B	**C**	**D**	**E%**	**B**	**C**	**D**	**E%**	**F%**
1	18	10	16	83.70	43	33	12	67.41	75.56
2	30	12	23	75.93	21	22	40	69.26	72.59
3	21	34	18	72.96	13	54	30	64.07	68.52
4	19	20	27	75.56	22	22	16	77.78	76.67

*Legend : A : Local server; B : Architecture switch count; C : Null switch count; D : Performance switch count; E : Σ(switch_count)/270*100; F : Σ(switch_count)/540*100*

5. Conclusion

The architecture of the IAQ sensing device was proposed to obtain the optimal WSN performance in IAQ. The testing done shows the performance and reliability of accuracy of the prototype. This can be an alternative solution that is cheaper and more affordable for very confined space owner.

Comparing to [5] test subject, the implementation of PDSAR prove the effectiveness of search based architecture, compared to a metaheuristic approach. This result also proves the freedom of architecture results in better output, and also is also suitable for sensing devices. The performance of the sensing device concludes that, similar applications need a search based architecture to enhance their reliability.

Acknowledgments

This work is sponsored by Universiti Malaysia Pahang Internal Research Grant (RDU170305) under the umbrella of HeDoo innovation product.

References

[1] Krawczyk DA and Wadolowska B 2018 Analysis of indoor air parameters in an education building *Energy Procedia* **147** 96–103

[2] Tronchin L, Fabbri K and Bertolli C 2018 Controlled mechanical ventilation in buildings: A comparison between energy use and primary energy among twenty different devices *Energies* **11** 8

[3] Yi WY, Lo KM, Mak T, Leung KS, Leung Y, and Meng ML 2015 A survey of wireless sensor network based air pollution monitoring systems *Sensors* **15** 12

[4] Schieweck A, Uhde E, Salthammer T, Salthammer LC, Morawska L, Mazaheri M and Kumar P 2018 Smart homes and the control of indoor air quality *Renew. Sustain. Energy Rev.* **94** 705–718

[5] Arcelli D, Cortellessa V, and Pompeo DD 2018 Performance-Driven Software Architecture Refactoring *Proc. - 2018 IEEE 15th Int. Conf. Softw. Archit. Companion* 2–3

[6] Romli A, Prickett P, Setchi R, and Soe S 2015 Integrated eco-design decision-making for sustainable product development *Int. J. Prod. Res.* **53** 2 549–571

[7] Karami M, McMorrow GV, and Wang L 2018 Continuous monitoring of indoor environmental quality using an Arduino-based data acquisition system *J. Build. Eng.***19** 412–419

[8] Cynthia BB, Priya BD, Nandhini R, Sindhuja P, Senthilkumar MA, and Raja S 2018 Proactive Indoor Air Quality Monitoring System *Int. J. Recent Inno. Trends Comput. Commun.* **6** 3 133–138

[9] Esquiagola J, Manini M, Aikawa A, Yoshioka L, and Zuffo M 2018 SPIRI: Low power IoT solution for monitoring indoor air quality *Proc.of the 3rd Int. Conf. on Intern.of Things, Big Data & Security* **2018–March** 285–290

[10] Parmar G, Lakhani S, and Chattopadhyay MK 2018 An IoT based low cost air pollution monitoring system *Int. Conf. Recent Inno. Signal Process. Embed. Syst.* **2018–Janua** 524–528

[11] Goh CC, Kamarudin LM, Sudin S, Visvanathan R, and Zakaria A 2018 Cloud-Based In-Vehicle Air Quality Monitoring System with GSM Module *J. Tele. Electron. Comput. Eng.* **10** 1 77–81

[12] Tijani IB, Almannaee AD, Alharthi AA, and Alremeithi AM 2018 Wireless sensor node for indoor air quality monitoring system *2018 Adv. Sci. Eng. Tech. Int. Conf.* 1–6

[13] Al-Sharafi MA, Arshah RA, and Abu-Shanab EA 2017 Factors affecting the continuous use of cloud computing services from expert's perspective *2017 IEEE Reg. 10 Conf.* 986–991

[14] Ahmed FD, Majid MA, Sharifuddin M, and Jaber AN 2016 Software Agent and Cloud Computing: A Brief Overview *Int. J. Softw. Eng. Comput.* **2** 108–113

IOP Conf. Series: Materials Science and Engineering **551** (2019) 012048 doi:10.1088/1757-899X/551/1/012048

Influence of Network Speed on Structured Query Language (SQL) Database Data Transmission Performance in an Internet of Things (IoT) Sensing Device on Single Board Computers (SBC)

A Nor Azhar[1*], S Nur Ayuni Hasna[2], AH Roslina[3], Z Fauziah[3] and M Rahmah[4]

[1]Department of Computer Science, [2]Department of Computer Systems & Networking, [3]Department of Software Engineering, [4]Department of Graphic & Multimedia Technology, Faculty of Computer Systems & Software Engineering, Universiti Malaysia Pahang, Lebuhraya Tun Razak, Gambang, 26300 Kuantan, Pahang, Malaysia

*Corresponding author: nazhar@ump.edu.my

Abstract. One of the crucial elements in an Internet of Things (IoT) environment is a database. IoT performance will be at stake if the wrong database is adopted. In this research, three Structured Query Language (SQL) databases were tested against multiple network speeds in an IoT device. Single board computers (SBC) were used as a media of testing instead of an ordinary computer. A controlled wireless sensor network (WSN) was developed with several uniform constants and variable network speeds. A new data transmission rate equation also was proposed and tested. From the result, it was found that not all SQL databases can cater IoT devices. Some of them showed that performance decreased along with network speed. The correlation, between 0.90 and 0.96, proved the strong influence between the research subjects. In addition, file-type database is the best option available for SQL-wise IoT storage.

1. Introduction

An emerging advancement in the Internet of Things (IoT) has the attention of the world and is quickly launching it into the 4th Industrial Revolution. Terms like 'data farming' and 'sensor networks' have become a frequent subject of discussion. Data transmission from sensors such as GPS, thermometers, or vital signs monitors need a robust communication network. Different applications and sensor types have different requirements for the transmission of data, but all must consider loss-tolerance, timeliness, priority and completeness.

The aim of this paper is to analyze the relationship between network speed and data transmission rates through wireless sensor networks (WSN). WSN are a distributed group of sensors able to secure data at a low cost that can be used for real-time monitoring of physical changes in multiple environments. This paper will focus on data transmission rates of different SQL databases by analyzing the latency of data between these WSN sensor nodes to the databases. Meanwhile, these selected SQL databases will be embedded in single board computers (SBC) which are quick becoming good mediums between sensors and farming servers.

IOP Conf. Series: Materials Science and Engineering **551** (2019) 012048 doi:10.1088/1757-899X/551/1/012048

1.1. SQL Databases

In recent development, alternative database systems called NoSQL databases have been introduced to overcome the problems of SQL databases. However, the non-relational concept in NoSQL databases make it hard to do reporting. So, even though the development of NoSQL databases is rapidly evolving, they cannot contend with the vast and easy adoption of SQL-based storage. In this paper, three known SQL databases will be tested on their latency in assisting fast data transmission in an IoT environment. With our simulation in mind, these databases must be able to be embedded in single board computers, which is the main factor for the implementation of IoT with an open-source license. The databases are:

(a) SQLite: a file-based database. It is extremely portable and reliable because it consists of a single file on the disk. It has an amazing library that gets implanted inside the application that uses its interface.

(b) MySQL: a full-featured open source relational database management system (RDBMS). The data stored are grouped using SQL to access data.

(c) PostgreSQL: a standard-compliant and extensible database. It is a high-performance, object-oriented and relational database. The support for concurrency is achieved with varied locking modes.

These relational databases are among the most famous data storage systems for an IoT environment. All databases should support the fast pace of data transmission, However, the implementation of them in single board computers is questionable; is the data transmission performance influenced by network speed or not?

1.2. Single Board Computers (SBC)

SBC are computers with small form factor. It is widely used in education, especially in microcontroller learning or programming. The need for small form factor processing devices in the IoT environment has contributed to its widespread adoption. Starting with SolidRun in early 1996 to the latest Wandboard in Spring 2018, the processors have become faster and device memory bigger. In IoT environments, these SBC can replace standard computers, being low cost and easily placed in remote areas. Several successful research papers have confirmed SBC capabilities, as reported by Stylia for DDoS attack [1], Mansoor in smart cities [2], and Guha Roy in message telemetry [3]. In this paper, we will use Raspberry Pi 3 B+ SBC as a server while Lolin NodeMCU as a client to perform the data transmission analysis in the SBC.

1.3. Related Works

Several researchers have done their work on comparing storage databases. In [4], SQL and NoSQL databases are compared. Experiments focused on concepts and characters of the databases instead of hardware. The comparison was made by determining the performance of the different databases. They found out that SQL databases are more stable in response, but NoSQL provided faster query response. Furthermore, [5] presented a discussion between the use of "without database approach" (WDA) and the SQL databases. Based on their result, the use of WDA gave a better performance result. Through this paper, we found that latency of data transmission was not well discussed as they focused more on data queuing. It is a loophole in a data transmission analysis, since data transmission performance in IoT have many parameters. Finally, as discussed in [6], there are three major metrics for evaluating data transmission performance in sensor devices: network delay, average CPU, and memory usage. In our paper, only network delay will be experimented by employing different limitations on network speed

IOP Conf. Series: Materials Science and Engineering **551** (2019) 012048 doi:10.1088/1757-899X/551/1/012048

2. Methodology

2.1 Hardware Preparation

Three Raspberry Pi 3 B+ and Lolin NodeMCU were chosen as the main SBC components of this research, whereas a DLink Wireless Router DIR-868L was employed as the WSN controller. For each SBC, different databases were embedded in its built-in storage. However, due to vast data, we moved these databases into separate USB drives. Figure 1 shows the WSN architecture and the embedded databases. Each node consisted of three sensors, one each for humidity/temperature, light, and carbon monoxide. These nodes sent the data from the sensors to the SBC through the router. This architecture follows the data integration in relational databases [7]. The hardware setup is shown in Figure 2.

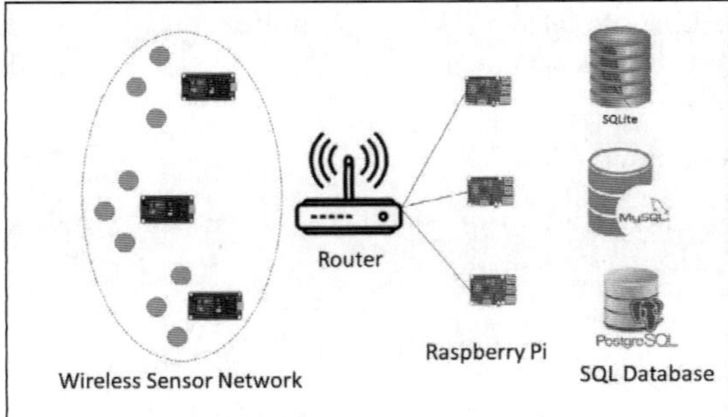

Figure 1. WSN architecture Mapping

Figure 2. Hardware setup

2.2 Software Preparation

To enable the WSN, Dynamic Host Configuration Protocol (DHCP) was set up. The sensor nodes were the client, while the SBC was the server. A router was connected to the laboratory network point to obtain real-time data from the Network Time Protocol (NTP) server. The nodes sensed change in the environment, captured the data and sent it to the respective SBCs with different databases. The time in milliseconds was taken when the packet of data was sent, and taken again when the packet of data arrived. This setting was constant for the three data capturing services. This semantic query concept is inspired by the query expansion technique [8]. Figure 3 explains how the data was sent back and forth within the simulation framework.

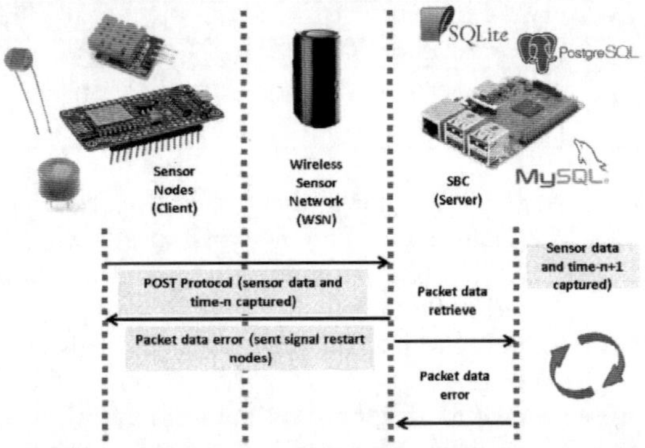

Figure 3. WSN with Hyper Text Transfer Protocol (HTTP) Architecture

2.3 Data Transmission Performance Rates Metrics

The WSN performance for sending and retrieving data can be measured by data latency. Isakovic had defined this term as a time delay between two machines when sending and receiving a packet of data [9]. In this paper, the equation for data latency is defined as

$$D_L = t_{(n+1)} - t_{(n)} \quad (1)$$

where DL = data latency in millisecond, t = timestamp, n = data in nodes and $n + 1$ = data in database

In this experiment, to determine the performance, packet data delay transmission was considered. Data stream classification [10] had been used to reduce the hardware load.

$$D_T = N / R \quad (2)$$

where D_T = transmission delay in millisecond, N = number of bits and R = bits per millisecond.

Therefore, the data transmission performance rates can be determined as

$$D_{TR} = D_L / D_T \quad (3)$$

where D_{TR} = data transmission performance rates.

Based on these equations, it is understood that the higher data transmission rates, the better the SQL database performed in the SBC.

3. Result and Discussion

The data transmission performance was observed using three different network speeds: 10bps, 100bps, and 1000bps. These speeds were throttled down using a network management program embedded in the router. The summary of results comparing the three significant network speeds is shown in Table 1. This IoT system was executed for 240 hours with a delay between readings being about five seconds (540 readings per database). For each data transmission, the latency rates were calculated using Formula 1 to 3 in real time. Then, the average of rate of data latency was calculated. We plotted the graph to visualise the relationship between network speed and data transmission performance in Figure 4.

Table 1. Summary of result for the data transmission performance of different SQL databases across three significant network speeds

Network Speed	**10 bps**			**100 bps**			**1000 bps**			**Average**
Metrics	D_L	D_T	D_{TR}	D_L	D_T	D_{TR}	D_L	D_T	D_{TR}	
SQLite	2.23		0.464	3.35		6.979	4.25		88.542	31.995
MySQL	5.65	4.8	1.177	2.75	0.48	5.729	2.50	0.048	52.083	19.663
PostgreSQL	2.98		0.620	2.78		5.791	3.00		62.500	229.70

In Figure 4(a), the plotted graph shows four further network speeds that was implemented: 200bps, 400bps, 600bps and 800bps. The inclusion of these network speeds was to clearer see the relationship between network speed and data transmission performance. The correlation between these two attributes are presented in Figure 4(b). All three databases had strong speed-transmission performance relationships, with PostgreSQL highly reliant on network speed. However, the other databases were also not far behind.

SQLite showed the highest average of data transmission rates in this experiment. However, based on network speed, PostgreSQL showed the best solution for any speed. On the other hand, MySQL database showed higher data transmission performance for small data, good for enterprise-level data storage. Comparing this to [4], [5] and [6], it is proven that SQL databases are reliable for IoT sensing devices embedded in SBC.

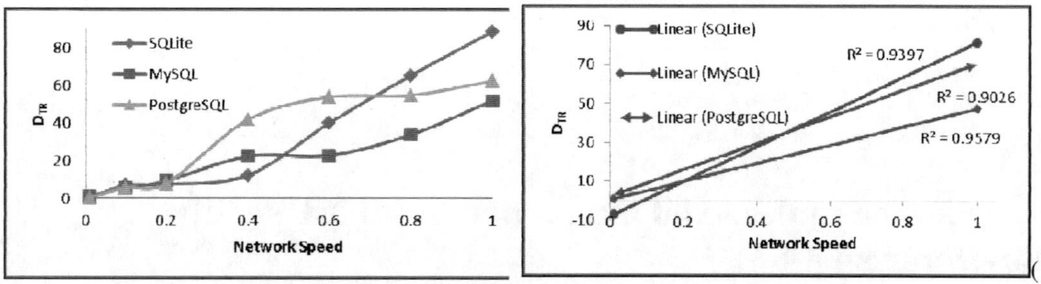

(a) b)

Figure 4. SQL-wise database in SBC data transmission (a) performance rates and (b) correlation analysis

4. Conclusion

From this research, we can conclude that not all databases are ready an IoT environment in SBC. Even with the same network infrastructure and hardware, data transmission rates are highly dependable to network speed. However, the best finding is that MySQL seems reliable in any computer architecture. It is robust database which does not rely too much on network architecture. On the other hand, PostgreSQL with its file-based architecture is reliable with higher network speeds; this performance is promising for newly-introduced 5G networks.

Acknowledgments
This work is sponsored by Universiti Malaysia Pahang Internal Research Grant (RDU170305) under the umbrella of HeDoo innovation product.

References

[1] Stylianopoulos C *et al* 2018 *Nordic Conf. on Secure IT Syst.* 187.
[2] Nasir M *et al* 2018 *J. Parallel Distrib. Comput.*
[3] Guha R D *et al* 2018 *Futur. Gener. Comput. Syst.* **89** 300.
[4] Rautmare S *et al* 2017 *IEEE Int. Conf. Adv. Comput. Appl.* 235.
[5] Zyrianoff I *et al* 2018 *IEEE Symp. Comput. Commum.* 688.
[6] Syafrudin M *et al Sensors* **18** 9.
[7] Maria G *et al* 2018 *Int. J. Softw. Eng. Comput. Syst.* **4** 95.
[8] Raza M A *et al* 2018 *Adv. Sci. Lett.* **24** 7490.
[9] Isakovic H *et al* 2018 *J. Comput. Syst. Softw. Eng.* 1153.
[10] Haneen A A *et al* 2018 *J. Phys. Conf. Ser.* **1018** 012019.

A review on multimodal interaction in Mixed Reality Environment

Aladin Mohamad Yahya Fekri [1,2] **, and Ismail Ajune Wanis**[1,2]

[1]Mixed and Virtual Environment Research Lab (mivielab), Vicubelab, Universiti Teknologi Malaysia, 81310 Johore, Malaysia

[2]School of Computing, Faculty of Engineering, Universiti Teknologi Malaysia, 81310 Johor, Malaysia

E-mail: yahyafekri@gmail.com, ajune@utm.my

Abstract. Mixed Reality (MR) environment proposes to blend together virtual and real environments where both world coexist and can respond to each other. As MR has the understanding of real environment, it is possible for user to use their surrounding as a means to interact with the virtual content. The user interaction in Augmented Reality (AR) and Virtual Reality (VR) are widely explored however very less in MR. Interaction must be intuitive as possible to produce natural user interaction. Multimodal interaction is an advanced interaction techniques when involve the combination of more than one input modalities such as speech with gesture, gaze and gesture, and gaze with speech. The inputs or sensory however are not restricted to just two modalities. Therefore this paper presents a review on multimodal interaction in MR environments.

1. Introduction

In the rapid advancements in computer vision, display technology, input systems and graphics processing, more researchers and large companies had their interest in Mixed Reality (MR). The term MR was originally discovered by Paul Milgram and Fumio Kishino [1], as a virtuality continuum where it extends from real world to completely virtual environments with Augmented Reality (AR) and Augmented Virtuality (AV) lies in between. AR lies close to real world continuum and it is an experience that overlay virtual imaginary onto the real world [2]. Virtual Reality (VR) on the hand is an experience that totally immerse users inside virtual imaginary world and cut off user's view from real world.

The emerging of many upgraded and new input systems invites more urges to make a natural user interaction. Interaction in AR usually obeys the VR principle. In MR however it looks different. For example, user can defend themselves from virtual enemy attack by using a physical book or bottle to act as a shield and use the same physical book to smash and damage the enemy. Multimodal interaction by combining several input modalities in order to achieve more natural user interaction in their application. As Bunt [3] and Quek [4] agreed that the interaction between this world and human is naturally multimodal. As to apply multimodal interaction in MR, Ismail [5] stated that it is a solution for more advanced interaction technique using gesture with speech inputs in AR. MR devices such as Microsoft HoloLens and Magic Leap did not cut users from the physical world, unlike immersive devices such as Oculus Rift and HTC Vive where these devices uses an opaque headset that blocks out the physical environment from user's view. Despite the differences, these available headset however offer user with full 6 Degree of Freedom (DOF), but both in rotational and translation movement within their respective environment. The user interaction in AR and VR are widely explored but very less in MR. Therefore, this paper presents the previous works on tangible interaction and multimodal interaction. The paper gives a review on multimodal interaction in MR environment.

2. User Interaction Techniques

There are many interaction techniques available and some of them are (1) Tangible User Interface (TUIs) and (2) Multimodal Interaction.

2.1 Tangible Interaction Techniques

Tangible interface invites different interaction techniques with real tangible objects are acting as a tool to perform 3D object manipulation. There several examples to present tangible interaction methods such as the Magic-Lens [6], Cubical Interface (CUI) [7], and the Magic-Cup [8]. As in Fig. 1 (a), the Magic Lens has used a physical handle to portray as a virtual lens. The interaction by moving the both hands to move the lens so it works to zoom or scaling. While the Magic-Cup (as in Fig. 1 (b)), uses the method of "covering" virtual objects with a real shape of cup, when the shape can actually hold a virtual object. A cube-based tangible AR interface (as in Fig. 1 (c)) when a user uses their both hands to manipulate and compose virtual objects. Fig. 1 (d) shows the EditDunia research project on Ancient Malacca [9] which has executed the paddle-based interaction method. The paddle is a single marker attached as a tool for user to perform virtual object manipulation and move the object around the ground marker.

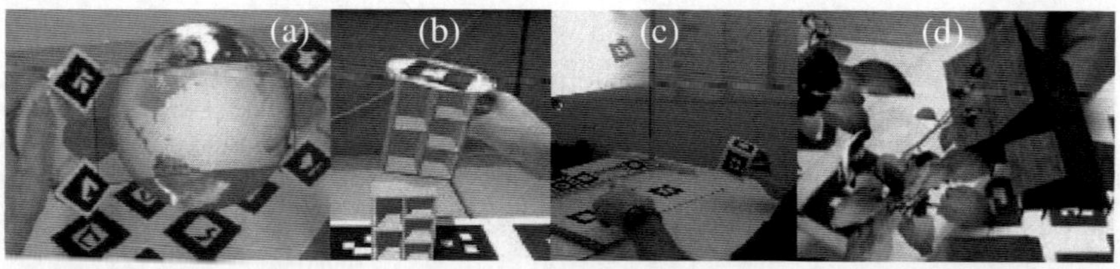

Figure 1. (a) CUI, (b) MagicLens, (c) MagicCup and (d) EditDunia, are the cases of tangible interface.

2.2 Multimodal Interaction Techniques

Multimodal interaction is an advanced interaction techniques when involve the combination of more than one input modalities such as speech with gesture, gaze and gesture, and gaze with speech. Multimodal interaction in AR has a great potential to enhance the user interaction and it is promising when more than input modalities have been invoked in the AR interface. For instance, G-SIAR [10] has introduced a natural interaction technique by two-handed free-hand gesture for object manipulation. It invites gesture and speech to execute a direct manipulation interaction technique to dynamically manipulate virtual content. As proposed by [9], the extended version of Ancient Malacca project [5], they also have created an advanced method to improve the user interaction. They have employed inputs modality, gesture and speech input to perform 6DOF virtual object manipulation. Compare to G-SIAR, [9] have advanced an intuitive and natural interaction by utilizing the sensor-based hand tracking when the motion data drawn as skeleton hands to represent the gesture input.

3. Multimodal Interactions in MR

Interaction methods in MR are based on the methods people have implemented in AR environment, researchers and experts are considering AR subsets to MR technology. Therefore, the methods in AR have been thoroughly studied recently the experts have begun to explore multimodal interaction in

IOP Conf. Series: Materials Science and Engineering **551** (2019) 012049 doi:10.1088/1757-899X/551/1/012049

MR. Table 1 shows the previous works with regards to multimodal interaction in MR spectrum. It describes and gives an overview of what the system is all about and the input modalities that are used.

Table 1. Previous works multimodal interaction in MR Spectrum

Year	Title	Input modalities
2014	Grasp-Shell vs gesture-speech: A comparison of direct and indirect natural interaction techniques in AR [10]	Speech, Gesture
2015	Vision-Based Technique and Issues for Multimodal Interaction in AR [11]	Facial Recognition, Gesture, Speech
2015	Multimodal Fusion: Gesture and Speech Input in AR Environment [9]	Speech, Gesture
2015	ARZombie: A Mobile AR Game with Multimodal Interaction [12]	Facial Recognition, Touchscreen
2016	Simplified Industrial Robot Programming: Effects of Errors on Multimodal Interaction in WoZ experiment [13]	Touch-Sensitive Table, Gesture, 6D Pointing Device, Touchscreen, Direct Robot Arm
2016	An Intelligent Multimodal Mixed Reality Real-Time Strategy Game [14]	Speech, Gesture
2016	Hybrid Team Interaction in the MR Continuum [15]	Eye, Head, Hand, Gesture
2017	Multimodal Interaction in AR [16]	Speech, Gesture
2018	Wall mounted level: a cooperative MR game about reconciliation [17]	Digital Input Device, Physical Touch
2018	Designing an Augmented Reality Multimodal Interface for 6DOF Manipulation Techniques [5]	Speech, Gesture

As presented in Table 1, in 2014 G-SIAR [10] has introduced natural interaction techniques which combines speech and gesture. These techniques support 6DOF movement, uniform scaling, as well as physics-based interaction. It be seen in 2018, [5] has proposed new multimodal interaction using gesture and speech that invites 6DOF with sensor-based interaction. In 2015, [12] has developed an AR game that uses facial recognition and detection to recognise players face and superimposed a 3D zombie face on top of player's face. Next in 2016, [14] has produced a role-playing MR table top game and the players give instructions via speech-gesture interaction. The game was to combine interaction styles with gameplay mechanics in order to control the character. MR game [17] was developed in 2018 which is a cooperative game that powers multimodal interactions to support its chronicle of 'reconciliation'. However in 2016, [13] found that gesture and 6D pointing are the fastest modalities in many interaction metaphors for robot-human. A robot-human system has been proposed by [15], it enable collaboration between a robot, a human, and a remote human where the remote human are connected via VR. This system leverage eyes, head, hand and gesture tracking as a mean to interact and communicate with and within the system.

4. Conclusion

Across the MR spectrum, each of the technology whether be AR, VR or MR, all of them has their own respective advantages and roles in enhancing our physical reality and making this world a better place to live in. In today's rapid technological advancement, more new devices will be announced and will push the limits and expand the range in MR spectrum. Interaction also need to be as intuitive as possible and natural that permits users to interact naturally and multimodality without the needs to wear the devices. It can also be seen that researchers nowadays are making a collaborative system to invites more users to use the system simultaneously whether it is currently present user or a remote user or even with a robot. Multimodal interaction is an advanced interaction techniques when involve the combination of more than one input modalities such as speech with gesture, gaze and gesture, and gaze with speech. The inputs or sensory however are not restricted to just two modalities. Interaction is one of the biggest issues as interaction is the component that allows user to engage and express themselves with the system. Lacking in interaction metaphor can make the system to be less intuitive and need more improvements. Hence this paper presents the works in multimodal interaction for MR environments. Based on this chronological timeline the experts and researchers are started to explore the remote collaborative works in MR while significantly grasping the multimodal interaction for MR.

Acknowledgments

We would like to express our appreciation to Mixed and Virtual Reality Laboratory (mivielab) at Universiti Teknologi Malaysia (UTM). This work was funded by UTM-HIR High Impact Research Grants Scheme (Q.J130000.2451.04G77).

References

[1] Milgram P *et al* 1994 *IEICE TRANS. Info. and Sys.* **77** 1321.
[2] Akçayır M *et al* 2017 *Edu. Res. Rev.* **20** 1.
[3] Bunt H *et al* 1998 *Springer Science & Bus. Media* **1374**.
[4] Quek F *et al* 2002 *ACM Trans. on Cop. Hum. Inter. (TOCHI)* **9** 171.
[5] Ismail A W *et al* 2018 *In Pro. SAI Intel.Sys. Conf.* 309.
[6] Lee H *et al* 2011*Virtual Reality* **15** 133..
[7] Looser J *et al* 2004 *ACM 2nd Int. Conf. Comp. Grap. & Inter. Tech.* 204.
[8] Kato H *et al* 2003 *IEEE Int. In Augm. Rea. Toolkit Work.* 75.
[9] Ismail A W & Sunar M S 2015 *Spr. In Comp. Intel, Inf. Sys.* 245.
[10] Piumsomboon T *et al* 2014 *IEEE Int. Sym. Mix. & Augm. Rea. (ISMAR)* 73.
[11] Ismail A W *et al* 2015 *ACM The 8th Int. Sym.Vis. Inf. Comm. & Inter.* 75.
[12] Cordeiro D *et al* 2015 *IEEE In 7th Int. Conf. on Inter. Entert. (INTETAIN)* 22.
[13] Materna Z *et al* 2016 *In RO-MAN* 200.
[14] Link S *et al* 2016 *IEEE In Vir. Real.* 223.
[15] Moniri M M *et al* 2016 *ACM 22nd Conf. on Vir. Real.Soft. & Tec.* 335.
[16] Chen Z *et al* 2017 *IEEE Int. Conf. on Sys. Man. & Cyb.* 206.
[17] Swearingen K L *et al* 2018 *ACM SIGGRAPH Posters* 20.

IOP Conf. Series: Materials Science and Engineering **551** (2019) 012050 doi:10.1088/1757-899X/551/1/012050

A Review on Collaborative Learning Environment across Virtual and Augmented Reality Technology

Nor'a Muhammad Nur Affendy [1,2], and Ismail Ajune Wanis[1,2]

[1] Mixed and Virtual Environment Research Lab (mivielab), Vicubelab, Universiti Teknologi Malaysia, 81310 Johore, Malaysia
[2] School of Computing, Faculty of Engineering, Universiti Teknologi Malaysia, 81310 Johore, Malaysia

Email: m.nuraffendy94@gmail.com, ajune@utm.my

Abstract. Collaborative Learning Environment (CLE) in various application settings are presented and describe in this paper. CLE focuses on facilitating learning process through collaboration task. This paper reviews the numerous Augmented Reality (AR) and Virtual Reality (VR) technology that adopt collaborative system. It provides for an insight view of the capabilities of collaborative learning through the adoption of collaborative system in AR and VR. The adoption perspective define the cruciality to adopt collaborative in AR and VR in order to support learning process of the user of the respective task in a specific subject domain.

1. Introduction

Augmented Reality (AR) is merging real and virtual objects together into its own setting and it be known as a technology that has the ability to superimpose or overlay virtual objects onto our real world in real time. Virtual Reality (VR) on the other hand is typically referred as a fully immersive computing technology that enables user to interact with digital workspace in a unique way [1]. VR technology focus on replacing the reality with a complete virtual environment. In order to manipulate the virtual content in both AR and VR workspace, a suitable interaction metaphor is needed.

As stated in [2], AR interaction needs to deliver its own way when it means to use the real objects to manipulate virtual objects. Interaction is one of research topic in VR and AR research as claimed in [3]. Moving from only a single user interaction recently looking forward to invite more than one user to manipulate virtual objects in real-time, the researcher and developer aims to support the idea of a collaborative environment. Over the last ten years, where designing user interfaces only focus on a single user at a time, an increasing number of paper focus on potential collaborative scenario arise [4].

The growing advancement of emerging technologies increase the potential of innovative environment that can promote and facilitate learning. The ability to create an immersive educational experience can be achieve through the adaptation of VR, AR and Mixed Reality (MR) in a various type of learning environments [5]. The ability to adapt multi-user interaction in the learning environment scaffolding the concept of Collaborative Learning Environment (CLE). Implementation of multi-user interaction promotes CLE with encouragement for creative learning technique that involves user collaboration on specific task of a particular subject domain. Subsequently, in CLE users are able to collaboratively interact with multi-user in a single shared space context [6].

IOP Conf. Series: Materials Science and Engineering **551** (2019) 012050 doi:10.1088/1757-899X/551/1/012050

2. Collaborative Learning Environment (CLE)

CLE is a virtual environment that allows learning process to happen where learners engage in a common task where each individual depends on and is accountable to each other. The following subsections discuss the features and interaction in CLE.

2.1 Features of CLE

As AR and VR technology allow user to interact with virtual content either 2D or 3D data, it can simply provide a collaborative interactive environment [7]. Users able to work towards specific task in CLE through the presence of 3D interaction to execute face-to-face and remote collaboration, support of virtual object manipulation and the ability to merge the reality and virtuality smoothly.

CLE invites multiple user to share the virtual working space with other users to their own working environment in AR/VR. There are five features included in collaborative environment [7]. The features as stated in [7] that can be apply in CLE are virtuality, augmentation, multi-user, independence, and individuality. Virtuality define as object that have the potential of being viewed or examined, either do not exist or unreachable in physical spaces can be projected as a virtual content in either AR or VR domain. Augmentation in CLE allows a smooth transition between real and virtual assets in delivering the task and purpose.

In order for collaboration to happen, multi-user features are essential to support the collaborative task in CLE. Multi-user allows users to perform discussion together or other cooperative task on collaboration. Higher intuitiveness among multi-user viably collaborates with themselves utilizing typical feeling of human associations, as verbal and gestures as AR and VR support this type of interaction. User independence authorize user to have control over its own independent viewpoint and interaction without interrupting other users action. The individuality of virtual content projected are shared among all users where they can observe the same exact virtual content, in a consistent manner of visibility.

2.2 3D Interaction in CLE.

3D Interaction are human-computer interaction where task are carried out by user in a 3D spatial context. It requires a 3D or 2D input devices that later were mapped into 3D. The use of metaphors is very essential for providing user a context of way for them to interact with virtual content. Interaction in CLE virtual environment are driven by the interface design to something the user are already familiar with. Metaphors employ user knowledge and help transitioning abstract concept more concrete. User familiarity with the kind of user interfaces that users used to, can convey a better understanding on how the system should work.

3D user interfaces that is effective for collaborative environment should take account the perceptual, cognitive, and social issue. Collaborative interface is beneficial as it provides individual users the perspective of the same shared workspace [8]. This can definitely enhance the user experience in CLE and increase the learning process while user collaboratively work towards the same shared task. Typical 3D interaction are relatively common in AR or VR domain. User perform the interaction through their understanding of the user interface to select and manipulate the properties of the virtual content.

3. Collaborative in AR and VR

AR and VR are well known technology that adopt collaborative as a medium of interaction that allow multiple user to collaboratively work on specific task. Learning process able to happen through collaboration of multiple user that work together in search for understanding, meaning, solution or product of the learning. Table 1 presents the papers ranging from year 2016 to 2018 that adopt collaborative concept on the respective domain and type of collaboration. The collaborative concept applies on various application setting that includes medical, engineering, education and more.

Table 1. Collaborative in AR/VR and Collaboration form.

Year	Research works	Domain		Collaboration Form	
		AR	VR	Co-located	Remote
2018	Collaborative Virtual Reality for Low-latency Interaction [11]		✓	✓	✓
2018	BoatAR: Multi-user Augmented Reality Platform for Boat [9]	✓		✓	
2018	Couples Designing their Living Room Together: A Study with Collaborative Handheld Augmented Reality [10]	✓		✓	
2018	SimCEC: A Collaborative VR-Based Simulator for Surgical Teamwork Education [12]		✓	✓	✓
2018	Emotion Sharing and Augmentation in Cooperative Virtual Reality Games [13]		✓	✓	✓
2017	CoVAR: A Collaborative Virtual and Augmented Reality System for Remote Collaboration [16]	✓	✓		✓
2017	Collaborative VR Painting in Web Browsers [14]		✓	✓	
2017	Remote Collaboration in AR and VR using Virtual Replicas [17]	✓	✓		✓
2016	Interactive Augmented Reality: A New Approach for Collaborative Learning [6]	✓			✓
2016	MedicalVR: Towards Medical Remote Collaboration using Virtual Reality [15]		✓		✓

Collaboration system is no longer a foreign matter in AR and VR domain. Table 1 shows 3 papers that implement collaborative system in AR domain [9, 10, 6], 5 papers in VR domain [11, 12, 13, 14, 15] and 2 papers that adopt both AR and VR domain [16, 17]. The table also present a number of 3 papers with co-located type of collaboration [9, 10, 14], 4 papers with remote collaboration [16, 17, 6, 15], and 3 papers that support both co-located and remote collaboration [11, 12, 13]. Among the various papers reviewed, there are several papers that apply the collaborative system in medical field settings. The VR simulator that aim to assist human motor rehabilitation in [11] focus on the CLE between therapist and patient respectively. Basic surgical routines of VR simulator in [12] considers task performed individually and collaboratively among a team of undergraduate students. MedicalVR as in [15] collaboratively shared the recorded digital breast imaging data to the possible remote collaborator in VR through network. The collaboration among users in the applied application settings allows information exchange hence establish a learning process.

In addition, three of the reviewed papers, [9, 10, 14] presents their applied application settings under the design and configuration theme. An AR boat configuration system [9] that allows multi-user to interact with the boat component in same shared space. The interior design AR application [10] allows two user to collaboratively design their living room through the virtual workspace. Multi-A-Painter [14] a multi-user VR collaborative system that works on a browser that supports WebGL and WebVR. Users in the same session work collaboratively design an empty space into a new virtual world give a sense of shared space. Knowledge and information transfer between users when exchanging roles promotes the establishment of CLE. On the other hand, two reviewed papers [13, 16], both focus to incorporate other input modalities such as facial expression and eye gaze tracking to increase social engagement and support collaboration awareness among users in collaborative environment. Higher level of social engagement are believe to benefit numbers of application such as education where collaborative learning can be establish [13]. The AR and VR collaborative system in [17] allows remote collaborator to work together with the local user in AR using virtual replicas representation. Remote user able to control the virtual replicas to refer parts of an aircraft engine combustion chamber. This approach are appropriate for training purposes. 3D Jigsaw AR puzzle application [6] allow users to collaboratively work together to complete the puzzle. The work aim to use the AR technology in order to establish collaborative learning. The users work in a different physical space but in the same network.

4. Conclusion

AR and VR as a mature technology able to adopt collaborative system in various application setting that can allow multiple user to work together working on a single shared space and task. The collaborative manner in the system allow learning process to happen when each user capitalize on one another's resources and skills. Knowledge created through collaborative learning are within populated members who are actively sharing experience and play asymmetric roles. Collaborative learning allows the others learn new things, understand clearly of certain task or able to make something faster through help of expertise [6]. Based on various paper of collaborative in AR and VR, it clearly shows that collaborative system applied in the technology can facilitate learning through various occasion and environment setting either in learning [6], art [14], medical [11, 12, 15], engineering [17], interior design [10] and conventional industry [9]. CLE should able to provide users with the product of the knowledge they gain through the collaborative task performed in either AR or VR environment.

Acknowledgements
We would like to express our appreciation to Mixed and Virtual Reality Laboratory (mivielab) at Universiti Teknologi Malaysia (UTM). This work was funded by UTM-HIR High Impact Research Grants Scheme (Q.J130000.2451.04G77).

References

[1] Berg L P *et al* 2017 *Virt. Real.* **21** 1.
[2] Zhou F *et al* 2008 *IEEE/ACM 7th Int. Sym. Mix. & Augm. Real.* 193.
[3] Kaiser E. *et al* 2003 *Int. Conf. Multi. Interfaces* 12.
[4] Kim K *et al* 2018 *IEEE Trans.Vis.& Comp.Grap.***24** 2947.
[5] Gautam A *et al* 2018) *TechTrends* **62** 119.
[6] Boonbrahm P *et al* 2016 *Int. Con. Lear.& Col. Tech.*115.
[7] Ismail A W *et al* 2009 *Int. Con. Tech. E-Learn.& Dig. Ent.* 444.
[8] Grandi J G 2017 *IEEE. In Virt. Real* 419.
[9] Liu Y *et al* 2018 *ACM 24th Sym. on Virt. Real. Soft.& Tech.* 74.
[10] Shin J G *et al* 2018 *ACM 9th Aug. Hum. Int. Conference* 3).
[11] Elvezio C *et al* 2018 *ACM Sym. U. Int. Soft. & Tech. Adj. Pro.* 179.
[12] Paiva P V *et al* 2018 *CIE* **16** 3.
[13] Hart J D *et al* 2018 *ACM Pro. A. Sym. Comp. Hum. Inter .P. Comp. Ext.* 453.
[14] Knispel J al *et al* 2017 *ACM SIGGRAPH VR* 4.
[15] Luxenburger A *et al* 2016 *ACM Int. J. Conf. Per. & Ubi. Comp.*321.
[16] Piumsomboon T *et al* 2017 *ACM SIGGRAPH Emer. Tech.* 3.
[17] Elvezio C *et al* 2017 *ACM SIGGRAPH VR* 13.

IOP Conf. Series: Materials Science and Engineering **551** (2019) 012051 doi:10.1088/1757-899X/551/1/012051

Application of Hybrid Predictive Tools for Prediction and Simulation in Supercritical Fluid Extraction – An Overview

Sitinoor Adeib Idris*[1,2,3] and Masturah Markom[2,3]

[1]Faculty of Chemical Engineering, Universiti Teknologi MARA, 40450, Shah Alam, Selangor, Malaysia.

[2]Chemical Engineering Program, Faculty of Engineering & Built Environment, 43600 UKM Bangi, Selangor, Malaysia

[3]Research Centre for Sustainable Process Technology (CESPRO), Faculty of Engineering & Built Environment, 43600 UKM Bangi, Selangor, Malaysia

*adeib@salam.uitm.edu.my

Abstract. Supercritical fluid technology (SFT) has been applied in many areas, such as in pharmaceutical and food sectors, due to its outstanding features. SFT is an efficient technology that performs extraction and leaves no or less organic residues compared to conventional processes. Recently, the simulation and prediction of the process output from supercritical fluid extraction was determined using intelligent system predictive tools. The prediction of the set of results from supercritical fluid extraction for designing and scale up purposes is crucial because it can not only reduce the usage of extraction solvent and the energy and time of the process but it can also solve the problem that the complex mathematical model cannot solve. A neural network is considered as one of the artificial intelligent systems and is a key technology in industry 4.0. The use of hybrid predictive tools is also a developing area in the prediction and simulation of supercritical fluid extraction and therefore will be further discussed in this paper.

1. Introduction

Supercritical fluid technology is now widely used in many areas such as in the chemical processing, food and pharmaceutical sectors. Supercritical fluid technology utilizes the unique properties of a supercritical fluid, namely, the pressure dependent dissolving power, to alter the process performance. Supercritical fluid offers the exclusive behaviour of possessing the dual characteristics of a fluid, which are gas and liquid. Among all the supercritical fluid available, carbon dioxide is a commonly used fluid because it is inexpensive and safe, it is available in high purity, and it can be used under mild conditions. Moreover, carbon dioxide is a small sized molecule with a linear structure that offers fast diffusion compared to other solvents. The critical temperature and pressure for carbon dioxide are 31.3°C and 71.9 bars; these mild conditions and this non-toxic behaviour of carbon dioxide have provided benefits to many food and health-related processes [1].

To apply supercritical fluid technology, fundamental knowledge, such as solubility, is crucial to control, design and initiate a more efficient and environmental friendly extraction process. For prediction purposes, in addition to having the appropriate mathematical model, the values of the experimental data and the thermodynamic properties of the sample and the extract are required. There are traditional mathematical modelling approaches for supercritical fluid extraction to determine solubility, as they can be developed from differential mass balance equations for the packed solid bed and, if possible, they can represent the overall extraction process. Some of the examples are hot ball diffusion, broken and intact cells, shrinking cores and desorption models. Assumptions and simplifications of a mathematical model are sometimes required, causing the model to not be reliable. However, many conventional mathematical models have been both challenging and impractical. Therefore, the prediction of the effects of temperature and pressure towards the yield and solubility of the supercritical fluid extraction process using tools is now developing because of time and cost concerns.

Predictive tools using intelligent systems are now increasingly being used for the prediction, to model and simulate a chemical process, as well as to compute, classify and optimize a process for process control. Intelligent systems apply the ability to learn, observe and memorize in a situation full of factors and important data. The main advantage of intelligent systems is that the predictions can be performed easily and quickly, as well as in an accurate manner, for systems that are difficult to simulate using physical models. Intelligent systems address imprecision, uncertainty, partial truth and approximation to achieve tractability, robustness and, more importantly, low solution cost. Examples of intelligent systems are neural networks, fuzzy systems and evolutionary algorithms. Thus, there is an increasing number of studies on the solubility of different compounds in supercritical fluid using intelligent systems. Nevertheless, this information is not well summarized. Therefore, this paper discusses several works involving the utilization of hybrid predictive tools based on intelligent systems for the simulation and prediction of the yield of plant extracts using the supercritical fluid extraction process.

2. Predictive Tools

Among all the predictive tools from intelligent systems that have been applied, the Artificial Neural Network (ANN) has been the common tool used for prediction because it is easily referenced and applied. ANN is based on the biological nervous system, which consists of a large number of neurons in an architecture inspired by the brain. Neural networks learn by example; therefore, they are trained by known examples of a problem to attain knowledge about it. After being properly trained, the network can be put to effective use in solving unknown parameters of the problem [1]. Over the years, neural network systems have evolved.

The architectures of neural networks can be identified by three types: single layer feedforward networks, multilayer feedforward networks and recurrent networks. There are various types of learning methods for neural networks; one of the most applied method is the backpropagation learning rule. Neural networks have been applied to several supercritical fluid extraction processes in predicting the yield of extraction [2]–[7], the initial slope [5] and the solubility [8]–[10].

3. Hybrid Predictive Tool Systems

Hybrid denotes the integration of two or more systems. Prediction using predictive tools with intelligent systems is a system integrating two or more technologies. The integration of technologies can effectively solved problems compared to using the technology individually.

3.1. Adaptive Neuro-Fuzzy Inference System (ANFIS)

Application of ANN individually leads to problems such as unsatisfactory extrapolation and also creating a 'black-box' problems. Thus, suggestion on hybrid models have been developed by combining ANN with simple models. These hybrid neural network models are supposed to perform better in terms of process identification tasks since extrapolation are confined only to the uncertain parts of the process [11].

The Adaptive Neuro-Fuzzy Inference System (ANFIS) is considered one type of hybrid model since it combines the Artificial Neural Network (ANN) and the Fuzzy Inference System (FIS) system. There are 5 layers that mapping the inputs through input membership functions (MF) and fuzzy rules in the ANFIS architecture. Similarly, output mapping is achieved through output membership functions along with its fuzzy rules. The number of membership functions assigned to each input variable is determined by trial and error. The hybrid of a neural network and the fuzzy logic in ANFIS allows it to have both the low-level learning and the computational power of neural networks as well as the advantages of the high-level human-like thinking of fuzzy systems and they can complement each other [12]. There are several studies using ANFIS to predict the solubility [13] and the mass of the extract [14], [15]. Some studies even compare the results of the application of ANN and ANFIS to identify the best system that can represent the data and that is reliable for optimization processes [16]–[18], [19]. A detailed description of the previous research applying ANFIS and the hybrid of it for yield and solubility predictions is shown in table 1.

Table 1. Compilation of research applying ANFIS tools for predicting yield and solubility of SFE.

Sample	Purpose	Remarks	References
***Pimpinella anisum L.* Seed**	Prediction mass of extract	• The input parameters: pressure, solvent mass flow rate, and extraction time • The output parameter: mass of extract	[12]
Pomegranate	Simulation of oil yield	• Comparing the data obtained between the ANN and ANFIS models	[16]
Glycyrrhiza glabra L.	Modelling the recovery of extraction	• Independent variables: dynamic time (t), pressure (P), temperature (T) and flow rate of $SCCO_2$ (Q) • Dependent variable: Recovery of extract	[17]
***Rosa damascene* Mill**	Modelling the recovery of extraction	• Independent variables: dynamic time (t), pressure (P), temperature (T) and flow rate of $SCCO_2$ (Q) • Dependent variable: Recovery of extract • Comparing the data obtained between ANFIS and RSM	[14]

3.2. Mathematical Model-Neural Network Hybrid

In 2003, a group of researchers from Canada proposed a hybrid model of the Radial Basis Function-Peng Robinson (RBF-PR) model. The hybrid model was developed by first developing a simple RBF model. The simple RBF model consists of three inputs of T, p and other factors, with one output, which is yield rate. This model has no knowledge of the whole process. To overcome the black box problem of the RBF model, the model is correlated with the Peng-Robinson equation to become a hybrid RBF-PR model. In the Peng-Robinson equation of state, there is an unknown interaction parameter k_{12}, for a binary mixture, whereby the predicted solubility is sensitive. The k_{12} can be obtained from the physical property data for the mixture, but it requires trial and error processes to obtain it. Moreover, the parameters are all temperature dependent, which does not fit the data compared to pressure. The proposed hybrid model by [20] can compensate for the black box problem. The proposed model can fit the experiment data well and can retain the physical meaning of the whole SFE process.

Ten years later, a parallel hybrid model was developed by researchers from Iran [15]. Combination of conventional mathematical modelling with ANFIS was used to predict the Epigallocatechin gallate (EGCG) recovery of supercritical extraction. The conventional model used was based on the differential mass balance in the solid and mobile phases. An analytical model, can improved the accuracy of ANFIS in non-training domains. Besides that, it can predict the unknowns of the analytical parameters. The hybrid model simulates the extraction system with any arbitrary value of adjustable parameters of mathematical modelling [21]. To date, only a few studies has approached the application of a hybrid model between neural networks and mathematical models. As a result, it is recommended to venture into this area of study in terms of the prediction of values other than the yield and recovery of extracts.

4. Analysis of Model

In both ANN and ANFIS, the correct and optimum structures were the most important problems during training. For ANN, the performance of the network will not be satisfied when the number of the neurons in the hidden layer is small or excessive. Too many neurons leads to a long length of training, and it may be compromised by over fitting. For ANFIS, whenever the number of input membership functions is increased, the structure becomes more complex, requiring more iterations to achieve the convergence to the target error. Thus, the training process is very long. Although, in the end, the predictive tools can solve the problem; the acceptance of the model highly depends on the values of the coefficient of determination (R^2) and the absolute average deviation (AAD). The correlation was meaningless because the model does not contain any meaning to the process.

The hybrid of the mathematical modelling and neural network was very promising because it overcomes the above-mentioned problem. The hybrid model combines the application of the mathematical model derived from the theory with the intelligent system. The limitation of solving the complex mathematical model can be resolved by applying the predictive tools, resulting in generating a much more meaningful model to be used in the prediction of yield and solubility values in supercritical fluid extraction processes. As seen in table 1, the prediction of the recovery and yield of the extract are the top applications of ANFIS.

In addition, a hybrid of the mathematical model with the neural network was mostly applied for the prediction of the yield of the extract. In ANN application, in addition to the recovery and yield, two other crucial predictions can be made: solubility and initial slope. Therefore, the hybrid between the mathematical model and the intelligent system may be applied for predictions other than the yield and recovery; such an approach will be an interesting effort because the correlation with solubility was different from that of the extraction yield mathematical model.

5. Conclusion

Overall, the application of an intelligent system to model and simulate a chemical process was achieved in the SFE process. Intelligent system models generalize the experimental results and present process behaviour, predicting and estimating problem. The model can be applied individually or by combining two or more models, called the hybrid model, to achieve the best estimation and prediction data. The appropriate model is beneficial in the development of new products by saving experimental time and costs. However, no research has been performed on the solubility of SFE using a hybrid model; thus, a comparison between a mathematical model and an intelligent model is recommended for further research.

Acknowledgements

The authors would like to thank the Ministry of Higher Education (MOHE) and Universiti Teknologi MARA for providing financial support. Moreover, the authors would like to thank Universiti Kebangsaan Malaysia for providing research facilities.

References

[1] S. Rajasekaran *et al* 2017 *Neural Networks, Fuzzy Systems and Evolutionary Algorithms : Synthesis and Applications*. PRENTICE-HALL OF INDIA.
[2] P. Davoodi *et al* 2016 *Korean J. Chem. Eng.* **32** 1.
[3] S. M. Ghoreishi *et al* 2013 *J. Supercrit. Fluids* **74** 128.
[4] P. Mitra *et al* 2011 *Food Bioprocess Technol.* **4** 737.
[5] B. Pavlic *et al* 2017 *J. Supercrit. Fluids* **130** 327.
[6] L. M. Salleh *et al* 2015 *J. Teknol.* **74** 79.
[7] J. Yin *et al* 2005 *Ind. Eng. Chem. Res* **44** 7420.

[8] A. M. Abbas Karim *et al* 2011 *Emirates J. Eng. Reseacrh* **16** 81.
[9] A. Aminian 2017 *J. Supercrit. Fluids* **125** 79.
[10] B. Vaferi *et al* 2013 *J. Supercrit. Fluids* **77** 44.
[11] D. C. Psichogios *et al* 1992 *AIChE J.* **38** 1499.
[12] V. Kumar *et al* 2017 *J. Food Meas. Charact.* **11** 704.
[13] M. R. Dadkhah *et al.* 2017 *J. Supercrit. Fluids* **120** 181.
[14] M. Davoody *et al* 2012 *J. Supercrit. Fluids* **72** 213.
[15] E. Heidari *et al* 2013 *J. Supercrit. Fluids* **82** 158.
[16] J. Sargolzaei *et al* 2013 *Front. Chem. Sci. Eng.* **7** 357.
[17] A. Hedayati *et al* 2016 *Chem. Prod. Process Model.* **11** 3.
[18] S. M. Ghoreishi *et al* 2016 *J. Supercrit. Fluids* **112** 57.
[19] J. Sargolzaei *et al* 2013 *Glob. Perspect. Artif. Intell.*, vol. 1, no. 2, pp. 15–22, 2013.
[20] J. Zeng *et al* 20013 *Proc. 2003 IEEE Int. Sym. Intelligent Control.*
[21] M. J. Kamali *et al* 2008 *J. Supercrit. Fluids* **47** 168.

A Concise Review of Named Entity Recognition System: Methods and Features

M. *Ikhwan Syafiq*[1,*], M. *Shukor Talib*[2], *Naomie Salim*[3], *Habibollah Haron*[4], and *Razana Alwee*[5]

[1,2,3,4,5]School of Computing, Faculty of Engineering, Universiti Teknologi Malaysia, 81310, Johor Bahru, Johor, Malaysia.

E-mail: shukor@utm.my

Abstract. Named Entity Recognition (NER) is an elementary tool for all application areas in Natural Language Processing (NLP) such as Automatic Summarization, Information Extraction, Information Retrieval, Text Mining, Machine Translation, Question Answering, and Genetics. NER is a task to discover and categorises the named entities ('atomic elements') in the text into predefined classes such as the names of persons, organizations, locations, terminologies of time, quantity and etc. Different languages may have different morphologies and thus involve dissimilar NER procedures. For example, an Arabic NER system cannot be practically used in processing Malay texts due to the different morphological features. The morphological features of every language are rich and complex and donates to the difficulties of implementing an actual method to develop the accurate NER system. In this paper, we review on three main techniques that commonly used to develop an NER system well-known as Rule-Based, Machine Learning, and Hybrid approach. This paper also highlights the features of each technique.

1 Introduction

Access to information extraction (IE) is a process to obtain relevant information from unstructured sources whether using classification or bridging methods. To get structured information on the article, the information process needs to through several phases, namely pre-processing of data, extraction and classification. One of the most widely used extraction techniques is identifying Names Entity (NE) based on unstructured text in the articles. However, texts are hard to understand by computers. A machine will not be capable to understand the content of texts. At the moment, an enormous capacity of documents can be easily regained and extracted

IOP Conf. Series: Materials Science and Engineering **551** (2019) 012052 doi:10.1088/1757-899X/551/1/012052

from internet. But, it would proceed a long time for people to physically process the documents in a short period time and the procedure of annotating texts manually frequently provides a partial results [1].

With a rapid growth of NER system, the capacity of data in many languages is accessible on the web, this make the procedure of extracting and classifying the NE is very useful [2]. However, domain specific, for example, DNA, biochemical, and etc., might not be valid for identifying named entities on other particular domains for example food guides. Also, identifying the categories of named entity for texts written in English language not applicable for other languages. Typically in the English language have proper nouns that normally begin with a capital letter. This proper nouns always been used to signify a unique named entity such as location, organization and person. Besides that, such methods might not be valid to be applied to texts written in Chinese and Arabic language as it does not hold capital letter that can be led to identify the NE. This is because most languages have rich morphological features. In short, the development of the NER system be subject upon on the domain of studies and also the type of languages used.

1.1 Named Entity Recognition

The word "Named Entity", nowadays extensively used in NLP field, was invented for the initial phase in the Sixth MUC conference (MUC-6) [3]. In other words, information extraction is a systematic procedure to solve the problem in the NLP area. The confusion of NE created the named entity system very challenging and complicated, and most of the NER fail to reach human level performance. Although, extracting information, for example classifying relationships between a single type of named entity; first need to require the extraction of entity's name in the text i.e., NER task. This process is important and its accomplishment defines the effectiveness of other information extraction tasks. NER focuses to detecting the rations of the text that state to specific entities. Due to the complexity of NER system, exists several steps to implement NER system;

i. **Pre-processing step**: To detect the boundaries of named entity in the text by tokenize and splitting the sentences. Larger chunks of article can be tokenized or splitting into one sentence and one sentence can be tokenized into a single word [4].
ii. **Feature processing step**: NLP has many complexities in language completion, generating a design to match the potential linguistic understandings of the single word facts involves the pre-processing on physical input, for example allocating Parts-of Speech (PoS) tagging and morphological features.
iii. **Named entity recognition step**: This step is to identify the NE and allocates it to a class or groups of entity.
iv. **Solving normalization or recognition errors step**: Normalization mapping NE and place all words on the same footing, and keep consistently [4]. This step to help in solving problems occur from distinctions in the synonymous word as well as the confusing abbreviations.

2 Methods

IOP Conf. Series: Materials Science and Engineering **551** (2019) 012052 doi:10.1088/1757-899X/551/1/012052

2.1 Rule Based Approach

Rule based are a method that relies on either heuristic or regular expression rules to classify the named entity. Also, rule based used in combination of language resources such as dictionaries, gazetteer lists, or list markers to classify the entity's name. This approach recognizes the NE by consuming a conservative of procedures and a list of dictionaries that are made by hand predefined by humans. The system using a rule based approach spreads over a customary of rules in order to extract design. The designs are generally made up from linguistic, syntactic and orthographic patterns. Also, rule-based approach is recognized in the form of either determinate state regular expressions or transducers. The followings are the main features and methods that can be performed in rule-based NER systems. Table 1 below illustrated the common features used in rule-based NER system.

Table-1. Common features used in rule-based NER system

Feature	Feature definition
Gazetteer	A gazetteer is a list of distinct NE, and it holds a particular list of named for the specific category of named entity class. For instance, Kuala Lumpur is in location class while Mansor is in a person names class with a gazetteer list. Gazetteers also made as dictionaries or whitelists word [5].
Blacklist (Filter)	Filters or blacklist is used as a rejection element for words or string of words that are acknowledged to be an unacceptable name entity and to refuse such incorrect named enity which could not be sentances/words. For instance, (A Singapore prime minister Lee Hsien Loong), here the phrase "prime minister" is unacceptable.
Trigger Words (Keywords)	Keywords or trigger words are regular words that usually come after or before named entity. These trigger words are used to recognize a named entity and can gross many forms such as a noun list or verb list, for instance word (drive) [6].

2.2 Machine Learning Technique

Machine learning technique (ML) in the viewpoint of text analytics is an established of statistical approaches for categorizing some characteristic of a document, for example, sentimentality, PoS tag, named entity, and etc. [7]. In NER system using machine learning techniques, there are several models and features that are used to identify the named entity. Features of ML NER system are important material or descriptors for words characteristics. Also, feature engineering is a leading critical task of NER for entirely classifiers. The common features of NER system using ML technique are illustrated in Table 2.

Table-2. Common features used in machine learning NER system

Feature	Feature definition
Word Length	Word Length can be used to determine NE, for example, if the word less than three or two words, it is not considered as NE.
Special Marker	To recognize some special marker/symbols in the text [8].
Word Suffix/Prefix	This feature used for pattern identical to capture word suffix and prefix of word length up to N [9]. However, prefix/suffix rarely comes as a named entity.
Stop word	This feature removes common words that cannot be part of NE.
Part-of-Speech (PoS)	PoS tagging features can be used to recognize part-of-speech class words (e.g. proper nouns, nouns, verbs, etc.).
Morphology-based	The languages extracted from a group of morphology-based features. This feature is significant and used extensively in NER.
Infrequent word	Infrequent words are gained by scheming the word frequency in the used corpus during the training segment and then choosing the cut-off frequency to form the binary method.
Word-Left/Right	Word-Left/Right feature used to investigate neighbour words (left/right) of length up to N, for example, NE tags or PoS tags from NER system that used as a feature [7].

ML technique can be specified as a standard that is well-designed to other text and article named as supervised approach, or probably will be a set of method that perform through a huge collection of files to annotated meaning that named unsupervised approach. Feature engineering in ML technique is a leading critical task of named entity recognition system for words categorizations [7]. Also, ML technique is more accomplished when compared to rule based approaches because the structure can be trained and also can perform in various domains. Machine learning can be allocated into three well-known methods: unsupervised learning, semi-supervised learning and supervised learning. The most regularly implemented are supervised learning (SL) method which characterize the NER problem as a

IOP Conf. Series: Materials Science and Engineering **551** (2019) 012052 doi:10.1088/1757-899X/551/1/012052

grouping task and involve the accessibility of huge annotated datasets. Table 3 illustrated the main model in ML NER systems.

Table-3. Common model in machine learning NER system

Learning Method	Model
Supervised	Conditional Random Field (CFR) Hidden Markov Model (HMM) Decision Tree (DT) Maximum Entropy (ME) Support Vector Machine (SVM) Artificial Neural Networks (ANNs)
Unsupervised-Learning	Clustering
Semi-Supervised	Bootstrapping

The Machine learning NER system aims to transform the identification problem into classification problem. In principle, the ML system identify and recognize NE into detailed NE classes such as persons, organization, locations, and etc. [10]. Most recent studies in NE for all major languages use a ML technique, also called statistical approach. Machine learning also have been widely used to identify named entity classes from annotated texts.

2.3 Hybrid approach

The hybrid approach is a technique used more than one named entity recognition approach in order to develop a good characteristics from each method. The term hybrid is concerned with the combination of rule based approach and dictionary-based approaches, this should not be confused with the hybrid technique used when two or more machine learning algorithms are combined to build the classification model. Such technique will produce an improved result and outcome [11]. On the other hand, the weaknesses of the rule based are still unescapable in the hybrid classification. A domain specific named entity system may possibly modify the usual of rules used to identify dissimilar forms of NE when the domain of trainings is reformed [12]. The way of the processing system may start from the rule-based approach to the machine learning approach or vice versa. Therefore, to solve the problem of identifying a precise entity terms, the rule based approach can be used to addressing the problem of overlapping entities, while, the machine learning techniques will measure strengthen of the co-occurrences between multi-word entities [13]. Hybrid NER systems also can be implemented using Local Grammars (LGs) that representing the contextual rules in linguistics approach by combining with any machine learning models [14]. This hybrid method seeks to discover the potential of the linguistics approaches to a specific type of language in NER systems.

3 Conclusions

NER is considered to be the most important task for many information extraction systems. Based on review above, we can conclude that the NER has been growing rapidly with various techniques and features. Many researchers have been proposed NER system for various domains using established methods such as rule-based, ML, and hybrid approaches, however, more investigation efforts are still needed to implement because we have a lack of resources on particular languages such as, Indian, Chinese, Malay, and many other languages, also on the specific domain, such as biomedical, DNA, criminal, food guide and etc. Hence, there is a necessity for researchers to employ a new brand handcrafted method that can utilize and perform for many languages and domains in the NER field.

Acknowledgements
This work was financially supported by a Tier 1 grant, reference number (PY/2017/01337), and Faculty of Engineering, Universiti Teknologi Malaysia (UTM), Skudai, Johor.

References
1. R. Alfred *et al* 2014 *Int. J. Mach. Learn. Comput.* **4** 300.
2. Ajees A. P. *et al* 2018 *Int. Conf. Data Sci. Eng.* 1.
3. K. B. Cohen *et al.* 2017 *BMC Bioinformatics* **18** 1.
4. S. Eltyeb *et al* 2014 *J. Cheminform.* **6** 1.
5. R. E. Salah *et al* 2017 *Int. J. Adv. Sci. Eng. Inf. Tech.* **7** 815.
6. W. Zaghouani *et al* 2010 *Adapting a resource-light highly multilingual* 563.
7. R. E. Salah *et al* 2017 *Int. J. Adv. Sci. Eng. Inf. Tech.* **7** 511.
8. A. Abdul-hamid *et al* 2010 *Proc. 2010 Named Entities Work* 110.
9. A. F. Wicaksono *et al* 2010 *Prooc. 4th Int. MALINDO (Malay Indones. Lang. Work.* 1.
10. T. Dasgupta *et al* 2017 *Proc. Int. Conf. Web Intell.* 541.
11. M. Basaldella *et al* 2017 *J. Biomed. Semantics* **8** 0.
12. M. Alruily *et al* 2009 *Proc. - Int. Conf. Dev. eSystems Eng. DeSE 2009* 153.
13. S. S. Balgasem *et al* 2018 *Proc. 2017 6th Int. Conf. Electr. Eng. Informatics Sustain. Soc. Through Digit. Innov.* 1.
14. J. P. C. Pirovani *et al* 2015 *Adv. Intell. Syst. Comput.* 4452.

uFA-FastSLAM: The New Hybrid of Firefly Algorithm and FastSLAM Algorithm

R J Musridho[1], H Hasan[1,2] and H Haron[1,2]

[1]Department of Computer Science, School of Computing, Faculty of Engineering, Universiti Teknologi Malaysia, Johor Bahru, Johor, Malaysia
[2]Applied Industrial Analytics Research Group, Johor Bahru, Johor, Malaysia

rajajoko@gmail.com

Abstract. Simultaneous Localization and Mapping (SLAM) is a challenging and fundamental issue in autonomous mobile robots. One of two main solutions of SLAM problem is FastSLAM. In terms of accuracy, FastSLAM is known to degenerate over time. This work hybridized Firefly Algorithm (FA) and FastSLAM to optimize FastSLAM. But in FA, there is a ranking process which ranked the fireflies based on their light intensity. It caused the output needed by particles in FastSLAM came in different sequence. Thus, the FA in this work is modified before it is applied to FastSLAM, the ranking process is deleted. The result has shown that the modified FA, that is unranked Firefly Algorithm (uFA) has been successfully hybridized with FastSLAM.

1. Introduction

Simultaneous Localization and Mapping (SLAM) is a challenging and fundamental issue in autonomous mobile robots. The problem of SLAM is state estimation, which is the robot and landmarks position estimation. Two main solutions to the SLAM problem are extended Kalman filter based SLAM (EKF-SLAM) and particle filter based SLAM (FastSLAM) [1].

In FastSLAM, robot position is estimated by the particle filter and landmarks position is estimated by EKF. Each particle represents the estimation of the robot position, where in each of it the estimations of the detected landmarks are saved. So decisions of data association is made on each particle. Therefore, different particles have different landmarks estimation, this is the key advantage of FastSLAM [2].

However, FastSLAM is known to have some shortcomings. It produces optimistic estimation of uncertainty in long-term [3], causing production of inconsistent estimation [2]. This happens because low weighted particles are replaced at the resampling phase. Meanwhile each particle carries their own collection of landmarks that represents the environment. So, the entire hypothesis of the environment in those replaced particles are lost, causing information depletion [4].

There are many algorithms [4–6] that have been proposed overcome the problem of information depletion. In [5], FastSLAM is optimized by using Particle Swarm Optimization (PSO). In [6], FastSLAM is optimized by using Genetic Algorithm (GA). And in [4], both GA and PSO are used to optimize FastSLAM.

In this work, FastSLAM is hybridized with a swarm algorithm, i.e. Firefly Algorithm (FA). It is used to update the estimation of robot and landmarks position estimation before the resampling phase

IOP Conf. Series: Materials Science and Engineering **551** (2019) 012053 doi:10.1088/1757-899X/551/1/012053

in FastSLAM. However, there is a ranking process in FA that caused the values processed by FA go to different particles. Thus, this work proposed a modified FA without the ranking process for its hybridization with FastSLAM.

The next section is methodology which explains about FastSLAM, FA and reason of ranking process deletion, and the proposed hybrid. Section 3 presents the results and discussion. Section 4 as the last section provides the conclusion of this work.

2. Methodology

2.1. FastSLAM

FastSLAM is one of two main solutions to the SLAM problem [1]. In FastSLAM, path is sampled by Particle Filter. In each particle, all detected landmarks which represent the map is stored [6]. The steps of FastSLAM algorithm can be described as follows [4]:

1. A new robot pose is drawn.
2. Update the map by Extended Kalman Filter (EKF) that associates observed landmarks in each particle with new detected landmarks.
3. Calculate the importance weight of each particle.
4. Draw a new particle set from the weighted particle set by resampling.

2.2. Firefly Algorithm (FA) and Reason of Ranking Process Deletion

Firefly's flashing behavior inspired [7] to develop an algorithm at Cambridge University in 2007. Three main rules in FA:

1. All fireflies are attracted to each other because they are unisex;
2. Attractiveness of the fireflies is related to their brightness, the brighter one attracts the less bright fireflies. And the attractiveness is inversely proportional to the distance between them. And if the brightness is the same for all fireflies, they will move randomly;
3. The brightness is determined by the objective function's landscape.

Objective function of FA is related to the brightness of the firefly. And FA is appropriate for multimodal optimization problems because they can divide themselves into smaller groups and those groups swarm around local models, this is due to their attractiveness [8].

In FA, there is a ranking process which ranked the fireflies based on their light intensity. It caused the output to be in different sequence. Meanwhile, the particles in FastSLAM need to use them. If the sequence of output (light intensity) from FA is different from output (weight of particles) that FastSLAM has previously sent to FA, it would cause the new values go to different particles. Thus, the FA in this work is modified, the ranking process is deleted and then it is called unranked Firefly Algorithm (uFA). The pseudocode of uFA can be seen in figure 1.

```
Objective function f(x), x = (x_1, ..., x_d)^T
Generate initial population of fireflies x_i (i = 1,2, ..., n)
Light intensity I_i at x_i is determined by f(x_i)
Define light absorption coefficient γ and randomization parameter α
while (t ≤ MaxGeneration)
    for i = 1:n all n fireflies
        for j = 1:n all n fireflies (inner loop)
            if (I_i < I_j), Move firefly i towards j; end if
            Vary attractiveness with distance r via exp[−γr]
            Evalute new solutions and update light intensity
        end for j
    end for i
end while
Postprocess results
```

Figure 1. Pseudocode of uFA, based on FA by [9] without the ranking process

IOP Conf. Series: Materials Science and Engineering **551** (2019) 012053 doi:10.1088/1757-899X/551/1/012053

2.3. The Proposed unranked Firefly Algorithm FastSLAM (uFA-FastSLAM)
The proposed uFA-FastSLAM is an optimized FastSLAM which has been hybridized with FA without the ranking process, which then is called uFA. The uFA is added into FastSLAM before the resampling phase to optimize the robot and landmarks position estimation. Weight of the particles from FastSLAM is the objective function, because the weight represents quality of the estimation [10]. Figure 2 shows the pseudocode of uFA-FastSLAM.

```
Start Algorithm
1)  Sample new robot pose (Predict mobile robot's pose by the motion model)
2)  Update the map by EKF with observed landmarks in each particle
3)  Calculate the weight of each particle
4)  Optimize robot and landmarks position estimation using uFA
    Objective function is the weight of the particles
    Define light absorption coefficient γ and randomization parameter α
4.1) Robot Position Estimation
    Initial population of fireflies is the current position of robot position estimation
    Light intensity I_i at x_i is determined by the weight of the particles
    while (t ≤ MaxGeneration)
       for i = 1: n all n fireflies
             for j = 1: n all n fireflies (inner loop)
                   if (I_i < I_j), Move firefly i towards j; end if
                   Vary attractiveness with distance r via exp[−γr]
                   Evalute new solutions and update light intensity
             end for j
       end for i
    end while
    Update robot position estimation
4.2) Landmarks Position Estimation
    for p = 1: dlm all detected landmarks
       Initial population of fireflies is the current position of landmarks position estimation
       Light intensity I_i at x_i is determined by the weight of the particles
       while (t ≤ MaxGeneration)
             for i = 1: n all n fireflies
                   for j = 1: n all n fireflies (inner loop)
                         if (I_i < I_j), Move firefly i towards j; end if
                         Vary attractiveness with distance r via exp[−γr]
                         Evalute new solutions and update light intensity
                   end for j
             end for i
       end while
    end for
    Update landmarks position estimation
5)  Draw a new particle set by resampling
```

Figure 2. Pseudo code of the uFA-FastSLAM.

3. Experiment Setup
The experiment is done in MATLAB, using the FastSLAM in the toolbox provided by [11] along with the environment which can be seen in figure 3. Appropriate setting of parameters is necessary to make sure that the experiment is done properly. The setup of parameters can be seen in table 1.

IOP Conf. Series: Materials Science and Engineering **551** (2019) 012053 doi:10.1088/1757-899X/551/1/012053

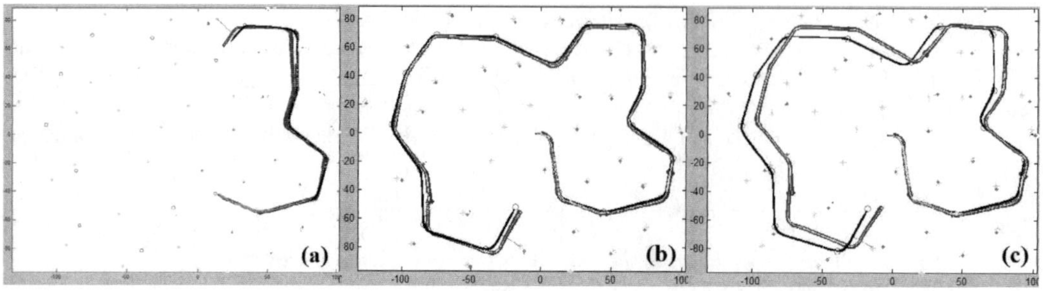

Figure 3. An Incomplete Run (a) and Two Samples of Different Run Results of uFA-FastSLAM (b and c).

Table 1. Setup of parameters.

No	Parameter	Value	Unit
1	Total of simulation	50	run
2	Velocity of the robot	3	m/s (meter per second)
3	Wheelbase size of the robot	4:2	m (meter)
4	Time interval (control signal)	0.05	s (second)
5	Time interval (observations)	0.2	s (second)
6	Number of particles	100	particles
7	Number of loops	1	loop
8	Max Generation	100	times
9	Population of firefly	100	fireflies

4. Results and Discussion

Proof that uFA is worked when hybridized with FastSLAM can be seen in figure 3. It has been proven to work successfully because there was no error happened in the simulations. The circles in figure 3 are the waypoints which connected by lines, this is the reason of showing a picture of an incomplete run, figure 3(a). The stars are true landmarks while landmarks estimations represented by the dots around them. The true robot is path is the thick line with sharper edge than the estimated robot path which is the line next to the true robot path. The sensor detection is represented by the line near the end of the true robot path.

5. Conclusion

This work explained the hybridization of FA and FastSLAM. But in FA there is a ranking process that caused the output for FastSLAM to be rearranged based on the light intensity. Thus, the FA in this work is modified, the ranking process is deleted and then it is called uFA. The goal is to successfully hybridize uFA with FastSLAM. A selected toolbox and an environment map are used for the experiment. The new proposed hybrid, uFA-FastSLAM, is proven to be successfully hybridized.

This work was supported by Tier 1 under grant number Q.J130000.2528.18H65 funded by Research University Grant for Universiti Teknologi Malaysia (UTM).

References

[1] Durrant-Whyte H *et al* 2006 *IEEE Robot. Autom. Mag.* **13** 99
[2] Bailey T *et al* 2006 *Proc. 2006 IEEE Int. Conf. Rob. Autom., 2006. ICRA 2006* 424
[3] Havangi R *et al* 2013 *Int. J. Humanoid Robot* **10** 1350018
[4] Khairuddin A R *et al* 2017 *Intelligent Systems Design and Applications ISDA 2016* **557** 57
[5] Lee H C *et al* 2009 *2009 IEEE Int. Conf. on Systems, Man and Cybernetics* 2763
[6] Xia Y M *et al* 2011 *Int. Symposium on Information and Automation* **86** 296
[7] Yang X S 2010 *Luniver press*
[8] Farahlina Johari N *et al* 2017 *J. Phys.: Conf. Series* **892** 012005
[9] Yang X S 2009 *Proc. of the 5th Symposium on Stochastic Algorithms, Foundations and Applications* **5792** 169
[10] Khairuddin A R 2016 *Universiti Teknologi Malaysia*
[11] Bailey T 2004 *SLAM Simulation Toolbox*

An overview of surface reconstruction using partial differential equation (PDE)

M N M Othman[1], Y Yusoff[1], H Haron[1], and L You[2]

[1]School of Computing, Faculty of Engineering, UTM Skudai, 81310, Johor, Malaysia
[2]National Center for Computer Animation, The Media School, Talbot Campus, Bournemouth University, Poole, BH12 5BB, United Kingdom

E-mail: yusliza@utm.my

Abstract. Surface reconstruction is the main process of reverse engineering where engineering model is reproduced in digital format. Surface reconstruction using PDE can be described as solving the PDE to generate the reconstructed surface of an object of interest. This paper provides a brief introduction to the process of surface reconstruction, partial differential equation and its application in surface reconstruction, as well as summarizing several works that utilize this approach. This paper also outlines the validation method used to assess a reconstruction model.

1. Introduction

Surface reconstruction, as a main part of reverse engineering, is the process of reproducing geometric-based model based on existing physical object [1]. Surface reconstruction allows recovery of engineering models from actual object as digital backup or in order to manipulate the object of interest virtually. Söderkvist [2] divides the process of surface reconstruction in to three separate steps, namely Data Acquisition, Data Organization and Surface Fitting. Further development in the field includes three additional steps in the process, which are Pre-processing, Post-processing and Visualization [3].

PDE is a mathematical tool used to describe a given physical phenomena, where the relation between rates of change with respect to continuous variables can be observed and calculated. Such phenomena that can be described by PDE include heat transfer, acoustics and fluid dynamics. A derivative is the rate of change of a quantity with respect to another variable. When the rate of change includes two or more variables, it is then known as partial derivative. Thus, PDE can be defined as the relations of partial derivatives that make up a given function [4].

Surface reconstruction with PDE can be described as solving the PDE with respect to the boundary curve extracted from an object of interest, to generate the surface that can represent that object. Bae and Weickert [5] describe Partial differential equation (PDE) as a good solution for image interpolation and compression. Ostrov [6] demonstrates the capability of PDE to decrease the time taken to calculate the solution of surface reconstruction. Osechinskiy and Kruggel [7] present an automatic reconstruction method using several PDE modelling stages, and demonstrate the capability of PDE method to handle biomedical image with their algorithm. Previous study shows that to represent a surface, when compared to spline-based techniques, PDE requires smaller amount of parameters, thus is easier to be manipulated. PDE generated surface are also guaranteed a basic degree of smoothness; where the higher the order of the PDE, the higher the smoothness degree of the generated surface would be [8].

IOP Conf. Series: Materials Science and Engineering **551** (2019) 012054 doi:10.1088/1757-899X/551/1/012054

The purpose of this paper is to give an overview on previous works on Surface Reconstruction that uses PDE. In this paper, previous works are summarized with the domain and the issues listed in a table, followed by a description on the method proposed and its result. The structure of this paper is as follows, Section 2 provides an overview of several works on PDE-based Surface Reconstruction, Section 3 gives a brief description on validation of reconstruction model, while Section 4 concludes this paper.

2. Overview of Previous Works

This section provides an overview on several works in surface reconstruction that uses Partial Differential Equation.

Table 1. Author, Domain and Issues of previous works on PDE-based Surface Reconstruction

Author	Domain	Issues
Rodrigues et al.	Reconstruction of realistic human faces from 3D facial scans data for facial recognition in airport security.	• Deals with vast amount of data and requires real-time processing. • Requires data encryption for secured transmission over network. • Compression of 3D data acquired from 3D scanners. • Reconstruction of the mesh from the compressed data whilst preserving the quality of the original mesh.
Ugail & Kirmani	Reconstruction of object with complex geometry.	• Reconstruction of complex geometric model using elliptic PDE formulation, where inclusion of only outer edge as boundary condition, may not produce close agreement between original and reconstructed shape.
Eyad & Ugail	Reconstruction of realistic human faces from 3D facial scans data	• Representation of human facial geometry through polygonal based approach require large amount of data to be saved. • Parametric patch approach, in particular spline techniques, requires extra storage, used to define complex geometry.
Linz et al.	Surface Reconstruction from unstructured point cloud data and multiple view 2D images	• Eulerian approach of deformable model requires post-processing to represent the object explicitly and does not guarantee to preserve the topology of the object over time. • Lagrangian approach of deformable model requires the topology information to be defined explicitly when the topology changes thus its implementation is computationally expensive.
Duan et al.	Shape reconstruction from volumetric data, unorganized point cloud and multi-view 2D images.	• PDE-Based surface evolution by applying Eulerian approach is computationally expensive. • Available methods are not flexible enough to handle all three types of data in one algorithm.
Osechinskiy & Kruggel	Reconstruction of cerebral cortex surface from MR Images	• MRI Scans produce images that have limited resolution, noisy, intensity inhomogeneities and partial volume effects. • Automated reconstruction of the cerebral cortex surface from such input can be topologically and geometrically inaccurate.

Pang et al.	Reconstruction and web visualization of complex geometric shapes	• Large polygon meshes are used to represent complex geometric shapes which require a significant amount of memory to handle. • Level of Detail (LOD) is predefined. • Web visualization using such shapes is impractical due to limited Internet bandwidth.

Rodrigues et al. [9] proposes Fourier-based data compression and PDE-based mesh/surface reconstruction. In preparation step consist of finding the minimum bounding box encompassing the whole mesh, and defining horizontal and vertical oriented cutting planes for vertex sampling. The compression step is comprised of Fourier analysis of data in each plane, saving the coefficients, boundaries and scale into ASCII file according to the prescribed format. The compressed data may be encrypted to ensure secured network transmission. Reconstruction step consists of solving the PDE of each pair of cutting planes with Laplace Equation acquiring patches, which is joined together to reconstruct the original mesh. The result shows data compression of over 96% with minimal loss of accuracy of the reconstructed surface against original mesh.

Ugail and Kirmani [10] use fourth-order elliptic PDE to generate the surface, where four curves acts as the boundary condition. Two curves can be extracted from the outer edges of the shape and is applied as the outer boundary condition, whereas Flexible parameterization is used to determine the position of two inner curves that acts as the inner boundary condition. The surface is reconstructed by solving the PDE with regards of the four previously extracted boundary conditions. Proposed scheme produces close agreement between original shape and reconstructed surface.

In [11], the method requires interaction with the user for initialization. User is prompted to highlight several key features of the facial data, such as tip of the nose, and corner of the eyes and mouth, which is then used to normalize and align the data. A series of plane which are parallel to the symmetry profile of the facial data are defined, where the distance between each two consecutive planes is dependent on the complexity of the topology of that section. The set of points that intersects each planes are identified, and following that, a series of profile curves are extracted from the facial data. These profile curves are used as the boundary curves. Solving the fourth order elliptic PDE with four consecutive boundary curves will generate a surface patch, and by combining these patches, the geometry of a human face may be adequately described. This work also looks into the importance of sampling ratio when extracting the profile curves. Significant saving of data storage whilst retaining the accuracy of the geometry can be observed. Higher sampling ratio guarantees smaller error rate of the generated surface, though depending on the application, such as animation, smaller sampling ratio may reduce the processing time.

Linz et al. [12] proposed a point-based approach to PDE-based surface reconstruction. The generated surface is made up of a set of oriented disks, deformed from initial point surface e.g. a sphere encompassing the object. In sampling, instead of considering the geodesic proximity between each point, spatial proximity, where the structure relies on the notion of k-nearest neighbour, is used for computation. Applying force to a point in a region affects every k-nearest neighbouring point as well, thus displacing the whole neighbourhood. By weighing the force applied to each point in the region with Gaussian kernel, the deformation in that region would still produce a smooth surface locally. The set of oriented disks are then ensured to produce a globally smooth surface by introducing Moving Least-Squares projection. To ensure the points are uniformly distributed, up-sampling scheme is introduced to under-sampled regions, and to control the scheme in order to avoid oversampling, point relaxation method is introduced next. When topology changes are involved, the point-based approach is more flexible and less expensive computationally compared to deformation of triangle meshes for surface reconstruction.

Duan et al. [13] works on the shape reconstruction from 3D and 2D Data using PDE-based deformable surfaces. PDE-based deformable surface model that can evolve from an initial shape to capture the geometric boundaries of the object of interest automatically is proposed. The PDE used is the general weighted minimal surface flow. Depending on the data types used for the reconstruction process, the stopping function that is used to stop the deformation of the model when the boundary of the object is reached would differ accordingly. The model only requires for the user to simply use the appropriate data interface function in the initialization phase, as the stopping function for each data type i.e. volumetric data, unstructured point cloud, multi-view 2D images, had already been defined beforehand. The model is capable of automatically processing all three types of data with minimal user interaction. In particular, for reconstruction of multi-view 2D images, the model allows additional images to be incorporated into the mesh representation for progressive refinement when required.

Osechinskiy & Kruggel [7] proposes reconstruction algorithm that is divided into four stages. First stage involves the computation of potential field, where gray matter is treated as dielectric layer that surrounds the conductive white matter, in an electrostatic model. Second stage requires the computation of the distance field from white matter along the streamlines of the potential field that solves the Eulerian framework PDE on a fixed grid. Third stage focuses on extracting the binary skeleton by finding shocks in the distance field, highlighting the areas where gray matter banks are present. Lastly, reconstruction of the cortical surface is performed by advection along the gradient of the potential field or the distance field, using a geometric deformable model which evolution is constrained by the binary skeleton computed previously. Comparison by visual inspection shows that the electric field model used can better describe the sulcal fold position than Laplacian field. The binary skeleton extracted by the method is consistent with the skeleton extracted using ACE algorithm. Overall, the method proposed is capable of producing outer cortical surface that is correct geometrically whilst preserving its initial topology.

Pang et al. [14] applies Patchwise PDE method, where the object is made up of a number of non-overlapping patches with its own parametric coordinate system. The number of patches required to represent an object is dependent on the complexity of said object. The original polygon mesh is first simplified using decimation algorithm to get the base mesh. Each facet of the base mesh is then used to define a surface patch, where the edges of this facet are used as reference to extract the outer boundary curve of each patch from the original object. The inner boundary curve of the patch can then be extracted, by mapping isometric polylines of triangular facet to the initial mesh and sampling accordingly. The boundary curves coordinates are finally stored in a file, which can later be processed by the PDE engine for rendering. Depending on the complexity of the shape and the Level of Details stated by the user, the proposed method requires significantly smaller storage size when compared to the original file.

3. Validation of Reconstruction Model

Validation of a reconstruction model involves assessing the quality of the reconstructed surface produced by the model; how well does the reconstructed surface can represent the original data. These assessments can be performed by either visually or statistically.

The most direct and meaningful assessment is through visual inspection, observation either by user or professional of the field, whether the reconstructed surface is a good fit or in close agreement to the original data. For application such as animation, where faster processing time is of more importance than accuracy, visual assessment is an adequate assessment.

When assessing a model that requires finesse, such as facial recognition, statistical assessment would be a better measurement tool. Through such assessment, the user can quantify the goodness of the fit. Some study compares the difference between each point in the generated surface against its closest point from the original data in Euclidean distance, averaging the calculated values and define it as the error rate; where smaller error rate corresponds to a better fit. Additional measurement approach includes comparing the running time of the model, features of the generated surface i.e. number of vertices, edges and faces, and size of the output file against other methods or models used in a similar domain.

IOP Conf. Series: Materials Science and Engineering **551** (2019) 012054 doi:10.1088/1757-899X/551/1/012054

4. Conclusion

Complex geometry shapes are generally represented using polygon meshes or set of points, which in turn requires large data storage. PDE-based approach that makes use of the boundary-value to represent an object is inherently more efficient. Depending on the complexity of the original data, several PDE patches can be blended together to represent a surface. Though proven to be quite expensive to solve and time consuming in computation, PDE surfaces preserve the topological and geometry of original object. It also requires small amount of parameters to define, thus is easier to modify or manipulate, adjust its Level of Detail (LOD), and produce smaller sized output file. These advantages encourages the utilization of PDE surfaces in application that emphasizes on topology and geometric accuracy of generated surface such as geometric modelling and facial recognition, flexible LOD for visualization such as animation, and small file size for transmission and sharing.

5. Acknowledgement

This research is supported by the PDE-GIR project which has received funding from the European Union's Horizon 2020 research and innovation programme under the Marie Sklodowska-Curie grant agreement No 778035.

References

[1] Guo, G X *et al* 2010 *Proceedings of the 2010 IEEE, Int. Conf. on Mechatronics and Automation* 1783.
[2] Soderkvist I 1999 *Technical Report 1999* **10**.
[3] Lim S P *et al* 2012 Artif Intell Rev. 1.
[4] Ugail H 2011 *Partial Differential Equations for Geometric Design*.
[5] Bae E *et al* 2010 *Dæhlen Metal (eds) MMCS 2008* 1.
[6] Ostrov D N 1999 IEEE Trans Geosci Remote Sens 37 **1** 335.
[7] Osechinskiy S *et al* 2010 *32nd Annual International Conf. of the IEEE EMBS*, 4278.
[8] González G C 2008 Visual Comput. **24** 213.
[9] Rodrigues M *et al* 2012 *Int. Conf. on Differential Equations, Difference Equations and Special Functions* 3.
[10] Ugail H *et al* 2006 *Proceedings of the 10th WSEAS Int. Conf on Comp.* ICCOMP'06 51.
[11] Elyan E *et al* 2007 JCP **2** 1.
[12] Linz C *et al* 2006 *Proceedings of the 28th Conf. on Pattern Recognition* DAGM'06 729.
[13] Duan, Y L *et al* 2004 ECCV.
[14] Pang M Y *et al* 2010 *Int. Conf. on Cyberworlds*.

Empirical Performance Evaluation of Imputation Techniques using Medical Dataset

O A Alade[1,2] R.Sallehuddin[1] and A.Selamat[1]

[1]School of Computing, Universiti Teknologi Malaysia, Skudai, Johor, 81310, Malaysia
[2]Computer Science Department, Federal Polytechnic, Bida, 912101, Nigeria

E-mail : kaleabel@gmail.com , roselina@utm.my and aselamat@utm.my

Abstract. This paper evaluates the error measures of missing value imputations in medical research. Several imputation techniques have been designed and implemented, however, the evaluation of the degree of deviation of the imputed values from the original values have not been given adequate attention. Predictive Mean Matching Imputation (PMMI) and K-Nearest Neighbour Imputation (KNNI) techniques were implemented on imputation of fertility dataset. The implementation was on three mechanisms of missing values: Missing At Random (MAR), Missing Completely At Random (MCAR) and Missing Not At Random (MNAR). The results were evaluated by mean square error (MSE), root mean square error (RMSE) and mean absolute error (MAE). PMMI performed better than KNNI in all the results. MSE for example, has the ratio of 0.0260/2.8555 (PMMI/KNNI) for 1-10% MAR – 99.09% reduced error rate; 0.1108/3.0120 (PMMI/KNNI) for 30-40% MCAR – 96.32 reduced error rate; and 0.0642/3.7187 (PMMI/KNNI) for 40-50% MNAR – 98.27% reduced error rate. MCAR was the most consistent missingness mechanism for the evaluations. Density distributions of the imputed dataset were compared with the original dataset. The distribution plots of the imputed missing data followed the curve of the original dataset.

1. Introduction

Data is the major operational facts of any organization including medical sector. However, medical datasets are blighted by missingness that characterize other data driven sectors. This makes discretization of medical data difficult for classifiers. Missing data comes from error of procedure, equipment error, measurement error, or respondents declining answer to queries of personal information. Most classifiers lack inbuilt routine to handle datasets with missing values, therefore there is need for explicit software for missing data imputation.

Several imputation techniques have been used to explicitly impute missing data. Zero substitution [1] has been proposed, but it is unacceptable by research community because it leads to wrong conclusion. Mean imputation is another approach [4], this approach is simple but it also undermines the variance and standard deviation of the imputed data. Support vector regression performed effectively well, but it was solely based on regression. Gaussian mixture model and extreme learning machines (GMM-ELM) was proposed as a reliable approximation technique for imputing missing data by Sovilj *et al.* in [5]. The result of their work improved imputation of missing values over mean imputation technique, however the evaluation of imputation result was not sufficiently investigated.

Pohar-Perme method [6] was used for the imputation of missing data in colorectal cancer dataset. They recommended sensitivity analysis of the results, but they did not evaluate the error measure of their technique. Hybrid of fuzzy c-means and GA was used in [7], but it is computationally complex. Monte Carlo simulation method was used to examine the performance of Bayesian imputation technique [8]. They concluded that the performance of Bayesian imputation technique depends on risk factors and the mechanism of missingness. This work of [8] motivated part of this work. KNNI is a state-of-the-art imputation technique. It is simple and works well with Gaussian distribution [9]. Also, PMMI imputes values that are like the original [10]. It is used to construct metrics matching of missing data instances with the observed instances. Despite all the advantages of KNNI and PMMI, there are less emphases on their evaluation of their performances [10].

The major missing patterns fall in three basic characteristics of missingness namely, Missing completely at random (MCAR), Missing at random (MAR) and Missing not at random (MNAR). Most researchers concentrate on MAR and MCAR, assuming that it is proper to find, and discard MNAR values in advance. During the analysis of incomplete datasets, it is necessary to consider the process that generates the missing values, and perform valid analyses based on that assumptions.

This work therefore posits to evaluate performance imputation of PMMI and KNNI based on the MAR, MCAR and MCAR.

2. Materials and Methods

Fertility medical dataset from UCI machine learning repository was used in this work. The dataset describes the quality of semen sample of 100 volunteers using artificial intelligence. The dataset consists of nine (9) continuous predictor variables and a binary class decision variable. The predictor variables are childish disease (C_disease), frequency of alcohol (Alcohol_Freq) and smoking habit (Smok_hab). The simulations were repeated for all the three mechanisms of missing values, that is MAR, MCAR and MNAR. All methods in this study were implemented in RStudio version 1.1.456 2018 on a HP computer with Core i3-4030U CPU @ 1.9Hz X64-based processor, RAM 6.00GB, TOSHIBA MQ01AB075 HDD and 64-bit Windows OS.

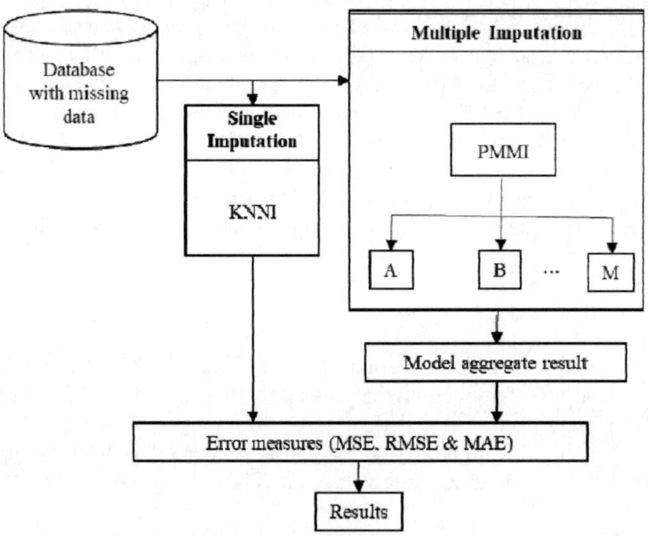

Figure 1: Framework of the proposed evaluation models for KNNI and PMMI imputation techniques.

2.1 Proposed evaluation models for KNNI and PMMI imputation techniques

Error! Reference source not found. shows the proposed framework of this study. Two methods of missing value imputations were considered – single imputation and multiple imputation. KNNI was used for single imputation, while PMMI was used for multiple imputation. The imputed datasets are labelled A, B, …, M. Multiple imputation algorithm generated five (5) imputed datasets for each run of 1-10%, 10-20%, 20-30%, 30-40%, and 40-50% of MAR, MCAR and MNAR. A sum of seventy five (75) imputed datasets were generated. The five imputed datasets for each run were aggregated; then, the error measures were used to evaluate the performance of the imputation models. The error performance for

PMMI and KNNI are compared and evaluated using MSE, MAE and RMSE. The method that produce small error indicates the better performance.

3. Result of the Evaluation

The evaluation results of the predictive mean matching imputation (PMMI) and k-nearest neighbour (KNNI) techniques are shown in Table 1. It shows different percentages of missing values ranging from 1 to 50% with and interval of 10%. The evaluations were based on the MAR, MCAR and MNAR using three error evaluation metrics that is MSE, RMSE and MAE. From the table, it shows that PMMI perform better than KNNI in all missing data mechanisms and in all missing values ratio. From the results in figures 2 – 4, the most obvious trend is the great disparity of error measures between PMMI and KNNI. PMMI has a sharp reduced error measures in the order of 10^{-1} - 10^{-2} than its counterpart KNNI. MSE for example, has the ratio of 0.0260/2.8555 (PMMI/KNNI) for 1-10% MAR – 99.09% reduced error rate; and 0.0642/3.7187 (PMMI/KNNI) for 40-50% MNAR – 98.27% reduced error rate. This high rate is observed for all the characterics of missing data under study. This shows that imputation accuracy of PMMI is higher than KNNI in all the experiments. This better performance of PMMI over KNNI may also be a proof of superiority of multiple imputation over single imputation model.

The error measures apparently increase for MAR, MCAR and MNAR as the percentage error increases for both PMMI and KNNI. In fact, looking at the figures 2-4, MCAR is relatively more consistent among the three mechanisms for every error measure, while MNAR is unstable for the two imputation algorithms with changes in the percentages of missing values. This may account for why MAR and MCAR are the choices of many researchers in the study of imputation of missing data. The slight fluctuations observed in MAR and MCAR may be due to the randomness of seed[3] which is subject to further study.

Table 1: MSE, RMSE, and MAE evaluations of missing data for MAR, MCAR and MNAR characteristics with 1 - 50% range of missingness at an interval of 10.

		MSE			RMSE			MAE		
% MISSING		MAR	MCAR	MNAR	MAR	MCAR	MNAR	MAR	MCAR	MNAR
1 - 10%	PMMI	**0.0260**	0.0299	0.1118	0.1613	0.1729	0.3343	0.1613	0.1729	0.2554
	KNNI	**2.8555**	2.8903	2.9119	1.6898	1.7001	1.7064	2.7164	2.2792	2.0889
10 - 20%	PMMI	0.0304	0.1303	0.0274	0.1744	0.3610	0.1655	0.1744	0.2684	0.1655
	KNNI	2.4492	2.5519	5.7325	1.5650	1.5975	2.3943	2.1981	2.1812	4.1111
20 - 30%	PMMI	0.0709	0.1686	0.0344	0.2663	0.4107	0.1854	0.2053	0.2943	0.1854
	KNNI	3.2843	2.6507	2.1947	1.8123	1.6281	1.4815	2.6447	2.5188	1.8922
30 - 40%	PMMI	0.1145	**0.1108**	0.0855	0.3384	0.3329	0.2925	0.2514	0.2417	0.2325
	KNNI	3.7773	**3.0120**	3.9638	1.9435	1.7355	1.9909	2.9120	2.4720	2.8062
40 - 50%	PMMI	0.0956	0.1439	**0.0642**	0.3092	0.3794	0.2534	0.2044	0.2772	0.2179
	KNNI	2.2018	3.7187	**3.7187**	1.4839	1.9284	1.9284	2.0802	2.8706	2.8706

NB: C_Disease – Childish disease, Alcohol_Freq – Alcohol frequency, and Smok_Hab – Smoking habit

MCAR was chosen to compare the distribution of multiple imputation with the original dataset for the percentage ranges of the missing values (1—10%, 10-20%, ..., 40-50%). This is because it was relatively consistency of as equally observed. Figures 5 represents the distribution charts used to compare imputed and the original dataset. The distributions follow the curve of the original data; i.e. the resultant imputation was very close to the original data, leading to a high degree of accuracy.

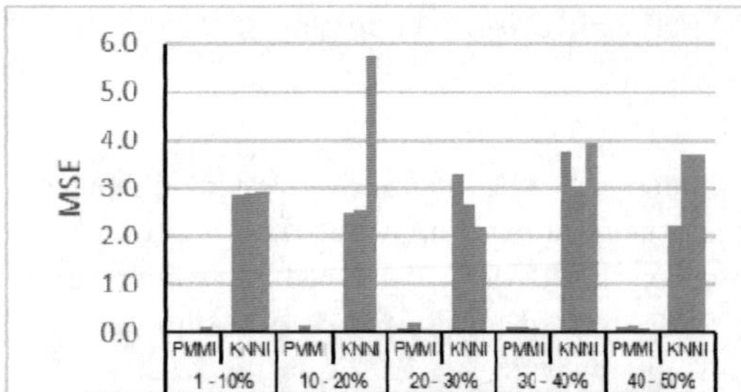

	PMMI 1 - 10%	KNNI 1 - 10%	PMM 10 - 20%	KNNI 10 - 20%	PMMI 20 - 30%	KNNI 20 - 30%	PMMI 30 - 40%	KNNI 30 - 40%	PMMI 40 - 50%	KNNI 40 - 50%
MAR	0.026	2.8555	0.0304	2.4492	0.0709	3.2843	0.1145	3.7773	0.0956	2.2018
MCAR	0.0299	2.8903	0.1303	2.5519	0.1686	2.6507	0.1108	3.012	0.1439	3.7187
MNAR	0.1118	2.9119	0.0274	5.7325	0.0344	2.1947	0.0855	3.9638	0.0642	3.7187

Figure 2: Comparative mean square error (MSE) of predictive mean matching imputation (PMMI) model and k-nearest neighbour imputation (KNNI) model for MAR, MCAR and MNAR with increase in percentage missingness.

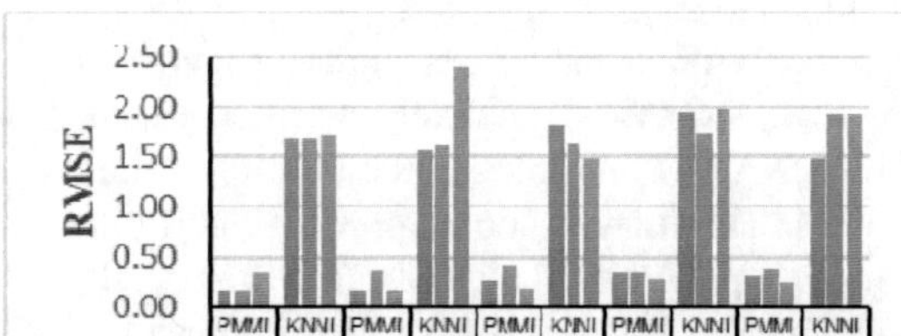

	PMMI 1 - 10%	KNNI 1 - 10%	PMMI 10 - 20%	KNNI 10 - 20%	PMMI 20 - 30%	KNNI 20 - 30%	PMMI 30 - 40%	KNNI 30 - 40%	PMMI 40 - 50%	KNNI 40 - 50%
MAR	0.1613	1.6898	0.1744	1.555	0.2563	1.8123	0.3384	1.9435	0.3092	1.4839
MCAR	0.1729	1.7001	0.361	1.5975	0.4107	1.6281	0.3329	1.7355	0.3794	1.9284
MNAR	0.3343	1.7064	0.1655	2.3943	0.1354	1.4815	0.2925	1.9909	0.2534	1.9284

Figure 3: Comparative root mean square error (RMSE) of predictive mean matching imputation (PMMI) model and k-nearest neighbour imputation (KNNI) model for MAR, MCAR and MNAR with increase in percentage missingness.

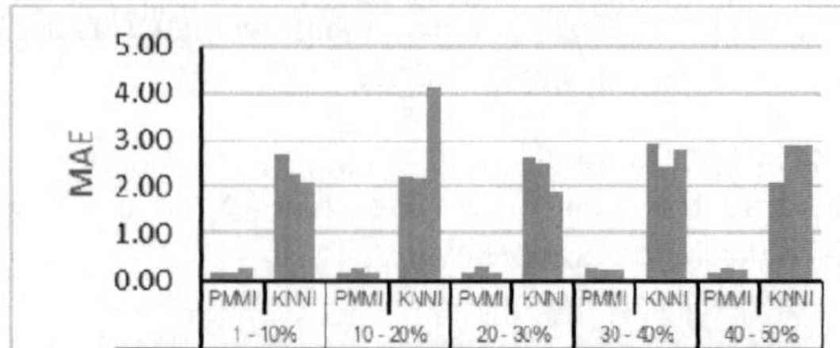

	PMMI 1 - 10%	KNNI 1 - 10%	PMMI 10 - 20%	KNNI 10 - 20%	PMMI 20 - 30%	KNNI 20 - 30%	PMMI 30 - 40%	KNNI 30 - 40%	PMMI 40 - 50%	KNNI 40 - 50%
MAR	0.1613	2.7164	0.1744	2.1981	0.2053	2.6447	0.2514	[illegible]	0.2044	2.0802
MCAR	0.1729	2.2792	0.2684	2.1812	0.2943	2.5188	0.2417	2.472	0.2772	2.8706
MNAR	0.2554	2.089	0.1655	4.1111	0.1854	1.8922	0.2325	2.8062	0.2179	2.8706

Figure 4: Comparative mean average error (MAE) of predictive mean matching imputation (PMMI) model and k-nearest neighbour imputation (KNNI) model for MAR, MCAR and MNAR with increase in percentage missingness.

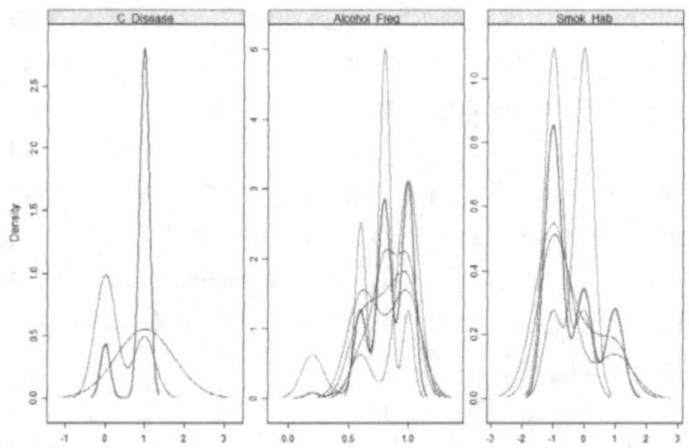

Figure 5: Density plot represents the missing value distributions for missingness in Fertility dataset. The blue line showing distribution of the original data plot, while the magenta lines shows the imputation

4. Conclusion

This work considered missing values as a serious problem in medical research that needs to be handled before classification, analysis and inferences for diagnosis. It focused on evaluation of error measure to determine the performance of imputations techniques in medical classification. The underlining causes of missingness were discussed, two state-of-the-art imputation techniques (PMMI and KNNI) were used to impute missing values in a dataset. The results of the imputations were empirically evaluated with three error metrics – MSE, RMSE and MAE – to measure and compare the degree of error in the imputed datasets under MAR, MCAR, and MNAR. The results showed that PMMI is more promising under all the mechanism of missing pattern than KNNI, and MCAR provide a better optimal missing pattern, while MNAR was the least. The density plots also showed that the distribution of the imputed datasets followed the distribution curve of the original dataset. However, the error measures of the imputation techniques were evaluated in their natural forms, not minding the correlation of the dataset. In future work, the effect of correlation on the error measures will be considered, this would be in a view to improve the algorithms' performances.

References

[1] Newgard CD *et al* 2015 *Jama* **314** 940.
[2] Garciarena U *et al* 2017 *Expert Syst Appl* **89** 52.
[3] Zeng D *et al IEEE 19th Int Conf e-Health Networking, Appl Serv* **1** 4.
[4] Liu Y *et al* 2017 *Data* **2** 8.

[5] Sovilj D *et al* 2015 *Neurocomputing* **174** 220.
[6] Falcaro M *et al* 2017 *Int J Cancer Epidemiol Detect Prev* **48** 16.
[7] Tang J *et al* 2015 *Transp Res Part C Emerg Technol* **51** 29.
[8] Austin PC *et al* 2005 *Comput Stat Data Anal* **49** 821.
[9] Wasito I *et al* 2006 *Comput Stat Data Anal* **50** 926.
[10] Gil D *et al* 2012 *Expert Syst Appl* **39** 12564.

A Comparative Study on Hoping Mechanism of LEACH Protocol in Wireless Sensor Networks: A Survey

M. *Mahmuddin*[1,*],W. A. *Alabadleh*[2], L. M. *Kamarudin*[3]
[1,2]Internetwork Research Lab, School of Computing, Universiti Utara Malaysia, 06010, UUM Sintok Kedah.
[3]School of Computer and Communication Engineering, Universiti Malaysia Perlis (UniMAP), First Floor, Pauh Putra Campus, 02600 Arau, Perlis.

E-mail:ady@uum.edu.my

Abstract. Low Energy Adaptive Clustering Hierarchy (LEACH) is a procedure, which is the most appropriate for the WSN routing. LEACH depends upon an adapted clustering procedure. The LEACH manipulates clusters and a single tier structure on the basis of a two-stage manoeuvre. LEACH functions in a single hop in such a way that every cluster head (CH) in WSNs my transfer its data to the BS and therefore the CH accepts the responsibility of sending information to the entire network. LEACH with Multi-Hop This procedure is the extension of single hop method, which works in a way that maximum CHs could send data to the BS.in this paper discussed the two extensions of LEACH in full detail with regard to their proposed schemes in terms of contributions and limitations.

1. Introduction

It is observed in the last few years that Wireless Sensor Networks (WSNs) have been utilized in various applications wirelessly. These applications include industrial applications such as mining, smart cities, smart emergency systems, and smart virtual power plants. WSNs may be categorized into infrastructure-less and infrastructure-based networks [1, 2]. The former one includes base station (BS) for sending data inside the network. Data networks that normally send data wirelessly are deployed either multi-hop or single-hop between the BS and all wireless nodes. Nevertheless, this category is unable to fulfil the demands of end users, even though the progress in this class has obtained a prodigious growth [3].

2. Discussion on LEACH Protocol

LEACH protocol is measured as the best procedure for WSN routing [4]. The idea of LEACH is being contemplated as an inventiveness for several routing procedures. The goal of this protocol is to select sensor nodes being the CHs in various cycles so as the outcome of extreme power excess (in communication with sinks) is obtained and then dispersed in the entire WSN.

3. LEACH Improvement

For the improvement of LEACH, several extensions were proposed in the last decade [5, 6]. In this section, we elaborate the features of those extensions.

3.1. Single-Hop LEACH Protocol

LEACH functions in a single hop in such a way that every CH in WSNs my transfer its data to the BS and therefore the CH accepts the responsibility of sending information to the entire network. Furthermore, a CH in this mode of transmission may receive information from WSN nodes and send it to the receiver (i.e., BS). A range of protocols have been designed for this category of LEACH, some of which are elaborated here in this section.

LEACH-C

In centralized LEACH, shortly called LEACH-C, a CH is chosen by the entire participating nodes in the network and therefore shapes a cluster [7]. Moreover, in the selection process of CH, the BS should be earlier aware of the level of energy in line with nodes placement. As a result, the BS must choose the most suitable sensor to become the CH while other nodes are assigned to CHs for shaping different clusters. The advantage of this procedure is such that a greater number of rounds may be accomplished in a little network portion. Nonetheless, this leads to additional operating cost because of transmission in a single hop and thus is not suitable for a network encompassing huge area.

V-LEACH

Vice-LEACH (V-LEACH) presents an associate sensor within the network to function being a CH if the central CH expires [8]. This is considered as a single hop due to the fact that it consists of merely one CH for the entire WSN. While an associate CH is chosen that begins working when the main CH expires, merely one node functions as a CH at one time. When a vice CH promotes to become the main CH, a third sensor is chosen to become the vice CH, etc. With this phenomenon, the selection of a new CH is not necessary every time when the main CH expires and thus the entire lifetime of WSN can be augmented. On the other hand, the vice CH is chosen on random basis, thus, this could be faraway, or would have a little amount of energy as well as battery life. Hence, that could expire immediately due to elongated distance data transmission.

Enhanced-LEACH (E-LEACH)

Enhanced LEACH or E-LEACH, smooth's the energy utilization of sensor nodes [9]. In the categorized routing procedure, numerous CHs are contemplated as the key component persuading the performance of routing protocols. When the number of CHs is less, all of them may need to cover a large area. It may encounter several challenges that some of the sensors would be placed nearby the CH that may utilize maximum energy. Data transmission or information sharing between the BS and CHs utilize maximum energy as compared to sharing sensors. A maximum level of CH would increase energy utilization and may ultimately minimize the lifetime of WSNs. Thus, this is primarily important to select the best node as a CH so that assist in consuming a little amount of energy. This scheme may expand the course to choose CHs by controlling the obligatory number of CHs, which ultimately minimizes the utilization of energy. Nonetheless, for getting information, a CH should retain its receptionist nodes ON every time. This help to minimize lifetime of the network.

Efficient Routing-LEACH (ER-LEACH)

In the Efficient Routing LEACH (ER-LEACH), the curiosity of a BS is pondered for gathering environmental information [10]. Therefore, sensors may collect information instead of forwarding to the sink. This protocol along with its process is presented. With this method, a notable execution is attained particularly when mobility is at the peak. On the other hand, since the selection of CH is done on the basis of sensor's life irrespective of distance, its energy could finish quickly because of data transfer in a void space.

3.2. Multi-Hop LEACH Protocol

This procedure is the extension of single hop method, which works in a way that maximum CHs could send data to the BS. In addition, a single CH in this scheme shares data with the remaining CHs and also with the entire nodes lying in that particular cluster. Various amendments of LEACH with multi hop have been proposed, where a few of which are presented in this section.

Two-Level Hierarchy LEACH (TL-LEACH)

A TL-LEACH present a Two-Level Hierarchy LEACH [11], i.e., the basic CH is called the major CH and the second level signifies subordinate CHs. This method includes 4 main phases, i.e., setup, advertisement cluster formation, and information sharing. In the setup phase, all nodes decide that in case a basic CH is formed then a secondary CH would also work the same as the primary CH. In case a sensor is selected as the CH, this would have the responsibility of advertising the nature of the network to all other devices. In the next stage, all secondary CHs decide their status, i.e., it would lie in which cluster. In the 3rd stage, all basic CHs design an agenda on the basis of TDMA that employs time frame for information sending for all nodes in the cluster. All basic CHs choose the code of CDMA and notify it to all sensors for utilizing it. In the last phase, clusters are formed, and all sensors forward data with respect to the timeframe designed by the basic CH.

Mobile-LEACH (M-LEACH)

Mobile-LEACH (M-LEACH) [12] allows the mobility relevant to non-CHs and the basic CH when the first phase, i.e., setup, begins. This procedure reflects the energy of devices allotted during CH selections. Initially, all sensors are treated the same with respect to the performance of antenna, sensors having knowledge about the placement with the assistance of GPS whereby the BS is considered as stationary. In this scheme, a CH is selected based on diminution development. The best CH is selected to minimize the influence of diminution. This scheme provisions the nodes' mobility and therefore assures the collaboration of nodes and CHs albeit the nodes is not mobbing at all. Yet, this utilizes maximum energy because of a large ratio of data loss if the CH travels prior to the election of a CH in the upcoming stage.

LEACH-D

LEACH-Density Distribution of Node (LEACH-D) determines load balancing through adapting the function of levels [13]. The basic feature of levels is connectivity density that determines delivery denseness of sensor nodes. In this scheme the radius of a CH is decided by the CH itself, i.e., it is placed how far from the BS. Lastly, the CH applies multi-hop steady state in order to send information to the sink (i.e., BS). This is believed that LEACH-D is used to minimize energy utilization of the whole network, however, the election of CH is done on the basis of distance wherein the closer sensor to the BS is selected as the CH that would have lower battery-life and therefore may expire quite early.

W-LEACH (Weighted Low Energy Adaptive Clustering Hierarchy Aggregation)

W-LEACH [14], is a decentralized form of LEACH designed to augment the entire lifetime of WSNs. The general procedure of this protocol is similar to the basic LEACH scheme, but W-LEACH includes rounds which are absent in the basic LEACH. Every round in W-LEACH begins by setting up the cluster formation and CH selection. The next phase is the steady state wherein it is utilized for data transmission between the BS and all network nodes. Before the information sharing begins, all nodes identify their adjacent nodes on the basis of a fixed distance that supervises the denseness of sensors in order to avoid adjacent nodes to send similar data. On the basis of various adjacent nodes, sensors make decision so that to be dynamic or persist in asleep phase in this round. Thus, a node that has various adjacent nodes works in the active status for as much time as required. This way, all sensors send their data to the CH and thus reduce redundancy of information for a particular CH. W-LEACH works efficiently as it consumes little energy.

Multi-Hop LEACH (MH-LEACH)

In this kind of procedure, data transmission between the BS and all sensors is achieved in such a fashion that requires little transmission cost. This is due to the fact that maximum data transfer cost is required for when it is sent form the CH to the BS MH-LEACH protocol [9, 15] ,which provides solutions to the mentioned problems in which sensors send information to the CH. The CH then collects and share this information with the BS either directly or via the BS. Similar to the basic LEACH, this procedure chooses a CH on random basis. A path, i.e., route best known for energy efficiency, is chosen for a CH placed quite data a distance from the BS. The distance is measured as the way for the selection of the middle CH.

A CH that is placed closer to the BS, obtains information from other CHs which is located at a distance from the BS. Through this way it assists to save CH's energy, which is portion of the cluster. Nevertheless, the election of CHs is done randomly, thus, a node which is chosen randomly would not essentially have high energy or power. Hence, the overall lifetime of WSN is minimized in this scheme, which is its main drawback.

The timeline of LEACH, their comparison, and its summary are given in table 1

Table 1. Timeline of LEACH and its Descendants

No	Type	CH Selection	Contributions	Limitations
1	LEACH	Random	Increases the network lifetime and achieves better energy efficiency	Does not guarantee good CH distribution , assumes uniform energy distribution
2	LEACH-C	Based on residual energy	A higher number of rounds is achieved in a small network area	Introduces extra overhead due to single-hop communication
3	TL-LEACH	Random	Brings better energy load distribution across the network.	Increases overhead during alternative CH selection and cluster formation.
4	M-LEACH	Based on residual energy	Supports the mobility of nodes and hence guarantees the CH and node's interaction even if node is static.	Consumes more energy due to huge packet loss ratio when the CH moves before the selection of a new CH for the next round.
5	LEACH-E	Based on residual energy	Enhances network lifetime by balancing energy consumption among all network nodes.	Selects CH based on energy which may have low battery and far from the sink, therefore could die soon.
6	MH-LEACH	Random	Helps in saving CHs' energy	Selects CH on random basis that may not necessarily be high in power and energy and hence decrease the network lifetime.
7	LEACH-V	Random	Prolongs the network lifetime	The CH may die soon due to long distance communication.
8	ER-LEACH	Based on residual energy	Good performance is achieved at the time of high mobility.	The CH is selected based on the node's energy regardless of the distance; therefore, its energy may vanish very soon due to long distance communication.
9	LEACH-D	Based on distance	Reduces the entire network energy consumption.	Selects CH based on distance, which may have low ba;ttery and thus could die soon.
10	E-LEACH	Based on residual energy	Improves the process of CH selection by determining the required number of CHs that may reduce the energy consumption.	For receiving data, the CHs keep their receivers turned on all the time, which reduces the battery lifetime.
11	LEACH-W	Random	Saves energy consumption.	Selects CH on random basis, therefore it is difficult to know which node has higher energy to be chosen as CH. Thus the overall network lifetime may be reduced.

4. Conclusions

In this research we surveyed deeply the comprehension of wireless networks in line with their applicability and next we elaborated the method that packets are forwarded in the wireless environment. In addition, we also presented the basic procedures of WSN routing, known as LEACH, in terms of its working and features. Additionally, we also put light on the two main extensions this scheme, called multi-hop and single-hop, in terms of their suggested procedures with regards to its advantage and disadvantage. Besides the presented protocols and procedures, LEACH has various extensions, but its comprehensive elaboration is required for discovering an accessible, vigorous and effective clustering mechanism so that to reduce the usage of energy and augment the lifetime of WSN in both large-scale and small network zones.

References

[1] Akyildiz I F *et al* 2002 *Comput. Netw. 38 393.*
[2] Yick J *et al* 2008 *Comput. Netw.* **52** 2292.
[3] Abbasi A A *et al* 2007 *Comput. Comm.* **30** 2826.
[4] Haro B *et al* 2014 *IEEE Trans. Signal Process.* **62** 492.
[5] Mehmood *et al* 2015 *Ad Hoc Sens. Wirel. Netw.* **28** 1.
[6] Zhao F *et al* 2012 *Int. J. Distrib. Sens. Netw.* **8** 649609 .
[7] Pantazis N A *et al* 2013. *IEEE Comm. Surv. Tut.* **15 155**.
[8] Yassein, M. et al,. 2009. Citeseer.
[9] Xiangning F *et al* 2007 SensorComm 260.
[10] Al-Refai H *et al* 2011. *Int. J. Aca. Res.* **3**.
[11] Hady A A *et al* 2013 *Egyptian Infor. J.* **14** 109.
[12] Abbasi A A *et al* 2007 *Compt. Comm.* **30** 2826.
[13] Liu M *et al* 2009 *Sens.* **9** 445.
[14] Abdulsalam H M *et al* 2013 *Int J Distrib. Sens. Netw* **9** 289527.
[15] Nikolodakis A S *et al* 2013 *Algorithms.*

WM-LEACH - An Improved Network Lifetime LEACH Protocol for Wireless Sensor Networks

M. *Mahmuddin*[1,*], W. A. *Alabadleh*[2], L. M. *Kamarudin*[3]

[1,2]Internetwork Research Lab, School of Computing, Universiti Utara Malaysia, 06010, UUM Sintok Kedah.
[3]School of Computer and Communication Engineering, Universiti Malaysia Perlis (UniMAP), First Floor, Pauh Putra Campus, 02600 Arau, Perlis.

E-mail:ady@uum.edu.my

Abstract. Ad hoc network is divided into subcategories in which wireless sensor network (WSN) is one of them, which include several small devices known as sensor nodes (SNs) connected to base stations (BSs). The network lifetime is one of the most vital features as a trustworthy transmission. The low energy adaptive clustering hierarchy (LEACH) is one of the prominent protocols for deploying in the WSN that depends upon a procedure of clustering. Nevertheless, as cluster heads (CHs) share information with a BS openly irrespective of their interior distance, a maximum energy is utilized if the distance between them is too long, which alternatively minimizes the lifetime of the network. To cope with these challenges, various extensions of the LEACH protocol have been designed that send data in two ways, i.e., either in a single hop or multi hop. However, most of these extensions bring extra computational complexity for CHs due to the usage of residual energies. This study proposes an adaptive routing multi-hop LEACH (WM-LEACH) that would minimize the utilization of energy and augment the entire lifetime of the WSN. This rationale of the WM-LEACH is examined in the MATLAB simulator and the obtained outcomes are associated with other routing protocols in terms of network lifetime.

1. Introduction

Routing in the WSN is a quite difficult task because of various features which distinguish this network from other wireless networks, for example, mobile network or cellular communications. A few common characteristics of WSNs that are vital for forwarding techniques design comprise of many-to-one (M21) transmission, location familiarity, fixed devices, strongly restricted power, partial managing capabilities, caching spaces, and information gathering data merging

1.1. Low Energy Adaptive Clustering Hierarchy (LEACH)

LEACH is a procedure, which is the most appropriate for the WSN routing [1]. LEACH depends upon an adapted clustering procedure. The LEACH manipulates clusters and a single tier structure on the basis of a two-stage manoeuvre. A CH in the LEACH is arbitrarily shaped and the CH gathers the sensed information in all clusters. This also has several features associated with supplementary protocols. This may attain some stable energy utilization because of three features. Principally, a CH is interchanged on random basis. Next, sensors identify the beginning of every novel cycle on the basis of coordinated clocks for the whole system. Further, the remaining devices inside the network (in any of the cluster), which do not perform as CHs may switch off the radio till the allotted time slots of devices. Besides, features of the LEACH are such that all sensors may send and receive data from/to others without the

IOP Conf. Series: Materials Science and Engineering **551** (2019) 012057 doi:10.1088/1757-899X/551/1/012057

demand of earlier data regarding locations. Lastly, the LEACH is able to improve the gathered information in all clusters with the intention that the information sent to the BS may be reduced to the lowermost level.

The LEACH method is founded on the idea of cycles (rounds) where it scuttles in several cycles [2]. Every cycle includes two conditions/states, i.e., the steady one and the setup of a cluster [3]. In the former one, the information is forwarded from the source to other network devices whereas in the latter one the cluster is shaped in a way called self-adaptation. The concept of selecting a CH is such that the LEACH protocol selects a CH in every cycle [4-5]. Therefore, some of the sensors would utilize energy quite soon when they are chosen as CHs many times.

Likewise, a CH in the LEACH sends data to the source directly, thus, the energy consumption between the CH and the source is more than the energy consumed between two CHs. As a result, energy is drained rapidly by the CH. In the multi-hop LEACH process, the entire network nodes may not die soon and therefore the network lifetime is prolonged in comparison with the normal LEACH process. This is done via specifying stability in the energy utilization among the entire sensor network.

Also, as a CH is connected to the source directly irrespective of their distance, majority of the energy is utilized if the distance is long and therefore it minimizes the whole lifetime of the network [6]. To surmount this problem, various editions to the multi-hop LEACH are available, however, most of these editions (extensions) such as MR-LEACH, LEACH-E, LEACHME, LEACH-C, and LEACH-B, produce extra communication overhead for a CH as these all utilize a kind of energy known as residual energy.

Some more extensions of the multi-hop LEACH, e.g., M-LEACH and LEACH-W, also follow a random selection process for the CH that is not an effective method to improve the lifetime of the network and minimize the utilization of energy. Thus, a stout, accessible, and effective method is desirable for the enhancement of network lifetime and minimization of energy utilization in both large and small WSNs [7-12]. Hence, this study proposes a multi-hop LEACH extension, named WM-LEACH, which would minimize the utilization of energy and upsurge the lifetime of WSN

2. Setup Phase

The setup phase is not identical for all communication rounds. It differs based on whether the network has clusters or not. When the network has zero clusters, each alive node forwards a control message to the BS that consists of the ID, location and energy information.

The main shortcoming of LEACH is the random selection of CH that is applied to all sensor nodes without taking into account any factor. In reality, to increase the lifetime of network and energy efficiency, we need to change the threshold of selecting CH. In other words, we must consider four essential factors: the distance between the nodes and the BS, the residual energy, RSSI and nodes degrees within the transmission range, to calculate the threshold, which are calculated presented in (1):

$$\cos t(i) = \frac{\alpha E_{res}(i) + \beta N_{nbr}(i) + \delta RSSI(i)}{\gamma d(i, bs))} \tag{1}$$

where, E_{res} is the initial energy, $N_{nb}(i)$ is the number of neighbours of node i, N_{alive} is the number of alive nodes, d (i, bs) is the distance between the node i and the bs. Then the threshold can be written as follows:

$$T(i) = \begin{cases} \cos t(i) & ; if \ i \in G \\ 0 & ; otherwise \end{cases} \tag{2}$$

IOP Conf. Series: Materials Science and Engineering **551** (2019) 012057 doi:10.1088/1757-899X/551/1/012057

3. Results and Discussion

3.1 Network Lifetime

The network lifetime of a sensing device means that a node may consume its energy in how much time that it dies after that. Thus, the device is counted as a dead node and is unable to function further in terms of information transmission. To elaborate it more, this is counted as in which round the first sensing device will lose its energy (i.e., die). Hence, it is considered as one of the most vital metrics to evaluate the capabilities of the network for prolonging sensing rounds of the network. A sensing device that dies after some time denotes that the system drops a routing choice and creates the lower and upper devices distressing from long-distance connections and thus misses some exposure of sensing.

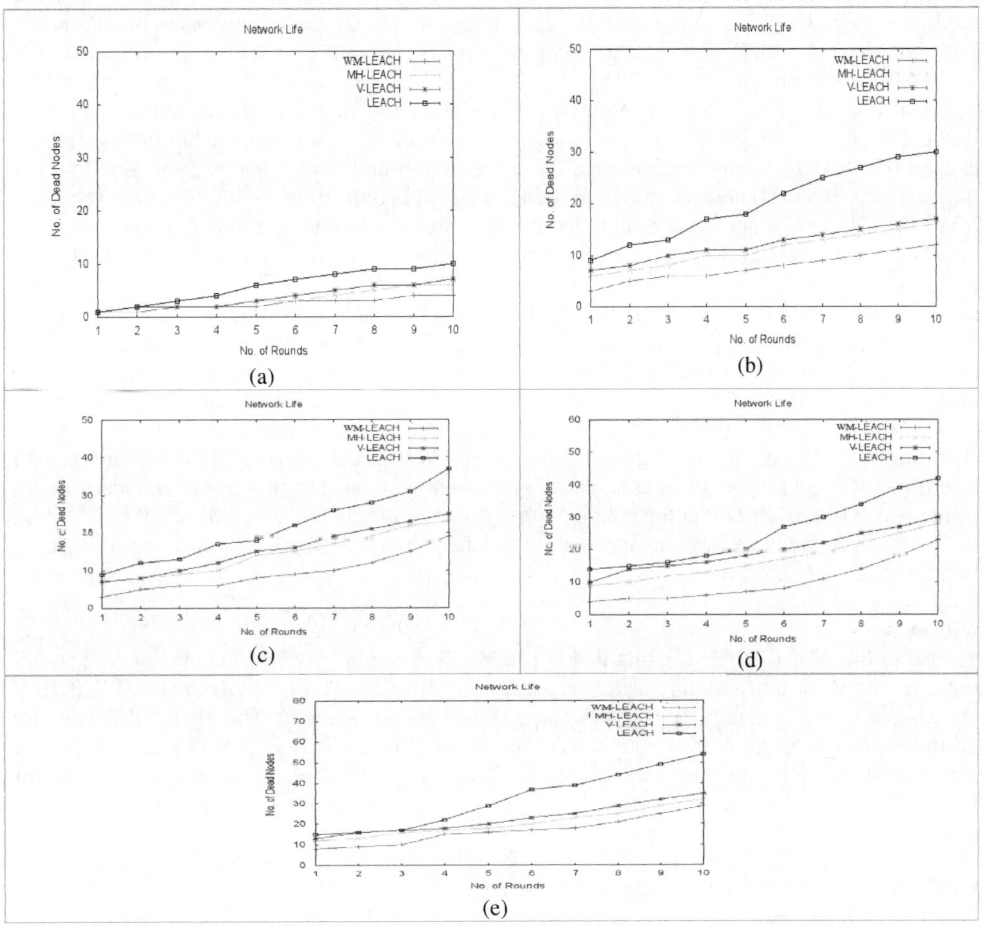

Figure 1. Network Lifetime with (a) 50, (b) 100, (c) 200, (d) 500 and (e) 1000 Nodes

The presented figure 1 demonstrates the lifetime the network for our proposed protocol in comparison with the basic LEACH, MH-LEACH, and V-LEACH having different number of nodes, i.e., 50 to 1,000 respectively, in 10 different rounds. The first figure, i.e.,1(a) shows that the number of nodes that have been died, (in a 50 nodes network) in the basic LEACH is one in the 1st round and 10

IOP Conf. Series: Materials Science and Engineering **551** (2019) 012057 doi:10.1088/1757-899X/551/1/012057

in the last round. In the MH-LEACH, dead nodes reached six from one in the 1st and last rounds, respectively. In the V-LEACH, the number of dead nodes reached seven in the 10th round as compared to one node in the 1st round. Whereas, in our proposed protocol, the number was recorded in such a fashion that it reached maxim four in the last round in comparison to one node in the 1st round.

Likewise, in the 100 nodes network, as shown in figure 1(b), the dead nodes were such that in the basic LEACH these were nine in the first round and then reached 30 in last round. In the MH-LEACH, dead nodes were six in the first and 16 in the last round, respectively. In the V-LEACH, the number reached 17 in the 10th round as compared to only seven nodes in the first round. Whereas, in the WM-LEACH, the numbers were such that in the first round these were only three and then reached 12 in the 10th one. Furthermore, when the network was extended to 200 sensing devices (nodes), the lifetime of the entire network was such that the basic LEACH lost nine nodes in the first round and 37 nodes in the final round (see figure 1(c)). The MH-LEACH lost 7 nodes in the first round as compared to 23 nodes in the last one. The V-LEACH was recorded such as it lost seven and 25 nodes in the first and 10th round, respectively. On the other hand, WM-LEACH lost only three nodes in the first round, which reached maxim 19 nodes in the 10th one. Moreover, when the network was deployed with a number of 500 nodes, LEACH got a number of dead nodes as 14 in the first round and 42 in the last one (see figure 1(d)). In the MH-LEACH, the number reached 28 in the final round, which was only nine in the first round. The V-LEACH lost 30 nodes in the final round, which was only 10 in the initial round. Whereas, in the WM-LEACH, the lost nodes were such that these it reached 23 in the last round, which were only four in the first round. In addition, these protocols were also tested with 1000 nodes in 10 different rounds.

This result was as follows: The basic LEACH lost 54 nodes in the last round, which were only 15 nodes in the first one. The MH-LEACH reached 32 lost nodes in the last round in comparison with 12 nodes in the first one. In the V-LEACH, the nodes reached 35 from 13 in the last round and first one, respectively. While our proposed protocol reached 29 nodes in the last round from eight nodes in the first one. Investigating the showing figures, this may be observed that the number of lost nodes is rationally minimized to a satisfactory degree in the proposed WM-LEACH protocol in comparison with other three protocols. Looking at all figures, i.e., (a) to (e), this can be seen that if the number of nodes is less or more, WM-LEACH works the best in comparison with basic LEACH, MH-LEACH, and V-LEACH. Therefore, this would have an appropriateness for deploying in the WSN so that to increase the entire lifetime of the network.

4. Conclusions

This rationale of the WM-LEACH is examined in the MATLAB simulator and the obtained outcomes are associated with three other routing schemes, i.e., basic LEACH, MH-LEACH, and V-LEACH in terms of network lifetime, energy consumptions, and network throughput. The obtained results are predicted to be momentous with regards to create the proposed WM-LEACH protocol satisfactory for subscribers where they may validate the rationale of WSNs afore choosing an appropriate scheme for routing.

References

[1] Abbasi A A *et al* 2007 *Comput. Comm.* **30** 2826.
[2] Kaur L *et al* 2014 Optim. **2**.
[3] Shan J *et al* 2013 Elektrotechniczny **89** 75.
[4] Bindal *et al* 2014. Energ. **1**.
[5] Vigneshkumar *et al* 2015 Energ. **4**.
[6] Vashist S *et al* 2014 *Int. J. Com. Sci. Mob. Com.* **3** 99.
[7] Abdulsalam *et al* 2013 *Int. J. Distr. SenNet* **9** 289527.
[8] Braman *et al* 2014 *Int. J. Adv. Res. Comm. Eng.* **3** 5683.
[9] Dhawan H *et al* 2014 *Int. J. Comp. Appl.* **95**.
[10] Gnanambigai *et al* 2012 *Int. J. Comm. Comp.* Tech. **1** 15.

[11] Kaur A *et al* 2015 *Int. J. Comp. Appl.* **116**.
[12] Akkaya et al, 2005 *Ad Hoc Netw.* **3** 325.

IOP Conf. Series: Materials Science and Engineering **551** (2019) 012058 doi:10.1088/1757-899X/551/1/012058

Throughput Analysis of WM-LEACH, LEACH, MH- LEACH and V-LEACH Protocol: A Simulation Based Approach

M. Mahmuddin[1],W. A. Alabadleh[2, *], L. M. Kamarudin[3]

[1]Internetwork Research Lab, School of Computing, Universiti Utara Malaysia, 06010, UUM Sintok Kedah.
[2]Internetwork Research Lab, School of Computing, Universiti Utara Malaysia, 06010, UUM Sintok Kedah.
[3]School of Computer and Communication Engineering, Universiti Malaysia Perlis (UniMAP), First Floor, Pauh Putra Campus, 02600 Arau, Perlis.

E-mail: ady@uum.edu.my

Abstract. This paper provides a passing description for some of routing protocols like LEACH, WM-LEACH, MH-LEACH and V-LEACH Protocol in Wireless Sensor Network (WSN) also a comparison study of these protocols based on one of performance matrices as throughput. Addition to this an attempt is done to calculate their throughput. The effect of these protocols are simulated through the MATLAB simulator. Finally, the throughput of the network increases with WM-LEACH with respect to send interval and number of bytes of data. In the WM-LEACH, the throughput is higher than in other protocols which are compared in this research. This is because the WM-LEACH protocol reduces the communication cost between the domains by allowing more data in less time.[-1].

1. Introduction

Wireless Sensor Networks (WSNs) have been utilized in various applications wirelessly. These applications include industrial applications such as mining, smart cities, smart emergency systems, and smart virtual power plants. WSNs may be categorized into infrastructure-less and infrastructure-based networks [1, 2]. The former one includes Base Station (BS) for sending data inside the network. Data networks that normally send data wirelessly are deployed either multi-hop or single-hop between the BS and all wireless nodes. Nevertheless, this category is unable to fulfill the demands of end users, even though the progress in this class has obtained a prodigious growth [3].

On the other hand, the latter class does not fix to a particular infrastructure and therefore known as ad hoc networks [4]. In this mode of transmission, the infrastructure is not involved as a BS, which connects other networks. Data transmission among contributing devices is achieved same as done in the infrastructure-less class, i.e., either through multi-hops or a single hop [5]. In this class of transmission, wireless devices share data with each other in a uniform communication range of an ad hoc network. Suppose, the required terminal is not in the range of the transmitting device, transmission is achieved indirectly, that is, in a multi-hop fashion [6, 7].

IOP Conf. Series: Materials Science and Engineering **551** (2019) 012058 doi:10.1088/1757-899X/551/1/012058

2. Related Works

For the improvement of LEACH, several extensions were proposed in the last decade [8, 9]. In this section, we elaborate the features of those extensions

2.1. Low Energy Adaptive Clustering Hierarchy (LEACH)

LEACH protocol is measured as the best procedure for WSN routing [10]. The idea of LEACH is being contemplated as an inventiveness for several routing procedures. The goal of this protocol is to select sensor nodes being the CHs in various cycles so as the outcome of extreme power excess (in communication with sinks) is obtained and then dispersed in the entire WSN.

This protocol includes the locations of CHs that are deemed of excessive power and random revolution as this switches among various nodes with the intention to stop a single node's battery to die. Primarily, rules are defined: a) sensor to reach sinks can utilize adequate energy for communications when it requires assistance from other nodes. b) Information that is inspected via a near device is measured the same and may be gathered. Sensors in this protocol administer itself with an indigenous block creating 1 device being a CH. The remaining sensors develop data transfer with the CH and the CH receives information from all nodes and exercise the task of data forwarding. Thus, the CH is believed powerful in comparison with other non-CH nodes.

3. Proposed Protocol

The proposed WM-LEACH protocol divides the communication process into several numbers of rounds. Each round comprised of a setup phase and data transmission phase. As previously mentioned, the setup phase is an overhead over actual data transmission. The primary focus of WM-LEACH is to reduce this overhead which causes improvement over the network lifetime. To accomplish this, WM-LEACH uses Alternative Cluster Head (ACH). This ACH hierarchy help to reduce the energy required to transmit necessary information for the setup phase to perform clustering operations. It also reduces the frequency of the setup phase using two schemes.

The first scheme is designed that relevant to multi-hop inter routing in which all CHs (include in clusters) are able to forward information to other CHs in other clusters. The second scheme copes with "multi-hop intra routing" in which sensors in the clusters are able to transmit data to all nodes for satisfying the idea of "multi-hop routing".

The designed two schemes focus on improving the lifetime of the network in line with minimizing usage of energy of the entire sensors in the system. Each of these phases is described in the following subsections comprehensively.

3.1. Initial Phase of WM-LEACH

This phase is not identical for all communication rounds. It differs based on whether the network has clusters or not. When the network has zero clusters, each alive node forwards a control message to the BS that consists of the ID, location and energy information.

The main shortcoming of LEACH is the random selection of CH that is applied to all sensor nodes without taking into account any factor. In reality, to increase the lifetime of network and energy efficiency, we need to change the threshold of selecting CH. In other words, we must consider four essential factors: the distance between the nodes and the bs, the residual energy, *RSSI* and nodes degrees within the transmission range, to calculate the threshold, which are calculated presented in (7):

$$\text{cost(i)} = \frac{\alpha E_{res}(\text{i}) + \beta N_{nbr}(\text{i}) + \delta RSSI(\text{i})}{\gamma d(\text{i}, bs)} \tag{1}$$

$$E_{res}(\text{i}) = \frac{E_{rem}(\text{i})}{E_{init}(\text{i})} \tag{2}$$

$$N_{nbr}(i)=\frac{N_{nb}(i)}{N_{alive}} \tag{3}$$

$$RSSI(\mathrm{i}) = \frac{1}{d^2} \tag{4}$$

$$\alpha+\beta+\delta=\gamma, \quad \frac{\alpha+\beta+\delta}{\gamma}=1 \tag{5}$$

$$\alpha,\beta,\gamma,\delta \in [0,1] \tag{6}$$

where, E_{init} is the initial energy, $N_{nb}(i)$ is the number of neighbors of node i, N_{alive} is the number of alive nodes, $d\ (i, bs)$ is the distance between the node i and the bs. Then the threshold can be written as follows:

$$T(i)=\begin{cases}\cos t(i) & ; if\ i\in G \\ 0 & ; otherwise\end{cases} \tag{7}$$

4. Simulation and Result

The following sections discuss the analytical result of the proposed WM-LEACH scheme compared to the primary LEACH and V-LEACH schemes. For comprehensive evaluation, it starts simulation environment then the evaluation with the network throughput

4.1. Analytical and Numerical Analysis

The effect of WM-LEACH is simulated through the MATLAB simulator in 5 diverse setups: The first scenario consists of 50 SNs, the second scenario includes 100 SNs, the third scenario comprises 200 SNs, while the fourth and last scenarios comprise of 500 and 1,000 SNs, respectively. The selected amount of SNs changes and is sufficient for the performance evaluation and representing different network conditions.

The functioning of the WM-LEACH is evaluated in comparison with the primary LEACH, MH-LEACH and V-LEACH schemes. The functionality of these schemes is evaluated in all scenarios with 100 rounds. Chosen metric is network throughput, In presented figures, the x-axis demonstrates the amount of rounds in such a way that 10 and 100 are represent;ted by 1 and 10, respectively.

The basic LEACH, MH-LEACH, and the V-LEACH procedures are quite limited in throughput. Thus, this may be an extremely essential to analyze the throughput at the time of developing a novel idea for routing. Here, five figures, i.e., (a) to (c) demonstrate the received amount of bits (on average basis) on the BS with 50 to 1000 sensing devices in in 10 different rounds.

IOP Conf. Series: Materials Science and Engineering **551** (2019) 012058 doi:10.1088/1757-899X/551/1/012058

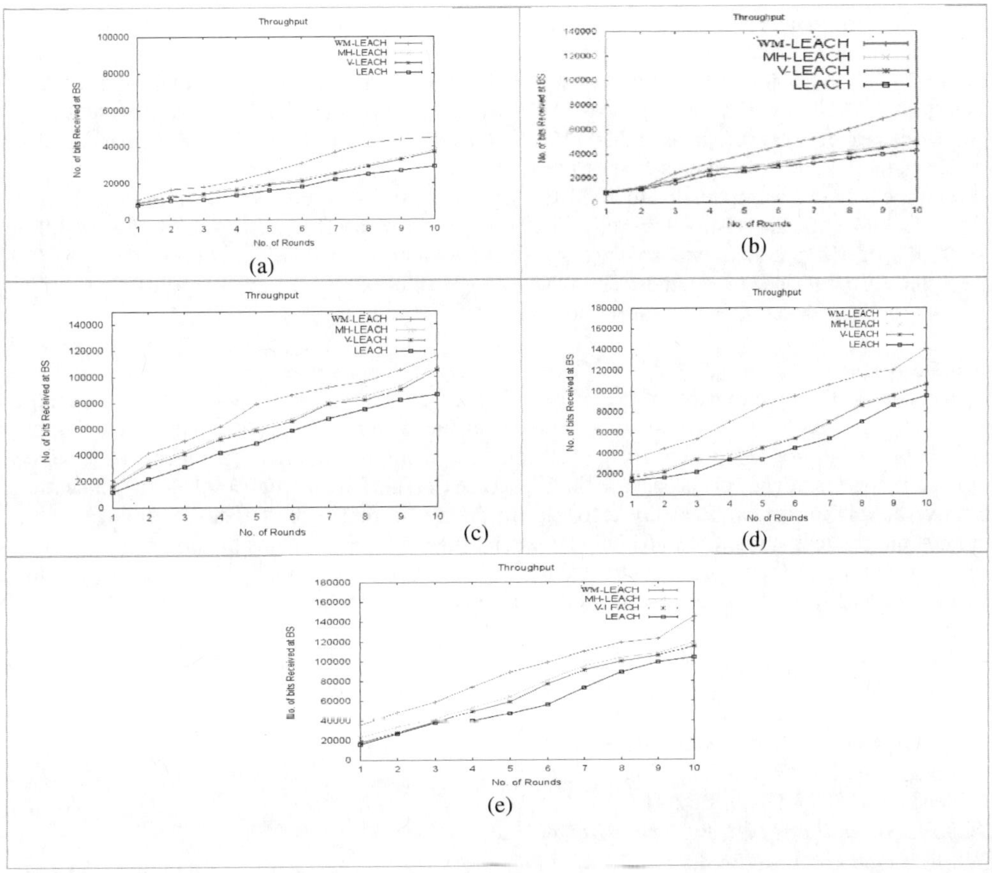

Figure 1. Obtained Throughput with (a) 50, (b) 100, (c) 200, (d) 500, and (e) 1000 Nodes Respectively

Our proposed protocol has the ability to improve the overall throughput of the network as this tries to reduce hops for a packet using splitting a network into different clusters. In addition, as compared to our proposed scheme, the remaining three protocols choose CHs on the basis of merely one parameter, i.e., either network lifetime, energy, or distance, thus, a multi-getaway to the network would aid to minimize the issue of blockage. Throughput, in terms of sent bits to the BS in all simulated schemes with only a 50 node network was such that the basic LEACH got 8000 initially and 29000 bits in the last round, as presented in figure 1. MH-LEACH = 9600 in the 1st round and 38000 bits in the final one. V-LEACH =9000 initially and 37000 finally. On the other hand, in our proposed protocol, in the first round the number of received bits was 11000 which then reached 45000 in the final round. Likewise, when the number of nodes was extended from 50 to 100 nodes, the proposed scheme received dominancy with 9000 bits initially which then finally reached 76000 bits, in comparison with LEACH, MH-LEACH, and V-LEACH that obtained 8000, 8600, and 8500 in the first round, and 4200 50000, 48000, in the final one, respectively.

Besides, the mentioned protocols were also examined when the network nodes reached 1000 from 200 nodes. That is, when the number of nodes was 200, LEACH obtained a throughput of 12000 bits initially which then reached 86000 in the final round. MH-LEACH received a maximum of 107000 bits in the final round as compared to 17,000 in the 1st one. V-LEACH obtained maximum 105000 bits in the 10th round in comparison with 16,000 initially obtained bits. Whereas, the WM-LEACH obtained the

following results: the number of received bits reached 116,000 in the final round which was 21,000 initially. When the number of network nodes was extended to 500, the recorded results were such that LEACH got 14,000 to 95000 in the 1st and 10th round, respectively. MH-LEACH attained 19,000 to 117,000 initially and finally, respectively. V-LEACH was recorded with 17,000 and 115,000 in the 1st and 10th rounds, respectively. Whereas in the WM-LEACH, the obtained results were as follows: 34,000 and 140,000 in the 1st and 10th rounds, respectively.
Lastly, when the network was extended to 1000 sensors, the recorded results were as follows; LEACH was observed with 16,000 and 104,000 in the 1st and 10th rounds, respectively. MH-LEACH was 23,000 and 119,000 in the mentioned two rounds. V-LEACH was recorded with 17,000 and 115,000, respectively, in the mentioned two rounds. The WM-LEACH achieved the following results: 36,000 bits in the initial round and 145,000 in the last round

5. Conclusions

This paper analyzed the functionality of our proposed scheme with regard to network throughput, the throughput of the network increases with proposed mechanisms with respect to send interval and number of bytes of data. In the proposed mechanism, the throughput is higher than in other mechanisms which are compared in this research. This is because the WM-LEACH mechanism reduces the communication cost (the overall amount of productively arrived information in one second. The determined stable throughput is the greatest amount of traffic in one second) between the domains by allowing more data in less time. In addition, WM-LEACH diverts the traffic in more than one path by choosing the path due to the multi-cluster approach. These are some of the reasons that help to increase the network throughput.

References

[1] Akyildiz, I F *et al* 2002 *Comput. Netw. 38 393.*
[2] Yick J *et al* 2008 *Comput. Netw.* **52** 2292.
[3] Abbasi A A *et al* 2007 *Comput. Comm.* **30** 2826.
[4] Lewis F *et al* 2004 *Wirel. Sens. Netw.* 11.
[5] Culler D *et al* 2004 *Comput. 37* 41.
[6] Akyildiz I F *et al* 2002 *IEEE Comm. Mag. 40* 102.
[7] Willig A *et al* 2005 *PIK 28* 86.
[8] Mehmood A *et al* 2015 *Ad Hoc Sens. Wirel. Netw.* **28** 1.
[9] Zhao F *et al* 2012 *Int. J. Distrib. Sens. Netw.* 649609.
[10] Pantazis N A *et al* 2013 *IEEE Comm. Surv. Tut.* **15** 551.

IOP Conf. Series: Materials Science and Engineering **551** (2019) 012059 doi:10.1088/1757-899X/551/1/012059

An Enhanced Energy Efficient Hybrid Asynchronous Duty Cycle MAC in Wireless Sensor

Maulidin[1], M. Mahmuddin[2]

[1,2]Internetwork Research Lab, School of Computing, Universiti Utara Malaysia, 06010, UUM Sintok Kedah.

Email: maulidin84@gmail.com

Abstract. Wireless sensor network (WSN) has countless potential application in many areas due to its easy deployment process, lower installation cost, less cabling required, and high mobility. However, limited energy residue (especially deposited in small size battery) is always a major obstacle in WSN. Improving mechanism on MAC layer is one of the potential research that could increase battery lifespan. In the active phase of the MAC layer, a node can receive and transmit packet when awake. Then, a node is in a sleep phase, it switches off the radio automatically to save energy and gives alert. While in idle listening phase, node still using the energy. Duty cycling performed in MAC layer don't involve turning off their radio and switching sleep mode but repeating off/on process. Therefore, this study proposed mechanism of WSN that used in order to enhance power energy using duty cycle in MAC layer. The proposed of asynchronous duty cycle MAC layer using a hybrid duty cycle MAC in implementation the performance of WSN. In addition, this paper also elaborated the design asynchronous hybrid duty cycle MAC with multi-hop broadcast.

1. Introduction

Nowadays, wireless system is becoming a popular system in industrial technology. It became popular due to their so many applications especially in wireless sensor network (WSN) that can have applied in any environment. The implementation of WSN is considered able to adopt in agriculture business [1], national defence [2], temperature monitoring [3], pollution [4], and so on. WSN provides some advantages including easy deployment process, lower installation cost, less cabling required, and high mobility [5]. It also offers expected transformation infrastructure for building automation, also process control applications [6]. WSN has expected to proper future in market prospective.

WSN is a regular of small networks component that contain of distributed particles called sensor nodes. Compared to mobile device, the coverage and communication range for sensor nodes compared is limited due to low power capacities. It consumes several powers to do the task as the laying of the sensor. Moreover, sensor node has some duties to collect and transmit data obtained by the sensor to other nodes or access point. WSN usually used a large number of nodes across the hostile and unstable environment.

Extending the lifetime of node and sustaining the proper processes of the network are becoming one of issue in this sensor networks. Furthermore, the energy efficiency becomes the first concern and main objective in designing this system, which can be useful if a user uses it properly and effectively. Those processes consist of sending, receiving, processing, forwarding data, and also processing or forwarding some requests. In other way, useless energy consumption consists of collision, overhearing, control overhead, idle listening, and over emitting [7].It used in order to enhance the power energy that used relevant technique Media Access Control (MAC) layer named duty cycle. Both synchronous and asynchronous are suitable for light traffic loads where frequent excessive traffic load and high traffic loads in wireless system network, such as broadcast [12] or converge cast [13-15] traffic could suddenly increase channel contention in a local neighbourhood. Broadcast process is commonly used for various network wide queries and updates [16], and the converge cast taken when various sensors can be observed under the same conditions by sending the report to the sink node or to the node that receives the data aggregation [17].

2. Duty Cycle on MAC

There are various low duty cycle protocols that is proposed for WSNs, which differ in aspects of synchronisation, the number of channels required, transmitter- or receiver-initiated operation etc. There are two different types of protocols, which control low duty cycle protocols: asynchronous and synchronous schemes [18-20].

Another variation of low duty cycle protocols such as Queue-MAC, a new traffic adaptive duty cycle MAC protocol considered for event-driven sensor networks [21] uses the synchronization based preamble sampling to deal with low traffic load in a long period and responds to burst events by dynamically dealing the TDMA slots. SVP sampling has high energy efficiency with short channel listening period and dynamically allocated TDMA slots could achieve low transmission delay by making full use of the communication bandwidth according to the traffic load. Moreover Distributed Slot Scheduling Algorithm for Hybrid CSMA/TDMA MAC [22] offered algorithm for hybrid mac to handle collision during communication. This protocol handle a schedule which bridges the gap between a feasible and an optimal schedule to handle the collision during transmission. First, find out two-hop neighbors of each node, then a particular slot is allotted to each node in order to prepare a feasible schedule using the RD-TDMA algorithm. Finally, the feasible schedule is adequate tuned in a novel way to improve the efficiency in handling the collision by reducing the number of allotted slots.

Energy Efficiency TDMA/CSMA Hybrid Protocol with Power Control (TCH-MAC) for WSN reduce the energy consumption and transmission throughput [23].This protocol takes advantages of TDMA and CSMA in the MAC layer by design a novel power control scheme to further reduce the energy consumption and optimize the transmission slots.

3. Proposed Hybrid Asynchronous MAC

Generally, this protocol is designed to achieve energy efficiency and delay WSN applications that combined technique time division multiple access (TDMA) with CSMA Asynchronous. Here TDMA concept used LMAC protocol [19] and then CSMA concept is used like WiseMAC Protocol [17].

The free timeslots of each channel are calculated by performing the OR operation between its local vector and the vector found from its neighbour. A node is allowed to transmit by using its controlled timeslots and receive from any timeslot of any channel by switching the interface between different free timeslot as well. Therefore, the transmitter has the option to choose the best timeslot between the two and hence the delay of packet advancing decreases dramatically. Figure 1 shown the idea of Hybrid Duty Cycle MAC WSN done in this paper.

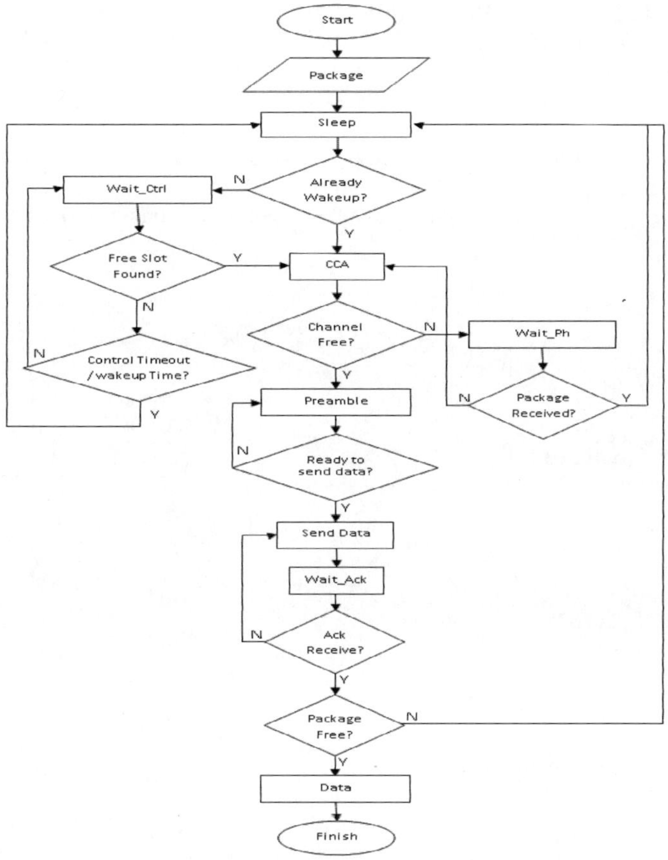

Figure. 1: Flowchart of Porposed Enhanced Hybrid Duty Cycle MAC

Based on Figure 1 is shown flowchart has been formulated to enhance the energy efficiency and reduce delay in this new model of Hybrid Duty Cycle MAC. When the package needs to send data, this model finds a free wakeup node. If all nodes are busy, then the package needs to wait in WAIT_CTL and schedule its turn. If the package cannot get their free slot in WAIT_CTL. Therefore, it expectedly goes back to sleep in order to save the energy.

Free wakeup node implemented to do clear channel assessment (CCA) to find free channel. If the channel busy, it uses to wait until there is a free channel do preamble. In preamble, package initiates the readiness of their neighbour to send the data. Then after sent the data, it waits for an acknowledgement to confirm that the data has been send successfully. If the package received the acknowledgement then the package become free, otherwise it will redo the sending process. Furthermore, in Fig. 2 shown that when the node is not wakeup yet, thus wait control will schedule the slot for package that being transmit, then continue to CCA. If the slot had been used by other package, then wait control will try to find another free slot.

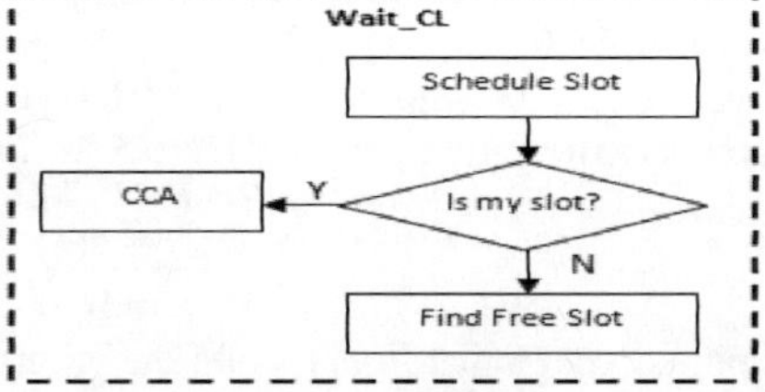

Figure 2: Wait control (WAIT_CTL) Mechanism

IOP Conf. Series: Materials Science and Engineering **551** (2019) 012059 doi:10.1088/1757-899X/551/1/012059

4. Operational Cycle of the Proposed Hybrid Protocol

This operational cycle in begin when a node wants to send the data. All nodes start from sleep state then it will indicate priority slot which already wakeup. If the wakeup slot is not for the node, then it will wait their scheduled slot in WAIT_CTL. If the wakeup slot is for the node, thus it will do clear CCA in order to find free channel. If the channel is busy, it will wait their turn in wait package then going back to sleep. If the channel free, then it will do preamble to send the data. Then it will check the acknowledge status. If the ack is true, means that the wait ack does not receive any acknowledgement then it will go back to sleep. If the ack is false means it will broadcast, then go to sleep. It shown in Figure 3 as follows:

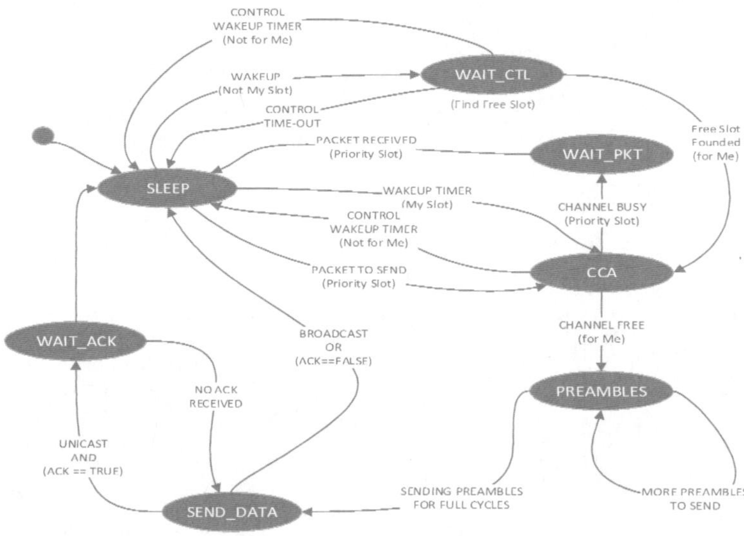

Figure 3: Overall Operational Cycle Hybrid Asynchronous Duty Cycle MAC

5. A Hybrid Asynchronous Duty Cycle MAC with Multi-hop Broadcast

Multi-hop broadcast is an important network service in WSNs, especially for applications such as code update, remote network configuration, route discovery, etc. Although the problem of broadcast that is studied in always-on networks such as wireless ad hoc networks, where neighbour connectivity is not a problem. The broadcast is more difficult in duty-cycled WSNs where each node stays awake only for a fraction of time slots and neighbourhood nodes are not simultaneously awake for receiving data. It becomes more difficult in asynchronous and heterogeneous duty-cycling scenarios.

EMBA [18] is the multi-hop broadcast protocol that uses asynchronous duty cycle. It allows nodes to periodically give-and-take link quality information with neighbour nodes. With this procedure, EMBA can significantly lessen duty cycle of nodes and also provide a considerable broadcast efficiency for networks with burst traffic. Hybrid Asynchronous Duty Cycle MAC integrated EMBA tehniques to reduce the number of transmissions, thus it minimizes the active time of nodes.

6. Conclusions

This paper presents about the development in designing model in hybrid asynchronous duty cycle MAC WSN as the guidance of the point. One of point in development is to measure the power energy in term improve or delay process. It elaborates operational cycle of hybrid mac protocol that consist with several stages such as; INIT, SLEEP, CCA, WAIT_PKT, WAIT_CTL, PREAMBLES, and SEND_DATA. It is condition that through the transmission in hybrid asynchronous duty MAC. In future work, the study about how to compare the protocol in asynchronous duty cycle MAC that used based on the proposed design mode.

Acknowledgements

This work was partly financially supported by from Ministry of Education (MOHE), Malaysia and Universiti Utara Malaysia (UUM), Shah Alam, Selangor.

References

[1] Abd El-kader S M *et al* 2013 Egyptian Informatics Jounal **14**(2) 109-115.
[2] Jokhio S H *et al* 2012 IET Wirel. Sens. Syst., **2** (3) 161-169
[3] Flammini A *et al* 2007 IEEE Int. Sym. on Industrial Electronics.
[4] Yi W Y *et al* 2015 Sensor (Basel) **15** (12).
[5] Karl H *et al* 2006 Protocols and Architectures for Wireless Sens Net. John Wiley & Sons, Ltd.
[6] Wheeler A 2007 IEEE Commun. Mag., **45**(4) 70-77
[7] Gopalan S A *et al* 2010 International Congress on Ultra Modern Telecommunications and Control Systems.
[8] Bashir Y *et al* 2009 Wireless Communication. Mob. Comput., **9**(12) 1572-1067
[9] Arfvidsson J *et al* 2006 Proceedings of the 4th International Conference on Embedded Networked Sensor Systems.
[10] Ni S Y *et al* 1999 Proceedings of the 5th Annual ACM/IEEE International Conference on Mobile Computing and Networking 151-162
[11] Zhang H *et al* 2007 Computing. Commun. **30**(13) 2560-2576
[12] Rodionov A S *et al* 2003 *Int.* J. Comput. Math. **80** (2) 181-191
[13] Shakhov V V *et al* 2004 Future Gener. Comput. Syst., **20**(2) 181-188
[14] Stann F *et al* 2006 Proceedings of the 4th Int. Conf. on Embedded Networked Sens. Syst.
[15] Estrin D *et al* 1999 Proceedings of the 5th Annual ACM/IEEE Int. Conf. on Mob. Comput. and Netw 263-270
[16] Cerpa A *et al* 2001 SIGCOMM Comput. Commun. Rev. **31.**
[17 El-Hoiydi A *et al* 2004 Proceedings. ISCC 2004. Ninth International Symp. on Comput. and Commun. (IEEE Cat. No.04TH8769), **1** 244-251
[18] Jang S *et al* 2013 IEEE Trans. Wirel. Commun., **12**(4) 1640-1650
[19] Lee A H *et al* 2008 International Symposium on Computer Science and its Applications.
[20] Wu L *et al* 2015 International Journal of Distributed Sensor Network, **2**.
[21] Lenka M R *et al* 2016 IEEE International Conference on Networking, Architecture and Storage (NAS).
[22] Wang X *et al* 2018 Wireless Communications and Mobile Computing 2018.

Evaluation Performance of Hybrid MAC Protocol in Wireless Sensor Network

Maulidin[1], M. Mahmuddin[2,*]

[1,2]Internetwork Research Lab, School of Computing, Universiti Utara Malaysia, 06010, UUM Sintok Kedah.

Email: maulidin84@gmail.com

Abstract. The implementation Wireless Sensor Network (WSN) in generally can be applied in any environment where the important MAC layer plays vital. Researchers discover that the major power consumption of wireless sensor network nodes are related to their radio transceiver's activities on MAC. This paper explains the implementation expected systeme that can be used in the WSN system. It focuses on the comparison some MAC protocol in order to discover the performance of each protocol. Therefore, this paper is benchmarking with other MAC protocol to identify the capability of the proposed hybrid protocol in term of some parameter. Then, this paper determines the proposed hybrid protocol that is camparised withother protocol based on evaluation of some parameter choosen .

1. Introduction

Wireless Sensor Network (WSN) is the most commonly well-known technology, especially in communication industry. It consists of tiny particles that consume some number of energies from irreplaceable batteries that is named sensor nodes. Sensor nodes gather packet data from worldwide networks, process it, and transmit the packet data via radio frequency transmitter to the sink node [1]. WSNs are usually static nodes that send data to a server or a sink node for processing.

Currently, WSNs are going into real deployments covering various areas, e.g., military industry [2], industrial application [3], environment monitoring [4] and health care [5, 6] for humanity. Moreover, sensor-networking communication manage is the main role in development of Internet of things (IoT) [5-8].

Wireless sensor network has limited battery. Most of researcher's efforts that discover the major power consumption of WSN nodes are related to their radio transceiver's activities have proved it. Then, running in the idle mode (switched on, or also called ready-to-receive mode) as when transmitting or receiving data is also significantly consume high energy consumption. Hence at the MAC layer that control the operation of radio, should avoid the frequent switching. Moreover, the problems of study about struggle to achieve power efficiency can be solved by the relevant technique in MAC layer called duty cycle.

Most of the existing Duty Cycle for MAC protocols in WSN use synchronous and asynchronous technique. In this paper, it proposes a hybrid asynchronous MAC protocol by combines the TDMA concept (L-MAC protocol) with the adaptive wake up preamble (WiseMac protocol). In order to improve end to end delay while ensuring energy efficiency, firts each node allows to choose different timeslot, which is not interfering with the communication between other nodes in the network. Then, after finished choosing slot and it will check the medium by using preamble sampling to alert the receiver node, after receiving the current frame. A wake-up preamble of size equal to the sampling period is transmitted, in front of every data frame. It ensures that the receiver will be awake when the data portion of the packet arrives. Therefore, by combine Scheduling mechanism (LMAC) and Preamble Sampling (WiseMac), it is possible to reduce the delay and energy of the node.

2. Proposed Hybrid MAC

There is a direct trade-off between energy efficiency and latency, it is because lowering the energy consumption may result in high latency. Conceptual model of this research is a hybrid MAC protocol combines the concept L-MAC protocol with the advantages of adaptive wake up preamble of WiseMAC protocol. In the first phase by use schedulling mechanism, each node allows to choose different timeslot, which is not interfering with the communication between other nodes in the network. The free timeslots of each channel are calculated by performing the OR operation between the s local vector and the vector found from its neighbor. A node is allowed to transmit by using its controlled timeslots and can receive from any timeslot of any channel by switching the interface between different free timeslot as well. In the next phase, after a node finished choosing slot and it will check the medium by using preamble sampling mechanism. It ensures that the receiver will be awake when the data portion of the packet arrives.

3. Performance Result

In this section is elaborated the result value of four protocols in term of energy consumption, throughput, latency and packet delivery ratio which formula as follow.

a. Energy consumption per bit: defined as the energy that being used by total number of bits transmitted. Energy efficiency unit is joules / bit. The formula for energy consumption per bit is given below:

$$\text{Energy Consumption per bit} = \frac{\sum energy\ consumed}{\sum bits\ transmit} \quad (1)$$

b. Average Network Throughput is throughput per unit of time (number of packets received by the sink node). The formula for Average Network Throughput is given below:

$$\text{Network Throughput} = \sum number\ of\ packet\ receive \quad (2)$$

c. Average Packet Latency is the amount of time delay between the sending of packet to the time when the packet reaches the sink node. The formula for the packet latency is given below:

$$\text{Packet Latency} = \frac{\sum(arrive\ time - send\ time)}{\sum Numberofconnections} \quad (3)$$

d. Average Delivery Ratio is the percentage of data packets received by the number of packets sent overall average node. The formula for packet delivery ratio is given below:

$$\text{Packet Delivery Ratio} = \frac{\sum Number\ of\ packet\ receive}{\sum Number\ of\ packet\ send} \quad (4)$$

3.1. Energy Consumption

Energy consumption values are the main focus in this simulation. In this simulation is calculated energy consumption value of the prototype realistic model. The result of energy consumption. The result of energy consumption is shown in Table 1.

In Table 1 showed the obtain result of energy consumption-based simulation implementation with different nodes. The highest achieve the energy consumptions in any nodes is L-MAC that reach 0.1626 in 10 nodes and in nodes 50 get until 0.2582. For second highest is B-MAC that reach until 0.1584 and in 50 nodes until achieve 0.2548 value. However, the lowest is different among each node between WiseMAC and Hybrid ADC MAC. In nodes 10 and 20 the lowest is WiseMac where for nodes 10 the value is 0.1551 and nodes 20 the value 0.1922. Meanwhile, in other nodes WiseMAC is higher energy consumption than Hybrid ADC MAC. The Hybrid ADC MAC has the lowest energy consumption in nodes 30 that reach 0.2128, nodes 40 with 0.2334 and nodes 50 with value 0.2497. In averages, the Hybrid ADC MAC has the lowest energy consumption compare with other:

Table 1. Result of Energy Consumption of Proposed MAC

MAC Protocol	Node(s)				
	10	**20**	**30**	**40**	**50**
WiseMAC	0.1551	0.1922	0.2148	0.2347	0.2507
Proposed MAC	0.1554	0.1930	0.2128	0.2334	0.2497
L-MAC	0.1626	0.2027	0.2230	0.2415	0.2582

3.2. Throughput
It calculated to get the value of network throughput. The simulation results are emphasized to get the value of the rate of successful message delivery over the prototype model realistic that used. The result of network throughput is measured in bits per second (bit/s or bps). Below is the value of network throughput of each MAC Protocol obtained from the simulation results in Table 2:

Table 2. Result of Network Throughput of Proposed MAC

MAC Protocol	Node(s)				
	10	**20**	**30**	**40**	**50**
WiseMAC	51946.14	110046.79	209743.23	251531.23	324813.45
Proposed MAC	52233.28	121261.44	219743.36	266535.89	334316.83
L-MAC	53573.57	132326.16	228243.74	287591.44	347253.14

In the Table 2 is described the result of throughput in each network. In throughput simulation L-MAC is the highest that has values around 53573.57 bps in nodes and in the highest nodes 50 L-MAC reach until347253.14 bps. The second highest throughput result in simulation is Hybrid ADC MAC that the lowest nodes (node 10) the result is 52233.28 bps and in nodes 50 the value is 334316.83 bps. Following the WiseMAC as the third highest with 51946.14 bps in nodes and in highest nodes is 347253.14 bps. Furthermore, the lowest of throughput simulation is B-MAC that achieved 51934.72 in nodes 10 until 314003.84 in nodes 50.

3.3. Latency
Latency measured from time interval between the stimulation and response, or, from a more general point of view. The result of energy consumption value is measured second (s). Below is the value of latency of each MAC Protocol obtained from the simulation results, as followed:

Table 3. Result of Latency of Proposed MAC

MAC Protocol	Node(s)				
	10	**20**	**30**	**40**	**50**
WiseMAC	0.0060	0.0144	0.0114	0.0110	0.0119
Proposed MAC	0.0046	0.0135	0.0111	0.0101	0.0113
L-MAC	0.0015	0.0120	0.0091	0.0082	0.0100

Table 3 showed the latency from simulation that B-MAC has the highest latency in any nodes from nodes 10 with value 0.0067 until nodes 50 with value 0.0127, following by WiseMAC that nodes 10 achieve 0.0060 and nodes 50 (0.0119). The third highest network is Hybrid ADC MAC value with nodes 10 is 0.0046 and in nodes 50 around 0.0113. Then, the lowest latency in this network is L-MAC with 0.0015 for nodes 10 and 0.0100 for nodes 50.

3.4 Packet Delivery Ratio (PDR)
The last is Packet Delivery Ratio (PDR) value also be taken from the result of this simulation. PDR defined as the ratio between the received packets by the destination and the generated packets by the source. The result of PDR value is measured in percentage (%). The PDR value of each MAC Protocol obtained from the simulation results, as follows:

Table 4. Result of PDR of Proposed MAC

MAC Protocol	Node(s)				
	10	20	30	40	50
WiseMAC	0.9998	0.9971	0.9993	0.9969	0.9977
Proposed MAC	0.9988	0.9928	0.9932	0.9937	0.9940
L-MAC	0.9955	0.9885	0.9832	0.9867	0.9882

In Table 4 the highest and the lowest are slightly different in packet delivery ratio (PDR). In nodes 10 the highest value is WiseMAC 0.9998 and the lowest is L-MAC with 0.9955. Then, in nodes 20 the highest is also WiseMAC with 0.9971 and the lowest in this node is L-MAC is 0.9885. Meanwhile, in nodes 30 the highest value is 0.9994 and the lowest also L-MAC with value 0.9832. Furthermore, in nodes 40 the highest PDR value is WiseMAC with 0.9969 and the lowest is also L-MAC with 0.9867. Then, in nodes 50 the highest PDR value is also WiseMAC 0.9977 and the lowest is L-MAC with value number 0.9882.

4. Conclusion

This paper is describing the evaluated of MAC protocols of WSN that used parameters in term of Energy Consumption, throughput, latency and PDR. Then, the MAC protocol that used compared the evaluation is WiseMAC, L-MAC, Hybrid ADC MAC and B-MAC.

The comparison values of the MAC protocol are different that achieved high rate based on each parameter. In energy consumption, the highest achieve the energy consumptions in any nodes is L-MAC. Hybrid ADC MAC has the lowest energy consumption compare with other. Meanwhile, in throughput performance, the highest value is L-MAC the lowest value simulation is B-MAC that achieved. Then, in latency the highest is also L-MAC and the lowest latency is B-MAC. Furthermore, in PDR parameter the average highest value is WiseMAC and the lowest is L-MAC.

The expected contribution on this paper is more towards of comparing the performance MAC protocol that can use to identify efficiency of WSN implementation. For future work, another parameter is anticipated to propose in order to obtain the effectiveness MAC protocol performance.

Acknowledgements
This work was partly financially supported by from Ministry of Education (MOHE), Malaysia and Universiti Utara Malaysia (UUM), Sintok, Kedah.

References

[1] Derdour Y *et al* 2014 Procedia Computer Science **32** 261-268
[2] Pan J J *et al* 2012 IEEE Transactions on Pattern Analysis and Machine Intelligence **34**(3) 587-600
[3] Baronti P *et al* 2007 Computer Communications **30** (7) 1655-1695
[4] Othman M F *et al* 2012 Procedia Engineering, **41** 1204-1210
[5] Al-Saud K A. *et al* 2011 3rd International Conference on Computing and Informatics, ICOCI 2011, Bandung, Indonesia
[6] Hu Y *et al* 2010 IEEE Potentials **29** (1) 19-23
[7] Gubbi J *et al* 2013, Future Generation Computer Syst. **29** (7) 1645-1660

IOP Conf. Series: Materials Science and Engineering **551** (2019) 012060 doi:10.1088/1757-899X/551/1/012060

[8] Yang C. *et al* 2016 Manufacturing Letters **7** 22-25
[9] El-Hoiydi A *et al* 2004. Proceedings ISCC 2004. Ninth International Symposium on Computers And Communications (IEEE Cat. No.04TH8769), **1**.
[10] Esparza L J R. *et al* 2011 Eleventh International Conference on Application of Concurrency to System Design

IOP Conf. Series: Materials Science and Engineering **551** (2019) 012061 doi:10.1088/1757-899X/551/1/012061

Educational Data Mining: Enhancement of Student Performance model using Ensemble Methods

Samuel-Soma M Ajibade[1], Nor Bahiah Binti Ahmad[2], Siti Mariyam Shamsuddin[3]

[1,2,3]Computer Science, Universiti Teknologi Malaysia, Johor Bahru, Johor, Malaysia.

E-mail: samuel.soma@yahoo.com, bahiah@utm.my, mariyam@utm.my

Abstract. Nowadays, Educational Data Mining (EDM), begun as a new research area due to the broadening of numerous statistical approaches that are used to perform data exploration in educational settings. One of the applications of EDM is the prediction of performance of students. In a web based education system, the behavioural features of learners are very significant in showing the interaction between students and the LMS. In this paper, our aim is to propose a new performance prediction model for students which is based on data mining methods which includes new features known as behavioural features of students and based on sequential feature selection which is used to identify most important features. The proposed performance model is evaluated using classifiers such as Support Vector Machine (SVM), K-Nearest Neighbour (KNN) and Decision Tree (DT). Furthermore, so as to enhance the classifiers performance, the ensemble methods such as Bagging, Boosting and Random Forest were applied. The achieved results show that there exists a strong relationship between behaviour of students and their academic performance. An accuracy of 91.5% was gotten when the ensemble methods were applied to the classifiers to improve the academic performance. Thus, the result gotten shows the reliability of the proposed model.

1. Introduction

EDM comprises of the utilization of machine learning techniques, Data Mining (DM) techniques and diverse educational statistical methods. EDM is a discipline that is taken after to extract significant knowledge from an educational setting. EDM applications such as model development helps to predict the performances of students. As a result, driving researchers to delve profound into different techniques of mining data to enhance existing techniques [1]. Originally, EDM and LA were significant areas of research in education, but gradually student performance prediction began to become prevalent as the main aim of the study is to examine and predict the performance of students for the students to obtain better results. This research work presents ensemble methods for enhancing various classification algorithm through which the student performance can be predicted by introducing a new feature category called behavioral features. The educational dataset is gotten from an eLearning system known as Kalboard 360 [2]. The gathering of the data carried out by making use of a learner activity tracker tool known as Experience API (XAPI). The obtained features are into three different categories: academic background features, demographic features and then behavioral features. The educational process consists of a new category which is the behavioral features and it is linked to the experience of learners.

In this research work, we utilized of educational data to predict the academic performances of students. This model has therefore evaluated the effect of the learning behavioral features of students on their academic performance. A data mining technique known as classification has been used to implement this work. We made use of three classifiers: Decision Tree (DT), K-Nearest Neighbor and

IOP Conf. Series: Materials Science and Engineering **551** (2019) 012061 doi:10.1088/1757-899X/551/1/012061

Support Vector Machine (SVM). In order to improve the performance of the classifiers, some ensemble techniques like Boosting, Bagging and Random Forest were used to enhance the performance model accuracy of the students. The remainder of this paper is as follows: Literature review is presented in Section 2 and the third Section describes the methodology while the fourth section presents the result and discussion of this research and Section 5 gives the conclusion and future work.

2. Experimental

To build a predictive model, there are several DM techniques used, which are classification, regression and clustering. Decision tree is a set of conditions arranged in a hierarchical frame. Most of researchers used this technique due to their simplicity, in which it can be transformed into a set of classification rules. Some of the famed DT algorithms are C4.5 [3], [4] and CART. In the work of [5] the objective of the author is to build the prediction model for students for first and second degree students of Computer Science & Engineering and Electronics & Communication streams by making use of two classifiers: Decision Tree(DT) and Fuzzy Genetic Algorithm. Parameters like internal marks, sessional marks and admission score were chosen to carry out this research. SVM is a learning algorithm developed by [6] to handle the challenges of pattern recognition and prediction and also for analysis and mapping of both linear and non-linear functions. A hyperplane or set of hyperplanes (classes)are being created in a high dimensional space which can then be used for classification [7], [8]. SVM was selected for the research because of its generic application nature and variety of applicability. KNN is a straightforward classification algorithm that stores every single accessible case and classifies new cases centered on a similarity measure (e.g., distance functions). The training samples are defined by n dimensional numeric qualities. Each sample signifies a point in an n-dimensional space. Along this lines, the majority of the training samples are kept in an n-dimensional pattern space. At a point, when an unknown sample is provided, a KNN classifier searches the pattern space for the k training samples that are closest to the unknown sample.

Ensemble method [9], [10], [11] combines two or more classification algorithm which is known as a based learner. The base learner can be identical or non-identical. Ensemble method organizes the predicted results of used base learners to achieve more strength in the system. In most cases, ensemble techniques such as Bagging, Boosting, and Random Forest are applied to improve the performance of classification algorithm [12]. Bagging ensemble method [9], [10] randomly takes data from training dataset and put into a bag. Several numbers of bags are created where each bag contains the subset of training data set. Then each bag is trained with a classification algorithm provides a model. In boosting ensemble method [10], [11] total training set used to train the first model and next model is trained from the performance of the previous model. At first, it gives equal weights to each instance of the dataset. If the class of an instance is misclassified, then gives it more weight to focus on this instance in the later model. This process is continued until the number of added model or accuracy is obtained. Random Forest algorithm [10], [12] is a large collection of Decision Tree algorithms which are not correlated. Random Forest creates a lot of Decision Trees from the subsets of training dataset where each subset provides a decision tree. Now each Decision Tree model classified an instance in a class. Then the majority voted class will be taken as the class of the instance. In [13], authors use some common ensemble techniques (such as Bagging, Adaboosting, and Random Forest) to predict student's academic performance more accurately where Adaboosting on Artificial Neural Network gives the best accuracy of 79.1 percent.

3. Methodology

The Educational dataset that was used in this paper is collected from a LMS called Kalboard 360. The Kalboard 360 is a multi-agent LMS that was developed to enhance learning via the utilization of leading-edge technology. The data was gathered by utilizing a learner activity tracker tool known as Experience API (XAPI). XAPI is a component of the Training and Learning Architecture (TLA) that allows the tracking of learning experiences and actions of learners like reading an article or watching an educational video. In this present study, the dataset lengthens into 500 students with 16 features. Table 1 displays

IOP Conf. Series: Materials Science and Engineering **551** (2019) 012061 doi:10.1088/1757-899X/551/1/012061

the attributes/features, data description and data type of the dataset. After the data preprocessing the dataset, the feature variable consists of 480 observations with 8 features. we used cvpartition to divide data into a training set of size 336 and a test set of size of size 144. Therefore, to evaluate the model, 10-fold cross validation was used. Forward sequential selection was used in a wrapper fashion to find important features.

Table 1. Student features, description and data type.

Feature Category	Features	Description	Data Type
Demographical Features	Nationality	Student nationality	Nominal
	Gender	The gender of the student (female or male)	Nominal
	Place of Birth	Student's place of birth (Kuwait, Jordan, Lebanon, Saudi Arabia, Iran, USA)	Nominal
	Parent responsible for student	Student's parent such as (father or mum)	Nominal
Academic Background Features	Stage ID	Stage Student belongs to such as (Low level, Middle level, High level)	Nominal
	Grade ID	Grade students belongs such as (G-01, G-02, G-03, G-04, G-05, G-06, G-07, G-08, G-09, G-10, G-11, G-12)	Nominal
	Section ID	Classroom student belongs to such as (A,B,C)	Nominal
	Semester	School year semester such as (First or second)	Nominal
	Topic	Course topic such as (Math, English, IT, Arabic, Science, Quran)	Nominal
	Student Absence Days	Student absence days (Above-7, Below-7)	Nominal
Parents Participation on learning process	Parent Answering Survey	Parent is answering the surveys that provided from school or not	Nominal
	Parent School Satisfaction	This attribute obtains the degree of parent satisfaction from school as follow (Good, Bad)	Nominal
Behavioural Features	Discussion groups	Student behavior during interaction with Kalboard 360 e-learning System.	Numeric
	Visited resources		Numeric
	Raised hand on class		Numeric
	Viewing announcements		Numeric

4. Results and Discussions

4.1. Evaluation results using traditional classifiers

In predicting the performances of students, numerous features have effect on the model. In this paper, the behavioral features have been considered as vital features that can have effect on the performance of students. As shown in table 2, we have shown the results through classification algorithms (DT, KNN and SVM) so as to demonstrate the impact of behavioral features. The classification results are fragmented into two different sections. Results of Classification with student's behavioral features (BF) and results of Classification without student's behavioral features (WBF). In the table, we can deduce that the DT model performs better than other data mining techniques. The DT model achieved 87.1% accuracy with BF and 84.4% accuracy WBF. For precision measure, the model achieved 85.2% with BF and 77.7% WBF. For recall measure, the results are 86.3% with BF and 78.1% WBF. For F-Measure, the results are86.0% with BF and 77.1% WBF. Hence from the above analysis, the results prove a strong effect of learner behavior on academic performance of students.

4.2. Evaluation results using ensemble methods

In this section, we made use of ensemble methods to enhance the accuracy of the evaluation results of the traditional Data Mining techniques. Table 3 shows the enhanced results using ensemble methods with three traditional classifiers (DT, KNN and SVM). Each ensemble trains the three classifiers and then now combine the results by a majority voting process in order to achieve the best prediction performance of students. The boosting techniques performs better than other ensemble methods in the case of DT, KNN and SVM, however DT gave the highest performance in which the accuracy of DT using boosting is enhanced from 0.87 to 0.89 while Precision result are improved from 0.85 to 0.87 and the Recall results are improved from 0.86 to 0.87 and F-measure result was enhanced from 0.86 to 0.89.

Table 2. Classification method results with BF and WBF.

Evaluation Measure	DT		KNN		SVM	
Behavioral Feature Extraction	BF	WBF	BF	WBF	BF	WBF
Accuracy	0.87	0.84	0.82	0.81	0.86	0.83
Precision	0.85	0.77	0.80	0.67	0.81	0.70
Recall	0.86	0.78	0.79	0.56	0.81	0.73
F-Measure	0.86	0.77	0.69	0.59	0.75	0.71

Table 3. Classification method results using ensemble methods.

Evaluation Measure	Classification Methods			Bagging			Boosting			Random Forest
Classifier	DT	KNN	SVM	DT	KNN	SVM	DT	KNN	SVM	DT
Accuracy	0.87	0.82	0.86	0.86	0.85	0.86	0.89	0.86	0.88	0.89
Precision	0.85	0.80	0.81	0.86	0.79	0.80	0.87	0.83	0.82	0.84
Recall	0.86	0.79	0.81	0.87	0.80	0.82	0.87	0.83	0.84	0.84
F-Measure	0.86	0.69	0.75	0.88	0.69	0.76	0.89	0.74	0.80	0.73

IOP Conf. Series: Materials Science and Engineering **551** (2019) 012061 doi:10.1088/1757-899X/551/1/012061

After the classification model has been trained using 10-folds cross validation, then the process of validation kicks-off. The process of validation is a very significant phase in the structuring of predictive models, it defines the accuracy of the predictive models. Table 4 displays the results of evaluation by the use of classification techniques (DT, KNN, and SVM) through the testing and validation phases.

Table 4. Classification method results through testing and validation.

Evaluation Measure	Testing Results			Validation Results		
Classifier	DT	KNN	SVM	DT	KNN	SVM
Accuracy	0.87	0.82	0.86	0.91	0. 85	0.88
Precision	0.85	0.80	0.81	0.87	0.83	0.84
Recall	0.86	0.79	0.81	0.89	0.82	0.84
F-Measure	0.86	0.69	0.75	0.89	0.75	0.82

As seen in Table 4, 91.5% accuracy is achieved in our proposed model through the validation phase. When compared to [15] et al which gave 82.2% accuracy, our model performed better. Hence, the result gotten from the validation phase proves the reliability of the proposed model.

5. Conclusion and future work

The prediction of student's academic performance has been a huge concern for higher institutions everywhere. The data gathered entails some hidden knowledge that are being used to improve the academic performance of students. In this research, a new performance prediction model for students was proposed which is based on various data mining methods which contains new features known as behavioral features. These attributes are associated with the interactivity of learners with the LMS. The predictive model is evaluated based on some classifiers like DT, KNN and SVM. Furthermore, we applied ensemble methods to enhance the performance of the classifiers. We made us of Bagging, Boosting and Random Forest. Based on forward sequential selection that was used to select the most important features, the accuracy of the predictive model is 91.5% which is a better performance than [13] that was 82.2%. In future works we will focus on analyzing the data of students to find other features that will identify the students that have weaker achievements and performances. Optimization techniques such as Differential Evolution, Genetic Algorithm and others could as well be applied to enhance the performance model of students in educational data mining.

References

[1] Amrieh E A *et al* 2015 *IEEE Jordan Conference* pg. 1-5.
[2] Quadri M M *et al* 2010 *Global Journal of Computer Science and Technology* **10** 2-5.
[3] Kumar M *et al* 2016 *International Journal of Computer Applications* **137** 2.
[4] Hamsa H *et al* 2016 *Procedia Technology* **25** 326-332.
[5] Costa E B *et al* 2017 *Computers in Human Behaviour* **73** 247-256.
[6] Subaira A *et al* 2014 *IEEE 8th Int. Conf. on Intelligent System and Control (ISCO)* **978** 274-280.
[7] Trstenjak B *et al* 2014 *IEEE Int. Conv. on Info. And Comm. Tech. Elect. And Microelectronics (MIPRO)* 1222-1227.
[8] Marquez-Vera C *et al* 2016 Expert Systems **33** 107-124.
[9] Dietterich T G *et al* 2000 *Int'l. Workshop on Multiple Classifier Systems.* 1-15.
[10] Sabzevari M *et al* 2018 *Cornell Uni. arXiv preprint arXiv:1802.07877.*
[11] Rahman M H *et al* 2017 *2nd ICEEE* 1-4.
[12] Amrieh E A *et al* 2016 *Int'l. Journal of Database Theory and Application* **9** 119-136.
[13] Moisa V *et al* 2013 *Journal of Mobile Embedded and Distributed Systems* **5** 70-77.

Utilization of Filter Feature Selection with Support Vector Machine for Tumours Classification

T A H Tengku Mazlin[1], R Sallehuddin[1] and M Y Zuriahati[1]

[1]Faculty of Engineering, School of Computing, Universiti Teknologi Malaysia, 81310, Skudai, Johor Bahru, Malaysia

Email : tgmazlynn@gmail.com, roselina@utm.my, zuriahati@utm.my

Abstract. Due to rapid technology advancement, machine learning has been widely used for solving cancer classification problem. Classification performance is highly depending on the quality of input features. With an explosive increase number of features of high dimensional data, the occurrence of ambiguous samples and data redundancy directly leads to poor classification accuracy. Therefore, this paper presents a utilization of filter feature selection using four filter methods such as Information Gain, Gain Ratio, Chi-Squared and Relief-F by performing attribute rankings to remove the irrelevant and redundant features and evaluate the significance and correlation of input data. Then, the classification will be performed using Support Vector Machine (SVM) to measure the accuracy performance based on the number of selected features. The performance measurement will be validated on standard Breast Cancer datasets consisting of 286 instances obtained from the UCI repository. Evaluation metrics such as accuracy, sensitivity, specificity and Area under Receiver Operating Characteristic Curve (AUC) will be used to assess the performance of the SVM classifier using four different filter methods. Experimental result shows that Gain ratio improves the accuracy of SVM classification compared to Information Gain, Chi-Squared and Relief-F in classifying breast cancer data with only small number of features selected.

1. Introduction

In recent years, the intelligence of machine learning approach such as classification has proven as one of the robust and reliable techniques in increasing the performance of cancer diagnosis. Despite the advantages provided by machine learning, several classification problems in cancer diagnosis have been reported recently. Majority of the classification issues are related to the amount of input variables in the training data or also known as the "curse of dimensionality" [1, 2]. In addition, the classification performance is highly depending on the number of features selected from the training data to train the classifiers [5]. An explosive increase of features in training data has caused a computation complexity where the classification become slower with the presence of noisy, redundant and irrelevant features that consequently degrade the performance of learning tasks [3]. These may cause difficulty to perform data interpretation due to the inconsistency and confusing data patterns. Regarding to the dimensionality issues, some irrelevant and redundant input features which occurred in high dimensional data should be discard in order to maintain or improve the accuracy performance. Hence, the classification model should able to produce an accurate and computationally fasters prediction with optimal number of input variables.

In this work, we present a utilization of four filter feature selection algorithms using Information Gain, Gain Ratio, Chi-square and Relief-F to determine the sets of relevant features. This research

IOP Conf. Series: Materials Science and Engineering **551** (2019) 012062 doi:10.1088/1757-899X/551/1/012062

aims to improve the classification accuracy while predicting the optimal number of significant features reliable for classification tasks using SVM classifier. Breast Cancer dataset obtained from UCI Library consisting 286 features were used for evaluating the performance of the proposed utilized method.

The organization of this paper are sorted as follows; section 2 presents the related work. The utilization of four filter feature selection algorithms is explained in section 3 while the implementation of classification algorithm is presented in section 4. All experimental results are discussed in section 5 and section 6 concludes the paper.

2. Related Work

An explosive number of features in high dimensional data have produce challenges towards the classification approach because the learning algorithms are unable to manage the overfitting data which resulting a poor classification accuracy due to the high computational time. As mentioned in reported study, many researchers have developed a classification model with feature selection approach to handle the explosive number of features in the training data for more accurate prediction.

The classification of breast cancer using SVM is proposed in [4] where SVM classifier is implemented with radial basis function (RBF) kernel function to improve the classification performance. In this context, the performance of classifier is highly depending on the quality of the training data which used to train the classifier. For that reasons, it is ultimately important to recognize and identify the doubtful features in the training data at the early phase of the classification. Besides classification techniques, feature selection is identified as one of the techniques used to address the issues. Since choosing the best features could be a challenging task, there are several solutions and methods have been proposed to remove the ambiguous and redundant features of training data.

Feature selection can be separated into three categories namely filter, wrapper and embedded techniques. According to studies on determining the significant features in [5] and [6], filter techniques outperformed wrapper and embedded techniques in terms of input space dimensionality reduction. The filter techniques have successfully improved the accuracy of the classifier and computationally faster when handling the high dimensional input variables. Unlike wrapper techniques, the classification tends to perform inefficiently if the input variables increase rapidly in the input space. As a result, the wrapper technique is more expensive when large input variables need to be used as the computational complexity increased with the dependency of classifiers. Even though the filter feature selection is more efficient when dealing with large dataset, the single filter method does not consider the correlation between each feature, which unable to produce an optimum selection of significant features [5,6]. Despite this, the filter techniques still proven the advantages of computationally faster and ability in handling the high dimensional datasets efficiently and highly recommended due to its simplicity and success performances [3,5,6].

Therefore, the classification in this research will be implemented using SVM classifier where its performance will be improved by utilizing four filter feature selection algorithms and each utilization will be compared with performance parameters such as accuracy, specificity, sensitivity and AUC.

3. Feature Selection using Filter Techniques

Basically, filter techniques are performed without including any learning algorithm or independent from classifiers. This caused the filter techniques to be computationally efficient as the time to compute the algorithm is reduced and ability in handling different dimension of datasets. The selection of features using filter approach is done by assigning a score or rank to each input features. Then, the ranked features are selected into the classifier or eliminated from the dataset according to the scored value. Due to the simplicity and attempts to consider the interaction between features, four filter feature selection algorithms such as Information Gain (IG), Gain Ratio (GR), Chi-Squared (CS) and Relief-F (RF) were implemented for selecting best features. Justification of choosing these four filter algorithms and the equations that implemented can be referred in [7], [8], [9] and [10] accordingly.

3.1 Utilization of Four Filter Feature Selection Algorithms with SVM

Overall, the utilization of four filter feature selection using Information Gain, Gain Ratio, Chi-Squared and Relief-F with SVM is illustrated in figure 1 and the algorithm to perform feature (attribute) ranking is presented as follows:

Step 1: Let X_i be the original feature set of UCI Breast Cancer dataset, where $X_i = \{X_1, X_2, ..., X_{286}\}$ and C_i represents the class of benign or malignant, where $C_i = \{C_1, C_2\}$.
Step 2: Rank and sort features X_i for each filter algorithm based on rank value in determining the output class C_i.
Step 3: Select the significant features based on threshold value and output as X'_i. The treshold value used for Information Gain, Gain Ratio, Chi-Squared and Relief-F are entropy value, gain ratio value, significance level and feature score which are set to 0.05 [7,8,9,10].
Step 4: Perform SVM classification on each X'_i and calculate the performance.

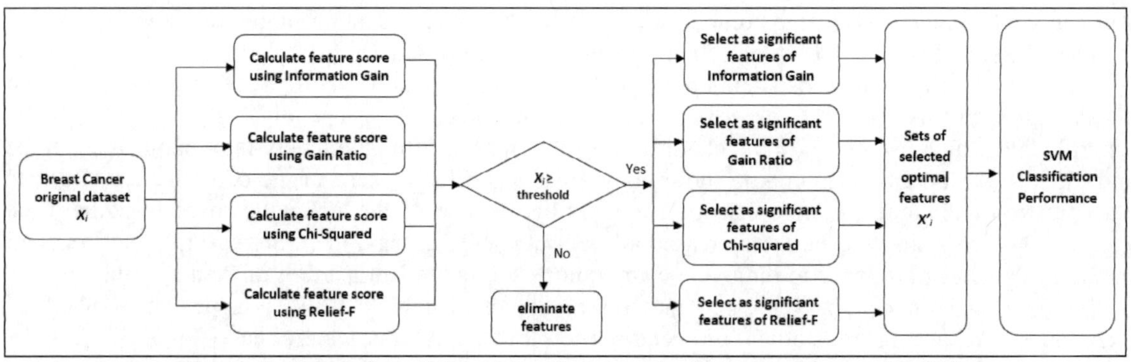

Figure 1. Utilization of four filter feature selection algorithms with SVM.

For the classification performance, SVM using Radial Basis Function (RBF) kernel function will be employed on the selected ranked features of each filter algorithms. Detail implementation of SVM using RBF kernel can be obtained in [11,12]. The SVM performance is measured using four statistical performance namely accuracy, specificity, sensitivity and AUC. Formulas for these statistical performances can be obtained in [2,11,12].

3.2 Experimental Data

UCI Breast Cancer dataset as described in table 1 is used to carry out the experiment. For performance evaluation, the classification dataset needs to be partitioned into training and test sets. Here, 10-fold cross validation is implemented.

Table 1. UCI Breast Cancer original dataset features.

Features	Age	Menopause	Tumour-size	Inv-nodes	Nodes-caps	Deg-malig	Breast	Breast-quad	Irradiat
Attributes	1	2	3	4	5	6	7	8	9
Notation	A1	A2	A3	A4	A5	A6	A7	A8	A9

4. Result and Discussion

This section presents the results of the utilization of four filter feature selection algorithms towards SVM classification. The results of selected features based on average ranking score are described in table 2 and figure 3 summarizes the overall SVM classification performance.

Based on analysis in table 2, four filter algorithms are defined as IG, GR, CS and RF. The information from the top side of the table indicates the features with highest significance and vice versa. For each filter method, the features that obtained values less than threshold value are discarded from the list as they are considered as not relevant or redundant. This also means that the information provided by these attributes may consists of redundant information which may slows down the SVM learning algorithm in the classifier. For example, Information Gain analysis, the attributes are ranked in descending order such as A6, A4, A3, A5, A9, A1, A8, A7 and A2. Based on the average merit score, attribute A6, A4, A3 and A5 are ranked with the highest average merit score higher than 0.05 among other attributes.

Table 2. Ranking score using Information Gain, Gain Ratio, Chi-Squared and Relief F[a].

	Attribute Ranking				**Average Merit Score** (score with standard deviation (+-))							
	IG	**GR**	**CS**	**RF**	**IG**		**GR**		**CS**		**RF**	
Descending Order	**A6***	**A5***	**A6***	**A6***	0.078	+- 0.011	0.071	+- 0.008	28.875	+- 4.215	0.093	+- 0.014
	A4*	**A4***	**A4***	**A8***	0.071	+- 0.01	0.054	+- 0.007	26.594	+- 3.857	0.062	+- 0.013
	A3*	**A6***	**A5***	**A2***	0.061	+- 0.008	0.051	+- 0.007	19.731	+- 2.515	0.057	+- 0.009
	A5*	A9	**A3***	**A1***	0.051	+- 0.007	0.033	+- 0.007	17.039	+- 1.913	0.051	+- 0.011
	A9	A3	**A9***	A7	0.026	+- 0.006	0.02	+- 0.003	9.792	+- 2.248	0.048	+- 0.005
	A1	A1	**A1***	A3	0.012	+- 0.003	0.006	+- 0.002	3.956	+- 1.339	0.005	+- 0.007
	A8	A8	A8	A9	0.01	+- 0.003	0.005	+- 0.002	3.462	+- 1.183	0.033	+- 0.008
	A7	A7	A7	A5	0.003	+- 0.001	0.002	+- 0.002	0.887	+- 0.6	0.026	+- 0.005
	A2	A2	A2	A4	0.003	+- 0.002	0.003	+- 0.001	0.94	+- 0.437	0.018	+- 0.01

[a] Hint: **X*** indicates the selected features (attributes).

From the analysis, we observed that certain features are similar on different filter techniques even though different ranking methods were applied in each filter. From the ranking score analysis, 4 features are selected by Information Gain, 3 features are selected by Gain Ratio, 6 features are selected by Chi-Squared and 4 features are selected by Relief-F. As an achievement from Algorithm 3.1, these selected significant features are the optimum features to be included into SVM for classification tasks.

4.1. *Performance of SVM Classification based on Selected Features*

The performance measures such as accuracy, sensitivity, specificity, and AUC are evaluated and averaged to obtain the overall SVM performance. Figure 2 presents the overall results obtained.

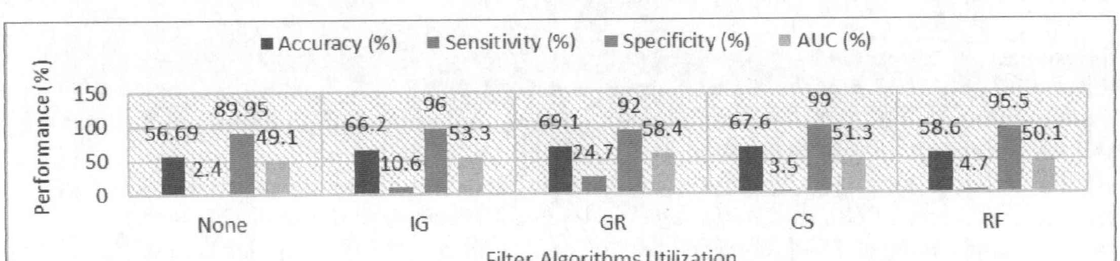

Figure 2. Overall SVM classification performance with four filter algorithms utilization.

Based on the results in figure 2, the classification performance achieved with feature selection utilization is better compared to the independent classifier. For example, the performance of SVM with Gain Ratio utilization perform better than others in terms of accuracy and sensitivity. However, in terms of specificity, SVM classification with Chi-Squared utilization achieved higher specificity (99.00%) compared to the utilization with Information Gain (96.00%), Gain Ratio (92.00%) and Relief-F (95.50%). AUC measures the effectiveness of classifier in terms of correctly classified the true positive (benign) and true negative (malignant) classes. It is observed that the AUC value of SVM with Gain Ratio utilization is better with highest percentage (58.40%) compared to the utilization with Information Gain (53.30%), Chi-Squared (51.30%) and Relief-F (50.10%). Based on the AUC performance obtained by SVM with four filter feature selection algorithms, SVM classification with

Gain Ratio utilization could identify the class of the data accurately than other three filter algorithms. By selecting 3 significant features (A5, A4, A6), the classification performance with Gain Ratio shows an improvement in terms of accuracy. This indicates that the other 6 features contain some redundancy, which have less contribution towards determining a certain class (benign) and (malignant). For overall effectiveness, it can be proved that filter feature selection using Gain Ratio can significantly improves the accuracy of SVM classification compared to Information Gain, Chi-Squared and Relief-F in classifying breast cancer data with only small number of features selected.

5. Conclusions

Handling massive increment of ambiguous and redundant samples in the training data has resulting a notable challenge faced by the current cancer diagnosis system in machine learning. Feature selection method is one of the solutions used to pre-process the datasets before proceeding it into the cancer classification process. In this paper, the performance of utilization of feature selection using four filter algorithms such as Information Gain, Gain Ratio, Chi-Squared and Relief-F has been examined on UCI Breast Cancer dataset to determine the significant features by ranking of attributes. Sets of threshold value are used to select the significant features from the ranked sample.

Experimental results show that SVM classification with gain ratio feature selection utilization contributes better performance than Information Gain, Chi-Squared and Relief-F utilization for breast cancer classification in terms of accuracy, sensitivity and AUC with percentage of 69.10%, 24.70% and 58.40%. This research indicates that SVM classification with gain ratio feature selection utilization could effectively improves the accuracy as well as helping medical experts in diagnosing breast cancer. For the future work, this research will be continued to optimize the number of significant features for optimal solution with different optimization approach while increasing the accuracy performance.

Acknowledgments

This research is supported by a Research University Grant, GUP TIER 1:16H57 and GUP TIER 2:14J13. The authors would like to thank the anonymous reviewer for providing constructive and generous feedback. The authors also would like to thank Research Management Centre (RMC) of Universiti Teknologi Malaysia, for the research activities and Applied Industrial Analytics (Ali@s) for the support and motivation in making this research a success.

References

[1] Wu J *et al* 2017 *J. of Biostatistics and Biometrics* **1(2)** 1-7
[2] Sharifah H *et al* 2013 *J. of Technology Science & Engineering* **65(1)** 73-81
[3] Miao *et al* 2016 *J. of Proc. Computer Science* **91** 919–926
[4] Liu B *et al* 2010 *J. of Pattern Recognition* **43(1)** 280–298
[5] Wang D *et al* 2018 *J. of Computing and Applied Mathematics* **329** 307- 321
[6] Sharifah H *et al* 2014 *Classification of Liver Cancer Using Artificial Neural Network and Support Vector Machine in Proc.of Int. Conf. on Adv. in Communication, Network and Computing CNC* February 21-22 Chennai, India pp 1-6
[7] Tresna M *et al* 2017 *Int. J. of Engineering Technology EMITTER* **5(1)** 36-71
[8] Dai J *et al* 2013 *J. of Appl. Soft Computing* **13** 211–221
[9] Seema A *et al* 2018 *Int. J. of Engineering and Technology* **4(2)** 222-226
[10] Ryan J U et al 2018 *J. of Biomedical Informatics* **85** 168–188
[11] Sallehuddin R *et al* 2016 *J. of Technology Science & Engineering* **78** 107–119
[12] Huang S *et al* 2018 *J. of Cancer Genomics & Proteomics* **15(1)** 41–51

A Review of Common Features inComputational Thinking Frameworks in K-12 Education

Filzah Zahilah Mohamed Zaki[1] , Su Luan Wong[2] and Mohd Ridzwan Yaakub[3]

[1,2]Department of Science and Technical Education, Faculty of Educational Studies, Universiti Putra Malaysia, 43400 Serdang, Selangor, Malaysia.

[3]Center for Artificial Intelligence Technology (CAIT), Faculty of Information Science and Technology, Universiti Kebangsaan Malaysia, 43650 UKM Bangi, Selangor, Malaysia.

Email : gs52106@student.upm.edu.my

Abstract. Computational Thinking (CT) is not just another application of computer science principles in humans' life. Computational Thinking has emerged as a systematic way of thinking, and problem-solving process. The awareness of inculcating CT into education, at different curricular level has started in various directions, and contexts. The article aimed to introduce the basic concepts of CT and highlight common key features within K-12 education frameworks, deployed by different bodies. Computational thinking is closely relatedto critical thinking, STEM (science, technology, engineering, and math) learning, as well as project-based learning. From the review, various computational thinking definitions and key features within education frameworks were identified, and this contributed in how computational thinking is being integrated in different context, respectively.

1. Introduction

Over the last decade, a growing body of research has provided evidence on the relevance of computational thinking in K-12 education. The concept of CT is gaining more attention due to the numerous benefits CT has to offer within different education level, for teaching and learning process across different subjects. Many study relates computational thinking as a great tool for harnessing interest in Science, Technology and Engineering (STEM) courses, for students at different level in education. In the context of K-12 education, it is highly valued as application of computer science principles within kindergarten, primary and secondary levels educational practice, by teachers and their pupils/students in classroom setting, or beyond. The National Research Council [13] emphasized the significance of introducing CT to students as early as possible, and assisting them to understand the application of these essential skills. The pervasive nature and potential of computational thinking is undeniable, as it permeates almost all diciplines, which includes both sciences and humanities [6].

Thus, the aim of this article is to highlight the common features available frameworks, or models which have been adopted in various K-12 educational context, as a useful guide for current and future endeavour to implement CT within various setting and objectives.

The Bureau of Labor Statistic [7] reported there were more computing jobs created annually. Nevertheless, there were insufficient number of graduates in computer science course, as substantial job opening for this qualification is estimatedin the future. Numerous studies have evidenced various advantagesof learning computational thinking, including improved student engagement,enthusiasms, with better confidence level, problem solving, communication, as well as a catalyst in STEM learning and performance [19]. A study by Israel, Pearson, Tapia, Wherfel and Reese [12] indicated even struggling learners, including students with disabilities and those living in poverty, also gain benefit from computing education through scaffolding, modeling, and peer collaboration.

As computing and the advances of artificial intelligence become more prevalent in human's life, computational thinking is emerging as very much neededcritical skill, and many international bodies are recognizing this fact. Support for teaching computing in K-12 schools is rising in the U.S. and other countries. Many developed nations such as England, Finland, South Korea, and Australia has made it compulsory for children education to learn computing or computational thinking [18].

2. History of Computational Thinking

In March 2006, Jeannette Wing released her seminal article titled "Computational Thinking" in the Journal of the Association for Computing Machinery. Wing suggested that "computational thinking involves solving problems, designing systems, and understanding human behaviour, by drawing on the concepts fundamental to computer science. It represents a universally applicable attitude and skill set everyone, not just computer scientists, would be eager to learn and use"[21]. Wing further said "to reading, writing, and arithmetic, we should add computational thinking to every child's analytical ability" [21]. She also asserted computational thinking as a critical skillset for everyone, not just for computer scientist nor exclusive to computer science field.

Piaget coined the term 'Constructivism' to reflect that children construct their own knowledge through experience rather than being taught by teachers [16]. Later, Papert and Harel [15] introduced 'constructionism' which extend the definition of constructivism by adding another useful dimension to it. Learning process was considered as developing knowledge structures regardless of learning conditions. The aim was to inculcate contentment during the conscious meaning making process in constructing new knowledge. This "learning by making" connotation echoes the concept of computational thinking, as the computational thinking core elements comprehends (at least with this four basic core elements), which requires learners to perform decomposition, recognizing available patterns, abstracting from existing knowledge and finally, designing algorithm, in problemsolving, and eventually developing new knowledge. Since then, computational thinkingteaching has conventionally been viewed as a primarily constructionist endeavour [4, 8].

3. Definitions of Computational Thinking

Computational thinking is a phrase frequently used in the literature, but there is little consensus on a common definition [1, 2, 13], although generally there are some agreement on the similarities and differences. Therefore, it is vital for researchers and educators to acknowledge the diverse version by different authors and organizations.

3.1 Operational definition

The International Society for Technology in Education (ISTE) and the Computer Science Teachers Association (CSTA) have collaborated with experts from K-12 education, higher education and industry, and developed an operational definition of CT. The definition has afforded a framework and some terminologies which are relevant to K-12 educators. ISTE and CSTA [11] regarded computational thinking as a problem-solving process whichcomprisesof (but is not limited to) all the below:

"Formulating problems in a way that enables us to use a computer and other tools to help solve them;Logically organizing and analyzing data; Representing data through abstractions such as models and simulations; Automating solutions through algorithmic thinking (a series of ordered steps); Identifying, analyzing, and implementing possible solutions with the goal of achieving the most efficient and effective combination of steps and resources; Generalizing and transferring this problem-solving process to a wide variety of problems."

IOP Conf. Series: Materials Science and Engineering **551** (2019) 012063 doi:10.1088/1757-899X/551/1/012063

Google For Education [10] defined CT as "problem solving process that includes a number of characteristics, such as logically ordering and analyzing data and creating solutions using a series of ordered steps (or algorithms), and dispositions, such as the ability to confidently deal with complexity and open-ended problems". In mathematics and science field, Weintrop et al [20] proposed a definition of computational thinking in the form of a taxonomy consisting of four main categories: data practices, modeling and simulation practices, computational problem-solving practices, and systems thinking practices.

4. Common features for Computational Thinking frameworks within K-12 Education: A work in progress

Collaboration with the computer science education community is essential to achieve the overarching objectives of computational thinking and "the process of increasing student exposure to computational thinking in K-12 is complex, requiring systemic change, teacher engagement, and development of significant resources" [2].Following the definition of computational thinking, CSTA and ISTE [11] presented a framework specifically for K-12 education, and nine major computational thinking concepts. Below table shows the summary of the framework:

Table 1.CT core concepts for Computational Thinking framework in K-12 education in different context

Author	**CT core concepts (or elements) introduced**	**K-12 Education framework / context of use**
Jeannette Wing	• problem-solving process, designing systems, and understanding human behavior [21] • Abstraction and automation [22]	Main reference to many CT framework in different countries
CSTA and ISTE [11]	• data collection, • data analysis, • data representation, • problem decomposition, abstraction, • algorithms and procedures, • automations, • parallelization • simulation	USA, Framework for K-12 Science Education (2012) [14] UK, Computing at School (2014) for primary and secondary school
Curzon, Dorling, Ng, Selby and Woollard [9]	Definition,Concepts,Classroom techniques and Assessment (Four interconnected stages of development)	UK's computational thinking framework
Selby and Woollard [19]	algorithmic thinking, evaluation, decomposition, abstraction and generalization	
Brennan and Resnick [5]	*computational concepts* (repetition, parallelism, etc) *computational practices* (debugging projects or reproducing) *computational perspectives* (the perspectives designers build about the world around them, and about themselves)	MIT Media Lab, using Scratch

The National Research Council [13] suggested mathematics and computational thinking to be part of the eight central practices for scientific and engineering development, as stated in "Framework for K-12 Science Education" [14]. In this framework, mathematics and computational thinking encompasses the use of computer tools to signify physical variables and the relationships among them.

MIT Media Lab, through its Lifelong Kindergarten research has developed Scratch, a computational authoring environment where learners would design their individual interactive media, which cover stories, games, animations, and simulations. This visual programming tool allow programming-like blocks to be incorporated during the process of designing, just like how LEGO bricks or puzzle pieces are snapped together [17]. Scratch is a visual programming environment that enables learners to develop interactive stories, games, and simulations, and then share it in an online community with others from around the world. By using Scratch as the context, Brennan and Resnick [5], promoted three key dimensions of computational thinking frameworks. The idea of the framework is in line with constructionism when learners participate as designer themselves whilst using interactive media with Scratch, to support the development of computational thinking among young learners.

In Malaysia, the Prime Minister has declared the integration of computational thinking skills into all subjects, starting from 2017 with Year 1 primary students [3]. Therefore, there is a crucial need for schools and especially teachers to be equipped with all the necessary skills to integrate computational thinking in their teaching and learning process. The newly revised curriculum was made available, with reference made from international bodies, and following the global trends, which also includes benchmarking.

5. Limitations

The scope of this article review is limited to only selected publications between 2006 and 2018, with focus on selected countries, not worldwide. This is mainly due to limited length of discussion allowed in this publication.Hence, more details can be addressed in a more comprehensive article in the future.

6. Conclusions

In summary, computational thinking is a thoughtful way of solving problems that is extensively appropriate throughout the K-12 curriculum and progressively relevant in the 21st century. Although there are varying definitions and multiple views of computational thinking frameworks in K-12 education. with some different core elements (constructs) being identified. This could be due to different epistemological perspective, and degree of application in various educational context. Integrating computational thinking into core and elective subject areas can help students to achieve better understanding and enable them to make imperative cross-curricular connections, which would improve their academic performance, and develop important skills for creating solutions in the wide variety of aptitudes. As computational thinking is gaining more popularity and significance, a lot of countries, states, and institutions are implementing it into their school curriculum. In general, it is important to acknowledge common key elements of CTs in ensuring successful computational thinking integration within K-12 education.

Acknowledgements

The authors gratefully acknowledge the Regional Cluster for Research and Publication (RCRP) grant scheme for supporting this research, through grant number RRCP-2016-002.

References

[1] Allan W *et al* 2010 White Paper for the ITEST Small Working Group on Computational Thinking.
[2] Barr V et al 2011 *ACM Inroads*, **2**(1) 48-54.
[3] BERNAMA 2016, August 11 Utusan Online retrieved from http://www.utusan.com.my/berita/nasional/pemikiran-komputasional-sains-komputer-akan-diajar-di-sekolah-tahun-depan-1.369190
[4] Bers M U *et al* 2014 Comp. & Educ. **72,** 145-157.
[5] Brennan K *et al* 2012 Proceedings of the 2012 annual meeting of the American Educational Research Assoc., Vancouver, Canada **1** 25.
[6] Bundy A *et al* 2007 *J. of Scientific & Practical Computing* **1**(2) 67-69
[7] Bureau of Labor Statistics 2018 Occupational outlook handbook.
[8] Buss A *et al* 2017 Emerging research, practice, and policy on computational thinking Springer 189-203
[9] Curzon P *et al* 2014 Monograph of Developing computational thinking in the classroom: A framework

[10] Google for Education 2016 Exploring Computational Thinking retrieved from https://edu.google.com/resources/programs/exploring-computational-thinking/
[11] International Society for Technology in Education and the Computer Science Teachers Association 2011 International Society for Technology in Education (ISTE) and the Computer Science Teachers Association (CSTA)
[12] Israel M *et al* 2015 Comp. & Educ. **82** 263-279.
[13] National Research Council 2010, *Report of a Workshop on the Scope and Nature of Computational Thinking National* Academies Press
[14] National Research Council 2012 *A framework for K-12 Science Education: Practices, Crosscutting Concept, and Core Ideas* National Academies Press
[15] Papert S *et al* 1991 *Constructionism Norwood NJ* **36**(2) 1-11
[16] Piaget J *et al* 1968 *Six Psychological Studies* **462** Vintage Books
[17] Resnick M *et al* 2009 Communications of the ACM **52**(11) 60-67
[18] Rich P J *et al* 2017 Int. J. of Comp. Sc. Educ. in Schools **1**(1) 1-20
[19] Selby C *et al* 2013 Monograph *Computational Thinking: The Developing Definition*
[20] Weintrop D et al 2015 Journal of Science Education and Technology **25**(1) 127-147
[21] Wing J M 2006 Communications of the ACM **49**(3) 33-35
[22] Wing J M 2008 Philosophical Transactions of the Royal Society of London A: Mathematical, Physical and Engineering Sciences, **366**(1881) 3717-3725

IOP Conf. Series: Materials Science and Engineering **551** (2019) 012064 doi:10.1088/1757-899X/551/1/012064

Integration of Statistical Method and Zernike Moment as Feature Extraction in Liveness Detection

A S Ahmad[1,*], R Hassan[2,*], Z Zakaria[3], R Ramlan[4]

[1,2]Software Engineering Department, School of Computing, Faculty of Engineering, Universiti Teknologi Malaysia, 81310, Johor Bharu, Malaysia.
[3]Artificial Intelligence and Bioinformatics Research Group, School of Computing, Faculty of Engineering, Universiti Teknologi Malaysia, 81310, Johor Bharu, Malaysia.
[4]Production and Operation Management Department, Faculty of Technology Management and Business, Universiti Tun Hussein Onn Malaysia, 86400 Batu Pahat, Johor, Malaysia.

Email: asrafulsyifaa.ahmad@gmail.com, rohayanti@utm.my

Abstract. Recently, fake fingerprints have been used to defeat fingerprint recognition systems. These fake fingerprints are created without the need for any expertise and use easily found materials. In this paper, a fake fingerprint detection method is proposed that employs a combination of eleven statistical methods and integrating them with Zernike Moment as the feature extractor. Based on the experimental results, the proposed method showed average classification accuracy, sensitivity and specificity of approximately 80% for all sensors used to capture fake fingerprint images fabricated by gelatine and latex materials.

1. Introduction

A fake fingerprint is an artificial finger that is used to fool a detection system. It is categorized into two groups; cooperative which means with the cooperation of the user, and non-cooperative i.e. without user consent. There are still some issues in the fake fingerprint images despite their ability to imitate real fingerprints and fool the detection system. A fake fingerprint deforms elastically as it makes contact with the plane surface [1]. Fake fingerprint images have noise due to the presence of organic molecules in the fabrication materials. Research has been carried out to calculate the image quality of fake fingerprint images and they concluded that silgum and wood glue resulted in very low quality images due to the noise present [2]. Meanwhile, a fake fingerprint fabricated using latex shows a result that is very similar to the original fingerprint sample. These results show that all materials that are used to fabricate the fake finger have imperfections which are hard to detect by human observers[3].

Fake fingerprint images are captured by using a sensor and can be digitally analysed. Digital images consist of pixel intensities which can be represented by grey-level distribution. Different images give different readings of histogram distribution. Therefore, the statistical method was chosen as the feature extractor in our method. However, to maintain the originality of the fake fingerprint images, no image enhancement is done. Thus, in order to ensure that the noises do not affect the extraction process by the statistical method, Zernike Moment is integrated as it can help in filtering the noises. There are several fake fingerprint detection approaches using machine learning which have

IOP Conf. Series: Materials Science and Engineering **551** (2019) 012064 doi:10.1088/1757-899X/551/1/012064

been employed in previous research. Among them, the Support Vector Machine (SVM) is the most widely used [4]. Therefore, we also select and use SVM as the classification method in our research.

Therefore, the work done in this paper has two objectives. The first is to analyse which datasets in the LivDet 2015 are suitable to use as the standard benchmark for performance classification of the fake fingerprint. The second is to manipulate the use of the statistical method and Zernike Moment in terms of feature extraction. The presented analysis and exploration will provide a simple understanding of fake fingerprints and the benefit of having the feature extraction in fingerprint liveness detection. The rest of this paper is organized as follows. The dataset and statistical equation used in this research are introduced in Section 2. The result of the performance of classification is presented in Section 3. Lastly, the conclusion is drawn in Section 4.

2. Methodology

Figure 1 below shows the flow of the proposed method.

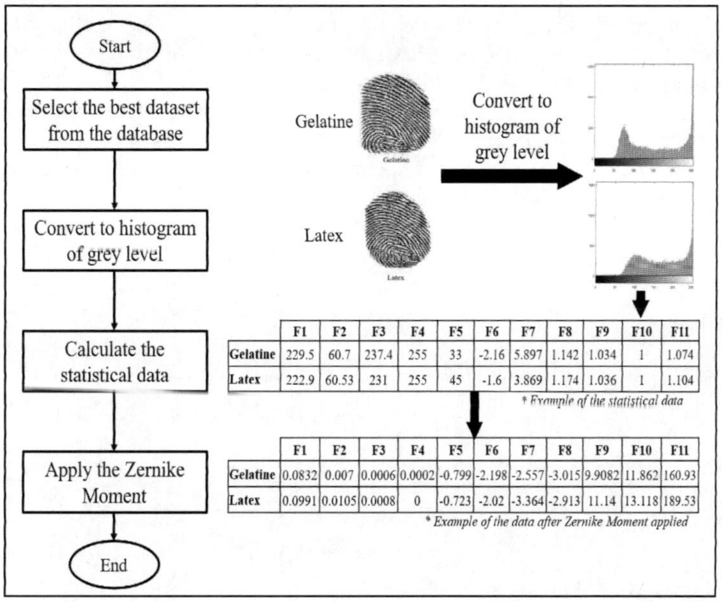

Figure 1. Flow chart of the research

2.1. *Dataset Selection*

The database chosen for this research is LivDet 2015. The first step in the fake fingerprint detection process is data acquisition from the databases. This step is essential as a failure in this step would lead to an inconsistency in the research objectives. The mathematical metrics approach is used in this research to calculate and evaluate the quality of the fake fingerprint images using Peak Signal to Noise Ratio (PSNR) and also Root-Mean-Square-Error (RMSE). The comparison is done by using the 250 live or known as the original fingerprint images as the references and comparing them against the 250 fake fingerprint images each made by gelatine, latex, ecoflex, and wood glue materials. The inverse relationship between PSNR and RMSE is where a lower RMSE and higher PSNR means less errors in the compared images. Table 1 shows that fake fingerprints made by latex and gelatine show the lowest RMSE values and high PSNR values. Therefore, this research will focus on latex and gelatine

IOP Conf. Series: Materials Science and Engineering **551** (2019) 012064 doi:10.1088/1757-899X/551/1/012064

materials in order to discover the characteristics that differentiates each material for a better classification process.

Table 1. Image quality assessment result

Materials	Ecoflex		Latex		Gelatine		Wood glue	
Camera	PSNR	RMSE	PSNR	RMSE	PSNR	RMSE	PSNR	RMSE
Digital	7.682	23.908	7.856	22.211	7.809	24.311	7.534	26.611
GreenBit	12.725	35.354	14.846	29.921	15.759	26.848	13.119	45.915
HiScan	10.771	32.747	11.747	30.526	10.104	21.253	12.286	41.329

2.2. Statistical Extraction Features

Generally, the extraction of features aims to reduce the dimension of the data and choose the most significant features. As the fingerprint pixels are associated with the grey-level, the statistical method is chosen to extract the meaningful information. This is due to the grey-level distribution of fingerprint images that changes according to the physical structure. Previous researchers used several statistical features to differentiate between real and fake fingerprint images [4], [5]. Thus, this research gathers eleven statistical data to be used as the features of a fake fingerprint. The features are Mean, Standard Deviation, Root Mean Square (RMS), Maximum Amplitude, Minimum Amplitude, Skewness, Kurtosis, Clearance Factor, Shape Factor, Impulse Factor, and Crest Factor. Then the images from both datasets, gelatine and latex, were converted into a histogram of grey-level distribution and the eleven statistical data were collected for each image. The equations are as follows where $H(n)$ is equalized and normalized histogram and N is the total number of bins in the histogram.

F1: Mean, μ

$$\mu = \frac{\sum_{n=1}^{N} H(n)}{N} \tag{1}$$

F2: Standard Deviation, σ

$$\sigma = \sqrt{\frac{\sum_{n=1}^{N}(H(n) - \mu)^2}{N-1}} \tag{2}$$

F3: Root Mean Square, RMS

$$RMS = \sqrt{\frac{\sum_{n=1}^{N}(H(n))^2}{N}} \tag{3}$$

F4: Maximum Amplitude, $getMax$

$$getMax = \max(H(n)) \tag{4}$$

F5: Minimum Amplitude, $getMin$

$$getMin = \min(H(n)) \tag{5}$$

F6: Skewness

$$skewness = \frac{1}{\sigma^3}\sum_{n=1}^{N-1}(n - \mu)^3 H(n) \tag{6}$$

F7: Kurtosis

$$kurtosis = \frac{1}{\sigma^4}\sum_{n=1}^{N-1}(n - \mu)^4 H(n) \tag{7}$$

F8: Clearance Factor, CLF

$$CLF = \frac{getMax}{\sum_{n=1}^{N} |H(n)|} \tag{8}$$

F9: Shape Factor, SF

$$SF = \frac{RMS}{\sum_{n=1}^{N} |H(n)|} \tag{9}$$

F10: Impulse Factor, IF

$$IF = \frac{\mu}{\sum_{n=1}^{N} |H(n)|} \tag{10}$$

F11: Crest Factor, CF

$$CF = \frac{getmax}{RMS} \tag{11}$$

2.3. Integrating Zernike Moment

To ensure the originality of the images, no enhancement is done on the images as applying image enhancement could result in many undefined or spurious minutiae being neglected and many genuine minutiae detected [6]. Thus, to handle the noise issues in the images, Zernike Moment is used as it is not sensitive at all towards the image noises. It does contain analytical invariances to rotation and luminance. Moreover, features extracted from fingerprint images are highly consistent in the change of translation since the centre of the unit disk is placed at the singular point of each image [7]. The eleven statistical data collected from each image is then applied to the Zernike Moment.

3. Performance of Classification

The table below presents the results of classification accuracy, specificity and sensitivity for the LivDet dataset based on the sensor and material fabricated. The proposed method of using the integration of Statistical Method and Zernike Moment is compared with the result of classification using Statistical Method only. Sensitivity and specificity are inversely proportional, meaning that as the sensitivity increases, the specificity decreases and vice versa [8]. Specificity is a test to define the ability of classification to detect the true negative classes while sensitivity is the ability to detect the true positive classes. In this experiment, a high specificity is needed since we want to test the specificity of the classification to specify fake classes.

Table 2 indicates that the accuracy of classification of the proposed method does show an average increment in score where the dataset of images captured by the HiScan sensor achieved a higher accuracy. It can be concluded that using a HiScan sensor that captures images in high resolution produces visible and clearer grey level distributions. This also helps in extracting the meaningful features that assist in the classification process and directly improves the accuracy score. Meanwhile, for the specificity of classification, a higher specificity means a lower sensitivity. Thus, we can see that the specificity of classification for a dataset captured using HiScan is higher compared to its sensitivity which has recorded a lower score. Meanwhile, a comparison of the accuracy of the fake fingerprint materials shows that all of them can achieve accuracy scores of more than 80%. The gelatine dataset achieves a better result when compared between the methods of extraction using the Statistical Method and the proposed method. Overall, we can see that the proposed method has a slight impact towards the classification. The presence of organic molecules affects the images captured by the sensor where each of the molecules will have their own pixel value and can be falsely detected as minutiae during the matching process. The average accuracy scores of more than 80% show that using the dataset with the most similar image quality compared to the original fingerprint images and with no image enhancement done, results in the high scores due to the usage of Zernike Moment to tackle the noise issues. Additionally, the dataset used in this research is the most similar to the original images in terms of quality. The above 80% average scores obtained in terms of accuracy and specificity respectively is therefore a good start for another detailed research in this area.

Table 2. Classification Accuracy, Specificity and Sensitivity

			Accuracy of Classification (%)		Specificity of Classification (%)		Sensitivity of Classification (%)	
Feature Extractor			Statistical Method	Proposed Method	Statistical Method	Proposed Method	Statistical Method	Proposed Method
Sensor	Digital	Gelatine	70.61	**71.19**	84.70	71.19	84.70	**71.19**
		Latex	78.98	76.22	81.19	**86.22**	81.19	**76.22**
	GreenBit	Gelatine	83.95	**86.39**	81.88	**86.39**	81.88	**76.39**
		Latex	80.15	77.78	83.57	77.78	83.57	**77.78**
	HiScan	Gelatine	84.96	**86.77**	82.65	**86.77**	82.65	86.77
		Latex	85.03	**85.18**	82.15	**85.18**	82.15	85.18

[a]The bold value is the highest score

4. Conclusions

Throughout this study, comprehensive explanations have been justified on the improvement of fingerprint recognition. Feature extraction is one of the crucial parts in the recognition. Therefore, a set of meaningful features need to be extracted. An enhancement of the feature extractor in terms of the framework does affect the result of the accuracy. Choosing the only meaningful statistical method as recommended by various literature also plays an important role. Using Zernike Moment as another feature extractor in combination with the Statistical Method shows a slight difference in terms of accuracy, sensitivity and specificity. This work provides insights into fake fingerprints and benefits the development of fake fingerprint liveness detection research. In future, employing the feature selection might produce a better impact on the classification. The usage of sensors also plays an important role and needs to be further researched.

Acknowledgement

This work was financially supported by FRGS, Vot No: 4F973 from Ministry of Education (MOHE), and also RUG, Vot No. 16H73 from Universiti Teknologi Malaysia.

References

[1] Arunalatha G, Ezhilarasan M 2015 Fingerprint spoof detection using quality features *Int J Secur Its Appl.* **9** 83–94
[2] Rattani A, Ross A 2014 Minimizing the impact of spoof fabrication material on fingerprint liveness detector 4992
[3] Kho JB, Lee W and Choi H, Kim J 2019 An incremental learning method for spoof fingerprint detection *Expert Syst Appl* **116** 52–64
[4] Huang Q, Chang S, Liu C, Niu B, Tang M and Zhou Z 2015 An evaluation of fake fingerprint databases utilizing SVM classification. *Pattern Recognit Lett* **60–61** 1–7
[5] Park Y, Jang U, Lee EC 2017 Statistical anti-spoofing method for fingerprint recognition *Soft Comput* **22** 1–10
[6] Hong L, Wan Y and Jain A 1998 Fingerprint image enhancement: algorithm and performance evaluation *IEEE Transactions on Pattern Analysis and Machine Intelligence* **20** 777-789
[7] Jang HU, Hyun DK, Jung DJ and Lee HK 2014 Fingerprint-PKI authentication using zernike moments 5022–6
[8] Trevethan R. 2017 Sensitivity, specificity, and predictive values: foundations, pliabilities, and pitfalls in research and practice. *Front public Heal* **5** 307

Improved Metaheuristic Algorithms for Metabolic Network Optimization

K Mohd Daud[1*], Z Zakaria[1], R Hassan[2], M S Mohamad[3,4] and Z Ali Shah[1]

[1] Artificial Intelligence and Bioinformatics Research Group, School of Computing, Faculty of Engineering, Universiti Teknologi Malaysia, 81310 Johor Bahru, Malaysia
[2] Software Engineering Department, School of Computing, Faculty of Engineering, Universiti Teknologi Malaysia, 81310 Johor Bahru, Malaysia
[3] Institute for Artificial Intelligence and Big Data, Universiti Malaysia Kelantan, City Campus, Pengkalan Chepa, 16100, Kota Bharu, Kelantan, Malaysia
[4] Faculty of Bioengineering and Technology, Universiti Malaysia Kelantan, Jeli Campus, Lock Bag 100, 17600, Jeli, Kelantan, Malaysia

Email: kautharmohdaud@yahoo.com

Abstract. Metaheuristic algorithms have been used in various domains to solve the optimization problem. In metabolic engineering, the problem of identifying near-optimal reactions knockout that can optimize the production rate of desired metabolites are hindered by the complexity of the metabolic networks. Through Flux Balance Analysis, different metaheuristics algorithms have been improved to optimize the desired phenotypes. In this paper, a comparative study of four metaheuristic algorithms have been proposed. Differential Search Algorithm (DSA), Particle Swarm Optimization (PSO), Artificial Bee Colony (ABC) and Genetic Algorithm (GA) are considered. These algorithms are tested on succinic acid production in *Escherichia coli*. The comparative performances are measured based on production rate, growth rate, and computational time. Hence, from the results, the best metaheuristic algorithms to solve the metabolic network optimization is identified.

1. Introduction
The metabolic network is defined as a series of biochemical reactions involving the transformation and modification of substrates into different products in which the enzymes act as catalysis agent. The metabolic network is important in assessing the properties of biochemical and physiological of a cell. Recently, advancements in genome sequencing have brought upon many developments that allow the biological researchers to have deeper knowledge and information of an organism. One of the development is the establishment of metabolic engineering, which allows the researchers to probe in detail the organizations of an organism including the reactions, pathways, metabolites, and genes as well as exploit the organisms for strains optimization by suggesting genetic alteration strategies that can optimize the desired objectives, such as growth rate or production rate.

A metabolic model may consist thousands or hundreds of reactions, genes, and metabolites that are interconnected among each other. This resulted in the complexity of the data. The complexity of metabolic model has resulted in the dimensions of solutions space being too large and thus the computational time increases exponentially [1], [2]. In addition, it is difficult to obtain a near-optimal set of reactions knockout in order to optimize the desired phenotypes. Therefore, with the principle of mass balances, various techniques and methods have been developed for modelling and optimizing

IOP Conf. Series: Materials Science and Engineering **551** (2019) 012065 doi:10.1088/1757-899X/551/1/012065

metabolic network. These methods apply constraint-based modelling and swarm-metaheuristic algorithms to identify a set of reactions that optimize the desired objective function.

Constraint-based modelling (CBM) is an approach to investigate the optimality of an organism by predicting and describing the metabolic phenotypes [3]. In CBM, constraints are applied to the systems, thus creates feasible flux distributions space. There are four approaches under constraint-based modelling methods – flux balance analysis (FBA), flux variability analysis (FVA), minimization of metabolic adjustment (MoMA), and regulatory on/off minimization (ROOM). Considering that a metabolic network comprises of vast information, FBA is able to handle large data compared to the three approaches by predicting the final higher steady-state of biological objective functions [4]. Furthermore, FBA successfully predicts the final steady-state growth rates or production rates of desired metabolites, without considering the regulatory and kinetic information that is difficult to obtain.

However, FBA alone could not optimize the capabilities of organisms by redirecting more fluxes toward the desired metabolites. Furthermore, the problem of finding combinations of reaction for knockout is considered as a combinatorial optimization problem, thus it is not practicable to test all the combinations of reaction knockout using an exhaustive brute force approach. Therefore, the use of metaheuristic algorithms have inspired the development of optimization-modelling methods for optimizing the production rate of desired metabolites by means of reaction knockout.

The advantages of metaheuristic algorithms that computationally less expensive have driven the progress of myriad strain design metaheuristic optimization-modelling methods. Although metaheuristic algorithms do not guarantee for finding optimal solutions, these algorithms allow for optimization of individual biological objective function and engineering objective function. Furthermore, metaheuristic uses multiple agents/individuals/particles to search for near-optimal solutions, thus, the search spaces are larger than the other frameworks [5]. In other words, metaheuristic allows a large number of genetic perturbations.

The focus of this paper is to evaluate and compare four metaheuristics algorithms in maximizing the production rate of succinic acid in Escherichia coli. These algorithms were improved with FBA as fitness function evaluation. The paper is organized as follows: Section 2 briefly describes the selected four metaheuristics algorithms. The following Section 3 presents the results and discussion. Lastly, Section 4 concludes the paper..

2. Metaheuristic Algorithms

2.1. Differential Search Algorithm
Differential Search Algorithm (DSA) is a swarm based metaheuristic algorithm that mimics the behavior of a group of animals that simulates the migration of a superorganism that consists of organisms, from unfruitful area to a more fruitful area [6]. The exploration in DSA is done by *Brownian-random walk*, while a greedy algorithm carries exploitation.

2.2. Artificial Bee Colony
Artificial Bee Colony (ABC) is inspired by the behavior of honey bee colonies in food foraging using waggle dance [7]. In ABC, the forages can be categories into two, namely employed and unemployed. For each employed bees, a new food source is generated which represent as a feasible solution. The information reside in food source is improved by the unemployed bees which is onlooker bees. On a certain limit of time, an employed bee may become a scout bee, such that the onlooker bees could not improve the food source

2.3. Particle Swarm Optimization
Particle Swarm Optimization (PSO) simulates the movement behavior of a group of animals [8]. PSO consists of a swarm of particles with different velocities and fitness. For every iteration, each particle will benefit from its own experience and global best experience of the swarm. Therefore, the

"communication" among particles are occurred and eventually, allow a unity among the particles in moving together as a swarm.

2.4. Genetic Algorithm

The earliest biological based evolutionary algorithm is a Genetic algorithm (GA) which involves exact biological mechanisms, namely mutation, crossover and selection [9]. GA uses chromosomes to exploit and explore the solution space in determining and enhancing the value of near-optimal solutions. GA is a well-known algorithm that has been improved in many ways.

Table 1 shows the advantages and disadvantages of the said algorithms.

Table 1. Summary of metaheuristic algorithms.

Name	Advantage (s)	Disadvantage (s)	Ref (s)
DSA	- The control parameters are not sensitive towards the problem. - Fastest CPU computational time. - Fewer parameters.	- Too many random numbers involved in generating stopover site. - The solutions may go outside the boundaries.	[6], [10]–[12]
ABC	- Better in global search - Fast convergence	- Require more control parameters - The solutions may trapped in local optima.	[11], [13]
PSO	- Has less parameter tuning - Efficient in global search	- Sensitivity of control parameters - The loss of inertia weight over time may cause the swarm to lose its ability to search.	[14]–[16]
GA	- Easy to implement - Fast convergence	- High computational complexity - The solutions may trapped in local optima.	[17], [18]

3. Results and Discussion

The problem in identifying reactions knockout for maximizing the production of desired metabolites can be formulated mathematically by representing the metabolic model in the stoichiometry matrix, *S* that consists of reactions and metabolites of size mxn. The stoichiometry matrix describes the dynamic mass balance equation of the metabolic model by a differential equation between the flux rates of the reactions (v) and concentrations of metabolites (c). The mass balance must be at a steady-state level that signifies there is no internal and external changes in the concentration of metabolites.

$$\frac{dx}{dt} = S \cdot v = 0 \tag{1}$$

Whereby

$$S_i = \sum_{j=1}^{r} n_{ij} \tag{2}$$

$$v_{lb} \le v \le v_{ub} \tag{3}$$

Thus, from the equations above, it can be concludes to

$$\max/\min\ Z = \sum_i c_i v_i \tag{4}$$

Where S is the stoichiometric matrix with size of r reactions and m metabolites, n is the coefficients of stoichiometric with the size of $i=1, 2...m$ and $j=1, 2...r$, v is the flux vector of size reactions, and Z is the objective function to be optimized that associated with a biological task such as product rate or growth rate. Herein, Z is refer to the production rate of succinic acid.

Table 2 shows the result comparison of different methods applied in this domain. The three methods were improved with FBA for fitness function evaluation, which is product rate. *E.coli* core model is tested and the target reaction to optimized is succinic acid.

Table 2. Result comparison of different evolutionary methods.

Method	Product rate (mmol/gDW/hr)	Growth Rate (hr-1)	Time (sec)	Knockout reactions
PSOFBA	15.4905	0.4769	25.8399	ACKr, GLUDy, AKGDH, ME1, ME2, PFL, PYK, RPE
GAFBA	15.5185	0.4621	45.4629	ACKr, GLUDy, GND, LDH_D, ME1, ME2, PYK, SUCOAS
ABCFBA	15.5185	0.4621	73.2244	ACALD, G6PDH2r, AKGDH, GLUDy, GND, ME1, ME2, PYK
DSAFBA	15.50	0.4836	11.9653	ACALD, G6PDH2r, GND, ME1, ME2, PFL, PYK, SUCOAS

Note: All the methods were developed and run again using the same dataset, with number of population is 100 and maximum iteration is 100.

As shown in **Error! Reference source not found.**, DSAFBA is compared with PSOFBA, GAFBA, and ABCFBA in terms of the production rate of succinic acid, growth rate after perturbations, and time. From the result comparison, GAFBA and ABCFBA are able to find the highest amount of product rate compared to DSAFBA and PSOFBA. Nevertheless, DSAFBA is able to identify the mutant with the highest growth rate. In addition, by looking at the result of GAFBA, ABCFBA, and DSAFBA; the difference in growth rate between DSAFBA and GAFBA-ABCFBA is 0.0215, whilst the difference in product rate between DSAFBA and GAFBA-ABCFBA is 0.0185. Furthermore, DSAFBA is able to find near-optimal combinations of reactions that optimize the production of succinic acid in a short amount of time, compare to PSOFBA, GAFBA and ABCFBA.

In terms of knockout reactions identified by the methods, DSAFBA and ABCFBA are able to find reactions that were predicted by other methods as well. From the total of 8 reactions being knockout, the first 2 of the suggested reactions identified by DSAFBA are found by ABCFBA, 3 of them are found by PSOFBA, GAFBA, and ABCFBA. Meanwhile, reactions PFL and PYK are found by other methods as well, whereas reaction SUCOAS is suggested by GAFBA. Therefore, from the results shown in **Error! Reference source not found.**, the searching strategy that uses *Brownian-random walk* movement and mutation structure, DSA are able to find the combination of reactions that improve the production rate of target metabolites in a short amount of time, while maintaining the viability of the mutant.

4. Conclusions

This paper focusing on modifying metaheuristic algorithms to solve the identification of near-optimal reactions knockout. From the four tested algorithms, DSA outperforms other metaheuristic algorithms in terms of CPU computational times, and performance. The discovery of the suggested reactions knockout are limited to the computational validation only. Thus, in future, these reactions can be serve as prior knowledge for in vivo implementation.

Acknowledgements
This work was financially supported by FRGS, Vot No: 4F973 from Ministry of Education (MOHE), and also RUG, Vot No: 16H73 from Universiti Teknologi Malaysia.

References

[1] Hansen A S L, Lennen R M, Sonnenschein N and Herrgård M J 2017 Systems biology solutions for biochemical production challenges *Curr. Opin. Biotechnol.* **45** 85–91

[2] Wang N, Zhao W J, Wu N and Wu D 2017 Multi-objective optimization: a method for selecting the optimal solution from Pareto non-inferior solutions *Expert Syst. Appl.* **74** 96–104

[3] Klamt S, Muller S, Regensburger G and Zanghellini J 2018 A mathematical framework for yield (vs. rate) optimization in constraint-based modeling and applications in metabolic engineering

Metab. Eng. **47** 153–169

[4] Bonarius H P J, Schmid G and Tramper J 1997 Flux analysis of underdetermined metabolic networks: The quest for the missing constraints *Trends Biotechnol.* **15** 308–314

[5] Maia P, Rocha M and Rocha I 2016 In Silico Constraint-Based Strain Optimization Methods: the Quest for Optimal Cell Factories *Microbiol. Mol. Biol. Rev.* **80** 45–67

[6] Civicioglu P 2012 Transforming geocentric cartesian coordinates to geodetic coordinates by using differential search algorithm *Comput. Geosci.* **46** 229–247

[7] Karaboga D 2005 An idea based on honey bee swarm for numerical optimization *Technical Report -TR06* **46** 185–189

[8] Kennedy J and Eberhart R 2011 Particle Swarm Optimization *Nat. Comput. Series,* Springer Boston, MA 760–766

[9] Mitchell M 1995 Genetic algorithms: An overview *Complexity* **1** 31–39

[10] Atay Y, Koc I, Babaoglu I and Kodaz H 2017 Community detection from biological and social networks : A comparative analysis of metaheuristic algorithms *Appl. Soft Comput. J.* **50** 194–211

[11] Kurban T, Civicioglu P, Kurban R and Besdok E 2014 Comparison of evolutionary and swarm based computational techniques for multilevel color image thresholding *Appl. Soft Comput. J.* **23** 128–143

[12] Liu J, Wu C, Cao J, Wang X and Teo K L 2016 A Binary differential search algorithm for the 0 – 1 multidimensional knapsack problem *Appl. Math. Model.* **40** 9788–9805

[13] Waghole V and Tiwari R 2014 Optimization of needle roller bearing design using novel hybrid methods *Mech. Mach. Theory.* **72** 71–85

[14] Yuce B, Packianather M S, Mastrocinque E, Pham D T and Lambiase A 2013 Honey bees inspired optimization method: The bees algorithm *Insects.* **4** 646–662

[15] Mora-Melià D, Martínez-Solano F J, Iglesias-Rey P L and Gutiérrez-Bahamondes J H 2017 Population Size Influence on the Efficiency of Evolutionary Algorithms to Design Water Networks *Procedia Eng.* **186** 341–348

[16] Kachitvichyanukul V 2012 Comparison of Three Evolutionary Algorithms : GA, PSO, and DE *Ind. Eng. Manag. Syst.* **11** 215–223

[17] Alam M N, Das B and Pant V 2015 A comparative study of metaheuristic optimization approaches for directional overcurrent relays coordination *Electr. Power Syst. Res.* **128** 39–52

[18] Goldberg D E and Holland J H 1998 Genetic algorithms and machine learning *Machine Learning* **3** 95–99

Audio Deformation based Data Augmentation for Convolution Neural Network in Vibration Analysis

M F M Esa[1] , N H Mustaffa[1] , N H M Radzi[1] and R Sallehuddin[1]

[1]School of Computing, Faculty of Engineering, Universiti Teknologi Malaysia 81310 UTM Johor Bahru, Johor, Malaysia

Email: pkharmony@yahoo.com

Abstract. Audio deformations in audio processing have proved ability in preserve semantic meaning for audio signal. Convolution Neural Network (CNN) is among deep learning model that requires huge dataset during training for excellence performance Thus, data augmentation (DA) method is used to overcome the problem of limited dataset number for vibration analysis. Several signal processing phases including segmentation and image converting need to be performed before the vibration signal can be used as input for CNN. In this research, audio-deformation based DA is proposed in generating the additional vibration signal dataset. The proses is start by encoding the raw vibration signal to audio signal format to enable the audio deformation process performing, then decoding back into new vibration signal. Speed and amplify transformation are selected for audio deformation process. The new vibration data set of bearing fault detection problem are used for training CNN to validate the proposed approach. The results obtained from 13 experiments setting have shown that the proposed DA able to increase the accuracy of training for CNN until 13% compared with the previous DA method.

1. Introduction

Bearing is significant mechanical component in rotary equipment and widely used in industrial application. The major fault in bearing may lead to the machinery malfunction and can interrupt the overall operation. The problem of bearing fault analysis has received increasing attention from the research community in recent years. The bearing fault analysis will identify the faulty bearing and its component in the system. Its applications range from rotating machine [1], helicopter gearbox and [2] electric locomotive [3] to wind turbine generator [4]. To date, variety of methods have been used for bearing fault analysis including the vibration analysis [5]. The vibration analysis is most commonly used due to its simple method of application [5].

In application of signal processing and machine learning, there are many techniques have been applied for vibration analysis and deep neural networks technique is the most recent technique used for the analysis [5]. In particular, Convolutional Neural Networks (CNN) is from the family of deep neural networks [6] and have advantages to solve the problem of bearing fault analysis. There are two significant characteristics of CNN that contributed to its successful. First, the capability of CNN in automatically extract complex feature representations [7] and second is the noise tolerance [8].

However CNN requires huge training dataset to be efficient, as they are composed of a lot of parameters [9]. Meanwhile, the collection of vibration signal data from defect bearing are usually limited because of expensive processing cost, laborious and depending to the device capabilities. These disadvantages will limit the application of CNN in vibration signal analysis for bearing fault

IOP Conf. Series: Materials Science and Engineering **551** (2019) 012066 doi:10.1088/1757-899X/551/1/012066

detection. Fortunately, data augmentation (DA) able to solve these problem and been extensively discussed in literature [10]. The term DA refers to methods for sampling algorithms via the producing the unobserved data or latent variables in population data [11]. It is also, a common strategy adopted to increase the quantity of training data. Therefore, DA is simultaneously can generate extra size dataset and variant data at the same time preserves the data label. With these abilities, DA can worked excellently for image recognition and speech recognition [12].

Previously, DA also is a useful strategy to reduce over-fitting in CNN training for the image recognition problem [13] and in the same time achieved superior results at the time. In image processing, DA can be classified into geometric and photometric transformations [14, 15]. A geometric transformation is altering the geometry of the image to generate variant in position and orientation. For instance, flipping, cropping, scaling and rotating. Meanwhile, a photometric transformation is modification of the colour channels to generate variant in lighting and colour. Among the technique in this category are Fancy Principle Component Analysis (PCA) and colour jittering [13, 7].

In other domain, the other type of data format such as time series data has inspired DA technique from image processing community such as window slicing [16], overlap window slicing [8] and window warping [9]. In speech recognition, audio deformation is commonly been used as DA. Audio deformation is a method that artificially modified the raw sound signal to different tone or harmony of sound but preserve the semantic meaning of signal. The example of audio deformation are including time stretching, pitch shifting, dynamic range compression [17] and speed perturbation [18].

Generally, DA should be used indiscriminately [19]. Each type data requires different technique to suit with solution. Commonly, vibration signal is in the form of time-series data and previously, the data segmentation techniques are used to create maximum number of data record. Despite many DA technique have been developed in others domain, only sliding windows and overlap sliding windows techniques have been commonly implemented in vibration signal processing.

In this paper, we propose the use of audio-deformation based DA to overcome the problem of limited dataset number of vibration signal and explore different types of audio deformations and their influence on the CNN's training performance.

2. Methodology

The traditional vibration analysis consisting several sequence processes which start with preparing the vibration raw signal dataset. The long length of vibration signal will be segmented into shorter length than original. Furthermore, after segmentation signal and domain transformation process, these sets of 1-D time series are converted to image. Thus, these images will be used as input to CNN model and produces the classification result.

2.1. CNN Model

CNN is a multi-layer artificial neural network (ANN) that is can be consisting with of several filter layer and ending with classification layer [20]. The filter layer is used to extract features representation from the inputs data, that is mainly contains of three basic operations of kinds of layers, the convolutional layer, activation layer and the pooling layer. The classification stage is a multi-layer perceptron, which is composed of fully connected layers.

The experiments implement the Matlab2018b software. The CNN training parameters setting in experiments are as follows. For the learning algorithm, the stochastic gradient descent with momentum algorithm is applied. For every 5 epochs, the learning rate will be reduced by a 0.2 factor and the epochs maximum number is 20 at each iteration. Meanwhile, the CNN architecture setting are as follows. A First convolution layer with single filter with size [256,1] is used, continued by size 2 of max pooling. Then, a second convolution layer is design of 2 filters with the size [32,1] and followed by max pooling of size 4 is used. Later, a third convolution layer is make by 3 filters with the size [16,1] and consecutively a max pooling of size 4 is used. All activation layer using RELU activation function.

IOP Conf. Series: Materials Science and Engineering **551** (2019) 012066 doi:10.1088/1757-899X/551/1/012066

2.2. Audio Deformation based Data Augmentation

This paper proposes an audio deformation approach to generate DA for CNN in vibration analysis. With the appealing capability preserve semantic meaning after the signal was modified by the approach, the limited of data sources signal can be solved. Transformation at data level is achieved by encode the raw data of vibration signal into an audio file format. Followed decode audio file back to vibration signal after performed the audio transformation. Figure 1 shows the flowchart of the proposed audio deformation for DA method at raw data level.

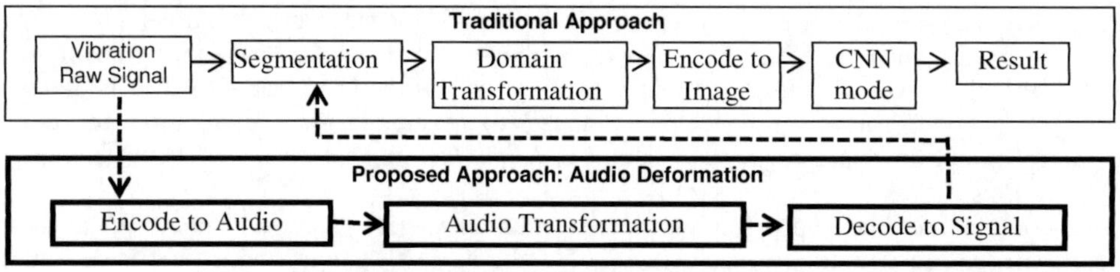

Figure 1. Flow chart of the proposed audio-deformation based DA.

Audacity® Software is used as tool to transform a raw signal in this study. Two audio effect function are selected to modify the raw signal are speed and amplify. Selection of the speed deformation due to its ability produce variant signal at frequency angle (axis x) meanwhile amplify deformation able to create variant signal at decibal(db) angle (axis y) of audio spectrum graph.

Basically, signal will be resampling in speed function meanwhile in amplify function will adjust the volume of audio signal. Five factor of deformation ware selected for each transformation in range 1% until 5%. Finally, five additional copies of the original training data were created by modifying each transformation function based on factor of the original rate.

2.3. Experimental Evaluation

To evaluate the effectiveness of the proposed approach for vibration analysis, the vibration signal dataset collected by Case Western Reserve University (CWRU) is analyzed. The dataset contains 9 defect and single good condition in various frequency, motor load, speed and bearing size. This study was selected a dataset with the properties are 48khz frequency, 1 hp load, 1772 rpm and 10 vibration signal that representing 10 defects condition bearing. From the raw signal, each signal will be executed the proposed technique and produces 12 additional signal where the first 10 additional are based on 5 different audio deformation factor and the rest are combination for all transformation factor for each audio deformation technique.

Then each signal included raw and additional signal are segmented using fix window technique with 2048 size window which was produced 200 sub signal for each signal. However, for mix transformation factor, total of sub signal for these signals are bigger 5 time than others.

All the generated signals are then divided into 2 part with ratios 70% training and 30% validation. Meanwhile, testing dataset are collected at 10% of an original raw data. Moreover, these setting are running 30 time each.

3. Results and Discussion

A similar CNN training and architecture parameter are used for all experiments in this paper. Figure 2 shows the result of 13 experiments of CNN training with speed deformation based DA and amplify deformation based DA in average.

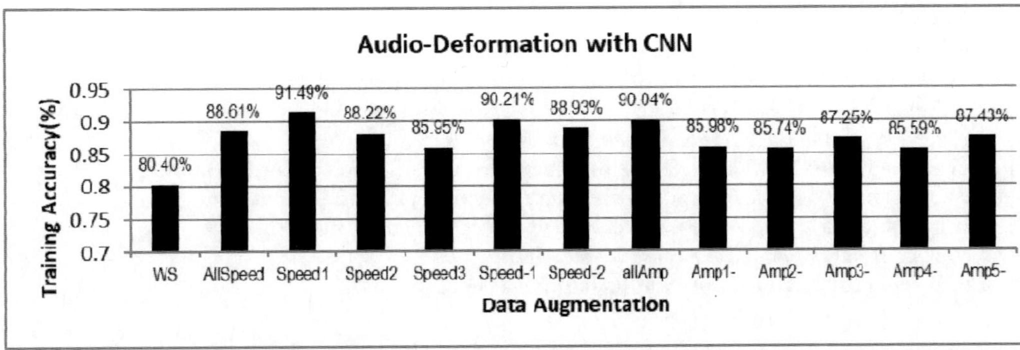

Figure 2. Comparison of CNN training accuracy with different factor of proposed approach.

The result shows that the proposed approach achieves higher training accuracy for all signals compared with window sliding (WS) DA due to ability of proposed approach to producing additional number and variant of dataset. The analysis has shown that the additional dataset using audio-deformation based DA has improved the efficiency of vibration analysis.

The result also shows that the fusion factor of speed deformation based DA is better than the fusion factor of amplify deformation based DA. However, a single speed deformation based signals perform better compared to the combination of all factor where the highest accuracy is achieved by speed with small factor either in positive or negative value. It is indicates that a fusion with various factor cannot guarantee a better result than the single factor.

4. Conclusion

In this paper, the audio deformation based DA for training CNN was proposed for vibration analysis in bearing fault detection. The idea is inspired from advantages of audio processing preserves sound meaning in various tone. Audio deformation is performed at raw data level to increase the CNN training accuracy by combining the raw signals to form the input of the CNN-based model.

Experimental studies on 10 kind of bearing defects verified the bearing fault analysis performance using vibration signal of the proposed approach for fault classification. The comparison between the proposed method and the traditional approaches shows that the proposed CNN-based method could achieve higher and more reliable diagnosis performance.

In addition, the capabilities of the proposed approach will allow for its extensive application in signal processing of various types of signals despite have limited in data sources and data generator methods. In the future, the fusion of different audio transformation will be considered in the analysis.

References

[1] Barszcz T *et al* (2018) *Advances in Condition Monitoring of Machinery in Non-Stationary Operations* (Springer,Cham) 235-224
[2] Zhou L *et al* (2019) *Appl. Acoustics* **147** 4-14
[3] Shao H *et al* (2018) *IEEE Trans. Ind. Electron.* ***65***(3) 2727
[4] Lu S *et al* (2017) *Mech. Sys. and Signal Process.* **85** 82
[5] Tandon N *et al* (1999) *Tribology Int.* **32** (8)
[6] Park C S (2013) *et al* (2013) *Mech. Syst. Signal Process.* **38**(2) 534–548
[7] Howard A (2013) Some improvements on deep CNN based image classification arXiv preprint arXiv:1312.5402
[8] Zhang W *et al* (2017) *Sensors* **17**(2) 425
[9] Guennec L A *et al* (2006) *ECML/PKDD workshop on advanced analytics and learning on temporal data* (Italy:HAL) 1-9
[10] LeCun Y *et al* (1990) *Adv. in Neural Inf. Process. Sys.* **2** 396–404

[11] Van Dyk D A *et al* (2001) *J. of Comp. and Graphical Stat.* **10**(1) 1-50.
[12] Rebai I *et al* (2017) *Procedia Comp. Sc.* **112** 316-322
[13] Russakovsky O *et al* (2016) *Adv. in Neural Inf. Process. Sys.* 3108
[14] He K *et al* (2014) *Comp. Vision ECCV 2014* (Berlin: Springer) 346
[15] Krizhevsky A *et al* (2012) *Adv. in Neural Inf. Process. Sys.* 1097
[16] Cui Z *et al* (2016) Multi-scale CNNs for time series classification arXiv:1603.06995
[17] Justin S *et al* (2017) *IEEE Signal Processing Letters* **24**(3) 279-283
[18] Ko T *et al* (2015) *16th Ann.l Conf. of the Int. Speech Comm. Ass.* 3586-3589
[19] Lemley J *et al* (2017) *IEEE Access* **5** 5858-5869.
[20] Lu W *et al* (2016) *IEEE Trans. Ind. Electron.* **64**(3) 2296-2305

IOP Conf. Series: Materials Science and Engineering **551** (2019) 012067 doi:10.1088/1757-899X/551/1/012067

Electromagnetic control of fluid with magnetic particles in the stenosed region

Salah Uddin[1,*], Mahathir Mohamad[1], Mghazali Kamardan[1], Mahmod Abd Hakim Mohamad[2], Suliadi Sufahani[1] and Roslan Rozaini[1]

[1]Department of Mathematics and Statistics, Faculty of Applied Sciences and Technology, Universiti Tun Hussein Onn Malaysia, Pagoh Educational Hub, 84600 Pagoh, Johor, Malaysia.
[2]Faculty of Mechnical and Manufacturing Engineering, Universiti Tun Hussein Onn Malaysia, Pagoh Educational Hub, 84600 Pagoh, Johor, Malaysia.

Email: salahuddinjalil@gmail.com

Abstract. Hydrodynamic flow of Newtonian blood mixed with magnetic particles (red blood cells) through stenosed region is studied in this paper. The fluid is acted upon by an oscillating pressure gradient in a perpendicular magnetic field. Governing fractional momentum equations are analytically solved for axial velocities. Graphical analysis of velocity profiles against various physiological parameters is developed by coding in Mathematica. Effect of Caputo fractional operator α on axial velocities is also considered. Results show that blood and magnetic particles has increasing effect with maximum height of the stenosis and electric field, whereas, it has a decreasing effect with perpendicular magnetic field. Hence, the flow of blood can be controlled by applying sufficient and strong electric and magnetic field. This is an important fact in medical diagnosis and clinical treatment of various cardiovascular diseases, hypertension and atherosclerosis.

1. Introduction

Vascular fluid dynamics assume a significant share in the advancement of blood vessel illnesses. Local narrowing/stiffening in the lumen of blood vessel section is usually alluded as stenosis. This happens due to deposition of various fatty materials like cholesterol on the endothelium of arterial wall. When a blockade is developed in an artery, one of the most severe consequences is the expanded obstruction and the related fall of the blood stream to the specific vascular bed provided by the artery.

In the comparative study of [1], various simulations were performed for Newtonian, non-Newtonian fluid and Reynolds number. In the stenotic region non-Newtonian blood flow model was found to be laminar and transient flow was observed for Newtonian blood flow model. By using certain assumptions in [2] and [3], the Caputo-Fabrizio fractional order $1<\beta\leq 2$, linear differential equations were transformed to differential equations with integer order and solved for explicit forms. Application towards the mass-spring-damper motion of the obtained result was found in the described literature. In [3], it was observed that with shape parameter n, volumetric flow rate increases. Moreover, in the diverging region, high variation in volumetric flow rate was detected for an inclined artery with magnetic field angle θ, in addition to it, opposite behavior was observed in the converging region.

Under the action of an external magnetic field and oscillating pressure gradient blood flow mixed with magnetic particles in [4] and [5] was reduced. [6] analyzed that for Reynolds number greater than

1000, flow moves towards turbulence downstream and frequency is connected with the periodic vortex formation. [7] used OpenFOAM programming to compare the influence of his elastic analog blood flow modeling with the analytic solution. [8] model is helpful for physicians in the severe stenosis case and can be extended to more severe case by introducing further rheological and physical parameters. In [9] mathematical simulation was effective in cardiovascular diseases and biomedical engineering.

In the current paper, new definition of Caputo-Fabrizo fractional derivative without kernel (NFD_t) is employed in the governing momentum equations and the resulting equations are obtained by coding in computer software Mathematica.

2. Mathematical Model

We study an in-compressible, laminar and unsteady flow of Newtonian fluid (human blood) mixed with magnetic particles (red blood cells being rich in iron) which is uniformly distributed in an axisymmetric but radially non-symmetric stenosed porous artery. Flow is driven by applying external pressure gradient in the electro-magnetic field.

Blood and magnetic particles are assumed to be flowing in the axial direction (z-axis) under the effect of an exterior electric field, whereas, the magnetic field acts upon perpendicularly to the circular channel of radius R_0 (as shown in the Fig. 1) and δ is the maximum height of the stenosis. Initially blood and magnetic particles are at rest. For $t > 0$ the electrolytic conducting solution is forced by an electromotive force, which depends on blood and magnetic particles flow and strength of electro-magnetic forces as well.

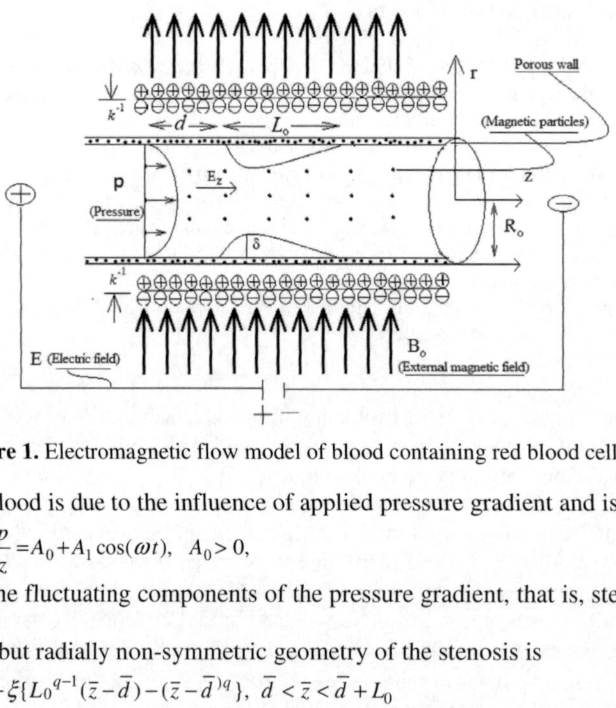

Figure 1. Electromagnetic flow model of blood containing red blood cells

Unsteady flow of blood is due to the influence of applied pressure gradient and is given by

$$-\frac{\partial p}{\partial z} = A_0 + A_1 \cos(\omega t), \quad A_0 > 0, \tag{1}$$

where A_0 and A_1 are the fluctuating components of the pressure gradient, that is, steady and amplitude pulsatile components.

Axially symmetric but radially non-symmetric geometry of the stenosis is

$$\frac{R_z}{R_0} = \begin{cases} 1 - \xi\{L_0^{q-1}(\bar{z} - \bar{d}) - (\bar{z} - \bar{d})^q\}, & \bar{d} < \bar{z} < \bar{d} + L_0, \\ 1, & otherwise \end{cases} \tag{2}$$

where $\xi = \dfrac{\delta q^{\frac{q}{q-1}}}{R_0 L_0{}^q (q-1)}, \dfrac{\delta}{R_0} < 1 \text{ and } \bar{z} = \dfrac{\bar{d} + L_0}{q^{\frac{1}{q-1}}}$. R_z and R_0 are the radius of the artery with and without stenosis respectively, L_0 is the length of the stenosis and $\bar{d}$ indicates its location, $q \geq 2$ is the stenosis shape parameter.

Introducing the following dimensionless variables with * notations,

$$A_0{}^* = \frac{\lambda A_0}{\rho u_0}, A_1{}^* = \frac{\lambda A_1}{\rho u_0}, M^* = \frac{u_0 \sigma B_0{}^2 \lambda}{\rho}, \mathrm{Re}^* = \frac{R_0{}^2}{\lambda \upsilon}, R^* = \frac{K N \lambda}{\rho}, G^* = \frac{m}{K \lambda}, K^{2*} = \frac{\varepsilon k^2}{\rho}. \quad (3)$$

Governing non-dimensional fractional momentum equations for the blood and magnetic particles flow under electro-magnetic forces and applied pressure gradient with porous stenotic region in cylindrical coordinates system, (r, θ, z) is given by (after dropping dashes)

$$D_t{}^{(\alpha)} u = A_0 + A_1 \cos(\omega t) + \frac{1}{\mathrm{Re}}\left(\frac{\partial^2 u}{\partial r^2} + \frac{1}{r}\frac{\partial u}{\partial r}\right) + R(v-u) - M\,u - K^2 \psi(r), \quad (4)$$

where $D_t{}^{(\alpha)} f(t) = \dfrac{M(\alpha)}{1-\alpha}\displaystyle\int_a^t \frac{\dot{f}(\tau)}{\exp\left(\frac{\alpha(t-\tau)}{1-\alpha}\right)} d\tau, \alpha \in [0,1] \text{ and } a \in [-\infty, t), f \in H^1(a,b), b > a \text{ and } M(0) = M(1) = 1$ (condition of normalization), m is the average mass of magnetic particles, $\lambda = \sqrt{\dfrac{R_0 \rho}{A_0}}$ has a dimension of time t, u_0 is the characteristic velocity, Re is the Reynolds number, R is the particle concentration parameter, G is the particle mass parameter, B_0 is the magnetic field strength, ρ is the fluid density, p is the fluid pressure, N is the number of magnetic particles per unit volume, u is the blood velocity, v is the magnetic particle's velocity, E_z is the electric field axial component, σ is the electrical conductivity, K is the stokes constant, υ is the kinematic viscosity, ρ_e is the net charge density.

According to Newton's second law of motion, magnetic particles movement can be govern by

$$G D_t{}^{(\alpha)} v = u - v. \quad (5)$$

Non-dimensional initial and boundary conditions are

$$\begin{cases} u(r,0) = 0 \text{ and } v(r,0) = 0 \text{ at } r = 1 \\ u(R_0,t) = 0 \text{ and } v(R_0,t) = 0, t > 0 \end{cases}. \quad (6)$$

Using Laplace and finite Hankel transforms on Eq. (3 & 5) together with Eq. (6), final velocity profiles for human blood and magnetic particles (red blood cells) has been obtained. Furthermore, analysis has also been made by coding in computer software Mathematica, where positive root of the Bessel function $J_0\left(r\dfrac{R_z}{R_0}\right)$ has been calculated.

3. Results and Discussion

The main aim of the study is to analyze the blood and iron-rich particles flow patterns due to externally applied electro-magnetic field through stenotic porous artery by means of the new approach of Caputo-Fabrizio fractional derivative (NFD_t) without singular kernel. For this purpose few simulations has been made in the local and non-local model against the fixed values given by $A_0 = 5, A_1 = 0.6, G = 0.5, \mathrm{Re} = 2, \omega = \frac{\pi}{4}, t = 0.2, r = 0.3$.

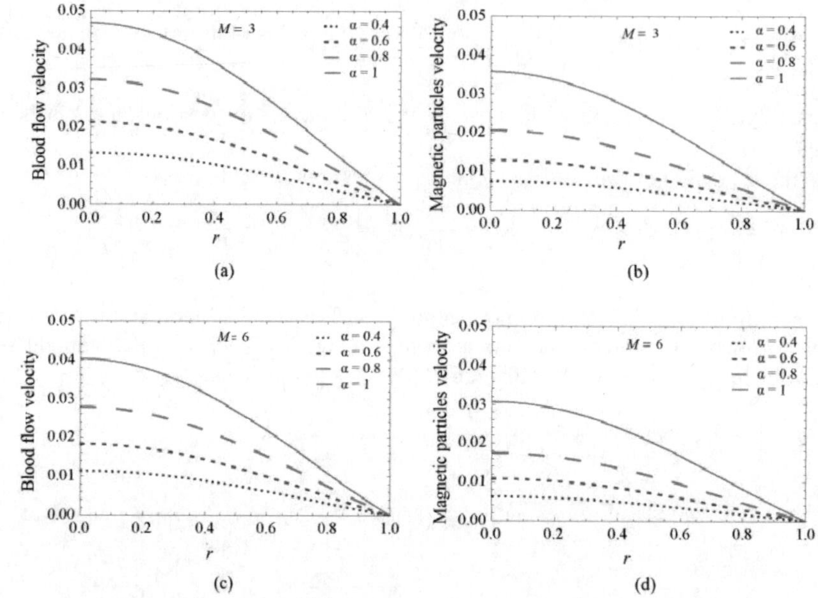

Figure 2. Velocities for different values of magnetic field M against r

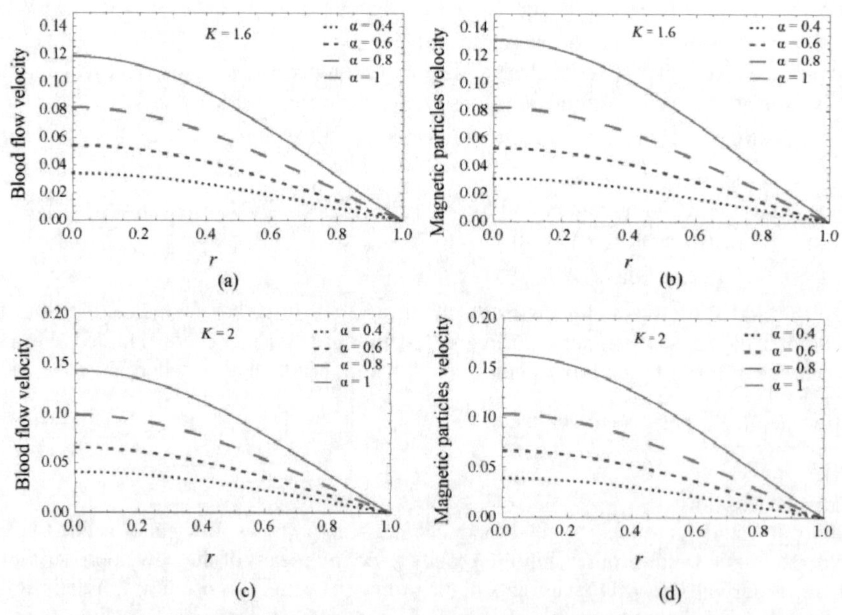

Figure 3. Velocities for different values of electric field K against r

In Fig. 2, axial velocity for blood and magnetic particles is decreased by enhancing the external magnetic field M. However, during the flow, blood moves faster than magnetic particles due to the drag and other retardation forces.

In Fig. 3, the external electric field at different strength is applied against *r*. Both the velocities has increasing effect with *K*. Moreover, particles flow is enhanced due to collision between the charged particles.

4. Conclusions
From the research, we reach at few conclusions which are;

- External electric field has an increasing effect on the velocity profiles, for $\alpha \leq 0.8$, particles velocity is found to be almost equal to blood.
- Magnetized solution decelerates with external magnetic field. Furthermore, magnetic particles move slower than blood.

5. Acknowledgements
The authors would like to acknowledge the financial support received from the Universiti Tun Hussein Onn Malaysia (Grant no: Tier 1/H158).

References
[1] Rabby A M G, Razzak A and Molla M 2013 *Procedia Engineering* 56.
[2] Al-salti N, Karimov E and Sadarangani K 2016 *Progr. Fract. Differ. Appl* 2.
[3] Uddin S, Mohamad M, Sufahani S, Kamardan M G, Mehmood O U, Roslan R 2018 *A letter on Applications of Mathematics and Statistics* 119-127.
[4] Srivastava N 2014 *Journal of Biophysics.*
[5] Ali N, Vieru D and Fetecau C 2016 *Journal of Magnetism and Magnetic Materials* 409.
[6] Uddin S, Mohamad M, Khalid K, Abdulhammed M, Rusiman M, Che-Him N and Roslan R 2018 *Journal of Physics: Conference Series* 995.
[7] Mittal R, Simmons S P and Najjar F 2003 *J. Fluid Mech* 485.
[8] Pinto S I S, Doutel E, Campos J B L M and Miranda J M 2013 *APCBEE Procedia* 7.
[9] Siddiqui S U and Shah S R 2015 *Applied Mathematics and Computation* 261.
[10] Rathod V P and Ravi M 2014 *International Journal of Research in Engineering and Technology.*

IOP Conf. Series: Materials Science and Engineering **551** (2019) 012068 doi:10.1088/1757-899X/551/1/012068

Support Vector Machine with Principle Component Analysis for Road Traffic Crash Severity Classification

N.H.M. Radzi[2], I.S.B. Gwari[2,] N.H. Mustaffa[3], R. Sallehuddin[4]

[1,3,4]School of Computing, Faculty of Engineering, Universiti Teknologi Malaysia.
[2]Department of Mathematical Sciences, Kaduna State University, Kaduna. Nigeria.

Email : haizan@utm.my

Abstract- Road traffic crash (RTC) is one among the leading causes of death in the world, including Nigeria. It also turns many victims completely disabled and generally affected the socio–economic development in the society. In this paper, we proposed to predict the road crash severity injuries in Nigeria by identifying the most significant contributory factors using Principal Component Analysis with Support Vector Machine (SVM) used for classification algorithm. Road crash data from year 2013-2015 obtained from Federal Road Safety Corps Nigeria is used in this study. The result shows that and increased to 87% compared to 82% without feature selection.

1. Introduction

Road Traffic Crash (RTC) is ranked as the ninth cause of death in the world [2]. RTC also gives significant impact to the socio-economic to the victim or the family as it incurs additional overhead cost for medical expenses and care taker. RTC is also contributing up to 3% lost in gross domestic product (GDP) in the country [6]. In Nigeria, there were 13,583 cases of RTC in 2013 with 6,544 people killed recorded by the Federal Road Safety of Nigeria [13]. Therefore, identification of significant contributory factors of RTC and the causes of each crash severe injury (fatal, serious and minor) is necessary to provide bases for rules and regulation enforcement [1].

The range of factors that contribute to RTC severity is very wide and different from one country to another. This is due to the different behaviors of the road safety awareness and also road infrastructure. Basically, the causes of RTC can be divided into three categories: human, environment and vehicle[3]. Further, each of these categories consists of several other factors. Therefore, identification and generalization of causes of RTC globally is very difficult [3]. In this study, we attempt to use principle component analysis (PCA)to identity the most significant factors and support vector machine (SVM) to classify the RTC severity in Nigeria.

2. Related Works

A considerable amount of studies on RTC severity classification has been conducted by many of researchers including in [3-8, 10]. Most of researchers used single contributory factor such as either features in human, environmental or vehicle. But, there are some researchers combine the features in environment and vehicle categories to classify the road crash and only a few [1,5, 8] combined the features in all three categories. Combination of all categories, however, may affect the accuracy of the classifier, as it will increase the dimension of the input. Therefore, researchers used various method to assess significant features contribute to RTC. In work [4], PART algorithm is used to identify the crash severity injuries significant features. Fuzzy granular decision tree was used in model proposed in [6]to classify and predict road crash into property damage, injury and fatal classes. The model in [6] used 7features from the environment category which are obtained from 1837 crash cases collected from 13 regions in California. Work in [7] classify the RTC into two classes: injury and fatal using crash data from Slovenia between year 2004 to 2009 with 9 features and used variables importance measure (VIM)to predict the most contributory factors to RTC. Work in [8] also used (VIM) in classification and regression tree (CART). Besides this, another method called Principle component analysis (PCA) is also promising for feature selection. Fengxi[11] used PCA to select significant features in face recognition. PCA also used in [12] and increase the accuracy in the classification algorithm. Therefore, this work attempt to use PCA to select most significant features contribute to RTC severity.

3. Methods

3.1. Principal Component Analysis

PCA learning algorithm will compute the covariance matrix and find all the eigenvalues and eigenvectors. The eigenvectors will be selected corresponded with the first m large eigenvalues as eigenvectors = $\{V_1 \cdots V_m\}$. Calculate the feature extraction result using the jth feature component, $C_j = \sum_{p=1}^{m} |V_{pj}|$, where $v = \{1,2 \cdots, N\}$, $p = \{1,2, \cdots m\}$, $|V_{pj}|$ is the absolute value of V_{pj}. The feature selection result will be the d_{1th}, d_{2th}, ... d_{nth} of the feature component[11]. The used of PCA feature selection will select the most important variables from all the large variables in the dataset.

3.2 Support Vector Machine

Support Vector Machine (SVM) is a non parametric machine learning algorithm proposed in 1990s by VN Vapnik. The input points are firstly maps into a high-dimensional feature space and then find a separating hyper-plane that maximizes the margin between two classes. The kernel functions are used in the training stage to select the support vectors with the surface of the function [9]. Given training data set (x_i, y_i), where x_i is a real valued n-dimension input vector and y_i is the corresponding target output. The SVM decision function learning algorithm is based on the following formula,

$$F(X) = \sum_{i=1}^{n} Y_I \alpha_i K(X_i, X) + b, \tag{1}$$

where ,

$K(X_i, X)$ - kernel function
n - number of input
b - bias

4. Data Description

This study used 641 crash data recorded from year 2013 to 2015. It was obtained from Federal Road Safety Corps of Nigeria Birnin Yero Unit Command (an agency that regulate safety, rescue and road crash activities) [1]. The data contains11 attributes from human, 6 from environment and 7 from vehicle categories. The target class defined in this study are:fatal, serious and minor[1]. Table 1 shows the description of each attribute. The data is divided into training and testing where 80% used for training and the remaining 20% for testing. The experiment is conducted using 10 – fold cross validation techniques In eachfold, the data set was split into 10 parts; 9 were used for training and 1 part for testing.

5. Experiments and Result

The PCA features selection algorithm in WEKA tools was used to extract the most relevant features in the classification of road traffic crash from the 24 attributes as describe in table 1. PCA firstly calculate the covariance matrix from the original data to find the eigenvalues and eigenvectors of each attributes and ordered them according to the eigenvalues. The eigenvector with highest eigenvalue will be the principal component of the new data set. PCA features selection generated 22 new attributes as in table 2 and ranked according to the eigenvalues.

These new22 attribute are then used as input features for SVM classifier. The RBF kernel functionis selected for training function in SVM with optimal values of gamma, γ and cost, C parameters obtained from experiment conducted in the [1]. We then run the experiment with various numbers of features. Table 3 shows the classifier accuracy with new attributes obtained from PCA with different number of features. The accuracy of the classifier is at the highest, which is 80%, whenever 15 features is chosen as the attributes.

Table 1: RTC contributory factors attributes

Attribute Name	Contributory Factors	Description
Crash Date	Environmental	The date that the crash occur
Crash Time	Environmental	An approximate time of the crash
Response Time	Human	The exact time the rescue teams attend to the crash victim.
Location	Environmental	The approximate location of the crash in the highway
Vehicle Type	Vehicle	The types of vehicle involved e.g. car, bus, lorry, etc.
Vehicle Category	Vehicle	categories of the vehicles e.g. Private, Commercial etc.
Vehicle Make	Vehicle	The vehicle make e.g. Honda, Toyota, etc.
Over Speeding	Human	The driver is over speeding
Under Alcohol/Drug	Human	The driver was under influences of alcohol/drug while driving
Dangerous Driving	Human	Dangerous driving of the vehicle by the driver.
Lost of Control	Human	The driver lost control, road crash occurs by hitting another vehicle or the vehicle move out of road.
Bad Road	Environmental	Bad road condition or a black spot, pothole, sharp bed etc.
Road Obstruction	Environmental	Obstructions on the highway
Poor weather	Environmental	Poor weather condition in the area like during rainy, haze, etc.
Sleeping on steering	Human	The driver is sleeping while driving.
Use phone	Human	Driver using phone while driving
Wrongful overtaking	Human	Overtakes in a corner, sharp bed without seeing his front.
Brake failure	Vehicle	Failure of the brake in the vehicle.
Mechanical deficiency	Vehicle	Vehicle has mechanical deficiency.
Overloading	Human	Overloading with either passenger or load of the vehicle
Sign light violation	Vehicle	Light sign in the vehicle is not proper working.
Route Violation	Human	Driver violate route on the highway.
Dangerous Overtaking	Human	Dangerous overtaking by the driver
Tire burst	Vehicle	Flat tires in the vehicle.

IOP Conf. Series: Materials Science and Engineering **551** (2019) 012068 doi:10.1088/1757-899X/551/1/012068

Table 2. New attributes

Sn	Eigenvalue	Proportion	Cumulative	New Attributes
1	1.57728	0.06572	0.06572	-0.578VEHICLE MAKE...
2	1.39531	0.05814	0.12386	0.509RESPONSE TIME ...
3	1.3049	0.05437	0.17823	-0.471SPEED VIOLATION...
4	1.22021	0.05084	0.22907	-0.527VEHICLE CATEGORY...
5	1.19383	0.04974	0.27881	0.42DANGEROUS DRIVING...
6	1.15863	0.04828	0.32709	-0.393TYRE BURST...
7	1.1287	0.04703	0.37412	-0.502LOST OF CONTROL...
8	1.08209	0.04509	0.41921	0.638ROUTE VIOLATION...
9	1.07585	0.04483	0.46403	-0.63WRONFUL OVERTAKING...
10	1.06432	0.04435	0.50838	-0.679MECHANICALLY DEFICIENT VEHICLE...
11	1.04029	0.04335	0.55172	0.486USE OF PHONE ON DRIVING...
12	1.03163	0.04298	0.59471	-0.409SIGN LIGHT VIOLATION...
13	1.02432	0.04268	0.63739	0.602 VEHICLE TYPE...
14	1.01856	0.04244	0.67983	-0.452 CRASH TIME...
15	1.01595	0.04233	0.72216	-0.595BAD ROAD....
16	1.01286	0.0422	0.76436	0.522POOR WEATHER...
17	0.95949	0.03998	0.80434	-0.527DRUNG ALCOHOL DRIVING...
18	0.92037	0.03835	0.84269	0.46 BRAKE FAILURE...
19	0.89747	0.03739	0.88008	-0.458VEHICLE CATEGORY...
20	0.86295	0.03596	0.91604	-0.57CRASH DATE...
21	0.7927	0.03303	0.94907	-0.59 DRUNG ALCOHOL DRIVING...
22	0.70243	0.02927	0.97834	-0.633LOCATION...

Table 3. New Attributes Accuracy

Number of Features	Accuracy
22	68%
20	69%
18	69%
16	71%
15	80%
14	69%
13	69%

We then used these 15 features for our PCA-SVM model and the result obtained as in Table 4. In the training phase the PCA-SVM has correctly classified 85% serious, 72% fatal and 36% minor while in testing dataset 96% serious, 85% fatal and 54% minor are correctly classified.The general performance prediction classification accuracy of the proposed build model is 77% for training set and 89% for testing.

Table 4. Classification Prediction Result

CLASS	Training Data		Testing Data	
	Correct Classified	Incorrectly Classified	Correctly Classified	Incorrectly Classified
MINOR	16	28	6	5
SERIOUS	284	47	80	3
FATAL	100	38	29	5
	400	113	115	13

The significant of SVM classifier performance in this paper was demonstrated using the commonly used measures in the literatures [1, 3, 5, 12] of accuracy, precision, recall and F-measure and compare with [1]. The general accuracy of the model classifier is 82%, serious class in the model has the highest values of precision, recall, f-measure; 77%, 86% and 81% respectively, while minor class accuracy outperform the other classes with 92%.

Table 5. Summary of experiment performance

	LIBSVM				PCA-LIBSVM			
	Precision	Recall	F – Measure	Accuracy	Precision	Recall	F – Measure	Accuracy
MINOR	65%	56%	60%	80%	78%	75%	76%	88%
SERIOUS	77%	86%	81%	75%	83%	88%	85%	81%
FATAL	56%	33%	41%	92%	55%	40%	46%	92%

Table 5 illustrates the comparative result of the proposed PCA model with the standard LIBSVM model with the 24 input attributes of road crash contributory factors it shows clearly that the proposed PCA –SVM with 15 most significant factors of road crash performed better in precision, recall and *f* measure for both fatal, serious and minor crash class. As early mention the accuracy rate of the proposed model presented are highly recommended for classification and prediction of road traffic crash severe injuries using principal component analysis PCA feature selection method to identify the significant contributory factors.

6. Conclusion

This research develops a PCA-SVM model that would provide an insights and helping road safety agency in Nigeria to predict and classify the RTC severity injuries. It also identified the most road contributory factors to help the policy makers in constructing a good roads with proper road sign and adequate regulation of highway rules to the road users for better traffic safety control policies.

After a thoroughly review of many literature in road traffic crash classification analysis a research gap was identified of using three main contributory factors of road crash data for severe injuries classification and prediction in Nigeria. In this study we collected a raw data from Federal Road Safety Corps Nigeria to clean, preprocess and construct a predictive model for crash severe injuries identification.

References

[1] Radzi N H M *et al* 2017 *International Journal of Innovative Computing, 7*(1) 15-18
[2] Organization, W.H. 2015 Global Status Report on Road Safety 2015 retrived from https://www.who.int/violence_injury_prevention/road_safety_status/2015/en/.
[3] Castro Y *et al* 2016 International Journal of Crash worthiness **21**(2) 104-111.
[4] Beshah T *et al* 2010 *AAAI Spring Symposium 2010: Artificial Intelligence for Development*
[5] Beshah T *et al* 2012 International Conference on Intelligent Networking and Collaborative Systems (INCoS)
[6] Moghaddam H K *et al* 2014 The 9th International Conference on Rough Sets and Current Trends in Computing Springer 280-287
[7] Rovšek V M 2017 Transport **32**(3) 272-281.
[8] Tavakoli K *et al* 2011 PROMET-Traffic&Transportation **23**(1) 11-17.
[9] Hsu C W *et al*, *Tech Rep* Department of Computer Science, National Taiwan University
[10] Beshah T *et al* 2013 *Int Journal of Tomography and Simulation* **22**(1) 73-86.
[11] Song F *et al* 2010 International Conference on System Science, Engineering Design and Manufacturing Information (ICSEM)
[12] Xu Y *et al* 2010 *Pattern recognition* **43**(3) 1106-1115
[13] Sheriff M 2009 Nitours Journal **2**.

Exclusion Radius Particle Swarm Optimizer

Alvin Soo Chun Kit[1], Haza Nuzly Abdull Hamed[1] , Mohd Adham Isa[2]

[1]Applied Industrial Analytics Research Group, Universiti Teknologi Malaysia, 81310 UTM Johor Bahru, Malaysia

[2]Software Engineering Research Group, Universiti Teknologi Malaysia, 81310 UTM Johor Bahru, Malaysia

Email: alvinsoock@gmail.com, haza@utm.my

Abstract. Function optimization is a problem that has existed since modern engineering began. With multiple variables, brute-forcing or simple regression models become unfeasible. This paper proposes a shrinking population control method for a dynamic population particle swarm optimizer. The proposed control mechanism creates an exclusion zone, where particles are periodically purged from the simulation. The selection of particles purged relies on their particle best i.e. the position where it found its personal best fitness. Particles with their particle best outside of the exclusion zone with a predetermined radius from the current global best are removed from the simulation. The purging is done periodically in stages of equal iteration counts, and the radius is shrunk by a factor after every stage. By testing 5 benchmark mathematical functions, the results show that the proposed population control mechanism achieves better final optima than only a basic particle swarm optimizer and a simple shrinking dynamic population particle swarm optimizer where the worst performing particle is removed every so often.

1. Introduction

Particle Swarm Optimization (PSO) has been one of the earliest optimization methods. It was first introduced by Kennedy and Eberhart [1] to solve a mathematical function which has now become ways to evaluate the performance of PSO. Many adjustments to the original algorithm has been made since then, such as the addition of ω as an inertial weight [2] and clamping of v_{max} [3] .Due to its simplicity in implementation, it is relatively fast to execute, and as such is rather popular as a way to solve optimization problems. It does have its drawbacks, however; as it cannot guarantee that a globally optimal solution is found. PSO is heavily dependent on the initial conditions, such as the search space and the control variables used to encourage exploration and exploitation respectively. Furthermore, PSO faces issues when used to solve higher dimensional problems [4]. As such, this work proposes a variant of Dynamic Population Particle Swarm Optimization, where a shrinking population control mechanism is used.

2. Experiments

The proposed approach still uses the basic Particle Swarm Optimizer algorithm as its base. As such, the following subsections discuss the parameters inherent to a basic PSO, and the modifications made in the proposed approach.

IOP Conf. Series: Materials Science and Engineering **551** (2019) 012069 doi:10.1088/1757-899X/551/1/012069

2.1. Base PSO Parameters

The base parameters for the PSO are as follows: c_1 and c_2 are set to 2.0, ω is set to scale linearly as the iteration count increases, from 0.9 at $t = 0$ to 0.7 at $t = t_{max}$ based on work by Shi and Eberhart [2] .The initial particle count is 20, while the maximum iteration count is set to 2000.

Velocity and position clamping is used in the proposed approach. The equations used are as detailed by Shi and Eberhart [2]. The position clamping is only used on problems with constrained search spaces. If the search space is deemed infinite, then $x_{min}$and x_{max} will be set to -IntegerMAX and IntegerMAX respectively.

2.2. Exclusion Radius PSO

This subsection details a shrinking population control mechanism, hereby referred to as Exclusion Radius Particle Swarm Optimizer (ERPSO). Exclusion Radius is used here to fence off an area that is safe from deletion. Particles outside of the exclusion radius at the time of purging (stage boundary).

The simulation is first separated into stages. For each stage, the following variables are set and initialized.

- Maximum deleted particles

$$pDel_{max} = \frac{p_{initial}}{stageCount} \tag{1}$$

- Exclusion radius (per dimension)

$$excludeRadius = 0.5 * shrinkFactor^{(phase_{count}-1)} * \frac{(searchRange_{max} - searchRange_{min})}{2} \tag{2}$$

- Shrinking factor, $shrinkFactor$

The exact values for used in the calculation of exclusion radius and shrinking factor may differ according to the problem. A grid search was performed during testing, and in general values lesser than 0.5 for the exclusion radius multiplier in Equation (2), and values greater than 0.5 for the shrinking factor itself were optimal.

A function to calculate the normalized distance between two particles is also defined:

$$\boldsymbol{distance} = \sqrt{\sum_{d=1}^{dim}\left(\frac{\boldsymbol{p_1.position[d] - p_2.position[d]}}{\boldsymbol{searchRange_{max}[d] - searchRange_{min}[d]}}\right)^2} \tag{3}$$

```
Initialize population
FOREACH iteration
     IF (iteration == stage boundary)
          FOR each particle in population
                Calculate pBest distance from gBest
          END FOR

          Remove numDel particles with distance > exclusion radius
     END

     Evaluate fitness of population
     IF (fitness > gBest)
         Update gBest
     END IF
     Update velocity of population
     Update position of population
END FOREACH
```

Figure 1. Pseudocode for the proposed ERPSO method.

IOP Conf. Series: Materials Science and Engineering **551** (2019) 012069 doi:10.1088/1757-899X/551/1/012069

In each stage, the simulation is allowed to run as is typical. At the end of each stage, the exclusion safety zone (shown as the shaded circle in Figure 2) is determined. This exclusion safety zone shrinks after each stage in accordance to the exclusion radius formula in Equation (2).

Figure 1 shows the general flow of the algorithm. At the stage boundary, the entire population is evaluated by calculating the distances of their p_{best} from the current global best (using Equation (3)), and removing those that are outside of the exclusion safety zone. If there are more particles outside of the exclusion safety zone than $pDel_{max}$, whichever particles the simulation finds first will be deleted. This is effectively a random removal method of the targeted particles as the particles are not ordered in any particular pattern.

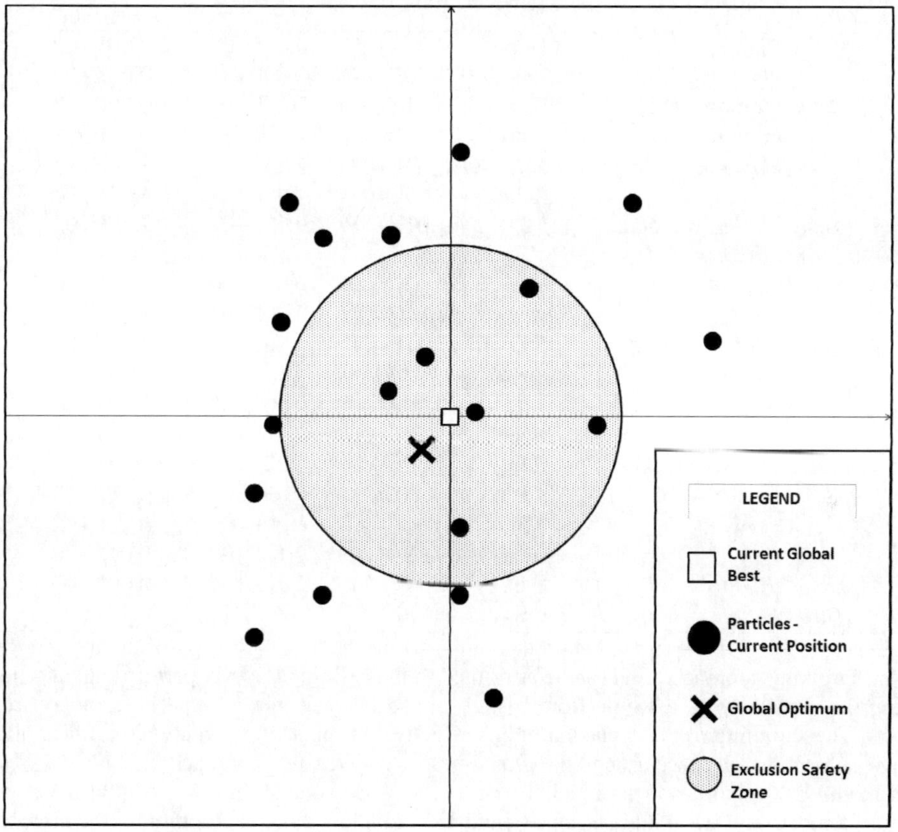

Figure 2. 2-dimensional representation of ERPSO in action.

3. Benchmark Evaluation

The proposed framework is evaluated based on the average (over 30 runs) best fitness value found. This is to mitigate the effect of "lucky" runs and is recommended in the work done by Uriarte, Melin [5]. A lower fitness value (closer to 0) indicates a better performing algorithm.

All of the evaluation functions used are benchmark optimization functions as seen in various other works concerning function optimization [4-8] and have multi-dimensional forms. This paper uses the Sphere, Rastrigin, Schwefel, Ackley and Griewank functions as benchmarks. These particular functions have been selected due to the ease of implementation and encompass a variety of shapes in the solution space, allowing for more varied testing conditions. The 30-dimension variants of these functions are used in this work, and are otherwise unchanged.

4. Results and Discussion

Table 1 shows the comparison between the 3 tested approaches. Base PSO refers to the basic PSO algorithm with linearly decreasing inertial weight as first proposed by Shi and Eberhart [2]. Remove Worst One refers to a shrinking population control mechanism, where the worst performing particle (as determined by current position from global best) is removed after each stage. Proposed - ERPSO is the proposed approach of this paper. Table 1 lists the comparison between a basic PSO approach, a shrinking (remove worst one) approach, and the proposed (ERPSO) approach.

Table 1. Results Comparison.

Problem	Metric	Base PSO	Remove Worst One	Proposed - ERPSO
Sphere	Min	0.00E+00	0.00E+00	0.00E+00
	Mean	1.50E+01	3.00E+01	0.00E+00
excludeRadius = 0.20	S.D.	5.71E+01	7.78E+01	0.00E+00
shrinkFactor = 0.33	Hit Rate	90%	76.67%	100%
Schwefel	Min	2.69E+03	2.79E+03	2.37E+03
	Mean	4.60E+03	4.84E+03	3.99E+03
excludeRadius = 0.50	S.D.	1.01E+03	9.62E+02	7.69E+02
shrinkFactor = 0.75	Hit Rate	-	-	-
Rastrigin	Min	0.00E+00	0.00E+00	0.00E+00
	Mean	4.80E+01	8.80E+01	5.51E+01
excludeRadius = 0.20	S.D.	3.41E+01	4.86E+01	4.29E+01
shrinkFactor = 0.75	Hit Rate	20%	3.33%	10%
Ackley	Min	0.00E+00	0.00E+00	0.00E+00
	Mean	1.59E+00	2.31E+00	9.59E-02
excludeRadius = 0.35	S.D.	4.60E+00	5.22E+00	3.71E-01
shrinkFactor = 0.60	Hit Rate	80%	80%	90%
Griewank	Min	0.00E+00	0.00E+00	0.00E+00
	Mean	9.08E+00	9.10E+00	0.00E+00
excludeRadius = 0.15	S.D.	2.77E+01	2.76E+01	0.00E+00
shrinkFactor = 0.50	Hit Rate	90%	86.67%	100%

As seen in Table 1, the proposed approach outperforms the basic PSO approach, and the Remove Worst One shrinking approach. This can be attributed to the culling of poorly performing particles, and preventing the simulation as a whole from being trapped in local minima that is far away from the global best. The shrinking of the exclusion radius effectively funnels the remaining particles into fully exploring the local search space around the global best. However, this approach may have a weakness in that it might lock-on too early to a global best and lead the simulation towards other local optima. This can be combated by tweaking the stage boundaries, to allow more exploration of the larger search space in the beginning.

5. Conclusion

From the research, we can conclude that the proposed approach performs better in terms of final fitness found compared to the Base PSO approach and a simple Remove Worst Performer approach. This is due to the effectiveness of culling particles outside of a predetermined radius around the current global best, rather than simply removing the worst performer.

Acknowledgements

This research was funded by the Ministry of Education Malaysia (MOE) through the Fundamental Research Grant Scheme FRGS/1/2018/ICT02/UTM/02/1, Cost Center No.: R.J130000.7851.5F069.

References

[1] Kennedy J *et al.* 1995 *International Conference on Neural Networks* 1942-1948.
[2] Shi Y *et al.* 1998 *IEEE Int'l Conf. on Evolutionary Computation Proceedings*. 69-73.
[3] Eberhart R *et al.* 1995 *MHS'95. Proc. of the Sixth Int'l Sym. on Micro Machine and Human Science* 39-43.
[4] Arasomwan M A *et al.* 2013 *IEEE Congress on Evolutionary Computation* 2352-2359.
[5] Uriarte A *et al.* 2016 *IEEE 8th International Conference on Intelligent Systems* 128-132.
[6] Jana N D *et al.* 2014 *IEEE Congress on Evolutionary Computation* 573-578.
[7] Li J *et al.* 2015 *IEEE Transactions on Cybernetics* **45** 2350-2363.
[8] Yu H *et al.* 2013 *Ninth International Conference on Natural Computation* 471-475.

IOP Conf. Series: Materials Science and Engineering **551** (2019) 012070 doi:10.1088/1757-899X/551/1/012070

A Review on Sentiment Analysis Techniques and Applications

Mohd Ridzwan Yaakub[1], Muhammad Iqbal Abu Latiffi[2] , Liyana Safra Zaabar[3]

[1,2,3]Center for Artificial Intelligence Technology, Faculty of Information Science and Technology, Universiti Kebangsaan Malaysia, 43600 UKM Bangi, Selangor, Malaysia.

E-mail: ridzwanyaakub@ukm.edu.my

Abstract. Nowadays, what user think is the most difficult and complicated task handled by organizations. The way to idetify the attitude of the speaker or a writer on some topics is to use sentiment analysis. The use of sentiment analysis is to identify user's opinion towards some topics whether it is positive or negative. This paper presents the techniques used by previous researchers in sentiment analysis which are Machine Learning and Natural Language Processing (NLP) in solving the classification task. The comparison among these two main approaches reveals that Machine Learning techniques can solve classification task with reasonable success and with very high accuracy compared to NLP-based techniques but it is depending on the training and test data with respect to the domain. This paper also presents the use of ontology in sentiment analysis that can help in achieving more high accuracy for the classification task.

1. Introduction

Nowadays, sharing opinion, sharing marketing, online ticket bookings, and online shopping are rising in daily life of people. In the decision-making process, we are always concerned about what other people think. Other people's opinions are the most important piece of information for most of us. Social networking sites such as Facebook allow users to post their comments or opinions about any issues and topics. Here, we will see the importance of reviews shared by the peoples.

Sentiment can be defined as a view or opinion that is held or expressed. Whereas, sentiment analysis (SA) is a process of computationally identifying and categorizing opinions expressed in a piece of textual content, particularly to decide the writer's attitude towards a particular topic, product or issue [1]. SA also known as opinion mining is widely used in many domains such as products, services, issues and politics to analyze user's behaviors or opinion regarding the related topics [2].

People's opinions or experiences influence users in making a decision. Typically, the use of SA is to extract what people feel or think about different things from the text. In the business industries, the use of SA is to analyze customer's opinions towards their product or services because user's satisfaction is the main attention in their industries. SA technology helps in improving businesses by analyzing user's opinion.

2. Techniques Used in Sentiment Analysis

SA aim is to identify the positive, negative or neutral sentiment in the set of text or documents. It is a very crucial task to build a system to analyze user's opinion in the form of review, social media, and microblogs. A lot of researches on SA have been done [3-5]. There are two main approaches used in SA which are Natural Language Processing (NLP) based and Machine Learning based.

NLP is commonly used for text mining, machine interpretation and mechanized inquiry. According to [6], NLP is a research and application area that explores how computers can be used to understand and manipulate natural language text or language for useful purposes. Furthermore, NLP researchers focus on trying to gather knowledge about how people understand and use language so that relevant mechanisms and techniques can be developed in allow computer systems to know as well as could manipulate natural languages to accomplish desired tasks. NLP method is broke down into basic methods and methods of modelling. The basic method in the NLP technique is to extract text features such as Part-of-Speech (POS) Tagging, frequency of documents, dictionary and weighting. In the meantime, the topic modelling method is a generic model of probability using distribution of vocabulary in searching for topics with text elements [7]. Both of these methods aim to identify the topic from a collection of documents. Based on the previous study, NLP approach uses variety of techniques in a different type of domain and most of the result is above 80% which can be say it is a quite impressive compared to Machine Learning approach. NLP approach had proven successful in [6,8-11] to get an excellent result.

Typically, NLP methods utilize statistical algorithms, which will eventually base or combine the algorithms with a classical Deep Learning Approach. Research in NLP is essential for opinion mining as it can monitor positive or negative human feelings based on the sentence or feature extraction in a user's comment.

Another method used in sentiment analysis is Machine Learning. Machine Learning is a set of statistical technique for analyze some form of text. The techniques can be indicate as a model that is then applied to other text (supervised), or could be a set of algorithms that work across large sets of data to extract meaning (unsupervised) [12]. Machine Learning approach has various of techniques, but there are a few techniques that most of the researchers frequently used.

Generally, among the popular Machine Learning methods, Support Vector Machine (SVM), Naïve Bayes and N-Gram are the most popular. SVM is a discriminative classifier formally defined by a separating hyperplane. In other words, having given labelled training data (supervised learning), the algorithm from SVM outputs an optimal hyperplane which categorizes new examples [13]. According to [14] The Naïve Bayesian classifier is based on Bayes' theorem with independence assumptions between predictors, and is easy to build, with no complicated iterative parameter estimation which makes it particularly useful for very large datasets. [14] also stated that despite its simplicity, the Naïve Bayes classifier often does surprisingly well and is widely used because it often outperforms more sophisticated classification methods. Additionally, N-gram is a simple model that assigns probabilities to sentences of words or of whole sequences. N-gram is widely used in text mining and NLP. It is one of the most important tools in speech and language processing. N-gram has been used for a variety of different tasks. These three techniques are the most used in SA because it is proven to achieve high accuracy in the previous researches such as in [15-24]. Machine learning method is able to eliminate overlapping and irrelevant features by using suitable feature selection method. However, according to [5], Machine Learning methods suffers a great deal when training on mixed data.

Machine Learning approach are the best technique among two approaches and able to achieve very excellent results. [25] is among the earlier research in SA that used machine learning approach in solving the classification task and is able to achieve very excellent results. Although this paper is in an early research, they managed to obtain the most satisfactory result up to 96% of accuracy compare to NLP approach where the highest accuracy is over 80% by [10]. Machine learning approach precedes NLP-based approach by 10% and based on these results, we can simply say that machine learning is the best method compared to NLP-based techniques according to the classification task.

3. Ontology in Sentiment Analysis

Ontologies are the key component in various fields such as information retrieval and extraction, knowledge management and organization. Ontologies has been used for modeling the terms in a domain of interest as well as the relations among these terms and are now applied in many fields [26]. The study by [25], is one of the early research in SA that use ontology in sentiment classification. In this paper they presents a method of ontology-based sentiment classification to classify and analyze online product reviews of consumers. In this research, the author used SVM text classification approach based on a lexical variation ontology. The use of ontology in this paper helped the classifier capture and analyzed the sentiments that use different words which may share the same meaning. The performance measure using precision, recall and F-measure allowed the author to be able classify with good accuracy after testing by F-measure with 96% accuracy. Many works have been done in ontology deployment in SA. Other than [25], researches by [27-31] are among the current researches in SA that uses ontology and managed to get a very satisfying results. The results in every experiment proved that the used of ontology can perform better than all baseline.

4. Conclusions

SA present a important role in decision making. Organizations will make decision about their product or service based on the reviews. There are a lot of approaches that can be used in SA. The Machine Learning techniques can solve classification task with reasonable success and with a very high accuracy compared to NLP- based techniques. From these results, we can conclude that even though Machine Learning obtained the highest results, but it is depending on the training and test data with respect to domain. This paper also presented the use of ontology in sentiment analysis which can obtain a very high accuracy.

Acknowledgements

The authors would like to thank Center for Artificial Intelligence Technology, Faculty of Information Science and Technology, Universiti Kebangsaan Malaysia by giving the authors an opportunity to conduct this research and funded by Universiti Kebangsaan Malaysia and Malaysia Government under the FRGS research grant scheme with code FRGS/1/2017/ICT02/UKM/02/4.

References

[1] Medhat W *et al* 2014 *Ain Shams Eng. J.* **5**(4) 1093–1113
[2] Kumar P N V S P *et al* 2018 *Proc. 2nd Int'l Conf. on Comp. Intelligence and Informatics* **712**
[3] Pang B *et al* 2002 *Proc. Conf. Empir. Methods Nat. Lang. Process.* 79–86
[4] Gamon M 2004 *Proc. 20th Int'l Conf. Comput. Linguist* **January 2004** 841–es
[5] Read J 2005 *ACL Student Res. Work.* **June** 43–48
[6] Jurek A *et al* 2015 *Secur. Inform.* **4**(1) 9
[7] Kim H *et al* 2011 *Comprehensive Review of Opinion Summarization* (Illinois Dig. Env. …)
[8] Khan A *et al* 2010 *Proc. 8th Int'l Conf. Front. Inf. Technol.* 1–6
[9] Palanisamy P *et al* 2013 *Proc – Int Workshop on Semantic Evaluation* 543-548
[10] Vu L 2017 *Int'l Conf. Inf. Knowl. Eng.* (July) 10–16
[11] Feng J *et al* 2018 *Wirel. Commun. Mob. Comput.* **2018**
[12] Abirami A *et al* 2016 *IEEE Eighth Int. Conf. Adv. Comput.* 72–76
[13] OpenCV 2017 *Introduction to Support Vector Machines — OpenCV 2*
[14] Sayad D S 2010 *Naive Bayesian* 1–4
[15] Dhaoui C *et al* 2017 *J. Consum. Mark.* **34**(6) 480–488
[16] Go A *et al* 2009 *Processing* **150**(12) 1–6
[17] Tripathy A *et al* 2015 *Procedia Comput. Sci.* **57** 821–829
[18] Xia R *et al* 2016 *Inf. Process. Manag.* **52**(1) 36–45
[19] Trivedi S K *et al* 2018 *Electron. Libr.* **36**(4) 590–606
[20] Al-Smadi M *et al* 2018 *J. Comput. Sci.* **27** 386–393
[21] Devika M D *et al* 2016 *IOSR J. Comput. Eng.*1–5, 2016.
[22] Rana S *et al* 2017 *Proc. 2016 2nd Int. Conf. Next Gener. Comput. Technol.* **October** 106–111
[23] Ibrahim M N M *et al* 2018 *2017 IEEE Conf. Open Syst.* **2018–January** 17–20
[24] Zvarevashe K *et al* 2018 *2018 Conf. Inf. Commun. Technol. Soc. - Proc.* 1–4
[25] Polpinij J *et al* 2008 *Proc. - 2008 IEEE/WIC/ACM Int. Conf. Web Intell. WI 2008* 518–524

[26] Gomez-Perez A *et al* 2002 *IEEE Intell. Syst.* **17**(1) 54–60
[27] Liu L *et al* 2012 *2012 5th Int. Congr. Image Signal Process.* (Cisp) 1620–1624
[28] Yaakub M R *et al* 2013 *Procedia Technol.* **11**(June 2015) 495–501
[29] Sam K M 2013 *Int. J. e-Education, e-Business, e-Management e-Learning* **3**(6)
[30] Schouten K *et al* 2017 *Proc - Int. Conf Web Eng* 302-320
[31] Ali F *et al* 2016 *Appl. Soft Comput. J.* **47**(November 2018) 235–250

A Review on Type of Attacks on Fingerprint Image and Watermarking Techniques

Mohamed Lebcir[1], Suryanti Awang[1]

[1]Soft Computing & Intelligent Systems Research Group (SPINT), Faculty of Computer Systems & Software Engineering, University Malaysia Pahang, UMP, Lebuhraya Tun Razak 26300, Kuantan, Pahang, Malaysia.

E-mail: mlebcir78@gmail.com, suryanti@ump.edu.my

Abstract. Fingerprint image watermarking techniques have several advantages over traditional security systems that based on token or knowledge. However, they are also vulnerable to attacks that can decrease their security. In this regard, this paper presents a study of a number of research studies on the attacks of fingerprint images watermarking techniques. The attacks can be categorized into intentional and unintentional. We study several attacks based on that category and several fingerprint image watermarking methods. We produce a comprehensive comparison of those methods to show how the methods able to prevent the attacks. We also discuss the advantages and disadvantages of the methods. Therefore, based on that, we can see what are the common attacks on the fingerprint image and what are the methods that efficient in preventing the attacks.

1. Introduction

Biometrics are an expanding area that focus on identifying and verifying individual identity using their anatomical (e.g. fingerprint) or behavioral (e.g. voice) characteristics. Since biometrics are permanently attached to the persons, they are more efficient than the normal authentication methods for example token-based or knowledge-based that can be forgotten, stolen or lost [1].

According to their uniqueness and immutability, fingerprint is the most widely used in biometric-based systems for user identification process that known as fingerprint recognition. Most of the fingerprint recognition systems are based on minutiae matching in the recognition step. Minutiae matching is based on minutiae points in the fingerprint pattern which is will be fingerprint features.

In order to increase the protection system, digital watermarking can be used to verify the authenticity of a fingerprint sample [2]. The basic idea of the digital watermarking approach is to embed watermark data into the original image fingerprint to protect fingerprint ownership. In addition, the watermark data can be encrypted before embedding the watermark, as a second layer of protection [3]. Therefore, the user's secret key which prevent the watermark from tempering or unauthorized access by attackers.

Although fingerprint image watermarking systems have several advantages over traditional protection systems, they also suffer from different attacks when they are used for achieving authentication. Those attacks can decrease their security considerably. Thus, it's needing to protect the transmitted content between the sender and receiver against those attacks and several measures that can be utilized to decrease the probability of such attacks. In this paper, we discuss about the attacks and the watermarking techniques that have been implemented to prevent the attacks.

IOP Conf. Series: Materials Science and Engineering **551** (2019) 012071 doi:10.1088/1757-899X/551/1/012071

2. Types of Attacks on Fingerprint Image Watermarking System

There are some attacks on fingerprint image watermarking systems that jeopardized the safety of that fingerprint image. Therefore, in this section we will discuss about the type of attacks based on a number of research studies that have analysed these attacks on fingerprint images watermarking techniques. In general, the attacks are broadly classified into two groups: unintentional and intentional attacks [4].

2.1. The unintentional attacks

the unintentional attacks can be defined as attacks that occur during normal fingerprint image processing operations such as printing, scanning, compression, transcoding, filtering, noise, gamma correction, geometric transforms and cropping. The attacks on a watermarked fingerprint image known as image processing attacks and geometric attacks. In the Geometric attacks like rotation, scaling, translation, cropping, row-column blanking, warping etc. attempt to destroy synchronization of detection, thus, making the detection process difficult and even impossible. For example, rotation or scaling can change pixel values and will damage the minutiae points and watermarks data.

2.2. The intentional attacks

The main threat to any assets protected by a biometric system is that of an impostor impersonating another person who is enrolled and gaining access to the protected assets[5]. The threats discussed by [6], include four major scenarios to attack a matching system, namely: use of a fingerprint dummy made e.g. from gelatin, use of latent fingerprint on sensor, use real finger of a biometric lookalike and use of the real finger of the victim.

fingerprints can be extracted from photos or copied from touched objects like coffee cups, keyboards, and other things. To test the device against this threat, [5] have developed two experiments.

Firstly, they reviewed the pictures taken and selected the most promising one in terms of image quality. Afterward, extracting a suitable fingerprint dummy using the following steps:

1. Crop the image area covering the whole fingerprint
2. Apply grayscale conversion, then color inversion
3. Crop and scale the relevant fingerprint area with respect to the physical and digital sensor size, e.g. 11×11mm, 192×192 pixel, 508dpi
4. Improve brightness, contrast, and gamma.

Using their tools, embedded the extracted fingerprint dummy into the recorded communication payload and uploaded this to the FPGA(Field-Programmable Gate Array). When activating the card, the FPGA injected the custom fingerprint created from the latent prints on the card. This way has successfully bypassed the authentication in a repeatable fashion.

Another, more passive way to extract the biometric data can be done by using pictures of the actual user covering his fingerprints. These pictures can be created by the attacker from a distance or can be found on the web as discussed by Starbug in Chaos Communication Congress 2014[7]. To evaluate this scenario, Fietkau et al. created multiple pictures covering the test person while showing his fingers and, set up an increasing target distance respectively to 3, 4, 5, 6 and 7 meters. The pictures were taken using a Canon EOS-D1 X with a 200mm lens in an outdoor daylight setting. After taking the pictures, starting the extraction process similar to the previous one. The main difference is in fact that we had to flip the image horizontally and further scale the area depending on the target distance.

Again, injected the obtained fingerprints into the recorded communication payload. defined a maximum amount of 3 attempts per image, which has given us the freedom to slightly improve the image gamma, brightness, and contrast. Under these constraints, 3 out of 5 dummies caused a valid authentication and could successfully bypass the matching algorithm

Due to these attacks, an efficient watermarking technique is needed to be implemented on a fingerprint image to prevent the attacks. We will discuss about the watermarking techniques that have been implemented by many researchers. The advantages and disadvantages of each watermarking technique is explained in that section as well.

3. Trend of preventing the attack on fingerprint image watermarking techniques

In fingerprint image watermarking techniques, the main consideration is the evaluation of the robustness and effectiveness of the watermarking method through measurement of the impact of different attacks upon the watermarked fingerprint image. The subsequent section will discuss on the fingerprint image watermarking techniques of preventing the attack.

Bousnina et al. introduced a new method of embedding watermark into a fingerprint image. This involves the use of a secretly generated key to identify which pixels to be watermarked. Thus, preserving the fingerprint minutia points and increase security when extraction step for watermark by attackers. After that, the face features (watermark) is then inserted into the fingerprint image (cover) such that the fingerprint minutia points are preserved through the help of Orthogonal Locality Preserving Projections (OLPP) method [8]. This proposed method of watermarking biometric fingerprint was verified mainly against conventional watermarking attacks notable for digital images. This includes median filter, Speckle noise, Gaussian noise and Poison noise. In addition, this proposed method succeeded in preventing the intentional attacks which attackers are needed of secret key to extract watermark. Hence, its resistance strength was not investigated against other potential attacks such as geometric attacks.

In a particular research, a technique for inserting two watermarks into fingerprint images was proposed by Alkhathami et al. The technique makes use of Discrete Cosine Transform (DCT) algorithm [9]. Although their approach was blind, means that the extraction stage of watermark does not require the original fingerprint image, but the robustness was tested using only noise attacks which may have a more vulnerable image while placing it on the sensor. It is however necessary that the robustness of the technique needs to be verified against different type of attacks especially geometric attacks. Although, this method is blind, robust for compression attacks and succeeded payload capacity which help to add more authentication factors based on the watermark messages and to protect the ownership of the fingerprint image against piracy. However, its low robustness against geometric attack such as rotation.

In another study, a technique which is based on DCT and a fuzzy-Particle Swarm Optimization was put forward by Bansal et al. for fingerprint watermarking. This technique was applied to secure the fingerprint image of an individual by watermarking its analogous face image [10]. Verification of the robustness of this approach was only confirmed for certain potential watermarked image attacks such as noise, JPEG compression and sharpening. However, it was not verified against other possible attacks especially the geometric ones such as rotations and translations which may lead to distortion of watermark or features for fingerprint.

Similarly, a technique for fingerprint watermarking was put forward by Cao et al. wherein (Contourlet Transform) and TC (Texture Complexity) is used for choosing the optimum blocks where the watermark may be embedded [11]. The watermarking algorithm is robust to JPEG compression, Gaussian noise and filtering attacks. However, it is generally known that there are possibilities for watermarked fingerprint images to be affected by several other attacks such as geometric attacks.

A blind algorithm in the Contourlet domain for watermarking was put forward by Kumar et al. and the algorithm does not depend on original template of the fingerprint at extraction. In their approach, Iris code was primarily used as the watermark in a fingerprint image [12]. However, the influence of the extraction of biometric features that are important to compare between them were not evaluated, and noise adding attacks were only used to verify the robustness of their algorithm. Therefore, testing of the robustness of their algorithm is considered incomplete.

In a different approach, a Particle Swarm Optimization (PSO) method was used by Bansal et al. [13]. The underlying principle behind their approach is the use of PSO to identify the best coefficients of DWT where there is possibility of embedding the facial image data so as to hide the face image's pixel data. However, this approach cannot be certified capable of retaining original fingerprint features as it was not verified against most of the conventional geometric and digital watermarking attacks.

An approach based on wavelet was introduced by Ghany et al. for embedding DNA data based multi-bit watermark into fingerprint images using DWT [14]. The method was robust and

IOP Conf. Series: Materials Science and Engineering **551** (2019) 012071 doi:10.1088/1757-899X/551/1/012071

straightforward, but it requires the original fingerprint template at the extraction stage (non-blind). Therefore, this method succeeded in imperceptibility and robustness but low security due to need the original fingerprint at the extraction step of watermark that may lead to the attackers can easily use the original secure data to attack the system.

Alkhathami et al. developed a watermarking algorithm to be embedded into fingerprint images with unique minutiae in order to prevent it from possible effects of the embedded watermark [4]. This technique uses a new Dual-tree Complex Wavelet Transform (DTCWT) and it presents desirable advantages especially for protecting the fingerprint minutiae from potential embedded watermark effects. However, this algorithm is non-blind. Therefore, the original image is needed to extract the watermarks. This exposes the original data to be easily accessed by potential attackers during process of transmitting the watermarked image from sender to the receiver.

A simple blind technique which uses DWT for watermarking and digital image detection was put forward by Abraham et al. [15]. The proposed method is blind as the original fingerprint template is not required at the extraction stage (blind). In addition, the robustness of embedded information in this approach is reasonable. However, it can only withstand certain type of attacks like noise addition and histogram equalization, but not tested against other possible attacks such as different other image processing or geometric attacks. Therefore, the robustness was not completely evaluated to justify its capability to better robust than other approaches.

A comparative study was carried out by Haddada et al. on the choice of domain whereby the watermark may be inserted such that the minutiae could be preserved both in terms of its number and its positions [16]. that mean this technique combined between robustness and imperceptibility. The approach presents a new biometric fingerprint watermarking technique that uses the facial Gabor characteristic features which are reduced by the Direct Linear Discriminant Analysis Technique (DLDA) method. However, the flaw of this proposed method is that it was not tested under rotation and other geometrical attacks. Therefore, the robustness was not completely evaluated to justify its capability to preserve the numbers and positions of the minutiae, better than other approaches.

In a separate research, a watermarking technique was proposed by Alkhathami et al. which combines the features of DWT and DCT techniques. These were combined incorporate the fingerprint owner's grayscale facial image and other identification details in the model of binary text image, into the fingerprint image [4]. The results obtained from this study indicated that incorporation of more than one watermark images of different sizes did not corrupting the minutiae. However, it is often necessary to evaluate the robustness of the proposed algorithm against various potential image attacker of such algorithm. Hence, the algorithm may be considered good if it achieves high payload without corrupting the minutiae points.

Based on aforementioned, most of the existing fingerprint image schemes demonstrate robustness against some categories of attacks. However, they failed to perform well against other types of attacks such as geometric transformations like translations or rotations. In addition, based on the results that were recorded in fingerprint image watermarking techniques, most of them cannot know its strong or weakness against intentional attacks but stay just analysis expected based on its algorithm and tools used in this technique.

4. Conclusion

In general, the attacks on watermarking fingerprint image are broadly classified into two groups: unintentional and intentional attacks.

Although, most of watermarking technique as show in previous section were able to achieve good robustness against some unintentional attacks such as image processing. However, they failed to perform well against other types of attacks such as geometric transformations like translations or rotations.

Based on the results that were recorded in fingerprint image watermarking techniques, Geometric attacks such as rotation, scaling and translation are the most common type of challenge attacks in digital watermarking. In intentional attack cannot know strong or weakness of this fingerprint image watermarking techniques as reviewed in this paper. This is still just analysis expected based on its proposed algorithm and tools used in this technique. Thus, the attacks are important to be evaluated to measure the robustness. This would help to provide higher security and robustness of watermarked fingerprint image.

Acknowledgements
We would like to show our gratitude to Universiti Malaysia Pahang (RDU vote number RDU180380) for supporting this study.

References

[1] Awang S *et al* 2013 *Proc. -2013 Int'l Conf. on Signal-Img. Tech. & Internet-Based Sys.* 706–713
[2] Noore A *et al* 2007 *Forensic Science International* **169** 188–194
[3] Mousavi S M *et al* 2014 *Journal of Digital Imaging* **27** 714–729
[4] Alkhathami M *Watermarking techniques for genuine fingerprint auth.* PhD (RMIT University)
[5] Fietkau J *et al* 2018 *WOOT'18 Proc. of the 12th USENIX Conf. on Offensive Tech.* 5
[6] Henniger O *et al* 2010 *Int'l Biometric Performance Testing Conf.* 1–10.
[7] Starbug 2014 *Ich sehe, also bin ich ... du* (31C3) [Online] https://events.ccc.de/congress/2014/Fahrplan/events/6450.html.
[8] Bousnina N *et al* 2016 *12th Int'l Conf on Innovations in Info. Tech.* 127–131.
[9] Alkhathami M *et al* 2013 *Proc. of the 2013 IEEE 8th Conf. on Ind. Elec.and App.* 1151–1155.
[10] Bansal R *et al* 2013 *IEEE Int'l Conf. on Fuzzy Systems*
[11] Cao Y *et al* 2010 *Int'l Symposium on Intelligent Signal Proc. and Comm. Sys., Proc.* 10–13.
[12] Kishore Kumar N K *et al* 2012 *3rd Int'l Conf. on Comp., Comm. and Networking Tech.*
[13] Bansal R *et al* 2012 *Information Security Journal* **21** 88–101.
[14] Ghany K K A *et al* 2014 *2014 International Conference on Informatics, Electronics & Vision*
[15] Abraham J *et al* 2016 *Data Mining and Advanced Computing Int'l Conf.* 145–149.
[16] Haddada L R *et al* 2017 *IPAS 2016 - 2nd Int'l Image Proc., App. and Systems Conf.* 1–6.

IOP Conf. Series: Materials Science and Engineering **551** (2019) 012072 doi:10.1088/1757-899X/551/1/012072

Flower Pollination Neural Network For Heart Disease Classification

Mohamad Haider Bin Abu Yazid[1], Mohamad Shukor Talib[1] , Muhammad Haikal Satria[2]

[1]School of Computing, Faculty of Engineering, Universiti Teknologi Malaysia, 81300 UTM Johor Bahru, Malaysia.

[2] School of Bioscience and Medical Engineering, Faculty of Engineering, Universiti Teknologi Malaysia, 81300 UTM Johor Bahru, Malaysia.

E-mail: mhaider04@gmail.com

Abstract. Heart Disease are among the leading cause of death worldwide. The application of artificial neural network as decision support tool for heart disease detection have been previously proposed. However, artificial neural network using conventional back propagation algorithm for error minimization and these algorithm tend to stuck at local minima. This paper proposed the use of flower pollination algorithm as a substitute to conventional back propagation algorithm for error minimization. Heart disease dataset obtain from UCI machine learning repository is used to evaluate the performance of the proposed framework. The results show that the proposed flower pollination neural network able to produce higher classification accuracy compared to the conventional back propagation neural network algorithm.

1. Introduction

Heart disease describes a range of conditions that affect your heart in which are the leading cause of death worldwide where more people die from cardiovascular diseases compared to any other causes annually. The early detection of heart disease can help the patient to adjust lifestyle and also help the medical professionals to prescribe appropriate medicine. However, the diagnosing patient with the presence of heart disease can be challenging where it depends on medical professional's experience and intuition [1,2]. It is imperative for medical professionals to have a system that can help them predict and classify the patient who have high risk of getting heart disease.

The implementation of machine learning algorithm can help medical professionals in diagnosing the presence of heart disease in the patient. Machine

IOP Conf. Series: Materials Science and Engineering **551** (2019) 012072 doi:10.1088/1757-899X/551/1/012072

learning algorithm have become very popular for solving classification problems where it is capable of mapping the relationship between variables or attributes with minimal human effort. Artificial Neural Network (ANN) are among the most popular machine learning algorithm where it proves to be powerful tools for mapping nonlinear data and are known to be useful in solving nonlinear problems where the rules to solve the problem is difficult to obtain or unknown [9,10]. However, artificial neural network using conventional back propagation algorithm for error minimization and these algorithm tend to stuck at local minima [6]. Among the the approach to avoid local minima is to train neural multiple times but it consume a lot of computational power [6]. This paper proposed the use of flower pollination algorithm which has been used to solve optimization problem for error minization in neural network. Statlog heart disease dataset, Cleveland heart disease dataset, Hungarian heart disease dataset and Switzerland heart disease dataset obtain from University of California, Irvine machine learning data repository are used to measure the performance of proposed flower pollination neural network [8].

2. Methodology

2.1 Flower Pollination Algorithm

Flower pollination algorithm has been proposed in 2013 by Xin-She Yang which inspired by characteristic of biological flower pollination in flowering plant. Essentially, flower pollination algorithm can be explained using following rules [4]:

1. Cross pollination and biotic is considered as global pollination process where polinators that carrying pollen performing random walk known as Levy Flight.
2. Local pollination is done by agent called abiotic
3. Flower constancy can be considered as the reproduction probability is proportional to the similarity of two flowers involved.
4. Switch probability $p \in [0,1]$ controlled local and global pollination. Local pollination can have significant fraction p in overall pollination activities because of physical proximity and factors such as local pollination and wind.

2.2 Flower Pollination Neural Network

In the proposed flower pollination neural network algorithm, each best pollen represents a possible solution (the initial weight space and corresponding biases for neural network optimization). The weight optimization problem and population size represent the quality of the solution. In the first epoch, the best initial weights and biases are initialized with Flower Pollination. Subsequently, those weights are passed on to the neural network. The weights in neural network are computed and compared with the best solution. In the second cycle, the flower pollination updates the weights with the best possible solution. The flower pollination continues searching the best weights until the last cycle/epoch of the neural network is reached or the target minimum error is reached. Figure 1 shows the simplified flow of flower pollination neural network algorithm.

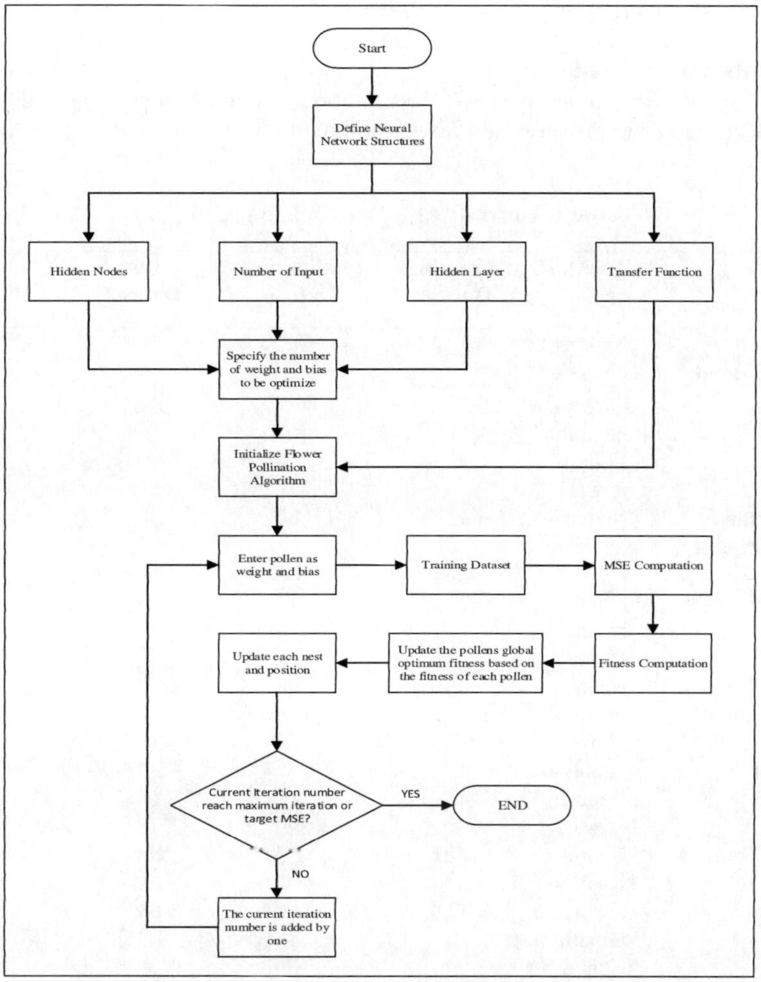

Figure 1 Flower Pollination Neural Network Flowchart

2.3 Experimental Setup

The proposed flower pollination neural network uses four sets of heart disease data obtain from machine learning repository of University of California, Irvine [8]. The Statlog dataset, Cleveland heart dataset, Hungarian heart dataset and Switzerland heart dataset is used to evaluate the performance of proposed framework. The result of simulation then compared to the performance of parameter tuned neural network that use conventional back propagation algorithm which has been proposed previously [3]. The previously proposed parameter tuned neural network using multi training approach to overcome local minima. Dataset is partition into

IOP Conf. Series: Materials Science and Engineering **551** (2019) 012072 doi:10.1088/1757-899X/551/1/012072

ratio of 70% for training, 15% for validation and 15% for testing. The proposed flower pollination neural network and parameter tuned neural network are simulated using MATLAB R2014b with the specifications of Intel core i-7, RAM of 12 GB and operating on Windows 10 64 bit.

3. Results and Discussion

Table 1 shows the comparison of classification result between standard neural network compared to the proposed flower pollination neural network.

Table 1. Heart disease classification accuracy

Dataset	Algorithm	Training Dataset	Validation Dataset	Test Dataset	Overall Accuracy
Cleveland Heart Dataset	Neural Network	91.4	86.7	71.7	86.2
	Flower Pollination Neural Network	93.2	86.7	88.3	90.9
Hungarian Heart Dataset	Neural Network	83.5	84.6	80	83.1
	Flower Pollination Neural Network	89	84.6	87.5	88.1
Switzerland Heart Dataset	Neural Network	92	93.3	80	90.5
	Flower Pollination Neural Network	97.3	93.3	100	97.1
Statlog Dataset	Neural Network	87.5	81.5	85.2	86.7
	Flower Pollination Neural Network	89.4	92.6	88.9	89.6

Flower pollination neural network able to produce higher accuracy compared to the previously proposed parameter tuned neural network using multi training approach. The result shows that the proposed flower pollination neural network able to achieve high classification accuracy with the overall accuracy of 90.9% for Cleveland dataset compared to 86.2% classification accuracy for neural network with multi training, 88.1% for Hungarian dataset compared to 83.1% classification accuracy for neural network with multi training, 97.1% for Switzerland dataset compared to 90.5% classification accuracy for neural network with multi training

IOP Conf. Series: Materials Science and Engineering **551** (2019) 012072 doi:10.1088/1757-899X/551/1/012072

and 89.6% of classification accuracy for Statlog dataset compared to 86.7% classification accuracy for neural network with multi training.

4 Conclusions

The flower pollination neural network for heart disease classification is proposed in this paper. The proposed flower pollination neural network classification accuracy is higher compared to the standard backpropagation neural network with multi training approach.

Acknowledgements

This research is funded by Universiti Teknologi Malaysia (UTM) via Research University Grant (RUG) Tier 1 with Vot No. (PY/2017/00781).

References

[1] Anooj, P. K. 2012 *J. King Saud Uni Comp and Inf Sys.* **1** 27.
[2] Hedeshi, *et al* 2014. *Comp Int and Neuro Sci* **6**.
[3] M. Haider Abu *et al* 2018. *Electric Eng, Comp Sci and Inf (EECSI) Int Con* **5**
[4] X.S. Yang, 2012, *Unconventional Comp and Natural Comp* **1** 240
[5] Amato *et al,* 2013, *J App Biomed,* **11** 47
[6] Bi *et al*, 2005, *Trans on Fund of Electro, Comm and Comp Sci,* **12** 3645
[7] Chakraborty *et al*, 2015 *ABLAZE Int Conf* **1** 261
[8] Blake *et al*, (1998), *Repo of Machine Learning Db*
[9] Er *et al*, (2010), *Expert Syst. Appl* **37** 7648
[10] Du *et al* (2012), *J of Med Sys* **36** 1283

An enhanced latent semantic indexing with term frequency-inverse document frequency variant for software traceability

R Tumeng[1], D N A Jawawi[2] and M A Isa[3]

[1,2,3]Software Engineering Research Group, School of Computing, Faculty of Engineering, Universiti Teknologi Malaysia, 81310, Johor Bahru, Johor, Malaysia

[1]rooster2@live.utm.my

Abstract. This paper proposes an improved Latent Semantic Indexing (LSI) with Term Frequency-Inverse Document Frequency (TFIDF) variant for software traceability. The main advantage of the method is the simplicity and accuracy of the algorithm such that it can produce high precision and recall. The accuracy of tracing precise links between software artefacts is achieved by feeding the LSI with TF-IFDF variant that halves the TF and IDF values, thus increases the probability of yielding precise links. To test the accuracy of the proposed method, two case studies were evaluated, namely Mushroom Management System (MMS) and Robotic Wheelchair System (RWS). The superiority of the proposed method over a conventional LSI is confirmed by 409.09%, 900.00%, and 620.00% improvements for precision, recall, and harmonic mean scores, respectively, in MMS case study, at cosine threshold of 0.94. It is anticipated that the method could be beneficial in the design of a practical and accurate system for retrieving precise software traceability links.

1. Introduction

In Software Development Life Cycle (SDLC), traceability plays a crucial role in analyzing the impact of alteration across different software artefacts. Traceability is described as the ability to trace links between source artifact and target artifact [1].

Successful traceability relies on the ability to concisely trace source artifact to target artifact. In real world scenarios, SDLC is often plagued by an overloading of information as software information is continuously generated. Researchers are in agreement that an immense wealth of information may be useful to assist in successful traceability across various SDLC phases via formulating traceability problem through the lens of an information retrieval (IR) [2-3]. Researchers have utilized and proposed various IR methods in literature. Several instances include algebraic models [2], probabilistic models [4], and statistical language models [5]. Widely used algebraic models include Vector Space Model (VSM) [6] and Latent Semantic Indexing (LSI) [7]. Fundamental understanding of algebraic models involves computing similarity between documents to determine the extent of document relevance. Researchers have argued that conventional LSI suffers poor accuracy (i.e. yielding low precision and recall scores) [2]. In addition, combination of conventional LSI with regular TF-IDF in retrieving latent meanings in software documents often have poor accuracy as shown by studies in [2, 8]. Thus, a new method is needed which could improve the accuracy in tracing software trace links between software artefacts.

In this study, the authors evaluate conventional LSI and the proposed LSI with TF-IDF variant and report the experiments with two case studies. The evaluation indicated that the proposed method improved precision and recall. The paper is structured as follows. Section 2 elaborates conventional

LSI. Section 3 elaborates TF-IDF. Section 4 describes the proposed TF-IDF variant. Section 5 describes the case studies. Section 6 describes the evaluation measures. Section 7 elaborate the experiment results. Section 8 concludes the paper.

2. Latent Semantic Indexing (LSI)

LSI is an extension of Vector Space Model (VSM) that was proposed by [7] to solve synonymy and polysemy issues. LSI operates by reducing document noise via vector space dimension reduction and utilizes single value decomposition (SVD) to determine new dimensions. The k-dimensional LSI is used to compute similarities between vectors. A query vector, q_k, is mapped into its representation in the LSI space by the transformation as in Equation 1;

$$\overrightarrow{q_k} = \Sigma_k^{-1} U_k^T \vec{q} \quad (1)$$

Where $\overrightarrow{q_k}$ is query vector with respect to coefficient of low-rank approximation dimension, k, Σ_k^{-1} is the inverse of diagonal matrix Σ_k, and U_k^T is transposed singular value decomposition term matrix with respect to *k*.

3. Term Frequency-Inverse Document Frequency (TF-IDF)

Term frequency, as expressed in Equation 2, refers to a normalized term weight occurring in document that is proportionate to term frequency [9];

$$tf(t,d) = \frac{f_{(t,d)}}{\max\{t' \epsilon d\} f_{t',d}} \quad (2)$$

Where $tf(t,d)$ is normalized term frequency of a term *t* that occurs in document *d*, $f_{t,d}$ is the raw count of frequency of term *t* in document *d*, and $\max\{t' \epsilon d\} f_{t',d}$

Meanwhile, inverse document frequency, as expressed in Equation 3, refers to specificity of terms quantifiable as inverse function of the number of documents in which the terms occur [10];

$$idf(t,D) = \log \frac{N}{|\{d \in D : t \in d\}|} \quad (3)$$

Where $idf(t,D)$ refers to inverse document frequency of a term *t* that occurs in corpus *D*, *N* is the total number of documents in corpus, and $|\{d \in D: t \in d\}|$ is the number of documents where term *t* appears.

4. TF-IDF Variant

In this paper, the authors proposed to halve TF-IDF through removing logarithm in IDF and replacing it with square root. Resulting TF-IDF variant is expressed in Equation 4.

$$tf(t,d) \cdot idf(t,D) = \sqrt{\frac{f_{(t,d)}}{\max\{t' \epsilon d\} f_{t',d}}} \cdot \sqrt{\frac{N}{|\{d \epsilon D : t \epsilon d\}|}} \quad (4)$$

Where $\sqrt{\frac{f_{(t,d)}}{\max\{t' \epsilon d\} f_{t',d}}}$ is the square root of normalized frequency and $\sqrt{\frac{N}{|\{d \epsilon D : t \epsilon d\}|}}$ is the square root of the inverse document frequency.

5. Case Studies

The authors utilize Mushroom Management System (MMS) and Robotic Wheelchair System (RWS). MMS is a general software developed by final year students from the School of Computing, Universiti Teknologi Malaysia for a local mushroom producer. RWS is a safety-critical, wheelchair equipped with AMD188ES microcontroller, parallel I/O, serial I/O, ADC/DAC, and an on/off switch [11]. The characteristics of case studies are listed in Table 1.

Table 1. Characteristics of case studies.

Characteristics	MMS	RWS
Domain	General	Safety-critical
Number of requirements	11	7
Number of test cases	38	29
Number of trace links	38	29

6. Evaluation Measures

Precision is defined in Equation 5, meanwhile Recall is defined in Equation 6. In addition, harmonic is also used, which is also referred to as F1 measure, as defined in Equation 7.

$$Precision = \frac{Correct}{Correct + Incorrect} \quad (5)$$

$$Recall = \frac{Correct}{Correct + Missed} \quad (6)$$

$$F1\ measure = 2 \bullet \frac{P \times R}{P + R} \quad (7)$$

Where *P* stands for *Precision,* while *R* stands *for Recall.*

7. Experiment Results

This section describes the results obtained from MMS and RWS case studies. Table 2 presents the results obtained from MMS case study. In general, the results in Table 2 revealed that the proposed LSI with TF-IDF variant had improved the precision, recall, and F1 measures significantly at cosine threshold of 0.94. Referring to Table 2, improvements of 409.09%, 900.00%, and 620.00% were observed for precision, recall, and F1 measures, respectively, for cosine threshold of 0.94, in comparison to conventional LSI. Meanwhile, a 100% improvement was observed for MMS' precision score at cosine threshold of 0.99. Such improvements indicate that the enhanced LSI with TF-IDF variant is capable of yielding higher accuracy at higher cosine thresholds. Meanwhile, at lower cosine thresholds (i.e. 0.80 and 0.60), no improvements but decrements of precision, recall, and F1 measures were observed for MMS, at cosine threshold of 0.60. This indicates that at lower cosine threshold, the proposed method was not able to return precise trace links for MMS case study.

Meanwhile, Table 3 presents the results obtained from RWS case study. Generally, precision and F1 measures improved significantly with values of 30.77% and 127.27%, respectively, when cosine threshold was 0.80. Meanwhile, high precision was maintained at cosine threshold of 0.90. The improvements at higher cosine thresholds were attributed to square-rooting the TF-IDF vector values, which yielded favorable results, particularly at cosine threshold of 0.94 for MMS case studies. Overall, in comparison with conventional LSI, the proposed LSI with TF-IDF variant produced unfavorable results for low cosine thresholds (i.e. 0.60). However, the findings could be inconclusive as the scale of the case studies were rather small.

In general, the results showed that the proposed LSI with TF-IDF variant is promising in improving precision to higher levels in all case studies, at higher cosine thresholds. In contrast, conventional LSI performed better at lower cosine threshold of 0.60, particularly in RWS. Based on this observation, the authors concluded that proposed LSI with TF-IDF variant has the most potential as a method to be used for automating software traceability at higher cosine thresholds.

Table 2. Characteristics of case studies.

MMS – Conventional LSI				
Cosine Threshold	0.99	0.94	0.80	0.60
Precision	0.50	0.11	0.00	0.57
Recall	1.00	0.10	0.00	0.25
F1	0.67	0.10	0.00	0.35
MMS – Proposed LSI with TF-IDF Variant				
Cosine Threshold	0.99	0.94	0.80	0.60
Precision	1.00	0.56	0.00	0.00
Recall	0.42	1.00	0.00	0.00
F1	0.59	0.72	0.00	0.00
Improvement				
Precision	100.00%	409.09%	N/A	-100%
Recall	-58.00%	900.00%	N/A	-100%
F1	-11.94%	620.00%	N/A	-100%

Table 3. Characteristics of case studies.

RWS – Conventional LSI				
Cosine Threshold	0.99	0.94	0.80	0.60
Precision	1.00	0.00	0.13	0.00
Recall	0.86	0.00	1.00	0.00
F1	0.92	0.00	0.11	0.00
RWS – Proposed LSI with TF-IDF Variant				
Cosine Threshold	0.99	0.94	0.80	0.60
Precision	1.00	0.55	0.17	0.17
Recall	0.31	0.55	0.44	1.00
F1	0.46	0.55	0.25	0.29
Improvement				
Precision	0.00%	N/A	30.77%	N/A
Recall	-63.95%	N/A	-56.00%	N/A
F1	-50.00%	N/A	127.27%	N/A

8. Conclusion

In this study, two IR methods have been investigated, conventional LSI and the proposed LSI with TF-IDF variant, across two case studies. The proposed method was targeted to solve the issue of poor accuracy of conventional LSI in capturing trace links between software requirements and test cases, encompassing precision, recall, and F1 scores. The findings of the experiment suggest that the proposed LSI with TF-IDF variant improved conventional LSI significantly at higher cosine thresholds (i.e. 0.99 and 0.94), particularly in MMS case study. Despite varied performances across the two case studies, the proposed method indicates a potential to improve the overall accuracy of IR-based software traceability systems.

IOP Conf. Series: Materials Science and Engineering **551** (2019) 012073 doi:10.1088/1757-899X/551/1/012073

As future works, further evaluation of benchmark datasets and larger case studies could be conducted. In addition, the future works could also consider comparing other IR-based techniques. Finally, hybridization of TF-IDF with other weighting schemes could also be investigated to produce precise IR-based software traceability systems.

References

[1] Center of Excellence for Software and Systems Traceability, [CoEST. Requirement Traceability. Retrieved 08 October 2018, 2018, from http://coest.org .

[2] Khatiwada S, Tushev M and Mahmoud A 2018 *Inform Software Tech* **93** 45-57

[3] Rempel P and Mäder P 2017 *IEEE TSE* **43** 777-797

[4] Turtle H and Croft W B 1990/2017 *SIGIR Forum* **51** 124-147

[5] Zhai C and Lafferty J 2017 *SIGIR Forum* **51** 268-276

[6] Salton G, Wong A and Yang C S 1975 *Commun ACM* **18**, 613-620

[7] Deerwester S, Dumais S T, Furnas G W, Landauer T K and Harshman R 1990 *J. Am. Soc. Inf. Sci.*, **41**, 391-407

[8] Mahmoud A and Niu N 2015 *RE*, **20**, 281-300

[9] Luhn H P 1957 *IBM J Res Dev.*, **1**, 309-317

[10] Spärck Jones K 1972 *JDoc.*, **28**, 11-21

[11] Jawawi D N A, Sabil S, Mamat R, Zaki M Z M, Talab M A S and Mohamad R A 2011 *In Mobile Robots-Control Architectures, Bio-Interfacing, Navigation, Multi Robot Motion Planning and Operator Training*

The authors fully acknowledge Universiti Teknologi Malaysia for Zamalah UTM Scholarship and UTM-TDR Grant Vot No. 06G23, and Ministry of Higher Education (MOHE) for FRGS 2019, which have made this research endeavor possible. Additionally, the authors would also like to express their sincere gratitude to members of Embedded Real Time Software Engineering lab for their continuous support and feedbacks.

IOP Conf. Series: Materials Science and Engineering **551** (2019) 012074 doi:10.1088/1757-899X/551/1/012074

Adoption of Machine Learning Techniques in Software Effort Estimation: An Overview

Siti Hajar Arbain[1], Nor Azizah Ali[1] and Noorfa Haszlinna Mustaffa[1]
[1]School of Computing, Faculty of Engineering, Universiti Teknologi Malaysia (UTM), 81310 Johor Bahru, Johor, Malaysia.

Email: shajar75@live.utm.my, nzah@utm.my

Abstract. Nowadays the significant trend of the effort estimation is in demand. It needs more data to be collected and the stakeholders require an effective and efficient software for processing, which makes the hardware and software cost development becomes steeply increasing. This scenario is true especially in the area of large industry, as the size of a software project is becoming more complex and bigger, the complexity of estimation is continuously increased. Effort estimation is part of the software engineering economic study on how to manage limited resources in a way a project could meet its target goal in a specified schedule, budget and scope. It is necessary to develop or adopt a useful software development process in executing a software development project by acting as a key constraint to the project. The accuracy of estimation is the main critical evaluation for every study. Recently, the machine learning techniques are becoming widely used in many effort estimation problems but there are limitations in some of the models and the variation research is still not enough. This paper presents an overview of the effort estimation using machine learning techniques and will be useful for researchers to provide future direction in the field of machine learning adoption in software effort estimation.

1. Introduction

Nowadays the software development system is becoming complicated. The usage of software arises in most companies. Depending on the organisation's size and accompanied tasks, each activity under a software project development should be updated regularly. They must deal in giving high-quality software with a low-cost budget. Therefore, more intelligent approaches are needed to solve the challenging problems in this domain. A software development project is one of the processes in the planning of software project management. It needs to be monitored by the manager to ensure a high-quality software can be produced at a low cost within a specified time and budget [6]. Software effort estimation (SEE) assumes an essential part of the improvement of software development. Lately, the product has turned into the costliest part of the software development efforts. The critical aspect of cost in software advancement is the human-effort, and most cost estimation techniques concentrate on this aspect and give estimates in regard to the individual [3][9]. During the early phase of software development project, there is a lot of work to be done. Software effort estimation is one of the steps in software development project which targets on the production of quality software, which can be delivered on time and within budget, and satisfying its requirements. It is also known as a feature of software engineering monetary on how to oversee restricted assets to meet the objectives and goals of the schedule, budget, and scope.

The growing concern by most developers is the complexity of estimation made in an early phase of the development process. Sometimes, the development of software product resulted differently due to uncertainties. It increases with the size of software project estimation mistakes that could cost a lot in terms of resource allocated to the project. Because software development problems have many dimensions, we need to investigate the use of several techniques to optimise these challenging issues, not only focusing on the software effort engineering approach, but also to include the incorporation of other methods that can contribute to the enhancement in effort estimation accuracy. The following sections present the overview of the techniques implemented in the development of the software effort estimation model. The methodology used for literature review is presented in Section 2. The background and the related theory of software effort estimation techniques are presented in Section 3, followed by the summary review techniques analysis and conclusions in Section 4.

2. Methodology

This paper presents the incorporation of machine learning and soft computing techniques with software effort estimation developed up to early 2007 publications. The publication search procedures are illustrated in Figure 1.

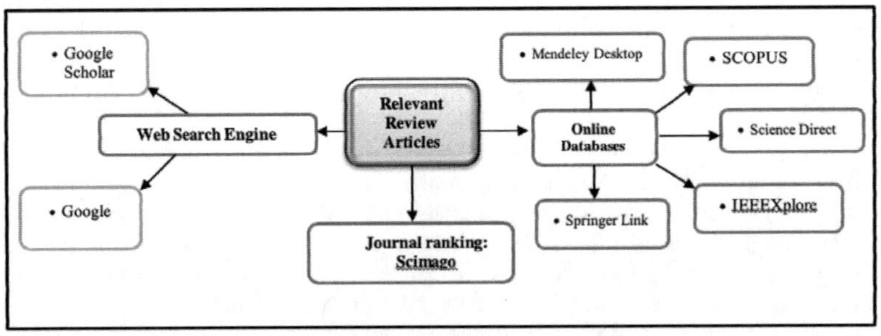

Figure 1: Publication Search

The searches include any keywords of research articles included in title, abstract and keywords. An example of the search queries used is *("software effort estimation") AND ("soft computing" OR "COCOMO II")*.

Upon completion of searching the publications, manual search was carried out to identify the redundant results. The redundancies in similar article publications were eliminated. Besides, we used the inclusion and exclusion characteristics to access the potential direction of the study as follows:

i. **Inclusion**: Publications with a clear focus on software effort estimation and machine learning.
ii. **Exclusion**: Publications of unrelated area of studies, or not peer-reviewed, for example, lecture notes and tutorials.

3. Background and Related Works

Software Effort Estimation (SEE) emphasises on how to estimate the effort, time and cost with the project activities plan. Generally, in SEE, two types of traditional approaches being implemented to calculate the effort estimation which are the non-algorithmic method, focusing on human expertise and the algorithmic approach which consists of mathematical modelling [1] [12].

The algorithmic process is produced by utilising some numerical demonstrations to carry out the software estimation. These scientific, mathematical models are based on the historical data and use the inputs such as size, number of attributes functions and other cost drivers. The Constructive Cost Model (COCOMO) model, Function Point Analysis (FPA) model, and Putnam Model are some of the models that use the algorithmic approach. A non-algorithmic method is used by an expert for software project estimation. In its process, the estimators must have the knowledge on an earlier completed project that is similar to the current project [4][5]. These estimation techniques are used to characterise the spending plan and the items are balanced with the goal in the spending figure. Table 1 shows the advantages and disadvantages of both the algorithmic and the non-algorithmic techniques in software effort estimation.

Table 1: Advantages and disadvantages of software effort estimation

Methods	Advantages	Disadvantages
Non-Algorithmic / Heuristic	Basic info parameter.	No formal/substantial motivation to judge that the estimation is adequate.
	Expert's understanding and learning give a high esteem.	Conflicting. For a situation, the blunder edge could be huge. Then again, it could likewise create a superior estimation than utilising the algorithmic technique.
	For the most part, estimation is made of the group accord.	
	Useful without measured, experimental information.	Estimate is just as great master's supposition.
	Can figure in contrasts between past venture encounters and necessities of the proposed undertaking.	Hard to archive the components utilised by the masters.
	Can calculate in effects brought on by new innovations, applications and dialects.	
Algorithmic	Objectively aligned to experience.	Confused information parameter.
	Reliable, since the conditions are consistent.	
	Utilising standard unit of estimation.	Requirements of having broad information as to each info parameter.
	Could be utilised as a part of each venture as long as the info parameter could be satisfied.	Unable to manage remarkable conditions.
	Generate repeatable estimations.	
	Easy to adjust information.	Some experience and components cannot be measured.

Studies show that both expert and formal methods have drawbacks and still, the best method is yet to be discovered. The efficiency of the processes depends on a specific context data, the techniques used and the problems domain to be solved. Software analysts and experts are giving numerous effort estimation procedures in quite a few years. The early software estimation models depend on relapse examination or numerical determinations. The present models depend on simulation, neural network, Genetic Algorithm (GA), fuzzy logic and so forth [2]. There are many implementation and improvement of effort estimation model using soft computing in order to overcome some limitations in the estimation accuracy which is applicable for every standard software development. Soft computing procedure is investigated to defeat the vulnerability and imprecision in estimation. In classical formulation, machine learning is one of the techniques under soft computing, included evolutionary computation and fuzzy logic. In previous research, Particle Swarm Optimization (PSO) method is used to tune function parameters *f(a, b)* consists in COCOMO II and take advantage of Fuzzy Logic to build a set of linear models over the domain of possible software Lines of Codes (LOC). The PSO algorithm starts by generating particle locations randomly throughout the designed space of the parameters, and the performance of the developed model was evaluated using NASA software projects dataset [10][13]. Moreover, some researchers suggest using the regression model to minimise the sum of relative errors and impose the non-negative coefficients, which is a favourable technique for calibrating the COCOMO model parameters [9].

The Artificial Neural Network (ANN), Support Vector Machine (SVM), Linear Regression (LR), K-Nearest Neighbour (KNN) are some of the data mining techniques combined with the algorithmic COCOMO II model in the quest for the precise estimation software development effort and time estimation [5]. Simulated annealing (SA) had been used to overcome the limitation of GA, which produced premature convergence population and was proposed to advance the COCOMO II PA model coefficients for achieving the accurate software effort estimation and to diminish the dubiousness of COCOMO II post architecture model. The study, however, suggested combining different techniques for calculating the best estimate for software engineering field. Recently, COCOMO II model is widely studied by most software researchers and practitioners in the enhancement of its ability to estimate software cost project precisely. COCOMO II is one of the numerous techniques in software effort estimation. It is widely accepted as an industry standard because of its applicability at diverging the stage of software engineering studies. The accuracy of COCOMO II is highly affected by of its input

IOP Conf. Series: Materials Science and Engineering **551** (2019) 012074 doi:10.1088/1757-899X/551/1/012074

parameters which are size of projects, coefficients and attributes cost drivers. The small changes in these parameters bring huge differences in their effort estimation [11].

Incorporating COCOMO II with novel soft computing and machine learning model becomes popular in research nowadays because it offers an added advantage with the ability to improve performance in tuning the parameters and to learn from experience data, which mainly focus on predictive accuracy [7]. Most of the research use many models with the combination with COCOMO II, to improve the signs of the particular model. However, the accuracy of the models is still questionable until now. The COCOMO model, either COCOMO I or COCOMO II is one of the established software effort estimation related to software engines, but it is not utilised in the practice even though it is widely known to the software industry. Table 2 shows the summary of the machine learning techniques that have been applied for ten years from 2007 until early 2018.

Table 2: Some of ML techniques for SEE for 10 years

YEAR	TECHNIQUES	DATASET	EVALUATION METHOD
2007, 2015	NeuroFuzzy	NASA &COCOMO	MMRE
2008	PSO	NASA	MMRE
2008	Regression Techniques	COCOMO	MMRE and PRED
2010, 2013,2015	Genetic Algorithm	-Turkish & Industry data set -COCOMO & NASA	Not specified
2011, 2014	ANN	COCOMO & NASA, PROMISE repository	Not specified
2011, 2018	Fuzzy Logic	NASA	MMRE, PRED
2013	PSO-ANN-COCOMO II	COCOMO I & NASA93	MRE and PRED (25%)
2014	Radial Basis function NN	COCOMO & NASA	MSE
2014	Bayesian with PSO	NASA 93	MMRE
2014	PCA with ANN	COCOMO 81	
2014	Simulated Annealing	NASA	
2014	Hybrid COCOMO and FP	NASA	MRE
2015	Bee Colony Optimization	Interactive Voice Response Software project dataset	MMRE
2016, 2017, 2018	Cuckoo, Hybrid Cuckoo Optimization	NASA60, NASA63, NASA93, MAXWELL, KEMERER, and MIYAZAKI	MRE, MMRE
2016	COCOMO TLBO	NASA	MMRE, PRED (30)
2016	Differential Evolution Method	NASA	MMRE, MRE
2016	Hybrid PSO-ANFIS	NASA	MSE, MMRE, PRED
2016	Bat Algorithm	NASA 93	MMRE
2016	COCOMO-GA	NASA	MMRE
2017	ANN-Dragonfly Algorithm	NASA	MSE, MMRE, MAE

The summary shows that the trends of estimation software effort are more on the exploration of enhancement techniques rather than proposing a new ML technique. The latest publication on 2017 enhances the previous ANN algorithm techniques using dragonfly to provide the optimal training of neural network, which offered more enhanced and accurate software effort estimation. In software effort estimation studies, there are several questions regularly asked; which data should be used? Which datasets are more reliable to be used for studies? How to choose a specific project to estimate a future project? These questions are important for the researchers in making their decision before analysing the historical data. Most researchers usually did not show the explicit ways on how to make a choice from the specific studies. The software process improvement is facing the difficulties with such effort estimation because of lack of data for analysis and lack of information regarding the data types.

The data quality concept is large. It does not only relate to consistency, but also the quality of completeness datasets used. There was a study warned about using the unbalance datasets in the research that it might cause the impact factor to be concealed. In software effort estimation, the whole data project effort is of utmost important to the developers to make reasonable plan or activities in project management. The occurrences of missing data in project data can bring significant impact on future estimation effort. There is an analysis showing that the pre-processing data analysis such as treatment of missing data can also influence the performance of the prediction model [8]. There are numbers of alternative ways of dealing with missing data. In some cases, deletion or elimination the missing variable is the default method for most procedures. However, there is only a few studies investigating the imputation of missing data in software effort estimation area. There should be an empirical investigation of the robustness and the accuracy in handling the missing data.

4. Conclusion and Future Research

There are still many problems regarding on how to compare and evaluate estimation methods. Most researchers either do not cooperate closely with the software industry and current issues or believe that it is better to focus on a replacement rather than improvement the of current approaches employed by the industry. Therefore, it is recommended that the future research works put a focus on the software enhancement as opposed to proposing new replacement techniques. It is good to note that some articles were based on datasets that are too old to be representative for the current or future projects. Future research should focus on understanding the relationship between project characteristics (dataset quality) and estimation evaluation. Most articles evaluated the estimation by employing historical datasets, but only a few evaluated based on the completed real-life estimation situation. Therefore, more studies on the methods used in real-life situation should be conducted.

Acknowledgements
This work was financially supported by RUG-Tier 2 Grant (Vot. No:15J16) from Universiti Teknologi Malaysia.

References

[1] Borade, J. G., & Khalkar, V. R. (2013). Software Project Effort and Cost Estimation Techniques. *International Journal of Advanced Research in Computer Science and Software Engineering, 3*(8), pg. 730–739.

[2] Choudhary, K. (2010). GA Based Optimization of Software Development Effort Estimation. *International Journal Of Computer Science And Technology,* Vol.(1), pg. 38–40.

[3] Choudhary, K. (2011). Parametric Estimation of Software Systems. *International Journal of Soft Computing and Engineering, 1*(2), pg. 17–20.

[4] Clause, J., Li, W., & Orso, A. (2007). Dytan: a generic dynamic taint analysis framework. In *Proceedings of International Symposium on Software Testing and Analysis* (pp. 196–206). ACM.

[5] Dan, Z. (2013). Improving the accuracy in software effort estimation: Using artificial neural network model based on particle swarm optimization. In *Proceedings of 2013 IEEE International Conference on Service Operations and Logistics, and Informatics, SOLI 2013*(pp. 180–185).

[6] Garbajosa, J. (2008). The emerging ISO International Standard for Certification of Software Engineering Professionals. In *IFIP International Federation for Information Processing* (Vol. 280, pp. 173–178).

[7] Maleki, I., Ghaffari, A., & Masdari, M. (2014). A New Approach for Software Cost Estimation with Hybrid Genetic Algorithm and Ant Colony Optimization. *International Journal of Innovation and Applied Studies, 5*(1), 72–81

[8] Missier, P., Lalk, G., Verykios, V., Grillo, F., Lorusso, T., & Angeletti, P. (2003). Improving data quality in practice: A case study in the italian public administration. *Distributed and Parallel Databases, 13*(2), 135–160.

[9] Rajper*, S., & and Zubair A. Shaikh. (2016). Software Development Cost Estimation Approaches - A Survey. *Indian Journal of Science and Technology, 9*((31)), pg.1–5.

[10] Sandhu, G. S. (2014). A Bayesian Network Model of the Particle Swarm Optimization for Software Effort Estimation tool. *International Journal of Computer Applications, 96*(4), 52–58.

[11] Vu Nguyen, Bert Steece, B. B., Nguyen, V., Steece, B., Boehm, B., & Vu Nguyen, Bert Steece, B. B. (2008). A Constrained Regression Technique for COCOMO Calibration. *Acm 978-1-59593-971-5/08/10*, 1–10.

[12] Zaid, A., Selamat, M. H., Azim, A., Ghani, A., Atan, R., Koh, T. W., … Albrecht, A. (2008). Issues in Software Cost Estimation. *International Journal of Computer Science and Network Security, 8*(11), 350–356.

[13] Zhang. "Improving the accuracy in software effort estimation: Using artificial neural network model based on particle swarm optimization." Service Operations and Logistics, and Informatics (SOLI), 2013 IEEE International Conference.

Hybrid Predictive Modelling for Motor Insurance Claim

Zuriahati Mohd Yunos[1] , Siti Mariyam Shamsuddin[2] , Roselina Sallehuddin[3] , and Razana Alwee[4]

[1,2,3,4]School of Computing, Faculty of Engineering, Universiti Teknologi Malaysia, 81310 Skudai, Johor, Malaysia.

Email : zuriahati@gmail.com

Abstract. The objective of this study is to propose a new hybrid model to predict the Malaysia motor insurance claim by estimating the two important components; claim frequency and claim severity. The proposed model are integrating between grey relational analysis and back propagation neural network. We proposed the hybrid model to handle the issue of the insurance data and the complexity of classical statistical technique. Moreover, the classic statistical techniques are incapable of handling huge information in the insurance data. Thus, hybrid model is proposed because it has a high learning ability and capability to learn. Finally, a comparative analysis is carried out to evaluate the predictive model performance between GRABPNN and BPNN. The results produce by MAPE show a small percentage and thus, show that GRABPNN model provides more accurate predictive results compared to BPNN model.

1. Introduction

Predictive modelling is a process that involves problem identification, analysis of data, model development, as well as prediction and validation [11]. Predictive modelling in the insurance industry helps actuaries and other insurance analysts employing predictive models to enhance business operations that previously used human expertise. Moreover, the influences of predictive modelling are also dependent on the quality of the data used to generate the model. Insurance is a unique type of agreement between the insurer or insurance company and the insured or client in which the insurers permit that upon the occurrence of specific events, whether to make a payment to clients or cover the specific costs. There are two types of insurance such as life insurance and general insurance. Among the primary products in general insurances are motor insurance, fire insurance, personal accident insurance, health care insurance, and traveling insurance. This study focuses on developing a predictive model for motor insurance claim by estimating the two important components, namely, claim frequency and claim severity [4, 6, 8]. Claim frequency is defined as the number of claims per exposure unit, whereas claim severity is defined as the average claim cost per claim [8]. The modelling of claim frequency and claim severity needs an information of exposure, number of claims and amount of claim (cost). The expected claim frequency and claim severity can be calculated through a process of identifying grouping risk, which has the same characteristics (combination of rating factors or variables).

2. Related Work

There are two important issues highlighted in motor insurance claims which are the data and techniques, and these issues are interrelated with one another. The first issue is related to the characteristics of insurance data, which contain huge information or large number of variables, as well as uncertainty, noisy and incomplete information as agreed by [5, 6, 7, 9].

This issued has been discussed in [20] on how to identify the rating factors and rating classes to build the grouping risk. Another study by [23] reported that the the grouping risk can be determined by a number of factors such as the vehicle cubic capacity, the geographic zone where the insured lives, age of the insured. This includes the data of gender criterion, the driver's age, the vehicle characteristics, and the driving behaviour [22, 24]. [3] also agreed that insurance data have complex data structures. Another problem is the existing of extreme values in the data which cannot be ignored or dealt as outliers [14]. The work of [18] has thoroughly investigated the insurance data consists a large number of variables and many important variables redundancies between them because they share the same information. The second issue is related to the complexity of statistical analysis that has become more apparent. Due to this, actuaries had to solve the problem of finding a model that can explain realistically the event of risk [3, 4] and a model that is able to handle complex problems in exploiting varying information [12]. There are some statistical methods and mathematical algorithms applied in insurance claim practices such as generalized linear model (GLM) [19], Almer's simple multiplicative model [15], an additive model in 1960 [16], log-linear model [17]. However, these techniques have limitations such as lack of discriminatory power, restrictions on the independent distribution, or poor ability to interpret [21].

Thus, we recommed a feature ranking using GRA as the technique attempt to maintain each feature which is most likely will be able to provide improvements in predicting accuracy and its not eliminate the features but ranking it. In addition, BPNN is choose as the technique use past experience to train the network to provide more consistent and reliable evaluations on claim frequency and claim severity.

3. Methodology

3.1 Experimental data

The data is provided by Insurance Services Malaysia Berhad (ISM), which is based on 1.2 million policies for the year 2001 until 2003 and are used to evaluate the proposed hybrid model, GRABPNN. The claim data motor insurance consist of two different types : third party property damage (TPPD), and third party bodily injury (TPBI). The input factors or rating factors used for predicting frequency claims and severity are coverage, vehicle cubic capacity, vehicle made, vehicle year and location. Any missing values are eliminated and data is normalized using min-max formula to smoothing data distribution for obtaining better data generalization. Then, the data is partitioned into two parts:training sets and testing sets based on 70%:30% ratio. Table 1 lists all the inputs and output used for claim frequency and claim severity. For claim frequency, each data is used to determine the number of claims made by the insured (clients); and for claim severity, the data are used to compute the amount claimed by the insured or amount paid by the insurer to insured.

Table 1. Input and output variables for TPPD and TPBI claims

Notation	Claim Frequency		Claim Severity	
	Input nodes	Output nodes	Input nodes	Output nodes
F1	Coverage		Coverage	
F2	Vehicle made		Vehicle made	
F3	Vehicle cc		Vehicle cc	
F4	Vehicle year	Claim frequency	Vehicle year	Claim severity
F5	Location		Location	
F6	Exposure		-	
F7	-		Number of claim	

3.2 Model Development

Figure 1 shows the research framework consist of two phases; feature ranking and model development. In phase 1, feature ranking using GRA is employed to determine the most and least significant features, and rank the six important features as listed in Table 1. The feature ranking is completed by rearranging the GRG value by put the greatest GRG value in the first position; the second largest is located in the second location, and subsequently. In phase 2, the output from the GRA feature ranking is employed and combined with the BPNN and known as, GRABPNN and compare with single model BPNN. The process of integrating the techniques is known as hybridization. The motivation of integration is to understand the underlying relationship between input nodes and hidden nodes, and to enhanced the prediction accuracy. There are nine network structure in GRABPNN are developed and to be tested on claim frequency and claim severity. There are six input nodes and one output node are used as listed in Table 1. The number of hidden nodes are determeine by using the number of input nodes (n) and used formula n, $2n$, $2n$+1 [12].

IOP Conf. Series: Materials Science and Engineering **551** (2019) 012075 doi:10.1088/1757-899X/551/1/012075

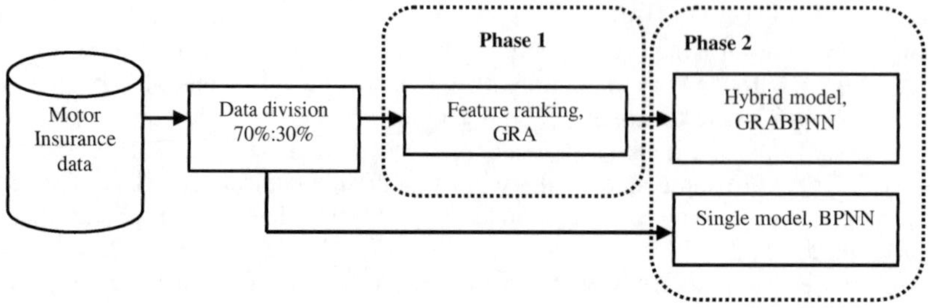

Figure 1. Research framework.

4. Experimental Results and Analysis

This section discusses the result obtained from the experiment. The discussion is divided into two parts. Part 1 discusses on how GRA rank the input and select the optimum features that will be fed into BPNN. The second part will discuss on the result obtained from BPNN using the optimum features. Then, the performance of the proposed model will be compared with the standard BPNN.

4.1 Feature Ranking using GRA

Based on Table 2 and 3, it was found that for each claim types, the most significant features for claim frequency are exposure (F6) and coverage (F1), and for claim severity are number of claim (F7) and coverage (F1) with a stronger correlation close to 1 which are greater than 0.5. The feature ranking result are based on GRG value in descending order and can be seen in the column 'rank' from the most to the less significant.

Table 2. The GRG value of claim frequency

Claim frequency				
Features	**TPPD claim**	**Features**	**TPBI claim**	**Rank**
F6	0.97229	F6	0.96695	1
F1	0.64829	F1	0.65069	2
F4	0.58924	F4	0.59232	3
F5	0.58831	F5	0.58212	4
F3	0.57892	F3	0.5803	5
F2	0.57855	F2	0.57686	6

Table 3. The GRG value of claim severity

Claim severity				
Features	**TPPD claim**	**Features**	**TPBI claim**	**Rank**
F7	0.90598	F7	0.90227	1
F1	0.61384	F1	0.61701	2
F4	0.60368	F4	0.60135	3
F5	0.58931	F3	0.59091	4
F3	0.58403	F5	0.59039	5
F2	0.58205	F2	0.58336	6

4.2 Hybrid Model using GRABPNN

In this study, the model development is done through trial-and-error with the aim to obtain the best predicting results. The number of input nodes is similar to a number of features. The maximum and minimum numbers of input nodes used are 6 and 4. The elimination number of input nodes is done one by one from 6 to 4. The output from the GRA was then fed into the network. Then, a comparative analysis was carried out to evaluate the predictive model performance between GRABPNN and BPNN. The best model is chosen based on four error measurements, mean squared of error (MSE), root mean square of error (RMSE), mean absolute error (MAE) and mean absolute percentage error (MAPE). The best model is chosen based on four error measurements, mean squared of error (MSE), root mean square of error (RMSE), mean absolute error (MAE) and mean absolute percentage error (MAPE).

Table 4. The best model for claim frequency

Predict Model	Claim types	N/W	Claim Frequency			
			MSE	RMSE	MAE	MAPE
GRABPNN	TPPD	6-13-1	310.5	17.6	10.1	20%
	TPBI	4-8-1	96.1	9.8	5.3	25.7%
BPNN	TPPD	6-13-1	369.50	19.22	10.73	21.9%
	TPBI	4-8-1	79.13	8.90	5.33	25.8%

Table 5. The best model for claim severity

Predict Model	Claim types	N/W	Claim Frequency			
			MSE	RMSE	MAE	MAPE
GRABPNN	TPPD	6-12-1	2713539.3	1647.28	958.99	48.1%
	TPBI	6-12-1	123158238.8	11097.67	6112.32	51.6%
BPNN	TPPD	6-12-1	2870793.89	1694.34	1105.20	65.2%
	TPBI	6-12-1	133139846.4	11538.62	5836.05	70.56%

Table 4 and 5 displays the best model with the best network structures for each claim type. It was found that for each claim type for claim frequency and claim severity, the best model is given by GRABPNN model. A small value of MAPE is acquired by GRABPNN compared to BPNN for claim frequency and claim severity (highlighted in grey). Therefore, the studies have proven that the proposed hybrid model is good and acceptable in predicting claim frequency and claim severity. Moreover, the application of GRA to determine and rank the most, and the least significant features have enhanced the accuracy of the proposed hybrid model.

5 Conclusions

This study proposes a new hybrid model for modelling the motor insurance claims. The proposed hybrid model was used and tested on two claims types, namely third party property damage (TPPD) and third party bodily injury (TPBI) claims. Each claim type has two elements, namely, claim frequency and claim severity, and the development of the predictive model has been applied on these two elements. Rating factors are used as an input factor in the proposed hybrid model. the implementation of GRA does not require any assumptions about the data and it is suitable in all situations, although the historical data are limited. The proposed hybrid model, GRABPNN has successfully been done and indeed improves the predictive accuracy. Moreover, the study has gained evidence that, given various numbers of features and hidden nodes, rank the informative features; GRABPNN obtained better performance in modelling claim frequency and claim severity for each claim type as compared to other models.

Acknowledgements
The authors gratefully acknowledge the Fundamental Research Grant Scheme, through grant number FRGS:4F825, Research Management Centre (RMC) of Universiti Teknologi Malaysia and Applied Industrial Analytics (Ali@s) for the support and motivation in making this research a success.

References

[1]Bahia H S I 2013 *Inter. J. Intell. Sci.* **3**136.
[2] Batty M, Tripathi A, Kroll A, Peter Wu C-S, Moore D, Stehno C, Lau L, Guszcza J and Katcher M 2010 *Predictive Modelling for Life Insurance* (Deloitte Consulting LLP).
[3] Christmann A 2004 *Allge. Stat. Archiv*, **88** 375.
[4] David M 2015 *Proc. Eco. Finance* **20 1**47.
[5] Dugas C, Chapados N, Ducharme R, Saint-Mleux X. and Vincent P 2011 *Inter. J. Bus. Intell. Data Mining,* **6** 237.
[6] Frees EW, Peng Shi P and Valdez EA 2008 *ASTIN Bulletin* **39** 165.
[7] Frees EW 2014 *Frequency and severity models.* In E Edward, G Meyers and R A Derrig Cambridge. Cambridge University Press.
[8] Ismail N and Jemain A A 2008 *Sains Malaysia* **37** 412.
[9] Kim J H T and Kim J 2014 *J. Appl. Math.& Info.* **32** 343.
[10] Kitchen F L 2009 *Inter. J. Electronic Finance* **3** 311.
[11] Rosenberg M A, Frees E W, Sun J, Johnson Jr P H and Robinson J 2007 *North American Acturial J.* **11** 54.
[12] Zhang G, Patuwo B E and Hu M Y 1998 *Inter. J. Forecasting* **14** 35.
[13] Shapiro A F 2009 *Ins. Math. & Eco.* **44** 307.
[14] Aftab S, Abbas W, Bilal M M, Hussain T, Shoaib M and Mehmood S H 2013 ICDIM p 1.
[15] Almer B 1957 *15th Inter. Congress of Actuaries* **2** 314.
[16] Bailey R A 1963 *Proc. of the Casualty Actuarial Society* **50** 4.
[17] Chang L and Farley W 1979 *J. of Risk and Insurance.* **46** 75.
[18] Chapados N, Dugas C, Vincent P and Ducharme R 2008 *ICDM Workshops* p 97.
[19] David M 2015*Proc. Eco.& Fin.* **20** 147.
[20] Laas D, Schmeiser H and Wagner J 2016 *The Geneva Papers on Risk & Ins.* **41** 398.
[21] Lindsey J K 2004 *Introduction to Applied Statistics* (Oxford University Press).
[22] Schmeiser H, St¨ormer T and Wagner J 2014 *The Geneva Papers on Risk & Ins.* **39** 322.
[23] Shi P and Valdez E A 2014 *Ins. Math. & Eco.* **55** 18.
[24] Wamwea C M, Muema B K and Mung'atu J K 2015 *American J. Theo. & Applied Stat.* **4** 527.

IOP Conf. Series: Materials Science and Engineering **551** (2019) 012076 doi:10.1088/1757-899X/551/1/012076

Overview of Machine Vision on Digital Imaging Approach for Automatic Tuna Length Measurement

Adam Ridhwan Sukiman, Haswadi Hassan, Azlan Mohd Zain, Alif Mukhtar and Kamarul Zaman Panatik[1]

Applied Industrial Analytics Research Group(ALIAS), School of Computing, Faculty of Engineering, Universiti Teknologi Malaysia(UTM), Skudai, Johor, Malaysia.

E-mail: aridhwan7@graduate.utm.my

Abstract. The demand of fish and marine product have increased significantly throughout the year. Therefore, overfishing had become major concern for various parties around the world especially the United Nation(UN). Collecting numerical data is a first step toward preservation and conservation of tuna and other fish population. This paper reviews automatic tuna length measurement by using machine vision techniques. This is one of the steps in collecting necessary data on size of fishes caught by fishermen. Previous research had been done on measuring the length of fish, but the accuracy rate of the length measurement is not satisfactory enough. Many factors affecting the accuracy of the length measurement processes such as lighting inequalities, blood on the platform and others. Machine vision techniques have a great potential to be used in automating the measurement of the fish length with high accuracy. Machine vision techniques could improve the accuracy and efficiency of the measurement of the fish length that would substitute the tedious manual work.

1. Introduction

The fisheries sector is one of the most important food sources throughout the world. Due to growth of human population, tones of fish have been catch throughout human history. Currently, overfishing has been a great concern due to the rapidly shrinking population of fish especially tuna. This issue is taken seriously by various parties including the United Nation Organization (UN). By 2020, the Sustainable Development Goals, a target from UN want to effectively regulate harvesting and end overfishing, illegal, unreported and unregulated fishing and destructive fishing practices and implement science-based management plans, in order to restore fish stocks in the shortest time feasible, at least to levels that can produce maximum sustainable yield as determined by their biological characteristics [1].

According to statistical data reported by Department of Fisheries, Ministry of Agriculture Malaysia, fish production in Malaysia increased from 2005 to 2017 with latest marine fish catches proportion are 1.46 million tons of fish in 2017. To avoid fish extinction, Malaysia government enforce Malaysian tuna fishermen to measure and record each catch's length and weight they caught. This enforcement is one of the government effort to preserve and sustain wild tuna stock in the ocean. Usually, tuna fishermen will measure their catch by using a measuring tape and weighing scale. This current method took the longest time and effort. To tackle this problem, several machine vision techniques had been proposed by researchers. Machine vision is the technology and techniques used to provide imaging-based automatic inspection and analysis.

IOP Conf. Series: Materials Science and Engineering **551** (2019) 012076 doi:10.1088/1757-899X/551/1/012076

In this paper, the author will review techniques of machine vision technology and the different method of measuring the length of the tuna on flat surface from 2D image. The machine vision technology can help fishermen to obtain the length information of that would overcome the limitation of the traditional methods as well as reducing the labor cost [2,3]. Criteria of this review is the accuracy of length measurement. Fish point identification will be discussed in section 2.1. then we discuss the length measurement method in section 2.2.

2. Fish Length Measurement

Fish length is one of the parameters of measurement needed to record the tuna's data. This section discusses automatic tuna length measurement from a 2D image. There are two steps of automatic tuna length measurement process which are; fish's snout and tail point identification (FSATPI), and length measurement method. Fish used in this review are tuna and the tuna will be treated as a linear measurement because the body of the fish is regarded as a straight line and not curve. Several types of fish length have been identified but the common length measurement used for tuna is fork length as in figure 1.

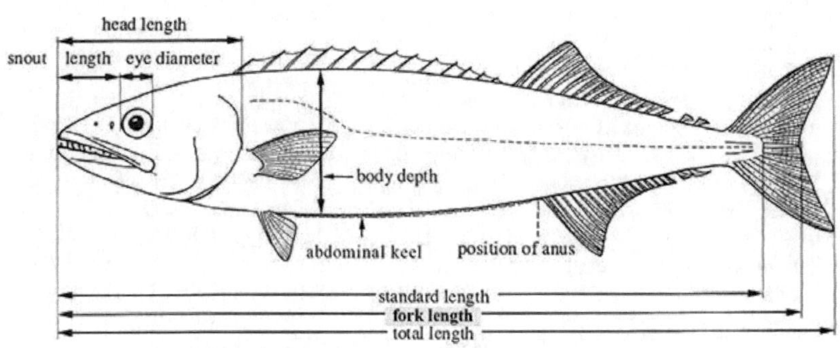

Figure 1. Different types of fish's length measurement.

2.1. *Fish Points Identification*

Past studies report on the method used to FSATPI. FSATPI play a major role on automatic length measurement. Without points identification, the length measurement cannot be done. Several methods are used for point identification such as; estimation by human eyes[3], Hsiu method[6], and Truss Morphometric Network (TMN) [8]. Estimation by human eyes were compared in several studies. From Table 1, human eyes are used to point the fish snout and tail before the points are used by the machine to automatically measure the fish length [3]. However, human eyes cannot be reliable on a long time task due to fatigue. There are result show approximately 10% of error [5]. While, Hsiu method[6] used in point identification has error rate less than 2.19%.

Simultaneously, TMN can also identify the fish snout and tail point. TMN constructed with the help of landmark points are one of the tools for fish body identification. Truss length measured between these landmark points should either lie on curved surfaces or be on straight lines lying on a flat plane[8]. Figure 2 show TMN constructed on a bluefin tuna body.

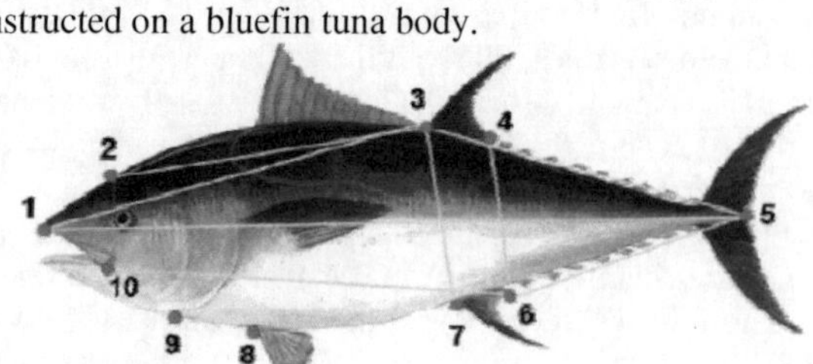

Figure 2. Truss Morphometric Network (TMN) of a Bluefin tuna.

Table 1. Detailed information on method of fish snout and tail point identification.

Method	Mode of measurement	Error rate	Strengths and weaknesses
Estimation by human eyes [5]	Manual	10%	- not very accurate - accuracy depend on the people
Hsiu method [6]	Automatic	<2.19%	+ Highly accurate - Accuracy depending on lighting qualities
TMN [10]	Automatic	-	- no result of accuracy rate show from previous research

2.2. Length Measurement Method

When the point of fish snout and tail have been identified, the length of the fish can be measured. The fish length can be calculated by using the relation of fish body pixel length and the image reference scale. The reference scale can be obtained through a square on the color plate. Currently, the widely used method used by fishermen are estimating by human eyes and measuring tape. However, estimation by human eyes is a hard labor job and concentration for a long time is needed. Less than 10% of error was shown by estimation by human eyes. From Table 2, Hough transform used by Hsieh [3] shown the error of measuring the length of tuna was less than 5%. But, the accuracy of the Hough transform method can be increased by using projective transform to reposition the reference plate before length measurement was calculated.

While, image thinning used by Jun showed a measurement error of less than 3% [9] as shown in Table 2. Man show that the Hsiu method used for measuring fish length where resulted less than 3% of measurement error [6]. Unfortunately, there were no past researches yet have been done to a bigger fish such as tuna because of the distance of targeted object and camera, and the angle of the targeted object and camera needed to be recorded for this method to calculated the length measurement of tuna [6]. The tuna caught by fishermen will be put into the freezer as fast as possible to retain the freshness of the fish and this will be a hindrance for researcher to record every single angle pf the fish during the photo taken by the camera.

Best fitting rectangle used by Hao for measuring fish length where resulted less than 0.5% of measurement error [2] as shown in Table 2. Unfortunately, same with Hsiu method, this method not yet been research on tuna fish which is much bigger than fish use in the research and tuna fin structure cause a significant factor of error for this method as best fitting rectangle always measure the total length of fish rather than fork length which are intended for measuring the length of a tuna. Though, the length measurement result may vary due to the different size of the subject [2].

Grade-3 polynomial regression method was used by Torres for measuring tuna where resulted less than 4% of measurement error [7] as shown in Table 2. Regression used in many research as it one of the simplest method used to measure the length of the object. Nevertheless, no research had been done yet to a bigger fish by using this method.

IOP Conf. Series: Materials Science and Engineering **551** (2019) 012076 doi:10.1088/1757-899X/551/1/012076

Table 2. Methods of fish length measurement.

Method	Object	Measurement error rate	Strength and Weakness
Hough transform [3]	Tuna	<5%	+ have relatively high accuracy of measuring tuna length + accuracy rate can be increase by using projective transform
Image thinning [9]	Tuna	<3%	+ have high accuracy on measuring tuna length
Hsiu Method [6]	Fish	<2.19%	+ have high accuracy on measuring smaller fish length - No past research have been done to measure tuna using this method
Best fitting rectangle [2]	Fish	<0.5%	+ have high accuracy on measuring smaller fish length - No past research have been done to measure tuna using this method
Grade-3 polynomial regression [7]	Fish	<4%	- No past research have been done to measure tuna using this method

3. Discussions

Automation of the tuna length measurement process is important because it will make the length measurement more accurate and faster compare to manual measurement. Thus, it is important to make the measurement process automatic to increase the efficiency of the measurement process and increase the effectiveness of tuna catch regulation.

There are other features that need to take into consideration in measuring tuna length using machine vision such as the visual perspective of the object, the angle of the camera to the object, and the distance between the camera and the object. The angle of the camera is very important as it will greatly affect the visual perspective of the object [3]. There are study had been made by taking in the one of the features for fish length measurement method [6,11] but no research had been done yet by using all of the features for fish length measurement method by considering all of the measurement environment factor. Usage of lightbox have higher accuracy in measuring the length of other smaller fish [2] but the disadvantage of using lightbox is it cannot fit the tuna body as tuna body is too big to be fit into the lightbox. There were a number of researches had been done with smaller fish species, but they were not yet tested on tuna fish [8].

Although the length measurement result has a low error rate, it is still not satisfactory. More research needed to be done in order to increase the accuracy rate of fish length measurement. Based on [2], although the method can be used on both smaller and bigger fish, the result of the length measurement very different from each other.

4. Conclusions

Problems such as image noise, lighting inequalities and poor contrast are obstacles to implementing effective and accurate machine vision techniques. Besides, there are factors that can affect the machine vision process such as blood stain on the reference plate, blood stain on the fish body and position of the fish. The approach of single image camera imaging system in a controlled environment was expected to improve image capture conditions, as backlighting techniques. Furthermore, segmenting fish before measurement made is a must do process as it eliminates foreign image noise and uncontrolled setting.

As for fish measurement, the fish was assumed to be positioned along the x-axis of the environment setup. The result from previous research shows that a relatively simple method such as polynomial regression can be used for calculating the fish length with remarkable accuracy [7].

5. Acknowledgements

The authors greatly acknowledge the Research Management Centre UTM for financial support through the Roboxa Technologies Sdn Bhd, private local company scheme VOT no. R.J130000.7651.4C208.

References

[1] Eric P 2015 *J. Global Ethics.* **11** 262-269.
[2] Hao M *et al* 2016 *Comp. & Comp. Technol. In Agri. IX* **479,** 15.
[3] Hsieh C L *et al* 2011 *Comp. & Electro. In Agri.* **75,** 44.
[4] Shah M *et al* 2015 *Inter. J. Env. Soc. Space* **3** 22.
[5] Ibrahim M Y *et al* 2006 *IEEE Inter. Conf. on Mechatro.* 462.
[6] Man M *et al* 2016 *J. Adv. Agri. Technol.* **3** 67.
[7] Torres G S *et al* 2018 *Eng. Lett.* **26:4** 461.
[8] Mojekwu T O *et al* 2015 *J. Aquac. Res. Dev.* **6** 354.
[9] Jun H *et al* 2009 *A. in Fishr. Sci.* **76** 1359.
[10] Addis P *et al* 2010 *Collect. Vol. Sci. Pap. ICCAT.* **65** 982.

IOP Conf. Series: Materials Science and Engineering **551** (2019) 012077 doi:10.1088/1757-899X/551/1/012077

Optimum Placement of Piezo-Polymer Power Harvester for Epicardial Pacemaker

Mohd. Hafizulhadi Bin Mohd. Asri[1] , Muhammad Haikal Satria[2] , Shukor Talib[3], Hasri Samion[4].

[1,2] Faculty of Biosciences and Medical Engineering, Universiti Teknologi Malaysia, 81310, Skudai, Johor, Malaysia.

[3] Faculty of Computing, Universiti Teknologi Malaysia, 81310, Skudai, Johor, Malaysia.

[4] Institut Jantung Negara Sdn Bhd, 50400, Kuala Lumpur, Malaysia.

E-mail: mhafizulhadi3@live.utm.my

Abstract. The major drawbacks of current pacemaker are the battery replacement. Patient will need additional surgery to replace the pacemaker unit with the new one. Utilization of rechargeable battery has been proposed to overcome this problem. However, recharging a battery within the body is not feasible due to tissue-heating and battery's charge-and-discharge life-time. By those reasons, the utilization of piezo-polymer is suitable for a self-powered pacemaker as energy harvester. Piezo-polymer has been widely used as energy harvesting but none for cardiothoracic implantable device. This study focuses on specific implementation of piezo-polymer for epicardial pacemaker. The proposed fundamental research aims to identify the optimum location on the heart to put the piezo-polymer. This research will be conducted by simulation of left ventricle of heart via ANSYS. Heart stress-strain Finite Element Analysis (FEA) will be employed to obtain the maximum harvested power. The result shows the location of myocardial contraction that produces sufficient kinetic energy for the placement of the pacemaker. The heart 3-dimensional images are taken from cardiac-CT or cardiac-MRI to search the optimum location on the heart for energy harvesting and minimize pacing energy. In conclusion, the left ventricular wall motion and deformation caused by the cardiac wall movements have been analysed based on the left ventricle model in the simulation to get the location of the maximum kinetic energy formed.

1. Introduction

Pacemaker is used to control an abnormal heart rhythm by providing and by inducing cardiac contractions with electrical stimulus to the heart [1]. Pacemaker is needed when the heart electrical system has a malfunction. In 2007, it is been estimated about 560,000 pacemaker implantation in North America and 680,000 in Europe [2]. Existing conventional and leadless pacemaker prototypes use non-rechargeable, long-life, and high-density battery to supply the power. One of the

IOP Conf. Series: Materials Science and Engineering **551** (2019) 012077 doi:10.1088/1757-899X/551/1/012077

drawbacks is, for example in paediatrics, the battery is still need to be replaced after couple of years, which means additional surgery to take out the pacemaker unit. To overcome this problem, utilization of rechargeable battery has been proposed.

One of challenges of the usage of rechargeable battery for pacemaker is recharging technique. Unlike the battery for lead-based pacemaker, which is located inside the generator exactly beneath the patient skin, battery of leadless pacemaker is located inside heart chamber; therefore, it is difficult to charge wirelessly. A technique has been proposed by using radio frequency (RF) [3]. However, RF needs space for antenna for power transmission, which is not possible for small size leadless pacemaker. Furthermore, efficiency of power transmission through RF is very low, with the mostly heat dissipation goes through antenna. Ultrasound based implanted battery recharging system has been demonstrated by [4]. Their in-vitro experiment successfully harvested 600mW through 10–15 mm of tissue depth. The drawback is power transmission using ultrasound can only be limited to a certain time due to heat effect on tissue and therefore, not suitable for leadless pacemaker.

The power may be obtained from several sources, such as: sound, solar, body heat, breathing, movement [5], muscle, joint movement, and glucose-based biofuel cell. Piezoelectric materials are one of the emerging concepts to provide self-energy harvested from human body. One of the promising materials for the pacemaker energy harvester is polymer piezoelectric (piezo-polymer). Although the electromechanical coupling factor is not as high as ceramic-based, piezo-polymer is biocompatible and more mechanically flexible; both are very important factors for implanted medical devices.

Energy harvesting by using piezoelectric materials may result in sufficient energy for pacemaker depending on the properties they have. One of the hypotheses provided by [6] has estimated that the power resulted from piezoelectric materials from heart movement could achieve hundreds of μJ.

2. Methodology

The methodology that will be applied by the study has been chosen in order to provide the useful information about the kinetic energy production at the left ventricle of heart. The idea of this research is to propose an ideal concept of leadless pacemaker which is "Implant and Forget". For this purpose, a continuous electrical power supply from energy harvester is an essential factor. The kinetic energy may supply the energy in order to make the pacemaker to be self-powered continuously. Figure 1 shows the steps involve in this methodology.

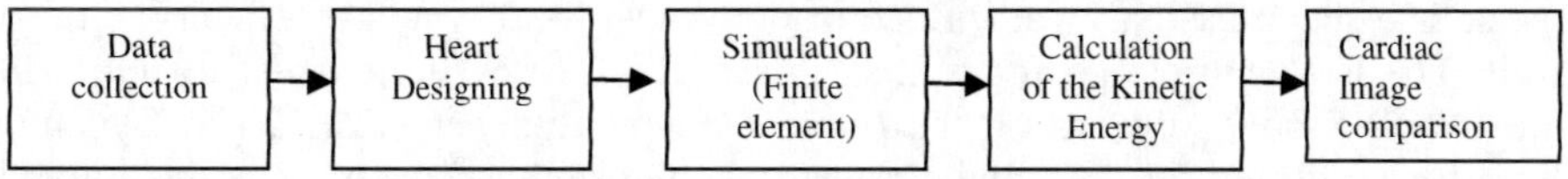

Figure 1. The flowchart of the methodology.

Based on Figure 1, the data is collected for designing and running the simulation. There are 2 types of data which is the pre-processing data and post-processing data. The pre-processing data is collected in order to design the structure of the heart, according to the real organ. The data includes the dimension of the heart, the density, Young Modulus and Poisson ratio. The collected data will be inserted as the engineering data inside the simulation software.

The heart is cone-shaped. The base positioned upwards and tapered down to the apex. The largest compartment of the heart is usually slightly left side of the chest even though it may infrequently be offset to the right. An adult heart generally has a mass in the rage of 250-350 grams. The heart size is usually described as the size of its individual fist, with typical values of 120mm in length, 80mm wide and 60mm in thickness. The dimension of heart in Figure 2 is derived from the cross diameter of each atrium and ventricle [7].

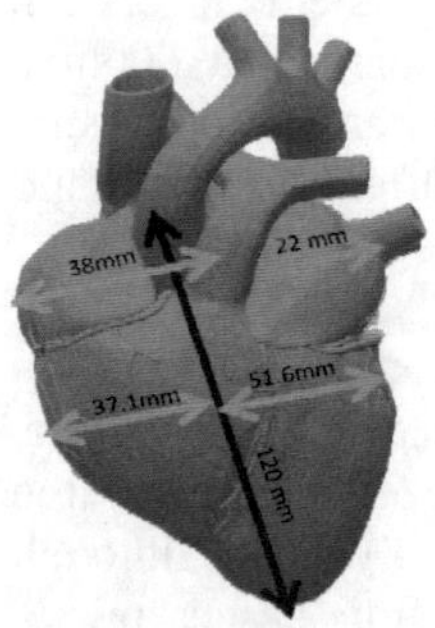

Figure 2. The dimension of heart.

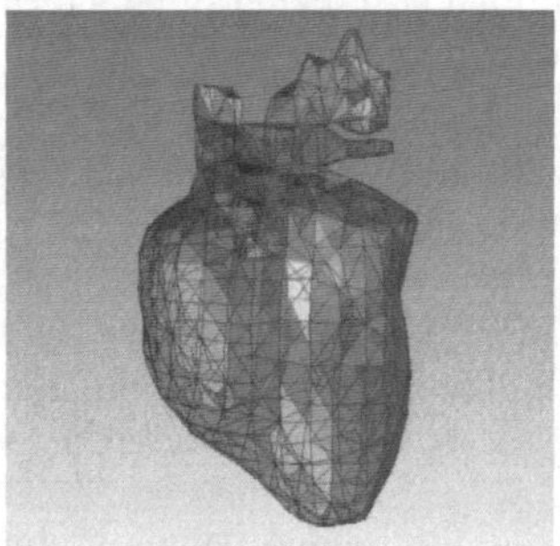

Figure 3. The design of heart

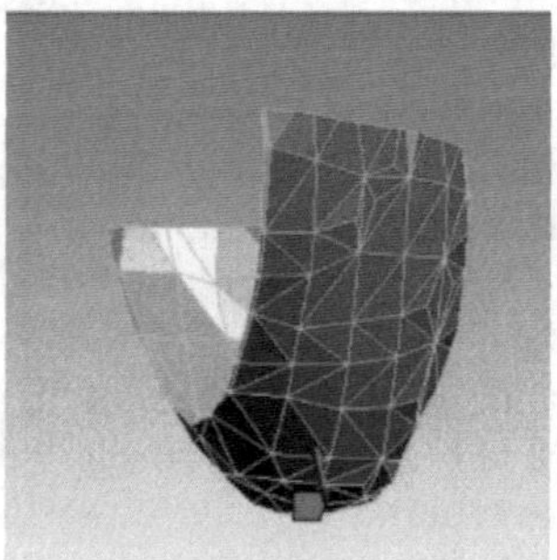

Figure 4. The design of the left ventricle.

Based on Figure 2, for the estimation of ventricular dilatation, the important female cross diameter of the left ventricle is 45.2 ± 3.4 mm diastolic and 30.5 ± 3.5 mm systolic. For the right ventricle, it is 30.7 ± 3.8 mm diastolic and 22.3 ± 3.8 mm systolic. For the determination of a left ventricular hypertrophy, relevant thickness of the septal wall measured in the short axis of the female left ventricle are approximately 8.0 ± 1.0 mm diastolic and 10.9 ± 1.4 mm systolic. Meanwhile, the measurement of male left ventricle cross diameter is 51.6 ± 4.6 mm diastolic and 33.8 ± 3.6 mm systolic. For the right ventricle, it is 37.1 ± 5.9 mm diastolic and 28.1 ± 4.4 mm systolic. For the left ventricular hypertrophy, the relevant septal wall thickness is 9.9 ± 1.2 mm diastolic and 13.6 ± 1.9 mm systolic [8].

Figure 3 and 4 show the geometries design of the heart and the left ventricle that is used for the simulation that have been constructed in 3-Dimension by using Solidworks software. The shape of heart is created by combining the triangular facet. The shape has only the surface of outermost layer which represent the heart wall. The left ventricle geometry then is saved as IGES file in order to import it into ANSYS 16.0 software for the Finite Element Analysis (FEA). The Dynamic Explicit is selected for the simulation because the study of the heart is undergoing the time-varying behaviour. In the software, it is compulsory to select the engineering data, trim the geometry, and set the boundary condition. Table 1 shows the engineering data that has been used in ANSYS.

IOP Conf. Series: Materials Science and Engineering **551** (2019) 012077 doi:10.1088/1757-899X/551/1/012077

Table 1. The Engineering Data for the simulation.

Property	Typical value
Modulus (MPa)	10-60
Density (kg/m³)	1037
Poisson ratio	0.4
Bulk Modulus (MPa)	25
Thickness (mm)	11

Based on Table 1, the engineering data is used as the input to simulate the geometry. The Post-processing data is the data that is the result provided after the simulation running. The provided data is the information regarding the location and the kinetic energy.

3. Results and Discussion

The results of finite element analysis are presented where the location of kinetic is formed. The result of the total deformation shows the colour contour that represents the movement of the left ventricle. The force from myocardial contraction and relaxation could provide piezo-polymer sufficient power to supply epicardial pacemaker with small dimension. This needs identification of the best harvesting placement and designing multilayer piezo-polymer. Figure 5 shows the total deformation of the left ventricle after the complete simulation and the location of the optimum kinetic energy produce for the placement of the pacemaker.

Figure 5. Total deformation of the left ventricle after the simulation complete.

Based on Figure 5, the optimum location of power harvesting for each pacing placement is located at the cardiac walls which have large movements (at the red contour). The kinetic energy that has been produced is greater than the other parts of the left ventricle. However, more rigorous conformation of the results is required in order to use this methodology for the calculation of actual kinetic energy values in the cardiac wall. Besides that, there are a few obstacles when the simulation is running. Most of them are the limitation of the hardware itself. The simulation takes too much time to complete and cannot accept the large scale design. The design need to be re-meshed and affecting the resolution of the design.

IOP Conf. Series: Materials Science and Engineering **551** (2019) 012077 doi:10.1088/1757-899X/551/1/012077

4. Conclusions

From the research, we can conclude several conclusions which are;

i. The myocardial contractility is able to produce sufficient force for piezo-polymer material to provide sufficient power for epicardial pacemaker pulse generator. It is a huge benefit towards self-powered pacemakers.
ii. The generated power from piezo-polymer could be optimized by providing best placement of the pacemaker on the heart epicardium. However, besides maximum power generated, the placement should also consider the location for effective pacing.
iii. The generated power from the piezo-polymer implanted on the heart epicardium could be improved by designing the piezo-polymer in multilayer topology and proper shape. The harvested power could be improved by properly tuning the Piezo-polymer resonance frequency that matches with the frequency of heart motion.

Acknowledgements
This research funded by University Technology of Malaysia (UTM) via Research University Grant (RUG) Tier 1 with vote no (PY/2017/00781).

References
[1] Das R *et al* 2015 *Biotelemetry and Wireless Powering for Leadless Pacemaker Sys. Microwave and Wireless Components Letters* **25** 262-264.
[2] Rotariu C *et al* 2012 *Wireless remote monitoring system for patients with cardiac pacemakers* 845-848.
[3] Ha B W *et al* 2015 *Energy Transfer and Harvesting for RF-Bio Applications* 54-55.
[4] Radziemski L *et al* 2015 *In Vivo Demonstration of Ultrasound Power Delivery to Charge Implanted Medical Devices via Acute and Survival Porcine Studies* **64** 1-9.
[5] Zurbuchen A *et al* 2013 *Energy Harvesting from the Beating Heart by a Mass Imbalance Oscillation Generator* **41**, 131-141.
[6] Deterre M 2013 *Toward an Energy Harvester for Leadless Pacemakers*
[7] Gordon Betts J *et al* 2013 *Anatomy & physiology*. 787–846.
[8] Hergan K *et al* 2004 *Normal Values of heart diameter in The Cine-MRT*, 176.
[9] Ampanozi *et al* 2018 *Comparing fist size to heart size is not a viable technique to assess cardiomegaly* **36** 1–5.
[10] Beurskens *et al* 2019 J. *Lead. Pace. imp. after exp. of infected conv. Pace. sys.* **1** 66-71.

Taguchi Method Used in Optimization of Plastic Injection Molding

Nur Fatihah Kamarudin, Suhaila Mohamad Yusuf and Azlan Mohd Zain

Applied Industrial Computing Research Group, School of Computing, Faculty of Engineering, Universiti Teknologi Malaysia, Malaysia

E-mail: nurfatihah3981@gmail.com

Abstract. Plastic Injection molding is the most important process of plastic product where it determines whether the product can be used or it has inadequacy. The inadequacy can make the product unusable or also can affect the quality of the product. So in order to avoid the inadequacy, optimization of the process parameter must be done. This is one of the steps to get the best quality of product. It will able to save cost as well as increase the quality of the product. Taguchi method is one of the methods that use to identify the parameter that can cause the defect of a product. This method is able to reduce the defect and also able to decrease the cost of creating the product.

1. Introduction

In each of the product, 70 % of element in consumer product is represent by mold manufacturing that is essential in all industry [1]. This process is difficult that enforced mold maker that has skill and qualification [1,2]. Today, injection molding is one of the crucial processing procedures of polymer in industry of plastic [3]. It is highly favored process in industry of manufacturing due to its availability to generate intricate-shape plastic part with high accuracy of dimension and shorter cycle times [4]. The process of molding required a proper design and material to avoid defects of the product [5]. The defect can affect the function and appearance of the product.

The defect occurs because of 3 main factors which are mold design, the chosen material and process parameter. Design of mold may cause cavities and flow lines occur on the product. Short shot is one of the defect causes by the unsuitable material. Process parameters play big roles in produce the high-quality product. Sink mark, weld lines, burn mark and warping are caused due to the process parameter. Shrinkage can be occurring on finished product that cause sink mark, short shot or warpage defect [6]. Figure 1 shows the roots causes of defect occurs on finished product.

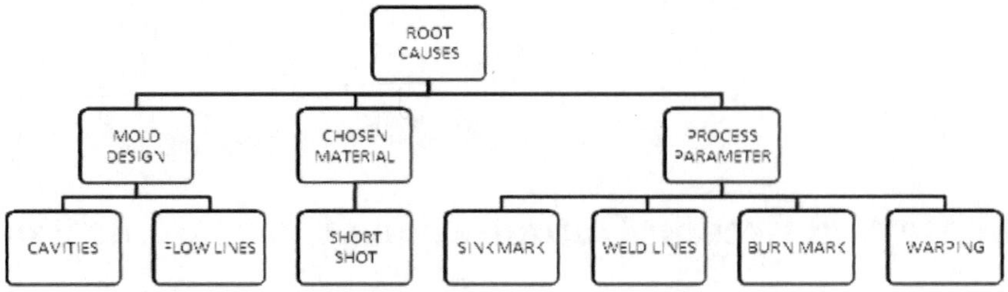

Figure 1. Root Causes and Defect

Shrinkage appears in most of the defect that caused by unoptimized process parameter. There are many method can be used to optimize the process parameter to minimize the defects such as Response Surface Methodology (RSM) and Taguchi method. Hatta et al [7] and Zakaria et al [8] applied the method together with ANOVA and able to get the factor that can be change to minimize the defect which are melt temperature, mold temperature, packing pressure and cooling time [7]. However, RSM need larger number of experiments in contra with the Taguchi method, which required less number of experiments. This paper will be discussed on use of Taguchi method for optimize the plastic injection molding parameter in order to reduce the percentage of shrinkage. Throughout the literature review, the general main factor that caused shrinkage is determined.

2. Taguchi Method

Dr Genichi Taguchi has developed a method to identify the best combination that used during the process in the manufacturing [9,10]. Taguchi method is a very effective tool that used to design and improve the high-quality system [11-15]. Industries able to decrease huge amount of time for the development of the product without adding any costs by applying this method [16,17]. Figure 2 is flow diagram that has been illustrated for a better understanding on Taguchi method.

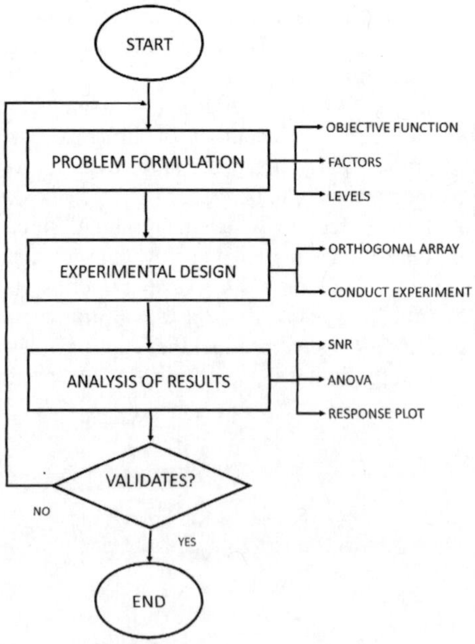

Figure 2. Taguchi flowchart

The first step is problem formulation. In this step the objective function, factors and levels is defined. It needs fully understanding of the chosen process. The second step is the process of design and conduct the experiment. Based on the number of defined factors and levels, a standard orthogonal array is created and used in experiment. The next step is result analysis. In Taguchi method, three main statistical method are used in analyze the design. The three statistical methods are Analysis of variance (ANOVA), signal-to-noise ratio (SNR) and Analysis of means (ANOM)[15-17]. The last step is performing validation of experiment. For each run in the external array, there are 27 experimental runs in the internal array. By design, only one run will yield the highest profit margin. Preliminary visual inspection of trends of each factor average contributions at all levels can be made through a response plot.

3. Discussion

3.1 Research using Taguchi method

Taguchi method has widely used as one of the optimization method in many areas. One of the areas is injections molding process. Each product has different process parameters that cause the defect due to different design of mold and different material. Each of products also has different value of process parameter depending on the design of product and material used in product. It is used the most in identifying the parameter that caused shrinkage and warpage [10,17 – 20]. This method used to reduce the number of experiment and investigate the factor that caused defect on the finished product [12].

Some researcher combines Taguchi with other method to increase the efficiency of the method. Moayyedian et al [11] implement Taguchi with FAHP and TOPSIS to increase the option in improving the quality of final product during injection molding. Khavekar et al [12] compared Taguchi with Shainin system and stated that Taguchi method consumed more number of experiment and time but its more trustworthy method in optimize the process parameter [20]. After that, the parameter must be optimizing to get the best settings. In the parameter optimization, some researcher has applied Artificial Intelligence techniques such as Artificial Neural Network [12] and some of them applied non – conventional technique which is less complicated and efficient. Most of the researcher choose naturally inspired algorithm to overcome the problem of optimizing warpage [23].

3.2 Process parameter chosen by taguchi method

Process parameter of plastic injection moulding is the parameter that involved before, during and after process of injection molding. The parameter is melt temperature, mold temperature, packing pressure, packing time, cooling time, injection pressure, injection time, cooling temperature and gate location. Based on previous research, melt temperature is the most chosen parameter that causes the shrinkage on the finished product. Mold temperature, melt temperature and packing pressure are the most chosen parameter that cause the defects on the product [2, 21, 23].

The present of weldiness can be reduce by increasing the injection speed and melt temperature and lowering the injection speed [24]. Taguchi based GRA gave an opportunity to increase and enhance the parameter of injection molding process. The weld-lines and sink mark defects successful reduce by obtain the best process parameter. Melt temperature, back pressure and holding time were rank as the top three parameters that contribute the defect [25].

Packing time and cooling time were choosing as parameters that cause the warpage of the final product [26]. Taguchi together with signal-to-noise ratio and variance analysis were used to get the consequence process parameter. The parameter then applied in Kriging surrogate model to get the mathematical relationship with warpage. As the method applied on front shell of lcd tv, packing pressure and packing time are the most important parameter [27]. The application of Taguchi with ANOVA and ANN analysis enable identification of best level for each factor. The most significant factor that has been identified by the method is melt temperature, coolant temperature, and packing time [23].

4. Conclusions

The Taguchi method is and old method yet it still popular in research of injection molding. It is used process of optimization of injection molding by choosing the most influence parameters which are mold temperature, melt temperature, packing pressure and packing time. This method is able to reduce the time taken in creating new product and testing it. It is able to reduce the cost and improve the quality of injection molding process. Taguchi method also provides robust design solutions. It can optimize several factors simultaneously. It can be hybrid with other method to get the best result.

References

[1] Tang SH *et al* 2007 *J. Mater. Processing Tech.* **182** 418.
[2] Wu CH *et al* 2006 *Sensors and Actuators A: Physical* **125** 367.
[3] Avery J 1998 Hanser Verlag 1.
[4] Bozdana A.T. *et al* 2002 *J. Mater. Processing Tech.* **128** 113.
[5] Kong JW *et al* 2003 *IEEE Trans. Elec. Pack. Man.* **26** 245.
[6] Chen WC *et al* 2014 *Int. J. Prec. Eng. and Man* **15** 1583.
[7] Hatta NM *et al* 2017 *AIP Conf. Proc.* **1885** 020152.
[8] Zakaria NA *et al* 2017 *AIP Conf. Proc.* **1885** 020169.
[9] Taguchi G *et al* 1987 *Amer Supplier Inst.* 1.
[10] Ozcelik B *et al* 2009 *Materials and Design* **30** 367.
[11] Moayyedian M *et al* 2018 *J. Man. Sc. and Tech.* **21** 150.
[12] Hatta NM *et al* 2018 *AIP Conf. Proc.* **2030** 020152.
[13] Shuaib NA *et al.* 2012 *International Review of Mechanical Engineering* **6.1** 11.
[14] Nasir SM *et al.* 2014 *Trans Tech Publications* **594** 842.
[15] Fathullah M *et al.* 2011 *International Review of Mechanical Engineering* **5.7** 1189.
[16] Hakimian E *et al* 2012 *Materials & Design* **42** 62.
[17] Sreedharan J *et al* 2018 *Materials Today: Proc.* **5** 12744.
[18] Fathullah M *et al.* 2011 *International Review of Mechanical Engineering* **5.7** 1278.
[19] Nasir SM *et al.* 2012 *International Review of Mechanical Engineering* **6.3** 372.
[20] Khavekar R *et al* 2017 *Mat. Sc. and Eng.* **225** 012183.
[21] Mehat NM *et al* 2017 *J. Physics: Conf. Ser.* **908** 012049.
[22] Shayfull Z *et al* 2014 *Key Eng. Mat.* **594** 593.
[23] Zakaria NA *et al* 2018 *AIP Conf. Proc.* **2030** 020154.
[24] Li H *et al* 2007 *Int. J. Adv. Man. Tech.* **32** 927.
[25] Sreedharan J *et al* 2018 *Materials Today: Proc.* **5** 12615.
[26] Chen WC *et al* 2016 *Int. J. Adv. Man. Tech.* **83** 1873.
[27] Zhao J *et al* 2016 *Adv. Polymer Tech.* **35** 288.

IOP Conf. Series: Materials Science and Engineering **551** (2019) 012079 doi:10.1088/1757-899X/551/1/012079

The Effect of Different Crumb Rubber Loading on the Properties of Fly Ash-Based Geopolymer Concrete

Ahmad Azrem Azmi[1,2], Mohd Mustafa Al Bakri Abdullah[1,2], Che Mohd Ruzaidi Ghazali[3], Romisuhani Ahmad[4], Luqman Musa[1,2] and Low Sheau Rou[1]

[1]School of Materials Engineering, Universiti Malaysia Perlis (UniMAP), Perlis, Malaysia
[2]Center of Excellence Geopolymer and Green Technology (CEGeoGTech), School of Materials Engineering, Universiti Malaysia Perlis (UniMAP), Perlis, Malaysia
[3]School of Fundamental Science, Universiti Malaysia Terengganu, Terengganu, Malaysia
[4]Faculty of Engineering Technology, Uniciti Alam Campus, Universiti Malaysia Perlis (UniMAP), Perlis, Malaysia

E-mail: ahmadazrem@unimap.edu .my, mustafa_albakri@unimap.edu.my

Abstract: This study aims at viability study the effects of addition of crumb rubber loading in fly ash-based geopolymer concrete for its properties. Crumb rubber is the recycled rubber from automotive scrap tires which can reduce scrap into uniform granules shapes that can absorb stress that reduces the reflective cracking because of its elastic properties. Geopolymer concrete includes an alternate material i.e Fly ash in replacement of cement, as a binding material. Mortar cubes of size 50mm x 50mm x 50mm were casted and curing at room temperature. X-Ray Fluorescence (XRF) and Scanning Electron Microscopy (SEM) were used to analyze the chemical composition of fly ash and the microstructure of the concrete. Moreover, compressive strength, water absorption capacity, and density of concrete are inter-related. Compression test, water absorption capacity test and density test were performed in hardened state, for different proportions of replacing the aggregate with crumb rubber i.e. 5%, 10%, 15%, 20%. Compressive strength test was performed at 7, 14 and 28 days. Results were obtained and compared. Decrease in strength was observed at 7, 14 and 28 days.

1. Introduction

The huge amount of waste tire becomes an emphasis topic that causes several environmental problems such as landfill pollution. An escalation in waste management problems was experienced in many countries [1]. To reduce the presence of non-biodegradable disposal materials in our surrounding, people can recycle disposal rubber tires by making use of pelletized cut rubber tire particles as a displacement of aggregates in a concrete mix. At the same time, it also helps to improve the supply demanding of coarse aggregates in low-strength concrete mixes as good strength building materials [2].

Environmental durability and sustainability characteristic is very important for a geopolymer material. A by-product of one industry use as source materials for other industrial application is a sustainable practice for industrial ecology demands. Geopolymer concrete is produced without the presence of Portland cement as a binder.

Fly ash which is rich in Silicon (Si) and Aluminium (Al) can be activated by an alkaline solution use as a binder in the concrete mixture instead of cement [3]. Low calcium fly ash is used as the major material in the mixture. The fly ash reacts with an alkaline solution to produce aluminosilicate gel that binds the aggregate to produce a geopolymer concrete.

The development of fly ash-based geopolymer concrete is an attempt to answer the challenge to produce more environmentally friendly concrete. The use of by-product material such as fly ash, as a base material for the concrete binder to exactly replace the use of Portland cement through a geopolymerisation process, has been attracting a lot of attention globally.

2. Experimental

2.1. Raw Materials

In this experiment, crumb rubber from tires used as a replacement of fine aggregates. Crumb rubber has a lower unit weight in between all the typically mineral aggregates. However, the addition of crumb rubber to the geopolymer concrete mixture can result a large reduction of unit weight of specimen. The addition of crumb rubber loading in the mixture by mass are 5%, 10%, 15% and 20%.

This study was using fly ash as the raw materials as cement replacement in concrete as geopolymer concrete. Fly ash was also gone through the XRF to analyze its chemical composition. Fly ash as the industrial by-product found that it can affect the mechanical properties of the concrete if the optimum amount of fly ash was used.

2.2. Mixing Method

The 400g of NaOH was dissolved in 1 L distilled water to produce 10M of sodium hydroxide solution. The hydroxide solution was left for one day prior. The alkaline activator was prepared by combining of a required amount of NaOH and sodium silicate solution, Na_2SiO_3. The ratio of Na_2SiO_3 to NaOH was kept constant at 2.0 for all mixing design. It must be one day prior to use.

Different loading of crumb rubber aggregate was first mixed by using hand mixer and fly ash was added to it in a dry state and mix for about 4 minutes. The ratio of fly ash to alkaline activator was fixed at 2.5. The alkaline activator solution was shaken properly before poured in the hand mixer. Then, continue stir for about 10 minutes until homogenous. In this way, the different loading of crumb rubber in the fly ash based geopolymer concrete mix was prepared.

2.3. Mix Design

In this study, the specimen cube is cast by using cube mold with dimension 50mm x 50mm x 50mm. Five different types of fly ash based geopolymer concrete mixtures sample were produced and five specimens for each type sample mixture. The mortar specimens were prepared with five different mix designs as shown in Table 1.

2.4. Casting and Curing

The mixing of fly ash and crumb rubber were mixed well with the alkaline activator solution for 15 minutes. The mixture was mix until homogenous mixture form by using a hand mixer. The geopolymer paste with addition of crumb rubber was poured into 50mmx 50mmx50mm cube molds. It was then compacted on a vibration table for few minutes. This step was accordance to ASTM C109 standard. The samples were sealed or covered with a thin plastic film to prevent contaminants and moisture loss. Then, the cube specimens left cured at room temperature.

IOP Conf. Series: Materials Science and Engineering **551** (2019) 012079 doi:10.1088/1757-899X/551/1/012079

Table 1: Mix proportion for the specimen.

Materials	Mix Proportion					
Crumb Rubber Aggregate (w %)	0%	5%	10%	15%	20%	0%
Ratio of Na_2SiO_3 to NaOH	2.0	2.0	2.0	2.0	2.0	2.0
Ratio of fly ash to alkaline activator	2.5	2.5	2.5	2.5	2.5	2.5

2.5. Testing

2.5.1. Compressive Properties Test

According to ASTM C 109/C, the compressive strength test of geopolymer cube specimens were performed using a GoTech Universal testing Machine. According to BS 1881: Part 116: 1983, the force applied on the cube was continuously increased with a nominal rate within the range of 0.2 to 0.4 $N/mm^2.s$ until no greater force can be sustained and the maximum force applied to the cube that was recorded as compressive strength.

2.5.2. Water Absorption Test

The water absorption test was carried out after the cubes were cured for 1, 7 and 28 days. The concrete cubes were left to dry for 24 hours. The weights of the dried cubes were measured before immersed in water for another 24 hours. The percentage increase in weight which is the result of the water absorption is then calculated by using the equation (1):

$$\text{Water absorption } \% = [(W_0 - W_i)/W_o] \text{ X } 100\% \quad (1)$$

Where W_o is the weight of the specimen at the absorbing equilibrium and W_i is the initial dry weight of the specimen.

2.5.3. Density Measurement

To determine the density of the fresh mixed geopolymer concrete, it gives formulas for calculating the unit weight, yield, and air content of the concrete. Yield means the volume of the concrete produced from a mixture of known quantities of the component materials. The density of the concrete cube was calculated by equation (2).

$$\text{Density} = \frac{\text{mass of the concrete cube after dried (kg)}}{\text{Volume of the concrete cube}} \quad (2)$$

2.5.4. Morphology Test

The SEM was used to carry out the morphology study in accordance with ASTM B748. The microscopic images of the fly ash based geopolymer concrete were captured using SEM on the surface of the geopolymer concrete after tested under the compression test. Then, the image was used to analyze the stress distribution and failure mechanism of the geopolymer concrete samples after the compression test.

2.5.5. Composition Test

XRF was used to analyze the composition of raw materials which such as fly ash before producing concrete cubes. Panalytical XRF was used in this study and the scan range used was 10° to 100° with a scan rate of 2° per minutes.

3. Result And Discussion

3.1. Compressive Strength

Based on figure 1, it shows that the average compressive strength for all five different mixes of mortar specimens cured for 7, 14 and 28 days under room temperature. These five specimens contain different loading of crumb rubber in fly ash-based geopolymer concrete. The control mix sample only contains fly ash without the addition of crumb rubber. The crumb rubber aggregate was used to replace sand aggregate in the fly ash geopolymer mortar at a replacement level of 5%, 10% 15% and 20% respectively by weight of the binder.

The compressive strength of the rubberized geopolymer mortar will increase with curing time for all different mixes. The longer the curing age resulted in higher compressive strength for the rubberized mortar. This might because longer curing time is required for optimum activation of pozzolanic reaction [4]. To produce a concrete with optimum potential performance, sufficient curing age is essential [5]. The curing is the most important step for concrete as it maintains the moisture content during its early stage in order to develop it properties like strength, porosity, shrinkage and durability of the concrete [6].

From the overall, the compressive strength of the geopolymer mortar reducing with addition crumb rubber loading. The increasing of crumb rubber loading in rubberized geopolymer mortar will gradually weakened its compressive strength. , the failure of the rubberized mortar specimens occur attributed to the more elastically deformable of crumb rubber than the matrix. The cracks begin at the softest areas when the mortar specimens were loaded. The site of the inclusion of rubber is where these sites appear [7]. In other words, the inclusion of crumb rubber implies defects in the internal structure of the concrete which resulting in a reduction in strength. This behaviour can be profitable when some ductility of the material is required [8].

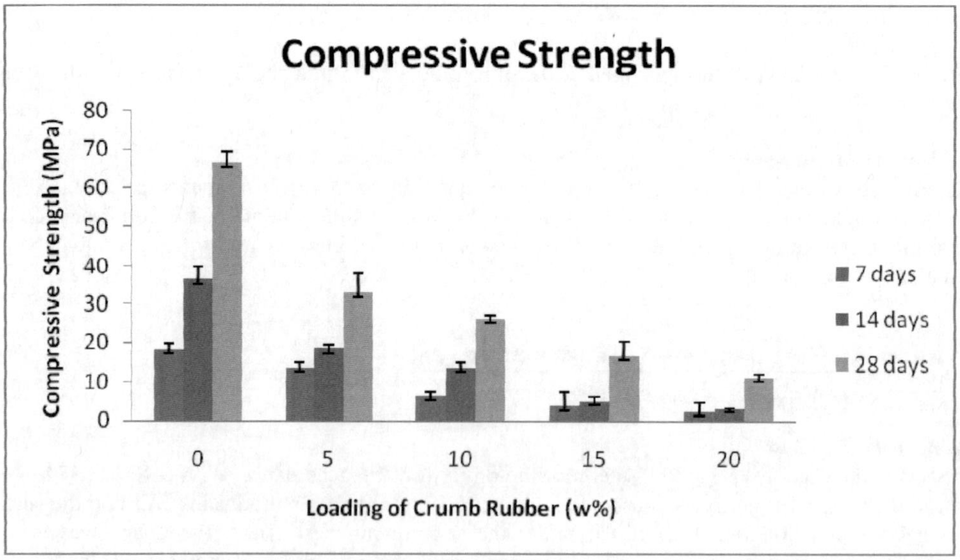

Figure 1. The compressive strength of the mortar specimens with different loading of crumb rubber for 7, 14 and 28 days of curing time.

3.2. Water Absorption Properties

Figure 2 showed the water absorption capacity of five different mixes of rubberized geopolymer mortar at three different curing periods. The water absorption test is correlation with the density test.

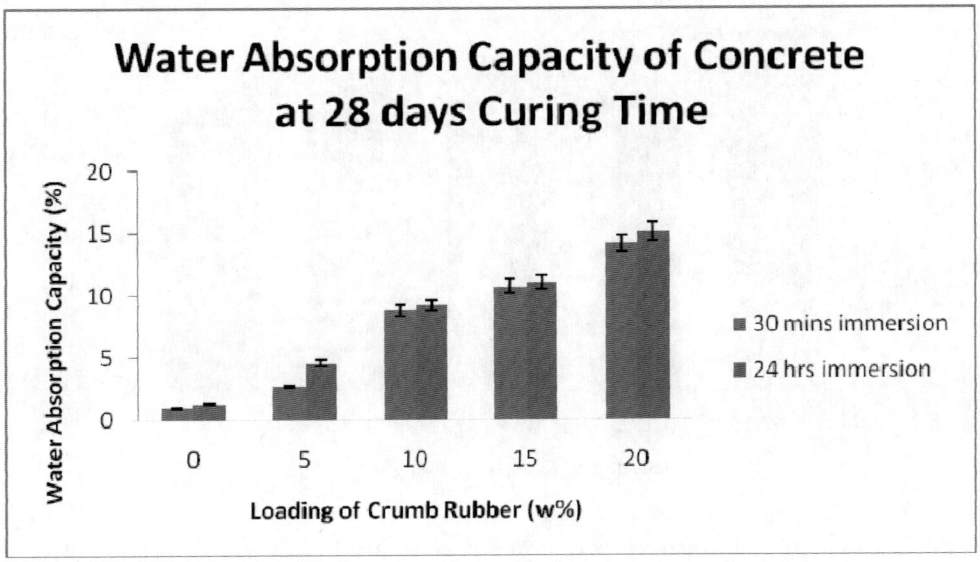

Figure 2. Effect of different loading of crumb rubber content on the water absorption capacity of rubberized geopolymer mortar at 28 days curing time with 30 minutes and 24 hours of immersion.

It can be concluded that the higher the water absorption capacity of the mortar, the lower density of the concrete. This is because of the presence of the amount of the voids in the mortar will decrease its unit weight.

Based on the data obtained from figure 2, the water absorption capacity of the rubberized geopolymer mortar shows an increasing trend as the crumb rubber content increases. The higher in water absorption capacity means that the density of the mortar is lower. This is because the amount of voids presence in the concrete is more than expected. It shows lowest water absorption capacity for plain geopolymer mortar among the five different mixes of rubberized geopolymer mortar, which means it has the best hydration characteristic. Hence, the low water absorption capacity and good hydration characteristic in plain geopolymer mortar results in highest compressive strength among all concrete mixes.

3.3. Density

Figure 3 shows the density of different mixes of hardened rubberized geopolymer concrete of curing day for 7, 14 and 28 days. The resulted obviously showed that the density of the rubberized geopolymer mortar reduce during the curing period from 7 to 28 days. This is because the rubberized mortars will loss of water during the curing period which contributes to a lower density. The loss of water may attribute to improper sealed of mortar specimens during the curing period [9]. But the loss of water do not contributed to a reduction of compressive strength on rubberized geopolymer mortar. This is because the main reaction product of the alkali activation of fly ash causes a zeolite-type phase [10].

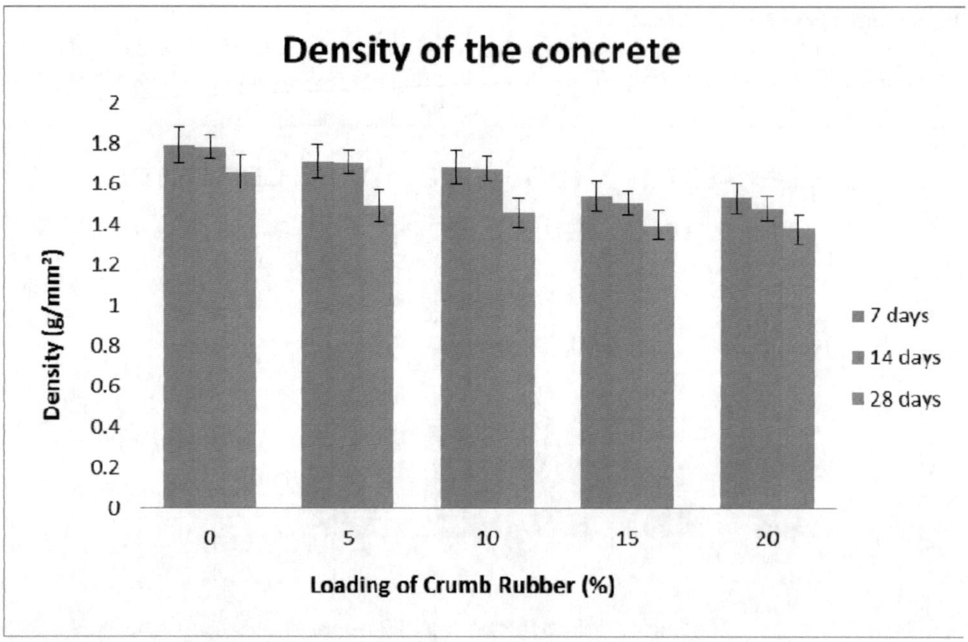

Figure 3. Effect of different crumb rubber content on the density of rubberized geopolymer mortar for 7, 14 and 28 days of curing time.

The density of rubberized concrete was found comparatively less than plaint geopolymer concrete. With the increase in the percentage of rubber in concrete the density of it decreases [11]. From the results obtained the unit weight of the rubberized geopolymer concretes ranged from 2.243 to 1.722 kg which depend on the crumb rubber contents. With the increasing crumb rubber loading, the unit weight of the rubberiezed geopolymer concrete was reduced which results into a lighter concretes [12]. The results reveal that rubberized geopolymer concrete mixtures showed lower unit weight compared to plain concrete was due to the low specific gravity of crumb rubber [13].

3.4. Scanning Electron Microscope (SEM)

The SEM micrograph for different loading of crumb rubber loading in the rubberized geopolymer mortar were compare under magnification of 200x as shown in figure 4.

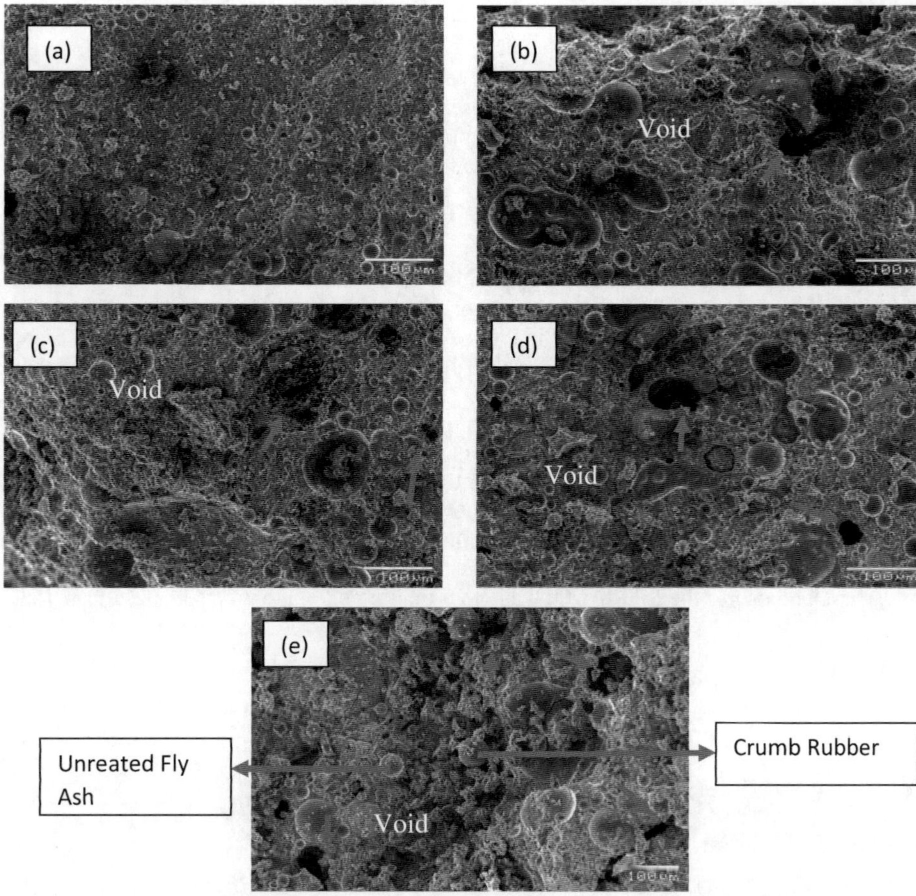

Figure 4. SEM micrographs of (a) Plain geopolymer mortar, (b) 5% of crumb rubber content, (c) 10% of crumb rubber content, (d) 15% of crumb rubber content, and (e) 20% of crumb rubber content under magnification of 200x.

These pictures can indicate the crumb rubber–mortar interfaces in all the rubberized geopolymer mortar specimens. In plain geopolymer mortar as shown in (a), the matrix has possess into a highly compacted paste. The non-reacted fly ash particles are embedded into the matrix, which bears the semispherical imprints of some of the grains that did. They are embedded into the sodium aluminosilicate gel produced during the reaction process. For the 5% crumb rubber content of rubberized geopolymer mortar as in (b), the interfacial zones are less dense with little micro-cracks. For the 10% crumb rubber content of rubberized geopolymer mortar as in (c), the interfacial zones become denser with increasing of micro-cracks. As the crumb rubber content up to 15% and 20% as in (d) and (e), the microstructure of the mortar shows the matrix appeared a lot of larger porosity and result in a relatively poor adhesion between the crumb rubbers and cement paste. These weak interfacial zones may have many serious influence on a range of properties of mortar.

The poor adhesion between the crumb rubber and the cement paste is very clear in these images. It was also observed that some rubber grains were buried in cement the matrix. Moreover, from SEM micrograph showed the porosity or void formed around the crumb rubber as a fine aggregate in the microstructure. These voids are susceptible to cracking when subjected to the influences of

IOP Conf. Series: Materials Science and Engineering **551** (2019) 012079 doi:10.1088/1757-899X/551/1/012079

compressive stress caused by the differential movements between hydrated cement paste and aggregate. Such movement can occur either during curing or drying of concrete. Furthermore, concrete can have microcracks in the interfacial transition zone even before a structure is loaded. This can be attributed to the content of rubber crumb loading present in the sample, which encourages poor bonding within the concrete structure. As a result, the application of compressive force puts a lot of pressure on the weak bonds. Hence, the mortar will be easy disintegration, which will lead to the large cracks in the sample.

3.5. Elemental Analysis

XRF analysis is used to determine the chemical composition of fly ash used in this study. For the use of fly ash in concrete, it can be classified into two categories, which are Class F and Class C [14]. In order to classify the fly ash as Class F, the sum if SiO_2, Al_2O_3, and Fe_2O_3, is not less than 70% and very low in CaO which lower than 10%, while Class C, the total of these constituent is not less than 70% and high content CaO which is more than 20% [15].

Therefore, based on the XRF results in table 2 obtained that the fly ash used in was belong to low calcium Class F- acidic ashes. This is because the sum of the main constituent such as SiO_2, Al_2O_3, and Fe_2O_3 was not less than 70% with less than 10% of CaO content. The summation of these three chemical constituents was 90.32% and the content of CaO is 5.06%, which is classified as Class F – acidic ashes according to the ASTM C-618 standard with McCarthy amendment [16]. According to the international system of classification of fly ash, the fly ash used in this study is belonging to aluminosilicate ashes. The due to the SiO_2/Al_2O_3 was more than 2 and content of CaO less than 15% [17].

Table 2. Chemical composition of fly ash

Chemical Constituents	% Mass of Chemical Constituents (%)
Al_2O_3	23.40
SiO_2	50.00
SO_3	0.08
K_2O	1.41
CaO	5.06
TiO_2	1.60
MnO	0.22
Fe_2O_3	16.92
BaO	0.37
SiO_2/Al_2O_3	2.1368
$SiO_2 + Al_2O_3 + Fe_2O_3$	90.32

4. Conclusions

From this study, it is concluded that the fine aggregate in geopolymer concrete was replaced by the addition different loading of crumb rubber with 0%, 5%, 10%, 15% and 20%. The different loading of crumb rubber content in the rubberized geopolymer concrete is highly influence the results of compressive strength, density, water absorption capacity and different SEM micrograph. From the overall of this research demonstrated that crumb rubber can be easily mixed in geopolymer paste without any difficulties. Potential use of crumb rubber cannot be used for the increasing compressive strength for concrete. This may be due to the improper bonding of crumb rubber with the mix and also due to the presence of voids in the concrete. The compressive strength of the rubberized geopolymer mortar reduces about 60% with 10% crumb rubber loading in the mixture. For large amount of crumb rubber in the mixture, the compressive strength gain rate is lower than plain concrete. Although the addition of crumb rubber in the mixture of rubberized geopolymer concrete does not show a significant improvement in the strength but these concrete still can be take advantage of many others engineering applications.

5. References

[1] Tomas U G Jr 2014 *International Journal of Advanced Science and Technology* **64** 21.
[2] Tomas U G Jr 2015 *International Journal of Advanced Science and Technology* **77** 7.
[3] Izzat A M *et al* 2013*Materiale Plastice* **3** 171.
[4] El-Dieb A S 2017 *Construction and Building Materials* **142** 444.
[5] Hassan K *et al* 1997 *In Proceedings of the Fifth International Conference* **16** 43.
[6] Kaloush K E *et al* 2004 *Arizona: Arizona State University* **32** 162.
[7] Emira N A *et al* 2012 *International Journal of Mechanic Systems Engineering* **2** 53.
[8] Albano C *et al* 2013 *Rev. de la Fac. de Ing. Uni.Cen. de Ven.* **28** 97.
[9] Lee S *et al* 2016 *Materials (Basel)* **9** 7.
[10] Bondar D et al 2007 *The Third International Conference on Concrete and Development* **8** 19.
[11] Michelle D 2006 *Purdue University Calumet* **4** 15.
[12] Güneyisi E *et al* 2004 *Cement and Concrete Research* **34** 2309.
[13] Aiello M A *et al* 2010 *Waste management* **30** 1696.
[14] Zailani W W A *et al* 2017 *Materials Science and Engineering Conference Series* **267** 012008.
[15] Bakri A M M *et al Revista de chimie* **4** 382.
[16] Lloyd *et al* 2010 *In second international conference on sustainable construction materials and technologies* **3** 1493.
[17] Miloš Š *et al* 2016 *Technical Gazette* **6** 1839.

IOP Conf. Series: Materials Science and Engineering **551** (2019) 012080 doi:10.1088/1757-899X/551/1/012080

Pull-Out Strength of Hooked Steel Fiber Reinforced Geopolymer Concrete

Meor Ahmad Faris[1], Mohd Mustafa Al Bakri Abdullah[1], Khairul Nizar Ismail[3], Nurul Aida Mohd Mortar[1], Mohammad Firdaus Abu Hashim[2, 1] & Abdul Hadi[1]

[1]Center of Excellence Geopolymer and Green Technology (CEGeoGTech), School of Materials Engineering, Universiti Malaysia Perlis (UniMAP), Perlis, Malaysia.
[2]Faculty of Engineering Technology, Universiti Malaysia Perlis (UniMAP), Perlis, Malaysia.
[3]School of Environmental Engineering, University Malaysia Perlis, Perlis, Malaysia.

E-mail: meorfaris88@gmail.com

Abstract. Pull-out strength of fly ash steel fiber reinforced geopolymer concrete was studied. Fly ash steel fiber reinforced geopolymer concrete was produced by mixing of Class F fly ash which taken from Manjung power station, Lumut, Perak, Malaysia with alkaline activator which are combination of sodium hydroxide and sodium silicate. Steel wool fiber were added into the geopolymer concrete as reinforcement with different weight percentage vary from 0 % - 5 %. Chemical compositions of Malaysian fly ash was first analyzed by using X-ray fluorescence. All geopolymer concrete samples with inclusion of hooked steel fiber with different volume percentage which are 0 %, 0.5 %, 1.0 %, 1.5 %, and 2.0 % were tested in terms of microstructure, flexural strength, flexural toughness, and pull-out strength. Microstructure of steel fiber reinforced geopolymer concrete shows a good degree of bonding between geopolymer binder and hooked steel fibers. Flexural strength of samples increased with the addition of hooked steel fibers until at 3 % and will slightly decrease when amount of fibers addition over than 3 %. Meanwhile, addition of hooked steel fibers shows significantly improvement to the toughness of samples. However, there is no improvement of toughness with addition of fibers over than 3 %. Pull-out strength of samples shows an excellent result where maximum result of 505 N can be achieved that indicates a good degree of bonding between geopolymer matrix and fibers.

1. Introduction

Currently, there are several types of fibers such steel fibers, polymer fibers, glass fibers, and natural fibers that have been used as micro reinforcement in Ordinary Portland (OPC) cement binder for a particular purposes. Steel is one of the most widely used material as reinforcement in cement concrete especially for certain structure that need a high flexural and toughness with high flexibility as well. This kind of micro fibers helps to improve post cracking behaviour through bridging the formation of micro-crack spot if the required amount of steel fibers addition into the cement concrete is sufficient [1].

Based on previous literature, the inclusion of steel fibers in OPC concrete is proved to form a better crack control [2-5]. In addition, tensile strength of OPC concrete before and after cracking was improved as well by addition of steel fibers [3-6]. There are a lot of studies have been made related to the potential of OPC concrete to be adjusted from brittle to ductile behaviour and many of previous studies shows positive results [3, 5, 7, 8].Currently, geopolymer is getting attention among researchers [9-12] in order to replace the using of OPC to reduce cost and carbon footprint

The addition of steel fibers in cement concrete has transformed the concrete into a reinforcement concrete called as steel fiber reinforced concrete (SFRC) and it is found to have an adequate for OPC structural [13]. However, reinforcement of fly ash geopolymer concrete by steel fibers is still questionable due to lack of collected data from previous studies. Therefore, special testing should be applied in order to measure the effectiveness of fibers addition on geopolymer. Pull-out test has been proposed by Leung & Chi where they proved the degree of bridging by fibers in OPC binder can be measured by this test [14]. This is due to the pull-out mechanism formed in SRFC is closely related to the function of fibers on a cracked surface in which the bridging effect formed. In the pull-out test, the effectiveness of fibers from randomly distributed fibers in SFRC has been tested where the force of pull out is plotted versus fibers slip [15-18].

2. Materials and Methodology

SFRGC was produced by mixing between geopolymer binder which is fly ash and alkaline activator with addition some amount of steel fibers. Fly ash used in this experiment was taken from electric power plant Manjung, Perak, Malaysia. Alkaline activator was first produced by the mixing between sodium silicate (Na2SiO3) and sodium hydroxide (NaOH). The Na2SiO3 was bought from South Pacific Chemicals Industry Sdn. Bhd. (SPCI), Malaysia with ratio between SiO and Na2O is equal to 3.2. Meanwhile, NaOH which is in the form of pallet was bought from Farmosa Plastic Corporation, Taiwan. The NaOH pellets was first diluted in distilled water and concentration of 12 M was produced. Chemical composition of fly ash used in this experiment was characterized by using X-Ray Fluorescence (XRF) as in Table 1. Ratio of fly ash to alkaline activator and Na2SiO3 to NaOH is equal to 2.0 and 2.5 respectively.

River sand and granite were sieved to ensure maximum size of fine aggregate and coarse aggregate are equal to 4.75 mm and 20 mm respectively. Details specification of steel fibers used in the production of SFRGC was summarized as in Table 2. Total amount of steel fibers addition into the geopolymer matrix was by the percentage of total volume of geopolymer concrete which are 0 %, 0.5 %, 1.0 %, 1.5 %, and 2.0 %.

Sample of SFRGC were cast in a mold size of 100 mm x 100 mm x 100 mm and 100 mm x 100 mm x 500 mm for pull out test and flexural test respectively. Samples were taken out from the mold after 24 hours and cured at room temperature.

After 28 days cured at room temperature, samples were tested for microstructure test, flexural test, and pull-out test. In microstructure test, samples were cutting into two pieces and cross-section image were captured. Flexural test were done by using universal testing machine (UTM) followed ASTM C1018 where 4 point bending were applied. Single fiber pull-out were set-up where 1 hooked steel fiber was embedded inside the geopolymer with depth of embedment equal to 6 mm, 10 mm, 20 mm, and 30 mm.

Table 1. Chemical composition of Malaysia fly ash characterized by using XRF

Element	Percentage (%)
Al_2O_3	14.70
SiO_2	38.80
CaO	18.10
Fe_2O_3	19.48
TiO_2	1.02
MgO	3.30
K_2O	1.79
SO_3	1.50

Table 2. Specifications of hooked steel fibers

Items	Specifications
Fiber type	Hooked
Diameter (mm)	0.60
Length (mm)	0.75
Aspect ratio	80
Young modulus (N/mm^2)	210
Tensile strength (N/mm^2)	1.225

3. Results and Discussions

Bonding of hooked steel fiber in geopolymer concrete can be observed under ordinary light microscope (OM). Hooked steel fibre and aggregates were seen having a good bonding with geopolymer binder as seen in Figure 1 where there is no gap between them. Meanwhile, surface roughness of steel fibers clearly seen has high degree of roughness where it helps to increase the bonding strength between steel fiber and geopolymer binder. This happened due to the increase of contact area between steel fiber and geopolymer binder where geopolymer binder filled the empty space from the surface roughness of steel fiber.

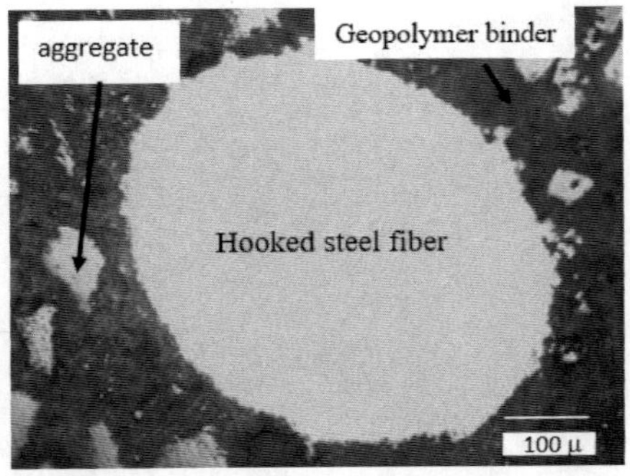

Figure 1. Cross-section image of SFRGC.

Flexural strength of geopolymer concrete with addition of hooked steel fibers is illustrated as in Figure 2. Result shows flexural of geopolymer concrete will increase with the increase of steel fibers addition until at maximum of 1.0 %. However, there is a reduction of flexural strength when steel fibers addition exceed 1.5 %. The increase of flexural strength by steel fibers addition is due to the hooked steel fibers helps to stop the crack propagation by create a bridge at the crack spot. The presence of hooked steel fibers may transform the behavior of geopolymer concrete from brittle to ductile.

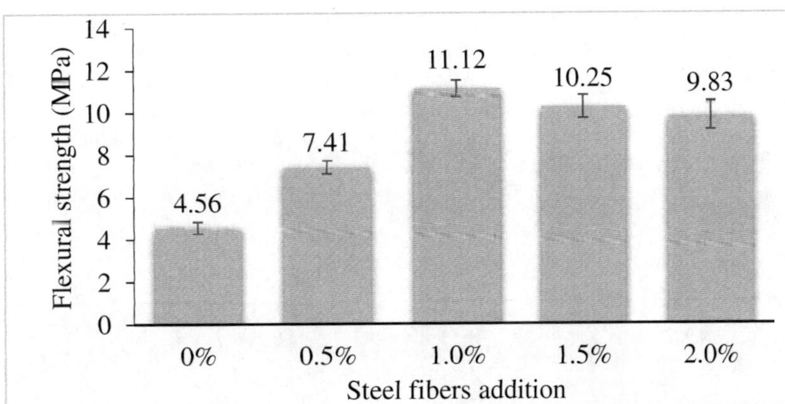

Figure 2. Flexural strength of hooked steel fibers.

Geopolymer concrete without reinforcement will failed immediately after first cracking. Meanwhile, sample of geopolymer concrete with steel fiber reinforcement undergo post-cracking load carrying capacity. There are no further improvement of flexural strength at fibers addition starting at 1.5 % where the reading was slightly decrease. This is due to the low workability of samples at fiber addition starting at 1.5 % which will cause a poor fibers distribution. This indicates the inclusion of hooked steel fibers is optimum at 1.0 % of addition.

Steel fibers inclusion shows significant improvement to the toughness of geopolymer concrete until maximum at 1.0 % of addition as shown in Figure 3. This is due to the addition of hooked steel fibers helps to increase the post-cracking strength of geopolymer concrete samples. Test results also shows reinforced geopolymer concrete with hooked steel fibers can absorb more energy compared to unreinforced sample. Fundamentally, the toughness of sample is very important measurement that indicates the energy absorption capacity of concrete and energy consumed of concrete to be failed.

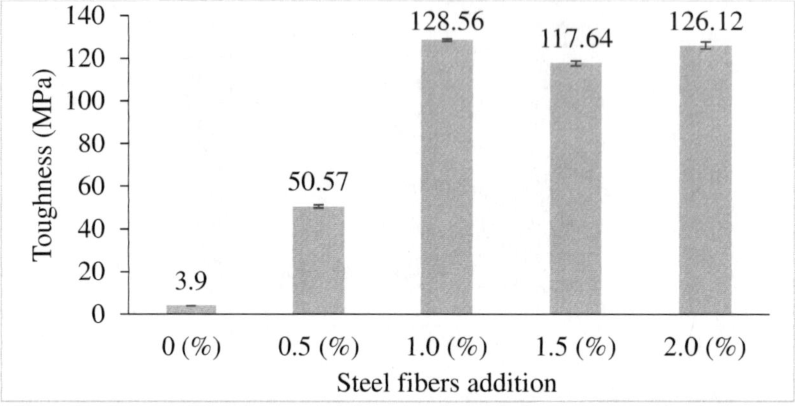

Figure 3. Toughness of hooked steel fibers.

Toughness of geopolymer concrete doesn't improve at hooked steel fibers addition exceed 1.5 % and become very slightly lower compared to the maximum toughness posted on fibers addition of 1.0 %. This also may due to the fibers distribution of hooked steel fibers starting at 1.5 % of fibers addition shows a low workability.

Pull-out strength of geopolymer concrete was measured versus embedment depth of hooked steel fiber and results is summarized as in Figure 4. Result shows the pull-out strength increase as the embedment depth was increased. The increase of pull-out strength by the increase of embedment depth is due to the contact area between the binder and the surface of hooked steel fiber getting higher. This will resulting in higher contact resistance between geopolymer binder hand the fiber when the fibers is pulled out from the geopolymer concrete by the UTM machine.

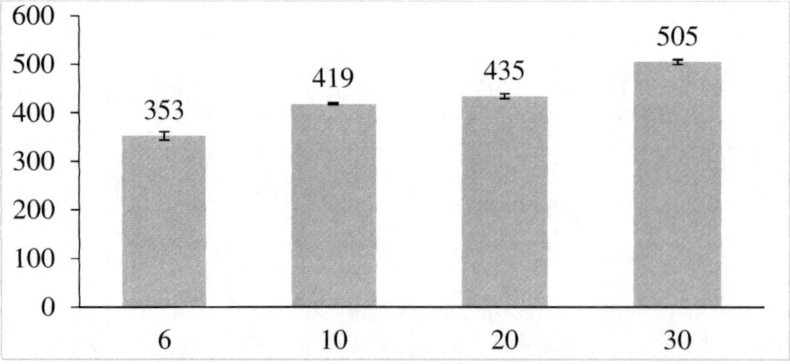

Figure 4. Pull-out strength of hooked steel fibers

Besides, the maximum pull-out of geopolymer concrete reinforced with hooked steel fibers shows an excellent reading which is 505 N compared to normal OPC concrete that have been done by previous researchers which are 400 N [17]. This is prove that interfacial bonding between steel fiber with geopolymer concrete is better than OPC concrete.

4. Conclusions

The reinforcement of geopolymer concrete by steel fibers give an improvement to the flexural strength and toughness. Plus, it is proven that hooked steel fiber and geopolymer concrete has an excellent interfacial bonding and better than OPC concrete.

References

[1] Cuenca, E J *et al* 2015 *Comp. Part B.* **75** 135.
[2] Cuenca E *et al* Cons. *Build. Mater.* 2013. **45** 145.
[3] Kim, K *et al* 2012 *Comp. Part* B **43** 2324.
[4] Tiberti G *et al* 2014 *Chem. Conc. Comp.* **45** 176.
[5] Chalioris C E 2013 *Const. and Buil. Mater.* **43** 317.
[6] Soroushian P *et al* 1991 *ACI Mater.* J. 88 129.
[7] Cuenca E *et al* 2010 *RIL. Book. Sprin.* **1** 429.
[8] Cuenca E *et al* Comp. Part B: Eng. **45** 952.
[9] Al Bakri Abdullah, *et al* 2015 *Revista de Chimie* **66** 1001
[10] Kamilah N *et al* 2018 *4th International Conference on Green Design and Manufacture* **2030** 67.
[11] Bakri, A.M.M *et al* 2013 *Revista de Chimie,* **64** 382.
[12] Izzat, A.M., *et al* 2013 *Materiale Plastice* **50** 171
[13] Brandt A M *et al* 2008 *Comp. Struc.* **86** 3.
[14] Leung K Y *et al* 1995 *J. of Eng. Mech.* 1995. **121** 131.
[15] Cunha V M C F *et al* 2010 *Dep. of Civ. Eng.* University of Minho.
[16] Wille K *et al* 2012 *ACI Mater. J.* **109** 479.
[17] Piazza G J Nuñez *et al* 2003 *J. of the Amer. Oil Chem. Soc.* **80** 901.
[18] Robins P S *et al Mater. Struc.* **35** 434.

IOP Conf. Series: Materials Science and Engineering **551** (2019) 012081 doi:10.1088/1757-899X/551/1/012081

The Effect of Geopolymer Ceramic Additions to The Wettability and Shear strength of Sn-Ag-Cu (SAC) Solder: A Preliminary Study

Nur Syahirah Mohamad Zaimi[1, a], Mohd Arif Anuar Mohd Salleh[1, b],Mohd Mustafa Al Bakri Abdullah[1,c], Marliza Mostapha[1,d], Romisuhani Ahmad[2]

[1]Centre of Excellence Geopolymer and Green Technology, School of Materials Engineering, Universiti Malaysia Perlis (UniMAP), Perlis, Malaysia
[2]Faculty of Engineering Technology, Uniciti Alam Campus, Universiti Malaysia Perlis (UniMAP), Perlis, Malaysia.

E-mail: arifanuar@unimap.edu.my

Abstract. The effect of geopolymer ceramics addition into Sn-3.0Ag-0.5Cu (SAC305) lead free solder was successfully investigated. The lead-free solder SAC305 with the addition of geopolymer ceramics was fabricated by using powder metallurgy microwave-sintering method in order to form a composite solder. The composite solder was analyzed based on wettability and shear strength. The wettability of SAC 305 solder was greatly enhanced with the decreasing in its contact angle. Then, the geopolymer ceramic addition enhanced the shear strength of SAC 305 solder.

1. Introduction

Solder interconnects are absolutely necessary parts in electronic devices which play vital roles in providing an electrical connection and mechanical support. Tin-lead (Sn-Pb) solders are widely used in electronic packaging industries back then since it gives outstanding reliability and solderability. However, the consumption of lead in the solders were banned by legal laws and regulations owing to the toxic nature of lead to human health and environments. Thus, the development of solders that are free from lead usage has been made by many researchers.

Among all the lead-free solders developed, the most promising and economically affordable approach is by introducing a composite solder. A composite solder is form when the reinforcing particles are introduced into the lead-free solder alloys [1]. The reinforcing particles can be in micrometer or nanometer size. Example of reinforcing material that have been used by previous researchers is ceramic particles. The ceramic particles can be in the form of oxide or non-oxide particles such as Silicon Nitride (Si_3N_4) [2-4], Silicon Carbide (SiC) [5, 6], Titanium oxide (TiO_2) [7-9], Aluminium Oxide (Al_2O_3) [10] and Zirconium dioxide (ZrO_2) [11, 12]. The addition of the reinforcement particles into lead free solder alloys are greatly improved and enhanced the solder performances. According to Mohd Salleh et.al., [1], he reported that the additions of TiO_2 into Sn-0.7Cu suppressed the growth of interfacial layer during soldering. Besides that, the addition of Si_3N_4 into Sn-0.7Cu as reported by Mohd Salleh et.al., [4] had improved the wettability of the solder by lowering the contact angle. A.K.Gain et.al., [11] studied the shear strength of Sn-Ag-Cu solder with the addition of nano-particles ZrO_2. He found that, the addition of ZrO_2 increase the shear strength of the solder due to the nano-particles (second phase) dispersion strengthening. Thus, it is believed that the addition of ceramic particles that acts as reinforcement materials in lead free solder alloys give positive effects to the solder performances.

In this study, Sn-3.0Ag-0.5Cu (SAC 305) solder will be reinforced with the geopolymer ceramic particles. To the best of author knowledge, there is no research on the effect of this type of ceramic to lead free solder alloys reported elsewhere. Therefore, in this research attempt has been made by incorporate geopolymer ceramic particles with 0.5 wt% into SAC 305 lead free solder.

2. Methodology

In this research, Sn-3.0Ag-0.5Cu (SAC305) lead free solder powders supplied by Sigma Aldrich with a size range of 25-45um was used as a base matrix. The micron size of geopolymer ceramics was fabricated and used as reinforcement materials in this study.

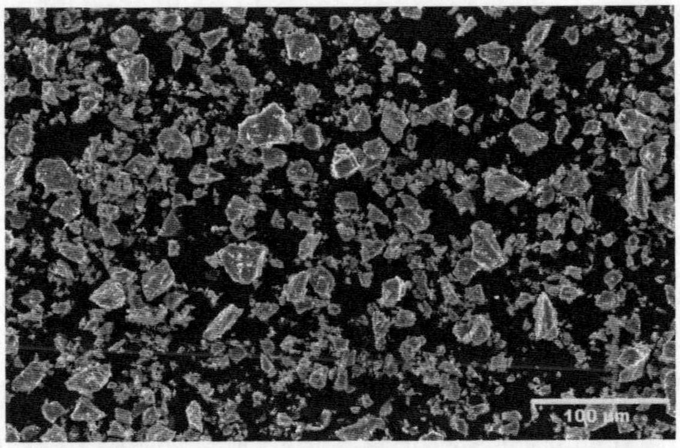

Fig. 1. Microstructure of geopolymer ceramics particles by using SEM

The fabrication of composite solder was carried out by using powder metallurgy (PM) microwave sintering method. The base material, SAC 305 lead free solder was mixed with 0.5 weight percentages (wt%) of geopolymer ceramics powder. The lead-free solder SAC 305 and geopolymer ceramics were pre weighted on a balance weight and homogeneously mixed in an airtight container by using a planetary mill. The rotation speed used was 200 rpm for one hour. The monolithic SAC305 was fabricated by using the same method. Then, each of the solder mixtures were pressed by using a Specac 15-ton Manual Hydraulic Press and uniaxially compacted in a 12 mm diameter mold at a pressure of 4.5 tonnes. The compacted cylinder pellets with about 1 mm thickness were sintered by using a microwave sintering method at a temperature of 185 °C under ambient conditions in an 800 W, 50 Hz Panasonic microwave oven.

Then, the solder wettability was determined by measuring its contact angle between solder and substrate as shown in Figure 2. The samples were placed on copper substrates and reflowed it in a reflow oven. About 10 samples of each composition were prepared to measure the contact angle in order to ensure the accuracy of the results. The contact angle was measured by using Optical Microscope and J-Image software. Besides that, a single-lap shear test was conducted by using Universal Testing Machine (UTM) INSTRON in order to determine the strength of the solder that are bonded with a copper substrate. Approximately 1.0 g compacted pellets were sandwiched between the copper substrates and were subsequently reflowed it. The test was done according to the ASTM D1002 standard with copper substrate specifications of 101.6mm x 25.4mm x 1.5 mm.

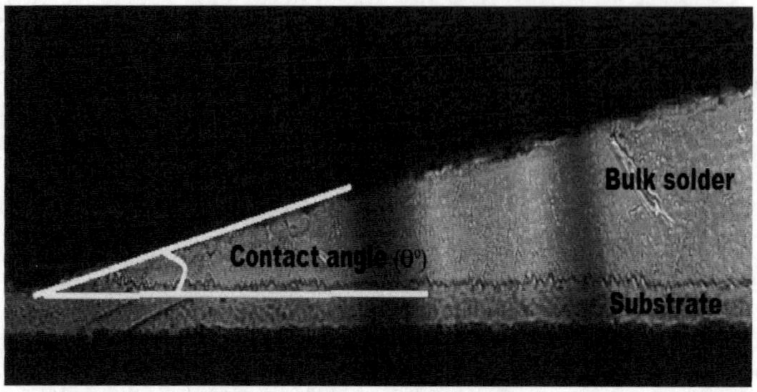

Fig. 2. Contact angle measurement

3. Result and Discussion

3.1 Wettability

Wetting is one of important part in soldering since it determines the good metallic bonding between the solder and the substrate. The wettability of the solder can be described through its contact angle (ϴ°) between the solder materials and the substrate. The smaller the contact angle give a better wettability results and it is desired in the process of soldering. This was owing to the fact that; the molten solder materials are able to spread out on the substrate in order to form a perfect solder joint and reliable solder bonding. As the contact angle value is less than 45°, it is considered to be accepted [8].

Based on the results in Figure 3, it is observed that the addition of 0.5wt% of geopolymer ceramics into SAC 305 lead free solder decreased the contact angle value. The contact angle for pure SAC 305 solder was 24.3°. Addition of 0.5 wt% geopolymer ceramics results in contact angle about 23.5°. The decreasing trend of contact angle in composite solder showing that the addition of geopolymer ceramics into lead free solder can improved the wettability of the solder. In addition, the lower contact angle of lead-free solder on substrate is favorable for a good solder bonding.

The possible reasons behind the improvement in wettability of SAC 305 composite solder is caused by the appropriate amount of geopolymer ceramics addition that acts as an agent in reducing the surface tension in a solder. As the geopolymer ceramics were added into SAC 305 solder, it tends to accumulate at the flux and molten solder interface during reflow soldering. Thus, the interfacial surface energy decreased and caused the interfacial tension between flux and solder to be lowered. As a result, a smaller contact angle formed.

Mohd Salleh et.al., [4] , discovered a similar phenomenon of the decrease in the wettability of composite solder. He reported that, the contact angle of Sn-0.7Cu/ Si_3N_4 was improved and the optimum contact angle achieved as 1.0 wt% of Si_3N_4 was added.

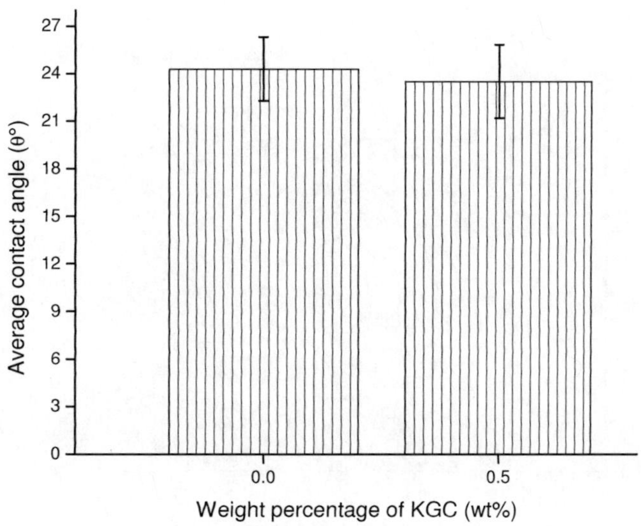

Fig. 3. Bar graph showing the influence of geopolymer ceramics addition to the wetting angle of SAC 305 lead free composite solder

3.2 Shear strength

Shear strength of the solder joint is the most relevant and reliable mechanical properties. It is important in a way that it provide a brief description about performance of new solder composition. In this study, a single-lap shear test was performed in order to mimic the actual solder joints in electronic and microelectronic industry. The results obtained for the single-lap shear test was presented in Figure 4.

Figure 4 shows the shear strength value that is obtained as the geopolymer ceramics added into SAC 305 lead free solder. Pure SAC 305 lead free solder has an average of shear strength of 6.518 MPa. As 0.5 wt% of the geopolymer ceramics added, the average shear strength is 8.098 MPa.

The improvements in the shear strength of the composite solder was attributed to the formation of IMCs that formed between the solder and the substrate [8]. As the geopolymer ceramics were added to the SAC 305 lead free solder, it will alter the IMCs morphology and its thickness and thus, improves the shear strength of the solder joint. Nai et.al., [13] working on 95.8Sn-3.5Ag-0.7Cu with the addition of various wt% of multiwalled carbon nanotube (MWCNTS) particles reported the similar phenomenon. The average shear strength of SAC lead free solder was improved with the addition of MWCNTS. The optimum value of shear strength is achieved when 0.01 wt% of MWCNTS was added to the lead-free solder.

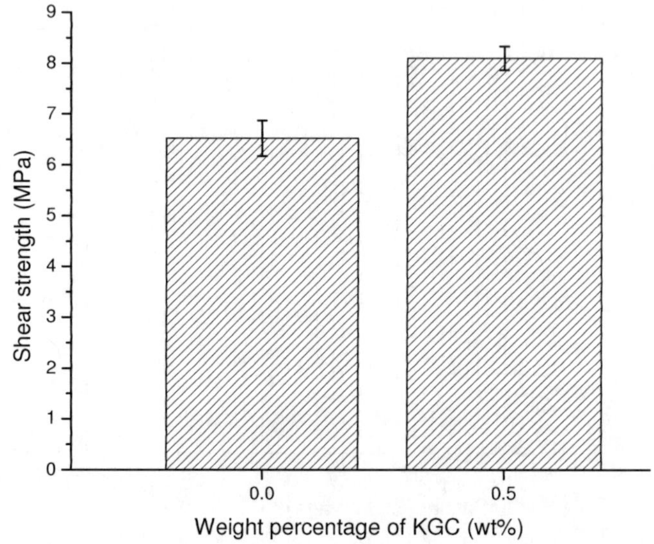

Fig. 4. Average shear strength of SAC 305 composite solder

4. Summary

A new lead-free composite solder reinforced with various wt% of geopolymer ceramics was successfully fabricated by using microwave-sintering powder metallurgy method. The addition of 0.5wt% of geopolymer ceramics improved the wettability of SAC 305 composite solder. Then, the shear strength of the SAC 305 solder also was improved with the addition of geopolymer ceramics. As overall, the addition of geopolymer ceramics gave positive effects to the lead-free solder SAC 305 and thus it can be applied for further analysis in order to elucidate and investigate in terms of its microstructures and IMCs formation.

Acknowledgement

The authors acknowledge the financial support through the Ministry of Education (MoE), Malaysia, under the Fundamental Research Grant Scheme (9003-00612), Center of Excellence Geopolymer & Green Technology (CEGeoGTech), School of Materials Engineering, Universiti Malaysia Perlis.

References

[1] Mohd Salleh M A A *et al* 2017 *J Mater Process Tech* **242** 235.
[2] Mohd Salleh M A A *et al* 2012 *Materials Science and Engineering: A* **556** 633.
[3] Mohd Salleh M A A *et al* 2011 *Adv. Mater. Res.* **277** 106.
[4] Mohd Salleh M A A et al *PP* 2011 **22** 299.
[5] Zaimi N S M *et al* 2018 *SSP* **280** 169.
[6] Said R M *et al* 2015 *AMM* **754** 546.
[7] Mohd Salleh M A A *et al* 2016 *Mater. Des.* **108** 418.
[8] Ramli M I I *et al* 2016 *Microelectron Reliab.* **65** 255.
[9] Mohd Salleh M A A *et al* 2015 *Mater. Des.* **82**, 136.
[10] Tsao L C *et al* 2013 *Mater. Des.* **50** 774.
[11] Gain A K *et al* 2011 *Microelectron Reliab* **51** 2306.
[12] Shen J *et al* 2009 *Alloy Compd* **477** 552.

[13] Nai S M L *et al* 2008 *J. Electron. Mater.* **37** 515.

IOP Conf. Series: Materials Science and Engineering **551** (2019) 012082 doi:10.1088/1757-899X/551/1/012082

Effects of Thermal Resistance to Fly Ash-Based Lightweight Geopolymer

Wan Mastura Wan Ibrahim[1,2], Mohd Mustafa Al Bakri Abdullah[1], Romisuhani Ahmad[1,2], Amir Naveed[3], Che Mohd Ruzaidi Ghazali[4] and Masdiyana Ibrahim[2]

[1]Center of Excellence Geopolymer and Green Technology (CEGeoGTech), School of Materials Engineering, Universiti Malaysia Perlis (UniMAP), Perlis, Malaysia.
[2]Faculty of Engineering Technology, Uniciti Alam Campus, Universiti Malaysia Perlis (UniMAP), Perlis, Malaysia
[3]Department of Chemical Engineering, University of Engineering & Technology Peshawar, Pakistan
[4]School of Fundamental Science, Universiti Malaysia Terengganu (UMT), Terengganu, Malaysia

Email: wanmastura@unimap.edu.my

Abstract. Geopolymers are inorganic polymeric materials and cementations materials that replace the Ordinary Portland Cement. This study aimed to determine the effect of thermal resistance of fly ash based lightweight geopolymer and to investigate the physical, mechanical and microstructure properties of lightweight geopolymer at different thermal resistance. Considering the fire endurance assessment of geopolymers, the evolution of geopolymer during thermal exposure is of interest. This paper presents a comparative study of the characteristic of unfoamed (control sample) and lightweight geopolymers after exposure to elevated temperatures which is 200 °C, 400 °C, 600 °C and 800 °C. Lightweight geopolymers were prepared by adding foaming agent which is polyoxyethylene alkyether sulphate with geopolymer paste. Fly ash, alkaline activator and foam were mixed to produce a homogeneous mixture, which was placed into a cube 50 mm x 50 mm x 50 mm mould and cured at 80 °C for 24 hours. The compressive strength, density, chemical composition, microstructure and functional group analyses were studied. Unfoamed geopolymer exhibit higher compressive strength at 47.04 MPa compared to the highest strength of lightweight geopolymer which is 33.26 MPa at 200 °C. However, the lightweight geopolymer produced low density in range 1200 kg/m³ to 1500 kg/m³ compared to the density of unfoamed is 1813.79 kg/m³.

1. Introduction

Geopolymers are inorganic polymeric materials, firstly developed by Joseph Davidovits in 1970s. Recently discovered geopolymers are the cementations materials to replace cement produced by adding pozzolanic compound or mineral rich alumino silicate (Si-Al) catalysts, including sodium silicate solution (Na_2SiO_3) and sodium composition hydroxide ceramic materials that are which are transformed into reaction product by polymerisation in a high pH environment. This geopolymerization reaction is exothermic and takes place under atmospheric pressure at relatively low temperatures which is up to 100 °C [1]. The attraction closer to the implementation of latest technology for the manufacture of extremely good prospects lightweight geopolymer based totally products has gradually grown [2]. Low strength of geopolymer are the requirements of manufacturing from common raw materials and their in flammability at high temperatures, these compounds are attracting increasing interest as ecologically friendly fireproof building materials, sound and warmth insulators and substances for encapsulating dangerous wastes for storage or disposal [3].

Lightweight geopolymers have good mechanical properties like excessive compressive electricity, low permeability, tremendous mechanical houses, consisting of low density and thermal stability [4]. The density of lightweight concrete normally tiers from 300 kg/m^3 to 1800 kg/m^3 whereas the density of ordinary concrete is about 2400 kg/m^3 [5] . Fire endurance of geopolymers will present as the most effective insulating lightweight geopolymer when fly ash are added compared with conventional concrete that have good thermal resistance behaviour and mechanical performance. Based on study from El-Gamal & Hashem [6] , the residual compressive strength value for OPC when exposed to elevated temperature is decreases by 98 %, 62 %, 46 % and 12 % after firing at 200 °C, 400 °C, 600 °C and 800 °C, respectively.

In this study, lightweight geopolymer materials were subjected to thermal resistance which is exposed to elevated temperatures from 200 °C to 800 °C. Geopolymer concretes experienced a lesser reduces in weight but greater energy loss with increasing exposing temperatures than geopolymer mortar. On the basis, geopolymers undergo phase changes like mechanical strength when exposed to fires. The responses of geopolymers towards the elevated temperature hereafter lead to microstructural stresses [7]. Previous studies claimed that geopolymers have the better thermal endurance to normal concrete or maybe some excessive overall performance concretes [8]. The concern on the thermal performance and fire resistance of building materials has become one of the research interests because the thermal stability is important to ensure that they are safe for usage at certain temperature range and improved environmental sustainability.

2. Methodology

2.1. Raw Materials

In this study, low calcium Class F [9] was collected from a coal combustion plant in Manjung, Perak, Malaysia. Class F Fly ash was used as the aluminosilicate source as the base material. Class F Fly ash was analysed by X-Ray Fluorescence (XRF).

A mixture of sodium hydroxide (NaOH) solution and liquid sodium silicate (Na_2SiO_3) was used as the alkaline activator solution to activate the fly ash. The NaOH was in pallet form with 99 % purity and it solution was prepared by dissolving NaOH pellets in distilled water to obtain 12 M of NaOH concentration based on previous study [10]. Sodium silicate solution (Vitrosol D - A53) was used. The chemical composition of sodium silicate solution contains Na_2O = 14.7 %, SiO_2 = 29.4 %, Water = 55.9 %, Specific gravity of 1.53 g/cm³ and Viscosity at 20 °C =400 cP.

The foaming agent used in this study is Polyoxy Ethylene Alkyether Sulfate. This type of foaming agent is known as surfactant which are utilized by entraining air during blending while the chemical products was mixed into the geopolymer paste to generate bubbles which produces a lightweight material.

2.2. Mix Design and Mixing Procedure

Lightweight geopolymer bricks were produced by mixing the foaming agent, Fly ash, and alkaline activator solution. The alkaline activator solution was prepared first by mixing a Na_2SiO_3 solution with NaOH solution and stir for at least 5 minutes. Then, the fly ash was added and continue mixing for 5 minutes until the mixture homogeneous to produce geopolymer paste. The foaming agent was added to the geopolymer paste mixture with ratio of foaming agent/water of 1/10 and foam/geopolymer paste of 1.0, by volume [11]. Afterwards, the pastes were poured into cube moulds (50 x 50 x 50 mm) and cured in the oven for 24 hours at 80°C. After 24 hours of curing, the samples were left at room temperature for testing at 7 days of ageing. Table 1 shows the mix design for this study.

Table 1. Vickers hardness properties of un-aged and aged Al-Cu with different TiB_2 contents

Mix Design	**Range Value**
Ratio of Na_2SiO_3/NaOH	2.5 (by mass)
Ratio of fly ash/alkaline solution	2.0 (by mass)
Ratio of foaming agent/water	1/10 (by volume)
Ratio of foam/geopolymer paste	1/1 (by volume)
Thermal resistance	200, 400, 600, 800 (°C)

2.3. Thermal Resistance to the Fly Ash Lightweight Geopolymer
For exposure to elevated temperature, the 7 days cured lightweight geopolymers were heated in a furnace at 200 °C, 400 °C, 600 °C, and 800 °C to determine the effect of different thermal resistance of fly ash based lightweight geopolymer with a heating rate of 10 °C/min and soaking time of 2 hours. Then, the specimens were left to cool inside the furnace to room temperature. For comparison, one set of the samples was kept at ambient temperature which is 29 °C. In general, all geopolymers samples exposed to heating retained in cubic shape up to 800 °C without showing any destruction or dimensional change.

3. Results and Discussion

3.1. Chemical Composition Analysis of Fly Ash and Foaming Agent
In this study, the chemical composition of fly ash and foaming agent were analyses using X-Ray Fluorescence (XRF) analysis. Each composition of Class F fly ash and foaming agent used are summarized in Table 2 and Table 3 respectively.

Table 2. Chemical composition of Class F Fly Ash

Compound	**Mass (wt%)**
SiO_2	55.9
$Al2O_3$	27.8
Fe_2O_3	7.093
TiO_2	2.25
CaO	3.95
V_2O_5	0.096
SO_3	0.33
K_2O	1.55

Table 2 shows that the main constituents of fly ash are SiO_2 and Al_2O_3 which is about 83.7 % of the total composition and this indicated that this fly ash was classified as class F fly ash according to the ASTM C 618 [9] standard specifications. Percentage of Fe_2O_3 also is higher than the other compound for Class F fly ash. Material with high silica and alumina content are good for binding action [10] . The result also shows that these material rich in silica and alumina which could be have a potential to produce geopolymer materials with good mechanical and thermal properties.

Table-3: Chemical composition of foaming agent
(Polyoxyethylene Alkyether Sulfate)

Compound	**Percentage (%)**
SiO_2	1.80
SO_3	78.60

K_2O	1.60
CaO	4.20
Sc_2O_3	0.10
CuO	2.20
PdO	11.0
OsO_2	0.90

Table 3 shows the major constituents of Polyoxyethylene Alkyether Sulfate are SO_3 and PdO with 78.60 % and 11.0 %, respectively. This type of foaming agent are called as synthetic foaming agent which are an amphiprotic substance that are strongly hydrophilic and easily dissolve in water yielding air bubbles [11].

3.2. Effect of ageing

The compressive strengths of geopolymer samples when exposing at different temperatures are summarized in Figure 1. The strength of unfoamed and unexposed fly ash geopolymer is 47.04 MPa. For the lightweight geopolymer specimens that exposed at 200 °C reached the highest compressive strength 33.26 MPa. The results obtained depicted that the compressive strength kept decreased to 27.72 MPa and 19.40 MPa when exposed to 400 °C and 600 °C respectively until it reach the lowest strength at 800 °C which is 18.62 MPa. This result represent the relationship between compressive strength and density of geopolymer. With decreases in density of geopolymer, compressive strength was also decreases. As stated above, the loss of water with increasing temperature led to producing of more pores in structure and consequently reduce the strength. However, in addition of foam, the merging of bubbles is seemed to produce larger voids that result in a wide distribution of void size and also lower the strength [11]. Besides, the microstructure shows in Figure 4 also can relate in decreasing of compressive strength as a more open microstructure like pores and cracks which is consistent with its lower strength [12].

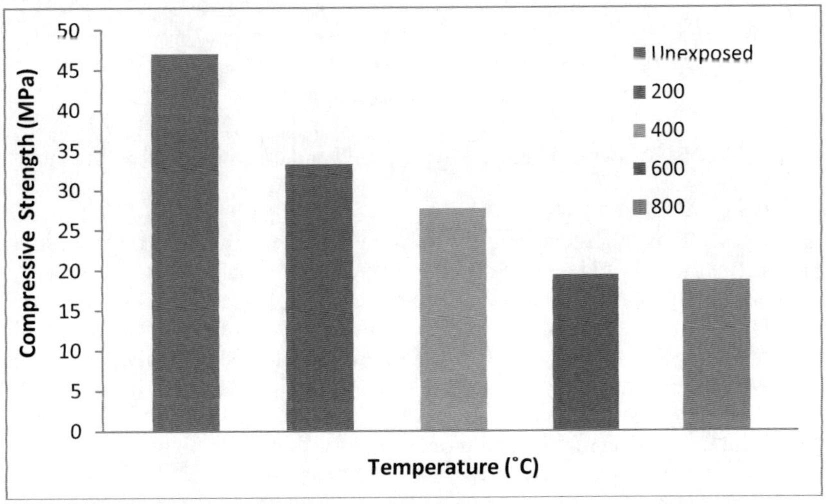

Figure 1. Compressive strength of the geopolymer when exposed and unexposed to elevated temperature.

3.3. Density Analysis

Figure 2 presents the test result on the density of lightweight geopolymer after exposure to the thermal resistance. The density data shown in Figure 2 is average value of three samples and the results reported the density values of the lightweight geopolymer are in the range between 1200 kg/m³ to 1500 kg/m³. As can be seen from Figure 2, the density of the unfoamed and unexposed geopolymer is 1813.79 kg/m³ and the highest from the exposed fly ash ash lightweight geopolymer is at 200 °C that presented 1493.65 kg/m³ density. The density keep decrease in the exposure to elavated temperatures to 1418.56 kg/m³ at 400 °C and 1314.35 kg/m³ at 600 °C. The lowest which is 1294 kg/m³ density value appeared at the sample with 800 °C. It was clearly shows that an addition of foaming agent and the increasing of the thermal resistance exposure decreased the density of lightweight geopolymer. There was a slight decrease in the density of lightweight geopolymer that affected from the presence of pore and crack to the surface of the sample [13]. In addition, the foaming agent caused the amount of voids inside the material with air contained could be generated, thus resulting in the reduction of density. This is due to the creation of tiny air bubbles in the geopolymer samples, which resulted in higher porosity and also reduced compressive strength of lightweight geopolymer [14] .

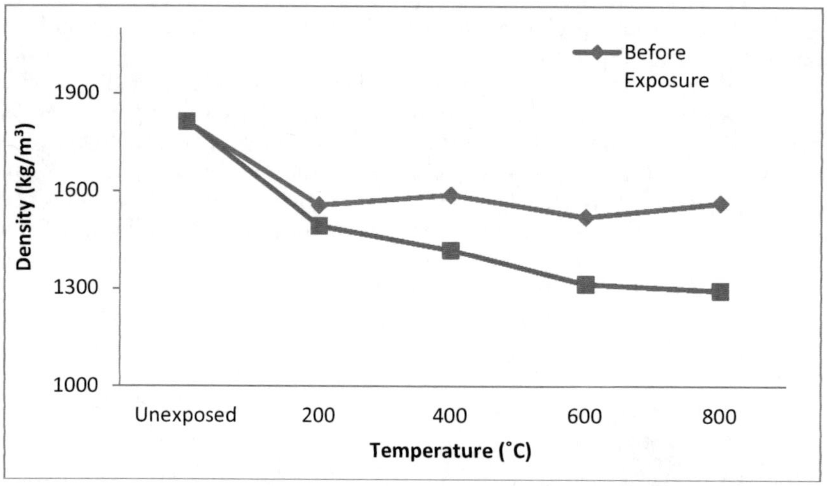

Figure 2. The density of lightweight geopolymer exposed and unexposed to elevated temperatures

3.4. Functional Group Analysis

Functional group analysis of high-temperature exposure can be analysed through Fourier transform infrared (FTIR) represented in Figure 3. Shifting of functional groups from fly ash toward fly ash geopolymers were identified using spectrum spectrometer. The specimen was powdered samples scanned from 4000 cm–¹ – 600 cm–¹ at the resolution of 4 cm–¹. Figure 3 shows the FTIR spectra of selected geopolymer samples. At 600 °C, 800 °C and unexposed fly ash geopolymer. The IR spectrum of fly ash geopolymer shows main absorption bands at 689 cm⁻¹, 981 cm⁻¹, 1453 cm⁻¹, 2348 cm⁻¹ and 3350 cm⁻¹. The O-H stretching bands also showed a well-defined trend at ~3300 cm⁻¹ -3610 cm⁻¹ for each of the sample. Deformation vibration of H-O-H for this lightweight geopolymer is at ~2348 cm⁻¹. It is noted also a small peak stretching of the C-O appears at ~1520 cm⁻¹. The bands become sharper and shift towards the lower frequencies ~689 cm⁻¹ which show the symmetric stretching of Al-O in Si-O-Al linkages. The entire sample at elevated temperature including the unexposed showed main broad absorption band at ~980 cm⁻¹ which corresponding to the asymmetrical stretching vibration of Si-O-Si and Si-O-Al. Table 4 shows the assignment of main FTIR bands.

With the concentration of network depolymerisation cations increasing, the vibrations attributed to stretching of Si–O and Al–O systematically shifts towards lower frequencies. The broadened peak around 3300 cm^{-1} - 3610 cm^{-1} is attributed to molecular water absorbed into the surface of the geopolymer [13]. At 2348 cm^{-1}, Ahmad et al., [15] stated deformation vibration of H-O-H groups from the weakly-bound water molecules that were trapped in the large cavities between the rings of the geopolymeric products. However, Zhang et al., [16] also defined 1420 cm^{-1} - 1660 cm^{-1} as the vibrational stretching frequency of the O-C-O in the carbonate, and caused by the unreacted alkali cations and the dissolved CO_2. The peak of band around 980 cm^{-1} exists in all mechanical activated materials prior to their geopolymerisation which formation of a new product called the amorphous aluminosilicate gel phase due to dissolution of fly in alkaline activator. This assigned to the Si (Al)–O–Si asymmetric stretch. This band is typical of the spectra of silicate glasses, and its frequency depends slightly on the state of hydration and Al content [17].

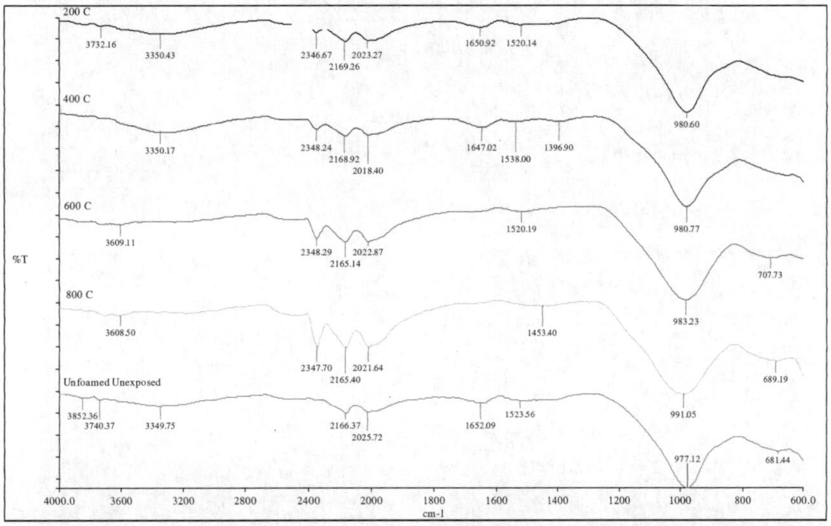

Figure 3. FTIR spectra of the geopolymer after expose to elevated temperatures.

Table 4. Shows the assignment of main FTIR bands.

Bands	Assignments
3350 cm^{-1}	Strong OH stretching vibration
2348 cm^{-1}	Bending vibration of H-O-H groups
1453 cm^{-1}	Stretching vibration of the O-C-O
980 cm^{-1}	Asymmetrical stretching vibration of Si-O-Si and Si-O-Al
689 cm^{-1}	Symmetric stretching of Al-O in Si-O-Al

3.5. Microstructure Analysis

Microstructure of fly ash based lightweight geopolymer with different exposed thermal resistance were examined by using Scanning Electron Microscope (SEM). For the exposure to elevated temperature of the lightweight geopolymer, it is clearly seen in Figure 4 that the number of pores distributed throughout the matrix are increases due to addition of foaming agent. There is fine pores and no crack at exposure to 200 ºC and 400 ºC. Comparatively, the micrographs of the sample heated at 600 °C and 800 °C shows large cracks and pores. However, micrograph exposure at 800 ºC shows the lightweight geopolymer matrix intervened with each other forming continuous matrix and also shows hollow cavities due to spaces left behind by dissolved fly ash particles. The results show that the increases of exposure to thermal resistance, the average pores size and cracks presence also increases. This change has occurred in the microstructure because of sintering and further geopolymerization with the increase in temperature [18]. Porosity is caused by different developments of microstructure skeletons in the porous geopolymer. The foam to geopolymer paste give large affects to the volume of voids and porosity in the lightweight geopolymer which directly influences the strength of geopolymer [19].

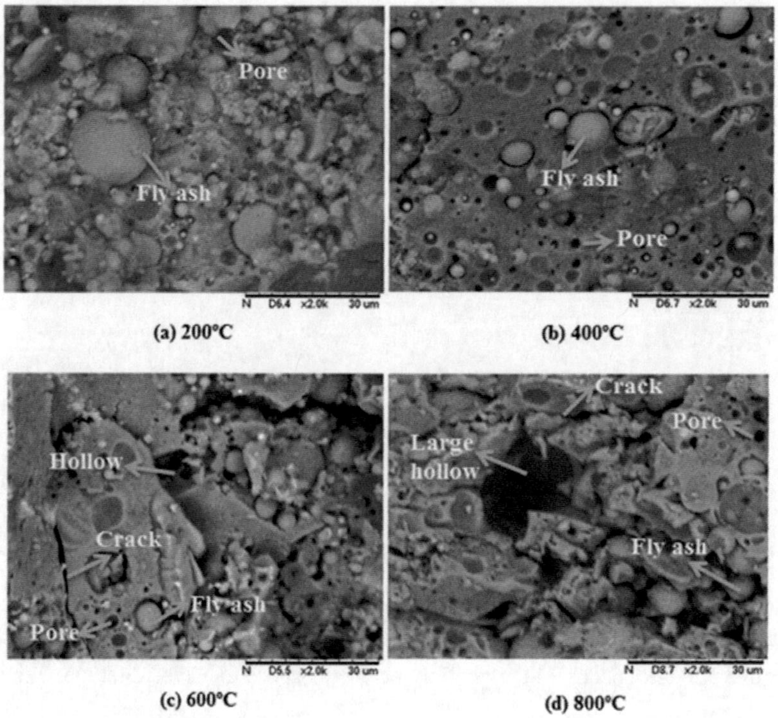

Figure 4. Micrographs of exposed to thermal resistance (a) 200 °C, (b) 400 °C, (c) 600 °C and (d) 800 °C

4. Conclusions

In this research presented the effect of fly ash unfoamed and lightweight geopolymer exposed to elevated temperatures. No disintegration and spalling of geopolymer samples occurred when subjected to high temperature. Unexposed unfoamed geopolymer exhibited excellent compressive strength which is 47.04 MPa with very high density 1813.79 kg/m³. With the addition of Polyoxyethylene Alkyether Sulphate as foaming agent, lightweight geopolymer was obtained. However, the strength was reduced from that of unfoamed geopolymer in range (18 MPa – 33 MPa) with lower density (1290 kg/m³-1500 kg/m³). When the geopolymers were heated at elevated temperature, the density and compressive strength decreased. It was deemed that the fast evaporation of water from the geopolymer structure induces internal damage to the structure within resulting in high water absorption. Foaming of geopolymers helped to minimize the disruption effect caused by thermal treatment as the pores provides room to counteract the damage by heat. It can be concluded that thermal resistance 400 °C to the fly ash lightweight geopolymer gives better properties in terms of density, compressive strength and water absorption compared to the others thermal resistance. However, it can be simplified that lightweight geopolymer have very excellent properties at exposing to the thermal resistance compared to the conventional concrete.

Acknowledgements
The authors would like to acknowledge the support of Center of Excellence Geopolymer and Green Technology (CEGeoGTECH), School of Materials Engineering, Universiti Malaysia Perlis.

References
[1] Živica V *et al* 2015 *Build. Res. J.* **61** 85.
[2] Roviello G *et al* 2017 *Compos. Part B Eng.* **128** 225.
[3] Wongsa A *et al* 2016 **111** 637.
[4] Messina F 2018 **86** 266.
[5] Abdullah M M A B *et al* 2012 *Int. J. Mol. Sci.* **13** 7186.
[6] El-Gamal S M A *et al* 2017 *J. Therm. Anal. Calorim.* **128**(1) 15.
[7] Shahedan N F *et al* 2017 *AIP Conf. Proc.* **020046** 020046.
[8] Cheng-yong H *et al* 2017 *Nat. Publ. Gr.* **March** 1.
[9] ASTM C 618 2012 44.
[10] Ibrahim W M W *et al* 2017 *AIP Conf. Proc.* **1885**.
[11] Mastura W A N *et al* 2013 *REV.CHIM.(Bucharest)* **9** 1.
[12] Kearsley E *et al* 2013 *Cem. Concr. Res.* **31**(1)105.
[13] Risdanareni P *et al* 2012 *AIP Conf. Proc.* **1835**.
[14] Yang T *et al* 2012 *Adv. Civil, Environ. Mater. Res.*905–914.
[15] Ahmad R et al 2017 *AIP Conf. Proc.* **1835**.
[16] Zhang Z *et al* 2014 *Constr. Build. Mater.* **56** 113.
[17] Bakharev T 2005 *Cem. Concr. Res.* **35**(6) 1233.
[18] Liu Z *et al.* 2014 **21**(1) 89.
[19] Sanjayan J G *et al* 2015 *Constr. Build. Mater.* **79** 236.

Influence of Solid-To-Liquid Ratio on Properties of Fly Ash Geopolymer Ceramics

Romisuhani Ahmad[1,2], Mohd Mustafa Al Bakri Abdullah[2], Wan Mastura Wan Ibrahim[1,2], Andrei Victor Sandu[3], Warid Wazien Ahmad Zailani[2], Sharun Nizam Shahrun Nahar[1]

[1]Faculty of Engineering Technology, Uniciti Alam Campus, Universiti Malaysia Perlis (UniMAP), Perlis, Malaysia.
[2]Center of Excellence Geopolymer and Green Technology (CEGeoGTech), School of Materials Engineering, Universiti Malaysia Perlis (UniMAP), Perlis, Malaysia.
[3]Faculty of Materials Science and Engineering, Gheorghe Asachi Technical University of Iasi, Romania

E-mail: romisuhani@unimap.edu.my

Abstract. Ceramics has come to be progressively very important in industry because of their exceptional mechanical and physical properties The motivation to develop fly ash geopolymer ceramics was to overcome the problem associated with the conventional technical ceramics like alumina, silicon, carbide, aluminium nitride that fracture easily under mechanical or thermo-mechanical loads. The novelties of this research were providing new material to fabricate geopolymer as ceramic precursor and move towards green technology. Geopolymer substrates were activate by adding the alkaline activator gradually with different ratios of solid to liquid ranged from 0.5, 1.0, 2.0 and 3.0 into fly ash at ambient temperature. The samples were cured before undergo powder metallurgy method. The best and optimum solid to liquid ratio is determined at ratio 2.0 based on flexural strength. The broad, amorphous peak showed by phase analysis indicates that geopolymerization has occurred. The formation of almost fully crystalline phase such as albite has been detected which contributes to the flexural strength of fly ash geopolymer ceramic. Microstructure image shows smoother and complete geopolymer matrix which gives denser structure as supported by the excellent density.

1. Introduction

Ceramic usually are from clay that widely used in the manufacture of many traditional ceramics. Each ceramic product requires clays having particular and appropriate characteristics, which must not contain a swelling phase, low loss of weight and low shrinkage after drying firing process [1]. Firing characteristics such as shrinkage, water absorption, and mechanical properties are very important characteristic to produce a good ceramic. Typically producing ceramic specifically used many types of fabricating method basically performs at high temperature which is complex and high cost process. A wide vary of process routes have indeed been proposed for the production of ceramics, using raw materials of varied nature and resulting in a range of morphologies that affected the properties [2].

Conventional ceramic materials fabrication require high sintering temperature up to 1600 °C (Kriven et al., 2013). The use of geopolymer technology is an alternate way in fabricating ceramic materials as the amorphous and semi-crystalline phase of geopolymer will convert into crystalline phases during sintering [4]. With the advantage of geopolymerization reactions, high temperature routes are no longer required to produce ceramic-like structure materials (Iwahiro et al., 2001). Besides, geopolymer also can be directly converted into final structural leucite, pollucite or nepheline ceramic part and thus the chemical compositions of final product also can be design (Peigang et al., 2010). Thus, geopolymer is parts of ceramic which are a well-known material for composites industry that become demands lately.

Inorganic polymers referred as geopolymers, are alumino-silicate materials which display superior physical and chemical properties with a diverse range of possible potential applications such as coating, high temperature ceramics, and cement for concrete and binder for fireplace resistant fiber composite [7]. Fly ash that contain a large amount of silica and alumina can be used as a source material for geopolymer [8]. Other than composition of the source material, several factors were identified as important parameters and solid to liquid ratio is one of the main factor as it rules the amount of solid and liquid concerning on homogeneous mixing, which specifically affect the dissolution, geopolymerization reaction, workability, and the strength of the final product. Therefore, this study is focusing on the effect of solid to liquid ratio on the mechanical and microstructural properties of the fly ash geopolymer ceramics.

2. Methodology

2.1. Materials

Fly ash is the main raw materials to produce the geopolymer ceramics. In general, the components of fly ash typically include SiO_2, Al_2O_3, CaO and Fe_2O_3, which exists in the form of amorphous and crystalline oxides or various minerals as shown in Table 1. The fly ash used in this study was obtained from Manjung power station, Lumut, Perak, Malaysia. The physical forms of fly ash used were in powder state with the mean particle size (X_{50}) of 46.22 µm.

The alkaline activators used in this research were sodium silicate (Na_2SiO_3) solution and sodium hydroxide (NaOH). NaOH in pellets form used was caustic soda micropearls with 99 % purity, Formosoda-P brand were made in Taiwan while Na_2SiO_3 with properties of Grade A53 with SiO_2 = 29.43 %, Na_2O = 14.26 % and water = 56.31 % were used.

Table 1. Chemical composition of Class F fly ash

Compound	Mass (wt. %)
SiO_2	55.90
Al_2O_3	27.80
Fe_2O_3	7.09
CaO	3.95
TiO_2	2.25
K_2O	1.55
SrO	0.37
SO_3	0.33
RuO_2	0.24
ZrO_2	0.13
V_2O_5	0.10
MnO	0.06
LOI	0.23

2.2. Sample Preparation

The alkali activator solution is prepared by mixing the 12M of NaOH solution with sodium silicate at alkali activator ratio of 2.5 which is fixed in this research until a clear solution is obtained to make sure the solution is well dissolved. The chosen ratio was based on previous work, as it certified for optimum properties. The geopolymer paste was prepared by mixing the fly ash with the alkali

IOP Conf. Series: Materials Science and Engineering **551** (2019) 012083 doi:10.1088/1757-899X/551/1/012083

activator at various solid to liquid ratio of 1.0, 2.0 and 3.0. The paste then were stirred well for a few minutes and poured into moulds. The samples were put into oven at 80 °C for 24 hour for the curing process.

To produce fine powder for powder metallurgy method, the fly ash geopolymer were crushed and sieved by using siever of 150μ. The fly ash geopolymer in powder form were then compacted using uniaxial compaction, hand press using 12 mm diameter mould under pressure of 4.5 ton for 2 minute. After being shaped into a pellet, the green bodies were sintered at a fixed temperature of 1000 °C with soaking time of 180 minutes and heating and cooling rate of 5 °C/minutes.

2.3. Testing

2.3.1. Flexural Strength
Flexural strength of fly ash geopolymer ceramic samples were measured using mechanical testing with Automatic Max (Instron, 5569 USA) according to the standard of ASTM C1161-02c. A minimum of five samples for each group were subjected to flexural strength testing were conducted on specimens (7 mm x 5 mm x 52 mm) using a three-point-bending fixture, with a span length of 30mm at a crosshead speed of 0.5 mm/minute.

2.3.2. Density
The theoretical density of fly ash ceramic and fly ash geopolymer ceramic was measured using densitometer model MD 300S that available in UniCITI Alam, UniMAP. A cylinder shaped sample with a diameter of 12 mm and a length of 5 mm were used in this experiment.

2.3.3. Morphology Analysis
JSM-6460LA model Scanning electron microscope (JEOL) was performed to reveal the microstructure of fly ash geopolymer ceramic at various solid to liquid ratio. The specimens were prepared and coated by using Auto Fine Coater; model JEOL JFC 1600 prior to examination.

2.3.4. Phase Analysis
The sample was characterized by using XRD 6000, SHIMADZU diffractometer. Samples were prepared by pressing the powder samples into aluminium holders. Operating conditions were 40 kV and 30 mA and performed using Cu-Kα radiation. The X'Pert High Score Plus software is used for peak identification and automated search match was used to analyse the results of diffraction patterns.

3. Results and Discussion

3.1. Flexural Strength
The flexural strength of fly ash geopolymer ceramics was assessed at three different solid to liquid ratio in order to determine the highest strength as shown in Figure 1. At 1.0 of solid to liquid ratio, the flexural strength was 19.64 MPa, whereas, the solid liquid ratio 2.0 achieved the highest strength which is 25.99 MPa, attributed to the increasing of fly ash content and activator solution content [9]. The strength of fly ash geopolymer ceramic is also depending on the workable mixture ratio. Less workable mixture causes difficulty in flexural testing where failure in providing good flexural properties and severely reduce the flexural strength of the samples. The flexural strength then dropped at the solid to liquid ratio of 3.0 which gives the lowest flexural strength of 18.92 MPa. The result clearly showed the best and workable solid to liquid ratio is at 2.0 where at this ratio, the most optimum ratio provided the highest flexural strength compared to other ratio which were less workable.

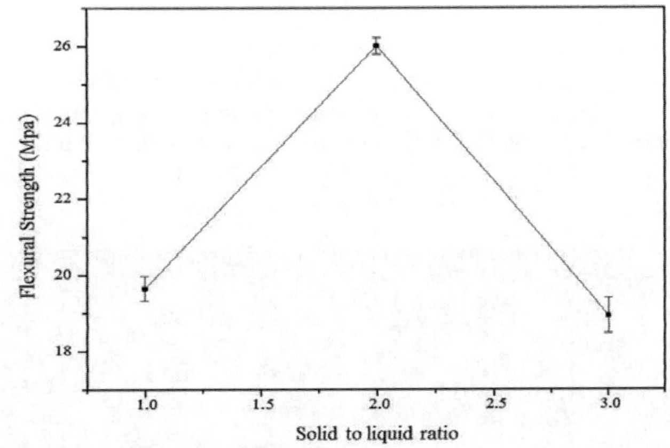

Figure 1. Flexural strength of fly ash geopolymer ceramic at different solid to liquid ratio.

3.2. Density

Figure 2 shows the density of the fly ash geopolymer ceramics as a function of solid to liquid ratio. It can be seen that the density of ceramic increased with the increasing of solid to liquid ratio from 1.0 to 2.0 ranging from 1.716 g/cm3 to 1.957 g/cm3. The results were contributed by the matrix of fly ash increased in the solid to liquid ratio 2.0. The increased amounts of fly ash matrix reduce the void in the structure of fly ash and denser the particle. However, the density of the geopolymer ceramic start to decrease at solid to liquid ratio of 3.0. The recorded density of fly ash geopolymer ceramics of solid liquid ratio 3.0 was 1.873 g/cm3. The density was slightly decreased due to the workable rate of the fly ash is not optimum and the exceeded amount of fly ash matrix in the structure make it less workable in the mixing process and leads to the decrease in the density [10].

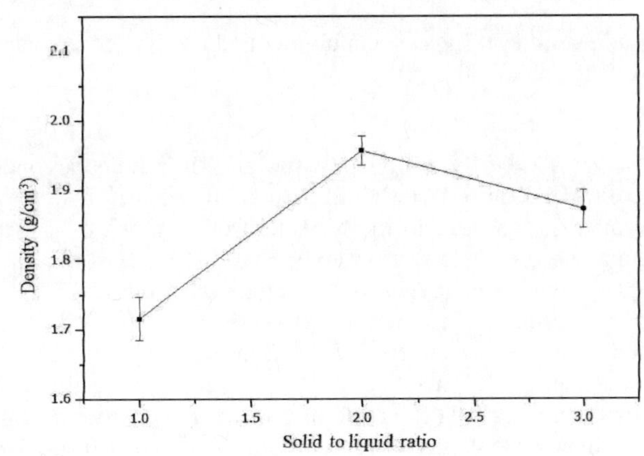

Figure 2. Density of fly ash geopolymer ceramic at different solid to liquid ratio.

3.3. Morphology Analysis

Figure 3 shows the microstructure evolution of the fly ash geopolymer and fly ash geopolymer ceramic at various solid to liquid ratio. The image of fly ash geopolymer sample surface showed heterogeneous elements and contained more unreacted particles. As the samples were sintered, at solid liquid to liquid ratios 1.0, large sintered area was formed due to the fly ash particle begin to fuse together where it demonstrates incomplete geopolymerization. For the solid to liquid ratio 2.0, the matrix of fly ash were started to fuse together and the appearance of pores started to presence. The samples appeared to have a smoother geopolymeric matrix and also denser as the concentration increased indicate that geopolymerization has occurred well and the effect of sintering [6]. However, at solid to liquid ratio 3.0 the pores started to increase and leave many pores in the structure. The solids to liquid ratio affect the amount of pores in the pastes which directly influences the strength of geopolymer ceramics.

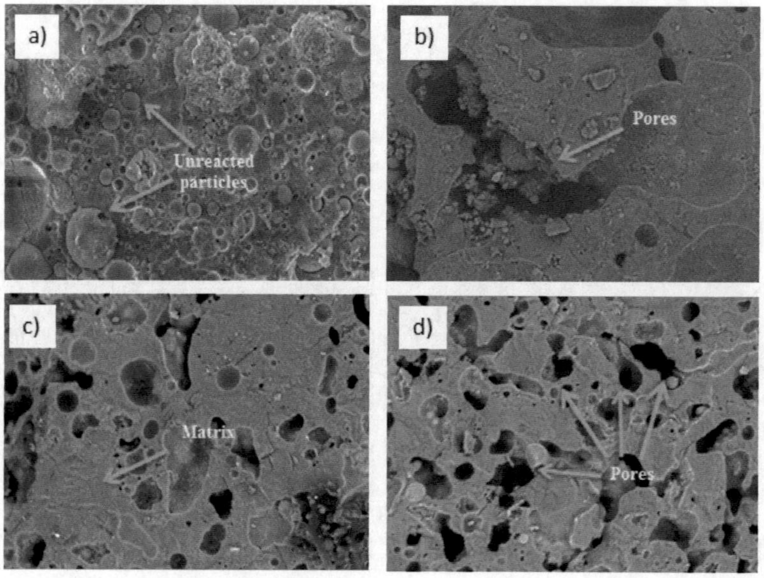

Figure 3. Microstructural images of a) fly ash geopolymer and fly ash geopolymer ceramics with solid to liquid ratio of b) 1.0, c) 2.0, and c) 3.0.

3.4. Phase Analysis
Figure 4 shows the phase analysis of fly ash geopolymer and fly ash geopolymer ceramic at various solid to liquid ratio. The difference in solid to liquid ratio and the effect of sintering produces a series of reactions and transformations that lead to the formation of new phases and the disappearance of humps in the diffractogram spectra. It can be observed that the peaks of quartz which contain SiO_2, disappeared as the samples were sintered. The broad reflection pattern of fly ash begins to show a sharp peak indicates the transformation of amorphous to crystalline phase upon sintering [11]. As the sample of solid to liquid ratio of 1.0 was sintered, the presence of nepheline phase can be found at the peak. However the peaks change into albite at solid to liquid ratio 2.0. Finally, analbite were obtained at the peak of solid to liquid ratio 3.0. The sample of fly ash geopolymer ceramic of solid to liquid ratio 2.0 shows the highest intensity compared to solid to liquid ratio of 1.0 and 3.0.

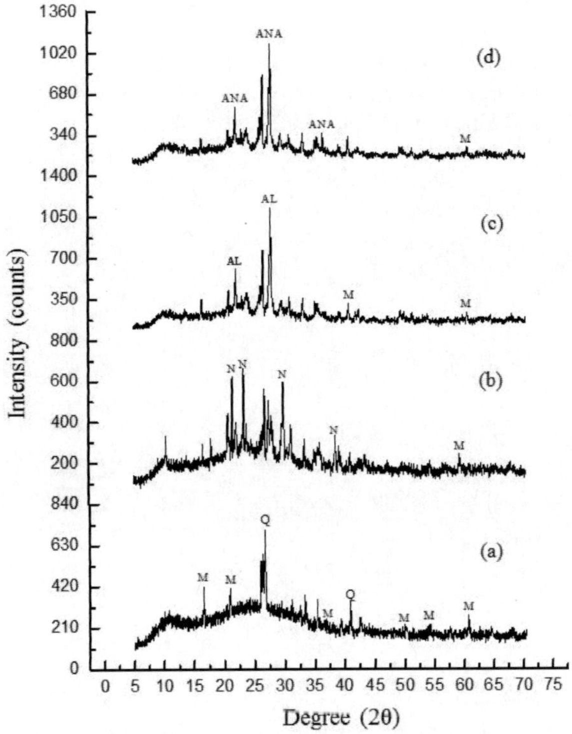

Figure 4. Phase analysis (a) fly ash geopolymer and fly ash geopolymer ceramics at solid to liquid ratio of (b) 1.0, (c) 2.0 and (d) 3.0 (N= Nepheline, AL= Albite, ANA= Analbite,Q= Quartz, M= Mullite,).

4. Conclusions

The aim of this paper is to study the effect of addition of Ultra High Molecular Weight Polyethylene (UHMWPE) on kaolin geopolymer ceramics. Results concluded that the addition of UHMWPE affected the strength of the kaolin geopolymer ceramics. The flexural strength of kaolin geopolymer ceramics with addition of UHMWPE increased compared to the kaolin geopolymer ceramics without addition of UHMWPE. Kaolin geopolymer ceramics with addition of 4 wt. % UHMWPE give the highest strength which is 92.1 Mpa. However, the strength started to decrease due to the excessive content of UHMWPE lead to the increasing of pores size appeared in microstructure of kaolin geopolymer ceramics.

Acknowledgements

The authors of the present work wish to dedicate their great thanks to Centre of Excellence Geopolymer System Research @UniMAP.

References

[1] Monteiro S N *et al* 2004 *Appl. Clay Sci.* **27** 229.
[2] Colombo P 2006.
[3] Kriven W M *et al* 2012 January 22.
[4] Peigang H *et al* 2013 *J. Mater. Sci.* **48** (2) 1812.
[5] Iwahiro T *et al* 2001 *J. Eur. Ceram. Soc.* **21**(14) 2515.
[6] Peigang H et al 2010 *Ceram. Int.* **36** 2395.

[7] Chen P *et al* 2007 *J. Mater. Process. Technol.* ***190***(1–3) 243.
[8] Wan Ibrahim W M *et al* 2015 *Key Eng. Mater.* **660** 9.
[9] Ozer I *et al* 2015 *Ceram. Int.* **41** 10192.
[10] Kong D L Y *et al* 2007 *Cem. Concr. Res.* **37**(12) 1583.
[11] Kuenzel C *et al* 2013 *J. Eur. Ceram. Soc.* **33**(2) 251.

IOP Conf. Series: Materials Science and Engineering **551** (2019) 012084 doi:10.1088/1757-899X/551/1/012084

A Review of Carbonate Minerals as an Additive to Geopolymer Materials

Ahmad Syauqi Sauffi[1], Wan Mastura Wan Ibrahim[1], Mohd Mustafa Al Bakri Abdulla[2], Romisuhani Ahmad[1], Fakhryna Ahmad Zaidi[1], Andrei Victor Sandu[3] & Subaer[4]

[1]Faculty of Engineering Technology, Universiti Malaysia Perlis (UniMAP), Perlis, Malaysia.
[2]Centre of Excellence Geopolymer and Green Technology (CEGeoGTech), School of Materials Engineering, Universiti Malaysia Perlis (UniMAP), Perlis, Malaysia.
[3]Faculty of Materials Science and Engineering, Gheorghe Asachi Technical University of Iasi, Romania
[4]Center of Excellence of Green Materials and Technology (CEoGMTech), Physics Department, FMIPA, Universitas Negeri Makassar, Makassar, Indonesia

E-mail: wanmastura@unimap.edu.my

Abstract. In Malaysia, several geopolymer materials were mainly from fly ashes, kaolin, metakaolin, and boiler ash which were mainly used to prosper sustainable development especially in the building and construction sector. Geopolymers can be classified as inorganic materials that were formed with the reaction between alminosilicate materials and alkaline solutions. This reaction causing a crystalline compounds mixture to form that toughen to become a novel compound through a process called geopolymerization process. These geopolymer materials have many admirable advantages, including high strength, small shrinkage, decent thermal resistance and good chemical resistance. Although they were used as a stand-alone material with great properties, combination with another material might be another way to improve their properties. A few studies had shown that the geopolymer material can achieve better properties with the addition of moderate amount of calcium-containing material to a geopolymer. The addition of carbonate materials can have insignificant effect on the geopolymer configuration and properties. Thus, this paper review the usage of carbonate materials for example Dolomite (CaMg(CO3)2) and Calcite (CaCO3) as an addition into these geopolymer material.

1. Introduction

Carbonate is a mineral that consists of CO3 component in its molecule structure. The molecular structure of CO3 consists of one carbon atom that is bonded with three oxygen atoms. It was built by the covalent bond between the atoms of carbon and oxygen [1]. One of the oxygen atoms shares two electrons with carbon, whereas the other one shares one electron for each oxygen atom. This structure leaves two negative charges for carbonate sub molecule. A carbonate molecule is built through ionic bond between this negative charge and the positive charge from certain metallic element [2].

Surrounded by all the carbonate minerals, calcite is categorized as the most mutual one that is often used conventionally. One of the usages is to make the composite for construction structures like roads and buildings when calcite is heated and assorted with additional materials [3]. Besides calcite, the carbonate mineral that frequently used for construction is dolomite. Dolomite is carbonate mineral that contains magnesium besides calcite. The existence of magnesium in dolomite make it is used as magnesium foundation for human, animal and plants, since magnesium is one of vital elements wanted by breathing mortals [4].

Moreover, dolomite is commonly used in road and building construction for foundation bases and as an aggregate for cement and asphalt concrete. Other than can reduce the mechanical properties of the product, dolomite at the same time can improve the formation of better contact zone for cement paste aggregate for the carbonate aggregate [5].

The benefits of using carbonate minerals in road construction are improve structural quality, greater load bearing, reduced maintenance cost and extended life [6]. Besides that, calcium carbonate materials such as calcite and dolomite are abundant and generally inexpensive natural minerals. It will be a great interest to be using this bulk of calcium carbonate materials to improve the properties of a geopolymer [7].

2. Chemical Composition of Dolomite and Calcite

Table 1 display the chemical composition of dolomite and calcite as determined by XRF. Both of the materials are from Clayworks, Australia. Based on the table, dolomite contains high calcium oxide (CaO) and magnesium oxide (MgO) composition. Besides that, dolomite also contains silicon dioxide (SiO_2) and alumina oxide (Al_2O_3). The silica and alumina are the core ingredients to produce geopolymer binder. Thus, it is possible to make dolomite as a raw material for geopolymer. Comparison between dolomite and calcite shows that calcite has higher CaO content meanwhile dolomite has higher MgO content. Other than that, dolomite has a small composition of K_2O Na_2O meanwhile calcite had none.

Table 1. Chemical composition of dolomite and calcite (% wt.) [7]

	CaO	SiO_2	Al_2O_3	Fe_2O_3	MgO	K_2O	Na_2O	LOI
Dolomite	33.4	2.5	0.7	0.3	17.1	0.07	0.02	42.9
Calcite	53.5	1.5	0.3	0.2	1.7	0	0	42.8

LOI: Loss on ignition at 1000°C

However, different sources of carbonate materials compose different composition. Based on XRF analysis, the dolomite from Jawa Timur, Indonesia composed 61.83 %wt. of CaO content is and MgO content is 25.75 %wt. [8]. Compared from table 1, the composition have huge different of CaO content. The dolomite form Indonesia has more CaO content compared to dolomite from Australia [9]. Another comparison is the dolomite composition content from Sverdlovsk Oblast, Russia. The CaO content is 23.16 %wt. and the MgO content is 15.6 %wt. Other than that, the SiO content is very high with 21.7 %wt. Thus, it can be conclude that, different raw materials form different country had various compositions and those compositions are different to one and another. The different of compositions can lead to dissimilar properties and reactions.

3. Geopolymer Concrete

Geopolymer concrete is a new eco green product synthesized from precursor and alkaline activator revealed as a promising construction material to reduce utilization of Portland cement [10]. Geopolymers is an inorganic material of cement where materials that are formed by reacting silica and alumina with a solution of alkaline solutions which consist of Na2SiO3 and NaOH resulting from a mixture of crystalline compounds that eventually harden into a new strong compound through a process called geopolymerization process [11]. The precursors for geopolymer are developed from various sources including the natural resource kaolin and industrial wastes or byproducts, such as fly ash and ground granulated blast furnace slag (GGBFS) [12].

Chemically, geopolymers contain of cross-linked units of AlO4 and SiO4 tetrahedra, where charge-balancing cations are provided by alkali metal cations such as Li+, Na+, K+, and Cs+. [13]. Duxson et.

IOP Conf. Series: Materials Science and Engineering **551** (2019) 012084 doi:10.1088/1757-899X/551/1/012084

al, reported that the geopolymer have many advantages which are high compressive strength, low shrinkage, fire resistance, acid resistance, and low thermal conductivity depending on materials used [14-15]. Additionally, the environmental impacts can be reduced by using the building material that was mixed with green building material for example, geopolymer. The environmental impacts that can be reduced are the low manufacturing energy consumption, low CO2 emission [16], and removal of these construction industry basis resources [17].

Metakaolin, a material produced from kaolin calcinations which has similar content to fly ash was considering as alternative precursor material besides fly ash [18]. Study from Pelisser et al. [19] stated that metakaolin based geopolymer showed high flexural strength and an excellent relationship between the flexural and compressive strength, which indicated good resistance to cracking while under stress.

Other than that, material that contains high calcium (Ca) and magnesium (Mg) also can be used as geopolymer materials. Materials that contain Ca and Mg can have certain effect to geopolymers properties. W.K.W. Lee and J.S.J. van Deventer [20] studied that it was found that calcium (Ca) and magnesium (Mg) shortened the setting time by providing heterogeneous nucleation centers in the initial paste solution. The Ca and Mg accelerated the setting in a similar fashion as all the other no treated systems.

4. Dolomite Addition into Geopolymer

A study conducted by Christina K. Yip et. al. [21] demonstrated 20% of dolomite additions into metakaolin increase its compressive strength of all the samples done. The strength was significantly increased up to 100 days with more than 50MPa. More than 20% of dolomite addition in the metakaolin was found to be detrimental, which is likely to be caused by a low degree of geopolymeric gel formation and connectivity. They also conclude that it is possible that Mg2+ plays a similar role to Ca2+.

E. A. Aizat [22] studied the compressive strength of the dolomite geopolymer with different NaOH molarities. The usage of solid/liquid ratio prepared is 2.5 and the Na2SiO3/NaOH is also 2.5 to this dolomite geopolymer. After that, they were cured at the temperature 80°C for 24 hours. After 7 day the test for compressive strength was done and the compressive strength of dolomite geopolymer increased as the NaOH increased. NaOH molarity up to 20M gave the highest compressive strength of all molarities. The usage of dolomite as the core material results that high concentration of NaOH and high curing temperature needed to be used. This is due to the low composition of Al and Si in dolomite that causes the less reactivity between dolomite and alkaline activator. Thus, a high NaOH concentratin is needed which is 20M to increase the dissolution process for ions Al3+ and SiO4+ as well as the development of geopolymer network systems.

The geopolymer material that had composition of high CaO will improve the geopolymer strength [23]. Although with the high content of CaO in the dolomite geopolymer, the compressive strength is still low compared to the fly ash geopolymer [24]. Thus, the dolomite geopolymer need a filler to increase its properties or it can be used as a filler to improve other properties.

Unfortunately, Study form Y. Zarina et al. [25] with the addition of dolomite into the boiler ash with up to 5% has decreased the strength of boiler ash based geopolymer. The NaOH molarity was fixed with 12M used for all the mix design. The solid/liquid ratio used is 1 and the Na2SiO3/NaOH used is 2.5 [26]. The boiler ash geopolymer without dolomite addition into it gave the highest compressive strength out of all compared to the dolomite addition into boiler ash. The compressive strength for boiler ash with no dolomite addition is 19.4 MPa. The comparison between the additions of dolomite showed that the addition of 4% dolomite has the best out of all addition compared to other composition with dolomite. The researchers also conclude that the addition of dolomite has disturbed the formation of three-dimensional geopolymeric aluminosilicate network which has influence the strength of geopolymer.

5. Calcite Addition into Geopolymer

C. K. Yip et al. also did the study on the calcite addition into metakaolin. Calcite content results also parallel to dolomite. Calcite addition tends to lead a stronger geopolymeric reaction compared to dolomite. It is due to simple a micro-aggregate effect which suggests that the minor quantity of dissolution of the carbonate minerals that is observed is enough to release enough Ca2+ or Mg2+ to have a noteworthy result on the geopolymer gel structure [21].

Another studies form C. K. Yip et al. which focuses on role of calcium in improving cement durability [27]. It is suggested that calcium provides a crucial link between the chemistries of geopolymerisation and the hydration of cement. Samples were synthesised using a metakaolin to calcite ratio of 0.8. The solid to liquid ratio was 0.69 for Ms = 2.3 and 0.61 for Ms = 1.5. These ratios were selected so as to maintain consistent water content in all synthesised material. Results showed that addition of calcite during Ms = 2.3 showed low strength at day 1 which was less than 5Mpa. But, on day 90, the strength increased to 30.1MPa. Meanwhile for Ms = 1.5, the strength shows steady increased from day 1 to day 90 which was from 42.1Mpa to 47.6Mpa. It is also clear that the type of calcium source added and the alkaline conditions used have a significant impact on the resultant binder.

Other than that, Patrick N. Lemougna et al. [28] reports the development of inorganic polymers from laterite, for a potential use at least as non-load bearing building materials The effect of blending laterite with Ground Granulated Blast Furnace Slag (GGBS) and calcite (CaCO3) was investigated. Laterite was substituted by calcite at 2 to 20% in mass. Sodium silicate solutions of modulus ranging from 1.6 to 2.2 were used to prepare inorganic polymers from laterite that had been thermally treated at 700 °C. The strength increased with the reduction of the modulus of the activating solution, achieving a 28 days maximum compressive strength of 36 MPa at 25 °C. The substitution of up to 20% laterite by calcite was found to have little effect on the compressive strength. At variance, slag/laterite combination presented better strength performance, mainly above 20% of slag, and achieved a 28 days compressive strength of 65 MPa with a composition made of 50% laterite and 50% slag.Conclusions

Epoxidized vegetable oils have been researched by many researchers in this past years because of they are derived from renewable and sustainable resources that can be used to replace petrochemical raw material in the future. In this study, palm olein from palm oil has been epoxidized using in-situ formed peracetic acid in order to produce epoxidized oleic acid. The DHSA can be synthesised from the epoxidized palm olein by reacting with water that can decreasing the RCO% and producing hydroxyl fatty acid that can also be used as ramaterial for industries.\

6. Conclusion

As conclusion, this paper summarized the usage of carbonate material which is dolomite and calcite in geopolymer. Carbonate material is a potential to become a precursor material for the replacement of used precursor material for geopolymer. Moreover, it was found that the CaO content in calcium sources had impact on the resultant compressive strength. On the contrary, the mineralogy of the calcium sources used and the concentration of sodium hydroxide present were found to have a significant influence on the resultant product. Besides, the use of dolomite which is abundant and potentially offer sustainable solution and environmental friendly for extending the service life of infrastructure and maintenance cost. Development of carbonate material geopolymer less been explored. It is hoped that future research in this field will drive a new era of greener materials in the geopolymer industry.

Acknowledgements

This study was supported by Center of Excellence Geopolymer and Green Technology (CEGeoGTECH) UniMAP and School of Materials Engineering, UniMAP. The authors of this research are very grateful to King Abdul Aziz City Science and Technology (KACST) for the financial support of this study.

References

[1] Omari M A *et al* 2016 31.
[2] Solihin 2018 *IOP Conference Series: Earth and Environmental Science* **118** 012065.
[3] James D *et al* 2012 *USGS Fact Sheet, United States of America* 1.
[4] Cygas D *et al* 2008 *The 25th International Symposium on Automation and Robotics in Construction. ISARC-2008.*
[5] Vaganov V *et al* 2016 *Construction Science* **19**.
[6] Rosli N A 2014 *Bachelor Thesis, Faculty of Engineering, University of Malaysia, Serawak, Malaysia* 1.
[7] Yip C K *et al* 2008 *Cement and Concrete Composites* **30** 979.
[8] Sulistiyono E *et al* 2017 *IOP Conference Series: Materials Science and Engineering* **176** 012041.
[9] Mikhailova O *et al* 2013 *Procedia Engineering* **57** 775.
[10] Bakri A M M *et al* 2013 *Revista de Chimie,* **64** 382.
[11] Muduli S *et al* **4** 032.
[12] Guo X *et al* 2010 **32** 142.
[13] Ahmad R *et al* 2017.
[14] Deraman L *et al* 2017.
[15] Izzat A M *et al* 2013 *Materiale Plastice* **50** 171.
[16] Al Bakri A M M *et al* 2012 *Applied Mechanics and Materials* **110** 734
[17] Zain H *et al* 2017 *MATEC Web of Conferences* **97** 01021.
[18] Risdanareni P *et al* 2017 *MATEC Web of Conferences* **101** 01021.
[19] Pelisser F *et al* 2013 *Construction and Building Material* **49** 547.
[20] Lee W *et al* 2002 *Cement and Concrete Research* **32** 577.
[21] Yip C K *et al* 2008 **30** 979.
[22] Aizat E A *et al* 2017 Chemical composition and strength of dolomite geopolymer composites.
[23] Yip C K *et al* 2005 **35** 1688.
[24] Al Bakri A M *et al* 2012 *Rev. Adv. Mater. Sci* **30** 90.
[26] Zailani W W A *et al* 2016 IOP Conference Series: Materials Science and Engineering **133** 012042.
[27] Yip C K *et al* 2008 *Cement and Concrete Research* **38** 554.
[28] Lemougna P N *et al* 2017 *Construction and Building Materials* **156** 486.

IOP Conf. Series: Materials Science and Engineering **551** (2019) 012085 doi:10.1088/1757-899X/551/1/012085

Tensile Toughness Improvement of Poly(ethylene-co-vinyl acetate)(PEVA) by Incorporation of Hybrid Pre-dispersed O-MMT and Destabilized Bentonite

Tuty Fareyhynn **Mohammed Fitri**[1] , *Azlin Fazlina* **Osman**[1*], *Rahimah* **Othman**[1,2], **and** *Khairul Anwar* **Abdul Halim**[1]

[1]Center of Excellence Geopolymer and Green Technology (CEGeoGTech), School of Materials Engineering, University Malaysia Perlis, 02600, Perlis, Malaysia
[2]School of Bioprocess, Universiti Malaysia Perlis, 02600 Arau, Perlis, Malaysia

E-mail: Tutyfareyhynn@gmail.com

Abstract. Poly (ethylene-co-vinyl acetate) (PEVA) is a biocompatible copolymer, easily processed and designed for intricate biomedical products. However, its toughness; which is the combination of high flexibility and high strength is quite low to resist tearing during long term use in human biological system. In this study, the possibility to enhance the tensile toughness of the PEVA was studied. Several ratios of organically modified MMT and bentonite were used as hybrid nanofillers in the PEVA matrix to form hybrid nanocomposites. Both nanoclays were pre-dispersed and destabilized prior to melt compounding with the PEVA. Results indicated that the use of O-MMT and bentonite in a ratio of 4:1 brought the best increment in the tensile toughness of the PEVA. This could be related to the improved interactions between PEVA with the hybrid nanofillers.

1. Introduction

Poly (ethylene-co-vinyl acetate)(PEVA) is a copolymer that exhibits flexibility similar to elastomeric materials, yet it can be processed like other thermoplastic. PEVA is biocompatible, offers excellent optical clarity, barrier properties, and stress-crack resistance [1, 2]. However, to be used as implantable biomedical device, further improvement in the PEVA toughness is needed because this property of PEVA is not sufficiently high. Therefore, this project aimed to employ organically modified montmorillonite (OMMT) and nano-bentonites (Bent) with different surface properties as hybrid nanofillers to reinforce the PEVA, thereby improving its tensile toughness [3]. Hybrid bentonites nanofillers with different polarity (hydrophobic versus hydrophilic) was hypothesized to result in more optimum nanofiller-PEVA interactions, as the PEVA copolymer contains the hydrophobic polyethylene chains and hydrophilic poly(vinyl acetate) chains in its structure [1,2,4]. Enhancement in nanofiller-PVAc interactions would also lead to improvement in overall hybrid PEVA nanocomposite's mechanical performance as biomaterial [4]. Recent works verified that incorporation of nanoclay into the PEVA copolymer can enhance its mechanical properties if only the nanofiller is well dispersed throughout the matrix [4-6]. In contrast, deterioration of mechanical properties of host polymer will occur if the nanoclay is inadequately dispersed and this has been reported by many researchers [5,7].

To produce plate- like nanoparticles with very high aspect ratio inside the polymer matrix, OMMT in stacking (tactoids) form can be exfoliated or delaminated into nanometer thick layers. Based on previous study, the pre-dispersing process of the OMMT in liquid medium can reduce its tactoid size and more facilitate its exfoliation and dispersion in the PEVA matrix during the melt compounding process [5]. In this research, the best parameters of the pre-dispersing process [6] were applied to prepare the OMMT as nanofiller in the PEVA. It is also worth mentioning that the use of unmodified bentonite as co-nanofiller in the PEVA could be challenging as it contains highly stacked platelets (tactoids) held tightly together by electrostatic forces [4]. Therefore, the destabilization of bentonite by pH control and salt addition was also done to break up these large tactoids. In this paper, the effect of hybrid OMMT/bent silicate nanofillers addition on the tensile toughness of the PEVA was studied. Different ratios of OMMT/Bent were used and the best ratio that resulted in the greatest enhancement in the tensile toughness of the hybrid nanocomposite was reported.

2. Methodology

2.1. Materials

Poly ethylene-co-vinyl acetate (PEVA) copolymer (COSMOTHENE EVA H2181) containing 18 wt% of vinyl acetate (VA) and 82 wt% of ethylene was supplied by the Polyolefin Company (Singapore) Pte.Ltd. Organo modified montmorillonite (organo-MMT), that comprising of 35-45% of dimethyl dialkyl (C14-C18) amine as an organic surfactant was employed. This organo-MMT was manufactured by Nanocore, Inc, and known as Nanomer (types 1.44P). The natural bentonite was supplied by Multifilla (M) Sdn. Bhd. The average particle size of this clay is ≤ 25 micron and density of the clay powder is in the range of 600-1100 kg/m^3. For destabilization process of bentonite, nitric acid (HNO_3) (65%) used was supplied by MERCK and sodium chloride (NaCl) used was supplied by Sigma Aldrich. For both nanofillers, dispersing medium used was distilled water.

2.2. Sample Preparation

2.2.1. Preparation of the pre-dispersed OMMT

The pre-dispersing method of the organo-MMT was prepared in water medium (distilled water: OMMT = 1:1). The suspension was stirred by using magnetic stirrer for 2 hours at room temperature. After that, it was ultrasonicated using Branson Digital Ultrasonic Disruptor/Homogenizer (Model 450 D) for 5 minutes. The suspension was placed onto the filter paper to filter out the distilled water and then was inserted into the oven for 48 hours at 50 °C. Lastly, the pre-dispersed OMMT powder was ground and sieved before being characterized by using fourier transform infrared (FTIR).

2.2.2. Preparation of the "destabilized" Bentonite
The bentonite was prepared as 20% suspension in water medium. After that, it was stirred by using ultrasonic probe for 15 minutes. Then, the pH of the suspension was adjusted by dropping HNO_3 until it reached pH 4 and stirred for 5 minutes by using ultrasonic probe. Next, NaCl (0.01M) was added and stirring was continued for 5 minutes. The suspension was filtered and underwent drying process for 24 hours at 80 °C. After that, the suspension was washed with hot water (80° C) and dried for 24 hours at 80° C. Lastly, the 'destabilized' bentonite were ground and sieved before being characterized by FTIR.

2.2.3. Preparation of PEVA hybrid nanocomposite samples
The PEVA hybrid nanocomposites were prepared based on the formulation summarized in Table 1. The hybrid nanocomposite samples were prepared by melt compounding the PEVA copolymer with the OMMT/Bent at 160 ° C using internal mixer (Brabender plasticoder) machine. The compounded materials were then compressed by using compression moulding machine (model GT-7014-H30C by GOTECH Co.) to obtain sheet form (~1mm thick). 5 minutes of pre-heat, 3 minutes of pressing and 10 minutes of cooling time at 130 ° C were employed in this process. Finally, the nanocomposite sheets were cut for testing.

Table 1. Formulation of nanocomposites.

Types of samples	Weight %			Acronym
	EVA	OMMT	Bentonite	
Neat PEVA	100	-	-	PEVA
PEVA/OMMT	96	4	-	PEVA-O
PEVA/Bentonite	-	-	4	PEVA-B
PEVA hybrid nanocomposite with OMMT/Bent ratio 4:1	96	3.2	0.8	PEVA-O_4B_1
PEVA hybrid nanocomposite with OMMT/Bent ratio 3:2	96	2.4	1.6	PEVA-O_3B_2
PEVA hybrid nanocomposite with OMMT/Bent ratio 2:3	96	1.6	2.4	PEVA-O_2B_3
PEVA hybrid nanocomposite with OMMT/Bent ratio 1:4	96	0.8	3.2	PEVA-EO_1B_4

2.3. Characterization of the nanofillers

IOP Conf. Series: Materials Science and Engineering **551** (2019) 012085 doi:10.1088/1757-899X/551/1/012085

2.3.1. Fourier transform infrared spectroscopy (FTIR)

Both nanofillers and hybrid nanocomposites samples were analyzed by Perkin Elmer Paragon FTIR spectrometer with Attenuated Total Reflectance (ATR) in the range of 4000-650cm^{-1} at 32cm^{-1} resolution and 32 scans.

3. Mechanical testing

3.1. Tensile test

Tensile test was carried out by using Instron machine model-552, on the neat PEVA, PEVA nanocomposite and PEVA hybrid nanocomposites according to the method ASTM D638. The testing was carried out by using dumbells samples punched from an ASTM D-638-M-5 die. The crosshead speed of 50mm/min was used. Tensile toughness was determined by calculating the area under the stress-strain curve. The mean values of tensile toughness was determined based on five replicates tested.

4. Results and Discussion

4.1. Morphology of pre-dispersed OMMT and destabilized Bentonite

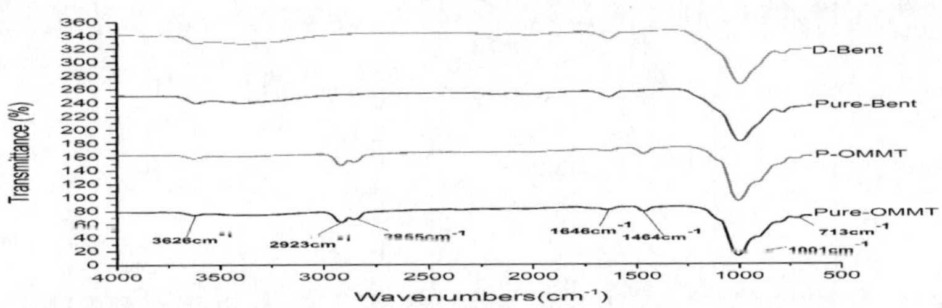

Figure 1. FTIR spectra of OMMT and Bentonite nanofillers.

Figure 1 displays the FTIR spectra of the pristine OMMT, pre-dispersed OMMT, pristine bentonite and destabilized bentonite. No significant difference in the feature of FTIR spectra between the pre-dispersed and destabilized nanoclays in comparison with their pristine form. This shows that upon the pre-dispersing and destabilization process, both nanofillers did not undergo significant structure /chemistry modification. At the band between~ 1464cm^{-1} to ~ 1646cm^{-1} and~ 3626cm^{-1}, the peak of bentonite is broader as compared to the OMMT due to the stretching and bending of hydroxyl group which is more prominent in the hydrophilic (unmodified) bentonite. The spectra peaks of OMMT at ~2855cm^{-1} and ~ 2923cm^{-1} represent the presence of both methylene groups and quartenary ammonium salt of the organic surfactant. In the spectra of Bentonite, these peaks were absent and this shows that the nanofiller was not surface modified with any organic compound [4]. Based on Figure 2, the incorporation of bentonite nanofillers (as single or hybrid form) into the PEVA matrix resulted in more intense spectra peak at ~3425cm^{-1} as compared to neat PEVA and PEVA-OMMT

IOP Conf. Series: Materials Science and Engineering **551** (2019) 012085 doi:10.1088/1757-899X/551/1/012085

nanocomposite. As mentioned earlier, this was due to the stretching and bending of hydroxyl groups, confirming the presence of bentonite in the PEVA matrix.

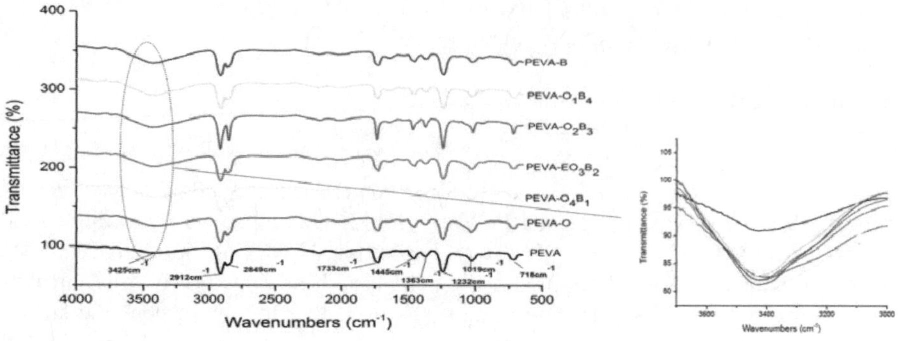

Figure 2. FTIR spectra of neat PEVA, PEVA nanocomposites and PEVA hybrid nanocomposites.

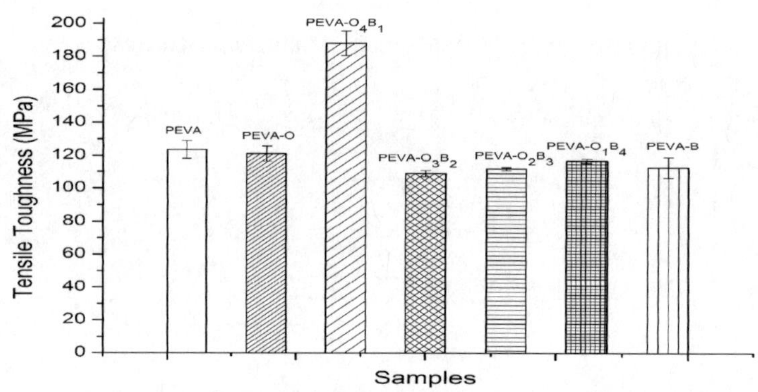

Figure 3. Tensile toughness of neat PEVA, PEVA nanocomposites and PEVA hybrid nanocomposites.

Figure 3 shows the comparison of tensile toughness of neat PEVA, PEVA nanocomposites and PEVA hybrid nanocomposites with different ratio of OMMT:Bent nanofillers. Apparently, PEVA-O_4B_1 sample shows the highest tensile toughness value, and reached about 52% increment when benchmarked with the neat PEVA. Other nanocomposites show lower or similar tensile toughness values with the neat PEVA. This suggests that the use of hybrid OMMT/Bent (4:1) nanofiller resulted in better degree of nanoclay exfoliation and dispersion inside the PEVA matrix due to the correct composition of hydropobic and hydrophilic nanofillers that would interact with the hydrophobic ethylene chains and hydrophilic vinyl acetate chains of the PEVA copolymer. Furthermore, the pre-dispersing process of the O-MMT and destabilization process of the bentonite might facilitate their dispersion throughout the PEVA matrix due to reduced tactoid size of the silicate platelets. As a result, the nanofillers could be better aligned in the direction of the stress to more efficiently absorbed energy during tensile deformation.

5. Conclusions

In summary, the use of OMMT and Bent nanofillers in hybrid form has provided greater toughening effect to the PEVA copolymer as compared to their single form (OMMT alone or Bent alone).The hybrid pre-dispersed OMMT/ destabilized Bent nanofillers have brought the largest increment in the tensile toughness of the PEVA when used in the ratio of 4:1 (PEVA-O_4B_1) .

References

[1] Fink J K 2010 187.
[2] Merinska *et al* 2013 35.
[3] Kotal M *et al* 2003 1539.
[4] Osman A F *et al* 2017 *J. Appl. Polym. Sci*. **134.**
[5] Osman A F *et al 2017 Mater. Sci. and Eng. C.* **74** 194.
[6] Abdul Hamid A R *et al* 2018 *Int. J. of Eng. and Tech.* **7** 288.
[7] Agubra V A *et al 2013 Nanomaterials* **3** 550.

IOP Conf. Series: Materials Science and Engineering **551** (2019) 012086 doi:10.1088/1757-899X/551/1/012086

Characterization of Fly Ash and Metakaolin Blend Geopolymers under Ambient Temperature Condition

Khairunnisa Zulkifly[1], Heah Cheng Yong[1], Mohd Mustafa Al Bakri Abdullah[1], Liew Yun Ming[1], Andrei Victor Sandu[2], Siti Fatimah Azzahran Abdullah[1]

[1]Centre of Excellence Geopolymer and Green Technology (CEGeoGTech), School of Materials Engineering, Universiti Malaysia Perlis (UniMAP), Perlis, Malaysia.
[2]Faculty of Materials Science and Engineering, Gheorghe Asachi Technical University of Iasi, Romania

E-mail: cyheah@unimap.edu.my

Abstract. This paper presents a characterization of fly ash and metakaolin blend geopolymers of the optimize mix proportion of 10 M NaOH concentration mixed with the solid-to-liquid ratio of 1.2 and sodium silicate and sodium hydroxide ratio of 2.6. The combination of amorphous and crystalline phase in the fly ash (FA) and metakaolin (MK) blend geopolymers showed a positive effect on the resultant geopolymers. The formation of new crystalline phases which is sillimanite as presented by XRD analysis. Fly ash and metakaolin blend geopolymers showed the presence of geopolymer bonding at around 1390 cm^{-1} and 960 cm^{-1}. This paper discloses effect of blending fly ash and metakaolin in the synthesis promotes the continual hydration of the geopolymers.

1. Introduction

For In 1978s, Joseph Davidovits named "geopolymer" as a new family of cementitious materials that are formed in direct synthesis of alkaline alumina–silicate materials at low temperature. On the basis, geopolymer can be regarded as a class of aluminosilicate materials polymerized by the geochemical action to form a three- dimensional structure "N-A-S-H" type gel. Besides, it is a network consisting of alternating SiO_4 and AlO_4 tetrahedra linked by shared O atoms [1].

The source materials play a major role in the geopolymerization reaction, controlling the chemical composition and microstructure of the final geopolymeric products [2]. Calcined source material such as fly ash and metakaolin display a higher reactivity during geopolymerization than non-calcined material. There is emerging trend of blending two or more source materials for the production of geopolymers for high performance applications.

Fly ash were found to be the most practical source material due to the plentiful worldwide raw material supply, derived from coal power plant [3]. As for metakaolin, it offers better strength and accelerated hydration reaction [4]. Metakaolin can attain high compressive strength at an early age due to the high pozzolanic reactivity and finer particles filling effect [5]. The main reaction product of fly ash and metakaolin is a N-A-S-H type gel (Na2O-Al2O3-SiO2-H2O) within a three-dimensional network [6]. The incorporation of blend fly ash and metakaolin is highly beneficial due to the difference in chemical composition overlook the handicaps of both starting material.

Thus, in this study, a study is carried out to investigate the characterization of fly ash and metakaolin blend geopolymers under ambient temperature curing.

2. Experimental

2.1. Material

Fly ash used in the present investigation is obtained from Tapah Power Plant, Perak, Malaysia. Whereas the metakaolin used was the calcined kaolin purchased from Associated Kaolin Industries Sdn. Bhd., Malaysia. Table 1 summarizes the chemical composition of the fly ash and metakaolin powder. The sodium hydroxide (NaOH) pellets used are the pellet type of Formosoda- P obtained from Formosa Plastic Corporation, Taiwan. Whereas the sodium silicate used in this study were supplied by South Pacific Chemical Industries Sdn. Bhd. (SPCI), Malaysia with chemical composition of 30.1% SiO_2, 9.4% Na_2O and 60.5% H_2O with a weight ratio SiO2/Na2O = 3.2, specific gravity at 20°C = 0.0014 g/cm^3 and viscosity at 20°C = 0.4 Pa·s.

Table 1. The composition of raw materials based on XRF analysis

Compound	SiO_2	Al_2O_3	Fe_2O_3	CaO	K_2O	TiO_3	Na_2O	ZrO_2
Fly ash, wt (%)	56.3	28.0	6.86	3.89	1.49	2.17	1.49	0.139
Metakaolin, wt (%)	55.7	38.6	2.03	-	2.43	0.78	2.43	0.035

2.2. Preparation of fly ash and metakaolin blend geopolymers

The NaOH concentration of 10 M solution was prepared and allowed to cool down to room temperature. Then as-received sodium silicate solution was added to the NaOH solution, followed by mixing it thoroughly. Dry fly ash and metakaolin was blend together and was added to this solution resulting the formation of geopolymer precursor in the form of slurry. The slurry was then transferred into 50 mm cubes, vibrated for 5 min and allowed to set at ambient temperature (30°C) for 24 h before being removed from the molds and kept at room temperature before it being tested for 28 days of ageing. The formulation of the fly and metakaolin blend geopolymer was the subject of an earlier study, which highlighted an optimal proportion regarding the compressive strength performances [7, 8].

2.3. Characterization

The chemical composition of both materials (fly ash and metakaolin) was determined by X-Ray Fluorescence (XRF) spectrometry with a PANanalytic, MiniPAL 4 spectrometer with PW4030 sample changer. Scanning electron microscopy (SEM) was conducted using a JSM- 6460LA model Scanning Electron Microscope (JEOL). The composition of the raw fly ash, metakaolin and the completely cured geopolymers was characterized by X-ray diffraction (XRD) using a XRD- 6000, Shimadzu x- ray diffractometer equipped with Cu X- ray tube operated at 40 kV. All XRD scans used Kα radiation, with a step size of 0.02° at a scan rate of 2° per minute. Shifting of functional groups from fly ash and metakaolin toward fly ash and metakaolin blend geopolymers were identified using spectrum RXI spectrometer. The specimen was powdered samples scanned from 4000–650 cm^{-1} at the resolution of 4 cm^{-1}.

3. Results and Discussions

3.1. Microstructural Analysis

Fig. 1 depict characteristic morphology of the fly ash and metakaolin particles, respectively. The original fly ash consists of a series of spherical particles of different sizes and relatively smooth. Moreover, with the rounder shape particles, it ensuring more promising workability with a low water demand [9]. The microstructure of metakaolin also reveals the irregular and sheet shape can be seen having the particle sizes in the range 0.5- 9 μm. It is crucial because the metakaolin particle size is reported to influence the mechanical properties of geopolymers [10]. The fly ash and metakaolin blend geopolymers in the micrograph below (Fig. 1) reflecting a gradual dissolution of geopolymers by the alkaline solution. The cracks and pores in these gel matrices may be introduced by the water evaporation and high amount of water that eliminate from the geopolymeric networks. Even so, other research study claimed that the formation of pores and crack were due to the fly ash components [11]. The proportion of 1:1 amount of fly ash and metakaolin components, on one hand, induces the formation of pores and crack. Yet, on the other hand, the effect of blend fly ash and metakaolin provoking a larger interconnecting and homogeneous geopolymers matrix as the alkaline activation proceed.

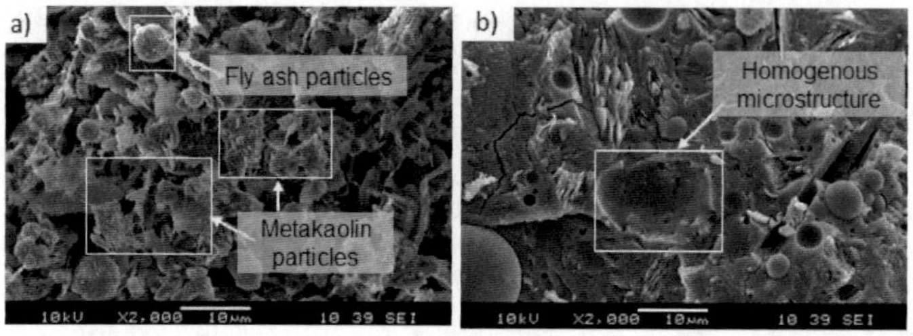

Figure 1. SEM micrograph of raw fly ash and metakaolin powder (a) and the fly ash- metakaolin blend geopolymer product.

3.2. Phase Analysis

Fig. 2 shows the XRD patterns of the raw fly ash, metakaolin and fly ash and metakaolin blend geopolymers. The XRD pattern of fly ash shows some sharp peaks indicating quartz (Q), mullite (T) and hematite (H). On the other hand, XRD pattern of metakaolin display some peaks attributable to quartz (Q), muscovite (M) and kaolinite (K). After the reaction with the alkaline activator solution, the diffuse hump of fly ash and metakaolin shifted slightly towards higher degree at 2θ ranging from 20° to 40°. This hump indicates that there is amorphous geopolymer matrix formed. This was the typical characteristic of aluminosilicates matrix and the shift indicated the formation of geopolymer matrix after the geopolymerization reaction [12].

Furthermore, a few sharp peaks are also observed in the fly ash and metakaolin blend geopolymers, such as sillimanite, quartz and mullite. Quartz and mullite were originally present in the starting materials, consequently it indicates that these phases are apparently not attacked during the alkaline activation process. This reflected that not all mineral phases are participating in the geopolymerization reaction towards the formation of geopolymer matrix. Thus, this system consisted of combination of amorphous geopolymer gels and crystalline phases.

3.3. Functional Analysis

FTIR spectra of raw fly ash, metakaolin and fly ash and metakaolin blend geopolymers are given in Fig. 2. The band positioned at around 3860 cm^{-1} is attributed to OH- stretching vibration. This band was associated with the residual OH- stretching of kaolin which indirectly associated by the partial transformation of kaolin into metakaolin [13]. Band located at 1521.21 cm^{-1} in the FTIR spectrum indicate the H-O-H bending vibration of water molecule bonds. The wavenumber corresponding about 1031 to 1035 cm^{-1} represents the Si-O/Al-O stretching vibration. The spectrum band approximately at 781 to 789 cm^{-1} are attributed to Si-O-Si symmetric stretching vibrations (quartz phase) [14]. This slight difference could be related to the low quartz content and high amorphous silica in the metakaolin.

The transformation that occurred during geopolymerization was indicated by the different absorption frequencies of the source materials and of the resultant geopolymers. The region of interest in all geopolymer product lies in the range between 800 and 1300 cm^{-1}. The broad absorption band between 3000- 3500 cm^{-1} and around 1640 cm^{-1} presented in all spectra are attributed to the bond vibration of water absorbed at surface or entrapped in cavities [15]. After geopolymerization, the Si- O- Si and Si- O- Al band shifts from 1031- 1035 cm^{-1} in fly ash and metakaolin to 968- 976 cm^{-1}, indicating the formation of geopolymer gel. These results can be explained by the formation of a new phase rich in sillimanite.

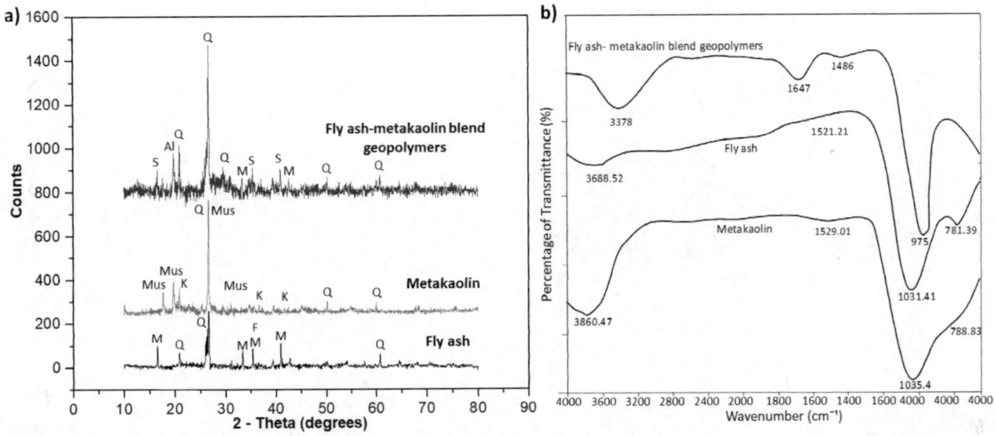

Figure 2: XRD patterns of raw fly ash, raw metakaolin (MK) and fly ash (FA) and metakaolin blend geopolymer samples (a) FTIR spectra of raw fly ash, metakaolin powder and fly ash- metakaolin blend geopolymers.

4. Conclusions

Based on the fly ash and metakaolin blend geopolymers produced, the following conclusion can be drawn:

1. The effect of fly ash and metakaolin blend geopolymers leads to larger formation of homogeneous geopolymer matrix.
2. Through phase analysis, it was observed that fly ash and metakaolin blend geopolymers show an appearance of new crystalline phase which is sillimanite.
3. The polycondensation of fly ash and metakaolin blend geopolymers network indicate by the formation of more geopolymer bonding as determined by functional group analysis.
4. Ambient temperature condition provides a continual ion mobility giving fly ash and metakaolin particles to react between each other to form a homogenous matrix.

Acknowledment

The authors would like to acknowledge the funding support from Fundamental Research Grant Scheme (FRGS/1/2015TK05/UNIMAP/02/1) by Ministry of Education, Malaysia.

References

[1] Li H *et al* 2010 *Cem. Concr. Res.* **40** 1349.
[2] Bakri A M M *et al* 2013 *Revista de Chimie,* **64** 382
[3] Al Bakri A M M *et al* 2012 *Applied Mechanics and Materials* **110** 734
[4] Izzat A M *et al* 2013 *Materiale Plastice* **50** 171
[5] Kelestemur O *et al* 2015 *Constr. Build. Mater.***81** 178.
[6] Al Bakri Abdullah *et al* 2015 *Revista de Chimie* **66** 1001
[7] Zulkifly K *et al* 2018 *AIP Conf. Pro.* 020028.
[8] Zulkifly K *et al* 2018 *AIP Conf. Pro.* 020100.
[9] Nath P *et al* 2014 *Constr. Build. Mater.* **66** 163.
[10] Duan C *et al* 2016 *Ceram. Int.* **42** 3504.
[11] Yunsheng Z *et al* 2008 *Constr. Build. Mater.* **22** 370.
[12] Boke N *et al* 2015 *Constr. Build. Mater.* **75** 189.
[13] Liew Y M *et al* 2012 *Constr. Build. Mater.* **30** 794.

[14] Musci G *et al* 2016 *J. Environ. Eng. Lands.* **24** 48.
[15] Somna K *et al* 2011 *J. Fuel* **90** 2118.

IOP Conf. Series: Materials Science and Engineering **551** (2019) 012087 doi:10.1088/1757-899X/551/1/012087

Correlation of Microstructure and Flexural Strength of Fly Ash Geopolymer of Varying Molarities

Ng Yong-Sing[1], Liew Yun-Ming[1], Heah Cheng-Yong[1], Mohd Mustafa Al Bakri Abdullahi[1,2], Ng Hui-Teng[1], Amir Naveed[2]

[1]Center of Excellence Geopolymer and Green Technology (CEGeoGTech), School of Materials Engineering, Universiti Malaysia Perlis (UniMAP), Perlis, Malaysia
[2]Department of Chemical Engineering, University of Engineering & Technology Peshawar, Pakistan

E-mail: ymliew@unimap.edu.my

Abstract. The present work studies and compares the flexural properties of fly ash geopolymer with varying molarity of sodium hydroxide (NaOH). Different molarities of NaOH solution (6M, 8M, 10M, 12M and 14M) are used to prepare the alkali activator (a mixture of NaOH and sodium silicate). The fly ash geopolymer were synthesised by using the solid/liquid ratio of 2.5 and sodium silicate/NaOH ratio of 2.5. The samples were cured at 60°C for 6 hours. The flexural test and microstructural analysis of fly ash geopolymer are performed after 28 days. Generally, the flexural strength of fly ash geopolymer increases when the molarity of NaOH increases. The 12M geopolymer has the highest flexural strength and a more densified microstructure is observed.

1. Introduction

Geopolymer is an inorganic aluminosilicate polymer where it is synthesised via the alkali activation process of mixing the solid aluminosilicate base materials with the mixture of an alkali metal hydroxide and silicate solution [1, 2]. The types of solid aluminosilicate base materials could be divided into natural materials such as kaolin and metakaolin [3] or industrial by-products such as ground-granulated blast furnace slag (GGBS) and fly ash [4]. Geopolymers are sustainable as they are ecological friendly with low carbon footprint and low energy consumption production process [5].

Geopolymers are amorphous in phase with the three-dimensional structure arranged in short-range order [6]. This three-dimensional framework of SiO4 and AlO4 gives geopolymer ceramic-like characteristics and exhibits a wide range of properties such as high mechanical strength, low shrinkage, low water absorption, fast setting time, acid resistance, fire resistance and low thermal conductivity [6, 7]. Thus, it has captured the interest of researchers in various application such as construction and building sector, fire resistance, toxic metal immobilization and waste management [8].

Geopolymerisation involves several stages including the alkali activation of SiO4 and AlO4 as precursor materials to yield geopolymer gel followed by hydrolysis where the bond breaking of SiO4 and AlO4 tetrahedral units by water molecules occurred [9]. These units reorganize and link to form the polymeric precursors. The geopolymer gel solidified and a three-dimensional aluminosilicate network was formed [10].

The mechanism of geopolymerisation for geopolymer formation is influenced by various parameters such as raw materials, solid/liquid ratio, alkali activator ratio, curing regime and molarity of alkali solution used. Consequently, this parameter would also indirectly affect the properties of the geopolymer. Among these parameters, the molarity of alkali solution plays a significant role as the cation (Na+, K+) contributed from the alkali solution balances the charge and maintains the neutrality of the framework of geopolymers [11]. Also, alkali solution promotes the dissolution of aluminosilicates and hence increases the degree of geopolymerisation reaction [12]. In this study, a relationship between the molarity of sodium hydroxide with the flexural properties and microstructure and fly ash geopolymer was investigated.

2. Experimental

2.1. Materials

Fly ash collected from Cement Industries of Malaysia Berhad (CIMA) Perlis Plant, Malaysia was used as the source of aluminosilicate. The chemical composition of the fly ash obtained through XRF analysis is tabulated in Table 1. Referring to Table 1, the fly ash used could be classified as Class F fly ash according to ASTM C618 as it contains total composition of SiO2 and Al2O3 of 84.30% and the CaO content is 3.89% which is less than 20% [13]. Figure 1 displayed the SEM micrograph of fly ash particle which was spherical in shape with smooth surfaces. The technical grade sodium silicate solution (Na2SiO3) has chemical composition of 30.1% SiO2, 9.4% Na2O and 60.5% H2O with modulus SiO2/Na2O of 3.2, specific gravity at 20°C = 1.4 and viscosity at 20°C = 0.4 Pa·s. The sodium hydroxide (NaOH) pellets used was classified as caustic soda micropearls with 99% purity.

Table 1. Chemical composition of fly ash.

Compound	SiO_2	Al_2O_3	Fe_2O_3	CaO	TiO_2	K_2O	Others
Mass (%)	56.30	28.00	6.86	3.89	2.17	1.49	1.29

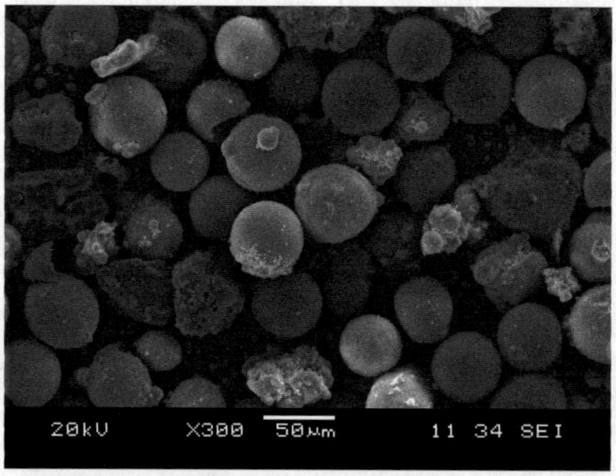

Figure 1. SEM micrograph of fly ash particles.

2.2. Geopolymer Formation

The NaOH solution with range of 6M-14M was prepared and then allowed to cool down to room temperature. The NaOH solution was mixed with liquid Na2SiO3 ratio of 1:2.5 to prepare the alkali activator. Then, the fly ash was mixed with the alkali activator at ratio of 1:2.5 until homogenous slurry was obtained. The fresh paste was then rapidly moulded and compacted into a 160 40 10 mm moulds. The moulded geopolymer samples were cured at 60°C in oven for 6 hours. The cured samples were kept at room temperature for 28 days before testing.

2.3. Testing and analysis

Flexural strength tests were performed according to ASTM C1505-15 by using the Instron machine series 5569 Mechanical Tester with a span length of 80 mm at a crosshead speed of 5 mm/min. JSM-6460LA model Scanning Electron Microscope (JEOL) was performed to reveal the microstructure of the fracture surface of fly ash geopolymer.

3. Results and Discussions

3.1. Flexural Strength

Based on the data obtained in [14], the flexural strength of fly ash geopolymer increased continuously when the molarity of NaOH increased from 6M to 12M but beyond of that the flexural strength reduced. The flexural strength and bulk density obtained is tabulated in Table 2. This flexural strength result complied with the bulk density of the fly ash geopolymer obtained where a more compact density indicates a higher flexural strength. This is further supported by the SEM micrographs shown in Figure 3. In this context, the geopolymerisation process becomes undesirable after reaching an optimum NaOH molarity [15]. The high viscosity of high molarity NaOH solution would hinder the solubility and deionisation of Si and Al ions from aluminosilicate sources. This statement is supported by He *et al.* [16] as they observed the occurrence of premature precipitation in geopolymer matrix and lead to deterioration in mechanical properties of the final product due to presence of remnant precursor material.

Table 2. Flexural strength of fly ash geopolymer with varying molarity of NaOH solution.

NaOH Molarity	6M	8M	10M	12M	14M
Flexural Strength (MPa)	2.35	1.95	3.31	4.80	3.53
Bulk Density (kg/m^3)	1753	1702	1852	1900	1856

3.2. Microstructural Analysis

Figure 2 displays the SEM micrograph of alkali activated fly ash particles. As compared to the fly ash micrograph in Figure 1, it could be observed that the surface of spherical fly ash particles opened up when the fly ash was mixed with the alkali activator. Fernandez-Jimenez *et al.* [17] reported that the dissolution process of Al and Si occurs when fly ash is added with alkali activator and followed by the condensation of higher molecules into a gel via nucleation and polymerisation. Then, the alkali attack opened up the spheres and exposed the internal small spheres. The small spheres inside would dissolve when in contact with alkali and the reaction products were formed inside and outside the sphere.

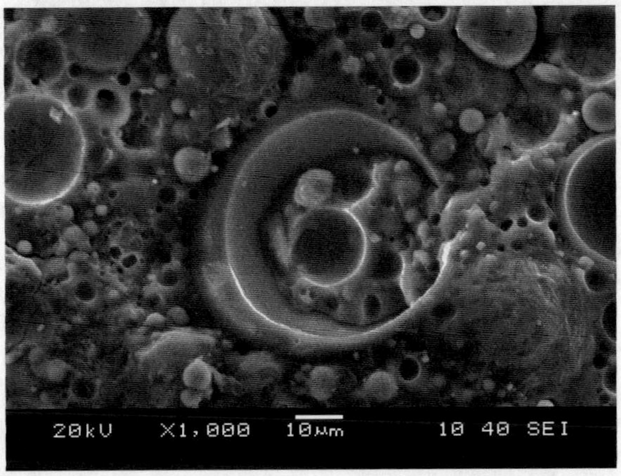

Figure 2. SEM micrograph of alkali activated fly ash particles.

The microstructures of the fracture surface of fly ash geopolymer with varying NaOH molarities are shown in Figure 3. It could be observed that there are large portion of unreacted or partially reacted fly ash particles present in the matrix of 6M and 8M NaOH-activated fly ash geopolymer (Figure 3a, 3b) which possessed a lower flexural strength. Similar observation was observed in Abdullah *et al.* [18] where the undissolved fly ash particles were loosely bonded to the geopolymer structure and interfacial boundaries could be observed in the structure. Besides, the presence of pore could be observed as well in the geopolymer matrix and these pores caused the imperfection in microstructure which result in poor mechanical properties [19].

On the other hand, fly ash geopolymer with a higher flexural strength (10M, 12M and 14M) have a smoother and more densified microstructure (Figure 3c, 3d and 3e). These densified and compacted structure once again complied with the flexural properties in Table 2 [14]. Hamidi *et al.* [20] reported that the geopolymeric gel formed from a greater extent of geopolymerisation was able to hold the constituent and fly ash particles well together to form a more compact structure. In this context, the increase in molarity of alkali solution could promote the formation of geopolymeric gel and hence results in a better mechanical properties [18].

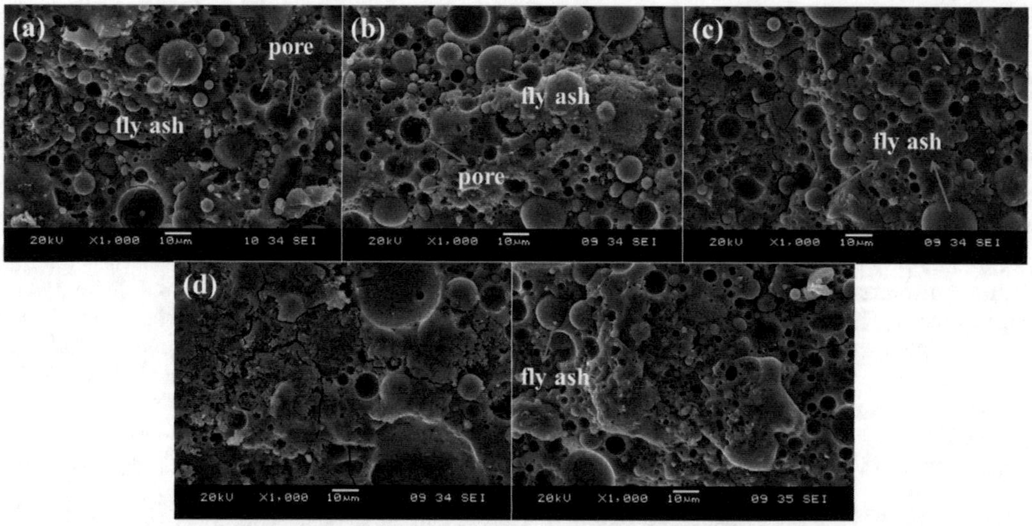

Figure 3. SEM micrographs of the fracture surface of fly ash geopolymer with NaOH molarity of (a) 6M; (b) 8M; (c) 10M; (d) 12M and (e) 14M

4. Conclusions

This paper studied the relationship between the molarity of sodium hydroxide with the flexural properties and microstructure and fly ash geopolymer. The optimum molarity of NaOH is 12M which gave better strength properties of fly ash geopolymer. A more compacted and densified microstructure is observed in which correlated with the flexural properties of 12M NaOH-activated fly ash geopolymer. However, further work should be implemented to boost the flexural strength of fly ash geopolymer.

Acknowledgements
The authors of the present work wish to acknowledge the funding by Fundamental Research Grant Scheme (FRGS/1/2015/TK05/UNIMAP/02/2) sponsored by Ministry of Education Malaysia.

References

[1] Heah, C.Y., *et al* 2013 *International Journal of Minerals, Metallurgy and Materials* **20** 313
[2] Heah, C.Y., *et al* 2011 *Australian Journal of Basic and Applied Sciences*, **5** 1026
[3] Yahya, Z., *et al* 2015 *Materials* **8** 2227
[4] Al Bakri Abdullah, *et al* 2015 *Revista de Chimie* **66** 1001
[5] Izzat, A.M., *et al* 2013 *Materiale Plastice* **50** 171
[6] Davidovits J 2011 Geopolymer Chemistry and Applications.
[7] Benhelal E *et al* 2013 *J. Cleaner Prod.* **51** 142.
[8] Zhang Z *et al Appl. Clay Sci.* **49(1)** 7.
[9] Yunsheng Z *et al* 2008 *Constr. Build. Mater.* **22(3)** 370.
[10] Hajimohammadi A *et al* 2010 *Chem. Mater.* **22(18)** 5199.
[11] Khale D *et al* 2007 *J. Mater. Sci.* **42(3)** 729.
[12] Yao X *et al* 2009 *Thermochim. Acta* **493(1-2)** 49.
[13] Antiohos S K *et al* 2007 *Waste Manage. (Oxford)* **27(5)** 675.
[14] Ng Y S *et al* 2018 AIP Conf. Proc. **2045(1)** 020101.
[15] Zuhua Z *et al* 2009 *Appl. Clay Sci.* **43(2)** 218.
[16] He J *et al* 2013 *Cem. Concr. Compos.* **37** 108.
[17] Fernández-Jiménez A *et al* 2005 *Cem. Concr. Res.* **35(6)** 1204.
[18] Abdullah M M A B *et al* 2012 *Adv. Mater. Res.* **476** 2173.
[19] Granizo M L *et al* 2007 *J. Mater. Sci.* **42(9)** 2934.
[20] Hamidi R M *et al* 2016 *Procedia Eng.* **148** 189.

Porous Metakaolin Geopolymers with Tailored Thermal Conductivity

Nur Ain Jaya, Liew Yun-Ming, Mohd Mustafa Al Bakri Abdullah, Heah Cheng-Yong

Center of Excellence Geopolymer & Green Technology (CEGeoGTech), School of Materials Engineering, Universiti Malaysia Perlis (UniMAP), Perlis, Malaysia.

E-mail: ymliew@unimap.edu.my

Abstract. In this work, a novel lightweight metakaolin geopolymers were produced by using hydrogen peroxide as foaming agent. The influence of the hydrogen peroxide incorporation content (0.25 – 1.25%) on density, compressive strength and thermal conductivity was evaluated. The experimental results show that the density and compressive strength were decrease with increasing content of hydrogen peroxide. However, the thermal insulating properties of the geopolymer foams were improved with increasing content of foaming agent. Geopolymers exhibiting thermal conductivity as low as 0.172 W/mK and a density of 0.5 g/cm^3 were produced, showing the potential of these materials in applications requiring lightweight and low thermal conductivity materials.

1. Introduction

The increase in world population has indirectly raised the needs for shelters in the means of the residential building leading to increased energy consumption and greenhouse gas emissions worldwide. In light to this matter, a key criterion for the selection of construction materials is their sustainability. Ordinary Portland cement (OPC) is the most commonly used cementitious building material where the manufacture of OPC releases a great amount of CO_2 into the atmosphere. Hence, the searching for alternative building materials with reduced environmental footprints is becoming important. Lightweight geopolymer appears to be an excellent alternative with good performance and environmental benefits.

Geopolymers are amorphous three-dimensional aluminosilicate binder materials named first by Professor Joseph Davidovits in 1978 [1]. Geopolymers are produced by geopolymerisation reaction that uses aluminosilicate materials and an alkaline activating solution as the primary raw materials [2-4]. The geopolymerisation is a reaction between amorphous silica and alumina rich solids with a highly alkaline solution to form amorphous to semi-crystalline aluminosilicate inorganic polymers.

Porous geopolymers have been the focus of attention in the field of thermal insulation materials because it has shown great potential to many practical applications as a desirable substitute to ordinary Portland cement due to low thermal conductivity. The production of porous geopolymers involves the addition of foaming agents such as aluminium powders and hydrogen peroxide to the geopolymer slurry [5, 6].

In this work, lightweight metakaolin geopolymers were produced using hydrogen peroxide (H_2O_2) as foaming agent. The influence of the foaming agent incorporation content on density, compressive strength and thermal conductivity were studied.

2. Experimental

2.1. Material

Metakaolin was used as the Si-Al source material in this work. The metakaolin was obtained by calcining the kaolin at 900 °C for 6 h in the furnace. The chemical composition of metakaolin determined by X-ray fluoroscene (XRF) spectrometer is shown in Table 1. The activator solution was a mixture of sodium hydroxide (NaOH) and liquid sodium silicate (Na_2SiO_3). The NaOH powder has 99% purity with a density of 2.13 g/cm^3 while the liquid Na_2SiO_3 contains 30.1% SiO_2, 9.4 Na_2O and 60.5% H_2O with a density of 2.4 g/cm^3. Hydrogen peroxide solution (diluted from 30 wt% H_2O_2, from Sigma-Aldrich) was selected as chemical foaming agent. Tween 80 (polyethylene glycol sorbitan monoleate, VWR BDH Prolabo, Briare, France) was added as surfactant.

Table 1. Chemical composition of metakaolin as determined by XRF analysis.

Chemical	**Wt%**
SiO_2	55.7
Al_2O_3	38.6
Fe_2O_3	2.03
TiO_2	0.78
CuO	0.03
ZrO_2	0.04
K_2O	2.43
MnO_2	0.04

2.2. Preparation of Metakaolin Geopolymer Foams (MkGps)

Geopolymer paste were prepared by mechanically mixing metakaolin and alkaline activator solution. The latter ones were obtained from a solution of 10M NaOH solution (prepared by dissolving sodium hydroxide pellets) and liquid sodium silicate. After that, wet foams were obtained by successively adding surfactant (500 rpm for 1 min) and H_2O_2 (500 rpm for 1 min) into the pastes. The mixing details is shown in Table 2. Immediately after mixing, the wet foams were poured into steel moulds with a dimensions of 50 x 50 x 50 mm. The moulded samples were sealed with a thin film to prevent moisture loss. The samples were cured subsequently in the oven at 60 °C for 24 h. The samples were demoulded and were left for 28 days at ambient temperature prior to compressive strength test and thermal conductivity.

Table 2. Foaming agent content of metakaolin geopolymers.

Sample name	Surfactant content (wt%)	H_2O_2 content (wt%)
G1	1	0.25
G2	1	0.50
G3	1	0.75
G4	1	1.00
G5	1	1.25

2.3. Test, Measurement and Characterization
The bulk density of the MkGPs was obtained as the ratio between the mass of the sample and its geometrical volume. The compressive strength of metakaolin geopolymer foam was measured in accordance to ASTMC109 by using Instron machine series 5569 Mechanical Tester. Room temperature thermal conductivity was measured using a Hot-Disk thermal constant analyzer (Hot-Disk AB Uppsala, Sweden). At least 3 measurements were performed to ensure reproducibility.

3. Results and Discussions
Fig. 1 presents the physical appearance of the geopolymer foams (sample G5). The structure of materials is homogeneous where the pores are evenly distributed and have almost the same size in the sample. However, the pore size is varied between different geopolymer foam.

Figure 1. The exemplary surface of the geopolymer foams (sample G5).

The influence of the foaming agent was investigated by applying five different concentrations of H_2O_2 to fixed amount of added surfactant, known as Tween 80. Table 3 presents the composition of the examined materials and their characteristics such as density, thermal conductivity and compressive strength (which, dependent on the sample, amounted to 0.36 – 6.00 MPa). As can be seen from the results, the measured density decreases with the increasing concentration of H_2O_2 content. With the increase amount of foaming agent added to the geopolymer mixture, the number of voids inside the material with air contained could be generated, thus resulting in increase of porosity and the reduction of density and thermal conductivity. Since the bulk density of geopolymer foams needs to be low, and is usually less than 1 g/cm^3, it can be concluded that the addition of only 0.25% of H_2O_2 is certainly too low. The compressive strength and the value of the thermal conductivity were dependent on the density of the geopolymer foams.

Table 3. Composition and physical properies of geopolymer foams.

Sample labels	H_2O_2 content (wt%)	Density (g/cm^3)	Thermal conductivity (W/mK)	Compression strength (MPa)
G1	0.25	1.212	0.298	6.00
G2	0.50	0.814	0.231	2.18
G3	0.75	0.720	0.203	0.66
G4	1.00	0.610	0.187	0.40
G5	1.25	0.500	0.172	0.36

As was expected, the compressive strength of the investigated samples were in correlation with their densities. For instance, the compressive strength of sample G1, which had a density of 1.212 g/cm^3, amounted to 6.00 MPa, whereas it was significantly lower, i.e. 0.36 MPa, in the case of sample G5, which had a density of 0.5 g/cm^3. The similar trend was reported by Feng, Zhang [7] whereby increasing H_2O_2 dosage of 0.03 to 0.45 wt% caused the density reduction of the foamed geopolymer from 0.33 to 0.23 g/cm^3.

Fig. 3 presents the relationships between the thermal conductivity and the density of the geopolymer foams. The relationship is linear where the lower the density, the lower the thermal conductivity. A research done by Korat and Ducman [8] also obtained similar results where lower density produces a better thermal insulating property of geopolymers.

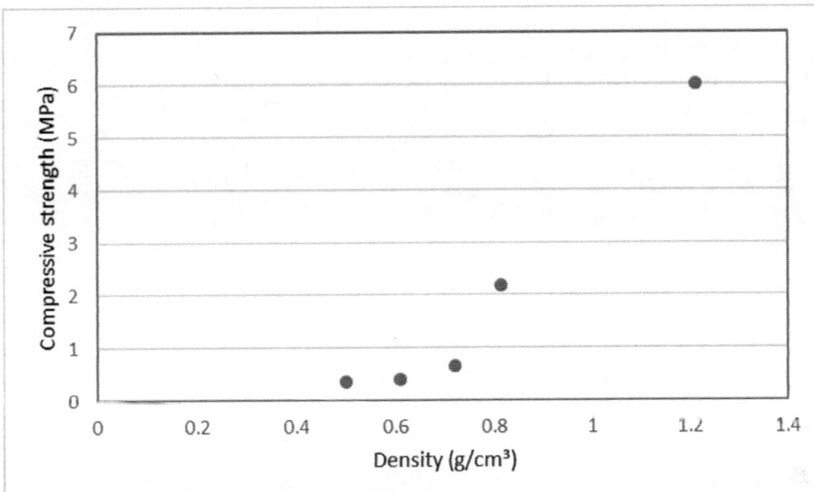

Figure 2. The relationship between the compressive strength and the density of the geopolymer.

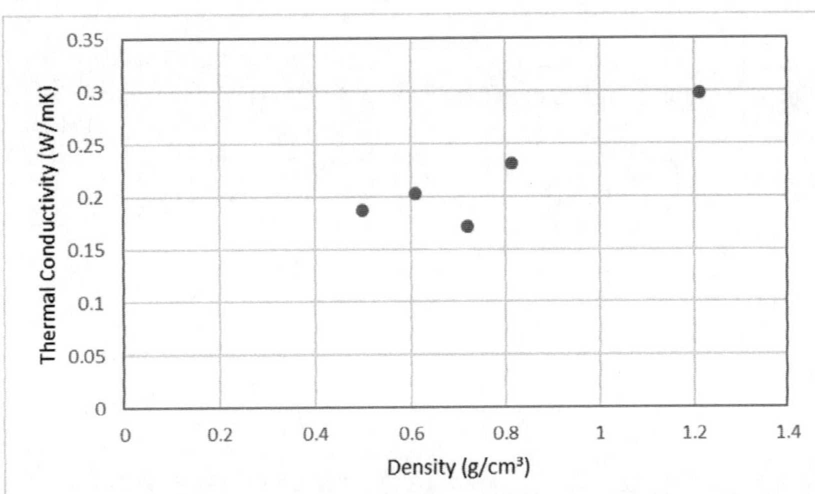

Figure 3. The relationship between the thermal conductivity and the density of the geopolymer.

4. Conclusions

From the research, we can conclude several conclusions which are;

i. The foamed geopolymers can be a promising alternative to other thermal insulating materials
ii. The increasing amount of hydrogen peroxide addition has decreased the density and compressive strength while improving the insulating properties of foamed geopolymers.

Acknowledgement
The authors wish to deliver their gratitude to the funding provided by Fundamental Research Grant Scheme (FRGS/1/2015/TK05/UNIMAP/02/2) sponsored by Ministry of Education Malaysia.

References

[1] Davidovits J 1991 *Therm Anal* **37** 1656.
[2] Bakri A M M *et al* 2013 *Revista de Chimie,* **64** 382
[3] Izzat A M *et al* 2013 *Materiale Plastice* **50** 171
[4] Al Bakri Abdullah, *et al* 2013 *Advanced Materials Research* **686** 227
[5] Singh N B 2018 *Mater Today* **5** 15252.
[6] Bai C *et al* 2018 *Ceram Int* **44** 16118.
[7] Korat L *et al* 2017 *Cement Concrete Comp* **80** 174.

Mechanical Properties and Thermal Conductivity of Lightweight Foamed Geopolymer Concretes

Siti Fatimah Azzahran Abdullah, Liew Yun-Ming, Mohd Mustafa Al Bakri Abdullah, Heah Cheng-Yong, Khairunnisa Zulkifly

Center of Excellence Geopolymer & Green Technology (CEGeoGTech), School of Materials Engineering, Universiti Malaysia Perlis (UniMAP), Perlis, Malaysia.

E-mail: ymliew@unimap.edu.my

Abstract. Foamed geopolymer concretes have a better performance in the thermal insulation properties compared to normal geopolymer concretes. In this research, lightweight aggregate geopolymer concretes was incorporated with different percentage of foaming agents (hydrogen peroxide). Compressive strength and thermal conductivity were measured. From results obtained, increased H_2O_2 contents will decrease the strength of lightweight foamed geopolymer concretes. Lightweight aggregate foamed geopolymer concretes (LWAFGC) with foaming agent content of 0.2wt.% obtained the highest strength of 19.601 MPa. Furthermore, Increased of H_2O_2 contents also will decrease the thermal conductivity of lightweight foamed geopolymer concretes. Lightweight foamed geopolymer concretes with 2wt.% H_2O_2 gave the good thermal insulating behavior when the thermal conductivity value recorded the lowest value compare to other wt.% of H_2O_2 content. The thermal conductivity value of lightweight foamed geopolymer concretes with 2wt.% of H_2O_2 was 0.072 W/ m. K while, the thermal conductivity of other mixtures ranged between 0.077 W/ m. K to 0.087 W/ m. K., respectively.

1. Introduction

In fact, geopolymers have potential in thermal application. There may be several ways to reach this goal and one of the major methods is to improve their thermal insulation properties. Insulation against environmental conditions and improving the energy efficiency of buildings is becoming increasingly important. The insulation properties of the buildings are one of the most important factors affecting both the operating cost and thermal comfort. Reduction of the heat loss in buildings decreases the consumption of energy, thus, reduces the cost of both heating and cooling. As a result of the lower use of energy, improvements in thermal insulation also affect sustainability [3].

Geopolymers are environmentally friendly concrete due to the low CO_2 emission during their production process [4,5,6]. Geopolymers, which have high strength, fire resistance, and chemical stability [7, 8], are formed by the condensation reaction between $Si(OH)_4$ and $Al(OH)^-_4$ after thermal curing. The curing temperature determines the strength development of geopolymers because temperature affects the condensation rate in the geopolymerization reaction [9, 10]. The sodium hydroxide (NaOH) and sodium silicate (Na_2SiO_3) have high alkalinity and are mostly used as activator solution to dissolve silica and alumina to $Si(OH)_4$ and $Al(OH)^-{}_4$, respectively.

Foaming agents generate a large volume of artificial pores in hardened pastes, leading to reduced specific gravities of the pastes. However, the foamed binder pastes are not suitable for structural purposes, due to their significantly low strength; thus, they are limited in their use for improvements in non-structural properties such as thermal insulation or sound-proofing capacity [11].

Abdollahnejad et al. [12] studied on the several mix parameters on the properties of foam geopolymers. In their study, the foaming agent used was hydrogen peroxide (H_2O_2) and sodium perborate ($NaBO_2$) with various mass ratio content of 1 %, 2 % and 3 %. They observed that low compressive strength below 3 MPa in geopolymer mixtures with 2 % and 3 % of H_2O_2. Masi et al. [13] studied a comparison between different foaming methods to produce lightweight geopolymers. Foaming agents reduced the density and compressive strength; however, it is effective in improving their insulating property. Increasing the concentration of the foaming agent (0.3 wt.%) reducing the density (1.12 g/cm^3 to 0.91 g/cm^3).

Furthermore, Liu et al. [14] studied on the thermal conductivity of oil palm shell foamed geopolymer concrete (OPSFGC) using a aluminosilicate source class F fly ash and palm oil fuel ash (POFA) as a raw material, and oil palm shell (OPS) as lightweight coarse aggregate (LWA). From their study, they observed that the foamed and non-foamed OPSGC obtained lower thermal conductivity compared to the conventional materials, which is less than 0.75 W/ m.K. This is due to the increase of void ratio thus decrease in thermal conductivity.

In this study, the effect of combination of lightweight aggregate and foam on geopolymer concrete samples was studied. The addition of foaming agent into lightweight geopolymer concretes would further reduce the thermal conductivity. It is important to properly tailoring the content of foaming agent and lightweight aggregate to achieve thermal insulation properties without seriously degrading the strength. Therefore, lightweight aggregate foamed geopolymer concretes are expected to have good properties such as lightness in weight, good thermal insulation and suitable as construction materials. In this study, physical, mechanical properties and thermal conductivity analysis of lightweight aggregate foamed geopolymer concretes are investigated.

2. Experimental

2.1. Material

Fly ash was used as the aluminosilicates for preparation of geopolymers. The alkaline activator used in this work was sodium hydroxide (NaOH) and sodium silicate (Na_2SiO_3) solution. Lightweight expanded clay aggregate (LECA) was used for the preparation of lightweight aggregate foamed geopolymer concretes (LWAFGC) as coarse aggregate. River sand was used as the fine aggregate in the preparation of LWAGC and LWAFGC. Hydrogen peroxide (H_2O_2) was used as the foaming agent in LWAFGC.

2.2. Preparation of Lightweight Aggregate Foamed Geopolymer Concretes (LWAFGC)

Fly ash and alkaline activator were first mixed together in mechanical mixer for 3 minutes. Before the alkaline activator was mixed with the fly ash, the foaming agent were firstly added into the alkaline solution. Then, the LWA and sand river were added and mixed for another 5 minutes. The fresh concrete was then poured into 100 mm × 100 mm × 100 mm cubic molds to produce LWAFGC. Hydrogen peroxide (H_2O_2) content used was set at 0.2%, 0.4%, 0.6%, 1.0%, 1.5% and 2.0%. The ratio of both S/L and Na_2SiO_3/NaOH were maintained at 2.0 and 2.5 with a NaOH concentration of 12 M for Mixture 1-6. The total aggregate was selected at 70%. The lightweight aggregate (LWA) to river sand was fixed at 2:3. Then, the samples of LWAFGC were cured at room temperature (29°C) for 1 day and 60°C for another 1 day. Thus, the samples were demoulded and kept at room temperature (29°) until the day of testing.

2.3. Testing and Analysis

Bulk Density Measurement

Bulk densities of LWAFGC was measured as according to BS EN12390-7. The bulk density measurement were taken after 7 and 28 days. Bulk density of samples were calculated using Equation (1) by measuring dimension and mass of the geopolymers.

$$\text{Bulk density, } \rho = m/V \tag{1}$$

Where m is the mass of specimen and V is the volume of specimen.

Compressive Strength Test

Lightweight aggregate foamed geopolymer concretes (LWAFGC) with dimension of 100 mm × 100 mm × 100 mm was prepared for compressive test. The compressive strength was tested using Mechanical Tester Shimadzu UH-1000 KNI as according to ASTM C 109/C 109M-08. Three samples were tested to obtain an average value of strength. The samples were loaded at a constant rate of 5 mm/min. The compressive test was performed after 7 and 28 days.

Thermal Conductivity Measurement

Lightweight aggregate foamed geopolymer concrete (LWAFGC) was measured at room temperature according to the ASTM D5334-14, using the transient heat method using the KD2 Pro. The samples of the LWAFGC were drilled with diameter 3.9 mm drill, to form a hole with length of 60 mm. In order to ensure good contact between the pilot pin and samples' hole, all samples were coated with the thermal grease on the pilot pin. After drilling the holes, all samples were cleaned from the dust and drill cuttings from the rotary hammer hole using a compressed air before inserting the pilot pin. Three measurement were taken to obtain the average values of thermal conductivity, thermal diffusivity and specific heat. The following Equation (2) was used to calculate the thermal conductivity, thermal diffusivity and specific heat of the samples, respectively in unit of W/ m. K, m^2/s and J/ kg. K:

$$k = \frac{\alpha}{\rho c_p} \tag{2}$$

Where, k is the thermal conductivity (W/ m. K), ρ is the density (kg/m^3)of the specimen, c_p is the specific heat capacity (J/ kg. K) and α is the thermal diffusivity (m^2/s).

3. Results and Discussion

3.1 Bulk Density Measurement

Figure 1 shows the bulk density of the lightweight aggregate foamed geopolymer concretes (LWAFGC) for Mixture 1 – 6 with different addition of foaming agent. In general, the bulk density of the lightweight aggregate foamed geopolymer concretes decreased with the increase in addition of the foaming agent from Mixture 1 (0.2%) to Mixture 6 (2.0%). The bulk density values of the lightweight aggregate foamed geopolymer concretes ranged between 1.523 g/cm^3 and 1.425 g/cm^3. Mixture 1 (0.2%) showed the higher bulk density meanwhile, Mixture 6 (2.0%) have the lowest bulk density. This is due to the increasing of foaming agent added reducing the bulk density.

According to the Feng et al. [14], the reduction of density has been identified due to the numbers of voids in the samples with air contained is generated. This is supported by the Demirboga and Gul [15].

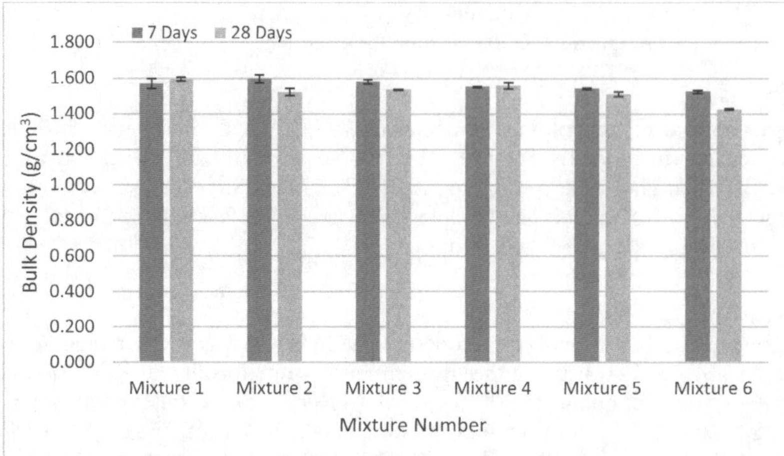

Figure 1. Bulk density of lightweight aggregate foamed geopolymer concretes (LWAFGC) for Mixture 1 – 6 after 7 and 28 days.

3.2 Compressive strength

Figure 2 shows the compressive strength of lightweight aggregate foamed geopolymer concretes (LWAFGC) with various hydrogen peroxide (H_2O_2) after 7 and 28 days. After 7 days compressive strength of lightweight aggregate foamed geopolymer concretes fluctuated with the increasing of foaming agent added. However, after 28 days, strength of LWAFGC recorded that these samples had strength values of 19.601 MPa (Mixture 1) with the lowest foaming agent percentage (0.2%) compared than other mixtures due to its less porous structure meanwhile Mixture 6 obtained 14.257 MPa, which is the lowest compressive strength values with the addition of foaming agent of 2.0% by weight. This might be due to the addition of the foaming agent depleted the compressive strength due to the foaming agent generally created air voids and resulted in a lower density [16,17]. The compressive strength result complied with the bulk density value as shown in Figure 1.

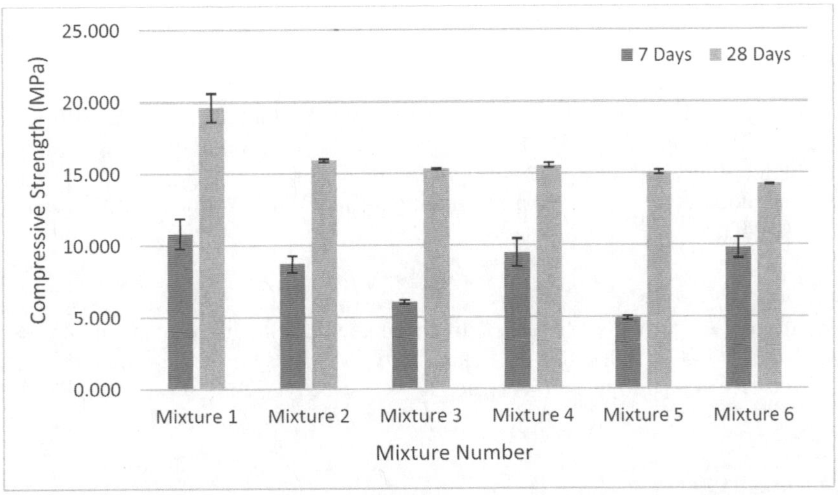

Figure 2. Compressive strength of lightweight aggregate foamed geopolymer concretes (LWAFGC) for Mixture 1 – 6 after 7 and 28 days.

3.3 Thermal Conductivity Measurement

Figure 3 presents the thermal conductivity values of lightweight aggregate foamed geopolymer concretes with varying of foaming agent addition for Mixture 1 (0.2wt.%) to Mixture 6 (2.0wt.%). There was positive correlation between thermal conductivity and foaming agent content. Mixture 6 (foaming agent of 2.0%) has lower thermal conductivity compared to other mixtures because of high foaming agent content and low of density value. Thermal conductivity value of Mixture 6 (foaming agent of 2.0wt.%) was 0.072 W/ m. K while, the thermal conductivity of other mixtures ranged between 0.077 W/ m. K to 0.087 W/ m. K. Moreover, moisture content and porosity structure affects the thermal conductivity values.

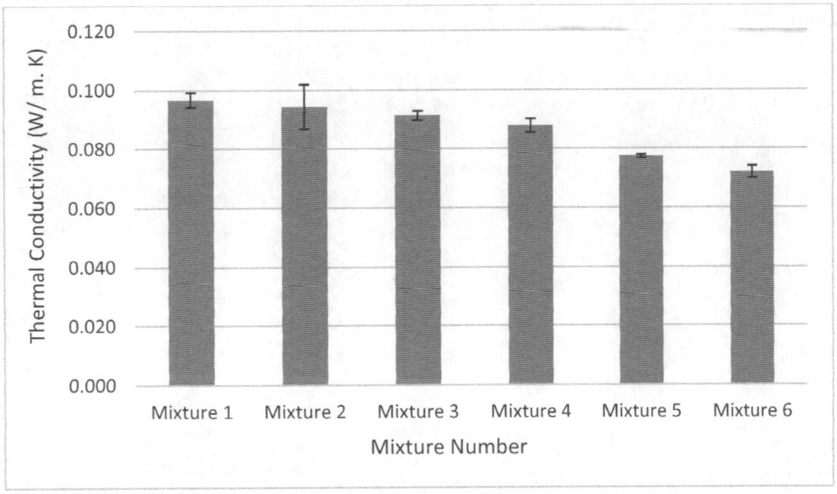

Figure 3. Thermal Conductivity of lightweight aggregate foamed geopolymer concretes (LWAFGC) for Mixture 1 – 6 after 28 days.

4. Conclusions

From the research, we can conclude several conclusions, which are;

1. Bulk density decreased with the increasing of addition of hydrogen peroxide (H_2O_2).
2. Strength of LWAFGC obtained is below than 20 MPa. The maximum strength at Mixture 1 (19 MPa) with foaming agent of 0.2% by weight in the mixtures.
3. Geopolymer concretes with addition of foaming agent could lowering thermal conductivity value (< 1.0 W/ m. K).

Acknowledgements

The authors of the present work wish to acknowledge the funding by Fundamental Research Grant Scheme (FRGS/1/2015/TK05/UNIMAP/02/2) sponsored by Ministry of Education Malaysia. Special thanks to those who contributed to this project directly or indirectly.

References

[1] Roodman *et al* 1995.
[2] Ismail *et al* 2012 *Procedia: Social and Behavioral Sciences* **35**.
[3] Tasdemir *et al* 2017 *Energy and Buildings* **151**.
[4] Heah C Y *et al* 2013 *International Journal of Minerals, Metallurgy and Materials* **20** 313.
[5] Al Bakri A M M *et al* 2012 *Applied Mechanics and Materials* **110** 734.
[6] Al Bakri Abdullah *et al* 2015 *Revista de Chimie* **66** 1001.
[7] Izzat A M *et al* 2013 *Materiale Plastice* **50** 171.
[8] Al Bakri Abdullah *et al* 2013 *Advanced Materials Research* **686** 227.
[9] Rovnaník P 2010 *Construction and Building Materials* **24.**
[10] Beiter *et al* 2016 *Renewable Energy Book.*
[11] Suh *et al* 2017 *Construction and Building Materials* **152.**
[12] Abdollahnejad *et al* 2015 *Construction and Building Materials* **80.**
[13] Masi *et al* 2014 *Ceramics International* **40.**
[14] Liu *et al* 2014 *Energy and Buildings* **72.**
[15] Feng *et al* 2015 *Materials & Design* **65.**
[16] Ahmad *et al* 2016 *IOP Conference Series: Materials Science and Engineering* **133** 012042.
[17] Demirboğa *et al* 2005 Cement and Concrete Research **33.**
[18] Nambiar *et al* 2006 Cement & Concrete Composites **28.**

Microstructural Analysis of Fly Ash-based Geopolymers with various Alkali Concentration

Ng Hui-Teng, Heah Cheng-Yong, Liew Yun-Ming, Mohd Mustafa Al Bakri Abdullah, Ng Yong-Sing

Center of Excellence Geopolymer & Green Technology (CEGeoGTech), School of Materials Engineering, Universiti Malaysia Perlis (UniMAP), Perlis, Malaysia.

E-mail: cyheah@unimap.edu.my

Abstract. In the present work, a comparative study on the effect of different concentration of sodium hydroxide (NaOH) on fly ash-based geopolymer was investigated. The geopolymer synthesis by mixing fly ash with alkali activator (a mixture of NaOH and sodium silicate) at solid/liquid ratio of 2.5. The NaOH were used 6M, 8M, 10M, 12M and 14M with constant sodium silicate/NaOH ratio of 2.5. The geopolymers were cured at room temperature (29°C) for 24 hours and 60°C in oven for another 24 hours. The testing and analysis of the fly ash-based geopolymers were performed after 28 days. The adequate Na^+ ions and densified microstructure were observed at optimum 8M-NaOH-activated fly ash-based geopolymers.

1. Introduction

There has been growing attention on new construction materials with low green house gas emissions and sustainability for the past few decades. Therefore, Davidovits [1] invented geopolymers as a new type of discovered materials with inorganic polymeric structure which can replace ordinary Portland cement. Geopolymer synthesis is considered as a sustainable binding material due to waste utilization and low greenhouse footprint [2].

Geopolymers are formed once the aluminosilicate material (fly ash and slag) is mixed with the alkaline solution through geopolymerization process. The alkaline solutionsact as a character to dissolve the active aluminosilicate species at the early stage of geopolymerization process [3]. There contain a few stages in geopolymerization [4]. First stage consist of starting materials (silicate (SiO_4) and aluminate (AlO_4)) which activated by alkali and geopolymer gel produced. Hydrolysis occured in the next stage when water molecules help to further break the bond and let SiO_4 and AlO_4 tetrahedral units to connect each other to yield polymeric precursors. Finally, polycondensation process occurred as geopolymer gel solidified and the three-dimensional aluminosilicate network is formed.

Different parameters such as raw materials, alkali concentration, solid/liquid ratio, alkaline activator ratio, curing time and any other factors could influence the mechanisms formation and properties of geopolymer synthesis. Among of the parameters, the sodium oxide concentration of the activator solution play an important role as it determines the degree of reaction and the densification of microstructure on geopolymer synthesis [5]. Several researchers were also studied on the consequence of using different NaOH content on geopolymer synthesis [6, 7]. Thus, the aim of this paper was to investigate the effect of NaOH concentration on fly ash-based geopolymers. The compressive strength, elemental composition and microstructural analysis of fly ash-based geopolymers with various NaOH concentration were examined.

2. Methodology

2.1. Material

Fly ash (FA) collected from Cement Industries of Malaysia Berhad (CIMA) Perlis Plant, Malaysia was used as the source of aluminosilicate materials. The chemical composition of FA is tabulated in Table 1, which was obtained from XRF analysis. Referring to Table 1, the lime (CaO) content was 3.89% which can be classified as Class F FA according to ASTM C618 due to less than 20%. Figure 1 displayed the SEM micrograph of FA particles which were sieved (75µm) with spherical in shape and smooth surfaces.

The alkaline solution used was a mixture of sodium hydroxide (NaOH) and liquid sodium silicate (Na_2SiO_3). The NaOH used was caustic soda flakes with the brand name Formosoda-P supplied by Formosa Plastic Corporation Taiwan. The liquid Na_2SiO_3 was supplied by South Pacific Chemicals Industries Sdn. Bhd. (SPCI), Malaysia with SiO_2/Na_2O modulus of 3.2.

Table 1. Chemical composition of fly ash as determined by XRF analysis.

Chemical	Wt%
SiO_2	56.3
Al_2O_3	28.0
Fe_2O_3	6.86
TiO_2	2.17
CaO	3.89
K_2O	1.49
Others	1.24

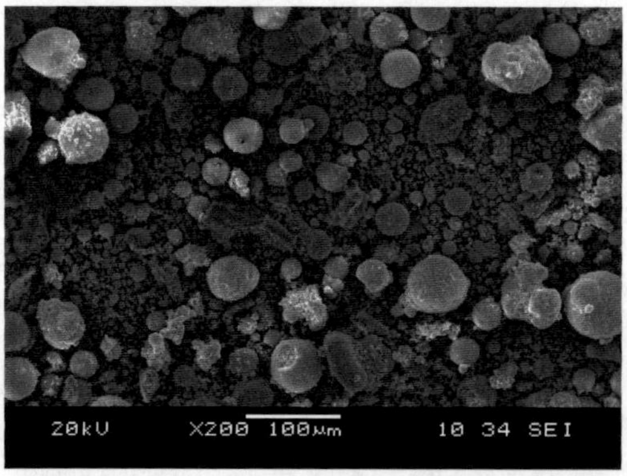

Figure 1. SEM micrograph of fly ash particles.

2.2. Samples Preparation

The Na_2SiO_3/NaOH ratio of 2.5 and solid to liquid ratio of 2.5 was fixed in FA-based geopolymers synthesis. Three sets of geopolymer samples were formed for each molarities of NaOH (6M, 8M, 10M, 12M and 14M). The geopolymer paste was moulded into 50mm × 50mm × 50mm plastic moulds. The moulded samples were cured at room temperature for 24 hours followed by curing at 60°C in oven for another 24 hours. The cured samples were sealed with plastic covering in order to avoid substantial loss of moisture and kept curing for 28 days in room temperature before testing.

2.3. Testing and Analysis

IOP Conf. Series: Materials Science and Engineering **551** (2019) 012090 doi:10.1088/1757-899X/551/1/012090

The compressive strength was carried out according to ASTM C109 by using the Instron machine series 5569 Mechanical Tester. Microstructural and elemental composition analysis were performed by using SEM machine with model JEOL JFC 1600. Before undergoing SEM characterization, specimens were cut in small pieces form coated with platinum using an Auto Fine Coater.

3. Results and Discussions

3.1. Compressive Strength

According to the previous studied on Ng et al. [8], the compressive strength (Figure 5) showed increasing trend as the NaOH concentration utilized in geopolymer synthesis increases from 6M to 8M (optimum strength) but beyond of that the compressive strength reduced. The highest strength obtained at 8M was due to the sufficient OH^- and Na^+ concentration as mention before. When NaOH concentration exceed 8M, a lower strength obtained in the FA geopolymer synthesis. It is reasonable suggest that the presence of excess OH^- and Na^+ ions content caused aluminosilicate gel precipitation at the very early stages [9]. It was consistence with the results on EDX (Table 2) where higher mass percentage of Na observed from 14M over 8M and 6M.

3.2. Microstructural Analysis

Figure 2 revealed the SEM micrographs of fly ash geopolymer synthesis with 6M, 8M and 14M-NaOH. After mixing the original FA and alkaline activator solution, there were undergoing geopolymerization reaction (arrow sign) as shown in Figure 2a and thus geopolymeric gel was formed, indicating that the structure experiences growth. In other words, the alkaline activator solution has dissolved the FA to release Al^{3+} and Si^{4+} for geopolymerization reaction to take place. However, the porous heterogeneous matrix formed at Figure 2a which does not exist in the original FA (Figure 1).

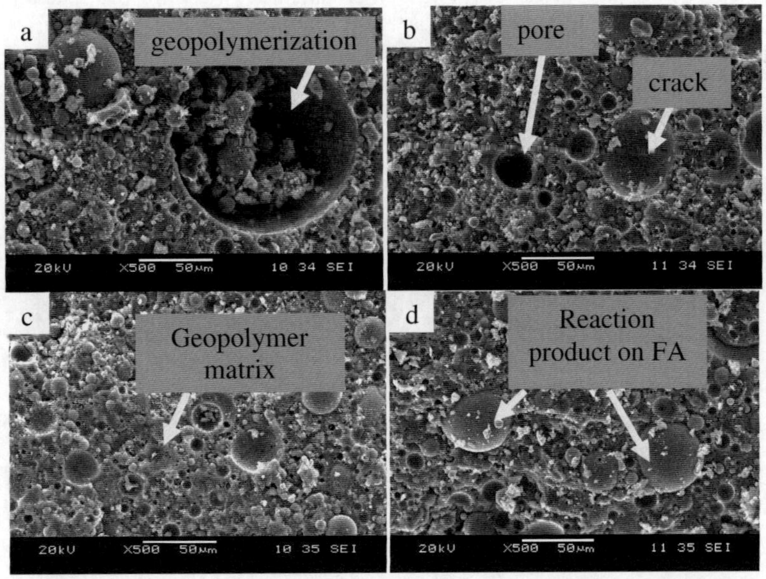

Figure. 2. SEM micrographs of fly ash geopolymer with a) geopolymerization reaction, b) 6M, c) 8M, and d) 14M-NaOH-activated.

IOP Conf. Series: Materials Science and Engineering **551** (2019) 012090 doi:10.1088/1757-899X/551/1/012090

Referring to Figure 2b, there showed a large proportion of un-reacted and/or partially reacted FA which did not completely dissolve with alkaline activator. A lot of FA particles were loosely bonded to the matrix and interfacial boundaries can be detected in the structure easily. The imperfect microstructures (cracks, pores) appear on FA geopolymers caused of poor compressive strength [10].

According to Figure 2c, it can be seen that the geopolymeric gel was enough to bind the constituents and some of the FA particles are firmly connected to the geopolymer structure, which indicates a greater extent of geopolymerization. The microstructures showed the more densified and homogeneous as compared to Figure 2b and 2d. The densified micro-structures (8M-NaOH) correlated well with the results on compressive strength [8], possessing a greater amount of geopolymeric gels. The results obtained suggested that the increase of alkali concentration contribute to higher geopolymerization rate and enhance the growth of geopolymeric gels, hence the compressive strength increased as observed for 8M-NaOH.

However, the excess content of alkali (14M) on Table 2 further increased the rate of geopolymerization caused the reaction products not enough time to disperse in the geopolymer matrix and it deposed on the surface of the FA particles (Figure 2d), as a result hindered further dissolution reactions. Therefore, Figure 2d produced lesser geopolymeric gel lead to strength reduction as compared to Figure 2c. These observations indicate that the geopolymeric gels formed by the reaction between FA and activator are rely on the reactivity and concentration of alkali activators solution [11].

3.3. Elemental Composition Analysis

Table 2 tabulated the elemental composition of 6M, 8M and 14M-NaOH-activated on FA geopolymers, as determined using EDX analysis. The Na^+ constituent of the activator solution was detected apart from the Si^{4+} and Al^{3+} constituent of FA. The presence of Na^+ cation was due to the formation of reactants via the condensation process caused by geopolymerization. These reactants combined with the Na^+ cation depart from the external NaOH, leading the reaction products to agglomerate and developing the strength at 8M-NaOH-activated geopolymers through the combinations of FA particles [12].

Table 2. Elemental composition of various NaOH concentration on fly ash geopolymers, as determined using EDX analysis.

Molarity of NaOH	Element with mass percentage			
	Si (wt%)	Al (wt%)	Na (wt%)	O (wt%)
6M	39.82	42.39	0.99	16.8
8M	47.21	35.3	3.57	13.92
14M	51.23	18.12	10.00	20.65

Geopolymeric system was under constant sodium silicate content but with the continuous addition of NaOH causes a reduction on Si/Na ratio. This reduced Si/Na ratio leading to the gradual change of the chemical system from the larger rings species, complex structures and polymers to monosilicates, chains and smaller rings species [13]. The predominance of monosilicates and species with smaller rings at high concentration of NaOH slows down the polycondensation process [13, 14] and causing the compressive strength reduced beyond 8M [8].

4. Conclusions

This paper presented the effect of different concentration of NaOH solution (6-14M) on fly ash-based geopolymers synthesis. The optimum NaOH concentration was revealed at 8M for the formation of FA geopolymers. It was supposed that there are sufficient of Na^+ ions for complete geopolymerization reaction which in agreement with the results on EDX analysis. There were also observed denser microstructure at 8M-NaOH with smoother surface than 6M and 14M-NaOH.

Acknowledgement
The authors wish to gratefully acknowledge the funding by Fundamental Research Grant Scheme (FRGS/1/2015/TK05/UNIMAP/02/1) sponsored by Ministry of Education Malaysia.

References
[1] Davidovits J 1989 *J Therm Anal* **35** 441.
[2] Bakri, A M M *et al* 2013 *Revista de Chimie,* **64** 382.
[3] Bakri A M M *et al* 2012 *Applied Mechanics and Materials* **110** 734.
[4] Izzat, A M *et al* 2013 *Materiale Plastice* **50** 171.
[5] Lloyd R R *et al* 2009 *Micropor Mesopor Mat* **126**, 32-39.
[6] Al Bakri Abdullah *et al* 2015 *Revista de Chimie* **66** 1001.
[7] Yahya Z *et al* 2015 *Materials* **8** 2227.
[8] Ng H T *et al* 2018 *AIP Conf Proc* **2045** 020098
[9] Lee W K W *et al* 2002 *Colloid Surface* A **211**, 115.
[10] Granizo M L *et al* 2007 *J Mater Sci.* **42** 2934.
[11] Duchesne J *et al* 2010 *Waste Biomass Valori* **1** 367-377.
[12] Xu H *et al* 2000 *Int. J. Miner Process* **59** 266.
[13] Panias D *et al* 2007 *Colloid Surface A* **301** 254.
[14] Zuhua Z *et al* 2009 *Appl. Clay. Sci.* **43** 223.

IOP Conf. Series: Materials Science and Engineering **551** (2019) 012091 doi:10.1088/1757-899X/551/1/012091

Fabrication of Novel Geopolymer Reinforced Tin Copper Solder in Suppressing Intermetallic Layer Growth

Nur Nadiah 'Izzati Binti Zulkifli[1], Mohd Mustafa Al Bakri Abdullah[2], Mohd Arif Anuar Mohd Salleh[3]

[1]Department of Mechanical Engineering Technology, Faculty of Engineering Technology, Universiti Malaysia Perlis (UniMAP), Sungai Chuchuh 02100, Padang Besar, Perlis.

E-mail: nadiahizzati.zulkifli@gmail.com

Abstract. Currently, lead-free composite solder technology system is the new approach that have been considered as a promising replacement for toxic SnPb solder alloy. Sn-0.7Cu was used as matrix solder alloy while fly ash geopolymer and kaolin geopolymer with different compositions (0.5 wt.%, 1.0 wt.%, and 1.5 wt.%) was added as a reinforcement to increase the properties of solder alloy. The solder composite was prepared by using powder metallurgy method which consist of mixing, compaction, and sintering process via microwave oven. Microwave sintering approach results in more uniform heating, lower pore coarsening with cost-effective and energy efficiency. The chemical composition and morphology of geopolymer reinforcement was analysed by using X-ray Flourescene and Scanning Electron Microscope. The influences of the geopolymer particulates in the monolithic matrix solder on the interfacial intermetallic compound thickness and wettability were investigated by using optical microscope, while the specific value was measured by using J-image software. The melting point temperature were investigated by using differential scanning calorimeter (DSC). The influences of adding types of geopolymer particulates into Sn-0.7Cu lead free solder were investigated in terms of interfacial intermetallic compounds (IMCs) and thermal properties. It is noted that several addition of fly ash geopolymer and kaolin geopolymer particles can remarkably supressed the thickness of intermetallic layer between composite solder and the substrate and thus gives higher melting temperature but somehow certain composition resulted low in mechanical properties. Overall, the addition of 0.5 wt.% of kaolin geopolymer reinforcements into Sn-0.7Cu lead-free improved all the listed properties of the solder materials.

1. Introduction

Nowadays, electronic packaging industries has become one of the most famous in the field of electrical engineering among the electronic industries and has grown rapidly becoming more sophisticated and advance in producing electrical device [1]. Some of the process occur in electronic packaging consist of soldering process which bond the solder materials together with the substrate forming one or more electrical connection by melting it at a controlled temperature. Two functions of soldering process are as electrically support; giving required electrical connection within the solder and circuit and mechanically support; holding the components to provide assembly together.

Over the past few years, Pb is typically used in soldering industries but nowadays, the uses of Pb as the solder alloy has been banned and restricted due to environmental issues. Therefore, other

IOP Conf. Series: Materials Science and Engineering **551** (2019) 012091 doi:10.1088/1757-899X/551/1/012091

alternatives is used to provide more safe and low cost with great performances effectiveness materials alloy such as Tin silver (SnAg), tin bismuth (SnBi), tin copper (SnCu), tin zinc (SnZn), tin rhenium (SnRE), tin zinc aluminium (SnZnAl), tin silver copper indium (SnAgCuIn), tin silver copper (SAC) etc to lower the consumption of Pb usage. Besides, the addition of other reinforcement such as from ceramic based are widely acknowledge by the researchers as it can increase the mechanical and thermal properties of the solder joint. Recently, many investigation about Sn-0.7Cu/ Si_3N_4, Sn-0.7Cu-0.05Ni/ TiO_2, Sn-3.0Ag-0.5Cu/ SiC, Sn-3.0Ag-0.5Cu/ NiO and other related to this study has been observe to have various properties for the matrix alloy solder [3,4].

Currently, some researchers had found that the addition of nano-ceramic particles will increase the mechanical and thermal behaviour of the solder joint. Since geopolymer powder is one of a type of reinforcement came from rare earth elements which also can be called as ceramic based, it can be added into the matrix of solder alloy. Geopolymer is getting a worldwide interest due to it low carbon dioxide emission [5]. Geopolymer materials comes from polymerization process which produce a chemical reaction between aluminosilicate materials such as alumina, silicon, fly ash, kaolin, slag with alkaline activators such as NaOH based, KOH based, or a mixture of Na_2O with SiO_2 based (Na_2SiO_3) forming a paste or binder. M. Albitar et. al. elucidates that fly ash geopolymer can withstand sulphate attack, corrosion and acid attack and therefore easy to get with a low cost. Fly ash geopolymer could give some advantages over metakaolin due to it is waste resource produced in huge quantities by coal-fired power stations making them environmentally friendly feedstock [6]. Previous research shows that NaOH solution above 12 M are more risky to handle compared to 12 M NaOH solution [7]. It also has high chance for the alkaline activator to undergo solidification faster compare to 12 M of NaOH solution. Mustafa Al Bakri et. al, 2011 also elucidate that fly ash geopolymer with 12 M NaOH concentration shows excellent result in compressive strength until 94.59 MPa when testing for seven days. This is because the microstructures appears to be homogenous and contains proportion of unreacted fly ash microspheres with continuous matrices of aluminosilicate and microcracks [8].

Nmiri et. al, also investigated the effect of different concentration of KOH and NaOH towards the compressive strength where they found that 13 M gives the highest compressive strength which is 33.5 MPa for KOH and 36.7 MPa for NaOH. Meanwhile, increasing molarity up to 15 M of NaOH and KOH concentration makes the compressive strength become lower. A new improved solders were fabricated by incorporating nanosized of geopolymer can suppress the grain growth and therefore contribute to small intermetallic layer growth which can give high surface energy to the solder-substrate [7].

2. Experimental

2.1. SnCu

In general, lead free solder cost almost two to three times higher than SnPb itself but Sn-0.7Cu solder usually cost only about 1.3 times higher than SnPb solder which explain why SnCu has been used in this study [9]. SnCu powder used was Sn-0.7Cu and was supplied by Nihon Superior, Japan with size range 25-40μm which act as the matrix alloy in the geopolymer particulates. Previous research elucidate that SnCu solder alloy can performed as the same level as SnPb solder alloy as they can give good fatigue, creep resistance, high melting point [10].

2.2. Fly Ash Geopolymer

Chemical composition analysis was undergoes to investigate the ASTM C618 standard. Based on ASTM C618 standard, there are two classes of fly ash which is Type C and Type F. In this study, the fly ash type F was used as the reinforcement in Sn-0.7Cu/fly ash geopolymer was obtained from Manjung Power Plant, Lumut, Perak, Malaysia. The alkaline activator used in this study was prepared by mixing 12M of NaOH solution with Na2SiO3 solution. The alkaline activator with the weight ratio

(Na2SiO3/NaOH) 2.5 was blended together with fly ash into weight ratio of S/L is 2.0. The solid/liquid which can be categorized as fly ash/alkaline activator and $NaSiO_3/NaOH$ ratio also influenced the workability of the mixture where less workability caused failure in providing good flexural properties. Some researcher agreed that 2.0 fly ash/alkaline activator ratio and 2.5 $NaSiO_3/NaOH$ ratio give the maximum strength up to 71 MPa [8]. Some of the researchers stated that the best flexural strength can be obtained when the ratio of fly ash/alkaline activator and $NaSiO_3/NaOH$ were 2.0 and 2.5 [11]. The mixture was then poured in a flat mould to make it easier during milling process to produce powdery reinforcement. After that, fly ash geopolymer was cured for 24 hours in ambient temperature. Huseien et. al, 2016 investigates that most fly ash geopolymer samples cured at ambient temperature shows lower water absorption compare when curing in oven and has resulted excellent compressive strength with low percentage of water absorption when curing fly ash geopolymer at ambient temperature [12]. Final fly ash geopolymer was milled to obtain powder reinforcement by using ball milling machine until achieve 63μm in size.

2.3. Kaolin Geopolymer

Kaolin refers to a family of kaolinitic clays that consist of 1:1-type clay mineral with one tetrahedral sheet and one octahedral sheet depending on the weathering and geological condition [13]. In this study, kaolin was obtained by Manjung Power Plant, Lumut, Perak, Malaysia. The same molarity of 12M NaOH was used with a ratio of alkaline activator (Na2SiO3/NaOH) 0.24. In terms of S/L, Heah et. al observed that kaolin geopolymer with S/L ratio of 0.60 disintegrated in water at 7 days while S/L 1.0 does not disintegrated in water even after 180 days [14]. The same observation was obtained when using $NaSiO_3/NaOH$ with ratio 0.16 and 0.32. Heah et. al (2012), concluded that as S/L and $NaSiO_3/NaOH$ ratio increase, the workability will decrease. As S/L ratio increase, the mix become stiffer where more solid content then fluid medium within the mixes. While as $NaSiO_3/NaOH$ ratio increase, viscous property of sodium silicate become higher than NaOH solution. Workability of kaolin geopolymers decreased when the ratio of solid/liquid and $NaSiO_3/NaOH$ increase. Therefore, workability decrease due to inter-particle friction and highly viscous properties [14]. In this study, the kaolin was then blended with alkaline activator with a ratio S/L is 0.8. After the mixture was well blend, the mixture was then poured in a flat mould and been cured in the oven temperature of 60°C. It was found that 60°C curing temperature gives the highest compressive strength [8]. Based on Joshi and Kadu, 2012, curing in the oven temperature of 60°C to 75°C for 24 hour can increase the strength as compared to room temperature [15]. Final kaolin geopolymer was milled to obtain powder reinforcement by using ball milling machine and sieve until achieve 63μm in size.

2.4. Sample Fabrication

Precise weighing of the solder alloy or solder material elements and the reinforcement materials are performed. Tin copper and geopolymer powder were weighted following the composition for 2 gram per sample. Several composition (0.5 wt.%, 1.0 wt.%, and 1.5 wt.%) of fly ash geopolymer and kaolin geopolymer were weighted. Sn-0.7Cu/fly ash geopolymer and Sn-0.7Cu/kaolin geopolymer powder were mixed together in a roller blender at speed of 250 rpm for 1 hour to ensure that the mixture were mixed homogenously. The compaction technique via powder metallurgy was used in the laboratory for this study was uniaxial compaction (hand press). The mould used was 12 mm diameter which the size consist of body base, two small plungers and main plungers. Then, the mixed powder were inserted into the mould and were compacted under pressure 4.5 ton for 1 minute. The mould was cleaned first by using WD40 liquid before undergo compaction process to ensure the mould clean from any contaminations.

The sintering of green body process was done by using a microwave oven. The temperature was set up to 175°C for 2 minutes. The sample were placed inside a ceramic crucible to make sure that the heat fully absorbed into the sample.

2.5. Analysis and Characterization

The microstructural elemental analysis of the metallographically polished samples of the composite materials were examined using optical microscope (OM model Olympus Optical Co. Ltd BX41RF) and SEM was utilized to determine the surface structure, thickness of intermetallic compound, and wettability formed in the solder matrix. This was aimed to investigate morphology, presence of intermetallic compounds (IMCs) formation of the solder-substrate and contact angle between solder and substrate from the cross-section area of the sample. Before the samples were observed under OM, the samples were mounted using a mixture of epoxy and hardener to make it easy to grind and polished the samples since the size of the sample is small and thin.

After the mounted samples solidify, the mounted part was removed from the mold. The surface area of the mounted samples were cleaned to remove foreign matters or unbalanced mounted part. The surface of the mounted sample were then grind and polished to ensure smooth and shinning surface area so that their microstructure can be view under a microscope.

Solder reflow process was the last process in soldering process where the solder paste was melted on the Cu substrate, mixed with active flux and wet the surfaces to be joined. This process produce the connection to form intermetallic compound when the matrix alloy was melted on Cu substrate. At the end, the solder solidified and form a strong metallurgical bond between them. F4N reflow oven used in solder reflow process. There are four zones involved in this process which are thermal soak zone, preheat zone, reflow zone and cooling zone. Thickness of intermetallic compound layer (IMC) and their wettability were determined by using OM. The IMC that formed at the Cu substrate layer of the cross-sectioned samples was measured by using J-Image software to obtain average result of IMCs thickness.

Differential scanning calorimeter (DSC model TA Instruments DSQ-10) was used to determine the melting point of the materials. The samples were weighed below 5 mg and were placed on aluminium pan and heated up to temperature 250°C with a heating rate 10°C/min under a protective nitrogen atmosphere. The solder samples were coated with small amount of rosin flux and attached on the Cu sheet before placed into aluminium pan.

3. Results and Discussions

3.1. Surface Microstructure

Microstructure of fly ash and kaolin powder, distribution of particles, and homogeneity degree were examined by using SEM and OM. There are several magnification involved in this analysis which are two magnification for SEM; 2000x and 5000x and three magnification for OM; 50x. SEM analysis revealed morphological of pure crystal in magnification 2000x and 5000x as in Figure 3.1. Figure 3.1 shows that fly ash are generally in glassy solid or hollow, spherical and finer in shape [16]. The micrograph image can be seen from the figure that fly ash consist of a series of spherical vitreous particles of different sizes. While usually hollow in shaped, some of the spheres almost appear within other interior small size sphere.

Generally, kaolin structure was flaky-like as can be observed on the surface of kaolin particles. Besides that, systematically arranged layered structure forming stacks of books can also revealed due to kaolin geopolymer composed sheets of silica and alumina held together by Van Der Waals and hydrogen bonding [17]. The nature kaolin structure contributed smaller surface area for geopolymerization process compared to materials with spherical-shaped particles such as fly ash. This will act as barrier that caused only minimum dissolution by alkaline solution and hence the geopolymerization reaction [17].

Based on Figure 3.2, clear grain boundary of Sn-0.7Cu without the addition of any geopolymer reinforcement can be observed. Increasing in weight percentage of fly ash geopolymer reinforcement making the black area become more obvious. By the addition of 0.5 wt.% and 1.0 wt.% fly ash geopolymer reinforcement, it shows that the fly ash particles does not uniformly distributed between the grain boundaries where at certain part the fly ash geopolymer particles agglomerate with each other compare to 1.5 wt.%. This shows that the fly ash geopolymer and the monolithic Sn-0.7Cu does not homogenously mixed during mixing process. Addition of kaolin geopolymer into the matrix alloy shows the grain boundary region had fairly distribution of kaolin geopolymer. As the weight percentage of kaolin geopolymer increases to 1.5 wt.%, the kaolin geopolymer particles which distributed along the grain boundaries of the bulk solder (black area) increases and produced a few agglomeration of kaolin geopolymer particles between the grain boundaries. This happened due to non-homogenous distribution of the Sn-0.7Cu/kaolin geopolymer between the grain boundaries during mixing process.

Figure 3.1: SEM analysis of a) fly ash, b) kaolin

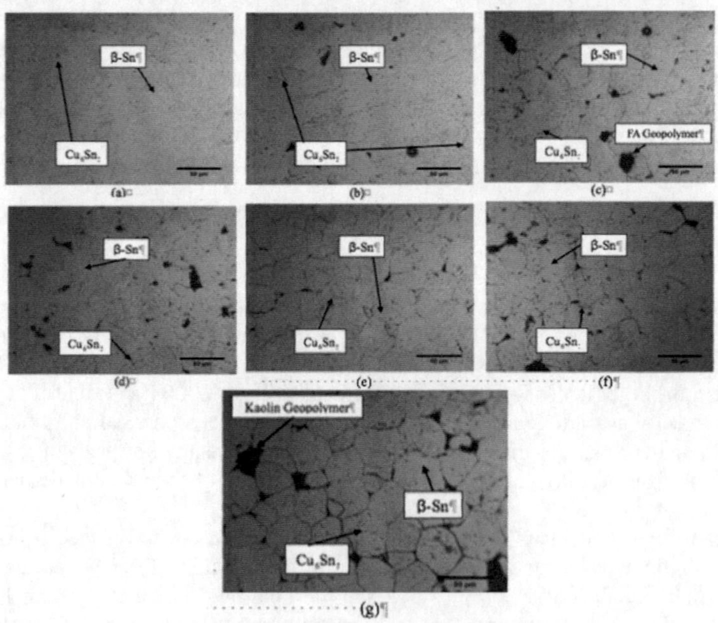

Figure 3.2: Optical micrographs of Sn-0.7Cu + geopolymer grains structure at 50x magnification with a) 0 wt.%, b) 0.5 wt.% FA geopolymer, c) 1.0 wt.% FA geopolymer, d) 1.5 wt.% FA geopolymer, e) 0.5 wt.% kaolin geopolymer, f) 1.0 wt.% kaolin geopolymer, g) 1.5 wt.% kaolin geopolymer.

Typically, microstructure of Sn-0.7Cu solder alloy that forms on the substrate consist of primary β-Sn cells or dendrites that surrounded by eutectic which is combination of β-Sn and intermetallic Cu_6Sn_5 [18]. Based on the figure above, the pure Sn-0.7Cu solder alloy consist of large number of primary β-Sn cells with small intermetallic compound Cu_6Sn_5. Besides, it also has uniform eutectic structure phase microstructure. Addition of 0.5 wt.% fly ash geopolymer and kaolin geopolymer reinforcement into Sn-0.7Cu solder shows that the primary β-Sn cells are still present in the microstructure and the microstructure become uniform with small β-Sn dendrites and small number of intermetallic compound Cu_6Sn_5. The microstructure of β-Sn dendrites was refined and has smaller intermetallic compound Cu_6Sn_5. Addition of 1.0 wt.% and 1.5 wt.% geopolymer reinforcement, the microstructure become non-uniform because of large β-Sn cells with larger number of intermetallic Cu_6Sn_5. This can be concluded that fly ash geopolymer did not completely dissolved homogenously in Sn-0.7Cu solder alloy.

The small changes in composition of reinforcement can affect the formation and properties of Cu_6Sn_5 while it involve actively during interface reaction [19]. According to Chellvarajoo et. al, many researchers proved that addition of ceramic reinforcement can affected the formation and properties of Cu_6Sn_5 while it involve in the interface reaction actively [20]. Ng et. al, elucidate that β-Sn dendrites are still present in the microstructure but the eutectic has been modified with the incorporation of Zn into the Cu_6Sn_5 crystals. The microstructures of the bulk alloys can be refines by added a small amount of Zn [19].

3.2. Intermetallic Compound Layer (IMC)
From the observation, the presence of fly ash geopolymer as reinforcement in Sn-0.7Cu solder alloy helped to change the thickness of IMC. By referring to Figure 3.3, the bar graph shows that 0.5 wt.% of fly ash geopolymer particles caused the intermetallic layer increases compared to pure Sn-0.7Cu solder. 1.0 wt.% of fly ash addition gives the best result in retarding the thickness of IMCs layer. It is noted that when fly ash geopolymer particles was added into solder and reflowed, the particles remain as stable solid particles. This shows that 1.0 wt.% was the optimum composition for Sn-0.7Cu/fly ash geopolymer to obtain good metallurgical bond for solder-substrate. While solder with 1.5 wt.% fly ash geopolymer begin to increase the thickness of IMCs near the value of pure Sn-0.7Cu IMCs. Figure 3.4 shows the bar graph of IMCs thickness for each composition of kaolin geopolymer added to Sn-0.7Cu solder. By adding 0.5 wt.% kaolin geopolymer can supress the intermetallic layer growth thus producing better performing of solder substrate. Addition of 1.0 wt.% kaolin geopolymer shows the highest IMCs thickness compare to pure Sn-0.7Cu and addition of 1.5 wt.% kaolin geopolymer.

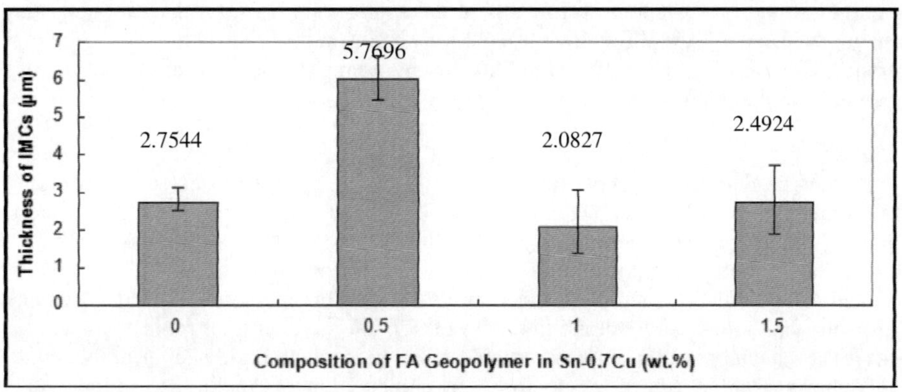

Figure 3.3: Thickness of IMCs with different composition of fly ash geopolymer

During reflow soldering process, the mobility of geopolymer particles caused the respective particles to embed in Sn-0.7Cu grain structure and Cu substrate [21]. Therefore, the free movements of Cu atoms from the Cu substrate and solder matrix slowly inhibited by the accumulation of these particles. IMC thickness will slowly reduce when a limited number of Cu atoms presence during IMC formation [18]. Previous researchers investigated that the pure Sn-0.7Cu solder indicated that the interfacial Cu_6Sn_5 were in needle like and scalloped structure compare to the shorter and more faceted structure when adding TiO_2 reinforcement [22].

The presence of other reinforcing elements appear to limit the growth of neighbouring grains making the IMC layer more stable [22]. But addition of 0.5 wt.% fly ash geopolymer resulted in more higher in thickness (5.7696 μm) compare to pure Sn-0.7Cu matrix solder (2.7544 μm) which is opposite from previous research [21]. This phenomenon occur due to addition of 0.5 wt.% cannot act as heterogenous nucleating agent to supress the IMC layer growth [23]. It also might occur due to too small of addition in the Sn-0.7Cu composite solder cannot be detected in the phase and therefore disturbed the IMCs formation [23].

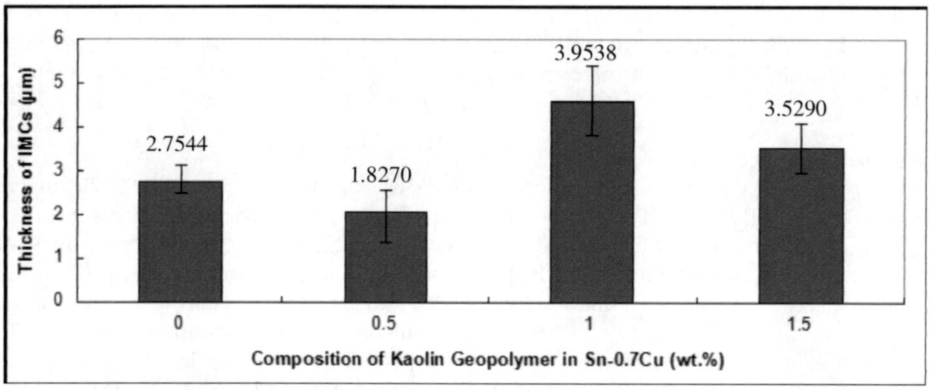

Figure 3.4: Thickness of IMCs with different composition of kaolin geopolymer.

But as soon as the kaolin geopolymer being increased, the thickness of IMCs increased more higher up to 3.9538μm and 3.5920μm compare to pure Sn-0.7Cu. Previous study by Kotadia et. al (2014), stated that reinforcing ceramic-based in matrix solder might be risk as the high amount of ceramic particles might contribute to brittleness of the solder joint. Besides that, an increasing IMCs layer also might cause by the difficulties to achieve homogeneous distribution of ceramic particles since ceramic is a hard particles [10]. Humidity in environment also can cause ceramic especially kaolin geopolymer to agglomerate again even after sieving.

3.3. Wettability

Figure 3.5 observed that contact angle for 0.5 wt.% fly ash geopolymer decrease until 30.57° from pure Sn-0.7Cu solder (32.026°) and increase to 36.363° after 1.0 wt.% addition of fly ash geopolymer. Addition of fly ash geopolymer more than 1.0 wt.% shows the contact angle decrease back to 35.946°. This shows that optimum composition of 0.5 wt.% is the optimum composition of geopolymer reinforcement can be added to increase the wettability. Wettability of solder alloys depends on surface properties and also solder alloy proportion. Generally, the smaller the contact angle, the better wettability of the interconnections. An acceptable range of contact angle should be below 45° [24]. Figure 3.6 shows average contact angle for each composition of kaolin geopolymer addition in monolithic Sn-0.7Cu solder alloy. The wettability shows at the best when 1.5 wt.% kaolin geopolymer was used in Sn-0.7Cu which it can obtained until 27.946°. The trend shows in the bar graph starting with high wettability angle which is pure Sn-0.7Cu up to 32.026°. But the addition of kaolin geopolymer contribute to lower the contact angle between the solder and substrate. The wettability of 1.0 wt.% kaolin geopolymer as a reinforcement in Sn-0.7Cu solder increased but it is still low than pure Sn-0.7Cu wetting angle.

Wetting can be relate with the thickness of intermetallic layer growth of the solder-substrate. When there is intermetallic formation, it means that there are contact angle (wetting) occurs between solder alloy and the substrate [25]. Some researchers stated that wettability angle of solder-substrate related to the formation of intermetallic compound where addition of 1.0 wt.% and 1.5 wt.% reinforcements resulted higher IMC thickness thus contributed to higher wetting angle [13]. As seen in Figure 3.5 above, 0.5 wt.% addition gives better contact angle. Generally, as composition of reinforcement in solder alloy increase, contact angle can be reduced [25]. But as the percentage of fly ash geopolymer increased up to 1.0 wt.% and 1.5 wt.%, the contact angle become larger. Based on Salleh et. al (2011), addition of reinforcement particulates will only improve the wettability until the optimum composition is achieved. Too higher of reinforcement percentage will further degrade the wettability of composite solder which can be worse than monolithic Sn-0.7Cu alloy since ceramic-based reinforcement gives brittle properties in nature [24].

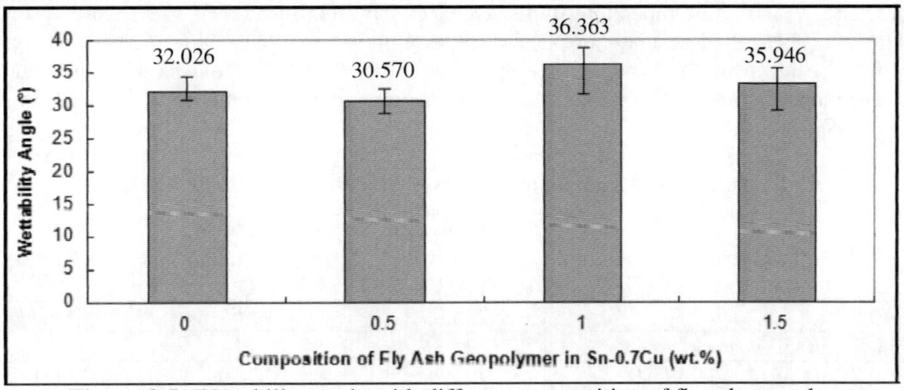

Figure 3.5: Wettability angle with different composition of fly ash geopolymer

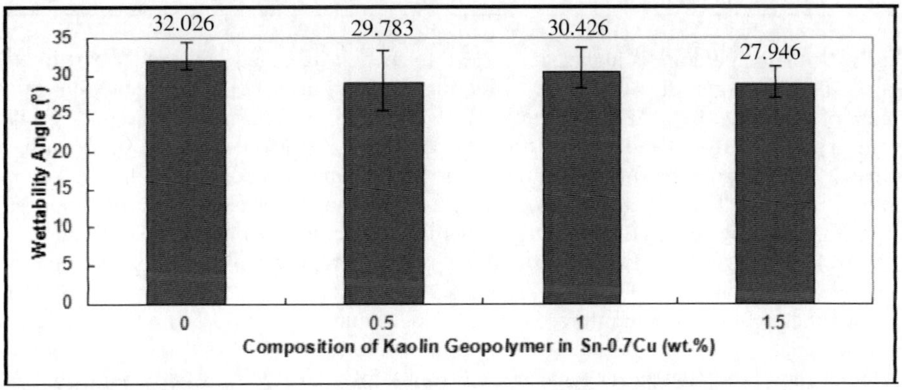

Figure 3.6: Wettability angle with different composition of kaolin geopolymer

Previous study state that formation of intermetallic layer can influenced the wettability of the solder (Zang et. al, 2009). When comparing with the IMC layer growth produced in Figure 3.6, the wettability of Sn-0.7Cu/0.5 wt.% kaolin geopolymer increased when the IMCs thickness decreased. This phenomenon has been supported by previous studies which agreed that addition Si_3N_4 particulates in molten Sn-0.7Cu solder act as an agent to increase the wettability between Cu substrate [24].

3.4. Melting Point

As seen in the Table 3.1, the result shows the reduction in melting temperature, TL of Sn-0.7Cu solder from 229.67°C to 228.15°C after addition of 0.5 wt.% fly ash geopolymer which is lower than the monolithic melting temperature of Sn-0.7Cu. This showed fly ash geopolymer particles has just slightly altered the melting temperature. But as the composition of fly ash geopolymer increase to 1.5 wt.%, the melting temperature start to slightly increase to 228.99°C but still lower than the monolithic solder alloy. Besides, the solidous temperature of pure Sn-0.7Cu solder and Sn-0.7Cu/fly ash geopolymer particles range from 225.09°C to 225.83°C shows increasing reading with increasing fly ash geopolymer content. Meanwhile, addition of kaolin geopolymer powder, the monolithic Sn-0.7Cu obtained up to 229.67°C while slowly decrease the melting point when 0.5 wt.% kaolin geopolymer particles is added. DSC results shows that increasing in composition of kaolin geopolymer up to 1.5 wt.% created low melting temperature until 228.83°C. The mixing of 1.5 wt.% kaolin geopolymer composite solder absorbs less amount of heat (228.83°C) compared than 0.5 wt.% kaolin geopolymer (228.99°C). Besides, the solidous temperature of pure Sn-0.7Cu solder and Sn-0.7Cu/kaolin geopolymer particles range from 225.09 to 226.47 shows increasing reading with increasing kaolin geopolymer content.

Table 3.1: Melting temperature of Sn-0.7Cu/geopolymer solder.

Materials	Solidous temperature, T_S (°C)	Liquidous temperature, T_L (°C)	ΔT (°C)
Sn-0.7Cu	225.09	229.67	4.58
Sn-0.7Cu/0.5FA geopolymer	225.83	228.15	2.32
Sn-0.7Cu/1.5FA geopolymer	225.79	228.99	3.20
Sn-0.7Cu/0.5Kaolin geopolymer	225.98	229.02	3.03
Sn-0.7Cu/1.5Kaolin geopolymer	226.47	228.83	2.36

If the melting temperature is higher, it can lead to several damage to the solder part in electronic components in packaging due to higher soldering temperature [12]. The reduction in melting temperature of Sn-0.7Cu.0.5 wt.% fly ash geopolymer was due to the addition of non-alloying element particles has just slightly altered the melting temperature [20]. More addition of fly ash ceramic reinforcement can cause increasing in melting temperature since geopolymer is highly resistance to high temperature [22]. This statement has been proved by this experiment since the addition of 1.5 wt.% fly ash geopolymer resulted increasing in melting temperature from 228.15° to 228.99°. Liu et. al (2008), explained higher addition of Bi particles in solder alloys could increase the melting range and therefore lead to initiation of solidification crack of the solder joints. This prove that addition of 0.5 wt.% of fly ash geopolymer reinforcement is the optimum composition which can produce better melting point [26].

Table 3.1 shows the melting range decreased from 4.58°C to 2.36°C. Chellvarajoo and Abdullah observed that addition of 0.5 wt.% NiO particles caused the melting range become lower than plain solder. This result indirectly saves the cost for thermal application for the composite solder to melt during reflow soldering process since high melting temperature increase the usage of energy [21].

4. Conclusions

The comparison between morphological, growth of interfacial intermetallic compound formation, wettability angle, and thermal properties of Sn-0.7Cu solder using different types and compositions of geopolymer reinforcement was examined through this project. The influences of adding fly ash geopolymer and kaolin geopolymer powder into Sn-0.7Cu solder alloy can refine the microstructures and IMCs as long as the Sn-0.7Cu/geopolymer reinforcement were homogenously mixed.

Based on this study, the addition of 1.0 wt.% fly ash geopolymer reinforcement into monolithic solder alloy Sn-0.7Cu can retard the formation of IMCs layer. In case for kaolin geopolymer reinforcement, thickness of IMCs layer become thinner at 0.5 wt.% kaolin geopolymer addition. Between both geopolymer, addition of 0.5 wt.% kaolin geopolymer reinforcement gives good result in IMCs thickness (1.827μm) compare to fly ash geopolymer reinforcment at composition 1.0 wt.% (2.0827μm).

The melting temperature become lower with the adition of geopolymer reinforcement. Addition of 0.5 wt.% fly ash geopolymer gives low melting temperature until 228.15°C while addition of 1.5 wt.% kaolin geopolymer can decrease until 228.83°C. Wettability of solder-substrate improves with the addition of 0.5 wt% fly ash geopolymer reinforcement and 1.5 wt.% kaolin geopolymer reinforcement. It should be the same observation with IMCs but the result were different. It can be concluded that solder substrate experience less wetting time or due to Tombstoning effect during reflow solder. Therefore, when the IMCs of the solder increases, the wettability also decreases. Overall, it can be concluded that 0.5 wt.% kaolin geopolymer reinforcement gives better properties in terms of interfacial intermetallic compound, wettability, and melting temperature compare to addition of fly ash geopolymer which shows inbalanced result.

References

[1] Mohd Salleh M.A.A et al 2015 *Materials & Design* **82** 136.
[2] Yoon J.W. et al 2004 *Journal of Alloys and Compounds* **381** 157.
[3] Mohd Salleh M.A.A et al 2015 *Scripta Materialia* **100** 20.
[4] Chellvarajoo S. et al 2015 *Materials and Design* **82** 215.
[5] Mehta A. et al 2016 *Construction and Building Materials* **127** 198.
[6] Bakri M. M. A. et al 2012 *Rev. Adv. Mater. Sci* **30** 97.
[7] Joshi S.V et al 2012 *International Journal of Environmental Science and Development* **3** 421.
[8] Abdullah M. M. 2012 *Rev. Adv. Mater. Sci* **30** 97.
[9] Yoon J.W. 2004. *Journal of Alloys and Compounds* **381** 157.
[10] Kotadia H. R. et al 2014 *Microelectron Reliability* ***54*** 1273.
[11] Liyana J et al 2014 *Engineering Materials* ***594-595*** 150.
[12] Huseien G. F. et al 2016. *Technology and Social Sciences* ***26*** 125
[13] Zang L. et al 2012 *Materials Letters* ***63*** 2069.
[14] Heah C. Y. et al 2012 *Construction and Building Materials* **35** 922.
[15] Joshi S.V. et al 2012 *International Journal of Environmental Science and Development* **3** 421.
[16] Mehta A. et al 2016 *Construction and Building Materials* **127** 198.
[17] Yunsheng Z. et al 2010 *Applied Clay Science* **47** 275.
[18] Harcuba P. et al 2010 *Journal of Electronic Materials* **39** 2553.
[19] Ng W. et al 2015. *Electronic Packaging and iMAPS All Asia Conference* **1** 809.
[20] Chellvarajoo S. et al (2015). *Materials and Design* **67** 208.
[21] Chellvarajoo S., 2015*Materials and Design* **82** 206.
[22] Mohd Salleh M.A.A. et al 2016 *Materials & Design* **108** 428.
[23] Ramli M.I.I. et al 2016 *Microelectronic Reliability* **65** 264.
[24] Salleh M. A. A. M. et al 2011 *Physics Procedia* **22** 304.
[25] Yuan Y. et al 2013 *Science Techniques* **1** 25.

[26] Leong L. C. et al 2011 *Journal of Material Science: Material Electron* ***22*** 1449.

IOP Conf. Series: Materials Science and Engineering **551** (2019) 012092 doi:10.1088/1757-899X/551/1/012092

Performance of Geopolymer Concrete when Exposed to Marine Environment

Fakhryna Hannanee Ahmad Zaidi[1], Romisuhani Ahmad[1], Mohd Mustafa Al Bakri Abdullah[2], Muhammad Faheem Mohd Tahir[2], Zarina Yahya[1], Wan Mastura Wan Ibrahim[1], Ahmad Syauqi Sauffi [1]

[1]Faculty of Engineering Technology, Uniciti Alam Campus, Universiti Malaysia Perlis (UniMAP), Perlis, Malaysia.
[2]Center of Excellence Geopolymer and Green Technology (CEGeoGTech), School of Materials Engineering, Universiti Malaysia Perlis (UniMAP), Perlis, Malaysia.

E-mail: romisuhani@unimap.edu.my

Abstract. Over the decades, concrete is widely known as the most resourceful construction material as it is suitable for many building applications. However, despite its ability to last hundreds of years in many applications, it has been proven that Portland cement concrete poses problems such as carbon dioxide emission and its durability when exposed to sea water, sulphuric soils or freezing weather. As a result, an alternative binder is in need to reduce these problem which nowadays, the uses of geopolymer concrete is promoted. Therefore, this research is intended to investigate the effect of marine environment towards geopolymer concrete and is compared to OPC concrete. By using the same curing conditions, both type of concrete is immersed in artificial seawater for total of 28 days before the properties such as water absorption, density and compressive strength is tested. As a result, despite the similar density, the GPC is found to have lower water absorption and higher compressive strength when compared to OPC concrete. As a conclusion, GPC exhibits a higher resistance to seawater thus, suitable for construction in marine environment when compared to OPC.

1. Introduction

Portland Cement is after water, the most utilized material in the world. It is largely used as the main material in producing concrete. The classic term of concrete refers to the mixture of Portland cement with the related materials such as water, aggregates and additives which when mixed together, are capable of producing concrete with desirable properties. Over the decades, concrete is widely known as the most resourceful construction material as it is suitable for many building applications. Previous studies has stated, from massive dams to the high skyscrapers all over the world, concrete was used as the main construction material [1].

However, despite its ability to last hundreds of years in many applications, it has been proven that Portland cement concrete poses problems such as carbon dioxide emission [2] and its durability when exposed to sea water, sulfuric soils or freezing weather. Major concerns is raised about its harmful impacts to the environment due to its significant contribution to the amount of greenhouse gas, resulting from the high volume of carbon dioxide emitted during its production [3].

As a result, an alternative binder are in need to reduce these problem which nowadays, the uses of geopolymer concrete is promoted. Geopolymer concrete is a new sustainable and environmental-friendly composite with great potential to replace conventional concrete mostly produced by ordinary Portland cement (OPC). The binder material used for geopolymer concrete such as fly ash and blast furnace are mostly the industrial waste or by-products containing high content of silica (Si) and aluminum (Al) which acted as precursor for geopolymerization [4]. Geopolymer technology also reveals the emission of carbon dioxide into the atmosphere is reduced by 80% as the result of aggregates and cement. Geopolymer also shows the ideal mechanical properties of concrete can be obtained such as high compressive strength, good acid resistance, low creep and low shrinkage [5].

2. Experimental Method

2.1. Raw Materials.

The fly ash used is attained from Manjung Power Plant, Lumut, Perak which is of low calcium, Class F is used as the base material of geopolymer and is equivalent to ASTM C618 while for Portland cement concrete, Blue Lion Cement is used. The chemical composition of fly ash and ordinary Portland cement obtained from the analysis as tabulated in Table 1.

Table 1: XRF analysis data of fly ash and ordinary portland cement

Component		SiO_2	Al_2O_3	Fe_2O_3	TiO_2	CaO	MgO	Na_2O	K_2O	P_2O_5	SO_3	MnO
Percentage (%)	FA	52.11	23.59	7.39	0.88	2.61	0.78	0.42	0.80	1.31	0.49	0.03
	OPC	21.5	3.8	3.8	-	64.3	0.8	0.1	1.1	-	2.6	-

Based on this result, it proves that the fly ash used is type F fly ash where the sum content of SiO_2, Al_2O_3 and Fe_2O_3 is greater than 70%. Besides that, the difference in composition of fly ash and ordinary Portland cement is compared. It is found that the most abundant component found in fly ash is SiO_2 where the percentage is 52.11% while for ordinary Portland cement is CaO with 64.3%. The SiO_2 and CaO component plays a similar role where it helps in providing strength towards the concrete. However, according to past research, fly ash is particularly useful in marine structures since fly ash concrete since fly ash satisfies the conflicts (requiring both high C_3A and low C_3A) to resist chlorides and sulfates [6].

2.2. Mix Proportions and Mixing Process.

Based on past research [7-8], the most suitable mix proportion for geopolymer concrete is chosen where molarity of sodium hydroxide solution is 12M, ratio of sodium silicate to sodium hyroxide is 2.5 and the ratio of alkaline activator to fly ash is 0.5. While for ordinary Portland cement concrete, the cement to water ratio used is 0.5 [9]. The mix proportion of both geopolymer concrete and Portland cement concrete are tabulated in Table 2 below.

Table 2: Mix proportions of geopolymer concrete and Portland cement concrete.

Concrete	Ingredient	Weight (g)
Geopolymer concrete	12M sodium hydroxide	85.0
	Sodium silicate	215.0
	Fly ash	600.0
Portland cement concrete	Ordinary Portland cement	600.0
	Water	300.0

2.3 Moulding, Curing and Testing.
The fresh geopolymer are poured into (50x50x50) mm size cubes. The samples are left for 24 hours in room temperature before it is demoulded. Similar steps are also done to the fresh Portland cement. After that, the samples are immersed in artificial seawater (ASTM 1141-98) for in total of 28 days. For each 7, 14 and 28 days of immersion, weight of the samples were taken to determine the density, water absorption and tested in compressive strength testing machine.

3. Results and Discussions

It is observed that after 7 day of exposing the concrete to artificial seawater, there is precipitation inside the container containing Portland cement concrete sample as shown in Figure 1. However, there is no precipitation can be found inside the container containing geopolymer concrete sample. This may occur because of when the chloride diffuses into the Portland cement concrete, the hydroxide ion leaches out to maintain the electro neutrality of paste, which results in the portlandite solubility and can also cause decrease in the concrete strength [10].

Figure 1: Sample observation 7 days after exposure to artificial seawater

Samples are the observed again after taken out from the artificial seawater solution where the Portland cement concrete samples was found to exhibits some defects that only take place after the exposure. As shown in Figure 2, the Portland cement concrete shows presence of porosity and cracks at the surface of the sample while the exposure of artificial seawater has no significant effect occurred towards the geopolymer concrete sample.

Figure 2: Surface of Portland cement concrete (left) and geopolymer concrete (right) after artificial seawater exposure

3.1 Water Absorption
The result of water absorption of geopolymer concrete and Portland cement concrete after 28 days of immersion are presented in Figure 3. Based on the result, it is identified that the water absorption of geopolymer concrete is lower than of Portland cement concrete where the percentage obtained is 4.58% and 11.33% respectively.

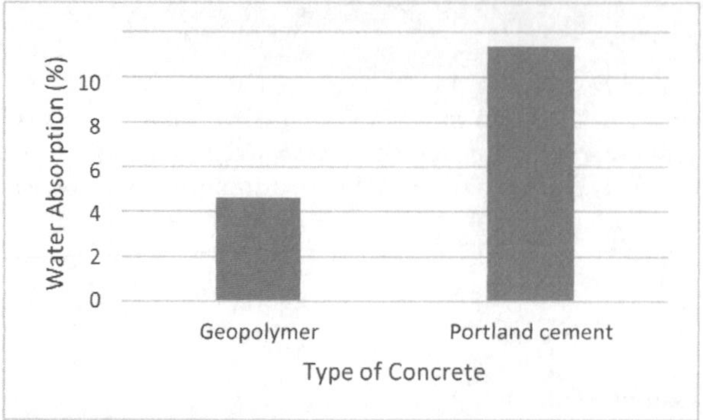

Figure 3: Water absorption result for both type of concrete

Water absorption with value ranged 3%-5% is classified as 'average' concrete. Low water absorption is a good indicator that there is limited open porosity that can inhibit high flow of water into the concrete [11]. However, the concrete will be classified as 'bad' if the value of water absorption is higher than 6% . Since Portland cement is a water-based concrete, it is expected that it will absorb more water if compared to geopolymer concrete [12].

3.2 Compressive Strength

The result on compressive strength of the geopolymer concrete and Portland cement concrete cube exposed to artificial seawater indicates that geopolymer concrete have better durability compared to Portland cement concrete as shown in schematics of Figure 5 where the compressive strength of geopolymer concrete at day 28 is 56MPa while Portland cement concrete is 46.86MPa.

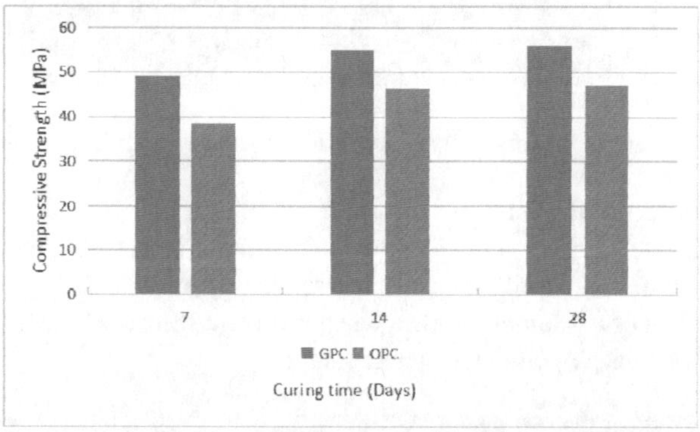

Figure 4: Compressive strength results

Since the reaction of pozzolanic takes time, it is able to be related to this result where as the curing time increases, the compressive strength also increases. The extended curing time intensify the geopolymerization mechanism thus increases the strength [13]. However, it is also observed that the difference of compressive strength between 7 day and 14 day of curing is higher when compared to the difference of compressive strength between 14 day and 28 day of curing. This shows that the concrete are steadily reaching its optimum compressive strength.

In addition, it is observed that there are no significant effect of the artificial seawater towards the compressive strength of the concrete. Sea water usually plays an important role in corrosion of steel. Past research had also noted that the increase time of curing in seawater had a massive impact on the deterioration of the reinforced steel inside the steel which will reduce the compressive strength of the concrete [14]. Since, there are no reinforcing steel used in this research, the effect of curing in seawater on compressive strength of the concrete are unable to be investigated properly.

4. Conclusion

The primary focus of this research has been to experimentally evaluate the durability of fly ash-based geopolymer concrete in marine environment and compared to ordinary Portland cement concrete. At early stage, research is done to identify the best formulation to produce fly ash-

IOP Conf. Series: Materials Science and Engineering **551** (2019) 012092 doi:10.1088/1757-899X/551/1/012092

based geopolymer concrete and ordinary Portland cement concrete. Using the knowledge obtained, product is produced starting from raw material into a sample product for testing according to standard.

The key outcome expected is the better durability performance of geopolymer concrete. Therefore, by analyzing and comparing the behavior and properties of both type of concrete, it was observed that the fly ash-based geopolymer concrete is more homogeneous and well bonded than ordinary Portland cement concrete. Consequently, lower water absorption and higher compressive strength are obtained with geopolymer concrete.

Reference

[1] Muttashar et al 2014 *Conference on the Mechanics of Structures and Materials* ***1*** 101

[2] Al Muhit, B.A. et al 2013 *Electronic Journal of Structural Engineering* ***13*** 15.

[3] Edouard, J. 2011 *Conference on the Mechanics of Structures and Materials* ***1*** 65

[4] Lloyd, N.A. 2010 International *Conference on Sustainable Construction Materials Technologies* ***1*** 13.

[5] Wallah S.E. 2006 *Research Report GC2* 133-138.

[6] Ken W.D. 2006 *3rd edition, Taylor & Francis, New York, NY, USA.*

[7] Mustafa Al Bakri A.M. et al 2011 *Journal of Engineering and Technology Research* ***3*** 49.

[8] Mustafa Al Bakri A.M. et al *2011 International Conference on Chemistry and Chemical Process* ***10*** 17.

[9] Singh, S.B. 2015 *Journal of Building Engineering* ***4*** 100.

[10] Olivia, M. et al 2013 *Conference on Structural Engineering and Construction* ***1*** *13.*

[11] Srinivas, T. et al 2014 *International Journal of Research in Chemical, Metallurgical and Civil Engg.* **4** 55.

[12] Huseien, G. F. et al 2016 *Journal of Science and Technology* ***9*** 48.

[13] Muhd F.N et al 2018 *Internatioanl Journal of Advance and Applied Science* ***5*** 35.

[14] Pradhan B. 2014 *Construction and Buildings Materials* ***72*** 410.

IOP Conf. Series: Materials Science and Engineering **551** (2019) 012093 doi:10.1088/1757-899X/551/1/012093

Investigation of Heat Released during Geopolymerization with Fly Ash based Geopolymer

Rosnita Mohamed[1], Rafiza Abd Razak[1,2], Mohd Mustafa Al Bakri Abdullah[2], Raa Khimi Shuib[3], Nurul Aida Mohd Mortar[2], Warid Wazien Ahmad Zailani[2]

[1]Faculty of Engineering Technology, Universiti Malaysia Perlis (UniMAP), Perlis, Malaysia.
[2]Center of Excellence Geopolymer and Green Technology (CEGeoGTech), School of Materials Engineering, Universiti Malaysia Perlis (UniMAP), Perlis, Malaysia.
[3]School of Materials & Minerals Resources Engineering, Universiti Sains Malaysia, Pulau Pinang, Malaysia

E-mail: rafizarazak@unimap.edu.my

Abstract. Geopolymerization consists of several chemical reactions that has been reported widely as an exothermic reaction. Previous studies on heat released during geopolymerization used metakaolin and fly ash of F class as a precursor for geopolymer. Meanwhile, in this study, fly ash of C class is used and a preliminary study has been conducted on determining the exothermic reaction of the geopolymerization with various ratios of solid-to-liquid ranging from 1.5 to 2.5. The amount of heat released was determined by using Differential Scanning Calorimeter. It was proven that different solid-to-liquid ratio affected the amount of heat released during geopolymerization as the highest amount of the heat released were recorded at the optimum solid-to-liquid ratio of 2.0.

1. Introduction

Production of OPC consumes energy during and emits high amount of carbon dioxide during its production. Alternatives researches have been reported widely on evaluating suitable methods and materials as a future replacement for OPC. Geopolymer was introduced by Davidovits around 1970s to describe the product that formed when any aluminosilicate materials react with alkaline solution. Geopolymer have been studied extensively particularly in terms of mechanical properties, microstructure and thermal properties.

Geopolymer have been accepted widely by researchers as one of alternative for Portland Cement, OPC. Any material with high aluminosilicate content has a great potential to be used as raw materials for geopolymers in which including waste or by-products such as slag, fly ash and rice husk [1]. The variability of chemical composition of fly ash, blast furnace slag or volcanic ash usually leads to geopolymers whose physical and mechanical properties vary from one aluminosilicate raw material to another [2].Geopolymer has empirical formula of $Mn\{-(SiO^2)z-AlO^2\}\cdot wH2O$ where M is a cation (K^+, Na^+, or Ca^{2+}), n is a degree of polycondensation, and z is 1, 2, and 3[3].

Geopolymerization is a process in which involves multiple reaction to produce a geopolymer. Generally, geopolymerization involve three steps which are dissolution of Si and Al atoms from materials, transportation or orientation of precursor's ions into monomers and polycondensation of monomers into polymeric structures. Geopolymerization had been reported widely as an exothermic reaction. Heat is evolved during the reaction and this heat released have been evaluated by previous studies. Zhang et al., reported that metakaolin geopolymerization consists of three exothermic reactions and the amount of heat released was affected by curing temperature [4].

In OPC, too much heat evolved will lead to temperature rise thus causing thermal cracking. As geopolymer have been studied widely as an alternative for OPC, therefore this study will be crucial in determining the amount of heat released during geopolymerization. Besides, the heat release profile during the setting and hardening reaction is also used as a highly sensitive measure of the kinetics of reaction of cement, including the balance between the reaction rates of different components present in the cement in which making it crucial to be evaluated for geopolymers too.

One of common method used for evaluating heat released during geopolymerization is calorimetric study. Isothermal conduction calorimeter (ICC) and Differential Scanning Calorimeter (DSC) are commonly used to monitor the peak formation and enthalpy of geopolymerization process isothermally. This study will focus on using Differential Scanning Calorimeter, DSC to monitor the reaction that occurs in class C fly ash geopolymerization. Various solid-to-liquid ratio will be applied in this study to identify its effect to the amount of heat released during the reaction. Class C fly ash was chosen as precursor since determination of heat released leading to kinetic study of class F fly ash have been reported briefly by [5].

2. Methodology

2.1. Materials

Class C fly ash was used in this study as a raw material for geopolymer and was obtained from Manjung Coal-Fired Power Station, Lumut, Perak Malaysia. The chemical composition of class C fly ash was determined by using X-Ray Fluorescence, XRF and listed in table 1. This fly ash should be achieved the standard specification of American Society for Testing and Materials (ASTM) C 618 in which stated that the sum of silicon dioxide (SiO_2), aluminum oxide (Al_2O_3) and iron oxide (Fe_2O_3) is equal to or greater than 50% by mass in order to be classified as class C fly ash. Morphology of the fly ash was also observed by using Scanning electron microscopy (SEM), SEM JSM-6460 LA Jeol Japan in School of Material Engineering, Universiti Malaysia Perlis (UniMAP) and the result is presented as in figure 1.

Table 1. Chemical Composition of Raw Class C Fly Ash

Chemical Compound	**Quantity (%)**
Silicon dioxide (SiO_2)	30.8
Calcium oxide (CaO)	22.3
Iron oxide (Fe_2O_3)	22.99
Aluminium oxide (Al_2O_3)	13.1
MgO	4.0
TiO_2	0.89
SO_3	2.67

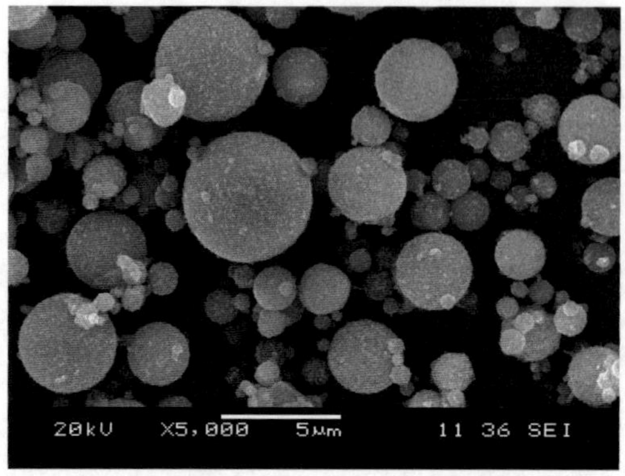

Figure 1. Morphology of Raw Class C Fly Ash

Alkaline activator used in this study was a mixture of sodium hydroxide, NaOH and sodium silicate, Na_2SiO_3. NaOH was required to be dissolved with distilled water to obtain desired molarity as it is in pellet form with 97% purity and must be prepared at least 24 hours' prior usage. Sodium silicate used was in form of solution with properties of Grade A53 with SiO_2 = 29.43 %, Na_2O = 14.26 % and water = 56.31 % were used.

2.2. Mix Design And Mixing Procedure

Solid-to-liquid ratio was the main parameter to be varied in this study meanwhile optimum molarity of NaOH and ratio of Na2SiO3/NaOH used were followed from previous work by Mustafa et al[6]., 12M and 2.5 respectively. The mix design was listed as in table 2 and was used for heat released testing and setting time.

Table 2. Mixing Design Procedure for Setting Time and Heat Released Testing

Mix Design	Solid-to-liquid ratio, (S/L ratio)	Raw material	Ratio of alkaline activator, Na_2SiO_3 : NaOH	Molarity of NaOH,M
1	1.5			
2	2.0	Fly Ash	2.5	12M
3	2.5			

2.3. Testing and Procedure Test

2.3.1. Setting Time Testing

By following the mix design, a freshly prepared geopolymer paste was tested for its setting time using Vicat apparatus as followed from American Society for Testing and Materials (ASTM) C191. This testing was conducted in room temperature to avoid any humidity disturbance to the paste.

2.3.2. Heat Released Testing using Differential Scanning Calorimeter, DSC

Differential scanning calorimeter, DSC was used to determine the amount of heat released during geopolymerization of fly ash based geopolymer. DSC used was Perkin Elmer DSC. Only small amount of geopolymer paste (less than 20 milligram) was required and it was directly inserted into a small aluminium crucible with lid in which also referred as sample pan. The weight of empty sample pan and the weight of sample pan with geopolymer paste was weighed to ensure no excessive weight of geopolymer paste contained in the pan as this will cause destruction to the DSC. The geopolymer paste was kept in DSC for duration of setting time period testing isothermally at 30°C as this study focuses on self-curing condition.

IOP Conf. Series: Materials Science and Engineering **551** (2019) 012093 doi:10.1088/1757-899X/551/1/012093

3. Results and Discussion

3.1. Effect of Different Solid-to-Liquid ratio on Setting Time of Geopolymer

Setting time can be defined as the transition of fresh concrete from liquid phase to solid phase[7]. Setting time is also can be indicated as a time required or available for transport, placing and compaction of paste. Class C fly ash based geopolymer has faster setting time compared to class F fly ash based geopolymer due to the calcium content[8]. Setting time also could be affected by different parameters including solid-to-liquid-ratio. As presented in figure 2, higher solid-to-liquid ratio gave faster setting time as increasing the ratio cause faster hardening or also called as gelation process and more rapid bonding to occcur. In this study, the highest ratio of solid-to-liquid which is the fastest setting time. However, fast setting time did not always have a good benefit as too fast setting time may result geopolymer to improper geopolymer bonding and the networking is low thus lead to the low compressive strength[9] and also setting time has to be suits with requirement for the application in construction .

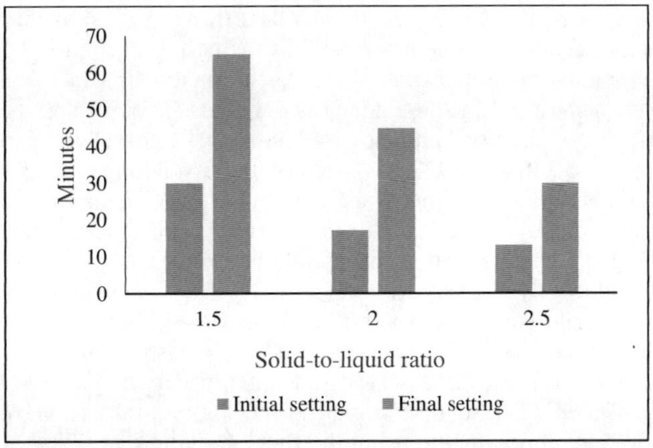

Figure 2. Setting time of Class C Fly Ash based Geopolymer

Table 3. Initial and Final Setting Time of Class C Fly Ash based Geopolymer

Solid-to-liquid ratio	Initial setting, min	Final setting, min
1.5	30	65
2.0	17.07	45
2.5	13.02	30

3.2. Enthalpy of class C Fly Ash based Geopolymer Determined by using Differential Scanning Calorimeter,DSC.

Geopolymerization is a chemical reaction in which heat is dissipated during the reaction. In this study, the heat released during geopolymerization was recorded by using Differential Scanning Calorimeter, DSC. The enthalpy of process that occured was calculated by using equation (1) and (2).

$$\left(\frac{dQ}{dt}\right)p = \left(\frac{dH}{dt}\right) \quad (1)$$

$$\Delta H = \int_{t1}^{t2} \frac{dH}{dt} dt \qquad (2)$$

Where $\frac{dQ}{dt}$ p is the amount of heat evolved and $\frac{dH}{dt}$ is the enthalpy change or can be represent as ΔH. t1 and t2 represent the time limit between the curve.

As in figure 3, there were two distinctive peaks appeared on the heat flow graph of the DSC in this study. The first narrow and sharp exothermic peak appeared immediately less than five minutes represented a dissolution peak in which is similar to the first peak formed when geopolymerization was investigated using isothermal conduction calorimeter,ICC[5][4][10]. This peak also indicates on the absorption of alkaline activator onto the fly ash particles in which leading to dissolution.

According to Table 4, peak 1 was higly exothermic and this increased as the solid-to-liquid ratio increased. The OH- anions attacks became more rapid as increasing in solid-to-liquid ratio due to the more surface available for it to attacks thus causing rapid gel formation in which leading to faster setting time. However, when the highest solid-to-liquid ratio was applied (2.5), the peak became less narrower and a bit broader in which having lowest enthalpy (-43.5067J/g) of peak 1 compared to solid-to-liquid ratio of 1.5 (-50.0302 J/g) and 2.0 (-52.3927 J/g). Besides, the first noting of the peak was endothermic or also known as an endothermic start-up hook and this was a normal behaviour in heat flow graph of DSC due to the differences in heat capacities of the pans (sample pan and reference pan) was not noted as part of the reaction in geopolymerization. Another endothermic peak formed after the first peak could be an induction period. Induction period is period of inactivity that separates the rapid initial dissolution that occurs when fly ash and alkaline activator first come into contact from the polymerization or gelation that leads to set. This period became narrower and shorter as the solid-to-liquid ratio increased.

Second exothermic peak(Peak 2) was observed and it is broader compared to the first peak (Peak 1).The amount of heat released was also lower compared to the first peak as some of the reaction could be occured directly after dissolution thus causing the first exothermic peak had higher heat released. This peak represented the polymerization and gelation process that occured during geopolymerization. Formation of alumina/silica-hydroxyl species and oligomers such as $OSi(OH)^3$–, $Al(OH)^4$ –, $(OH)^3$–Si–O–Al–$(OH)^3$ occured rapidly during this process in which leading to set of the geopolymers[10].

By corelating the setting time and the DSC curve obtained, it could be seen that increased in solid-to-liquid ratio making the setting time became faster but the peak 2 appeared later when using the highest solid-to-liquid ratio (2.5) compared to solid-to-liquid ratio 1,5 and 2.0. Although the second peak was included in setting time of the geopolymer when using solid-to-liquid ratio of 2.5, the late appearance of the peak could be due to the rapid dissolution in which leading to gelation and polymerization occured faster making it less detectable in the peak 2 of DSC curve.

Other than that, since this study used DSC, the third peak observed in previous studies using isothermal conduction calorimeter could not be determined. This could be due to the low sensitivity of DSC or the peak merged together with the second peak as class C fly ash geopolymer hardened faster comparing to the geopolymer that using other materials as precursor. Therefore, the amount of heat released was assumed by total sum of the two peaks formed in the DSC curve.

In this study, solid-to-liquid ratio had been applied in order to observed the heat released during geopolymerization. Increased in solid-to-liquid ratio thus not only causing faster setting time but also higher in amount of heat released. However, same as mechanical properties researches reported by previous studies, an optimum ratio of solid-to-liquid was selected. This study obtained an optimum ratio of solid-to-liquid ratio at 2.0 since it has the highest amount of heat released. At 2.5, the amount of heat released became decreased. This could be occured due to the heat released quicker during dissolution comparing to during polymerization. Solid-to-liquid ratio of 2.0 also was proven in this study to have an optimum setting time when comparing to ratio of 1.5 and 2.5.

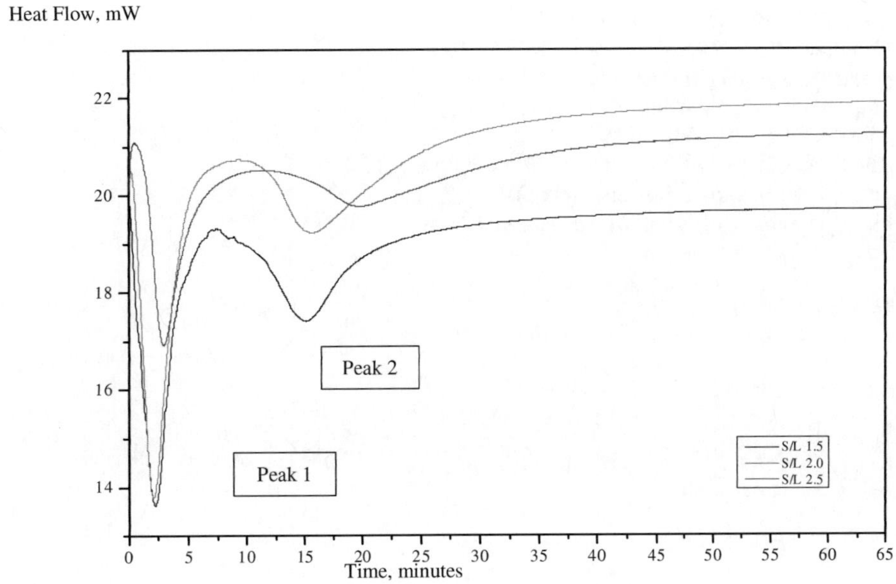

Figure 3. DSC Curve of Class C Fly Ash based Geopolymer with Various Solid-to-Liquid Ratio (S/L ratio)

Table 4. Enthalpy of Fly Ash Geopolymer with Different Solid-to-Liquid ratio

Solid-to-Liquid Ratio	Enthalpy of Peak 1(ΔH),J/g	Enthalpy of Peak 2(ΔH),J/g
1.5	-50.0302	-31.0944
2.0	-52.3927	-43.8944
2.5	-43.5067	-26.5124

4. Conclusions

In this study, class C fly ash was used as a starting materials for geopolymer. The raw fly ash was characterized and the chemical composition determined by XRF is equivalent to the standard specification of ASTM C 618. Sample was prepared with different solid-to-liquid ratio at room temperature as to evaluate the self-curing of the geopolymerization. Based on the setting time and enthalpy of the geopolymerization, it could be concluded that the optimum ratio of solid-to-liquid ratio applied in this study was 2.0 in which has the maximum amount of heat released, -96.2871 J/g. When ratio of 2.5 was applied, the enthalpy became decreased which was 70.0191 J/g. The influence of solid to liquid ratio were also evaluated based on the time taken for the formation of the exothermic peaks observed as well as the setting time of the geopolymerization higher ratio used lead to faster formation of exothermic peaks (exceptional when solid-to-liquid ratio of 2.5 was applied).

Acknowledgements

The authors acknowledge Centre of Excellence Geopolymer and Green Technology(CeGeoGTech), School of Material Engineering, Universiti Malaysia Perlis and School of Materials and Mineral Sources Engineering of Universiti Sains Malaysia for lab facilities. Special thanks to those who contributed to this project directly or indirectly.

IOP Conf. Series: Materials Science and Engineering **551** (2019) 012093 doi:10.1088/1757-899X/551/1/012093

References

[1] Bakri A M M *et al* 2013 *Revista de Chimie* **64** 382
[2] Izzat A M *et al* 2013 *Material Plastice* **50** 171
[3] Al Bakri Abdullah *et al* 2015 *Revista de Chimie* **66** 1001
[4] Zhang Z *et al* 2012 *Thermochim Acta* **539** 23
[5] Nath S K *et al* 2017 *J. Therm. Anal. Calorim.* **127** 1953
[6] Abdullah M M A. *et al* 2011 **330** 1475
[7] Piyasena R R C *et al* 2013 *Natl. Eng. Conf. 19th Eru Symp.* **19** 47
[8] K. Ramujee 2016 *Int. J. Earth. Sci. Eng.* **09** 360
[9] Abdul Rahim R H *et al* 2014 *Appl. Mech. Mater.* **625** 46
[10] Yao X. *et al* 2009 *Thermochim Acta* **493** 49

IOP Conf. Series: Materials Science and Engineering **551** (2019) 012094 doi:10.1088/1757-899X/551/1/012094

Solderability of Sn-0.7Cu-0.05Ni-xZn Solder Ball on Sn-0.7Cu and Sn-0.7Cu-0.05Ni Solder Coating

M. I. I. Ramli[1,*] and M. A. A. Mohd Salleh[1]

[1]Center of Excellence Geopolymer & Green Technology (CeGeoGTech), School of Materials Engineering, Universiti Malaysia Perlis (UniMAP), Taman Muhibbah, 02600 Jejawi, Perlis, Malaysia

E-mail: mohdizrulizwan@gmail.com

Abstract. In this study, the solderability of the Sn-0.7Cu-0.05Ni-xZn solder ball on solder coating were investigated by using Gen3 wetting balance test machine. By using wetting balance test in globule mode, the solderability of solder ball was determined by the maximum force and wetting time. The result of a good solderability is generally referred to short wetting time and high wetting force. It was found that by increasing amount of Zn in Sn-0.7Cu-0.05Ni solder ball, the maximum force was decrease and wetting time was increase. This result indicates that by increasing amount of Zn in Sn-0.7Cu-0.05Ni solder ball, it will decrease the solderability of solder ball to the solder coating.

1. Introduction

Surface mount technology (SMT) is an area of electronic assembly used to mount electronic components to the surface of printed circuit board (PCB). This surface mount technology is important with the growth of printed circuit board in electronic industry. This advantage of SMT is it has lower

IOP Conf. Series: Materials Science and Engineering **551** (2019) 012094 doi:10.1088/1757-899X/551/1/012094

cost and can placed closer to the board. In surface mount devices, the surface finish has an important role. This surface finish has several functions, to protect surface impurities such oil and dust, a barrier to prevent oxidation and to be solderable. There are several surface finishes that currently used in PCB manufacture such as immersion tin (ImSn), immersion silver (ImAg), Electroless Nickel immersion Gold (ENIG) and Hot Air Solder Levelling (HASL). However, HASL surface finish has been a popular surface finish for PCB used for circuitry with large components used in heavy-duty applications in harsh environments. In HASL surface finish, the copper was coated with solder as a coating. The traditional Sn-Pb solder has been widely uses in the solder coating. However, due to health issue and negative effect to the environment, the use of Sn-Pb solder alloy is banned in electronic packaging industry [1-3]. The studies on new lead-free solder have been studied to replace the Sn-Pb solder alloys as solder coating [2, 4]. Among the lead-free solder, the Sn-0.7Cu and Sn-0.7Cu-0.05Ni solder is the most commonly used as a solder coating for hot air solder leveling (HASL).

Solderability of solder can be defined as the ability of the molten solder to spread over on a substrate during the reflow process. In the HASL process, a PCB is fluxed and dipped into a molten solder bath so that entire copper surfaces are wetted. The result of a good solderability is generally refer to short wetting time and high wetting force. Recently, Zn addition has found to be able reduce the nucleation undercooling of β-Sn in Sn based Pb free solder alloy [5]. The formation of Cu-Zn intermetallics can be collerated with an enhancement of mechanical properties. However, the effect of Zn in Sn-0.7Cu-0.05Ni to solderability is rarely studied.

In this study, we investigate the IMC thickness of Sn-0.7Cu and Sn-0.7Cu-0.05Ni solder coating by aging method to mimic the storage condition of surface finish. The influence of Zn in Sn-0.7Cu-0.05Ni solder ball for solderability properties of Sn-0.7Cu and Sn-0.7Cu-0.05Ni solder coating were also investigated.

2. Experimental

In this study, the Sn-0.7Cu and Sn-0.7Cu-0.05Ni solder alloy was supplied by Nihon Superior Co., Japan as a base solder for dipping method. The dipping process was simulated by coating copper strips with solder by using wetting balance machine. The copper strips with a dimension of 10mm x 3mm x 0.3 mm was cleaning by acid to remove the surface oxides. The copper strips were then fluxed and mounted on the clip holder then immersed in the molten solder with withdrawal speeds of 30 mm/s and immersion time of 20 s using the dipping method. The coated samples were then aged at 180°C for 24 hours, 120 hours and 240 hours. The samples were then cross sectioned and observed the IMC of solder coating. The interfacial IMC thickness were then measured by using J-Image software. The solderability testing were performed by using wetting balance machine in globule mode. In this mode, the sample were fixed at an angle in the grip and the solder ball raised until it touches the surface coated copper strip. The result of this testing is evaluated by wetting time and maximum force.

For solder ball fabrication, the base Sn-0.7Cu-0.05Ni were mixed with three different compositions of Zn (0.5, 1.0, and 1.5 wt.%) by using conventional casting process and have been namely Sn-0.7Cu-0.05Ni-0.5Zn, Sn-0.7Cu-0.05Ni-1.0Zn, and Sn-0.7Cu-0.05Ni-1.5Zn. The casting sample were takeout and poured onto the stainless-steel plate and rolled into 30μm foils thickness. The sheet solder was melted in a reflow oven and form a solder ball by the action of surface tension. Figure 1(a) show a schematic diagram of dipping process and Figure 1(b) explain the solderability test in globule mode. In this study, the effect of aging to the interfacial IMC has been investigated by cross sectioned and the solderability performance for the coated sample with influence of Zn in Sn-0.7Cu-0.05Ni solder alloys has been investigated.

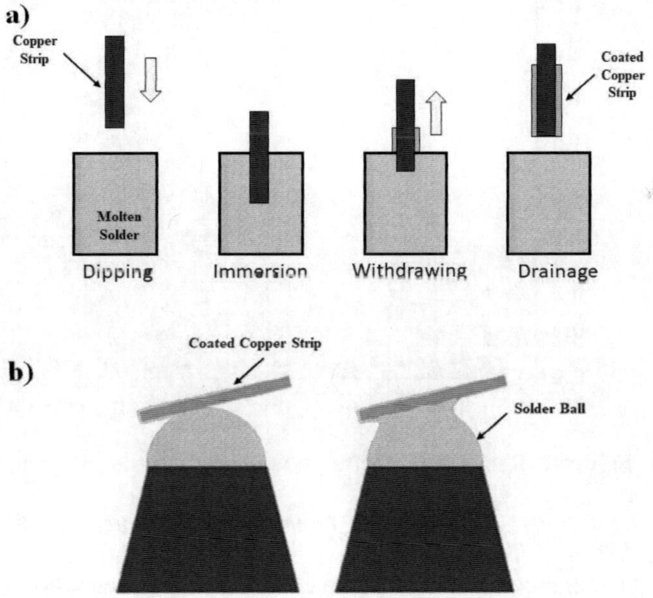

Figure 1.(a) show a schematic diagram of dipping process and (b) explain the solderability test in globule mode.

3. Results and Discussion

3.1. Wettability of Sn-0.7Cu and Sn-0.7Cu-0.05Ni solder coating

Figure 2 show the wettability of Sn-0.7Cu and Sn-0.7Cu-0.05Ni solder coating on copper strip. The wettability performance of copper strip was evaluated by wetting time and maximum force that carried out using dipping wettability balance test. This result show that the Sn-0.7Cu-0.05Ni solder coating show better wettability compared to Sn-0.7Cu solder coating. The highest maximum force and lowest wetting time show a better wettability for coating solder. The above result indicate that the addition of Ni may influence in the surface tension of the solder which eventually change the wetting time and maximum force. Wang et al. also [6] reported that the influence of Ni addition on Sn-0.7Cu have influenced the surface tension between solder and flux interface in the molten state.

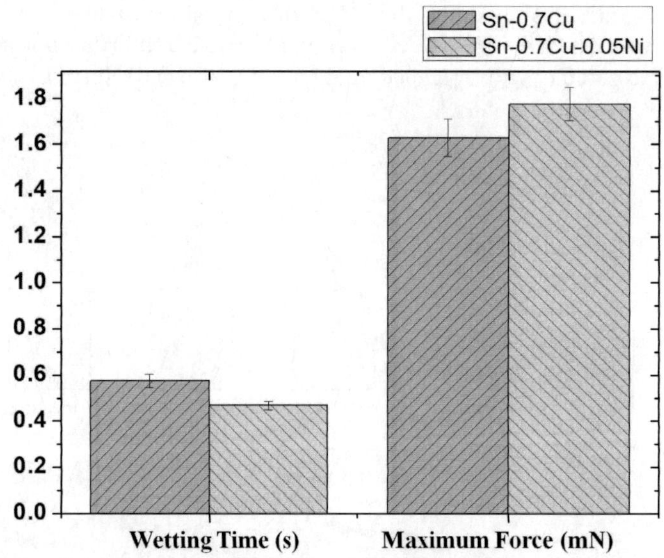

Figure 2. Relationship between wetting time and maximum force.

3.2. Effect of aging sample to the interfacial intermetallic of Sn-0.7Cu and Sn-0.7Cu-0.05Ni solder coating

After the aging process, the sample were cross sectioned to investigate the interfacial IMC of Sn-0.7Cu and Sn-0.7Cu-0.05Ni solder coating as shown in Figure 3. The interfacial IMC of Sn-0.7Cu is increase from 2.45 μm to 3.85 μm, 7.70 μm, and 8.78 μm after 24 hours, 120 hours and 240 hours. For

IOP Conf. Series: Materials Science and Engineering **551** (2019) 012094 doi:10.1088/1757-899X/551/1/012094

Sn-0.7Cu-0.05Ni with 3.54 μm, and increases to 5.02 μm, 10.9 μm and 11.94 μm at 24 hours, 120 hours and 240 hours. This indicated that the IMC thickness will increase with increasing of aging time. The IMC growth are occur by the dissolution of element in the substrate to the solder and interdiffusion kinetics.

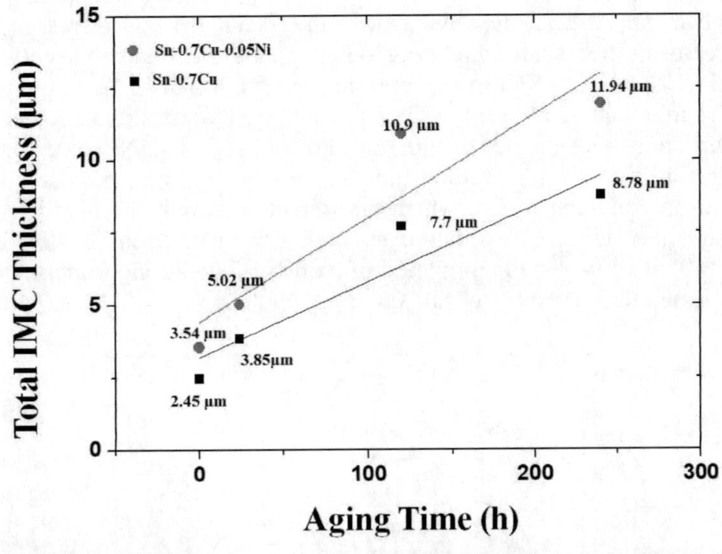

Figure 3. Total IMC thickness of Sn-0.7Cu and Sn-0.7Cu-0.05Ni solder coating as a function of aging time.

3.3. Effect of interfacial IMC to the solderability of Sn-0.7Cu and Sn-0.7Cu-0.05Ni solder coating

Solderability is crucial in soldering process and plays an essential role to determine a good bonding formed between solder materials and substrate. It is also an important issue in the reliability of electronic products. Figure 4 show the effect of Zn addition on Sn-0.7Cu-0.05Ni solder ball to the solderability of solder coating in term of wetting time and maximum force. After the aging process, the interfacial IMC are measured and has been found that with the increasing of aging time, the thickness of the IMC layer were increase. It clearly shown that the wetting time has been increased with increases of Zn addition on Sn-0.7Cu-0.05Ni solder ball. The maximum force of Sn-0.7Cu and Sn-0.7Cu-0.05Ni solder coating also decrease with increases of Zn element in Sn-0.7Cu-0.05Ni solder ball. It can be concluded that when interfacial IMC increases by aging process, it also increase the wetting time and decrease the maximum force. The shorter wetting time and higher maximum force will indicates a good solderability.

With increasing the interfacial IMC, it were found that the solderability of solder coating will decrease. This finding related to the increases of interfacial IMC will decrease the free solder thickness. The increases of IMC also reduce the free solder thickness were suggested by Ramli et al.[7]. They reported that if interfacial IMC is increase, it will decrease the free solder thickness thus decrease the solderability of solder coating. Noh et al. [8] also reported that, by the increasing maximum force and shorter wetting time are resulted from a better wettability of solder alloy. In this study,with the addition of Zn into Sn-0.7Cu-0.05Ni solder ball, the solderability in term of wetting time and maximum become worse. This is due to the addition of Zn, which can directly prevent the wetting on Cu substrate because of the oxidation and high activity of Zn in Sn alloy. Kamal et al. [9] also suggest the formation of oxide residue during soldering which may lessen the wettability of Sn-Ag-Cu solder.

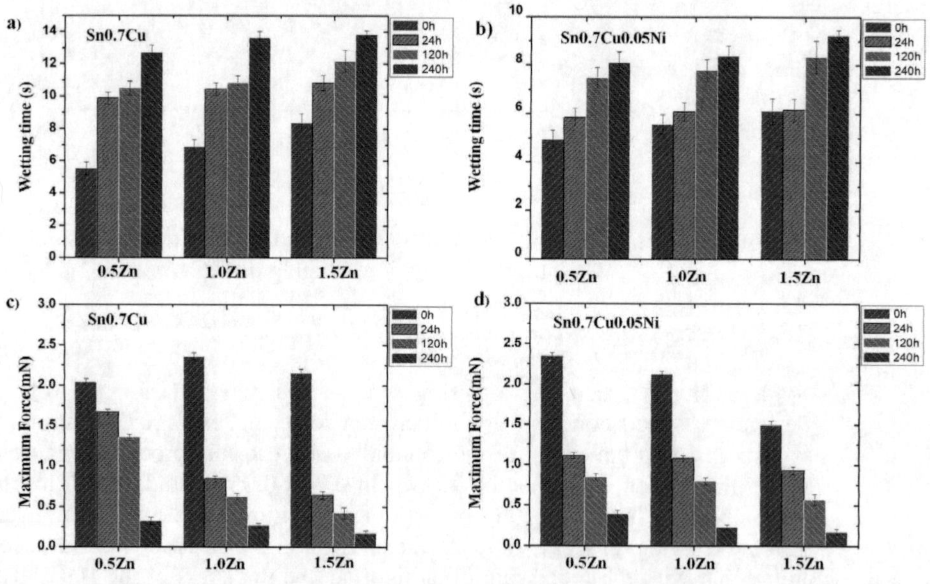

Figure 4. Effect of Sn-0.7Cu-0.05Ni-xZn solder ball on wetting time, a) Sn-0.7Cu coating, b) Sn-0.7Cu-0.05Ni coating and on maximum force, (c) Sn-0.7Cu coating and (d) Sn-0.7Cu-0.05Ni coating.

4. Conclusions

This study investigated the solderability of Sn-0.7Cu-0.05Ni-xZn solder ball to the Sn-0.7Cu and Sn-0.7Cu-0.05Ni solder coating that influenced by the aging process. The conclusion was summarized as follows:

i. The Sn-0.7Cu-0.05Ni solder coating show better wettability compared to Sn-0.7Cu solder coating in dipping test.
ii. The IMC thickness of Sn-0.7Cu-0.05Ni were higher than Sn-0.7Cu solder coating IMC thickness.
iii. The addition of Zn into Sn-0.7Cu-0.05Ni solder ball reduced the solderability of solder coating for both solder coating.

Acknowledgements

The work was financially supported by fundamental research grant scheme (FRGS) (FRGS/1/2017/TK05/UNIMAP/02/7)(9003-00635) from the Ministry of Higher Education, Malaysia. We also acknowledge Universiti Malaysia Perlis (UniMAP) and Nihon Superior Co. Ltd. collaboration research project for the financial support throughout the research project.

References

[1] Nasir N M *et al* 2015 *Mater Sci Forum.* **803**
[2] Ramli M I I *et al* 2016 *Key Eng. Mater* **700**
[3] Ramli M I I *et al* 2018 *Solid State Phenom.* **273**
[4] Salleh M A A M et al 2017 *Sci Rep.* **7**
[5] Zeng G *et al* 2016 *J Alloy Comp* **685**
[6] Wang Y *et al* 2017 *Mater Des* **119**
[7] Ramli M I I *et al* 2019 *J Mater Sci Mater Electron*
[8] Noh B I *et al* 2010 *Alloy Comp.* **499**
[9] Kamal M *et al* 2008 *J Mater Sci Mater Electron* **19**

Morphology Analysis of Spontaneous Tin (Sn) Whisker Growth on Lead-Free Solder

Aimi Noorliyana Hashim, Mohd Arif Anuar Mohd Salleh and Noor Zaimah Mohd Mokhtar

Centre of Excellence Geopolymer and Green Technology, School of Materials Engineering, Universiti Malaysia Perlis (UniMAP), Taman Muhibah 02600, Jejawi, Arau, Perlis, Malaysia

Abstract. Due to the transition of the electronics industry to lead-free solders, the issue of tin (Sn) whisker has become one of the most imperative issues that need to be resolved. Moreover, with the increasing demand for electronics miniaturization, Sn whisker growth is a severe threat to the reliability of microelectronic devices. Sn whisker is a conductive tin structure that grows spontaneously from tin finished surfaces, which can lead to well-documented system failures in electronics industries. Assessment of the Sn whisker growth is difficult because of the slow and unpredictable nature of the Sn whisker formation. This paper presents the results of morphology analysis of Sn whisker growth in terms of Sn whiskers shape, average density and maximum length of Sn whisker on Sn-Cu solder finished using scanning electron microscope (SEM). Micro-indentation test was used to accelerate the growth of Sn whisker at room temperature storage.

1. Introduction

Due to the implementation of the Restriction of Hazardous Substances Directive 2002/95/EC (RoHS) in 2006 prohibiting the use of lead in electronics, the issue of tin (Sn) whisker has become one of the most imperative issues that need to be resolved [1]. Moreover, with the increasing demand for electronics miniaturization, Sn whisker growth is a severe threat

IOP Conf. Series: Materials Science and Engineering **551** (2019) 012095 doi:10.1088/1757-899X/551/1/012095

to the reliability of microelectronic devices [2]. Sn whisker is a conductive tin structure that grows spontaneously from tin finished surfaces, which can lead to well-documented system failures in electronics industries [3-5].

It has been believed that compressive stress in Sn solder coating is a necessary factor for tin whisker growth [6, 7]. Many factors may contribute to the compressive stress level in Sn solder finished, including residual stresses generated during plating process, formation of intermetallic compound (IMC) due to interfacial reaction between Sn solder and copper substrate, mechanical loading, surface damage, and mismatch of coefficient of thermal expansion (CTE) between solder and substrate, or under layer [6].

Whiskers have been observed in a variety of shapes including filaments, nodules and hillocks [8]. Filament whiskers are mostly monocrystalline, needle-like have been observed to grow with lengths up to millimetres but lengths in the 10–100 μm range are observed commonly. Column and spiral whiskers are quite understandable. Nodule whiskers are normally defined by their diameter and are often referred to as mounds or hillocks [5].

Assessment of the Sn whisker growth is difficult because of the slow and unpredictable nature of the Sn whisker formation. It is very hard to estimate the period for the growth of whiskers, which can vary from hours to years. Effects of applied mechanical stress or mechanical deformation on tin whisker growth have been investigated in a few studies. Although a greater extent of mechanical loading or deformation on tin plating could enhance tin whisker growth, little quantitative work has been conducted to correlate the relationship between tin whisker growth and continuously applied mechanical stress [6]. The approach by Doudrick et al. [9] has been successfully used to accelerate whisker growth using microscale indenters. Micro-indentation, is a simple and cost effective method which consist of a ball shaped indenter, has been revealed to induce potential growth of whiskers beyond the directly damaged area and increases the overall stress. As stresses are induced by the IMC and indenter, such as a small ball-bearing, whiskers and hillocks will form [9].

This paper presents the morphology analysis of Sn whisker growth in terms of Sn whiskers shape, average density and maximum length of Sn whisker on Sn-Cu solder finished using scanning electron microscope (SEM) and Image-J Software according to JEDEC standard JESD22-A121. A simple micro-indentation test was used to accelerate continuously the growth of Sn whisker at room temperature storage for 24 h.

2. Methodology

In this experiment, Cu substrates (>99.9% Cu) were used from sheared 1.5 cm x 1.5 cm squares of 1.0 cm in thickness. The Cu substrates were

IOP Conf. Series: Materials Science and Engineering **551** (2019) 012095 doi:10.1088/1757-899X/551/1/012095

immersed into acid liquid bath that contain of 5 g (35%) of hydrochloric acid with 95 g of deionized water (1.75%) for 5 s to remove surface oxides, rinsed with acetone followed by distilled water and dried. Flux solution was applied to the Cu substrates before solder-dip process. The flux used was a standard B type (JIS Z3198- 4), which is mixtures of rosin, 2-propanol and diethylamine hydrochloride to remove the oxidation from the solder as well as from the surface of the substrates. The Cu substrate was then mounted on the clip holder of a solder-dip machine followed by immersed in the Sn-0.7Cu molten solder bath with withdrawal speeds of 10 mm/s and an immersion time of 5 s using the dipping method. Table I shows the temperature profile parameters of the Sn-Cu solder-dip process. The solder bath is held at a peak temperature of 250 °C. After the solder-dip, the samples were cooled down before the water rinse in ultrasonic bath.

Table-1. Sn-Cu solder-dip temperature profile

Solder-Dip Profile Data	Value
Preheat Temperature (°C)	200
Preheat Time (s)	60
Preheat Rate (°C/s)	2.5
Peak Temperature (°C)	250
Cooling Rate (°C/s)	3.5

Micro-indentation test consisted of a 200 g weight was used to accelerate the growth of Sn whisker at room temperature storage. After the samples were indented for 24 hours, morphology analysis of Sn whisker growth in terms of Sn whiskers shape, average density and maximum length of Sn whisker was done using scanning electron microscope (SEM) and image-J Software according to JEDEC standard JESD22-A121.

3. Results and Discussion

Sn whiskers growth on Sn-0.7Cu solder samples were observed after 24 h continuously accelerated with indentation compressive stress. The properties of Sn whiskers were shown in Table 2. The average of Sn whisker length was about 430 μm and the longest Sn whisker was about 920.02 μm. The average density of Sn whiskers was noticed on the surface was about 28 mm^{-2} with the average diameters of Sn whiskers were 4.6 μm.

IOP Conf. Series: Materials Science and Engineering **551** (2019) 012095 doi:10.1088/1757-899X/551/1/012095

Table-2. The properties of Sn whiskers after 24 h indentation test.

Sample (indented 48h)	Average whisker length (µm)	Max whisker length (µm)	Average Whisker Diameter (µm)	Whisker Density (mm^{-2})
Sn-Cu	430.61	920.02	4.6	28

Fig. 1 was showed SEM observations that whiskers formation was greatly enhanced by induced mechanical compressive stress. All whiskers found were filament-like whiskers, hillocks, and nodule-like whiskers. In Fig. 1(b), the long and spiral whisker with length was 294.47 µm apparently grew from a defect on the tin oxide surface. According to Lin and Lin [6], defects, cracks or weak spots, either pre-existent before mechanical testing or induced by mechanical bending, could provide a path for whiskers to initiate and grow for relaxation of the built-up compressive stress. The most well-known as shown in Fig. 1(c) is the filament or needle type whisker: long and straight ones with small diameter. This type

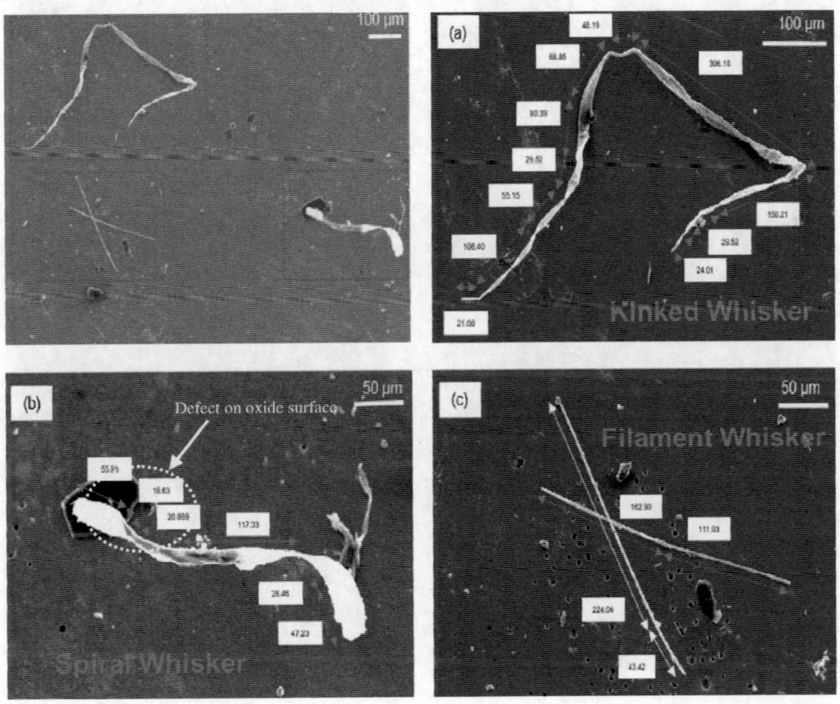

IOP Conf. Series: Materials Science and Engineering **551** (2019) 012095 doi:10.1088/1757-899X/551/1/012095

Fig. 1 . SEM images of the types of Sn whiskers growth on Sn-Cu solder after micro-indentation test at room temperature for 24h. (a) Kinked whisker (b) Spiral whisker (c) Filament whisker

of whisker significantly is the most dangerous to circuit reliability as the conductive tin whiskers can grow long enough to cause electrical shorts circuit by bridging neighbouring conductors or may break off and cause damage elsewhere in the device.

After indentation test, whiskers and hillocks formed around the indentation area. The morphology of Sn whiskers was shown in Fig. 2. Obviously, the hillock and the nodule type of whiskers can be seen a lot in this sample. Even though hillock are short, thick objects protruding not so high from the surface but these types of whisker can be regarded as whisker roots, because if it has the time and conditions, it is likely to grow longer and form other types of Sn whisker [10].

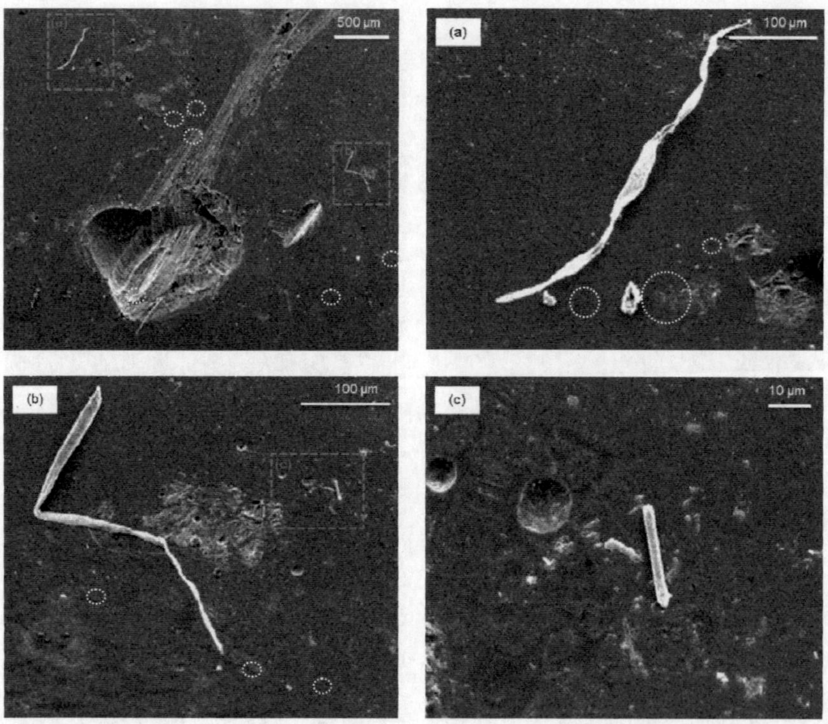

IOP Conf. Series: Materials Science and Engineering **551** (2019) 012095 doi:10.1088/1757-899X/551/1/012095

Fig. 2 . SEM images of the Sn whiskers morphology at indentation area of Sn-Cu solder (yellow line area showed distribution of hillock (a) Spiral whisker (b) Kinked whisker (c) Filament whisker.

4. Conclusions

From the research, the following conclusions have been obtained;

i. After 24 h continuously indented by mechanical compressive stress, The Sn-0.7Cu samples started extreme whiskering, with the average of Sn whisker length was about 430 μm and the longest Sn whisker was about 920.02 μm.

ii. Sn whiskers nucleate and grow critically after accelerated by micro-indentation test on Sn-Cu solder joint. Sn whiskers found were filament-like whiskers, hillocks, spiral, kink and nodule-like whiskers.

iii. Sn whiskers significantly is the most dangerous to circuit reliability as the conductive Sn whiskers can grow to several hundred microns, which are long enough to cause electrical shorts circuit by bridging neighbouring conductors or may break off and cause damage elsewhere in the device.

Acknowledgements

This work was financially supported by Unimap-Nihon Superior Research Colaboration (Grant No: 2016/10/0001) and Tin Solder Technology Researchs Group (TSTRG) Research Grant 2019.

References

[1] Ashworth M A *et al* 2014 *J. Elec. Mater.* **43** 1005.
[2] Sun M *et al* 2017 *Mater. Charac.* **134**, 354.
[3] Jadhav N *et al* 2013 *J. Elec. Mater.* **42** 312.
[4] Skwarek A *et al* 2011 *Mater. Sc. Eng.* **176**, 352.
[5] EChason N *et al* 2013 *Prog. Surf. Sc.* **88** 103.
[6] Lin C K *et al* 2008 *Microelec. Reliab.* **48** 1737.
[7] Mathew S *et al* 2011 *Packag. Manufac. Tech.* **1** 957.
[8] Hashim A N *et al* 2018 *Sol. State. Phen.* **280**, 151.
[9] Doudrick K *et al* 2015 *Microelec. Reliab.* **55** 832.
[10] Horváth B *et al* 2012 *Thin Solid Films* **520** 5733.

IOP Conf. Series: Materials Science and Engineering **551** (2019) 012096 doi:10.1088/1757-899X/551/1/012096

Preliminary Study of Electrical Current Stressing on Tin Whisker Formation

Noor Zaimah Mohd Mokhtar[1,*], Mohd Arif Anuar Mohd Salleh[2], Aimi Noorliyana Hashim[3]

Center of Excellence Geopolymer and Green Technology (CEGeoGTech), School of Materials Engineering, Universiti Malaysia Perlis (UniMAP), Perlis, Malaysia.

E-mail: arifanuar@unimap.edu.my

Abstract. The growth of Sn whiskers on Cu substrate Pb-free solder is considered as one of the emerging problems in electric and electronic devices as well as in aerospace interconnects. Electrical current stressing have been identified as a major driving force for Sn whisker formation. In this paper, we investigate the mechanism of Sn whisker growth through the effect of electromigration of Sn on copper substrate. Test samples were prepared by dipping method before proceed with current testing, and the characterization of whisker growths were evaluated using the Scanning Electron Microscopy (SEM) on the deposits near the anode and cathode areas. Results demonstrate that the accumulation took place at the anode side forming whiskers and depletion occured at the cathode side leading to the formation of void with the effect of current stressing associated with different exposure of time.

1. Introduction

Tin whiskers are tiny electrically conductive filaments, which continuously formed on pure tin or tin-alloy finished surfaces. Numerous failures due to shorting from whiskers have been reported such as medical devices applications, military and aerospace [1][2][5][6]. Tin whiskers are usually obtained by various methods such as electroplating, soldering or hot dip. They can grow up to several millimeters in length with a diameter of a few hundred micrometers which caused a lot of damage, costs and reduced the lifespan of many electronics components. Sn-Pb solders and plating were used extensively for a long time to mitigate and reduce the growth of whiskers. However, due to the restriction of hazardous substances for environmental concerns, the materials containing lead (Pb) has been banned in electric and electronic interconnects [2][7][8].

Generally, compressive stress has been accepted as a primary driving force for tin whisker growth [3]. The mechanism of how the whiskers form is still limited. Nevertheless, there is some literature on other driving force for the whiskers to growth. In previos studies, the whiskers formed on the thickness below than <10μm with the environmental condition by electroplating [3][9]. The straight whiskers also formed on the Sn-Cu solder as illustrated in Figure 1.In this paper, we investigate the current stressing effect to the whiskers growth The effects of current stressing on pure tin was investigated at the increasing exposure time.

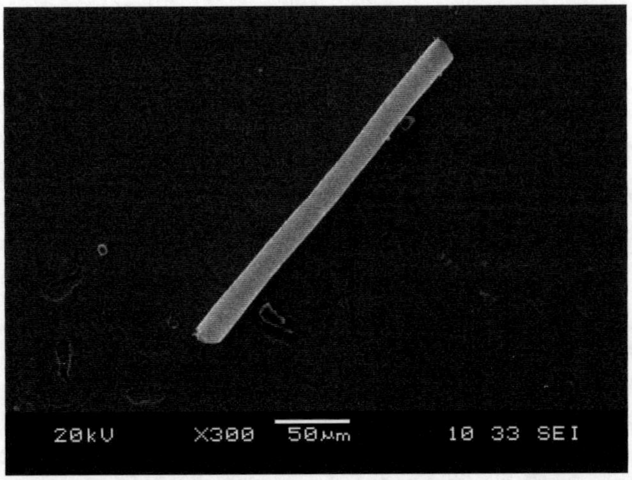

Figure 1. SEM of straight shaped whiskers of Sn-Cu solders.

2. Experimental

2.1. Material Preparation

First, a 200g granular shape of pure tin was prepared using melting method at a temperature of 350°C in a solder pot. A flat square-shaped copper shim with dimensions of 15 mm x 15 mm size and 1 mm thickness was used as a substrate. The substrates were cleaned using acid cleaning liquid that contain of 5g (35%) of hydrochloric acid with 95g of deionized water (1.75%) to remove surface oxides and contaminations.

2.2. Dipping Procedure

The dipping process was carried out using a hot soldering machine. Hot dipping process has been chosen compared to the electroplating technique since it can save cost and reduce time consuming [3][4]. The preparation of the solder also will be easier compared with the electroplating technique. Each coupon was hang using a crocodile clip at the dipping machine as shown in the Figure 2 below. It is equipped with a direct current (DC) motor with 12V power supply and 531 rpm speed. The coupon was first covered with flux before dipped in the solder pot containing molten bath for 3 seconds and blown at a pressure of 300 Mpa to get an even thickness. The thickness for each samples were 20μm each. We prepared three coupon for each solder coating. Then, the samples were cleaned with acetone for 3 min using ultrasonic cleaner.

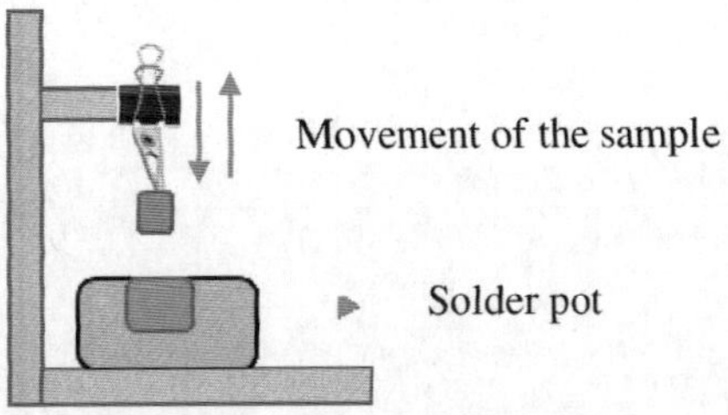

Figure 2. Illustration of dipping test.

IOP Conf. Series: Materials Science and Engineering **551** (2019) 012096 doi:10.1088/1757-899X/551/1/012096

2.3. Electrical Current stressing
In order to accelerate whiskers, electrical current test was conducted at different time period and current stressing as listed in Table 1.

Table 1. The test condition.

Parameter	Condition
Solder	Tin
Substrate	Copper
Dipping time	1s **Plating**
thickness	20 μm **Current**
stressing	0.0215A
Current stressing	0h, 24hrs, 72hrs and 120hrs

A 5V DC power supply was used at 0.0215A to conduct the electromigration test. Both sides of the coupon are connected with crocodile clips at the anode and cathode area. The schematic diagram of the interconnection of coupon for the electromigration test is shown in Figure 3.

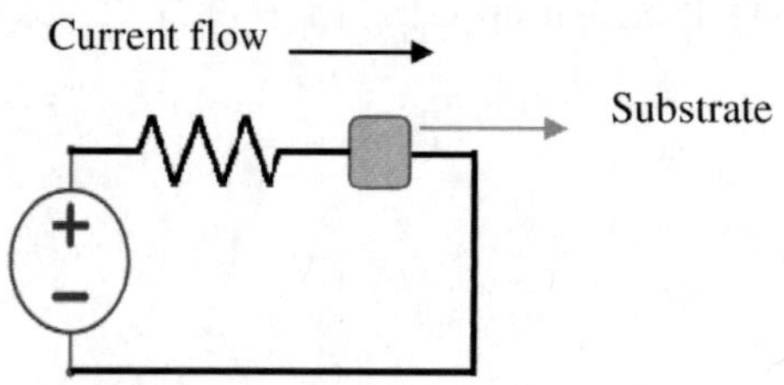

Figure 3. Schematic drawing of EMI test.

During the inspection, the area Of 0.09mm^2 has been screening for minimum 9 areas of coupon with secondary electron microscope (SEM). The maximum length and the whisker density has been recorded and shown in Table 2.

2.4. Whiskers Measuring
During the inspection, the area Of 0.09mm^2 has been screened for minimum 9 areas of coupon with secondary electron microscope (SEM). The maximum length and the whisker density has been recorded and shown in Table 2.

Table 2. Whiskers characterization of pure tin solder after 0.0215A current stressing.

Time (hrs)	Whiskers number [μm]	Whiskers length [μm]	IMCs thickness
24	4	2	2
72	22	8	3
120	48	13	5

IOP Conf. Series: Materials Science and Engineering **551** (2019) 012096 doi:10.1088/1757-899X/551/1/012096

3. Results and Discussions

3.1. Effects of current stressing on the anode and cathode surface after 0.0215A electrical testing.
Figure 4 presents a secondary electron image that shows the surface morphology of tin on copper substrate. By comparing figure 4 a,b and c, it can be seen that whiskers has been formed when it goes to anode side. Moreover, densities of whiskers getting lower from anode to cathode area and there are voids being observed on the cathode area. The morphologies of samples at x2000 magnification shows whiskers in straight and kinked shape grew up to 10micrometer and diameter is about 2micrometer of current exposure. From the cross section in figure 5, it has been shown that, the layer of intermetallic compound (IMCs) is slightly thicker compare with the cathode side [8]. The thicker of IMCs layer shows that the formation of whiskers on anode has effects to the IMCs layer. Although the thicker layer of IMCs on anode, the IMCs layer on the cathode side seems no different after 5 days test.

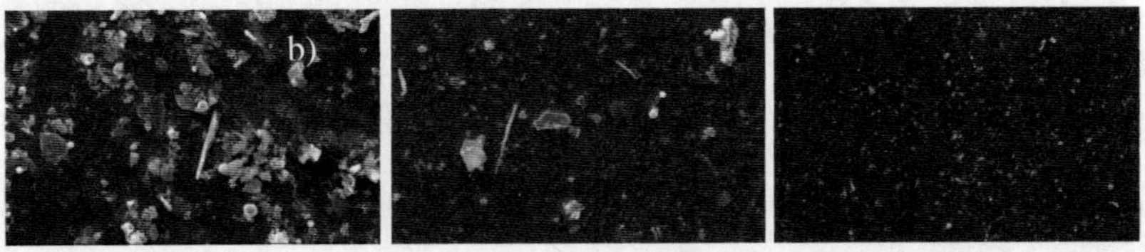

Figure 4. SEM images of 120hrs of current stressing shows the growth of whiskers on different area; a) anode b) between anode and cathode c) cathode.

3.2. Effects of current stressing within increasing time
Figure 5 shows the whiskers formation of pure tin on copper substrate under controlled environment. It can be seen that the whiskers have grew longer when the time of current applied increased. It can be shown also that the densities of whiskers were higher over exposure of time.

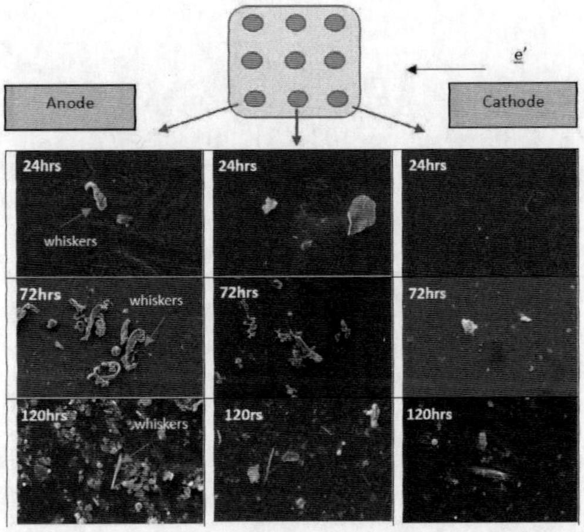

Figure 5. SEM images of current stressing under different exposure of time.

Whisker growth by electrical current stressing comes from the bombardment of electrons moving in the electric field from the cathode to the anode, pushing Sn atoms towards the anode. This phenomeno creates voids on cathode and as a result, compressive stress happens within tin coating, which is relieved by whiskers formation through the layer near the anode region [9][10]. The schematic diagram in Figure 6 shows the whisker formation under the effects of electrical current stressing.

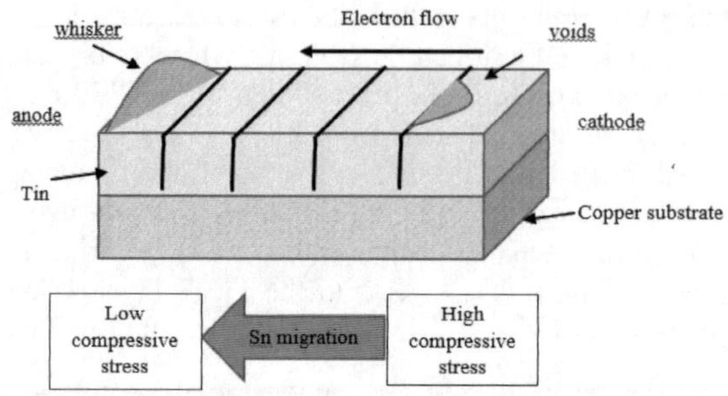

Figure 6. Schematic diagram of tin whiskers growth.

4. Conclusions

The formation of whiskers on electrical current stressing shows the least formation of whiskers on cathode while more whiskers formed on anode region. The whisker densities show higher under controlled environment over exposure of time.

5. Acknowledgment

This work was financially supported by the CREST R&D (P14C1-17/001) grant of University Malaysia Perlis and Nihon Superior Co. Ltd.

References

[1] K. N. M.A.A. Mohd Salleh *et al* 2017*Sci. Rep.*, vol. 7, 2017.
[2] E. Chason *et al* 2013 vol. 88, no. 2. pp. 103–131, 2013.
[3] N. M. Mokhtar *et al* 2018, vol. 280, pp. 175–180.
[4] A. N. Hashim *et al* 2018, vol. 280, pp. 151–156.
[5] K. N. M. A. A. Mohd Salleh, *et al* 2013 *Appl. Mech. Mater.*, vol. 421, pp. 260–266.
[6] K. N. M. A. A. Mohd Salleh *et al* 2015 vol. 82, pp.136–147, 2015.
[7] K. N. M.A.A. Mohd Salleh *et al* 2017 *Sci. Rep.*, vol. 7, 2017.
[8] L. Reinbold *et al* 2009 *J. Mater. Res.*, vol.24, no. 12, pp. 3583–3589, 2009.
[9] C. Su, H. A. O. Chen *et al* 2014 vol. 43, no. 9, pp.3290–3295.
[10] K. N. Tu *et al 2005 Mater. Sci.Eng. A*, vol. 409, no. 1–2, pp. 131–139.

Influence of Expandable Graphite and Aluminium Hydroxide on Mechanical Properties of Polyurethane Foam

Rozyanty A.R1,*, Mofeed D.M.S 1 , Firuz Z 2, Amini M.H.M 3

[1]Center Of Excellence Geopolymer & Green Technology (CEGeoGTech), School of Materials Engineering, University Malaysia Perlis, Malaysia

[2]School of Materials Engineering, Universiti Malaysia Perlis.

[3]Faculty of Earth Science, Universiti Malaysia Kelantan, Jeli Campus, 17600, Jeli, Kelantan, Malaysia

E-mail: rozyanty@unimap.edu.my

Abstract. In this research, the influence of expendable graphite (EG) and aluminium hydroxide (ATH) on mechanical properties of polyurethane composite were investigated. Polyurethane foam was prepared by using oil palm based polyurethane with ratio 1:1.1 and mechanical properties was evaluated by compression test. Results shows that the addition of expendable graphite and aluminium hydroxide into polyurethane had improved the compression strength compared to polyurethane composite without additif. SEM micrograph shows the present of expendable graphite in the polyurethane cell wall which help to improve the mechanical properties. Especially, the compression strength at

IOP Conf. Series: Materials Science and Engineering **551** (2019) 012097 doi:10.1088/1757-899X/551/1/012097

10% EG/2% ATH loading sharply inproved by 66% over polyurethane foam.

1. Introduction

Polyurethane (Pus) are polymers which have traditionally been produced by petroleum [1]. PU foam covers a wide range of applications including insulators, automotive interior, bedding, constructions, packaging, furniture, etc. Most of the rigid PU foam that currently available on the market are made from petroleum based polyester or polyether polyol. However, due to concern on environmental problems, researchers were driving the search of biodegradable and renewable materials as alternative to petroleum based materials. PU can be produced by reacting isocyanate (NCO) with polyol (OH). Palm based polyol and soy-based polyol known as natural oil based polyol are widely used to replace petroleum based polyol recently [2].

The other concern about that still not resolved is the polyurethane foams flammability-despite the very good thermal insulation properties. Foamed materials have a highly developed pore surface thus facilitate the access of oxygen to the materials which lead to easier combustion [3]. Typically, the main flame retardants of PU foam used are compounds containing halogen as well as antimony trioxide [4]. However due to problems related to large amount of toxic and corrosive gasses and smoke released during PU decomposition, researchers have found the halogen-free flame retardants. These include, among others, phosphines and their oxides, phosphonates, phosphates and phosphites, characterized by low toxicity, lack of release of toxic gases and production of small amounts of fumes during combustion. Furthermore, through the use of synergistic effect, flame retardant which contain, in addition to the phosphorus atom, additionally nitrogen, expandable graphite, inorganic compounds or halogen were used [5-11].

Expandable graphite (EG) is one of intumescent additive known to impart fire retardant to various materials and in particular to PU foam [12]. Weiguo et. al, has studied the synergetic flame retardant effect of EG in water-blow PU foam. They had found that EG together with ammonium polyphosphate has effectively improved the flame retardancy of semi rigid polyurethane foam. Furthermore, EG shows negative effect on the mechanical property which is decrease in compression modulus [13]. Several investigations have been devoted in recent years to incorporate fillers and additive in polymer matrix with purpose increase its mechanical properties. In this research, attempt has been made to use aluminium hydroxide (ATH) to increase the compression strength of EG/ polyurethane foam composite.

2. Methodology

2.1 Materials

Materials. Polyether polyol with amine catalyst (Maskimifoam 9935B/35) is palm oil based polyol while isocyanate (Maskiminate 80) contains the mixture of Polymeric MDI with a concentration of 55% and Methylene Diphenyldiisocyanate with 45% concentration. Both materials were supplied by Maskimi Polyol Sdn. Bhd. Additives used for polymerization of PUF are aluminum hydroxide (ATH) and expandable graphite (EG). Both of the additives were manufactured by Sigma-Aldrich

2.2 Preparation of composites

PU foam composites were prepared according to the formulation in Table.1. The palm oil based polyol was stirred at 2000 rpm with a high speed mechanical stirrer and allowed to degass for 2 minutes. Then, EG and ATH were added into the polyol. The amount of EG and ATH were varied as shown in the Table.1. The mixture was stirred continuously until a homogenous. Later, the mixture was reacted with isocyanate and continuously stirred for 20 second. The mixture then was casted into a mold, covered and allowed to cure. The cured foam was removed from the mold after 24 hours. Then, the samples were placed in the oven for 2 hours at 80°C for post cure.

Table.1: Formulation for polyurethane foam composite based on ratio 1:1.1.

Expandable graphite (EG) (wt%)	Aluminum hydroxide (ATH) (wt%)			Polyol (g)	Isocyanate (g)
	1st series	2nd series	3rd series		
5	0	2	4	23.66	26.03
10	0	2	4	23.66	26.03
15	0	2	4	23.66	26.03

2.3 Mechanical test

IOP Conf. Series: Materials Science and Engineering **551** (2019) 012097 doi:10.1088/1757-899X/551/1/012097

Compression test for rigid cellular plastic was conducted under ASTM D1621-10 by using Instron 4206 machine. The cure PUF composites were cut by using a band saw machine with dimensions of 50 mm x 50 mm x 50 mm (length x width x thickness). Compression test was performed at room temperature with a constant crosshead speed at 10mm/min and the load was applied until the foam was compressed to approximately 80% of its original thickness (height). The samples were compressed in the direction parallel to foam rise. The strength and modulus of five specimens for a sample were measured and averaged.

2.4 Morphology
Scanning electron microscope (SEM) was conducted using JEOL JSM 6460 model. This analysis was done to determine the effect of additives and filler loading on foam cell size and distribution. It also reveals information about the sample surface morphology and mechanical properties. The samples were cut into 5 mm x 5 mm x 5 mm (length x width x thickness) and were coated with a very thin layer of platinum by using Auto Fined Coater to attain observation on the samples. Observation was made at 100 and 500 x magnifications.

3 Results and Discussion

3.1 Mechanical Properties
Figure 1 and Figure 2 represent the compression strength and compression modulus of PUF composites with varying amount of EG and ATH in polyurethane foam. Generally, compression strength shows significant increment as the EG and ATH loading in PU foam composite increase compared to pure PU foam composite. Compression strength increase by 41% for (5%EG/2% ATH), 66% for (10%EG/2%ATH) and 60% for (15%EG/2%ATH). Generally, the achievement of good mechanical properties in filled PUF system was determined by the distribution and dispersion of filler in the polymer matrix [14]. It can be seen from the SEM micrograph that EG distributed well and located mostly on the foam cell struts as shown in Figure.3 (C) which can contribute to increases the strength and modulus of PUF composites.

Compression modulus shows increment trend as EG and ATH loading increased (Figure 3). Incorporation of EG and ATH will affect the formation of cell and cell wall in PU foam composite. PU foam composite with high loading percentage of EG and ATH will have ticker cell wall, smaller and more uniform cell size compared pure and lower percentage of EG and ATH. It was believed that ticker cell walls in PU foam composite

IOP Conf. Series: Materials Science and Engineering **551** (2019) 012097 doi:10.1088/1757-899X/551/1/012097

contributed to the foam stiffness and higher stress was required to impose cell bending, buckling and collapsed.

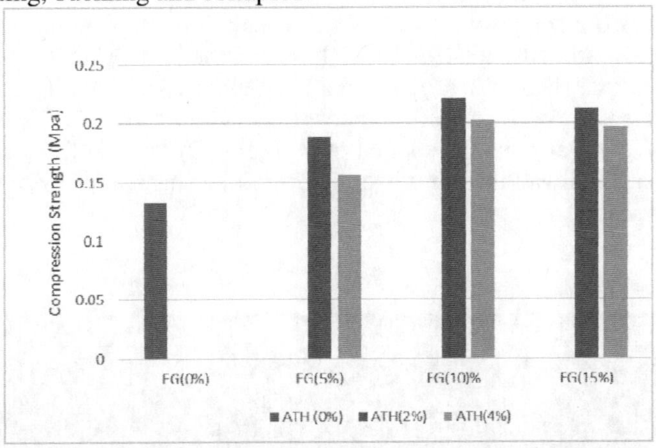

Figure 1: Compression strength of PU foam composites

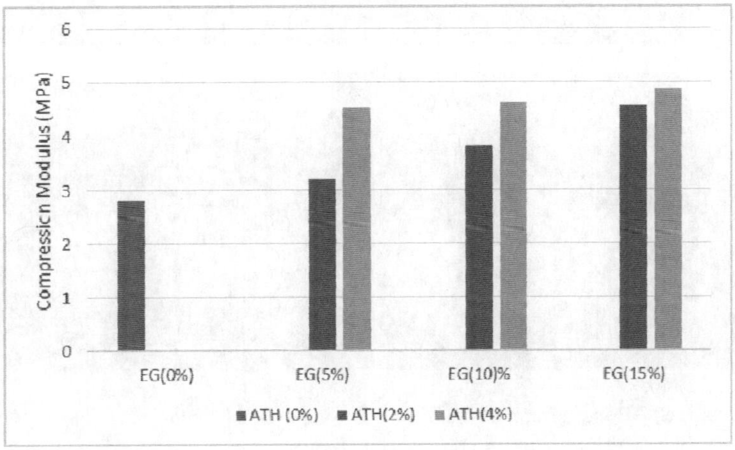

Figure 2: Compression modulus of PU foam composites

3.2 Morphological properties

Figure. 3 represent morphology of PU foam composite. It can be seen that the addition of EG and ATH has affect the cell size and cell distribution in overall PU foam composite. The cell size become smaller and more uniform for samples with 10%EG/4% ATH (Figure 3 (B)) compared to pure PU foam composite (Figure 3(A)). The wall thickness which highlighted in red circle, increased for PU foam composite sample with 15% EG/4% ATH (Figure 3(C)). It also can be spotted in Figure 3 (D) that the EG particles at the PU foam cell wall thus improve the overall mechanical properties of PU foam composite.

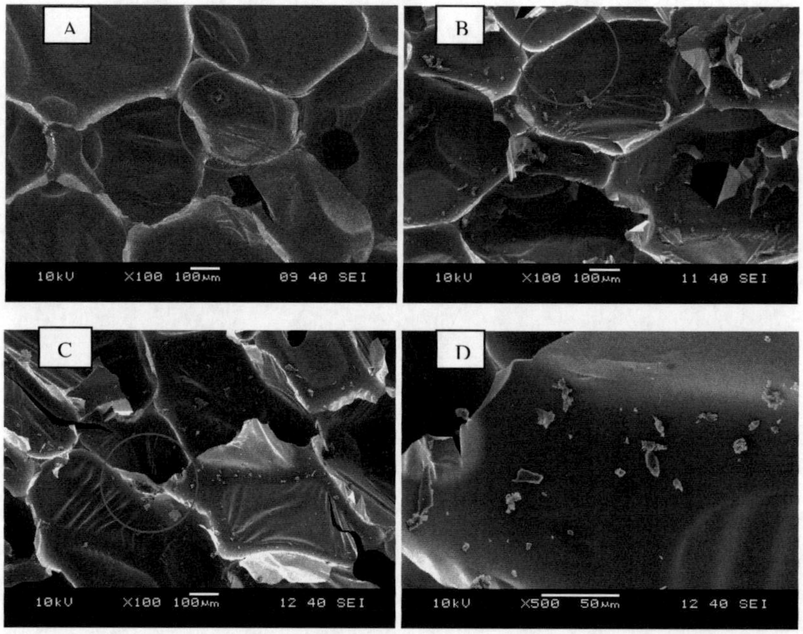

Figure 3. SEM micrograph of (A) Pure PU foam, (B) 10%EG/2% ATH PU foam, (C) and (D) 15%EG/4%ATH PU foam.

4. Conclusion

The incorporation of EG and ATH into palm oil based polyurethane foam has been carried out and Compression test shows that 10%EG/2% ATH shows the highest compression strength and 15% EG/2% ATH shows the highest compression. This proved EG and addition of ATH will improve the mechanical properties PU foam composite.

References

1. Avar G *et al* 2012 **10** 411
2. Saifuddin *et al* 2010 *Proceedings of International Conference on Advances in Renewable Energy Technologies.*
3. Michałowski *et al* 2018 *J Therm. Analy. Calorimetry.*
4. La M. G. *et al* 2015 *Environment International* **79** 106
5. Liu L *et al* 2017 *J Appl Polym Sci* 134
6. Li L *et al* 2018 *J Appl Polym Sci* 135
7. Bian X-C *et al* 2009 *J Appl Polym Sci* **109** 1935
8. Gao L *et al* 2013 *Ind Crops Prod* **50** 638
9. Lorenzetti A *et al* 2011 *Polym Degrad Stab* **96** 1455
10. Zhang M *et al* 2014 *Polym Degrad Stab* **110** 27
11. Zheng X *et al* 2014 *Polym Degrad Stab* **101** 32
12. Ming *et al* 2017 *Comb. Sc. Tech.* **189** 793
13. Weiguo Yao *et al* 2019 *Adv. Mat. Sc. Eng.*
14. Thirumal M *et al* 2008 *J Appl. Pol. Sc* **108** 1810
15. Zunaida Z *et al* 2018 *J Cell. Plas.* **54**

IOP Conf. Series: Materials Science and Engineering **551** (2019) 012098 doi:10.1088/1757-899X/551/1/012098

Mesoporous Structure of Doped and Undoped PEG on Ag/TiO_2 Thin Film

Kamrosni Abdul Razak[1], Dewi Suriyani Che Halin[1], Azliza Azani[1], Mohd Mustafa Al Bakri Abdullah[1], Mohd Arif Anuar Mohd Salleh[1], Norsuria Mahmed[1], Muhammad Mahyiddin Ramli[2], Suhaila Sepeai[3]

[1]Center of Excellence Geopolymer & Green Technology (CEGeoGTech), School of Materials Engineering, Universiti Malaysia Perlis (UniMAP), Perlis, Malaysia
[2]School of Microelectronic Engineering, Universiti Malaysia Perlis (UniMAP), Perlis,Malaysia
[3]Solar Energy Research Institute, The National University of Malaysia, 43600 UKM, Bangi, Selangor, Malaysia

Email: kamrosni@unimap.edu.my

Abstract. In this reaserch, photocatalyst silver titanium dioxide was doped and modified by Polyethylene Glycol (PEG). The purpose of the present study was to analyse the synthesized Ag/TiO_2 thin film doped and undoped PEG. Ag/TiO_2 thin films on silicon wafer have been prepared by sol–gel spin coating. The samples were characterized by Grazing Incidence X-ray diffraction (GIXRD), Field Emission Scanning Electron Microscopy (FESEM) and Atomic Force Microscope (AFM). The doped and undoped PEG Ag/TiO_2 thin films showed a mesoporous TiO_2 matrix which includes TiO_2 crystallites of 10-20 nm in size and small Ag nanoparticles (white spots) with various sizes ranging from 10 to 30 nm. However, doped PEG Ag/TiO_2 thin film showed the Ag nanoparticles became agglomerates but still remained roughly uniform on the surface.

1. Introduction

Thin films of titania (TiO_2) composed of nanosized particles in the anatase form show a very high photocatalytic activity due to their large internal surface. Both the optical properties and the photocatalytic activity of TiO_2 coatings depend on the crystallite size, the phase, and porosity of the coatings. Hence, the micro- or nanostructure of the TiO_2 films and their strict control through the preparation methodology of paramount importance. Recently, significant effort has been directed towards the low-temperature transformation of TiO_2 by using the sol-gel process. In a previous work, they have found that nanostructured anatase phase TiO_2 thin films can be fabricated by a sol-gel method at temperatures as low as 100 °C. These films, with nanometric (5 nm) grain size, were obtained by using a structure-directing agent such as poly(ethylene glycol) (PEG) [9].

Main advantages of this photocatalyst reside in its strong resistance to a wide variety of chemicals, as well as to the photocorrosion, besides its low cost. However, a drawback to this material is that it has a forbidden band value (band gap) in the range of the ultraviolet light (3.2 eV), which within the solar spectrum extents to only 4%. Therefore, its activity is negligible under visible light irradiation and consequently its application is limited [1]. For this reason, studies have been conducted to reduce the bandwidth of TiO_2 into the visible light range through its doping. So far, there is an important amount of literature related to Ag doped TiO_2 for the decomposition of organic molecules [2,3]. However, reports about Ag doped TiO_2 for water splitting hydrogen production under visible light are scarce [4,5]. Recently, researchers have concentrated in modifying TiO_2 with Ag to achieve a high-efficiency electronehole separation by forming a Schottky barrier at the AgeTiO_2, thus enhancing its photocatalytic activity [6,7]. Therefore, metal Ag, which is cheaper than other precious noble metals like Pt, Au and Pd, present a greater opportunity to modify TiO_2 [8].

Porous films have been obtained by using PEG as a chelating agent. Some researchers used PEG in order to control the porosity of Fe^{3+} doped TiO_2 films starting with $Ti(OC_2H_5)_4$ as TiO_2 precursor and PEG with molecular weight of 600. The porosity increases with the PEG amount introduced in the film. Other researchers prepared TiO_2 films starting with $Ti(OC_3H_7)_4$ and PEG with different molecular weights and amounts. The amount of PEG used in the papers mentioned above varied between 1 and 9% [9].

In the present work, we prepared porous Ag/TiO_2 thin films by a sol- gel spin-coating process using titanium alkoxide with and without organic or polymer compound additives. The phase analysis was done using GIXRD. The morphology of the sample were observed using FESEM and the surface topography and roughness of the thin films were characterized and analyzed with atomic force microscopy (AFM).

2. Methodology

All the chemicals used in the study were of analytical reagent grade, or the highest purity available from several suppliers and were used as received. Titanium tetraisopropoxide (TTIP) was purchased from Acros, Silver Nitrate (AgNO3) from Bendosen, and Acetic Acid (37%) were from J. T. Baker. Other reagents used were ethanol and 2- propanol from Qrec. The additive agent used in this experiment is Polyethylene Glycol 2000 (PEG 2000), from Merck, Germany. When precursors with strong reactivity towards water (like titanium alkoxides) are used, the hydrolysis reaction can result in precipitation. The problem can be overcome using non-aqueous sol–gel techniques with titanium chloride or titanium alkoxides as a precursor. In case of the reactions using TiCl4 as a precursor, TiO2 forms under release of a large amount of HCl gas. Because HCl gas is toxic and chlorine impurities often remain in the final oxide material, we preferred titanium isopropoxide as precursor. In order to increase the stability of the solutions from titanium alkoxide precursors, it is not sufficient to switch to a non-aqueous sol–gel route. Indeed a chelating agent is still needed in a non-aqueous sol–gel route to obtain highly stable sols [10].

Procedures of solution preparation can be referred to our previous study [11] For the doped PEG Ag/TiO2 sols, 0.3g of PEG 2000 was added and the solution were kept stirring until it was completely dissolved.

The PEG doped and undoped Ag/TiO2 thin films were characterised structurally through GIXRD, morphologically by FESEM and AFM. The phase composition of the synthesized thin film were determined by using an Grazing Incident X-ray diffraction (XRD) over a 2θ range 10°-80° using Cu Kα (λ=1.5046). Analysis of the XRD pattern will be carried out using Diffract Eva Software. Systematic information on the thin film morphology was obtained by (FESEM). The topographical 2D and 3D AFM images were taken over area of 5 x 5μm , for displaying purpose and subsequent statistical data analysis.

3. Results and Discussion

3.1 Phase Analysis

Measurements were performed in the 10-80° range using for the PEG doped and undoped Ag/TiO_2 thin films as shown in Figure 1. The undoped Ag/TiO_2 thin film shows a sharp and intense peak in spectra are indication of the presence of highly crystalline films. The observed diffraction peaks matched the characteristic peaks for the crystal structure of anatase coincide with the JCPDS Card No: 01-078-2486. The diffraction peaks shows the 2θ at 25.31°, 35.57°, 48.04°, 55.07° and 62.69° which were corresponding to (101), (112), (200), (211) and (204) plane, respectively. However, the X-ray pattern for the doped PEG Ag/TiO_2 thin film showed a shape and intensity of the diffraction lines is not representative for well-crystallized, but for nanosized, poorly crystallized materials. The addition of the PEG 2000 in the Ag/TiO_2 influenced on the unit cell parameters of anatase [10]. Both doped and undoped PEG 2000 Ag/TiO_2 thin films shows the presence of Ag nanoparticles with the characteristic peak at 37.04° with the plane of (101).

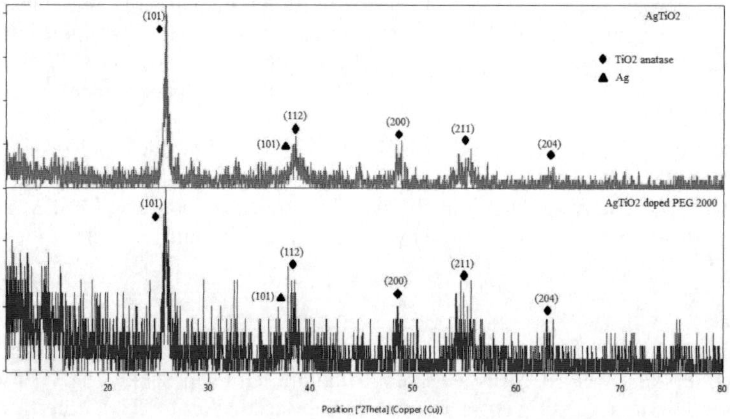

Figure 1: XRD patterns of doped and undoped PEG Ag/TiO_2 thin films.

3.2 Morphology and Topography Observations

Field Emission Scanning Electron Microscope (FESEM) was utilized to observe the surface morphology of doped and undoped PEG on Ag/TiO_2 thin film. As shown in Figure 2 (a), the undoped Ag/TiO_2 thin film exhibit the forming of wormlike TiO_2 grains with white spots uniformly distributed throughout the samples. Considering that heavy elements (e.g. Ag) backscatter electrons more strongly than light elements (e.g. O, Ti), the metallic silver appears brighter in the image. Instead of having a smooth surface, the Ag/TiO_2 composite films displayed a rough surface morphology. The addition of silver salt results in a more mesoporous TiO_2 matrix, which includes TiO_2 crystallites of 10–20 nm in size and small Ag nanoparticles (white spots) with various sizes ranging from 10 to 30 nm. For the doped PEG Ag/TiO_2 thin film, it can be detected that the Ag nanoparticles became gradually aggregated but still remained roughly uniform on the surface, as shown in Figure 2 (b).

Although the size of the silver particles is smaller with diameter around 10 nm, there are more nano-pores detected in the thin film. It is reported that heating promotes aggregation of Ag atoms. As radius of Ag^+ ions (ca. 126 pm) is much larger than that of Ti^{4+} (ca. 68 pm), the Ag^+ ions introduced by the sol–gel process would not enter into the lattice of the TiO_2 anatase phase. When the samples were heat-treated to a certain temperature, the reduction of Ag^+ to Ag^0 took place as the $AgNO_3$ decomposed. During the annealing process, these uniformly dispersed Ag^+ ions would gradually migrate along with the anatase grain boundaries to the surface of the TiO_2 film, while TiO_2 anatase grains would grow at the same time. Finally, the Ag+ ions probably exist on the surface of the anatase grains by forming Ag–O–Ti bonds. In literature study, the movement of the resulting nanoparticles in the TiO_2 matrix could also be restricted by tuning the molar ratio of water to Ti when adding the Ag^+ preparing sol, and hence particle aggregation is controlled. The anatase grain growth is thereby depressed and the specific surface area increases. The porous feature of the film indicates that it should have a rough surface [12].

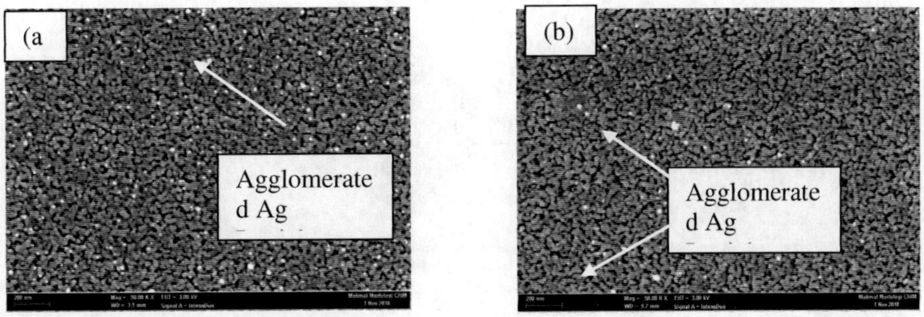

Figure 2: FESEM micrograph of (a) undoped Ag/TiO_2 thin film, (b) doped PEG Ag/TiO_2 thin film

AFM has been demonstrated to be a very versatile and powerful tool for surface imaging at the submicrometer level and the revelation of the surface characteristics of the films. AFM was used to characterize the morphology and surface roughness of the samples. The darkest and brightest regions of all AFM images in this paper correspond to those regions with the largest height gradients. The brightness of a region corresponds directly to its height. Figure 3 shows the representative top view images and angle view images of the surface morphology of a scan area 5 μm × 5 μm of the doped and undoped PEG Ag/TiO_2 composite films on silicon wafers by three spin-coating cycles. It was observed that the surface morphology and roughness of the nanocrystalline thin films reveal a significant difference between doped and undoped PEG Ag/TiO_2. The film prepared without PEG is covered with homogenly dispersed silver particles and has a surface roughness of 3.020 nm and statistical mean roughness 7.700 nm. The particle size and average roughness of the film decreased with the addition of PEG into the sol, which is 2.300 nm and statistical mean roughness is 5.238 nm. The Ag particles were dispersed uniformly, throughout the samples (white dot), but in the doped PEG Ag/TiO_2, there were more larger size of Ag particles, shows that the Ag particles are agglomerates, in agreement with FESEM results. The PEG present in Ag/TiO_2 leads to the agglomeration of Ag particles, hence the TiO_2 matrix with less Ag particles distribution become smooth, and the surface roughness value is lower than Ag/TiO_2 particles without PEG.

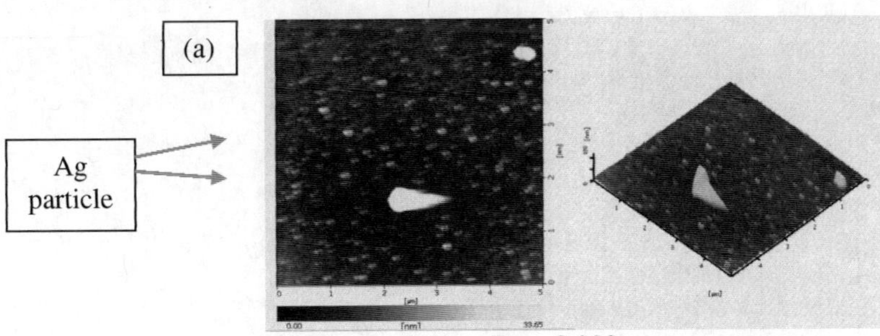

Ra = 3.020 nm

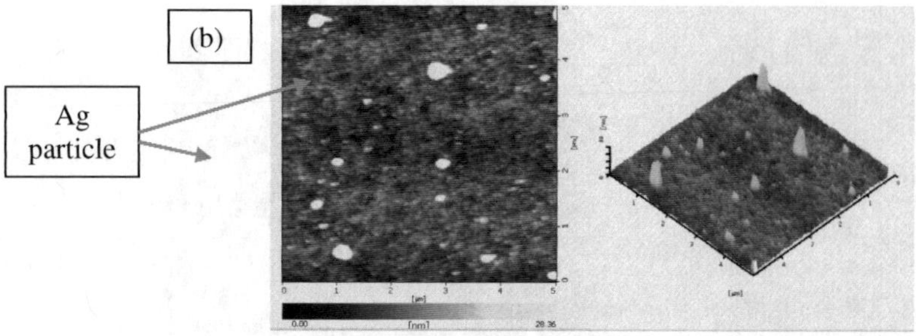

Ra = 2.322 nm

Figure 3: AFM micrograph of (a) undoped Ag/TiO_2 thin film (b) doped PEG Ag/TiO_2 thin film

4. Conclusions

This work observes the effect of doped and undoped PEG on the growth of Ag/TiO_2 thin films synthesized via sol-gel spin coating method. X-Ray diffraction shows both samples, doped and undoped PEG Ag/TiO_2 exhibit peaks of anatase TiO_2 and Ag particles. FESEM observation shows some agglomeration of Ag particles on the sample of doped PEG Ag/TiO_2. As for the topography analysis, the surface roughness of the sample decreased for the doped PEG Ag/TiO_2.

Acknowledgements

The author would like to acknowledge the support from the Fundamental Research Grant Scheme (FRGS) under a grant number of FRGS/1/2017/TK07/UNIMAP/02/6 from the Ministry of Education Malaysia and under grant number of 03-01-02-SF1322 from Ministry of Energy, Science, Technology, Environment and Climate Change (MESTECC), Malaysia. The authors wish to thank the Center of Excellence Geopolymer & Green Technology (CEGeoGTech), School of Materials, Engineering, Universiti Malaysia Perlis, UniMAP for their partial support.

References

[1] Ashkarran A A *et al* 2014 *Appl. Surf. Sci.* **301** 338.
[2] Klabunde K J *et al* 2007 *Interf. Sci.* **31** 1514.
[3] Mogyorosi K *et al* 2009 *A React Kinet Catal Lett* **98** 215.
[4] Ni M *et al* 2007 *Renew Sustain Energy* **11** 401.
[5] Zaleska A et al 2008 *Recent Pat Eng* **2** 157.
[6] Hu C *et al* 2006 *J Phys Chem B* **110** 4066.
[7] Lopez Ortiz A *et al* 2015 *Int. J. Hydrog. Energy* **40** 17308.
[8] Djaqued Y *et al* 2005 *Materials Science-Poland* **23** 1.
[9] Calderon-Moreno J M *et al* 2014 *Ceram. Int.* **40** 2209.
[10] Vetrivel V *et al* 2014 *UGC Sponsored National Seminar on Emerging Trends in Plasma Technology and Its Applications* (ETPTA-2014).
[11] Abdul Razak K *et al* 2018 *Solid State Phenomena* **280** 26.
[12] Yu B *et al* 2011 *Nanotechnology* **22** 115603.
[13] Halin D S C et al 2017 *IOP Conf. Series: Materials Science and Engineering* **209** 012002.

Microstructure and Wettability of Graphene Oxide/TiO_2 Thin Film Prepared via Sol-gel Method

Azliza Azani[1,*], Dewi Suriyani Che Halin[1], Kamrosni Abdul Razak[1], Mohd Mustafa Al Bakri Abdullah[1], Mohd Arif Anuar Mohd Salleh[1], Norsuria Mahmed[1], Muhammad Mahyiddin Ramli[1,2], Suhaila Sepeai[3]

[1]Center of Excellence Geopolymer & Green Technology (CEGeoGTech), School of Materials Engineering, Universiti Malaysia Perlis (UniMAP), Perlis, Malaysia
[2]School of Microelectronic Engineering, Universiti Malaysia Perlis (UniMAP), Perlis, Malaysia.
[3]Solar Energy Research Institute, The National University of Malaysia, Selangor, Malaysia.

Email: dewisuriyani@unimap.edu.my

Abstract. The microstructure of graphene oxide TiO_2 (GO/TiO_2) films on wettability and morphology properties were studied at different amount of Titanium isopropoxide with fixed amount of GO. GO/TiO_2 thin films were prepared by using sol-gel method and deposited on glass substrate by spin-coating technique. The formation of GO/TiO_2 was confirmed by using Fourier Transform Infrared Spectroscopy (FTIR) study while the morphology of GO/TiO_2 was observed by Scanning Electron Microscopy (SEM) analysis. The hydrophilic/hydrophobic interactions of these films were examined by wettability test by measuring the water contact angle of the water drop on the surface of the coating. The wettability tests display that the contact angle of GO/TiO_2 was increased from 90.0° to 118.7° after reduce the amount of titanium isopropoxide (TTIP). A small peaks of graphene and Ti were observed in the coated films and more pronounce peaks occur after the films were annealed. However, it was found that the annealing process does not improve the contact angle of the films as compared to the non-annealed films.

1. Introduction

Self-cleaning technology is among the technologies existing in today's world. Self-cleaning coatings can be classified into two major categories which are hydrophilic and hydrophobic. Both of the categories involve water to self-clean themselves. In a hydrophilic coating, the water is made to spread surface in which a liquid on a solid surface makes a film over the surfaces. This will carry the dirt and other impurities, whereas in the hydrophobic coating, the water droplets slide and roll over the surfaces in which a water drops remain spherical on the solid surface thereby cleaning them [1]. Self-cleaning action of TiO_2 over the past years that mainly discuss photocatalytic activity of TiO_2.

Titanium dioxide (TiO_2) also known as titania is a white inorganic solid substance and also a most investigated materials in photocatalysis [2]. Pure TiO_2 is obtained from ilmenite or leuxocene ore [3]. TiO_2 holds many holds many possible applications including environmental and energy applications due to its excellent properties such as nontoxic, chemically stable, cheap, and easily available material, having a bandgap of 3.2 eV in the ultraviolet (UV) region [4]. However, there are limitataion in the photocatalytic e ciency of pure TiO_2 due to rapid recombination of photogenerated electron–hole pairs within TiO_2 particles and lack of visiblelight absorption. Nobel metal deposition is one of the technique to overcome the limitation [5]. Suitable metal oxides deposition can enhance the hydrophilic and hydrophobic porperties.

Graphene oxide is the appropriate nobel metal to overcome the limitation due to its high surface area, high electrical conductivity (106 S^{cm-1}), high carrier mobility (200,000 cm^2 V^{-1} S^{-1}) and e cient electron (e^-) transfer from TiO_2 to graphene as its redox potential is just below the conduction band (CB) edge TiO_2 [6]. Graphene film can provide improved surface properties such as friction coefficient [7], optical transmittance [8], and thermal conductivity [9].

IOP Conf. Series: Materials Science and Engineering **551** (2019) 012099 doi:10.1088/1757-899X/551/1/012099

In this study, we have investigated the effects of doping with graphene oxide on the hydrophilic property and hydrophobic property at different concentration of TiO_2 as well as annealing process on the hydrophobicity of films. Sol-gel method is used to fabricate thin films due to its advantages such as easy, good homogeneity, ease of composition control and good optical properties. In particular, the sol-gel process is very efficient in producing thin and transparent multi-component oxide layers on various substrates such as stainless steel plates, alumina plates, glass panels and ceramic tiles. This work contributes to the comprehension of using modified TiO_2 as self-cleaning materials.

2. Methodology

2.1 Materials

All the chemicals used in the study were of analytical reagent grade available from several suppliers and were used as received without further purification. Graphene Oxide powder (4-10% edge-oxidized) and titanium tetraisopropoxide (TTIP) 97% was purchased from Sigma Aldrich, 2-propanol AR from QReC and Acetic Acid (99.5%) was from Daejung Reagent Chemicals. Hydrolysis reaction will result in precipitation by using precursors with strong reactivity towards water (eg. Titanium alkoxide). To overcome the percipitation, non-aqueous sol–gel techniques was used with titanium chloride or titanium alkoxides as a precursor. In case of the reactions using TiCl4 as a precursor, TiO2 forms under release of a large amount of HCl gas. Because HCl gas is toxic and chlorine impurities often remain in the final oxide material, titanium tetraisopropoxide are preferred as a precursor. In order to increase the stability of the solutions from titanium alkoxide precursors, it is not sufficient to switch to a non-aqueous sol–gel route. Indeed a chelating agent is still needed in a non-aqueous sol–gel route to obtain highly stable sols [10].

2.2 Substrate Preparation

In this study, the glass slide with dimension size of 10 mm × 10 mm are used as substrate. The substrate was cleaned by using acetone and was placed in ultrasonic cleaner for 20 minutes in order to remove the contaminants on the glass surface. Then, the glass slides were rinsed with distilled water to remove the hydroxyl contaminants, and these glasses were then dried under room temperature for about 30 minutes.

2.3 Preparation of Graphene Oxide / TiO_2

Firstly, titanium tetraisopropoxide (TTIP) was mixed with 2-propanol with two different ratios (TTIP: 2-propanol) which is 1:9 and 1:20 ratio with fixed loading of 1 mg of graphene oxide was added to the solution together. This resultant titanium precursor solution was stirred vigorously for 20 minutes. Then, 0.10 mL acetic acid was added to the solution. Continue vigorous stirring for 1 hour to obtain the final solution for the preparation of graphene oxide and TiO2 thin films. After the substrate was properly cleaned, the solution was deposited dropwise on the substrate at a speed of 2000 rpm for 30s by spin coating technique. The wet films were dried at 60°C for 10 minutes, it was then annealed at 350°C in the muffle furnace for 1 h. The heat treated films were stored in Petri dishes with lid for further use.

2.4 Characterization of Graphene Oxide/ TiO_2 thin films

The structure analysis was determined by FTIR spectra using Jasco FTIR 4200 spectrometer (Jasco International Co. Ltd., Japan) at the room temperature in the spectral range of 600–4000 nm. The surface morphology of the layers was observed and the microstructure of the films was determined by EDX technique by Surface Electron Microscopy (SEM) JSM-6460 LA Jeol Japan in School of Material Engineering, Universiti Malaysia Perlis (UniMAP). While all contact angle measurements were conducted using self made water contact angle apparatus with image capture by optical microscope and analyze by J Image software to obtain the angle.

3. Results and Discussion

3.1 Functional Group of GO/TiO_2

The FTIR spectra of glass substrate and GO/TiO_2 are shown in Figure. 1. Figure.1(a) shows the peak of Si-Si interaction at 2300 cm^{-1} and 2123 cm^{-1}. As compared blank glass substrate with GO/TiO_2, Fig.1(b) shows that an obvious peak was observed at 1399 nm^{-1} corresponded to C=O, indicating that the existed of GO in the films. The characteristic peaks of graphene oxide can also be found at 2900 cm^{-1} and 2700 cm^{-1} confirming that successfully prepared GO-TiO_2. Moreover, Figure.1(b) shows the broad peak the broad absorptions at low frequencies below 1000 cm^{-1} were ascribed to the stretching vibrations of Ti–O–Ti bonds in TiO_2. The peak in FTIR spectra observe similar to previous study by Martins et.al [11].

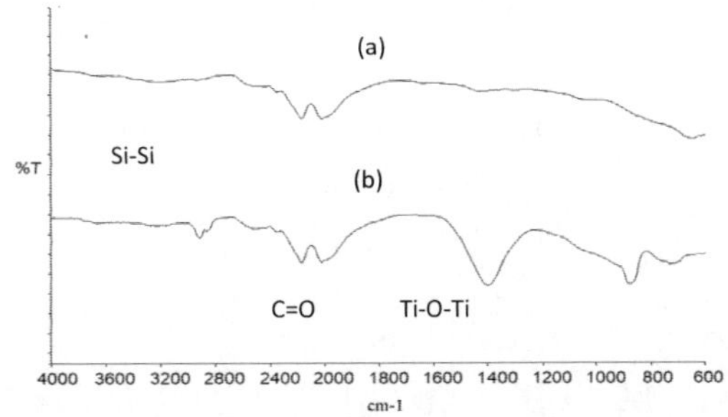

Figure. 1. FTIR spectra of (a) glass substrate and (b) GO/TiO_2 film.

3.2 Surface Morphology of Graphene Oxide/TiO_2

Figure 2 shows micrographs of two different ratio of GO/TiO_2 thin films which are 1:9(GO/TiO_2) and 1:20(GO/TiO_2) microstructure of the dispersion-distribution of graphene nanosheets within the matrix TiO_2. Fig. 2(a) shows a non-annealed film at low ratio of 1:9(GO/TiO_2) while Fig. 2(b) and 2(c) were films for non- annealed and annealed at high ratio 1:20(GO/TiO_2) respectively. It can be seen that the morphology 1:9(GO/TiO_2) showed spongy morphology with most of the TiO_2 particles were combined together with each other. Interestingly, Fig. 2(b) shows the film was denser and compact at high ratio 1:20(GO/TiO_2). However, Fig. 2(c) exhibit that annealing process resulted of smaller particles with an even surface. It was clearly observed that TiO_2 particles were agglomerated at higher ratio by supported of graphene oxide sheets. The similar micrographs of GO/TiO_2 can also be observed in previous study by Abdelmajid et.al [12].

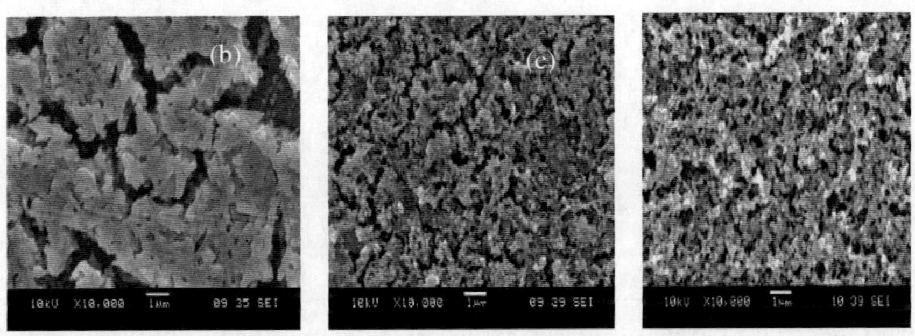

Figure 2. Scanning Electron Micrscopy microgarphs of GO/TiO_2 films at two different ratios, (a) 1:9 without annealing) , (b) 1:20(without annealing) and (c) 1:20 (annealing at 350ºC)

3.3 Chemical Analysis

The chemical composition analysis was carried out on GO/TiO_2 thin film sample by Energy Dispersive X-ray (EDX) analysis. The representative (EDX) pattern of a ratio 1:20(GO/TiO_2) layer presented in Table 1, confirms the presence of C, O and Ti, indicating the formation of nanocomposite with high purity. The signal for C should mainly originate from the GO sheets, while those for Ti are from the TiO_2 nanoparticles. The signal of O could contribute by TiO_2 nanoparticles and a small amount of oxygen-containing groups on GO sheets.

3.4 Water Contact Angle

The water contact angle was carried out to analyze the hydrophilicity of TiO2 composite film. Fig. 4 shows the water contact angle of the prepared films with a significant decrement after addition of graphene oxide due to alteration of interaction of hydrophilic/ hydrophobic of TiO2. Figure. 4(a) shows that the photograph of water contact angle on the blank substrate. It was found that the wettability tests display that the contact angle of GO/TiO2 non-anneal was increased from 90.0º to 118.7º after reduce the amount of TTIP as shown in Figure. 4(b) and Figure. 4(c) respectively. Table 2 shows all the water contact angle of GO/TiO2 thin film at different ratio of TTIP with and without annealing process. This result shows that the films was subjected to high wettability due to addition of GO particles.

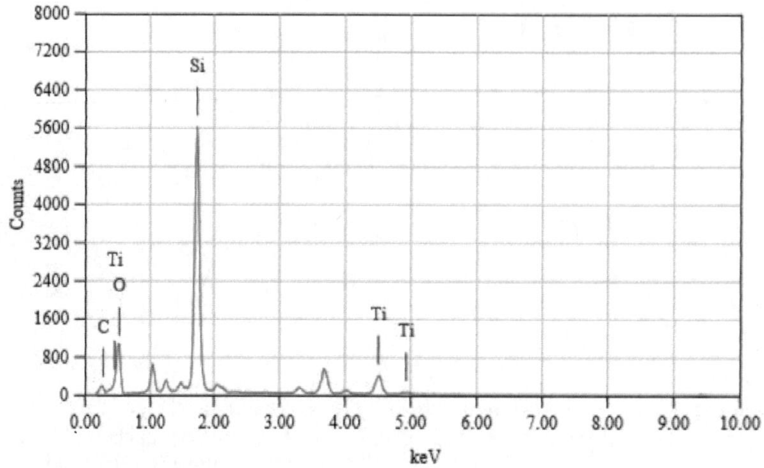

Figure 3. EDX diagram of ratio 1:20(GO/TiO2) film without annealing.

Table 1. Table of EDX element of ratio 1:20(GO/TiO_2) film without annealing

Element	Mass %	Atomic %
C K	8.62	14.94
O K	37.28	48.50
Si K	42.58	31.55
Ti K	11.51	39.39
Total	100.00	100.00

This can be confirmed in previous study by Narayanam et.al [13]. Furthermore, this finding was correlated to the amount of C, Ti and O_2 as reported in the EDX result. Nevertheless, it was found that annealing process does not improve the water contact angle of the film as observed in Figure 4(d).

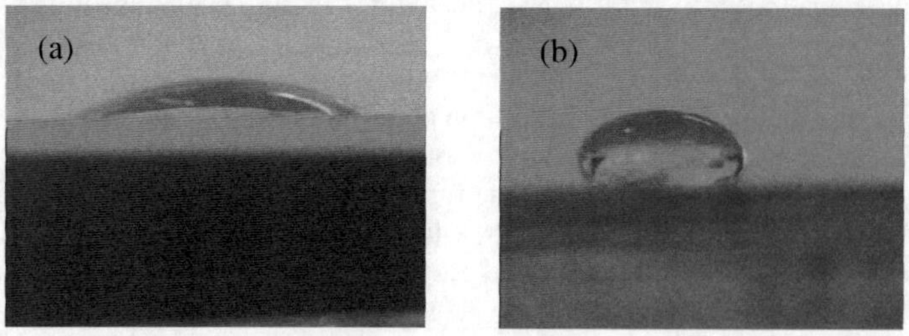

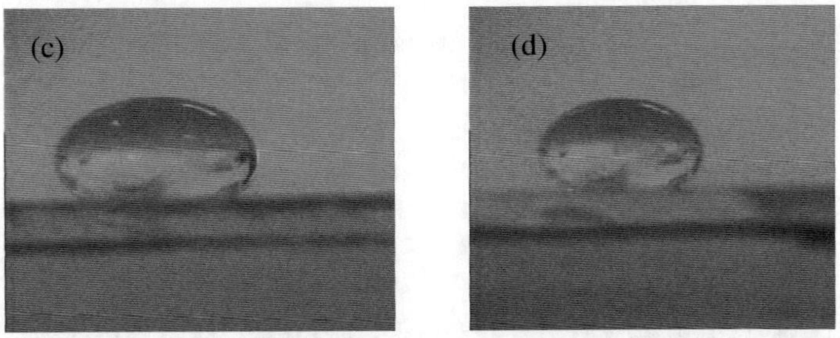

Figure. 4. Photograph of water contact angle of GO/TiO_2 at different ratio of TTIP.

Table 2. Water contact angle of GO/TiO_2 at different ratio of TTIP

Thin Film	Water Contact Angle (°)
Glass substrate without thin film	60.17
1:9 (GO/TiO_2) - without annealing	90.00
1:20 (GO/TiO_2) – annealing at 350°	114.44
1:20 (GO/TiO_2) - without annealing	118.70

4. Conclusions

From the research, GO/TiO_2 thin film were prepared in two different ratios of titanium isopropoxide with fixed amount of GO. The thin film was successfully prepared by sol-gel method using spin-coating deposition technique. Functional group analysis were obtained from FTIR spectra which indicated the presence of GO and TiO_2. While for morphology analysis by SEM analysis shows that at lower ratio (1:9) showed spongy morphology with most of the TiO_2 particles were combined together whereas at higher ratio (1:20) shows that TiO_2 particles was denser and compact. The hydrophilic and hydrophobic properties are reflected from wettability test which shows that the thin film will become hydrophobic at the higher ration of TiO_2 from 90.0° to 118.7°. From the wettability test also it was found that annealing process does not improve the contact angle of the film from hydrophilic to hydrophobic.

Acknowledgements

The author would like to acknowledge the support from the Fundamental Research Grant Scheme (FRGS) under a grant number of FRGS/1/2017/TK07/UNIMAP/02/6 from the Ministry of Education Malaysia and under grant number of 03-01-02-SF1322 from Ministry of Energy, Science, Technology, Environment and Climate Change (MESTECC), Malaysia. The authors wish to thank the Center of Excellence Geopolymer & Green Technology (CEGeoGTech), School of Materials, Engineering, Universiti Malaysia Perlis, UniMAP for their partial support.

References

[1] Anandan S *et al* 2012 *Applications 1.*
[2] Kusiak-Nejman E *et al* 2017 *Catal. Today* **287** 189.
[3] Damsyik N *et al* 2016 *Politek. Kolej Komuniti J. Eng. Technol.* **1** 77.
[4] Saleem A *et al* 2018 *J. Electron. Mater.*
[5] Anandan S *et a 2013 ACS Appl. Mater. Interfaces* **5**, 207.
[6] Prabhu S *et al* 2017 *Sol. Energy Mater. Sol. Cells* **169** 304.
[7] Berman D *et al* 2014 **17** 31.
[8] Kim K K *et al* 2010 Enhancing the conductivity of transparent graphene films via doping.
[9] Zhang C *et al* 2017 *Nat. Publ. Gr* 1.
[10] Abdul Razak K *et al* 2018 *Solid State Phenomena* **280** 26.
[11] Martins P M *et al* 2018 *Compos. Part B Eng.* **145** 39.
[12] Timoumi A *et al* 2018 **11**. 46.
[13] Narayanam P K *et al* 2018 *ACS Appl. Nano Mater.* **1** 5691.

IOP Conf. Series: Materials Science and Engineering **551** (2019) 012100 doi:10.1088/1757-899X/551/1/012100

Assessment of Surface Ozone Concentration in Northern Peninsular Malaysia

Nur Izzah Mohd Hashim[1], Nur Alia Izzati Mohd Yusoff[1], Norazian Mohamed Noor[1], Ahmad Zia Ul-Saufie[2]

[1]School of Environmental Engineering, Universiti Malaysia Perlis (UniMAP), Perlis, Malaysia
[2]Faculty of Computer and Mathematical Sciences, Universiti Teknologi MARA, Pulau Pinang, Malaysia

Email: norazian@unimap.edu.my

Abstract. This study investigate the surface ozone variations in northern Peninsular Malaysia which is Penang and Ipoh. A database with hourly air quality data and meteorological parameters covering the period of 2013 – 2016 obtained from Air Quality Division, the Department of Environment (DOE) were analysed. The association between the O_3 concentration and meteorological parameter as well as the other air pollutants were identified by using Pearson Correlation Analysis and Principle Component Analysis. In order to predict the next three hour of O_3 concentration, two models were developed which is MLR and PCA-MLR. The results indicate that Ipoh has higher ozone concentration with maximum concentration of 0.124 ppm. The diurnal pattern of ozone displayed a peak during afternoon and low concentration during night-time until morning. On the other hand, the NO and NO_2 concentration shows opposite pattern from ozone concentration. Maximum concentration of O_3 reached the peak at late afternoon at approximately 2 p.m. in Penang and Ipoh while the lowest reading was recorded was at 8 a.m. A significant correlation was observed between O_3 concentration with temperature and wind speed in both urban areas. MLR was chosen as best predictive and more accurate model in predicting O_3 concentration during the peak hour compared to PCA-MLR.

1. Introduction

Ozone is a combination of three oxygen atoms which majority found in lower stratosphere, to maintain the thermal structure of the atmosphere and absorption of ultraviolet rays (UV) [1]. Due to rapid increasing in industries and anthropogenic activities in the ground-level atmosphere lead to higher concentration of the nitrogen oxides (NOx) and volatile organic compounds (VOC) which is the ozone precursors. When the nitrogen oxides (NOx) interact with the VOCs under the action of UV radiation at the earth surface, it will result to the form of secondary pollutant, which is ozone, O_3 [2].

Surface ozone is one of the major causes of air pollution in urban areas around the world. Breathing ozone can trigger to respiratory problems, lung diseases such as asthma and also can give harm on vegetation and environment [3,4].

There are several factors that can affect human health including the exposure duration, concentration of surface ozone in atmosphere, length of time between short-term exposures and average amount of air inhaled per minute (ventilation rate) [5-7].

Hence, surface ozone should be firstly identify and analyses. The understanding of the conditions that contribute to its formation [8] and the influence of meteorological parameters such as temperature and humidity will help to thoroughly understand the photochemical reactions and the relationship of O_3 concentration with the weather parameters. Prediction of ground-level ozone concentration can be done with the good understanding of the pollutants. Thus, this paper evaluate the relationship of O_3 with its precursors and weather parameters in the norther areas of Peninsular Malaysia. Then, 2 models were developed and the performances were compared.

2. Methodology

2.1 Study Area

The chosen areas in this study are Seberang Perai, Penang and Ipoh,Perak. These two areas are known as urbanized area and was chosen due to lack of studies were conducted in these two selected areas. Seberang Perai and Ipoh had a tropical rainforest climate also borders on tropical monsoon climate [9]. Seberang Perai is located in the mainland of Penang. The data for air quality and meteorological parameters used in this this study were collected from the monitoring station at Sekolah Seberang Perai II located at latitude 5.3975, 100.4048. Seberang Perai is served by a major highways, which experience heavy traffic in the morning and peak hours. Besides that, this city also was surrounded with a heavy industry area with high pollution.Meanwhile,the monitoring station at Sekolah Menengah Kebangsaan Jalan Tasek, Ipoh (4.6305, 101.1156) were used to collected the air quality and meteorological parameters data which will used in this study. This city consist a lot of industrial areas located at nearby the urban area which may lead to formation of pollution, especially ozone.

2.2 Statistically Analysis

Descriptive statistics is used to present the data in simple and easier way to interpret. It include the descriptive statistic table and boxplot generated by using Statistical Package for Social Science (SPSS) software version 23. Process of collecting, organizing, summarizing and presenting the data also part of the descriptive statistic.

2.3 Diurnal Plot

It is use to identify the behaviour, plotting and average of hourly value on chosen time of 24-h scale. It can show the relationship and effect of ozone precursors towards the surface ozone concentration, and show the variations were in 24-h scale.

2.4 Contour Plot

Contour plot is used to show three dimensional surface on two dimensional plane. The line are drawn for connecting the x and y axis while the contour value represents as z. In this study, the contour plot is used to show the monthly variation of surface O_3 concentration by using hourly data. The contour was plotted by using open software, Plot.ly.

2.5 Pearson Correlation Analysis

Pearson correlation analysis use to measure the strength of the relationship between two continuous variable X and Y. In Pearson Correlation, the letter r stand for a value from +1 and -1 where 1 is the positive correlation while 0 means no correlation and -1 is negative correlation. It can show the strength of linear relationship between surface ozone concentration and meteorological parameter.

2.6 Principle Component Analysis (PCA)

PCA was used to find a small sets of linear combinations of covariates that not correlated with one another. This will avoid multicollinearity problems. Additionally, it can also ensure that the selected linear combinations have the maximum variance [10].

Principal Component Analysis is generally expressed as follows:

$$PC_i = l_{1i}X_i + l_{2i}X_i + \dots l_{ni}X_n \quad (1)$$

Where:
PCi = i^{th} principal component
1_{ji} =loading of observed variable X_j

2.7 Multiple Linear Regression (MLR)
MLR have capabilities to predict the contribution of selected variables to the variations of O_3 concentration. This was an attempts to model O_3 concentration the relationship between two or more explanatory variables such as the traces gases and a response variable surface ozone concentration with meteorological factors by fitting a linear equation to predict the next hour of O_3 concentration. 80 percent of the data will be used to develop the model while 20 percent will be used to test the performance of the model developed using performance indicator [11].

2.8 PCA-MLR
The original variables (dataset of NO_2, NOx, PM_{10} and other) and PCA (PC1, PC2, PC3 and PC4) results were used as the input in MLR analysis. Mostly PCA is used to reduce the various dimensions associated with MLR which then creates a new variable called the principle component (PC). The PCs has been extracted so that the first principle component (PC1) accounts for the largest number of variations in the data set, while the following components contribute to variations that have not been considered on PC1.

2.9 Model Performance Measure
Performance measure are used to evaluate the predictive model between Multiple Linear Regression (MLR) and Principle Component Analysis (PCA). This will ensure that the model developed for prediction of ozone concentration in the studied areas were comparable and had reliable performances. The performance measures used in this study were Mean Absolute Error (MAE), Root Mean Squared Error (RMSE), Normalised Absolute Error (NAE), Index of Agreement (IA) and Prediction Accuracy (PA) and coefficient of determination (R^2). The equations for all the performance measures applied in this study is shown in Table 1 [12].

Table 1. Performance Indicators

No	Performance Indicator	Formula			Better Predictability if
1.	Mean Absolute Error	MAE	=	$\frac{1}{n-1}\sum_{i=1}^{n}(P_i - O_i)$	Closer to 0
2	Normalized Absolute Error	NAE	=	$\frac{\sum_{i=1}^{n}\lvert P_i - O_i\rvert}{\sum_{i-1}^{n} O_i}$	Closer to 0
3	Root Mean Square Error	RMSE	=	$\frac{1}{n-1}\sum_{i=1}^{n}(P_i - O_i)^2$	Closer to 0
4	Prediction Accuracy	PA	=	$\frac{\sum_{i-1}^{n}(P_i - \bar{O})^2}{\sum_{i-1}^{n}(O_i - \bar{O})^2}$	Closer to 1

5	Coefficient of Determination	R2	=	$\left(\frac{\sum_{i-1}^{n}(P_i - \bar{P})(O_i - \bar{O})}{n \cdot S_{pred} \cdot S_{obs}}\right)^2$	Closer to 1
6	Index of Agreement	IA	=	$1 - \left[\frac{\sum_{i-1}^{n}(P - O_i)^2}{\sum_{i}^{n}{}_{1}(\lvert P_i - \bar{O}\rvert + \lvert O_i - \bar{O}\rvert)^2}\right]$	Closer to 1

3. Results and Discussion

3.1 Boxplot of Ozone Concentration

Figure 1 represents the box plot of hourly ozone concentration in both studied area, Penang and Ipoh from 2013-2016. Based on the plot, O_3 concentration in Ipoh was higher than Penang in 2013 and 2014, and the median in Ipoh is the most highest compared to other year for Penang and Ipoh. Other than that, the dataset in Ipoh was more varies than Penang since the boxplot for Ipoh on 2013 and 2014 have the longer distribution ozone concentration due to high value of standard deviation. Then, the boxplot for ozone concentration in both Penang and Ipoh for all years are skewed to the right. Hence, it shows that the O_3 concentration is quite high. However, the distributions in Penang on 2015 and 2016 were more varies because the concentration of O_3 in Ipoh was droped from 2014 to 2016. This is due to the decreasing in O_3 precursor concentration, NO_2 in that years.

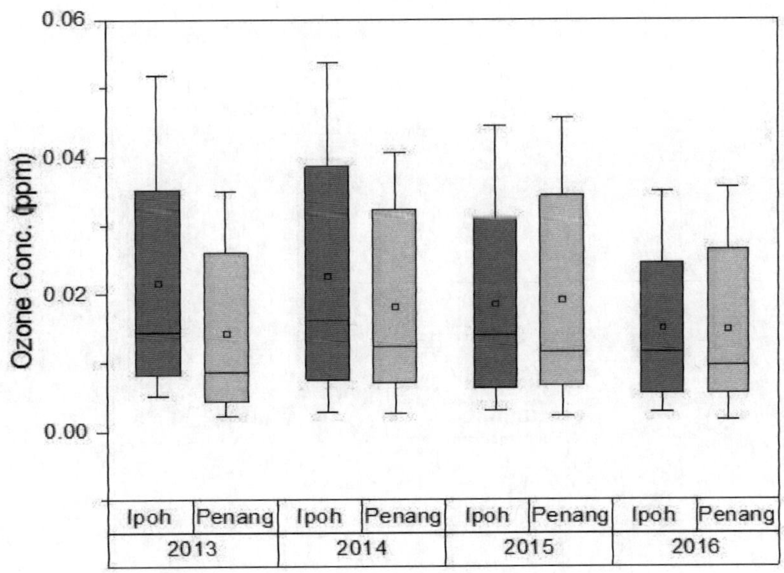

Figure 1. Boxplot of Surface Ozone Concentration

3.2 Boxplot of Other Pollutant and Weather Parameter

The boxplot of traces gases (NOx, NO_2, CO, SO_2, PM_{10}) and meteorological parameter (temperature, humidity, wind speed) in Penang and Ipoh were displayed in Figure 2. The boxplots for NOx concentration in Ipoh has longer distribution than Penang because monitoring station in Ipoh surrounded with high volume of vehicles. The variation of dataset for NO_2 concentration and temperature for both studies area are just similar. Furthermore, boxplot for SO_2 and PM_{10} concentration show very small distribution in both area, even though the concentration in Penang for

IOP Conf. Series: Materials Science and Engineering **551** (2019) 012100 doi:10.1088/1757-899X/551/1/012100

both parameter was higher than Ipoh except for PM_{10} concentration in Ipoh for 2014. This is due to the location of monitoring station in Penang which surrounded by heavy industrial areas.

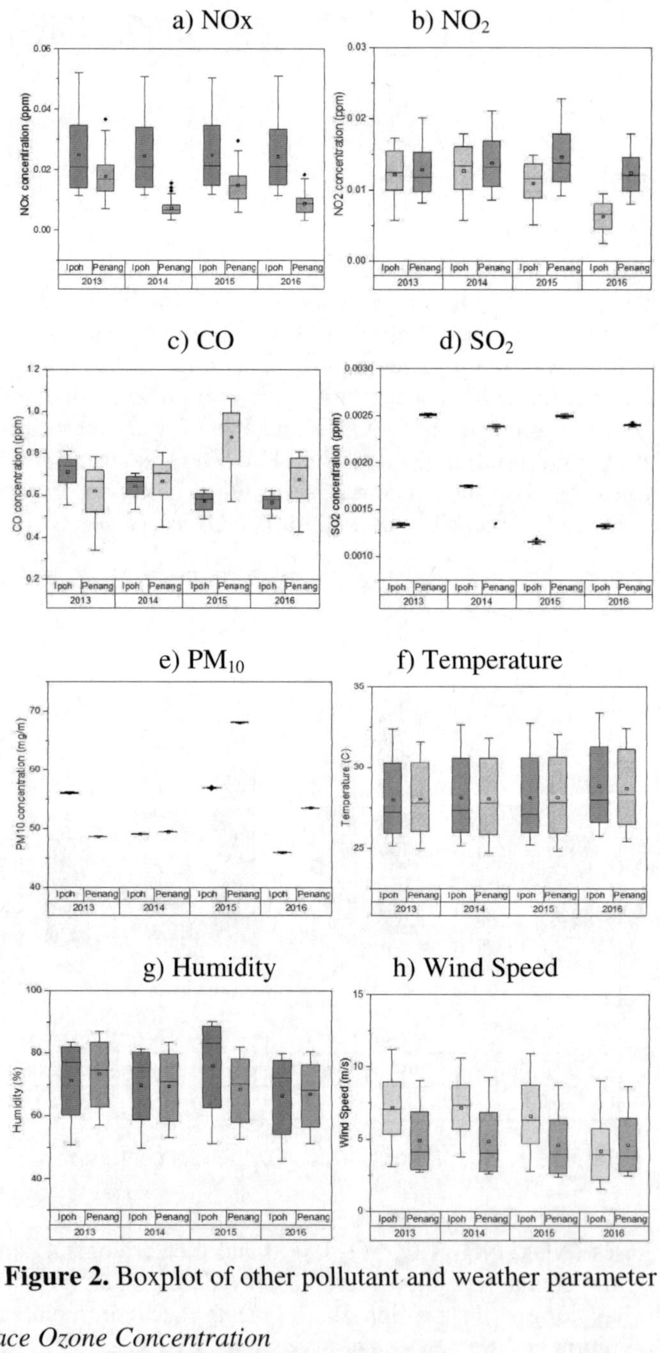

Figure 2. Boxplot of other pollutant and weather parameter

3.3 Variation of Surface Ozone Concentration

The surface ozone concentration was analysed diurnally and the result was presented in in the Figure 3. The ozone peak was observed in both sites and higher ozone concentration was observed in Ipoh compared to Penang. This is because the location of the monitoring station in Ipoh situated nearby

IOP Conf. Series: Materials Science and Engineering **551** (2019) 012100 doi:10.1088/1757-899X/551/1/012100

schools and surrounded with high traffic flow. The minimum ozone concentration was observed during the night-time and early morning hours. Urbanization, industrialization, high-speed propulsion, and atmospheric emissions of a large numbers of active substances such as carbon dioxide and aerosol are some of the major sources that contribute the increase of ozone concentration. The lowest ozone concentration consistently measured at 0800 and reached the peak around 1300-1500 during midday and appear to reduce in the night at the both site. The value is reached at approximately 2 p.m. in Penang and Ipoh.

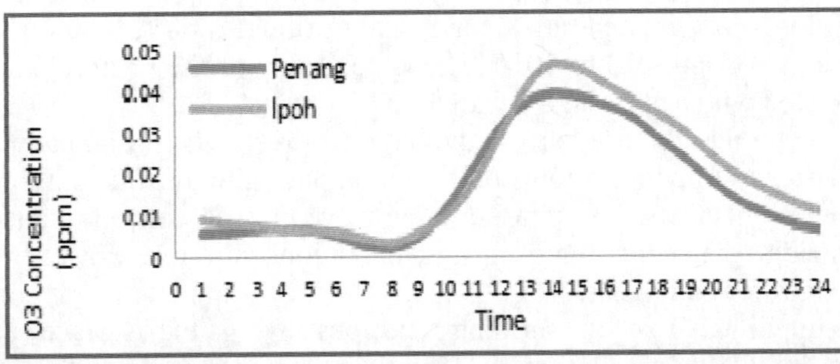

Figure 3. Diurnal concentration of surface ozone in Penang and Ipoh

3.4 Monthly Variation of Ozone Concentration

The surface ozone concentration was analysed by using average hourly data for every month to identify the monthly variation of ozone concentration. Figure 4 presented the contour plot of surface O_3 concentration for Penang and Ipoh. Based on the plot, the O_3 concentration in Penang was higher in March and April, meanwhile in Ipoh, the highest O_3 concentration was in February. This is because, Peninsular Malaysia especially northern area were experiencing the effects of the El Nino phenomenon and northeastern monsoon winds before the monsoon wind shifts to wet weather and thunderstorms from April to middle May [10].

a) Penang

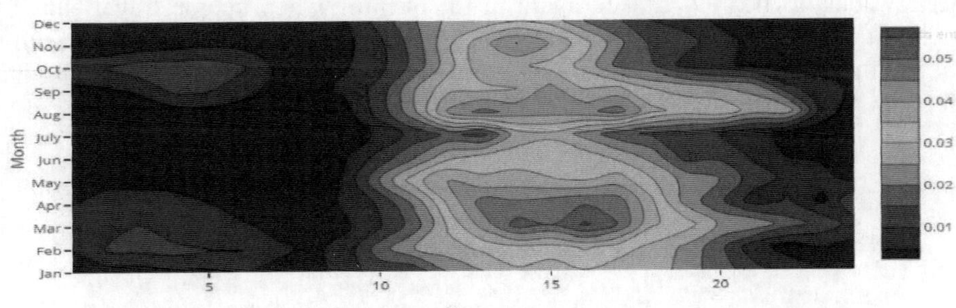

b) Ipoh

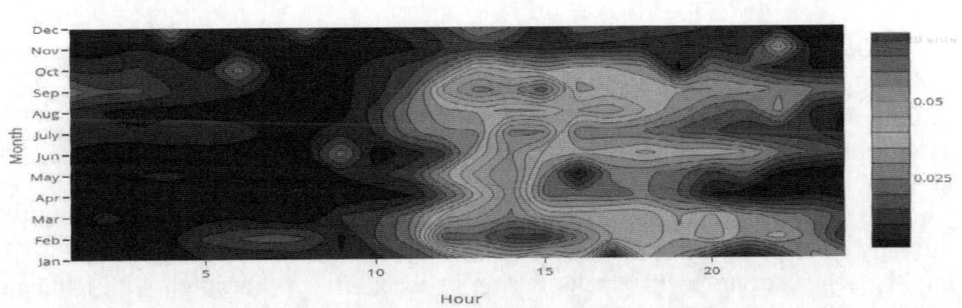

Figure 4. Contour plot of monthly O_3 concentration

This phenomenon causes the current temperature to rises and enhance the production rate of surface ozone concentration. The concentration of surface ozone are lower after July until end of December in both studies area especially Ipoh. This regarding to heavy rain and thunderstorm between the months which causes low temperature and wind speed to increase [14]. This will affect the photochemical reaction between O_3 precursors.

3.5 Diurnal Fluctuation of Ozone and Its Precursor

Figure 5 (a-b) compares the diurnal variation of surface ozone and its precursor which are NO_2, NOx and CO. The diurnal plots of O_3 concentration for Penang (a) and Ipoh (b) were plotted according to the average hourly data set from 2013 to 2016. The dependency of the O_3 formation to the precursor concentration is associated with a clear hourly variation.

The diurnal variation for all urbanized area around the world shows similar pattern which displayed peak concentration during afternoon and low concentration during night-time until morning. While the NOx and NO_2 concentration shows the opposite pattern from ozone concentration. Increasing ozone concentration correspond to the decreasing of its precursors concentration. However, CO concentration shows only slight changes for both areas.

NOx played an important role in forming and destroying the surface ozone. When NO concentrations is high, it will effectively titrate O_3 in urban areas. During rush hour between 8.00 until 10.00, the NO_2 concentration was recorded at the highest levels. O_3 precursors such as NOx and CO are originated from the sources like motor vehicles and industrial activities [15]. In addition, high concentrations of NO_2 will produce especially in areas with high number of vehicles since high emission of NO levels. Other than that, VOCs also a precursor to O_3, which is released from anthropogenic activity and natural resources as well [16].

The concentration of surface O_3 was highest during midday and reached the peak around 13.00-15.00 and appear to reduce in the night. This pattern happen when the intensity of sunlight was at the maximal level and it will promote the formation of O radical, which then responsible for the formation of O_3 from its precursors [11,17]. Meanwhile, in the absence of sunlight during evening until early morning, the surface O_3 concentration reached minimum value since the UV intensity was very small, hence the photochemical reaction is stopped.

The concentrations of the NO_2 peaked again in the evening when people travel back home from work. The concentrations of O_3 precursors showed a less steep drop during the night due to the higher atmospheric stability and the absence of sunlight, which reducing the scavenging reaction of photochemical oxidants [16].

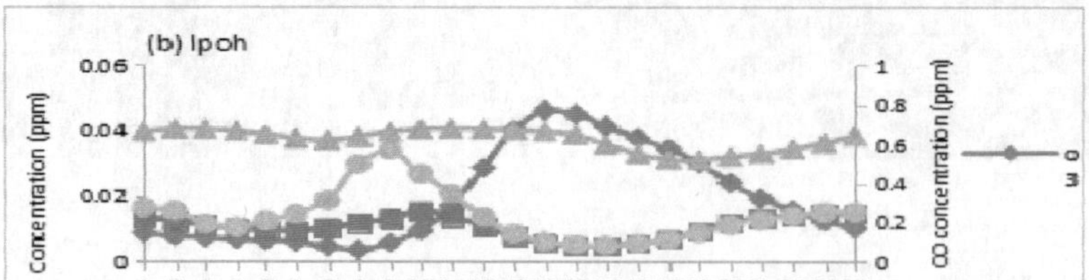

Figure 5. Diurnal plot of O_3 and its precursors in (a) Penang, (b) Ipoh

3.6 Pearson Correlation Analysis

Figure 6 show linear relationship between the two variables in this two urbanized areas which are Penang and Ipoh. In this analysis, the surface ozone concentration were correlated between its precursors and the meteorological parameter. Based on this result, ozone concentration in Penang and Ipoh were strongly correlated with temperature, humidity and wind speed.

In Penang, there is a strong positive relationship between O_3 concentration with temperature with the r value of 0.970 and negative correlated with the wind speed with r = -0.962 respectively.

IOP Conf. Series: Materials Science and Engineering **551** (2019) 012100 doi:10.1088/1757-899X/551/1/012100

Meanwhile, in Ipoh, the most correlated meteorological parameter were temperature and wind speed with the r value of 0.959 and r = -0.850. This result shows that formation of surface ozone are affected by temperature, as well as the wind speed, which both affect the rate of chemical reaction of O_3 production [18].

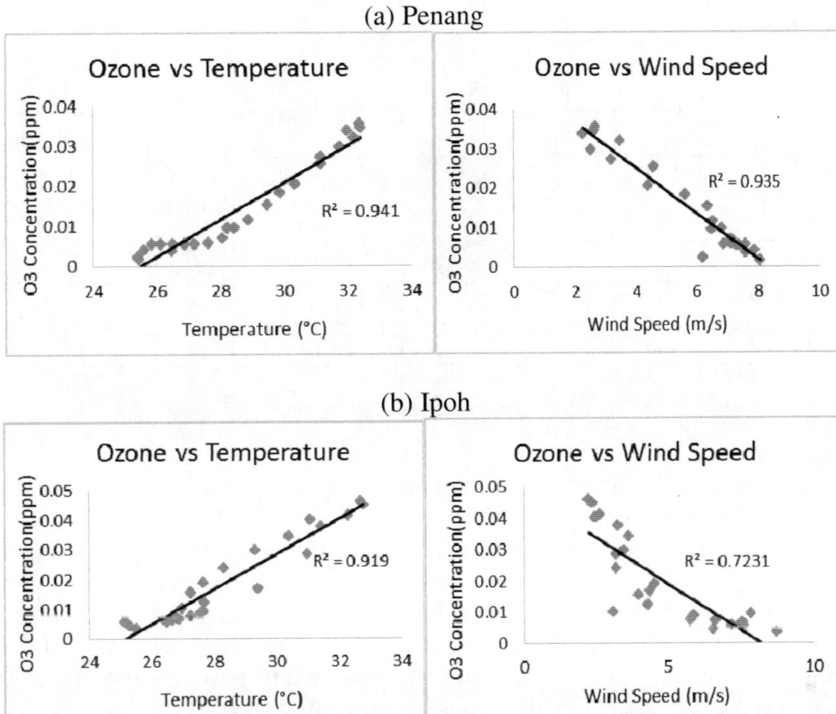

Figure 6: Relationship of O_3 concentration with temperature and wind speed in (a) Penang (b) Ipoh.

3.7 Principle Component Analysis (PCA)

There are three principal component (PCs) in Ipoh where the total variance is 31.35% which mainly comprised of O_3, temperature and relative humidity same as in Penang for the first component, thus was classified as the meteorological factor. The second component (PC2) comprises of NO_2 and NOx (16.96% of total variance) since the monitoring station is surrounded by densely residential areas and schools which, contributed to high number of vehicle.

Besides that, the third component (PC3) which represented by 14.25% of the total variance shows the positive loading of CO (0.872) and PM_{10} (0.686) as the main source from diesel vehicles such as lorries and bus [17]. While the fourth component (PC4) as the meteorological parameter which represented by wind speed with total variance of 11.19%. In short, first principle component (PC) group in Penang and Ipoh could be renamed as meteorological parameters and the second component could be appointed as the O_3 precursors which was similar to the finding from [10] that conducted PCA for the same 9 parameters in Shah Alam.

Table 2. Rotated Component Matrix

Air pollutant	Principle Component			Principle Component			
	Penang			Ipoh			
	PC1	PC2	PC3	PC1	PC2	PC3	PC4
O_3	0.671			0.725			
NO_2		0.847			0.861		
NOx		0.775			0.872		
SO_2			0.642		-0.475	0.358	
CO			0.599			0.863	
PM_{10}	0.312		0.730	0.442		0.686	
Temperature	0.865			0.839			
Humidity	-0.835			-0.840			
Wind speed	-0.723						-0.875

3.8 Model Development

The original air quality data in Penang and Ipoh was further analysed by development of model by using Multiple Linear Regression for prediction of O_3 concentration for the next hour, next two hour and next three hour during the peak hours (1300-1500). Then the hybrid model was also develop by using Principle Component (PCs) as the input in order to enhance the predictability of MLR model. The data for three peak hours which are from 13.00 – 15.00 were selected for both model. The model for O_3 prediction for both method was presented in Table 3. The input for MLR model include O_3, NO_2, NOx, CO, SO_2, PM_{10}, temperature, humidity and wind speed. While for PCA-MLR, the significant variable which was greater than 0.5 was used as input.

Table 3 : Summary model for O_3 concentration predictions

Method	Prediction Day	Model
MLR	Next hour	$O_{3,\ t+1,\ PNG} = 0.067 - 0.001T - 0.141NOx + 0.494SO_2 + 0.307NO_2 + 0.289O_3 + 0.003CO$
		$O_{3,\ t+1,\ IPH} = 0.058 + 2.123*10^{-6}WS - 0.001T + 0.09NOx - 0.249SO_2 - 0.33NO_2 + 0.374O_3 - 0.002CO$
	Next two hour	$O_{3,\ t+2,\ PNG} = 0.032 - 0.132NOx - 0.003SO_2 + 0.045NO_2 + 0.204O_3 - 0.002\ CO$
		$O_{3,\ t+2,\ IPH} = 0.046 + 4.233*10^{-6}WS - 9.781*10^{-5}H + 0.087NOx - 1.461SO_2 - 0.296NO_2 + 0.287O_3 - 0.001CO + 8.826*10^{-5}PM_{10}$
	Next three hour	$O_{3,\ t+3,\ PNG} = 0.039 - 9.065*10^{-5}T + 0.018NOx + 0.218SO_2 - 0.053NO_2 + 0.186O_3 - 0.002CO + 9.366*10^{-5}PM_{10}$
		$O_{3,\ t+3,\ IPH} = 0.079 + 3.449*10^{-6}WS - 0.001T + 0.075NOx - 0.643SO_2 - 0.293NO_2 + 0.239O_3 - 0.001CO$
PCA-MLR	Next hour	$O_{3,\ t+1,\ PNG} = 0.026 + 0.001PC1 + 0.001PC2 - 9.837*10^{-5}PC3$
		$O_{3,\ t+1,\ IPH} = 0.043 - 6.926*10^{-5}PC1 - 0.248PC2 + 8.346*10^{-6}PC3 - 4.24*10^{-6}PC4$
	Next two hour	$O_{3,\ t+2,\ PNG} = 0.039 + 0.001PC2$
		$O_{3,\ t+2,\ IPH} = 0.043 + 9.183*10^{-5}PC1 - 0.153PC2 + 1.064*10^{-5}PC3 - 4.809*10^{-6}PC4$
	Next three hour	$O_{3,\ t+3,\ PNG} = 0.039 + 5.195*10^{-5}PC1\ ^{-}7.764*10^{-5}PC2 + 7.151*10^{-5}PC3$
		$O_{3,\ t+3,\ IPH} = 0.043 - 0.196PC2 + 2.916*10^{-5}PC3 - 3.071*10^{-6}PC4$

3.9 Performance Index

The developed MLR and PCA-MLR model were assessed by using a few performance indicators.The calculated performance indicators for both urbanized areas were summarized in Table 3. The MLR was chosen as the most reliable method compared to PCA-MLR with smaller error (MAE=0.0191: NAE=0.0562: RMSE=0.0236) and higher accuracy (PA=0.4675: R2=0.4351: IA=0.5092) for the next hour in Penang. In general, both models give low R^2 values indicating that the performance of the models in prediction of O_3 concentration (next hour, next two hour and next three hour) are not very good. However, MLR shows better performance than PCA-MLR. This is because MLR models is better in prediction of O_3 levels for time periods of greater than 1 hour. It has also successfully used in predictions of O_3 concentrations one day ahead as proved by previous researches [19].

In this study, Multiple Linear Regression give the best performance of O_3 concentration prediction for the next hour, next two hour and next three hour in both area. Most of the models studied on ozone concentration have shown that PCA-MLR has higher accuracy than MLR except for the model studied by [20] on the variation of O_3 concentration in Dilovasi, Turkey which had similar results with this study. It is due to the dataset used in this study was minimal which the three peak hours (1300 - 1500) of dataset were used. Thus, the usage of PCs as input in MLR could not improve the performance of the model to study the variations of O_3 concentrations.

Table 4. Performance Comparison of O_3 Prediction Models

Study Area		Method	MAE	NAE	RMSE	PA	R^2	IA
Penang	Next hour	MLR	0.0191	0.0562	0.0236	0.4675	0.4351	0.5092
		PCA-MLR	0.0254	0.0643	0.0467	0.4361	0.3164	0.4672
	Next two hour	MLR	0.0221	0.0648	0.0264	0.4675	0.4351	0.4703
		PCA-MLR	0.0276	0.0704	0.0436	0.4105	0.3132	0.4361
	Next three hour	MLR	0.0253	0.0708	0.0316	0.4443	0.4177	0.4033
		PCA-MLR	0.0297	0.0738	0.0395	0.3848	0.2961	0.3942
Ipoh	Next hour	MLR	0.0158	0.0463	0.0200	0.3812	0.2447	0.3827
		PCA-MLR	0.0239	0.0510	0.0259	0.3170	0.2212	0.3594
	Next two hour	MLR	0.0220	0.0542	0.0269	0.3497	0.2224	0.3784
		PCA-MLR	0.0294	0.0548	0.0297	0.3043	0.2203	0.3236
	Next three hour	MLR	0.0228	0.0610	0.0276	0.3029	0.2186	0.3526
		PCA-MLR	0.0301	0.0654	0.0306	0.2831	0.2135	0.3192

The relationship between the prediction and actual for the next hour O_{3+1} concentration for both studies area is presented by using scatter plot as shown in Figure 7. The figure shows that all plots have positive trend and showed weak association between the predicted and the actual values. The MLR model for both Penang and Ipoh has higher R^2 values than the PCA-MLR model with the value of 0.4351 and 0.2447.

(a) Penang

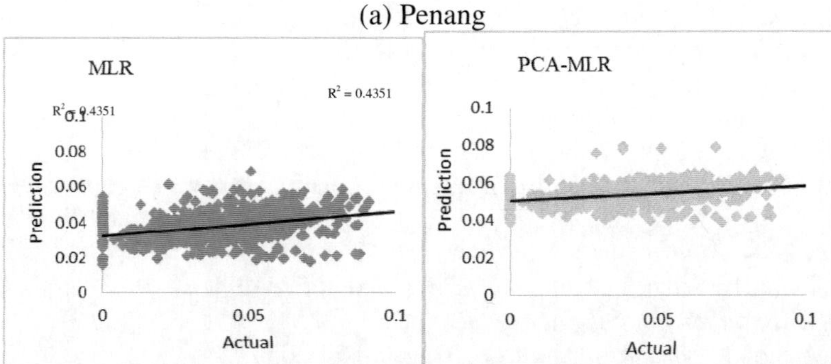

(b) Ipoh

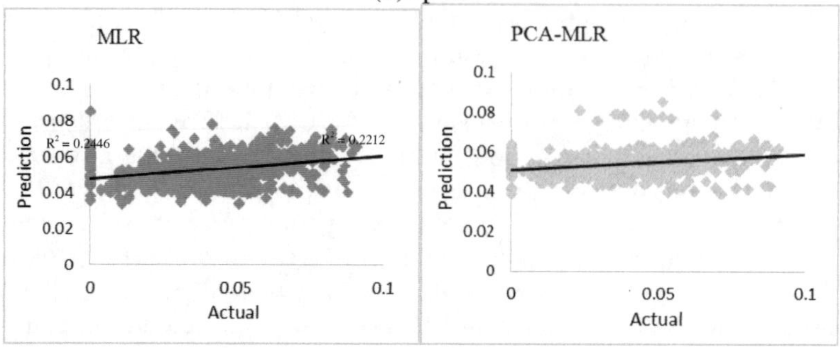

Figure 7. Relationship between prediction and actual O_3 concentration

4. Conclusions

In this study, MLR model was developed and modified MLR model was also develop using Principle Component (PCs) as the input to enhance the predictability of Multiple Linear Regression model. Then, the performance indicator were used to find the goodness of the models. The result shows that the MLR model was more accurate model in predicting O_3 concentration prediction during the peak hour compared to PCA-MLR in both study areas.

By conducting this study, a sufficient database of ozone concentration, can be used as reference for future development since the weather parameter at the study area has been achieved. Other than that, a prediction of O_3 by using MLR is capable and a really useful tool in providing early information to the public in an urban area.

Acknowledgements

Author would like to thank Department of Environment, Malaysia for the air pollutant dataset and Ministry of Education Malaysia for the Fundamental Reserach Grant Scheme (FRGS 9003-00508).

References

[1] Burghardt T E *et al* 2016 **14** 714.
[2] Abdul-Wahab S A *et al* 2002 *Environmental Modelling & Software* **17** 219.
[3] Alkasassbeh M *et al* 2013 *International Journal of Advanced Science and Technology* **55** 1.
[4] Noor N M *et al* 2015 *Revista de Chimie* **66** 1443.
[5] Noor N M et al 2014 *Key Engineering Materials* **594-595** 902.
[6] Sara Y Y et al 2013 *Advanced Materials Research* **620** 502.
[7] Karlsson P E *et al* 2017 *Science of total Environment* **576** 22.
[8] Kumar N *et al* 2017 *Urban Climate* **20** 148.
[9] Department os Statistics Malaysia 2010 *https://www.dosm.gov.my/v1* (Retrived on 15 Jan 2018).
[10] Awang N R *et al* 2015 *Atmospheric Pollution Research* **6** 726.
[11] Zhao W *et al* 2016 *Atmospheric Environment* **144** 182.
[12] Noor N M *et al Autralian Journal of Basic and Applied Sciences* **5** 2796.
[13] Utusan Online *http://www.utusan.com.my/berita/nasional/negeri-di-utara-berhadapan-cuaca-panas-lebih-tinggi-sehingga-mac-15-* (Retrieve on 5 May 2018).
[14] Department of Environment 2015 Environmental Quality Report. Department of Environment: Putrajaya.
[15] Vingarzan R 2004 *Atmospherics Environment* **38** 3431.
[16] Latif M T *et al* 2016 *Science of the Total Environment* **573** 494.
[17] Xie M *et al* 2016 *Science Total Environment* 551.
[18] Alghamdi M *et al* 2014 *Atmospherics Environment* **94** 205.
[19] Sousa S I V *et al* 2006 *Chemosphere* **64** 1141.
[20] Ozbay *et al* 2011 *Atmospheric Research* **102** 57.

Utilization of rice straw ash during composting of food waste at different initial C/N ratios for compost quality

Siti Noor Baya Binti Khalib[1], Irnis Azura Binti Zakarya[1,2,3] and Tengku Nuraiti Binti Tengku Izhar[1,2]

[1]School of Environmental Engineering, Universiti Malaysia Perlis, 02600, Arau, Perlis, Malaysia

[2]Centre of Excellence Geopolymer and Green Technology, Universiti Malaysia Perlis.

[3]Centre of Excellence for Biomass Utilization, Universiti Malaysia Perlis.

E-mail: irnis@unimap.edu.my

Abstract. The increases in rice productivity and size of paddy areas have led to the huge excess of rice straw in paddy field after the harvesting season. Rice straw can be used as a soil conditioner by composting process instead of on-farm burning that been practice by farmer which can caused pollution. Nevertheless, it required a long time for the decomposition process to happen since the main fraction of rice straw is lignocellulose which was difficult to degrade. Thus, the objective of this study was to evaluate the effect of different initial C:N ratio used (20,25,30) during composting of rice straw ash and food waste towards compost quality in term of pH, temperature, moisture content, C:N ratio and germination index. A laboratory-scale bin composter reactor was used during 30 days of composting period. The composting mixture with initial C:N ratio of 30 maintained the temperature exceeding 55°C for more than 3 days but the composting mixtures with initial C:N ratios of 20 and 25 did not meet the requirement of pathogen destruction. The maximum final C:N ratio and germination index obtained at a composting mixture with initial C:N ratios of 30 compared to the other two composting mixtures. The pH and moisture content value in three composting mixture was recorded in a range of 7-9 and 40-60% for mature compost to establish.

1. Introduction

The increases in rice productivity and size of paddy areas have led to the huge excess of rice straw in paddy field after the harvesting season [1]. The production of paddy was increases years to years, from more than 2.5 million tonnes of paddy was produced in year 2010 into more than 2.7 million tonnes of paddy in year 2014. This means that, the same amount of rice straw residue also was produced in that year [2]. It is a non-edible product, which often left in the paddy field after harvesting process [3]. The elimination of rice straw is frequently carried out by the farmers all over the world by burning it in the field [1,4,5].

The other method to avoid the rice straw from on-farm burning was removing the rice straw from paddy field for others environmental application like composting to produce a good compost used as a soil enhancer back towards the paddy field. However, the main fraction of rice straw is lignocellulose which was difficult to degrade and make the efficient composting performance generally hard to be achieved or required a longer time for the decomposition process [6]. Regarding to the issue of lignocelluloses, rice straw will be burned before being used to help in accelerating the composting process, which is known as rice straw ash (RSA). The product from RSA composting known as soil enhancer will be applied back to paddy field soil to recover the nutrient content loss from removing the rice straw for composting process.

Food waste was used to be composted with rice straw ash during this study because of its high moisture content. Besides that, in Malaysia, food wastes are one of the largest components of municipal solid waste produced [7]. Although considerable research has been conducted on composting of rice straw with various organic materials [6, 8, 9, 10, 11, 12], but no information available on the effect of composting rice straw ash with food waste was done.

The objective of this study was to evaluate the effect of different initial C:N ratio used (20,25,30) during composting of rice straw ash and food waste towards compost quality in term of pH, temperature, moisture content, C:N ratio and germination index. The compost quality which is stability and maturity compost cannot be well described by a single property or parameters [13,14]. Thus, the stability of compost during this study will be refers to the changes of pH, temperature, moisture content and C:N ratio. Meanwhile, the maturity of compost will refer to the germination test.

2. Materials and method

2.1. Source materials

Food waste (FW) was collected from stalls near School of Environmental Engineering, UniMAP. Rice straw (RS) was collected from a local farmer of paddy field at Arau, Perlis. RS was shredded to 2-3cm to allow a uniform size before the burning process using a shredder machine. RS was burned in a muffle furnace at 400°C for 30 minutes to produce rice straw ash (RSA). Goat dung used as a source of nitrogen for the composting mixture of RSA and was collected from goat farm at Padang Siding, Pauh, Perlis.

Besides that, the prepared liquid effective microorganisms (EM) will be poured into the mixture of compost as a source of microorganism to help in accelerating the decomposition process [20]. EM also functions to reduce the odour emission during the process. EM will be prepared according to the recipe obtained from Solid Waste Corporation (SWCorp), Perlis. One piece of *tempe* will mix with 250g granulated brown sugar and 3liter (fermented the mixture for 1 week before used) for 3kg of waste. The ratio of liquid EM (liter) to compost mixture (kilogram) is 1:1. RS, RSA, FW and GD was characterized for pH, moisture content, C:N ratio, total carbon and nutrient content (nitrogen, phosphorus, potassium) meanwhile the liquid EM was characterize for pH only.

2.2. Composting experiment

A laboratory-scale of bin composter reactor used for this study was modified from other research [9,15] as shown in Fig.1. The reactor is a cone-shape. The dimension of bin composter is 50 cm height and 40 cm and 30 cm in diameter (top and bottom, respectively). The reactor had a removable lid to be opened each times when the temperature inside the reactor was dropped for sample collection and also for mixing and turning process. A hole was made on the lid of bin composter for gas measurement and air outlet. A rubber stoppers was used as a sealer to the hole during the composting process.

The reactor was equipped with a fine-masked net, 8 cm from the bottom to place the composting materials. An aeration air flow tube was installed 40 cm from the bottom of the reactor to maintain an aerobic condition during the composting process. The air was supply by using an air pump at a controlled rate. A hole also made at the center of the bottom reactor for leachate collection.

Three different initial C:N ratios of 20, 25 and 30 was used during the composting process at an aeration rate of 0.6L/(ming.kg) OM. The mixture of rice straw ash and food waste with goat dung and EM as organic accelerator are manually mixed in a ratio of 1kg RSA: 3.4kg FW: 2kg GD: 1L EM for C:N 30, 1kg RSA: 3kg FW: 2kg GD: 1L EM for C:N 25 and 1kg RSA: 1.6kg FW: 2kg GD: 1L EM for C:N 20. Each composting process lasted 30 days. The characteristics of the raw materials are presented in table 1.

The air pump run continuously, but it turned off about 15 minutes for sample collection and also for mixing and turning process once every three days. About 50 gram of sample was withdrawn after each turning once every three days until the end. The sample was divided into two parts. One part was immediately analyzed for pH, total carbon and phytotoxicity evaluation. The other part was air dried to a constant weight at 60°C for 2 days for chemical analysis and then ground the sample to pass through 1-2mm sieve and stored in a dessicator.

2.3. Compost analysis

The temperature of the compost pile was measured by a digital thermometer recorded daily. About 1g of sample was placed into 10ml de-ionized water, stirred up and then left the mixture to settle before measured the value of pH by using a pH meter electrode [6]. The moisture content was determined by oven drying 5g fresh sample at 105°C for a period of 24 hours [16]. The carbon to nitrogen ratio was obtained by dividing the value of total carbon to the value of total nitrogen. The seed germination technique was used to evaluate the phytotoxicity of compost extracts [17].

Table 1. Characteristics of the raw composting materials

Materials	pH	Moisture content (%)	Total Carbon (%)	Total Nitrogen (%)	C:N	Phosphorus (%)	Potassium (%)
Rice straw ash (400°C)	9.859	5.44	37.25	2.00	18.63	0.09	0.49
Food waste		81.92	40.00	2.50	16.00	-	-
Goat dung	7.556 7.997	11.55	18.22	1.01	18.04	-	-
Effective microorganisms	8.021	-	-	-	-	-	-

3. Result and discussion

3.1. pH

The changes of pH in composting of rice straw ash and food waste at different initial C:N ratio was shown in figure 1. The pH values for composting mixture of initial C:N ratio of 20, 25 and 30 increases from initial pH 7.826, 7.428 and 7.508 on day 3 to a maximum pH 9.462, 8.566 and 8.657 on day 24. The releasing of ammonia from ammonification and mineralization of organic nitrogen during the initial phase of composting causing the increases of these pH values [18,19]. At the end of the composting process on day 30, the pH values for composting mixture of initial C:N ratios 20, 25 and 30 decrease to 8.776, 8.332 and 8.165 due to the production of organic and inorganic acids, the decomposition of organic matter and the release of carbon dioxide during the composting process [18]. The pH value of the composting mixture with an initial C:N ratio 20 was slightly higher than the other two composting mixture was possibly due to the lower amount of food waste contained in that composting mixture.

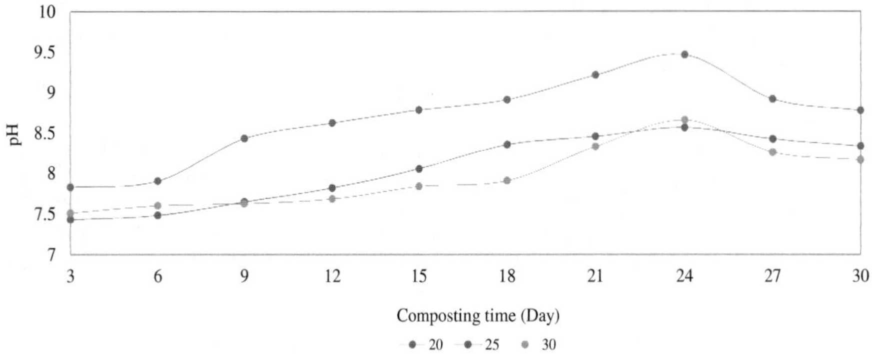

Figure 1. pH variation over composting time

3.2. Temperature

Temperature is a direct parameter to measure the status of the process. It is an indicator of the microbial activities during the decomposition process [13,20]. There are three phases of temperature during the composting process; mesophilic, thermophilic and curing phase. Figure 2, shows the changes of temperature during the composting mixture with initial C:N ratio 20,25 and 30. The temperature in three composting mixture increases rapidly to reach a maximum temperature of 56°C for initial C:N ratio 20,25 and 58°C for initial C:N ratio 30. The increased of temperature was due to the rapid microbial activities that breakdown the available organic matter and nitrogenous compounds in the compost pile [19]. When the organic matter become more stabilize and the easily degradable compound was depleted, the microbial activities in compost pile and the organic matter decomposition rate starting to slowdown. At this state, the composting mixture entering the curing phase causing the temperature slowly to dropped to reach ambient temperature [19,21]. The composting mixture with initial C:N ratio 30 had a longest thermophilic phase for 10 days with maintained the temperature exceeding 55°C for 5 days. It ensured the maximum pathogen reduction and the stabilization of organic matter. To destroy the phatogen and weed seed, all the composting mixture must exceed 55°C for at least for three consecutive days [22]. Although the composting mixture with initial C:N ratio 20 and 25 had a thermophilic phase for 8 and 9 days, but they did not meet the requirement for pathogen destruction because only reach temperature more than 55°C for two days. Beside, composting mixture with low initial C:N ratio of 20 and 25 resulting in a shorter thermophilic phase compared to the initial C:N ratio of 30 caused by the insufficient of carbon source for low C:N [19].

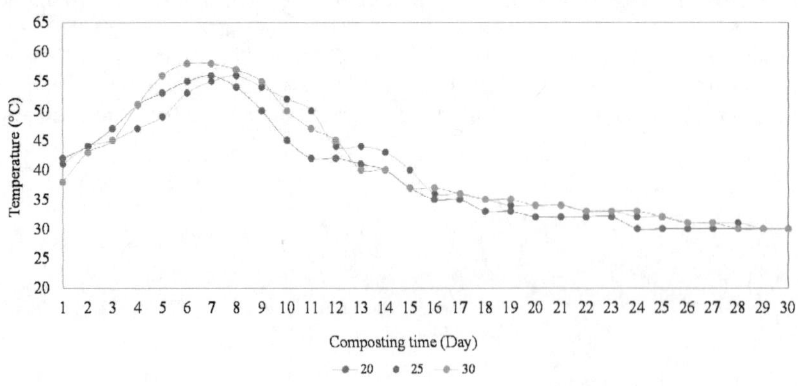

Figure 2. Temperature variation over composting time

IOP Conf. Series: Materials Science and Engineering **551** (2019) 012101 doi:10.1088/1757-899X/551/1/012101

3.3. Moisture content

The moisture content during the composting process plays an important role as a medium for transporting dissolved nutrients that are required for the physiological and metabolic activities of microorganisms [23]. Figure 3 shown the changes values of moisture content that increasing at the early stage of composting process before decreased at the later stage for all composting mixture with initial C:N ratio 20,25 and 30. Based on the values of moisture content obtained, the initial C:N ratio has no significant influence on moisture content. The high amount of moisture content at the end of the composting mixture with initial C:N ratio 20 restricted the decomposition of toxic compounds causing the compost produce in this compost pile slightly toxic and required to extend to a bit longer to form a mature compost [21]. However, according to George, 2010 [24] and Tchobanoglous et al., 1993 [25], the values of moisture content obtained for all the composting mixture was in a range of 40-60-%, which had sufficient to support the growth of microorganisms during the decomposition process.

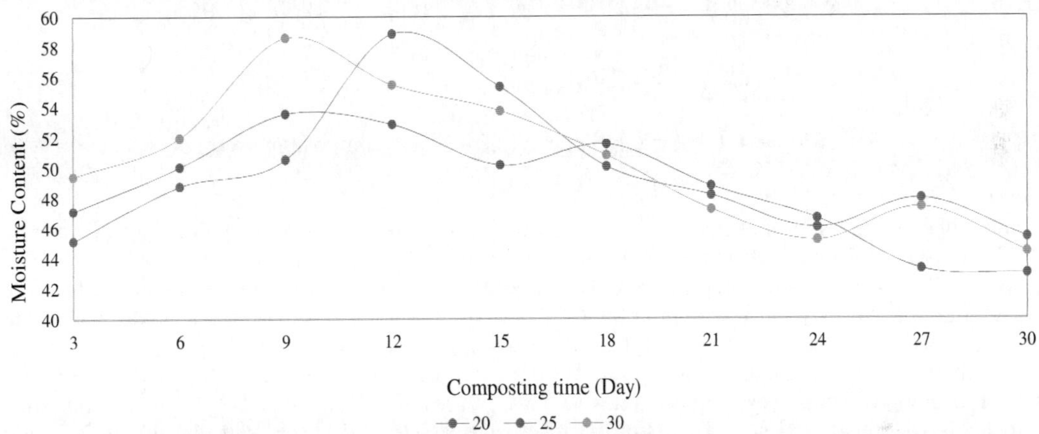

Figure 3. Moisture content variation over composting time

3.4 Germination Index

According to Gariglio et al., (2002) [26] and Brewer and Sullivan, (2003) [27], germination test is a sensitive indicator of maturity and often conducting to evaluate the phytotoxicity of compost produced. If the GI value obtain more than 80%, it can be considered that the compost has reach the level of maturity and practically the compost was free from phytotoxins substances and mature [28,29]. Figure 4 demonstrates the changes of germination index throughout the composting mixture with initial C:N ratio 20,25 and 30 for 30 days composting period. The GI values was increased for all composting mixture with increase the composting time resulting that all the toxic materials like volatile fatty acids mainly acetic acid and ammonia was decomposed completely [30]. At the end of the composting process, the GI for composting mixture with initial C:N ratio 30 and 25 were higher than initial C:N ratio 20 with the values over 80% suggesting that the compost produce was free from toxic substance and enough mature. Composting mixture of initial C:N ratio 20 was never over 80% at day 30 which only reach 70.02% of GI value. It showed that composting mixture with initial C:N ratio 20 was slightly contain toxic materials due to the incomplete decomposition of volatile fatty acids and ammonia. Thus, a longer time was required to form mature compost when a low C:N ratio was used. Similar suggestions were also reported by others researcher during co-composting of spent pig-manure sawdust litter and pig sludge [31], composting of pig manure by two static aerobic piles with initial C:N ratios of 15 and 30 and co-composting of pig feces and corn stalks with initial C:N ratio 15, 18, 21 [21].

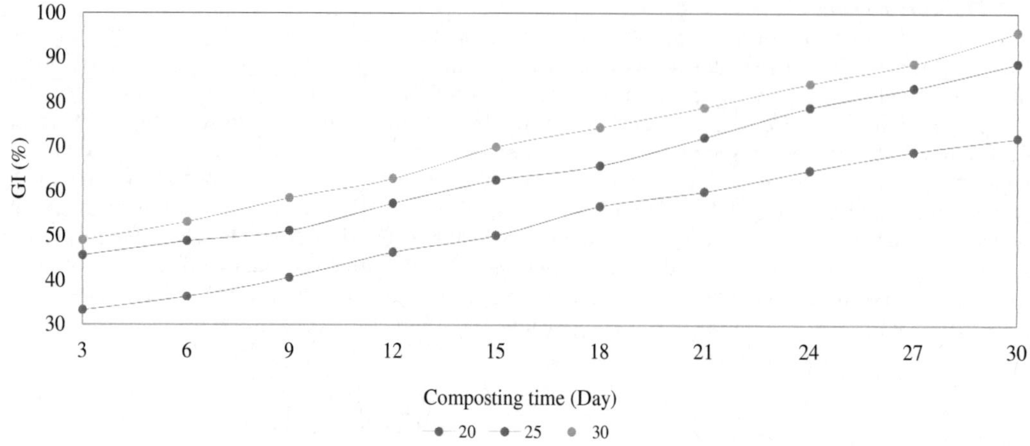

Figure 4. Germination index variation over composting time

3.5 C:N ratio

Table 2 shown the result of composting mixture with different initial C:N ratio of 20,25 and 30 and end C:N ratio on day 30 of the composting process with constant aeration rate of 0.6L/(min.kg). The end C:N ratio for all composting mixture were decreased due to the mineralization of organic matter [33]. Saha et al., (2010) [34] stated that an initial C:N ratio of 30:1 is ideal for feedstock mixtures. It then, decreases to less than 20:1 at the end of the composting process as obtained by the composting mixture with initial C:N ratio 30 in this study. Composting mixture with initial C:N ratio 20 and 25 shown a bit poorer decomposition process because of a smaller decreases of C:N at the end of the composting time especially for initial C:N ratio 20. However, Bernal et al., (2009) [13] stated that when C:N ratio decreased below 15, the compost had satisfied an acceptable standard of maturation. Thus, all the composting mixture with initial C:N ratio 20,25 and 30 can be defined mature, but compost produce from composting mixture with initial C:N ratio 20 and 25 must be used with precaution because of the bit poorer decomposition process.

Table 2. The results of end C:N ratio for the composting mixtures

Aeration rate (L/kg.min)	Initial C:N ratio	End C:N ratio
0.6	20	13
0.6	25	11
0.6	30	12

4 Conclusions

The compost quality during the composting of rice straw ash and food waste have been investigated at initial C:N ratio of 20, 25 and 30 in term of pH, temperature, moisture content, germination index and C:N ratio. The composting mixture of initial C:N ratio 30 was maintained the temperature exceeding 55°C for 5 days but the composting mixtures with initial C:N ratios of 20 and 25 did not meet this requirement of pathogen destruction. The final germination index in composting mixture of 30 and 25 was over 80% with 95.56% and 88.79% compared to the initial C:N ratio 20 that was not reach over 80% of GI value. The maximum final C:N ratio was recorded at the composting mixture of 30 due to a larger decrease of C:N ratio rather than initial C:N ratio 20 and 25 that had a smaller decreases of C:N at the of the composting process. The value of pH and moisture content, in a range of 7-9 and 40-60% for all composting mixture obtained in a range that acceptable for mature compost to establish. To

IOP Conf. Series: Materials Science and Engineering **551** (2019) 012101 doi:10.1088/1757-899X/551/1/012101

economically manage the increasingly quantities of rice straw ash and food waste produced each year, composting of these two waste materials with initial C:N ratio 30 is recommended with aeration rate of 0.6L/(min.kg) and burning the rice straw at 400°C for 30 minutes.

Acknowledgement
The authors gratefully acknowledge the supporting staff in the Laboratory School of Environmental Engineering, UniMAP, Centre of Excellence Geopolymer and Green Technology and Centre of Excellence Biomass Utilization. for the good support in this research.

References
[1] John A 2013 *Agricultural and Food Policy Studies Institute, Malaysia* **18**.
[2] FOASTAT 2017 http://www.fao.org/faostat/en/#data/QC
[3] Lal R 2005 *Environ. Ent.* **31** 575.
[4] Gadde B. *et al Environ. Pollut.,* **157** 1554.
[5] Rosmiza M Z *et al* 2014 *Malaysian Journal of Society and Space,* **10** 5.
[6] Zhou C *et al* 2015 *Waste. Manage.* **40** 150.
[7] Kathirvale S *et al* 2003 *Renew. Energ.* **29** 559.
[8] Zhu N 2007 *Bioresource Technol.* **98** 9.
[9] Li X *et al* 2008 *Bioresource Technol.* **99** 6271.
[10] Roca-Perez L *et al* 2009 *Chemosphere,* **75** 781 .
[11] Rashad F M *et al* 2010 *Bioresource Technol.* **101** 5952 .
[12] Qian X *et al* 2014 *Waste Manage.,* **34** 901.
[13] Bernal M P *et al* 2009 *Bioresource Technol.* **100** 5444.
[14] Nolan T *et al* 2011 *Bioresource Technol.* **102** 9425.
[15] Andersen J K *et al* 2010 *Waste Manage.* **30** 2475 .
[16] International A 2003 *Standard Test Methods for Moisture, Ash and Organic Matter of Pest and Other Organic Soils.*
[17] Zucconi F *et al* 1981 *Biocycle* **22** 2.
[18] Wong J W C *et al* 2001 *Bioresource Technol.* **76** 99.
[19] Gao M *et al* 2010 *Bioresource Technol.* **101** 3808.
[20] Li Z *et al* 2013 *Chemosphere.* **93** 7.
[21] Guo R *et al* 2012 *Bioresource Technol.* **112** 171
[22] Zhang Y *et al* 2006 *Bioresource Technol.* **97** 1061
[23] Kumar M *et al* 2010 *Waste Manage.* **30** 4
[24] George W D 2010 *Extension Horticulture Specialist*
[25] Tchobanoglous G *et al* 1993 *Engineering Principles and Management Issues*
[26] Gariglio N F *et al* 2002 *J. Crop. Hort.* **30** 135
[27] Brewer L J *et al* 2003 *Compost Sci. Util.* **11** 2
[28] Tiquia S M *et al* 2010 *Environ. Pollut.* **110** 535
[29] Wei Y S *et al* 2000 *Resour Conserv and Recy.* **30** 203
[30] Fang M *et al* 1999 *Bioresource Technol.* **67** 19
[31] Tiquia S M *et al* 1998 *Bioresurce Technol* **65** 43
[32] Huang G F *et al* 2004 *Waste Manage.* **24** 8
[33] Rasapoor M *et al* 2009 *Waste Manage.* **29** 2
[34] Saha J K *et al* 2010 *Waste Manage.* **30** 2

IOP Conf. Series: Materials Science and Engineering **551** (2019) 012102 doi:10.1088/1757-899X/551/1/012102

Design and Development of Tape Applicator For Managing Wire Purposes

Mohamad Syafiq A.K[1,2*], Nur Amirah Adila H[1], Muhammad Ikman Ishak[1], C.Y. Khor[1], M.U. Rosli1, Mohd Riduan Jamalludin[1], M.A.M. Nawi[1]

[1] Faculty of Mechanical Engineering Technology, Universiti Malaysia Perlis, 02100 Sungai Chuchuh, Padang Besar, Perlis, Malaysia
[2] Centre of Excellence for Unmanned Aerial Systems (COEUAS), UniMAP, Kangar, Perlis

E-mail: msyafiq@unimap.edu.my

Abstract. The project titled Design and Development of Tape Applicator for Managing Wire Purpose is a project that needs to be achieved by evaluate the users' performance while using manual method when taping wires and design and developing a tape applicator. This project is focused on to make it easier for the workers to tape wires on the floor without have to bend their bodies. Primary and Secondary data is gathered for completing this project. Interview is done to determine the usual method applied by users in taping wires on the floor and the difficulties that users faced when using manual method. After data is collected, idea generation of concept and sketches are made manually by sketching. Several aspects are being considered to choose the final design which is the most suitable. The final design chosen for the tape applicator is designed by Computer Aided Drawing (CAD) using Solidwork Software. At the end of the project, fabrication process is done before the product is testing. As for the result, time consumed for taping wire cables by using product is shorter than the time consumed when manual method is applied. All of the objectives are achieved.

1. Introduction

In 2006, based on Bureau of Labor Statistics, United States Department of Labor, 250 workers involved in deaths caused by too exposed with the electric current. Half of the employees were injured in fires and explosions and from this information, it is well known that electricity can be conducted through the body. In a condition when the body is in contacted directly towards the surface of equipment or any tools, and at the same time the equipment is also in contacted with any surface that is electrically conducted under, current flow will occur and enters the body at the point that is in contact and traverse through the body then exits at contact point at the other side which is usually the ground [1].
From here, it can be concluded that electrical accidents that commonly happened due to conditions that is not safe can be in term of no rubber insulator around the wire, how the cord is placed while not in use and managing the scattered cord when the main plug is plugged in to the main power. The maintenance that is insufficient causes the tools, equipment and installations that at first is considered as secure and harmless, to decrease the performance and results for unsafe condition [2].
Thus, based on Occupational Safety and Health Act of 1970, it is the responsibility of the workers to provide more secure and workplace with safety features to the workers. OSHA's main function is to ensure the safety and health of America's working men and women by outlining and implementing

specification which includes providing training, outreach, and education. They introducing the partnerships and encouraging the continual process of improvement in workplace safety and health [3].

2. Methodology

2.1. Interview
In approaching the first objective, method of interviews is used. This method is used to identify closely towards the respondent's feelings, emotion and opinions in respect to this research study. In using this method, it gives benefit to researcher as the respondent includes their personal and direct contact to the researcher. It results for the process of gathering information can be done more smoothly. Researcher needs to have some skills in order to carry the interview successfully.
An interview is made around Perlis towards the individuals that involve in managing wires in their job scope and also to identify the method that they applied in the process of taping wire cables. The target place at first is the hall where events are usually held. Due to the hall that is not in operation for a few months, researchers focused on the second target place which is cybercafés. Some questions are prepared before interview session to get the details of the workers and gain information from them.

2.2. Questionnaire
Questionnaires are contributed to 20 people including the employees in the scope of wedding planning service, and individual that involves in photography career, and business site such as the night market. The female participated in this questionnaires were 9 people while the male respondents were 11 people. Most of the respondents were among the age range from 29 to 39 years old. Most of the respondents involves in night market seller which is 8 respondents, 6 respondents in scope of wedding planning service and 5 respondents from photography career.

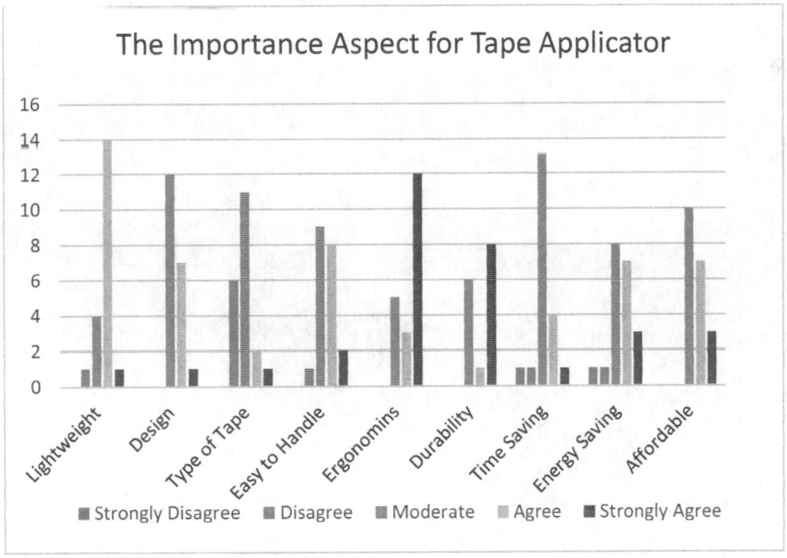

Figure 1. The questionnaire result for the important aspect for tape applicator.

The questionnaire also investigated the time consume when they tape the wires manually on the floor. It is also to identify level of importance in the aspects from the users' perspective in order to apply it in designing tape applicator. The aspects that has been listed are lightweight, type of tape used, save users' energy, save users' time, must be affordable, aesthetic value, easy to handle, easy to handle, and high durability. The range of importance rate from strongly disagree, disagree, moderate, agree, and strongly

IOP Conf. Series: Materials Science and Engineering **551** (2019) 012102 doi:10.1088/1757-899X/551/1/012102

agree. It was also use to help generate the customer needs and help in designing the final design of tape applicator.

Figure 1 shows the aspect level of importance based on user needs. These aspect level of importance helps in determining the customer needs for the tape applicator. The most important aspect that needs to be considered is the ergonomic of the product which it strongly agreed by 12 respondents. It is followed by the product that is durability and easy to handle which means it can be carried easily.

2.3. Prototyping Fabrication

All the obtained aspects were analysed and being transformed into the product prototyping. The aspect with the highest score became the major reference and consideration in designing a prototyping. The design process started with several sketches by using hand drawing. After the evaluation, only the best design was chosen to proced with CAD drawing. For this product, the CAD software used is Solidworks.

All the detail drawing was designed including the joint components and for assembly process. The joint components must be considered thoroungly since to easy the assemble and disassemble process. If the assemble process takes the longer time, time saving aspect was not achieved. Apart from that, for the stick or handle, which acts as a guide to push during taping process, the suitable height was evaluated based on ergonomics charts[4]. By considering average height of male and female Malaysians, the suitable height, 100 cm of stick was chosen. This is important to avoid the users need to bend their body during using this tape applicator. By using the suitable height, the user did not need to bend their body during taping process and the risk of back pain can be reduced [5].

The chosen material for the product is 1 kg ABS filament. This is because of abs plastic has good mechanical properties and can wistand high force [6]. All the CAD drawing data was converted and imported to 3D Printing Flashforge. The printing process takes 48 hours, for 4 parts before being assembled as final product.

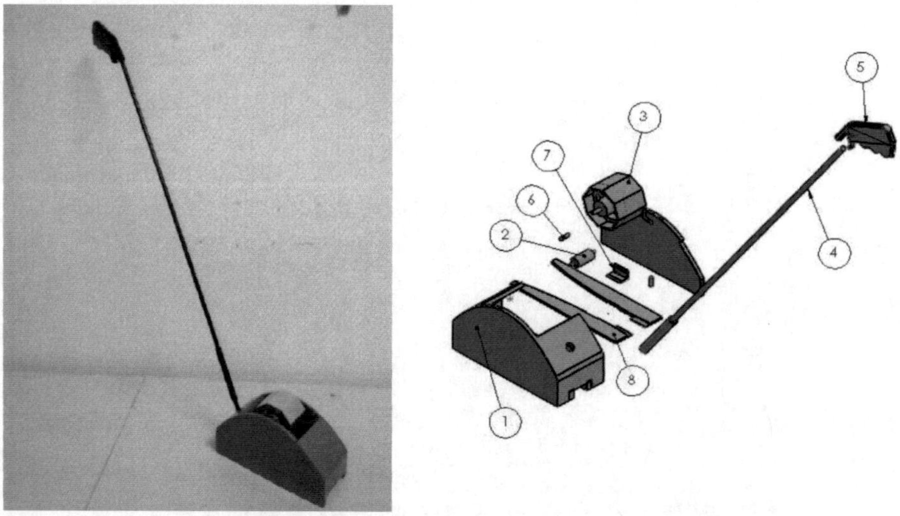

Figure 2. The drawing and the actual product of tape applicator.

IOP Conf. Series: Materials Science and Engineering **551** (2019) 012102 doi:10.1088/1757-899X/551/1/012102

3. Results and Discussions

3.1. Product Testing

For product testing, an experiment for the product testing has been done for user that works with wires. Three of them is chosen for the user testing. Based on the questionnaire result, they usually worked with wires that is 3 to 5 meters length. For the experiment, 1 meter of wire is chosen for the user testing to make the process of experiment easier.

The experiment is done using masking tape because of the same properties as the actual product, Gaffer tape that it does not left residue behind on the wire cables and it can be easily tear after taping process to make the process easier. Table 2 shows the results of the product testing.

RCO% is calculated from the oxirane oxygen content (OOC) that can be obtained from the direct titration of hydrogen bromide. The RCO% is calculated for the reaction process with different temperature and presence of catalyst. The percentage RCO calculated for epoxidation without presence of catalyst with different reaction temperature shown in table 1.

Table 1. Result of the Product Testing

User	Time taken when using manual method (s)	Time taken when using the product (s)
A	103.3	31.4
B	110.2	32.6
C	104.8	35.2

Based on the results of testing, it shows that the difference of the users' height proves the length of the handle is suitable with any height of user. The aspect has been considered during research to ensure it can be used by users in any job scope including household usage.

From this experiment, average time taken for both manual method and using prototype is taken. The average time taken when product is used is 33.1 seconds while the average time taken when manual method is applied is 106.1 seconds. Graph for both of the time taken by users is shown in figure 3.

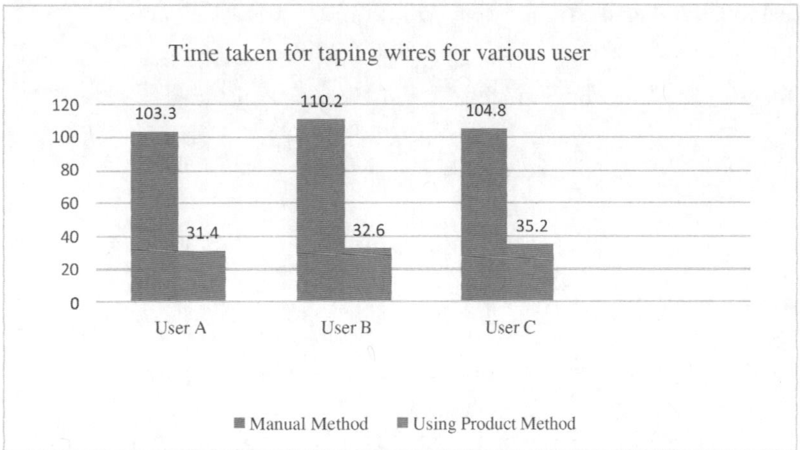

Figure 3. Time taken for taping wire for various users.

Based on the graph in figure 3, it is proven that the average time taken consumed when the tape is laid manually is longer than the time taken when product is used. Based on survey that has been done

IOP Conf. Series: Materials Science and Engineering **551** (2019) 012102 doi:10.1088/1757-899X/551/1/012102

during interview session, the time that usually consumed by users when taping 4 meter wires is in range 5 to 10 minutes. By means, it is equivalent to the time taken for 1 meter of wire during experiment is done.

One of the aspect that incorporate in wire taping process is the roller that act as the suppressor of the product. Manual method is a process that required more force to be applied on which it required user to press the tape harder while bending their body [6]. Product testing process required less force since the process is more user- friendly and the wire cables can be tape easily by the roller. Thus, the less force required for the wire cables to be laid on the floor, the less time taken is consumed to complete the process [7]. Based on the graph of Figure 2, the efficiency of the product is 73%, which means the time consuming for taping process for 1 meter of wire can be reduced up to 73%. Therefore, the productivity of the process can be enhanced besides the risk of back pain for the user can be improved.

4. Conclusions

The purpose of this project is to evaluate the user's performance in applying manual method in taping wire and to design and develop a tape applicator of wire cables. It can be concluded that the first objective was successfully done and achieved. All the methods done such as the interviews, questionnaires, watching You Tube videos, and doing experiment towards user helps researcher to analyze the accidents that usually happened among the employees in the job scope involving wire cables. Other than that, the questionnaires that has been distributed towards 20 respondents succeed to evaluate the users' performance in taping wire cables by using manual method. The common side effects towards the users' body has been identified due to the non- ergonomic method that they applied while doing their job.

References

[1] Bures M 2015 *Procedia - Soc. Behav. Sci.* **174** 3204.
[2] Elena M *et al* 2015 *Procedia Econ. Financ* **23** 54.
[3] Reese C D *et al* 2006 *Handbook of OSHA Construction Safety and Health* **2**.
[4] Shahriman A B *et al* 2017 *J. Phy Con. Ser* **908** 501.
[5] Ishak M K *et al* 2017 *MATEC Web Conf* **97** 01040.
[6] Abdullah A H *et al* 2015 *Adv. Mater. Res.* **1102** 139.
[7] Shahriman A B et al 2017 *AIP Conf Proc* 20120.

IOP Conf. Series: Materials Science and Engineering **551** (2019) 012103 doi:10.1088/1757-899X/551/1/012103

An Experimental Investigation of Humidity Effect on fabrication of Green Ceramic Membrane Prepared Via Dry Phase Inversion

Mohd Riduan Jamalludin[1,*], Siti Khadijah Hubadillah[2], Zawati Harun[3], Mohd Hafiz Dzarfan Othman[3] Muhammad Ikman Ishak[1], C.Y. Khor[1], M.U. Rosli[1], M.A.M Nawi[1], Mohamad Syafiq A.K.[1], Lailina N.M[1], Ras izzati Ismail[1] and Lee Yit Leng[1]

[1] Faculty of Engineering Technology, Universiti Malaysia Perlis (UniMAP), Kampus UniCITI Alam, Sungai Chuchuh, Padang Besar 02100 Perlis Malaysia.
[2]Advanced Membrane Technology Research Centre (AMTEC), Faculty of Petroleum and Renewable Energy Engineering, Universiti Teknologi Malaysia, 81310 UTM Johor Bahru,Johor Darul Takzim, Malaysia.
[3]Advanced Materials and Manufacturing Centre (AMMC), Faculty of Mechanical and Manufacturing Engineering, Universiti Tun Hussein Onn Malaysia, 86400 Parit Raja, Batu Pahat, Johor Darul Takzim, Malaysia.

Email: riduanj@unimap.edu.my

Abstract. Ceramic membranes are gaining a market demand nowadays due to its characterization and performance. However, in the preparation of ceramic membrane for water and wastewater separation, the processes involved are taking a higher cost. In this work, ceramic membranes were prepared via phase inversion within controlled humidity by using different composition of low cost and green ceramic material, kaolin. The preparation of ceramic membranes was conducted via phase inversion in high humidity; i) 3°C and, ii) room temperature. In this paper, we report on the high humidity affected the viscosity of membrane suspension. Thus, the viscosity of ceramic suspension was giving integration towards membrane characterization and performance. A dense membrane structure was observed when the ceramic membranes were prepared in high humidity and kaolin composition. The morphology of overall, porous and dense structure of ceramic membranes was characterized by field emission scanning electron microscope (FESEM) at x50, x500 and x5000 magnification respectively.

1.Introduction

Recently the use of membrane technology is widely employed into water treatment processes and industrial wastewater treatment due to their low cost production [1]. In fact, stringent regulations in industrialized countries and increasing water problem have promoted the use of membrane for water treatment. For two decades experience, water companies now operate thousands of membrane plants of treated water every day using membrane processes such as microfiltration (MF), ultrafiltration (UF), nanofiltration (NF), and reverse osmosis (RO). Compared with conventional water treatment process, such as clarification and sand filtration, membrane technology is an absolute barrier and therefore offers the advantage of selectively removing contaminants based on their sizes.

In membrane technology, polymeric membranes have been almost exclusively dealt with; however, alternative membrane materials have been getting attraction nowadays. Ceramic membranes have much greater advantages, which, because of its endurance to high temperature, high chemical resistance as well as high pressure [2-4]. In selectivity of raw materials, several studies on ceramic membrane preparation using ceramic materials such as Alumina (Al_2O_3), Zirconia (ZrO_2), metal oxide or metal alloys were carried out by Liu et al. [5]. However, in water and wastewater treatment, the production of ceramic membrane has been limited due to high cost production. Therefore, the low cost ceramic such as kaolin, raw clay, moroccan clay, Tunisian clay, sepiolite clay, Algerian clay and dolomite was intensively studied by many researchers in the preparation of low cost ceramic membrane. Kaolin is one of the cheapest membrane raw materials and easily available [6-8]. In addition, kaolin is hydrophilic and can be dispersed in water and in some other systems. Because of the nature of the chemistry of its surface, kaolin can be chemically modified so that it will become hydrophobic or organophilic, or both.

As far as we are concerned, the investigations in preparation of ceramic membrane with low cost ceramic materials (kaolin) composition via dry phase inversion system were rarely studied. Thus, in this work, the effect of different composition with low cost ceramic, kaolin was studied by phase inversion system within the controlled higher humidity involved. The prepared membranes were characterized via scanning electron microscopy (SEM) while the performance of the ceramic membrane was tested by flux within the pressure of gravity.

2. Methodology

2.1 Materials

Kaolin powder was purchased from QReC as the ceramic material. 1-Methyl-2-pyrrolidinone (NMP) was purchased from Merck. Polyether Sulfone (PESf, UDEL) was provided by BGOIL CHEM. The kaolin powder and the PESf were stored in 100°C in the oven before use to remove adsorbed moisture, and the other materials were used without any purification.

2.2 Fabrication of Green Ceramic Membrane via Dry Phase Inversion

To fabricate flat-sheet ceramic membranes within controlled humidity, the ceramic membranes were casted on the glass. The masking tape with 1.5mm was used as to control the thickness of flat-sheet ceramic membrane. The membrane was placed in the ice box set up within 3°C in order to let the phase inversion occur at high humidity. In order to investigate the preparation of flat-sheet ceramic membrane via high humidity, the membrane was differentiated with the fabrication of flat-sheet ceramic membrane in room temperature. The ceramic membranes casted on the glass were placed in desired controlled humidity for 24 hours. The phase inversion occurred at this time. After that, the glass with suspension is immersed into the non-solvent bath horizontally (Fig 1) for another 24 hours. The non-solvent used was distilled water. This immersion was done in order to remove NMP from the ceramic membrane and solidified the membrane completely.

The membrane was dried at room temperature for 24 hours. Then, the membrane was cut into the circle shape in order to record the diameter and thickness before and after sinter. This step was repeated for different composition of kaolin.

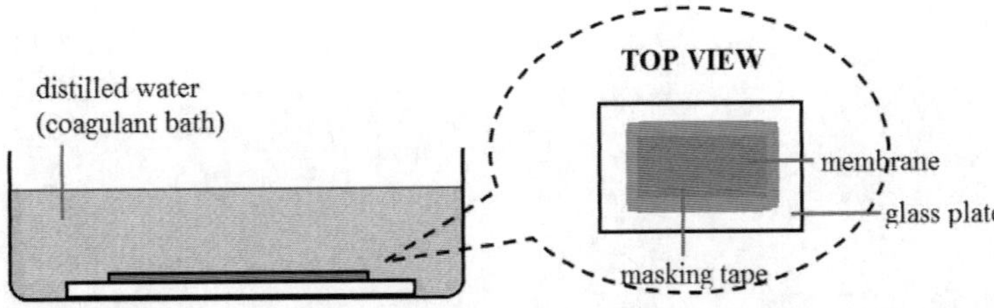

Figure 1. Schematic diagram for flat sheet ceramic membrane fabrication into coagulant bath via immersion method

2.3 Characterization of Green Ceramic Membrane
Observation of the membrane structure was conducted using a scanning electron microscope (JEOL JSM6380LA, Germany). The diameter and thickness of ceramic membrane before and after sinter were recorded to study the effect of humidity on the shrinkage of ceramic membrane prepared by low cost ceramic material. The samples were cut into the circle shape before sinter. In order to study the difference of structure formed by the effect of different humidity of the phase inversion method, the membrane were prepared in the 2x2 cm shape for porosity and density of ceramic membrane test. The test was carried by Mettler Toledo density test.

3. Results and Discussion

3.1 Morphology of Green Ceramic Membrane
In this study, the macrostructure of membrane cross section do not showed any finger-like formation of macrovoid (Figure. 2). This is due to the structure have been solidified at top of the flat-sheet ceramic membrane during phase inversion method at high humidity. Previously, during the phase inversion system, the NMP will move out from the bottom of the membrane while the droplet of water formed the fingerlike structure on the surface of the ceramic membrane. In this work, the phase inversion system is controlled within the high humidity in the dry condition. The phase inversion in this work was studied by drying method in order to avoid the unbalanced membrane solidified. This method was also successfully prepared in the formation of the dense layer at the top of ceramic membrane.

Ceramic membranes are usually asymmetric or composite type consisting of several layers of one or more different ceramic materials. In this study, the structure morphology was significantly showed that ceramic membrane that fabricated at higher humidity was more porous compare to membrane fabricated at room temperature. Previous study has shown that low humidity shrink the surface pore size. In the study, the effect of high temperature water have been studied towards the structure of pore on Cellulose Acetate (CA) membrane and found that low humidity made the pore size getting smaller [9].

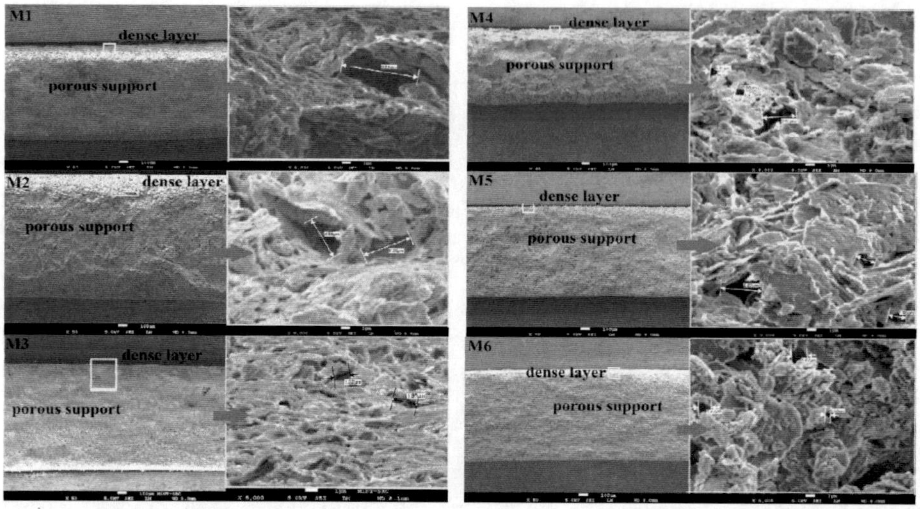

Figure 2. FESEM images of prepared ceramic membranes: (a) whole view x50 (b) dense layer cross section x5000 at (M1) 60g of kaolin at 3°C; (M2) 70g of kaolin at 3°C; (M3) 80g of kaolin at 3°C; (M4) 60g of kaolin at room temperature; (M5) 70g of kaolin at room temperature; (M6) 80g of kaolin at room temperature.

3.2 Porosity and Density of Green Ceramic Membrane

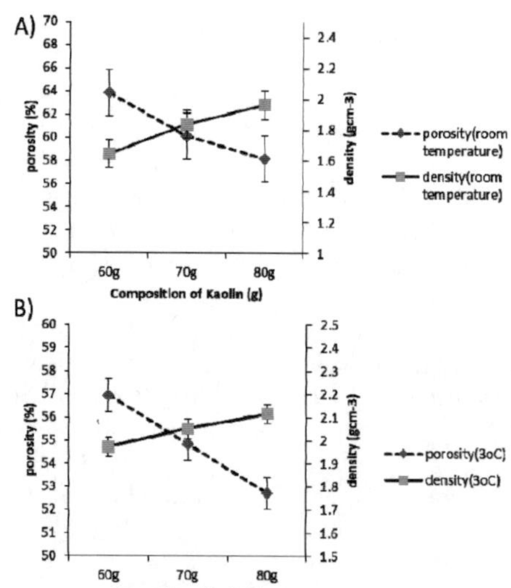

Figure 3. Experimental porosity decline and density incline within the composition of membrane and at temperature of A) room temperature; and B) 3°C

Figure 3 above presents the porosity and density of green ceramic membrane prepared at two different humidity. Accordingly, sintering temperature at 1100 °C lead the porosity surface appeared. Theoretically, when the porosity is high, the density opposed the value of porosity. Water absorption from the high porosity is high due to number of pore consist in the specimen. P. Hristov et al. prove

IOP Conf. Series: Materials Science and Engineering **551** (2019) 012103 doi:10.1088/1757-899X/551/1/012103

the same result as plotted at the porosity and density graph. In their work, they studied about the porosity decrease with the increasing sinter temperature. In meantime, when the porosity declined, the density inclined as well [10]. In this work, we observed that the same trend plotted for both condition. However, in comparing the porosity of the ceramic membrane fabricated by different temperature, the porosity of ceramic membrane fabricated in low temperature (3°C) was higher than room temperature. This was supported by the morphology structure (Fig.2). The ceramic membranes fabricated at high humidity (3°C) had bigger pores sizes.

4. Conclusions

This investigation verified the following two distinguishing characteristics of green ceramic membrane suspension:

1) The structural morphology of membrane's dense and porous layer are affected by the factor involved during phase inversion system. In this work, we are focused on the composition of the kaolin and temperature of the phase inversion system. We attribute this system produced a dense membrane which hence more suitable in the gas separation application.

2) By controlled the humidity of the phase inversion system, the unsolidified of the membrane can be solved. As, expected, the humidity will affect the viscosity of the ceramic suspension, hence, develop the non-fingerlike structure of membrane.

References

1. Yunos M Z *et al* 2013 *Jurnal Teknologi* **65** 111.
2. Wang H *et al* 2010 *Mater. Sci. Eng.* **527** 2881.
3. Ishak M I *et al* 2017 *MATEC Web of Conferences* **97** 01040.
4. Shohur M F *et al* 2013 *Jurnal Teknologi* **65** 59.
5. Ishak M I *et al* 2017 *Journal of Medical and Biological Engineering* **37** 336.
6. Liu L *et al* 2010 *Ceramics International* **36** 2087.
7. Hubadillah S K *et al* 2018 *Ceramics International* **44** 10498.
8. Harun Z *et al* 2014 *Applied Mechanics and Materials* **575** 31.
9. Jamalludin M R *et al* 2016 *Chemical Engineering Research and Design* **114** 268.
10. Ghaseminezhad S M *et al* 2019 *Composites Part B: Engineering* **161** 320.
11. Hristov P *et al* 2012 *Journal of the University of Chemical Technology and Metallurgy* **47** 476.

IOP Conf. Series: Materials Science and Engineering **551** (2019) 012104 doi:10.1088/1757-899X/551/1/012104

Characteristics and Heating Values of Pellets Made from Raw and Charred Rice Residues-Rubber Tree Twigs Mixture

Lee Yit Leng[1,2], Osumanu Haruna Ahmed[2,3,4], Samsuri Abdul Wahid[5], Norawanis Abdul Razak[1], Nurul Nadzirah Zulkeple[1],Ras Izzati Ismail[1], Nurlailina Makhtar[1]

[1]Department of Mechanical Engineering Technology, Faculty of Engineering Technology, Universiti Malaysia Perlis, Kampus Uniciti Alam, Padang Besar, 02100 Perlis, Malaysia
[2]Department of Crop Science, [3]Agriculture and Environment, Borneo Eco-Science Research Center, Faculty of Agriculture and Food Sciences, Universiti Putra Malaysia Bintulu Sarawak Campus, 97008 Bintulu, Sarawak, Malaysia
[4]Institute of Tropical Agriculture and Food Security, Universiti Putra Malaysia, 43400 Serdang, Selangor, Malaysia
[5]Department of Land Management, Faculty of Agriculture, Universiti Putra Malaysia, 43400 Serdang, Selangor, Malaysia

E-mail: yllee@unimap.edu.my

Abstract. In this study, the characteristics and heating values of raw pellet (RP) and charred pellet (CP) made from rice residues-rubber tree twigs mixture were compared. The carbonization process to produce the charred plant residues was conducted in a locally fabricated stainless steel kiln. Both the RP and CP were evaluated for their heating values, pellet durability, carbon and inorganic ash content. The CP had higher heating value and carbon content than the RP. However, the ash content of CP was higher. The results indicate that the pellet made from the charred rice residues-rubber tree twigs mixture could be a suitable source of renewable fuel.

1. Introduction

Malaysia produces a large amount of waste from paddy and rubber tree plantation. The rubber tree twigs from the pruning are generally left to decompose in the plantation for mulching and nutrient recycling purposes. About 3 million tons of rice straw and 0.48 million tons of rice husk are produced in Malaysia annually [1]. Open burning is a common practice to manage the paddy residues as it is the most convenient and cheapest way to get rid of them. However, open burning releases pollutants to the atmosphere and contributes to the global climate change. These crop residues should be valorized and utilized as an energy to become less dependent on fossil fuels.

Jablonsky [2] discovered that a highly significant linear correlation between the heating value of biomass and its C content. Nevertheless, content of volatile matter in rice husk is higher than in wood and coal, whereas carbon is much lower than in coal [3]. High ash content in rice husk cause barriers for energy conversion [3].

The carbon content in biochar has been reported in the range of 54 – 84% depending on the pyrolysis temperature [4]. Although biomass are decomposed during carbonization, it retains a large part of its carbon content. Biochar is produced by thermal decomposition of organic material under limited supply of oxygen and at relatively low temperatures (<700 °C) [5]. According to the literature, the biochar sometimes has slightly higher energy content than bio-oil [6] and can be used as a replacement of coal as well as carbon storage and also as a soil amendment. However, the low density of the biochar makes it difficult to handle and to transport. Densification of biochar into pellets could reduce material waste and improve its handling, transportation and storage.

Densification of biomass is one of the methods that can be used to increase the energy density and to overcome the handling difficulties [7]. While there are reports in the literature on the quality of pellets from other sources of biomass, little information is available on the quality of pellets generated from charred rice residues-rubber tree twigs mixture. This study was carried out to determine whether charring can improve the heating value and the physiochemical properties of pellets generated from rice residues-rubber tree twigs mixture as an energy source.

2. Experimental

2.1 Samples Collection and Preparation

The rice residues were collected from a paddy field in Perlis while the rubber tree twigs were collected from a rubber tree plantation at Sg. Chuchuh Campus, UniMAP. The samples were sun-dried for 3 days. The rubber tree twigs were chopped into 2 cm sections. The rice straws were cut into 15 cm size and kept in a gunny sack while the rice husks were ground and sieved into 1 mm and kept in a sealed container. For the raw pellet (RP) production, all the three types of plant residues were ground and sieved through a 1 mm sieve prior to pellet production. The rubber tree twigs, rice straws and rice husks were charred separately prior to the production of the charred pellet (CP).

2.2 The Charring Process

The charring process was done using a stainless steel kiln fabricated locally at the Department of Mechanical Engineering Technology, Universiti Malaysia Perlis. During the charring process, a maximum temperature of 500 °C was attained after 1 hour of ignition. The average temperature of the kiln for the entire charring process was about 340 °C. The charring process was completed in about 3 hours after which the charred product was collected. The charred materials were ground and passed through a 1 mm sieve prior to the CP production.

2.3 Pelletization

For both RP and CP pellets production, the same proportion of rice husk, rice straw and rubber tree twigs were used; 50% rice husk, 30% rice straw and 20% rubber tree twigs. The plant residues were blended using a mixer with 20 % (w/w) starch and 30 % (w/w) water added as binding agents. Afterwards, the mixture was densified using a Model PC-30 hydraulic cold press with a punch weight of 2.5 tonnes. The dimensions of the RP and CP produced were 14 mm in length and 12 mm in diameter.

2.4 Pellet durability

Pellet durability was determined by mass loss upon vibration. The pellet durability was calculated using the following equation:

$$Pd = 100 - ((mi - mf)/mi) \times 100\%$$

where, Pd is tablet durability (%), mi is initial mass of samples (g), and mf is final mass of the samples (g).

2.5 Inorganic Ash

Inorganic ash contents of the pellets were determined by calculation from the inorganic ash remaining in pellets after the samples were heated in a DAIHAN WiseTherm digital muffle furnace at 600°C for 2 hours. About 2 g of the pellets were weighed on an analytical digital balance (0.0001 Resolution) and then placed in a crucible at 600 °C for 2 hours. The crucible was removed from the muffle furnace

IOP Conf. Series: Materials Science and Engineering **551** (2019) 012104 doi:10.1088/1757-899X/551/1/012104

and cooled in a desiccator. Then, the crucible was weighed on a precision digital balance and the inorganic ash content was calculated based on the following equation:

$$\text{Inorganic ash, Ia (\%)} = \frac{\text{weight of inorganic ash remaining}}{\text{initial weight of pellet}} \times 100\%$$

2.6 Carbon Content

Total carbon (TC) of the RP and CP were determined by a LECO CNS elemental analyzer.

2.7 Heating Value

The caloric value of the RP and CP were measured using IKA model C200 Bomb Calorimeter. About 0.005 g RP or CP sample was introduced into the calorimeter and burnt under oxygen about 5 minutes and the reading were recorded.

3. Results and Discussions

3.1. Pellet durability

According to Adapa [8], durability is high when the computed value is above 80%, medium when the value is between 70% and 80%, and low when the value is below 70%. The raw biomass (RP) shows a significantly higher (96.18%) than the charred biomass (CP) (76.56%). A lower durability of CP could be attributed to destruction of some hydroxyl groups of the biomass materials during the carbonization [9] and reduce binding sites for hydrogen bonding to occur between particles [10]. High durability of RP (96.18%) met the lowest limit standard for both industrial and domestic uses [11].

3.2 Total carbon content

The TC of CP (46.13%) was significantly higher than the RP (39.10%). This difference could be attributed to the carbonization process which condensed the carbon of the rice residues-rubber tree twigs mixture, thereby increasing the carbon content after carbonization. The energy content in C=C bonds is high therefore more energy per unit mass will be observed in pellets with higher TC content [12].

3.3 Inorganic ash content and heating value

As shown in Table 1, the inorganic ash of RP was significantly lower (8.54%) than that of CP (30.25%). The inorganic ash of RP pellet was consistent with the ash content of wood-straw residue reported in the literature [13], while the ash content of CP pellet (30.25%) fell within the range of ash content in coal compiled by Vassilev [13]. Similar observation for high ash content of biomass after thermal treatment have been revealed in the literature [13]. This is most likely due to the pellets consist of paddy residues are rich in silica (Si) [14] and subsequently formation and condensation of aromatic Si-C occur during carbonization process [15]. The CP pellet with higher inorganic ash content seems to less desirable as fuel because it will have lower bioenergy conversion efficiency.

Table 1. Inorganic ash content and heating values of pellets made from raw and charred rice residues-rubber tree twigs mixture

Types of pellet	Inorganic ash content (%)	Heating value (MJ/kg)
Raw pellet (RP)	$8.54^{*b} \pm 2.03$	$15.00^{b} \pm 0.04$
Charred pellet (CP)	$30.25^{a} \pm 0.35$	$15.32^{a} \pm 0.03$

The heating value is the amount of energy that a mass unit releases when upon combustion. The heating values of the RP and CP pellets were greater than the heating value of rice husk (12.87 MJ/kg) and rice straw (14.87 MJ/kg) reported by Biswas [16]. High heating value of RP and CP pellets suggests addition of rubber twigs together with densification increased the heating value as compared to the rice husk or rice straw separately. The higher heating value of the CP can be attributed to its higher TC as C-C bond has higher energy content if compared to O-H, C-H or C-O bonds [12]. Similar observation was obtained by Biswas [16] who found the agricultural residues with high C content released more energy per unit mass.

4. Conclusions

The effects of carbonization on pellet durability, carbon content, inorganic ash content and heating value were statistically significant. After being carbonized, the carbon content and heating value of the rice residues-rubber tree twigs pellets were improved. However, the carbonization process increased the ash content and reduced durability of the pellets. The carbonization process parameters should be varied and investigated in further improving the quality of agricultural residues derived pellets.

5. Acknowledgements

We express our gratitude to Universiti Malaysia Perlis and Universiti Putra Malaysia for providing the facilities to support our research work related to the biomass and bioenergy. This research did not receive any specific grant from funding agencies in the public, commercial, or not-for-profit sectors.

References

[1] *Malaysia economics statistics* 2011 Kuala Lumpur Department of Statistics Malaysia.
[2] Jablonsky M *et al* 2013 4th *International Conference Renewable Energy Sources, High Tartas, Slovak Rebublic.*
[3] Jenkins B M 1998 Biomass Handbook. Gordon and Breach.
[4] Ngatia L W *et al* 2017 *Chemosphere* **208** 211.
[5] Tripathi M *et al* 2016 *Renew. Sust. Energ. Rev.* **55** 481.
[6] McHenry M P 2009 *Agric Ecosyst. Enviro.* **129** 7.
[7] Tumuluru J S *et al* 2011 *Biofuels Bioprod. Biorefin.* **5** 707.
[8] Adapa P K et al 2003 ASABE Pap. No. 036069, ASABE, St. Joseph, MI.
[9] Whittaker C *et al* 2017 *Renew. Sust. Energ. Rev.* **71** 11.
[10] Stelte W *et al* 2011 *Biomass Bioenergy.* **35** 918.
[11] Verhoest C *et al* 2012 http://www.enplus-pellets.eu/wp-content/uploads /2012/04/Industrial-pellets-report_PellCert_2012_secured.pdf.
[12] Pahla G *et al* 2017. *Afr. J. Chem.* **24** 175.
[13] Vassilev S V *et al* 2017 *Fuel* **208** 409.
[14] Mukome F N D *et al* 2013 *J. Agric. Food Chem.* **61** 2204.
[15] Guo J *et al* 2014 *Enviro. Sci. Technol.* **48** 9101.
[16] Biswas B *et al* 2017 *Bioresour. Technol.* **237** 63.

IOP Conf. Series: Materials Science and Engineering **551** (2019) 012105 doi:10.1088/1757-899X/551/1/012105

Effect of Solder Joint Width to the Mechanical Aspect in Thermal Stress Analysis

J. S. Tan[1], C. Y. Khor[1], Muhammad Ikman Ishak[1], M. U. Rosli[1], Mohd Riduan Jamalludin[1], M. A. M. Nawi[1], Mohamad Syafiq A.K. [1] and M.S. Abdul Aziz[2]

[1]Faculty of Engineering Technology, Universiti Malaysia Perlis (UniMAP), Aras 1, Blok S2, Kampus UniCITI Alam, Sungai Chuchuh, Padang Besar, 02100 Perlis Indera Kayangan, Malaysia.
[2]School of Mechanical Engineering, Universiti Sains Malaysia, Engineering Campus, 14300 Nibong Tebal, Penang, Malaysia.

E-mail: cykhor@unimap.edu.my

Abstract. The solder joint has a different dimension which depends on the component or package size. The thermal induced stress, strain and displacement is difficult to visualize due to small size during the component assembly process. Thus, this study is aimed to investigate those mechanical aspects via thermal stress analysis using finite element method. A model of surface mount capacitor was created in the simulation software. The influence of the different widths to the stress, strain and displacement were investigated. The simulation results revealed that the solder width significantly affects the stress, strain and displacement of the solder joint. The increment of the width decreases the strain and stress of the solder joint. The lowest stress and strain were found on the 3.1 mm of solder joint. Thus, the simulation results are expected to provide the understanding of mechanical aspects of solder joint in the thermal stress analysis.

1. Introduction

In the electronic assembly industry, surface mount technology (SMT) [1-2] is widely used in the assembly process. The surface mount component is assembled to the printed circuit board (PCB) via the solder paste material. The solidified solder material forms a solder joint or interconnector between the component and PCB. The interconnector provides the mechanical support, electrical connections to the component and also signal as well [3-6], which ensure the component could function properly in the entire electronic system. Therefore, the quality and reliability of the solder joint are important in the electronic assembly process.

The assessment of the solder joint reliability can be achieved via the thermal stress analysis at a fixed rate of temperature change or at the constant temperature [7]. The thermal analysis technique can be applied to assess the thermal endurance of PCB FR4 epoxy laminates [8] by ranging the temperatures range from 170 to 200 °C. Moreover, the thermal analysis can be performed by using simulation methods. Zhang and Bagnoli [9] used COMSOL solver to validate the methodology through different PCB layouts. They reported that Mesh density and the number of eigenvalues were the main influence factor to the simulation results.

In the simulation analysis, the accurate data input yields reliable results in the thermal stress simulation [10]. For example, the input data of coefficient of thermal expansion (CTE) is crucial, in which induced by the surrounding temperature variations during the reflow and wave soldering process [11-13]. As the extension of our previous work [14], the effect of the width to the mechanical aspects was considered in the current study. Moreover, finite element (FE) based software [15-19] was employed to create the model and perform the thermal stress analysis.

2. Methodology

Figure 1 illustrates the three dimensional model of capacitor with solder joint on the printed circuit board. The dimensions of the 3D model are 6.21 mm × 3.1 mm (length ×width) for the capacitor. However, the solder joint width was ranging from 2.3 mm to 3.1 mm. The length of the solder joint is 3 mm. The dimensions of the PCB are 16 mm × 8 mm × 1.6 mm (length × width × height) [14]. In the simulation analysis, the interfacial copper layer between joint and the PCB was neglected.

A similar simulation setting and the material properties of different components were applied from our previous work [14]. Table 1 summarizes the value of the mechanical and thermal properties used in the current simulation, which are Young modulus, Poisson's ratio, thermal conductivity, density and coefficient of thermal expansion (CTE) of the 3D model. The 3D model was meshed by hexahedral element with 0.4 of mesh sizes and no error was found for the quality of the mesh.

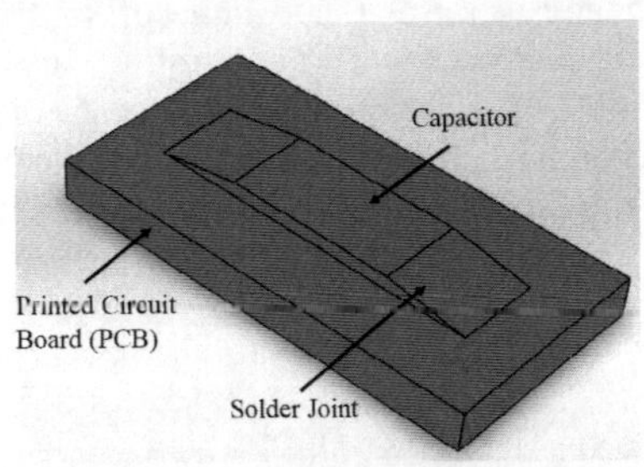

Figure 1. 3D model.

Table 1. Material properties for various components [20].

Component	Material	Young's Modulus (GPa)	Poisson's Ratio	CTE (pmm/ °C)	Density (g/mm^3)	Thermal conductivity (W/mmK)
Capacitor	Aluminium	282.7	0.2222	6e-6	2.7e-3	205e-3
PCB	FR4 polyclad	27	0.17	14e-6	1.85e-3	0.81e-3
Solder	95.5Sn3.9Ag0.6Cu	43	0.3	23.2e-6	7.44e-3	60e-3

3. Results and Discussion

The simulation results in term of mechanical aspects (i.e., stress, strain and displacement) were studied. The relationship between the solder with and the mechanical aspects were analyzed and presented in this section. The width of solder joint used are 2.3 mm, 2.5 mm, 2.7 mm, 2.9 mm and 3.1 mm. Figure 2 shows the distribution of von Mises stress in the thermal stress analysis. It is noticed that the maximum stress was concentrated around the interfacial region between solder joint and capacitor. This situation may attribute by the variation of thermal properties between these components, which induce the thermal mismatch between the component and the solder joint. Hence, the expansion of solder joint lead to the stress generated around the interfacial region between solder joint and capacitor. The stress imposed on the solder joint may also cause by other phenomena such as mechanical load and during the fluid structure interaction [21].

IOP Conf. Series: Materials Science and Engineering **551** (2019) 012105 doi:10.1088/1757-899X/551/1/012105

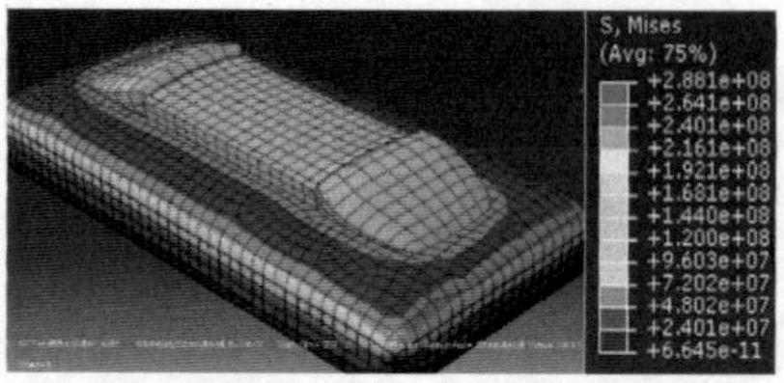

Figure 2. Von Mises stress distribution.

The correlation between the solder joint width and the mechanical aspects are presented in Figures 3-5. The results revealed that stress, strain and displacement of the solder joint exhibits in polynomial behavior to the solder width. The increment in solder width resulted in the decrement of the strain and stress as clearly shown in Figure 4 and Figure 5. This phenomenon may be attributed to the increase of the contact area of solder with PCB. The expansion of the solder joint is decreased due to the increasing of area or space. The expansion of the solder joint is not restricted in a confined space resulted the decreasing of strain and stress.

Figure 6 illustrates the correlation between the displacement and the solder width. The displacement of the solder joint increases gradually when the width increases from 2.3 mm to 3.1 mm. The increment of width simultaneously raises the volume of solder and lead to the greater volume expansion. Thus, the displacement increases when the width is increased. A similar phenomenon was also noticed in our previous work [14] in which the increase of solder length also contributes to the greater solder displacement. Therefore, the optimization of the process and design [22] of the electronic package is vital for the engineer to ensure the high reliability of the solder joint.

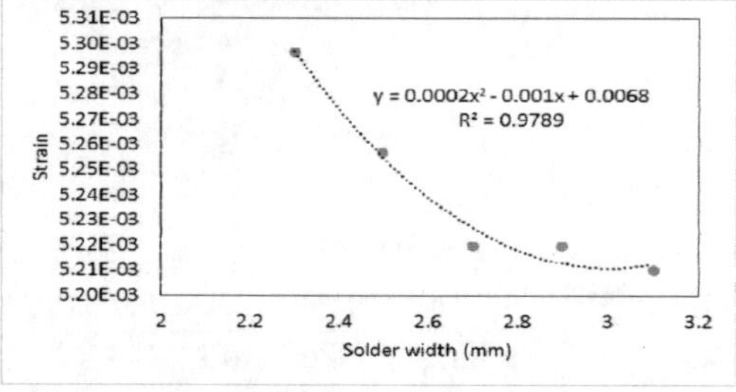

Figure 3. Maximum strain.

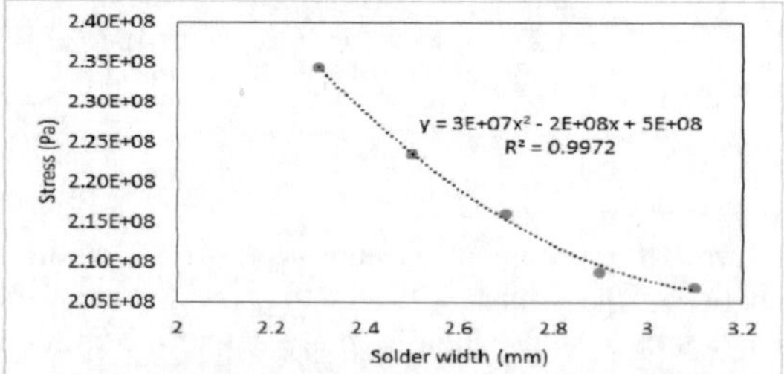

Figure 4. Maximum stress.

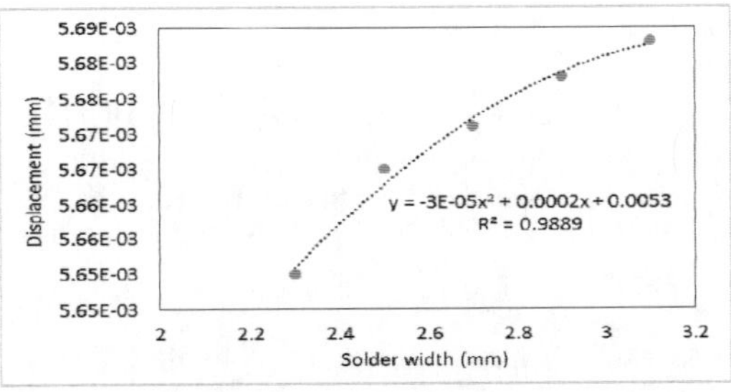

Figure 5. Maximum displacement.

4. Conclusion

The investigation of the solder joint width in the thermal stress analysis has been carried out via finite element based simulation. The influence of the solder width towards the mechanical aspects (i.e., stress, strain and displacement) has been studied. The distribution of the thermal induced stress of the component was visualized in three-dimensional model. The correlation between the width and the mechanical aspects was also discussed. The results revealed that solder width significantly affects the strain, stress and displacement in the polynomial behavior. Increase of solder width leads to the greater volume expansion and resulted in the increasing of displacement. Moreover, extremely high stress and displacement of the solder joint will lead to the unintended malfunction of the electronic component. Thus, the current results are expected to provide better understanding in term of mechanical aspects and the visualization of stress distribution in the 3D model. This study will be extended to focus on other parameters (e.g., PCB thickness and component size and shape) in the thermos-mechanical analysis.

Acknowledgements

The author would like to acknowledge the support from the Fundamental Research Grant Scheme (FRGS) under a grant number of **FRGS/1/2016/STG07/UNIMAP/02/4** (No: 9003-00600) from the **Ministry of Education Malaysia**.

References

[1] Lau C-S *et al* 2013 *Microelectronics International* **30**(3) 151.
[2] Najib A M *et al* 2015 *International Journal of Heat and Mass Transfer* **87** 49.
[3] Ong E E S. *et al* 2012 *International Communications in Heat and Mass Transfer* **39**(10) 1616.
[4] Ong E E S *et al* 2014 *Microelectronic Engineering* **113** 40.
[5] Khor C Y *et al* 2014 *Microelectronics reliability* **54**(4) 796.
[6] Khor C Y *et al* 2013 *Microelectronics Reliability* **53**(2) 334.
[7] Haines P *et al* 1995 *Thermal method of analysis: Principles, application and problem.*
[8] Polanský R *et al* 2014 *Polymer Degradation and Stability* **105** 107.
[9] Zhang Y *et al* E 2014 *Microelectronics Journal* **45**(8) 1033.
[10] Abdul Aziz M S *et al* 2014 *Advances in Mechanical Engineering* **6** 275735.
[11] Khor C Y *et al* 2014 *Microelectronics Reliability* **54 (8)** 1511.
[12] Abdul Aziz M S *et al* 2014 *Soldering & Surface Mount Technology* **26**(3) 97.
[13] Lau C-S *et al* 2016 *Soldering & Surface Mount Technology* **28**(2) 41.
[14] Tan J S *et al* 2017 AIP Conference Proceedings **1885** 020063.
[15] Ramdan D *et al* 2012 *IEEE Transactions on Components, Packaging and Manufacturing Technology* **2**(11) 1786.
[16] Abdul Aziz M S *et al* 2013 *International Communications in Heat and Mass Transfer* **48** 116.
[17] Leong W C *et al* 2012 *Transactions on Components, Packaging and Manufacturing Technology* **2(8)** 1335.

IOP Conf. Series: Materials Science and Engineering **551** (2019) 012105 doi:10.1088/1757-899X/551/1/012105

[18] Omar A *et al* 2012 *Applied Mechanics and Materials,* **121** 1214.
[19] Jizat N M *et al* 2017 *AIP Conference Proceedings,* **1885**(1) 020126.
[20] Amalu E H *et al* 2015 *Finite Elements in Analysis and Design* **107** 13.
[21] Khor C Y *et al* 2014 *Microelectronics Reliability* **54**(4) 796.
[22] Khor C Y *et al* 2012 *Simulation Modelling Practice and Theory* **29** 109.

The Influence of Spiral Blade Distributor on Pressure Drop in a Swirling Fluidized Bed

M. A. M. *Nawi*[1*], *Mohd Razman* Amin[1], M. S. *Kasim*[2], R. *Izamshah*[2], *Muhammad Ikman* Ishak[1], C. Y. *Khor*[1], M. U. *Rosli*[1], *Mohd Riduan* Jamalludin[1] and *Mohamad Syafiq* A. K.[1]

[1]Faculty of Engineering Technology, Universiti Malaysia Perlis (UniMAP), Aras 1, Blok S2,Kampus UniCITI Alam, Sungai Chuchuh, 02100 Padang Besar, Perlis, Malaysia
[2]Precision Machining Group, Faculty of Manufacturing Engineering, Universiti Teknikal Malaysia Melaka, Hang Tuah Jaya, 76100 Durian Tunggal, Melaka, Malaysia.

E-mail: nawi@gmail.com

Abstract. This work reports studies on the aerodynamics of air flow in the plenum chamber of a swirling fluidized bed. The study focused on the effect of annular spiral blade distributor configuration whereby the effect of various pitch length (60 mm, 80 mm and 100 mm) via various horizontal inclination angle (0°, 12° and 15°). The CFD method is used to compute and obtain the performance results as well as pressure drop data. Moreover, high pressure drop is undesirable since high pumping power is needed to sustain the pressure needed to run the swirling fluidized bed processes. It should be as low as possible to reduce the power being wasted during the processes. The numerical result at the parameter of horizontal blade inclination of 15° via pitch length of spiral blade distributor of 100 mm shows that effect on large overlapping length of blade (trapezoidal) area consume less pressure drop.

1. Introduction

Fluidized beds are widely used in many industrial especially in chemical process industries. The mechanism of a fluidized beds can be seen when gas passes through the particle and experiencing sufficient drag force from the air or fluid [1]. Compare to the conventional fluidization, the current design of swirling fluidized bed had an array inclination distributor which froms annular opening as shown in Fig. 1 [2]. The current study of spiral distributor was carried out in order to seen the same effectiveness of air flow of the conventional fluidized bed. The main objective of this study is to determine the pressure drop in a swirling fluidized bed through to the various pitch length of annular spiral blade distributor.

The idea of spiral blade distributor was come up from the screw conveyors design. The screw conveyors was most reliable and cost-effective methods for conveying some amount materials. Moreover, the screw conveyors can be operated at any angle position, from vertical to horizontal condition. Futhermore, the design was base on the objective, the suitable design and development will bring a long times undisrupted service and productivity [3].

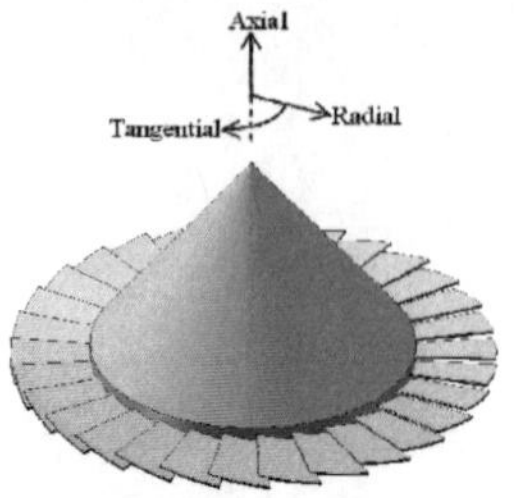

Figure 1. Distributor blade with a conical centre body.

2. Methodology

2.1 Numerical Methodology

Investigation of the air flow distribution in a SFB was conducted using commercial CFD software – FLUENT 6.3. The computation domain and grid generation was developed via GAMBIT 2.4.6. The two parameters were changed to obtain the correlation between the spiral blade length and blade horizontal inclination angle as shown in Table 1. The horizontal inclination has been selected based on previous studies by [4].

Table 1. Spiral blade distributor configuration in a swirling fluidized bed.

Case	Cylinder Base Diameter (mm)	Pitch Length (mm)	Horizontal Inclination Angle (°)
1	8	60	0
2		80	
3		100	
4		60	12
5		80	
6		100	
7		60	15
8		80	
9		100	

2.2 Boundary Condition

The air inlet was modeled as velocity inlet boundary condition of 2.25 m/s. The velocity and other parameters were based on the actual SFB system in the Faculty of Mechanical and Manufacturing Engineering, UTHM, which is currently used for drying of biomass. Same setting with previous researcher [4] of numerical simulation in CFD software in swirling fluidized bed has been applied on current study. The air outlet was modeled as a pressure outlet of 1.01325 bar (atmospheric pressure). Meshing of the SFB system was generated in such a way that it may capture airflow taking place through narrow opening of the distributor blades. Grid sensitivity was done prior to actual investigation to ensure the independence of grid size on the flow and to reduce possible numerical errors. The suitable grid size is chosen in such a way that they total number of elements are less (to reduce compunational costs) without compromising simulation results. Hence, the Tri:Pave Meshing Scheme was applied to the surface and it allowed GAMBIT 2.4.6 to create a face mesh consisting of irregular triangular mesh elements. The Tet/Hybrid parameter type that specifies tetrahedral, hexahedral, pyramidal and wedge element were defined to the meshing algorithm. The mesh elements in the computation domain as well as the spiral blade distributor is presented in Fig. 2.

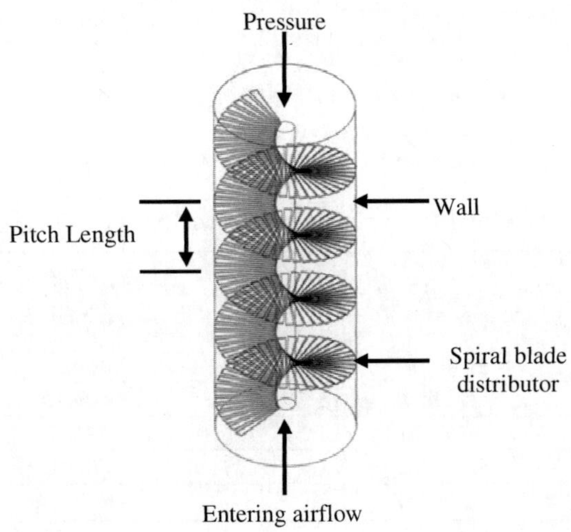

(a)

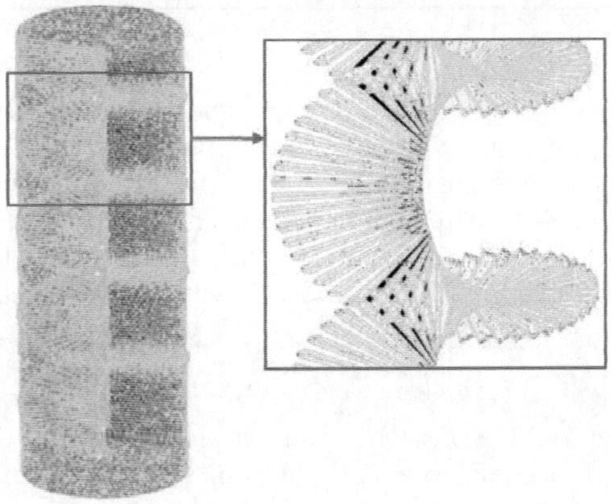

(b)

Figure 2. (a) Domain ; (b) meshes on the spiral blade distributor.

2.3 Solver Solution Techniques

The Reynolds Averaged Navier Stokes (RANS) turbulence Equation Models that being considered for usage in this study is RNG k-ε RANS turbulence of RNG k-ε model with enhanced wall treatment is used as this type of equation gives the closed results when being compared to the experimental value. The RNG k-ε is similar in form to the standard k-ε model, but has additional term in its ε equation that improves the accuracy significantly for rapidly strained flows and provides and analytical formula for turbulence Prandtl numbers and also the effect of swirl on turbulence is included in the RNG model [6] and [4].

3. Results and Discussion

3.1 Low Pressure Drop

The pressure drop was obtained by computing the pressure differences amongst the studied planes which located at 28 mm and 32 mm above the inlet. Effect on axial entry type plenum chamber with various number of blades (depends on pitch length and the 10 mm spacing for each designated configurations) and the trapezoidal area opening of blades would determine the pressure drop an each parameter like as shown in Table 2. The data that was obtained have shows that the pressure drop is significantly differ to its pitch length with horizontal inclination angle of blade distributor. The blade angle of 0° at pitch length of 60 mm has the highest pressure amongst the other parameters for every section of the system. It has the total of

1674.16 Pa of pressure drop overall for the system. Whilst, when the blade angle at 15° and the pitch length in 100 mm, the lowest pressure drop has been acquired on the system compared to other study parameters. The pressure drop and the condition was coincided with the previous statement [1].

Table 2. Static pressure data on blade 0°, 12° and 15° horizontal inclination.

Horizontal Inclination Angle	Pitch Length (mm)	Average Pressure at 28 mm Above Inlet (Pa)	Average Pressure at 32 mm above Inlet (Pa)	Pressure Drop (Pa)
0°	**60**	**10743.39**	**9069.24**	**1674.16**
	80	7723.97	6609.15	1114.81
	100	6267.26	5481.63	785.62
12°	60	15756.22	14784.19	972.02
	80	11953.37	11218.26	725.11
	100	6812.04	6256.71	555.33
15°	60	13010.14	12213.95	796.18
	80	8813.20	8189.91	623.28
	100	**5661.70**	**5182.62**	**479.08**

4. Conclusions

From the research, we can conclude that the pitch length of spiral blades distributor in a swirling fluidized bed (SFB) affects the pressure drop values. The findings from numerical study have shows that when the air flow go past at high horizontal inclination angle along with large pitch length of spiral blade the value of pressure drop tends to fall down. Moreover, the large area of trapezoidal have also effected on the pressure drop decline. At the horizontal inclination angle of 0° with 60 mm pitch length has the highest pressure drop (1674.16 Pa) compared to the lower pressure drop (479.08 Pa) at 15° with 100 mm pitch length parameter. The result was occurred due to the effect of trapezoidal area (spaces between blades) and the pitch length itself that allows air flow to travel more.

Acknowledgements

A sincere thanks also goes to Dr. Mohd Faizal Mohideen Batcha from Universiti Tun Hussein Onn Malaysia (UTHM) for his contribution in excellents ideas and guidance on fluidization systems research.

References

[1.] Batcha M F M *et al* 2011 *J. Applied Sci* **11** 1980-1986
[2.] Faizal M *et al* 2012 *Materials Research* **468-471** 25-29
[3.] Mecke B 2019 *Designing Inclined Scre Conveyors for Industrial Applications.*
[4.] Al Hafiz M *et al* 2013 IOP *Conf. Series: Materials Science and Engineering* **50**
[5.] Batcha M F M *et al* 2013 *Journal of Scientific Research* **6** 157-66
[6.] Safiah O *et al* 2008 *Proceedings of International Conference on Mechanical & Manufacturing Engineering*

IOP Conf. Series: Materials Science and Engineering **551** (2019) 012107 doi:10.1088/1757-899X/551/1/012107

Investigation on the Mechanical Properties of Banana Trunk Fibre–Reinforced Polymer Composites for Furniture Making Application

Muhammad Ikman Ishak[1,*], C.N. Ismail[1], C.Y. Khor[1], M.U. Rosli[1], Mohd Riduan Jamalludin[1], M.H.M. Hazwan[1], M.A.M. Nawi[1], and Mohamad Syafiq A.K.[1]

[1]Faculty of Engineering Technology, Universiti Malaysia Perlis, Level 1, Block S2, UniCITI Alam Campus, Sungai Chuchuh, 02100 Padang Besar, Perlis, Malaysia.

Email: ikman@unimap.edu.my

Abstract. The banana plant is unproductive after it has produced fruits once in its lifetime and the trunk commonly be disposed. Among the disposing methods used are by herbicide injection and burning using kerosene in which are unfavourable due to bad effects to the environment. Thus, the application of the unused banana trunk could be advantageous in saving the environment through its fibre properties as composite. Three tensile test specimens were prepared with each has different weight of banana fibres – 0, 0.45, and 0.9 g. The results demonstrated that the composite with high amount of fibre promoted superior maximum and break stress values. Also, the properties of the composite are convenient to be applied in furniture making as adequate stress level recorded in the numerical analysis.

1. Introduction

Banana plant is classified as a gigantic herb that springs from an underground stem to form a trunk with 3- to 6-meters high. It consists of several main parts such as stem or trunk, leaves, and fruits. The banana trunk which also known as pseudostem, formed by the tightly packed overlapping leaf sheaths. It continues to grow as the leaves emerge one after the other, and reaches its maximum height. Besides the fruit of banana plant can be eaten, its trunk may be used to make garment and natural craft materials [1]. Also, the banana trunk promotes several advantages in medical treatment application, as examples for digestion, detoxification, blood pressure, weight loss, and healing acidity and gastric problems.

The banana plant would only be productive once in its lifetime before it dies. It usually takes about 13 months for the plant to achieve maturity and the fruits are ready to be plucked. The dead plants are usually disposed through few methods. Among the methods are by herbicide injection, burning using kerosene, and cutting down using mechanical tools. The former methods are considerably unfavourable due to bad effects to the environment such as air pollution and toxicity. Therefore, the unproductive banana plant or particularly its trunk can alternatively be utilised for related potential applications owing to its fibre properties. To date, natural fibres are extensively used in many applications for the factors of low cost, low density, light weight, less pollution, and eco-friendly nature. Natural fibre is defined as substances that produced by the plants or animals which can be spun into filament, thread, or rope [2]. Banana trunk fibre is similar to that of natural bamboo, however, it has better fineness and spin ability.

Composite is described as a material made of two or more constituent materials (matrix and reinforcement) with significantly different physical or chemical properties, to produce a new material with characteristics different from the individual component. Typical engineered composite materials are reinforced concrete, composite wood, reinforced plastics, ceramic matrix composite, and metal matrix composite [3]. This study only focuses on the use of common organic matrix which is polymer, to be examined.

The present study, therefore, places a major emphasis on the investigation on the mechanical properties of banana trunk fibre–reinforced polymer composite for furniture making application. There is no study, as yet, found with regards to the use of the proposed composite in the making of furniture. Based on past findings, it was observed that rattan is commonly used as main parts of furniture since 1970s. The technique of weaving reeds and other natural fibres is one of the popular methods considered and used in constructing relevant furniture parts.

2. Materials and Methods

2.1 Banana Trunk Fibres Preparation

A banana trunk from an unproductive matured banana tree was collected from a farm. It was then cleaned and washed with water before cut into several small segments using a cutter, with each 50-mm thick. Only one segment was used and it was chopped into smaller pieces. The fibre was placed under the direct sunlight for 36 hours until it was completely dry or all moisture content is removed.

In this study, the effect of banana trunk fibre quantity (weight) embedded in the polymer matrix on the mechanical properties was a major concern. Thus, it is noteworthy that the size of the fibre was set similar (0.5 mm) among the specimens. For the matrix's material, epoxy resin was chosen to bind all the fibres fixedly together. In total, there were three specimens prepared; 1) Specimen 1: Without fibre (pure epoxy resin matrix), 2) Specimen 2: 0.45 g fibre (14.6 ml epoxy resin matrix), and 3) Specimen 3: 0.9 g fibre (14.6 ml epoxy resin matrix). A manual sifter with the hole size of 0.5 mm was used to filter out all the required fibre size.

2.2 Specimen Mould Preparation

Before proceeding to the testing specimen fabrication, it is a necessary to create its mould first. Hence, a three-dimensional (3-D) model of dumbbell shape representing the specimen's design and dimension, was developed using SolidWorks software and then printed using a rapid prototyping machine. The dimensions of the model are 150, 20, and 4 mm for the length, width, and thickness, respectively. The model acts as the basic shape used to cast the specimen's mould. Silicon rubber and hardener are the type of materials utilised to build the mould with the ratio of 25:1. The silicon rubber offers several advantages over any other materials included its good elasticity and heat resistance. As the mixture of both materials were prepared, it was poured into a container. The printed specimen models were then placed into the solution and the mould was let dry and harden. Figure 1 depicts the dried banana trunk fibres and the hardened silicon rubber mould.

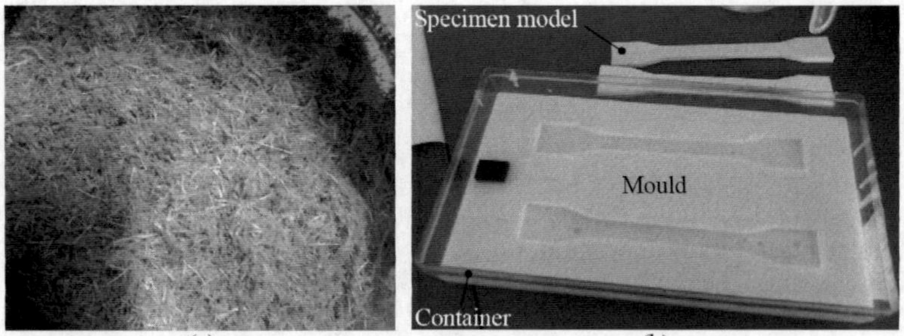

(a) (b)

Figure 1. (a) The dried banana trunk fibres. (b) The hardened silicon rubber mould.

2.3 Tensile Test Specimens Fabrication

To fabricate the testing specimens, a mixture of epoxy resin and hardener was prepared at a ratio of 2:1 to be as the composite matrix, and it was poured into three different containers. Different weights of the fibre – 0, 0.45, and 0.9 g – were inserted into each container separately and stirred evenly. Then, the composite mixture was flowed into the mould and dried at room temperature for about 24 hours. For testing, the tensile test machine was initially calibrated to avoid any errors which could jeopardise the results and the speed of the stroke was set to 30 mm/min.

IOP Conf. Series: Materials Science and Engineering **551** (2019) 012107 doi:10.1088/1757-899X/551/1/012107

2.4 Computational Analysis of Furniture
A numerical analysis was performed on a furniture model to investigate its performance with regards to the properties of the composite obtained from the experimental testing. The 3-D model of a coffee table was therefore designed using SolidWorks software and exported into Abaqus software for the finite element analysis (FEA). Finite element analysis plays an important role in solving mathematical modelling problems in various fields of science, technology, and industry [4-15].

3. Results and Discussion

3.1 Maximum and Break Stress Values
Figure 2 exhibits the specimens condition after the tensile test was performed as well as the comparison of the maximum and break stress values recorded. It was evident that the maximum stresses were significantly increased as the amount of fibre increased. The maximum stress magnitude was slightly lower for both Specimen 1 (0 g fibre) and Specimen 2 (0.45 g fibre) as compared to that for Specimen 3 (0.9 g fibre) with differences of 23.5 and 12.4%, respectively. Similar observations were seen for the break stress results wherein Specimen 3 (29.8 MPa) still superior with approximately 1.3- and 1.1-fold greater break stress value than Specimen 1 (22.1 MPa) and Specimen 2 (26.1 MPa), respectively.

A possible explanation for the higher stress levels generated in the specimen with increased amount of fibres is due to the high fibre-to-matrix contact area. The increase in fibre numbers has increased the contact area accordingly. In comparison, it was noteworthy that there was no difference observed between the maximum and break stress magnitudes for Specimen 2 and Specimen 3. In contrast for Specimen 1, the maximum stress was slightly greater than the break stress level by about 3%. This could be owing to the brittleness of the materials possessed which has resulted in no plastic deformation or strain hardening phenomenon occurrence before fracture. The results were in agreement with the type of specimen fracture shown in Specimen 2 and Specimen 3 by which no constriction or necking found to indicate the elongation under the application of tensile load.

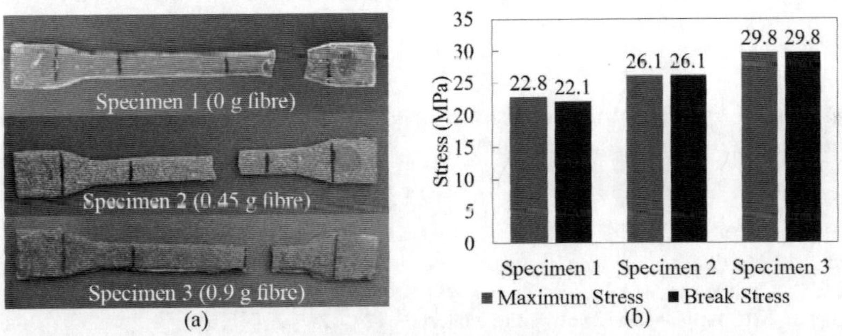

Figure 2. (a) The condition of specimens after tensile test. (b) The comparison of maximum and break stresses among all specimens.

3.2 Maximum von Mises Stress Distribution in Furniture Model
The results of FEA in form of von Mises stress dispersion within the coffee table model are illustrated in Figure 3. Based on the tensile test outcomes, it was recorded that the elastic modulus, E of the tested composite is about 2 GPa with Poisson's ratio, v of 0.3 assumed. These material properties had been considered in the pre-processing settings of the analysis which were assigned for the table legs. The model was meshed with four nodes solid tetrahedral elements and a load of 981 N was applied onto the top surface of the table.

IOP Conf. Series: Materials Science and Engineering **551** (2019) 012107 doi:10.1088/1757-899X/551/1/012107

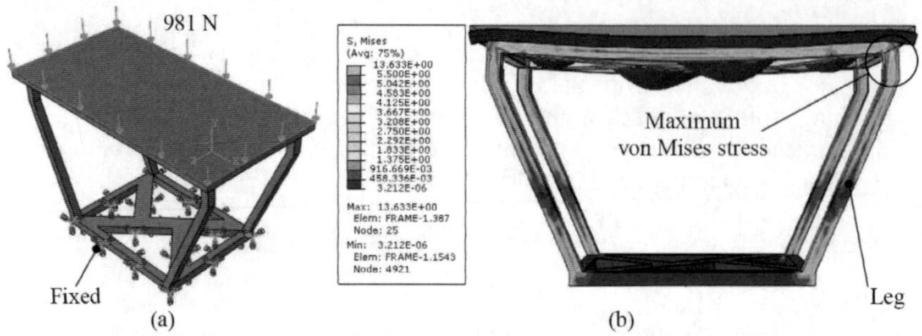

Figure 3. (a) The applied loading and boundary conditions on the FEA model. (b) Von Mises stress plot for the table model that assigned with the composite properties.

The findings demonstrated that the load was evenly distributed through out the legs with high stress concentration discovered at the top regions. The maximum von Mises stress value generated in the table was 13.6 MPa shown in grey colour contour plot. The value seems considerably way lower than the typical range of ultimate tensile stress value of the natural fibre–reinforced polymer composite. It may therefore be claimed that the properties of banana fibre–reinforced epoxy resin composite are favourable in producing encouraging stress level within the furniture.

4. Conclusions

The results of this study support the following conclusions. The mechanical properties of banana fibre–reinforced epoxy resin composite was remarked to be satisfactory in furniture making application as it predicted promising mechanical stress level through computational analysis. Moreover, the properties of the composite were evident to significantly be improved with the increase in the amount of fibre embedded within the polymer matrix.

Acknowledgements

Appreciation is given to Faculty of Engineering Technology, Universiti Malaysia Perlis. The authors reported no conflicts of interest related to this study.

References

1. R. Violette, *Ten Uses of Banana Trees*, (Dengarden, 2016)
2. Rosli M U *et al* 2017 *MATEC Web of Conferences* **97**
3. John M J *et al* 2008 Carbohydrate Polymers **71** 343-364
4. Omar A. Omar *et al* 2012 Applied Mechanics and Materials **121-126** 1214-1219
5. Ishak M I *et al* 2013 *SpringerBriefs in Applied Sciences and Technology* 9-26
6. Ishak M I *et al* 2010 *Conference on Biomedical Engineering and Sciences* 210-215
7. Jizat N M *et al* 2017 *AIP Conference Proceedings* **1885**
8. Ishak M *et al* 2017 *MATEC Web of Conferences* **97**
9. Ishak M I *et al* 2017 *Journal of Medical and Biological Engineering* **37** 336-344
10. Ishak M I *et al* 2013 *SpringerBriefs in Applied Sciences and Technology* **37-52**
11. Khor C Y *et al* 2017 *Microelectronics Reliability* **54** 796-807
12. Khor C Y *et al* 2017 *MATEC Web of Conferences* **97**
13. Shahriman A B *et al* 2017 *Journal of Physics: Conference Series* **908**
14. Tan J S *et al* 2017 *AIP Conference Proceedings*, **1885**
15. Yunos M Z *et al* 2013 *Jurnal Teknologi* **65** 111-115

Simulation-Based Optimization of Plastic Injection Molding Parameter for Aircraft Part Fabrication Using Response Surface Methodology (RSM)

M.U. Rosli[1,*], Muhammad Ikman Ishak[1], Mohd Riduan Jamalludin[1], C.Y. Khor[1], M.A.M. Nawi[1], Mohamad Syafiq A.K.[1]

[1]Faculty of Engineering Technology, Universiti Malaysia Perlis (UniMAP), Level 1, Block S2, UniCITI Alam Campus, Sungai Chuchuh, 02100, Padang Besar, Perlis, Malaysia.

Email: uzair@unimap.edu.my

Abstract. The small size and thin part characteristics of the honeycomb floor panel for the aircraft part product may cause problems in injection molding fabrication process. The objectives of this study is to find the most appropriate parameter in injection molding process for honeycomb floor panel by simulation using MoldFlow software and optimize the quality of injection molding process parameter by using Response Surface Methodology (RSM) as to obtain an optimal response and meets the requirement specification in the aviation scope. The crucial responses are shrinkage and warpage. Melt temperature, filling pressure and injection time are selected as the most influential factor for shrinkage and warpage. Model response was fitted by quadratic model. As the results, the optimum value suggested by the software were melt temperature of 360.02c, 60 MPa filling pressure and 4.70s injection time. With small differences error value between solution and simulation, 0.3% for shrinkage and 0.6% for warpage, the results was acceptable.

1. Introduction

In vehicles engineering, mainly in aircraft industries, floor panel is made up from honeycomb. Honeycomb for flooring in commercial aircraft product fabrication is the process of producing the component or part in light weight for requirement in specific strength with high accuracy and precision. Light in weight is one of the most important characteristic in the aerospace industry, parallel with the current trend in the automotive industry and manufacturing [1,5,8,10]. In the field of aerospace, the honeycomb panel is optimized by characterizing in light structure with the good stiffness, strength and good energy absorption properties [2]. The basic method of manufacture of the honeycomb involves printing stripes of adhesive onto flat foils which are subsequently stacked, bonded to each other and then expanded to form a series of hexagonal cells making up the honeycomb blocks. The type of the process that has been chosen to produce honeycomb for the floor panel product is injection molding. In injection molding, there are various material can be used that meet the requirement in aircraft for light weight [3,6,7]. Injection molding is a machining process to produce a complex shape product and mostly used because injection molding can produce a product in short cycle time, high product accuracy and can be fabricated in large quantities.

Based on the previous research for honeycomb fabrication, there are numbers of the researcher has used the invention of honeycomb in injection molding process but not in aircraft fabrication. From a research by Scharkowski, an injection molding process was used in a honeycomb-shaped hollow body plastic, preferably polyolefins [4]. This simulation based research used polyetheretherketone (PEEK) as the material selection to produce the honeycomb for floor panel. The type of the product is in light weight and for requirement in specific strength with the high accuracy and precision. Simulation based optimization is the most effective way to study the strength of plastic honeycomb floor panel before the actual fabrication processes begin

[9,12]. The injection molding parameters are considered in this project to determine the most significant factor and accurate molding process parameter setting [10].

2. Methodology

2.1 Response Surface Methodology (RSM)
In the honeycomb floor panel fabrication process, the most crucial aspect is the accuracy and precision without changing the lightweight aspect ratio. By using the Response Surface Methodology (RSM) method, the most suitable parameters can be defined to control the warpage and shrinkage. Melt temperature, filling pressure, injection time is the process factor that is the most significant parameters that effecting the warpage and shrinkage of an injection molded part. The different value order of processing parameter that generates from the Design Expert software was completed in 60 times of numerical run. Table 1 shows the summary of the design expert software which evaluates the minimum and maximum of the two responses that would like to be controlled in this project that is warpage and volumetric shrinkage. The two values were generated from Moldflow software. The minimum value for a warpage is 4.08 and maximum is 18.53. For shrinkage, the minimum and maximum value are 15.56 and 24.95, respectively.

Table 1. Design of experiment.

Response	Name		Obs	Min	Max	Trans	Model
Y1	Warpage		60	4.08	18.53	None	Quadratic
Y2	Shrinkage		60	15.56	24.95	None	Quadratic
Factor	**Name**	**Unit**	**Type**	**Low Actual**	**High Actual**	**Low Coded**	**High Coded**
A	Melt temperature	C	Numeric	360.00	400.00	-1	1
B	Filling pressure	Mpa	Numeric	60.00	100.00	-1	1
C	Injection time	s	Numeric	2.00	5.00	-1	1

3. Results and Discussion

3.1 3D Surface
In this study, the important goals that need to achieve are the minimum value of the response volumetric shrinkage and warpage. The 3D surface response and contour plots of the warpage and shrinkage were plotted using Design Expert software to study the interactive relationship between each factor and responses. The maximum or minimum value of the most optimal result for shrinkage and warpage can be displayed in the 3D surface plot. In the 3D surface response, the selection of two variables and a constant variable was determined according to the level of sensitivity towards the responses that depended on the perturbation plots. Filling pressure and melt temperature are the most sensitive factor toward warpage. Figure 1 shows the most optimal warpage in 30% of fiberglass at 4.23. The optimized (lowest value) warpage can be achieved at higher filling pressure 100Mpa and higher melt temperature 380 °C.

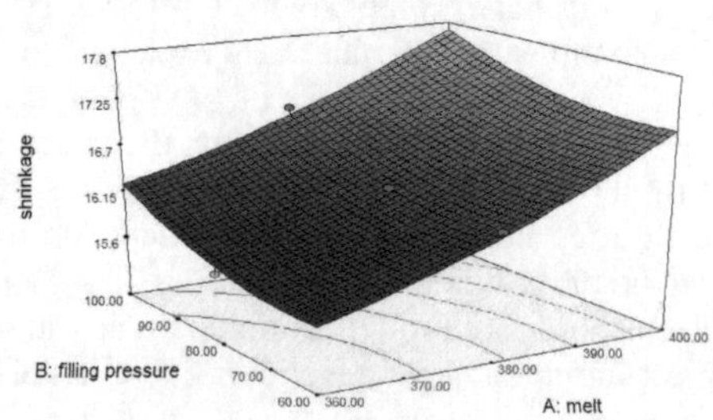

Figure 1. 3D Plot for warpage in 30% fiber glass material.

The most influential factor for shrinkage was filling pressure and melt temperature. For shrinkage, to find the most optimize result (the lowest value) of shrinkage must at higher filling pressure and lower melt temperature. In Figure 2, the most optimal result for shrinkage is 16.38 at 100 MPa and 380 °C in 30% fiber glass.

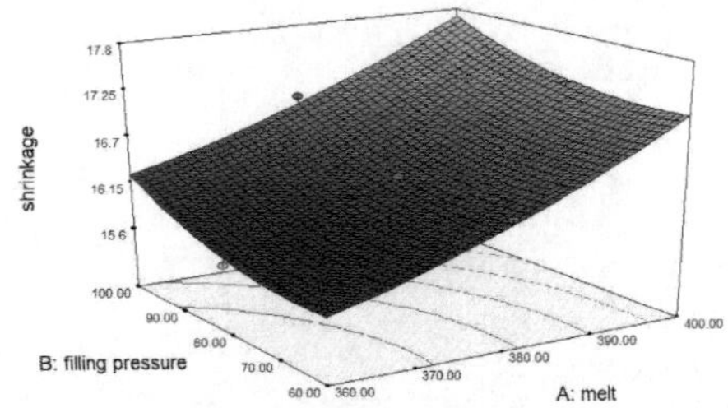

Figure 2. 3D Plot for shrinkage in 30% fiber glass material.

3.2 Optimization of Design Parameters

From the design expert analysis, there are 56 solutions were generated by RSM. The recommended solution with the highest desirability in Table 2 is solution number one. The warpage and shrinkage value of a selected solution are 4.1884 and shrinkage 15.56. The solution for the optimized factors in order to minimize the responses as suggested by the Design Expert software was examined through the Moldflow simulation at melt temperature 360.02, filling pressure 60 MPa and injection time 4.70s.

Table 2. Recommend solution.

No	Melt temp	Filling pressure	Injection time	Warpage	Shrinkage	Desirability
1	360.02	60.00	4.70	4.1884	15.56	0.996
2	360.04	60.00	4.75	4.18856	15.5572	0.996
3	360.01	60.03	4.76	4.18873	15.5553	0.996
4	360.17	60.00	4.87	4.18917	15.5525	0.996
5	360.49	60.01	4.89	4.1896	15.5598	0.996
6	360.00	60.01	4.93	4.18937	15.5433	0.996
7	360.63	60.00	4.95	4.19008	15.559	0.996
8	360.11	60.31	4.72	4.19033	15.5772	0.996
9	360.00	68.44	4.83	4.22979	15.56	0.995

3.2 Validation of Simulation Results

The comparison results between model response and simulation are listed in Table 3. The predicted value from the selected solution and the result run from Moldflow software based on recommending processing parameter as shown in Table 3. From this result, the percentage error can be calculated by using the formula. The percentage of error for shrinkage was only 0.3 % and warpage 0.6 %, which the value difference between the solution and simulation in small difference and acceptable.

Table 3. Result and percentage of error.

	Predicted	Simulation	% Differences
Warpage	4.1884	4.217	0.6
Shrinkage	15.56	15.51	0.3

Figure 3 and Figure 4 show the result in Moldflow software by using the selected solution for three parameters (i.e., melt temperature, filling pressure and injection time). The suggested optimize factor value was examined in Moldflow software, and the result was compared to the predicted value for response in the design expert software. The value for volumetric shrinkage is 15.61 and 4.217 for the warpage. In summary, the optimum values for each factor were successfully determined by RSM.

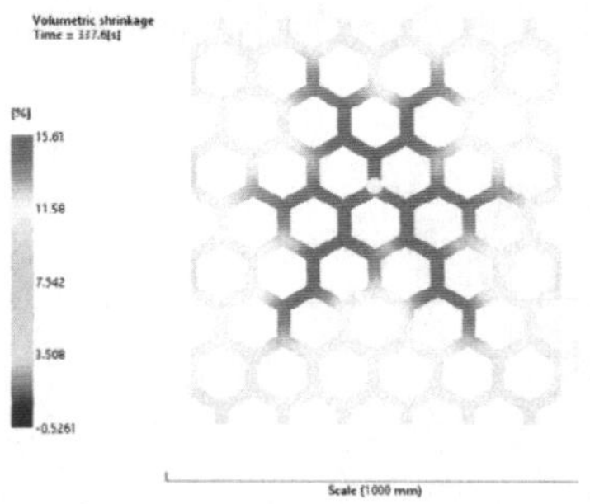

Figure 3. Result for warpage.

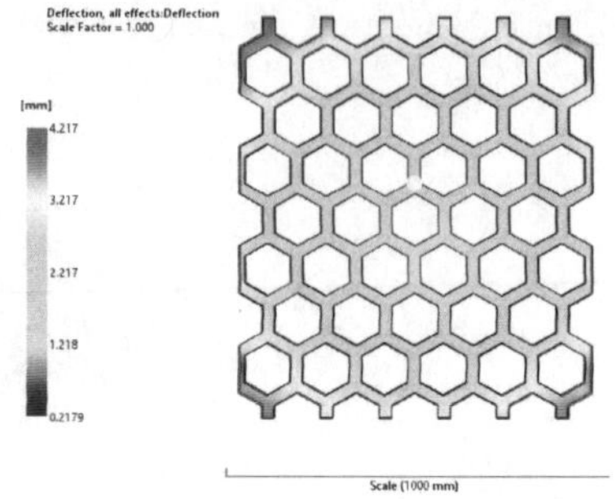

Figure 4. Result for shrinkage.

4. Conclusions

In conclusion, factor of melt temperature, filling pressure and injection time are selected as the most influential factor for shrinkage and warpage. As the results, the optimum value suggested by the software were melt temperature of 360.02°C, 60MPa filling pressure and 4.70s injection time. With small differences error value, 0.3% for shrinkage and 0.6% for warpage, the results was acceptable. It is significant to optimise parameters setting for injection molding in order to reduce the material waste and cost of a high accuracy finishing product.

Acknowledgements

The research was supported by Faculty of Engineering Technology, Universiti Malaysia Perlis (UniMAP) for experiment facilities and finically supported under Short Term Grant (STG) through project number 9001-00503.

References

[1] Rosli M.U *et al* 2013 *International Journal of Engineering and Technology* **5** 3158-3167.
[2] Giglio M *et al* 2012 *Computational Materials Science* **56** 69.
[3] Rosli M U *et al* 2017 *MATEC Web of Conferences* **97** 01039.
[4] Scharkowski J 1999 U.S. Patent No. 5,922,438. Washington, DC: U.S. Patent and Trademark Office.
[5] Rosli M U *et al* 2014 *International Journal of Mechanical & Mechatronics Engineering* **14** 62.
[6] Huang R *et al 2016 Journal of Cleaner Production* **135** 1559
[7] Ganesh B *et al* 2015 *International Journal of Advanced Engineering and Global Technology* I **3** 1451.
[8] Rosli M U *et al* 2013 *International Journal of Materials, Mechanics and Manufacturing* **1** 32
[9] Jizat N M *et al* 2017 *AIP Conference Proceedings* 1885
[10] Jou Y T *et al* 2014 *Applied Mathematics & Information Sciences* **8** 1277
[11] Ishak M I *et al* 2017 *MATEC Web of Conferences* 97
[12] Ramdan D *et al* 2012 *Packaging and Manufacturing Technology* **2** 1786

Theoretical framework of universal design principle (UDP) approach considering haptic exploratory procedure (HEP) of visually impaired person in using TRIZ integration.

S. Shahrin[1,2], KAAA. Rahman[1], KM. Kamarudin[1], and R. Che Me[1]

[1]Faculty of Design and Architecture, Universiti Putra Malaysia, Malaysia
[2]Faculty of Engineering Technology, Universiti Malaysia Perlis, Malaysia

Email: drkhairulazlin@upm.edu.my

Abstract. Visually impaired person has a difficulty in identifying the product function and usability. There is a need to integrate haptic exploratory procedure (HEP) of visually impaired person with universal design principle (UDP) in order to construct structural design approach. The purpose of this study is to construct theoretical framework for universal design approach considering haptic exploratory procedure (HEP). This study uses literature on visually impaired person haptic exploratory procedure (HEP) in order to identify the object properties translated by visually impaired person. Then, it uses literature of universal design to identify universal design approach in concept design process. Then, approach of TRIZ is explored in proposing systematic approach. Finally, the results are expected to lead towards model of theoretical framework of universal design approach using TRIZ and HEP. It can provide a helpful foundation for structuring the experiences to develop further the knowledge and understanding of objects for universal design. This theoretical framework contributes in designing structural UDP approach with consideration of visually impaired person HEP. It improves product usability in designing a universal design for visually impaired person.

1. Introduction

Recent statistic published by World Health Organization (WHO) in October 2018, estimate 1.3 billion people live with some form of vision impairment. With regards to distance vision, 188.5 million people have mild vision impairment, 217 million have moderate to severe vision impairment, and 36 million people are blind. Visual sense is important in guiding humans to move around and navigate through path in known and unknown environments regardless indoor or outdoor. Unfortunately, visually impaired person often faces a challenge in performing daily life routine task. In order to support spatial mapping of visually impaired person, two level of orientation skill which are perceptual and conceptual are widely studied [1], [2]. Visually impaired person relies on other normal functioning senses in order to navigate them in daily life. Majority of visually impaired person relies on the auditory and haptic sensory in order to help them in daily life [3]

Haptic exploratory can be described as perception that relates to the sense of touch, particularly to the ways in which it is possible to discriminate and recognize objects from handling them as opposed to looking at them (Bushnell & Boudreau, 1993). Haptics is commonly viewed as perceptual system that most typically involves active manual exploration. While vision furthermore perceived for giving exceedingly exact spatial and temporal data, the haptic exploratory framework is particularly powerful at processing the material qualities of surfaces and objects [4].

IOP Conf. Series: Materials Science and Engineering **551** (2019) 012109 doi:10.1088/1757-899X/551/1/012109

Good design is represented by the human factors before design specifications are made (Nielson, 1994). It is important for designer to consider basic human processing requirements in order to produce "user friendly" or universal designed products. Universal design has unique challenges because it aims to improve usability for a wide range of users, Universal design philosophy is a transfer towards inclusion promising improved solutions for people with diverse abilities. As stated by Singh & Tandon, (2016) universal design is important in order to aims to integrate various user groups within the design space to benefit the user of all ages with distinct abilities.

2. Review of Literature

2.1. Spatial Perception

In order to understand the needs of visually impaired person need in using product, a review on literature related to spatial perception of visually impaired person is reviewed. Majerova, (2015) highlights the important role of compensatory function of hearing, touch, smell and taste involved in perception of information process by the mental to evaluate. Smitsman & Schellingerhout, (2000) state that key to understand the exploration of blind infant is through kinesthesis and cutaneous sense. Brain will construct conscious image through stimulation of sensory system. This involves the recognition that stimulation has occurred and the ability to discriminate various aspects of the stimulus. Rhodes and Castel, (2008) state that the relation between cognition and perception to show that the size of a stimulus influences its perceived memorability.

Haptic sensory is specific to touch stimuli. Haptic memory is used regularly when assessing the necessary forces for gripping and interacting with familiar objects (Johansson et al., 1993). R. L. Klatzky & Lederman, (1992) state that human explored through touching in two stages, stage 1 is grasp/lifting then executing further exploratory procedures. In stage 1 of exploratory, sufficient information about multiple object properties is extracted, whereas in stage 2 was directed toward precise information about particularly diagnostic properties.

2.2. Haptic Exploratory Procedure

Human has use common behaviour in using their senses in order to explore or identify any objects they held. Klatzky, Lederman, & Matula, (1993) found that human performed highly stereotypical movements patterns using their hand, which possessed both necessary and typical features. Generally, this movement patterns are done unconsciously by human. The movement patterns are called exploratory procedure by Lederman & Klatzky, (1987) in their study. Eight exploratory procedure highlighted by Klatzky & Lederman, (1992, 1993).

Human also strategize the exploratory procedure sequence in order to identify or discriminate objects they currently explore [10]–[12]. Human normally combine different exploratory procedure during exploration to enhance the exploration speed. This statement is supported by study by Sommer & Billard, (2016), haptic exploration on tactile sensor offer more efficient when applied on multiple contact point. Therefore, haptic sensory is important in developing good design for visually impaired person.

2.3. Universal Design

Universal design (also known as inclusive design in United Kingdom and Design for All in Europe) is not a special requirement, for the benefit of only a minority of the population [14], [15]. It is a fundamental condition of good design. However, product design is often base on subjective understanding of designers and does not involve accurate conversion in design needs. Universal design principles are a transfer towards inclusion promising improved solutions for people with diverse abilities. Nevertheless, often in design practice universal design becomes only a formulation among designers. This claim is supported by statement from Mustaquim, (2015) claims an existing theoretical guidelines as design principles often fail to result in real universal design. In addition, Mustaquim, (2015) also claim there is no standard ways to evaluate the impacts of customer needs in universal design principles exist in product performance.

In order to develop a structural standard to evaluate and improve universal design principles implementation in design process. A systematic approach is needed in order to develop a concept meet the customer need. Bertoncelli et al., (2016) state that TRIZ offers a systematic approach for problem solving requires creativity abilities when translating the recommendations offered by inventive principles and standard inventive solutions into the specific domain of the problem.

2.4. TRIZ Application in Universal Design

There is lack of evidence showing TRIZ application for universal design in design practice. A study by C. M. Yang et al., (2010) proposed a TRIZ-based innovative product design process that incorporates universal design principles (UDP). This study highlights the importance of approach in order to develop concrete product concept throughout the product design and development with creative and systematic problem-solving procedure. However, broad application of TRIZ has attracted numbers of researcher to integrate other method with TRIZ. Most of them are in new product development (NPD) and innovation area.

Ko, (2017) state that abstract conceptual design is a critical activity during the early phases of product development, most creative ideas will be generated in this process. Deimel, (2011) found that TRIZ has more similarities than differences with classical design methods. Yang, Kao, & Liu, (2012) stated that an innovative problem-solving method should be the one that can eliminate conflict or contradiction in a problem effectively and efficiently and help generate innovative solution. And TRIZ is the one to serve this purpose well. Study by Chang et al., (2016) found that TRIZ has a strongly positive effect on an ability to analyse problems, and to generate, select, and execute a strategy. TRIZ also increased the creativity in designed products, including ability to develop and implement novel ideas.

3. Results and Discussion

3.1. Summary of Literature Review

In summary, visually impaired person relies on haptic sensory in order to develop spatial perception. Perceptual information spatial perception is important for brain during interpreting and recognizing using sensory naturally. However, for visually impaired person, other senses will be more dominance than visual. Haptic sensory trigger the haptic memory when assessing the necessary forces during interacting with familiar objects. That particular behaviour is identified as haptic exploratory procedure. In order to meet diversity of user needs factor by difference in age, capabilities, social and cultural background, an approach of universal design is introduced. Unfortunately, there are some drawbacks shown by the existing universal design practice. Designer mostly use subjective approach in proposing universal design during conceptual development. There is no structural design guidelines or framework in universal design approach.

3.2. Summary of Literature Review

Therefore, a theoretical framework is constructed base on the summary of the literature review, as per shown in Figure 3 below. There is a gap for potential research of HEP consideration towards UDP in universal design practice. From literature, it shown that there is no standard design practice available for designers to applied since most of the guidelines merely theoretical and shown no positive results in actual practice [16]. Therefore, a potential TRIZ integration with UDP is propose in order to develop a structural approach of universal design. However, in order to achieve universal design among visually impaired person, HEP consideration is needed as knowledge support system for TRIZ – UDP as problem solving tools for universal design.

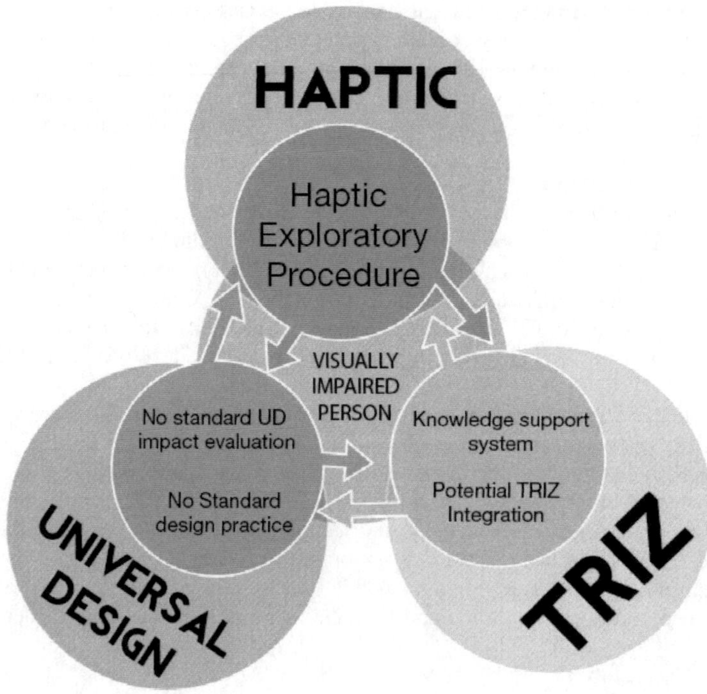

Figure 1. Theoretical framework.

4. Conclusions

For further study, the focus of the research in constructing contradiction matrix from properties of object interpreted by haptic exploratory procedure (HEP) and developing correlation matrix between universal design principles and 40 TRIZ inventive principles. A validation through concept design is needed in order to verify the HEP contradiction matrix and UDP-TRIZ sheet in order to prove this theoretical framework. Findings from this study will contribute a structural universal design tools to develop design with consideration of visual impaired person haptic exploratory procedure. HEP contradiction matrix and universal design tools integrated with TRIZ 40 inventive principles is proposed to replace existing theoretical guidelines as design principles often fail to result in universal design.

References

[1] Majerova H 2015 *Procedia - Soc. Behav. Sci.* **174** 3278
[2] Lahav O *et al* 2005 *Int. J. Disabil. Hum. Dev.* **4** 231
[3] Hecht D *et al* 2009 *Exp. Brain Res* **193** 307
[4] Lederman S J *et al* 1993 *Acta Psychol. (Amst).* **84**, 29
[5] Singh R *et al* 2016 *Int. J. Ind. Ergon.* **55** 46
[6] Smitsman A W *et al* 2000 *Infant Behav. Dev.* **23** 485
[7] Klatzky R L *et al* 1993 *Journal of experimental psychology.* **19** 726
[8] Lederman S J *et al* 1987 *Cogn. Psychol* **19** 342
[9] Klatzky R L *et al* 1992 *Percept. Psychophys.* **52** 661
[10] Klatzky R L *et al* 1993 *Acta Psychol. (Amst).* **84** 29
[11] Kalagher H, 2015 *Cogn. Dev.* **35** 111
[12] Baumgartner E *et al* 2015 *Vision Res.* **115**, 238
[13] Sommer N *et al* 2016 *Auton. Syst.* **85** 48

[14] John Clarkson P *et al* 2015 *Appl. Ergon.* **46** 235
[15] Waller S *et al* 2015 *Appl. Ergon.* **46** 297
[16] Mustaquim M M, 2015 *Procedia Comput. Sci.* **67** 57
[17] Bertoncelli T *et al* 2016 *Procedia CIRP* **39** 191
[18] Yang C M *et al* 2010 *Int. J. Syst. Innov.* **1** 49
[19] Ko Y T 2017 *Comput. Ind. Eng.* **107** 345
[20] Deimel M 2011 *Procedia Eng.* **9** 512
[21] Yang C *et al* 2012 *Syst. Innov.* **2** 1
[22] Chang Y S *et al* 2016 *Think. Ski. Creat.* **19** 112

IOP Conf. Series: Materials Science and Engineering **551** (2019) 012110 doi:10.1088/1757-899X/551/1/012110

Optimization of Laminated Composites Characteristics via integration of Chamis Equation, Taguchi method and Principal Component Analysis

Syafiq Syarafuddin Salam[1], Nik Mizamzul Mehat[1] and Shahrul Kamaruddin[2]

[1]Department of Mechanical Engineering Technology, Faculty of Engineering Technology, Uniciti Alam Campus, Universiti Malaysia Perlis, Sungai Chuchuh, Padang Besar, 02100, Perlis, Malaysia.
[2]Mechanical Engineering Department, Universiti Teknologi PETRONAS, 32610 Seri Iskandar, Perak Darul Ridzuan, Malaysia

E-mail: nikmizamzul@unimap.edu.my

Abstract. To date, composite material has drawn significant attention due to its extended properties in various application. Many factors need to be considered in designing the composites which might leads to complex decision making. Therefore, this study focuses on analyzing and optimizing various factors including combination of material, volume fraction of fiber, ply angle, ply quantity, ply thickness and load applied via integration of the Taguchi method / Principal component analysis along with analytical analysis of Chamis equation. The finite element was adopted in conducting the analysis. Three responses were considered for the laminated composites namely displacement, stress and strain. The findings from the main effects analysis showed that the set of optimum parameters was identified as load applied of 2000N, ply thickness of 0.08 mm, ply quantity of 12 plies, ply angle of 0,90,0 °, material combination of boron/epoxy as well as volume fraction of 65%.

1. Introduction

Laminated composite have been used increasingly in a variety of industrial areas due to their high stiffness and strength to weight ratio, long fatigue life, resistance to electro chemical corrosion, and other superior material properties of composite [1]. A true understanding of their structural behavior is required, such as the deflections, buckling loads and extreme importance for obtaining strong, reliable multi-layered structure failure characteristics [2], [3]. Kelkar et al., [4] for example studied the characteristics of tensile strength, tensile modulus and thermal conductivity of laminated composites. Meanwhile, Fan et al., [5] investigated the fracture toughness, bending strength and work of fracture of the alumina/nickel laminated composites. Many researches were conducted in evaluating the various factors that affecting the properties of laminated composite including combination of material [6], volume fraction of fiber, angle of fiber [7], ply angle, ply quantity and ply thickness [8] that relate with each other. Final design of laminated composite is limited to some extent, which results in complex decision making. This develops dilemma in modelling the relation between properties and the characteristics of the laminated composite material. According to Thiago et al., [9] the lamination parameters play an important role as design variables in composite laminates layer optimization. All these influential parameters that affected the quality of the laminated composite need to be systematically controlled. Therefore, there is a needed to have a proper and analytically optimization method. There are many optimization methods that can be implemented for example Taguchi method [10]. However, by adopting the Taguchi method alone will only applicable for single response, whereby in real industry application, the requirement to have a material with multiple properties such as strength, stiffness, lightweight, rigidity which depending on the material functionality. Therefore, in this study, the optimization method of the Taguchi method will be integrated with the Principal Component Analysis (PCA) along with the analytical Chamis equation to investigate the multi response of laminated different composite simultaneously.

2.Methodology

In this study, the finite element simulation was used to conduct the numerical simulation. The robust parameter design of Taguchi method and principal component analysis (PCA) were integrated in conducting the simulation and the overall procedures is elaborated in the following section.
Selection of parameters

Several parameters with four levels of each were selected in this study including load applied, ply thickness, ply quantity, ply angle, combination of materials and volume fraction. Table 1 shows the parameters control and their levels.

Table 1. Parameters and level studied

Column	Parameter	Level 1	Level 2	Level 3	Level 4
A	Load applied (N)	1000	2000	3000	4000
B	Ply thickness (mm)	0.04	0.08	0.12	0.16
C	Ply quantity	3 plies	6 plies	9 plies	12 plies
E	Ply angle (°)	0, 30, -30	0, 45, -45	0, 60, -60	0, 90, 0
D	Material combination	Kevlar + Epoxy	Boron + Aluminium	E-glass + Polyester	Boron + Epoxy
E	Volume fraction (%)	50	55	60	65

2.1 Selection of Orthogonal Array (OA)

After determining the number of control parameters and their levels, an appropriate OA was established for laying out the design of experiment in the finite element analysis. The selection of an appropriate OA depends on the total degree of freedom (DOF) of the control parameters. In this study, there were six control parameters with four levels of each. As DOF was calculated by subtracting the number of levels with one (DOF = number of levels-1), therefore for six control parameters with four levels each, the total DOF is equal to 18. The total DOF of selected OA must be greater than or equal to the total DOF required for the experiment. Table 2 shows the OA that was used in the simulation.

Analytical calculation by using Chamis equation.

The material properties of Young modulus, Poisson ratio and shear modulus were determined with respected volume fraction as follows:

Young's modulus $$E_1 = E_f V_f + E_m V_m \quad (1)$$

$$E_2 = E_2 = \frac{E_m}{1-\left\{\sqrt{V_f}\left(1-\frac{E_m}{E_f}\right)\right\}} \quad (2)$$

Poisson ratio $$v_{12} = v_{13} = v_f V_f + v_m V_m \quad (3)$$

$$v_{23} = \frac{E_2}{2(G_{23}-1)} \quad (4)$$

Shear modulus $$G_{12} = G_{13} = \frac{G_m}{1-\left\{\sqrt{V_f}\left(1-\frac{G_m}{G_f}\right)\right\}} \quad (5)$$

$$G_{23} = \frac{E_2}{2(1+v_{23})} \quad (6)$$

TABLE 2. OA L_{32} of simulation runs

Trial No	A	B	C	D	E	F
1	1	1	1	1	1	1
2	2	2	2	2	2	2
3	3	3	3	3	3	3
4	4	4	4	4	4	4
5	1	1	2	2	3	3
6	2	2	1	1	4	4
7	3	3	4	4	1	1
8	4	4	3	3	2	2
9	1	2	3	4	1	2
10	2	1	4	3	2	1
11	3	4	1	2	3	4
12	4	3	2	1	4	3
13	1	2	4	3	3	4
14	2	1	3	4	4	3
15	3	4	2	1	1	2
16	4	3	1	2	2	1
17	1	4	1	4	2	3
18	2	3	2	3	1	4
19	3	2	3	2	4	1
20	4	1	4	1	3	2
21	1	4	2	3	4	1
22	2	3	1	4	3	2
23	3	2	4	1	2	3
24	4	1	3	2	1	4
25	1	3	3	1	2	4
26	2	4	4	2	1	3
27	3	1	1	3	4	2
28	4	2	2	4	3	1
29	1	3	4	2	4	2
30	2	4	3	1	3	1
31	3	1	2	4	2	4
32	4	2	1	3	1	3

2.2 Analysis on results by PCA

To overcome the limitation of Taguchi method, the PCA is integrated with Taguchi method. By using PCA, a set of original responses is transformed into a set of uncorrelated components to find the optimal factor or level combination. The application of PCA involves a series of steps that are capable of solving the weakness of the standalone Taguchi method, which requires engineering judgement to handle multiple quality characteristics because the judgement of an engineer increases uncertainty during the decision making. The procedures of implementing PCA in analyzing the results are as follows:

Step 1. Find the normalization data, to avoid discrimination of variables in calculation results

$$X'_i(j) = \frac{X_i(j) - min_i X_i(j)}{max_i X_i(j) - min_i X_i(j)} \tag{7}$$

Step 2. Arrange the data in matrix form. The data taken from step 1 was arranged in matrix form

$$X' = \begin{bmatrix} x'_1 & x'_2 & \dots & \dots & x'_1 \\ x'_2 & x'_2 & \dots & \dots & x'_2 \\ \vdots & \vdots & \dots & \dots & \vdots \\ x'_m & x'_m & \dots & \dots & x'_m \end{bmatrix} \tag{8}$$

Step 3. Calculate the correlation coefficient array of the normalized response.

$$R = \frac{Cov\,(x'_i(j), x'_i(l))}{\sigma_{i(j)} \times \sigma_{i(l)}} \tag{9}$$

Where, $(j = 1,2 \ldots . n; \quad l = 1,2 \ldots .. m)$

Step 4. Determination of eigenvalues and eigenvectors. Eigenvalue is the original total variance while eigenvector is the list of coefficient of the original variables.

$$(R - \lambda_k I)\,(V_k) = 0 \tag{10}$$

Where, $\Sigma\lambda_k = n; \quad k = 1,2, \ldots, n \quad V_k$: Eigenvector, $V = V_1, V_2 \ldots, n$

Step 5. Evaluating the Principle Components (PC) Score. The PC scores can be obtained as linear combination of the original variable and the weighted.

$$Y_k = \Sigma\, x'_m(i). V_k \tag{11}$$

Step 6. Multiple Quality Characteristic Index (MQCI) to represent all responses of quality characteristic.

$$MQCI = \textstyle\sum_{i=1}^{n} W_i \times Y_{ij} \tag{12}$$

3. Results and Discussion

3.1 Analytical results by using Chamis equation.

In this study, Young's modulus (E) , Poisson ratio (υ) and shear modulus for the different composites material as in Table 1 are manually calculated by using the Equation 1 to Equation 6 with respected of

IOP Conf. Series: Materials Science and Engineering **551** (2019) 012110 doi:10.1088/1757-899X/551/1/012110

four selection of volume fraction where are 50%, 55%, 60% and 65%. Young's modulus (E) is used to measure the ability of material to withstand changes in length when under lengthwise tension or compression. Poisson's ratio (υ) is used to determine ratio transverse strain of longitudinal extension strain in the direction of stretching force. Meanwhile, shear modulus (G) is used to define as the ratio of shear stress to the shear strain, where the shear stress is the force which acts is the area on which the force acts to shear strain. The results are tabulated in Table 3.

Table 3. Young's modulus (E) , Poisson ratio (υ) and shear modulus for the different composites material

Composite material	Volume Fraction (%)	E1 (GPa)	E2 (GPa)	E3 (GPa)	υ12	υ13	υ23	G12 (GPa)	G13 (GPa)	G23 (GPa)
Kevlar + Epoxy	50	67.81	14.54	14.54	0.34	0.34	0.33	4.61	4.61	21.80
	55	74.13	16.24	16.24	0.34	0.34	0.33	5.08	5.08	24.36
	60	80.45	18.28	18.28	0.34	0.34	0.33	5.63	5.63	27.42
	65	86.77	20.79	20.79	0.35	0.35	0.33	6.27	6.27	31.19
Boron + Aluminium	50	244.50	168.68	168.68	0.26	0.26	0.33	64.83	64.83	253.02
	55	262.05	181.47	181.47	0.25	0.25	0.33	69.27	69.27	272.21
	60	279.60	195.66	195.66	0.25	0.25	0.33	75.61	75.61	293.48
	65	297.15	211.51	211.51	0.24	0.24	0.33	82.00	82.00	317.27
E-Glass + Polyester	50	37.60	9.87	9.87	0.29	0.29	0.33	9.98	9.98	14.80
	55	41.04	10.98	10.98	0.28	0.28	0.33	4.09	4.09	16.47
	60	44.48	12.32	12.32	0.28	0.28	0.33	4.59	4.59	18.47
	65	47.92	13.94	13.94	0.27	0.27	0.33	5.21	5.21	20.90
Boron + Epoxy	50	212.31	5.32	5.32	0.26	0.26	0.33	5.34	5.34	7.98
	55	233.08	17.33	17.33	0.25	0.25	0.33	6.03	6.03	26.00
	60	253.85	19.75	19.75	0.25	0.25	0.33	6.88	6.88	29.63
	65	274.62	22.80	22.80	0.24	0.24	0.33	7.95	7.95	34.20

3.2 Results of Finite Element Analysis

The stress, strain and displacement values for each trial conducted by using the selected OA (refer Table 2) obtained from the finite element analysis is normalized to a unit value, ranging from 0 to 1. The normalized values are summed up to represent the stress, strain and displacement simultaneously as shown in Table 4.

Table 4. The result of before and after normalization and MCQI

No Triall	Before normalization			After normalization			MQCI
	Displacement	Stress	Strain	Displacement	Stress	Strain	
1	161.0 X 10^{-6}	16.3 X 10^{+3}	244.0 X 10^{-9}	0.448	0.321	0.349	1.406
2	1.5 X 10^{-6}	5.5 X 10^{+3}	20.6 X 10^{-9}	0.003	0.085	0.022	-0.639
3	10.0 X 10^{-6}	5.7 X 10^{+3}	128.0 X 10^{-9}	0.027	0.091	0.179	-0.224
4	857.0 X 10^{-9}	3.0 X 10^{+3}	11.0 X 10^{-9}	0.001	0.031	0.008	-0.799
5	9.9 X 10^{-6}	9.3 X 10^{+3}	135.0 X 10^{-9}	0.026	0.168	0.190	-0.022
6	2.8 X 10^{-6}	10.0 X 10^{+3}	37.8 X 10^{-9}	0.006	0.183	0.047	-0.351
7	4.8 X 10^{-6}	2.7 X 10^{+3}	62.3 X 10^{-9}	0.012	0.026	0.083	-0.618
8	2.0 X 10^{-6}	7.6 X 10^{+3}	27.5 X 10-^{9}	0.004	0.131	0.032	-0.509
9	9.4 X 10^{-6}	9.5 X 10^{+3}	128.0 X 10^{-9}	0.025	0.172	0.179	-0.039
10	478.0 X 10^{-9}	1.6 X 10^{+3}	6.5 X 10^{-9}	0.000	0.001	0.001	-0.885
11	12.5 X 10^{-6}	8.3 X 10^{+3}	172.0 X 10^{-9}	0.034	0.147	0.244	0.064
12	26.6 X 10^{-6}	18.1 X 10^{+3}	71.8 X 10^{-9}	0.073	0.360	0.097	0.283

13	$7.7 X 10^{-6}$	$5.0 X 10^{+3}$	$68.8 X 10^{-9}$	0.020	0.075	0.092	-0.471
14	$624.0 X 10^{-9}$	$2.1 X 10^{+3}$	$8.5 X 10^{-9}$	0.001	0.012	0.004	-0.852
15	$359.0 X 10^{-6}$	$32.3 X 10^{+3}$	$687.0 X 10^{-9}$	1.000	0.668	1.000	4.631
16	$3.2 X 10^{-6}$	$4.1 X 10^{+3}$	$16.4 X 10^{-9}$	0.008	0.055	0.015	-0.714
17	$4.2 X 10^{-6}$	$5.1 X 10^{+3}$	$57.5 X 10^{-9}$	0.011	0.077	0.076	-0.522
18	$3.8 X 10^{-6}$	$14.4 X 10^{+3}$	$51.1 X 10^{-9}$	0.009	0.279	0.066	-0.079
19	$3.8 X 10^{-6}$	$2.2 X 10^{+3}$	$52.6 X 10^{-9}$	0.009	0.014	0.069	-0.685
20	$3.1 X 10^{-6}$	$8.2 X 10^{+3}$	$38.8 X 10^{-9}$	0.008	0.145	0.048	-0.432
21	$1.3 X 10^{-6}$	$1.5 X 10^{+3}$	$17.3 X 10^{-9}$	0.002	0.000	0.017	-0.848
22	$435.0 X 10^{-9}$	$1.7 X 10^{+3}$	$5.8 X 10^{-9}$	0.000	0.003	0.000	-0.884
23	$30.2 X 10^{-6}$	$16.5 X 10^{+3}$	$408.0 X 10^{-9}$	0.083	0.324	0.591	1.358
24	$18.2 X 10^{-6}$	$47.7 X 10^{+3}$	$229.0 X 10^{-9}$	0.050	1.000	0.328	2.249
25	$61.0 X 10^{-6}$	$37.3 X 10^{+3}$	$482.0 X 10^{-9}$	0.169	0.776	0.699	2.789
26	$3.5 X 10^{-6}$	$14.1 X 10^{+3}$	$47.7 X 10^{-9}$	0.009	0.273	0.062	-0.107
27	$3.2 X 10^{-6}$	$1.6 X 10^{+3}$	$43.5 X 10^{-9}$	0.008	0.002	0.055	-0.745
28	$560.0 X 10^{-9}$	$1.9 X 10^{+3}$	$8.1 X 10^{-9}$	0.000	0.008	0.003	-0.864
29	$2.7 X 10^{-6}$	$2.8 X 10^{+3}$	$35.6 X 10^{-9}$	0.006	0.028	0.044	-0.716
30	$895.0 X 10^{-9}$	$3.6 X 10^{+3}$	$12.2 X 10^{-9}$	0.001	0.046	0.009	-0.762
31	$11.3 X 10^{-6}$	$5.7 X 10^{+3}$	$152.0 X 10^{-9}$	0.030	0.090	0.215	-0.136
32	$12.8 X 10^{-6}$	$16.4 X 10^{+3}$	$69.1 X 10^{-9}$	0.035	0.322	0.093	0.123

3.3 Analysis of the normalized data by PCA

By using PCA, the set of normalized responses is transformed into a set of uncorrelated component, so that the optimal factor or level combination can be found. Table 5 and Table 6 display the eigenvalues and eigenvectors obtained from the PCA, which was conducted by utilizing the statistical software package. The new position coordinate system of the principal component was referred by the principal scores in this analysis. Besides, the linear combination of the original variable and the weighted were obtained by the PC scores. As studied, all the responses composites analysis which include the internal responses (stress, strain, and displacement) were represented by the used of multiple quality characteristic indexes (MQCI). The MQCI was obtained by integrating the total PCs scores via linear combination method correspond to their explanatory power for the total variance in ordered to facilitate the optimization analysis. The results of MQCI are tabulated in Table 4.

Table 5. The eigenvalues and explained percentage of variation for principal component

Component	Eigenvalue	Variability (%)	Cumulative %
1	2.360	78.652	78.652
2	0.504	16.797	95.449
3	0.137	4.551	100.000

Table 6. The eigenvectors for the principal components

Component	Displacement	Stress	Strain
1	0.565	0.541	0.622
2	-0.645	0.761	-0.076
3	-0.515	-0.358	0.779

3.4 Main effects analysis

The main effect analysis is the effect of an independent variable, averaging the levels of any other independent variables. The mean response at each level of the parameters was found out by calculating the average, based on the total MQCI (refer Table 4) for each parameter at different levels. Figure 1 illustrates the main effects analysis for the six parameters studied in this research.

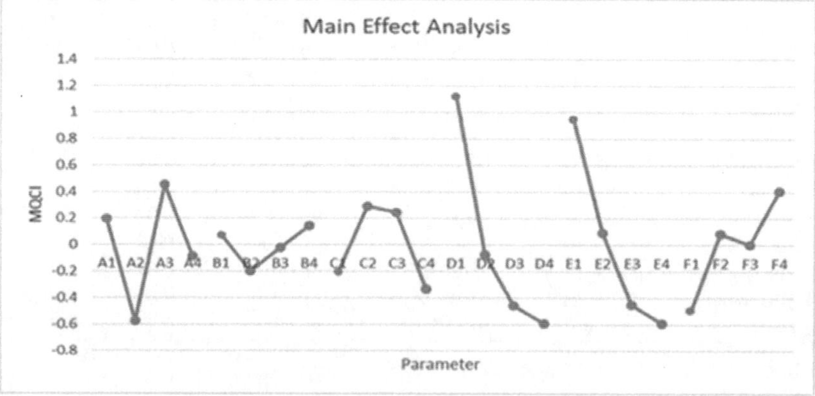

Figure 1. Main effects plot for MQCI

Figure 1 shows that the displacement, stress and strain which are represented by the used of MQCI are affected significantly by variations in the parameters. The lesser magnitude of MQCI will result in better parameter characteristics. The increment of load applied (A) from 1000 N to 2000N decreases the MQCI, yet then rapidly increased at 3000 N of load applied 3. By contrast, the ply thickness (B), an opposite trend is observed, shows the value of average MQCI slightly decreases at Level 1 but greatly increases from Level 2 to Level 4. Besides, the average of MQCI for ply quantities (C) is decreased gradually from Level 2 until Level 4. For ply angle (D) and material combination (E), similar trends can be observed for the decrement of the MQCI value as the levels of each increase. On the contrary, an opposite trend occurs on volume fraction (F) where the increasing of the fraction values apparently increases the value of MQCI in general.

From the result of main effect analysis, the best combination of parameters and levels could easily obtain by selecting the level of each parameter with the lowest average of MQCI. Referring to Figure 1, A2, B2, C4, D4, E4 and F4, show the lowest values of average MQCI for parameters A, B, C, D, E and F respectively. As a result, the optimal parameter setting which statistically results in stress and strain components as well as spatial displacement, are predicted to be A2, B2, C4, D4, E4 and F4. The set of optimum parameters is identified as load applied of 2000N, ply thickness of 0.08 mm, ply quantity of 12 plies, ply angle of 0,90,0 °, material combination of boron/epoxy as well as volume fraction of 65%.

4.Conclusion

Based on the findings of the research, it can be concluded that, by using the integration of the Taguchi method and PCA, the significant mathematical model function can be obtained in order to predict the optimal parameters in designing the different composites with variation of parameters. The set of optimum parameters is identified as load applied of 2000N, ply thickness of 0.08 mm, ply quantity of 12 plies, ply angle of 0,90,0 °, material combination of boron/epoxy as well as volume fraction of 65%.

Acknowledgments

The author would like to acknowledge support from the short term grant under a grant number of 9001-00523, Universiti Malaysia Perlis (Unimap) and Universiti Teknologi Petronas.

References

[1] Anish A.K *et al* 2018 *Structures*, 95-110.
[2] Rousseau M *et al* 2005 *Biomaterials* **26(31)** 6254-6262.
[3] Nikbakt S *et al* 2018 *Composite Structures* **195** 158-185.
[4] Kelkar A. *et al* 2016 *Materials Chemistry and Physics* **176** 136-142.
[5] Fan H *et al* 2017 *Ceramics International* **43 (13)** 10632-10636.
[6] Fiore V *et al* 2018 *Polymer Testing* **69** 310-319.
[7] Luis N *et al* 2017 *Composites Part B: Engineering* **128** 53-66.
[8] Chen X *et al* 2018 *Composite Structures* **192** 336-346
[9] Dutra T.A *et al* 2015 *Composite Structures* **133** 166-177.
[10] Mehat N.M *et al Journal of Physics: Conference Series* , **908 (1)** 012009
[11] Pattanaik A *et al* 2016 *Engineering Science and Technology, an International Journal* 710-716.

IOP Conf. Series: Materials Science and Engineering **551** (2019) 012111 doi:10.1088/1757-899X/551/1/012111

Effect of Extraction and Fractionation on Antioxidant Activity of Extract and Fraction *Crescentia Cujete* L Leaves

Hartati[1], Abd Muis[1], Narhaeda[1], Hasmida Mohd Nasir[2] and Nur Salsabila Md Norodin[3]

[1]Biology Department, Faculty of Mathematics and Natural Sciences, Universitas Negeri Makassar, Indonesia
[2]Centre of Lipids Engineering & Applied Research (CLEAR), Ibnu Sina Institute for Scientific & Industrial Research, Universiti Teknologi Malaysia, 81310 UTM Johor Bahru, Johor, Malaysia
[3]School of Chemical and Energy Engineering, Faculty Engineering, Universiti Teknologi Malaysia, 81310 Johor

E-mail: hartati@unm.ac.id

Abstract. Antioxidant compunds like phenols and flavonoids scavenge free radicals and thus inhibit the oxidative mechanisms that lead to control degenerative and other diseases. The aim of this study was to investigate the antioxidant activity in extracts and fractions of *Crescentia cujete* leaves. 1,1-difenil-2-pikrilhidrazil (DPPH) assay was performed to measure the free-radical scavenging activity of extracts. The results of the study show that leaves of *C. cujete* possesses significant free radical scavenging properties. All the ethanol extract, etyl acetat extract and fractions exhibited antioxidant activities, however, ethanol fractions of leaves showing the highest antioxidant activity based on the results of DPPH tests.

1. Introduction
Natural antioxidants occur in all parts of plants. These antioxidants include vitamins, carotenoids, flavonoids, phenols, dietary glutathionine, and endogenous metabolites. Plant-derived antioxidants have been shown to function as singlet and triplet oxygen quenchers, free radical scavengers, peroxide decomposers, enzyme inhibitors, and synergists [1]. The phytochemicals in plant tissues responsible for the antioxidant capacity can largely be attributed to the phenolics, anthocyanins, and other flavonoid compounds [2]

Crescentia cujete L Family of Bignoniaceae is commonly known as calabash tree. It is widely distributed in the Caribbean region, Mexico, Northern and Southern American and later introduced to tropical Africa from Senegal to Cameroon then to other parts of Africa [3]. In Indonesia, it is known as Maja or Bila. Maja leaves in traditional medicine are used to treat new wounds and reduce hypertension. The powdered leaves are used for headaches, and internally as a diuretic and in the treatment of tumors and hematomas [4]. This plant contains active compounds namely tartrate acid, stenohidric, citric acid, stearic acid, palmitic acid, flavonoid and quercetin [5]. The aim of this study was to investigate the antioxidant activity in extracts and fractions of *Crescentia cujete* leaves.

2. Experimental

IOP Conf. Series: Materials Science and Engineering **551** (2019) 012111 doi:10.1088/1757-899X/551/1/012111

2.1. *Preparation Extracts*
The leaves of the *C. cujete* was made into coarse powder. Five hundred gram of *C. cujete* was macerated in 70% ethanol for 3 days. The liquid component was filtered through whatman no.1 filter paper and evaporated to dryness under vacuum at 40°C using a rotary evaporator. All the step will be repeated using ethyl acetate solvent. The sample was stored under refrigeration (-20°C) condition for further analysis.

2.2. *Fractionation of Active Extracts*
The extract which showed the highest antioxidant activity was carried out fractionation. Fractionation is done by using a vacuum chromatography column to separate the compounds contained in it. The column was inserted PF254 silica gel as a stationary phase, then eluted using a mobile phase system using solvents which began with the lowest to high polarity level. Each fraction obtained was visualized to determine its profile using TLC aluminum plate GF254 (E-Merck) to determine the fraction to be combined. The combined fractions obtained were tested for antioxidant activity by using DPPH test.

2.3. *Antioxidant activity of C. cujete*
Antioxidant activity was measured by using DPPH assay. This assay was carried out according to the method [6] with a slight modification. Extract solution was prepared by dissolving 0.025 g of dry extract in 10 ml of methanol to give final concentration at 2.5 mg/ml. Then, 77 μL of the extract solution was mixed with 3 ml of 6 x 10^{-5} M methanolic solution of DPPH. After that, the mixture was placed in the dark for 30 minutes at room temperature and the decrease in the absorption was measured at 517 nm by using spectrophotometer. The DPPH radical concentration was calculated by using the following equation:

$$\text{DPPH radical concentration } (\%) = \frac{\text{A Control} - \text{A Sample}}{\text{A Control}} \times 100 \quad (1)$$

Where $A_{Control}$ is the absorbance value of the control reaction and A_{Sample} is the absorbance value with the presence of the tested extracts in the sample.

2.4. *Analysis of Data*

The results of these experiments are expressed as mean ± S.E, of three animals in each group. The data were evaluated by one-way ANOVA followed by Tukey's pair-wise comparison test. The values of $p < 0.05$ were considered as statistically significant.

3. Results and Discussions

3.1. *Yield extract of C. cujete*
Figure 1 shows that the result of yield on *C. Cujete* extract. There are two different solvents were used to extract the *C. cujete* which are 70% ethanol and acetate ethyl.

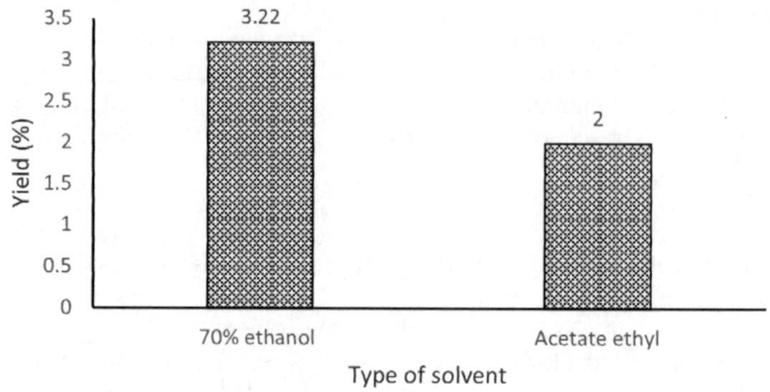

Figure 1. Percentage of extraction yield on different solvent

The figure 1 shows that the highest extraction yield was found with 70% ethanol solvent. On the other hand, acetate ethyl resulted in lowest extraction yield. Extracts using 70% ethanol solvent were higher than ethyl acetate solvents. The yield extract value was related to the number of secondary metabolites, the greater the yield value the more the content of secondary metabolites [7].

3.2. Activity antioxidant of C. cujete

The presented figure 2 shows that the DPPH free radical scavenging assay result of five extract from *C.cujete* leaves. The figure 2 show that the 70% ethanol fraction (75.43 %) gives highest DPPH radical scavenging activity compared to the other. This indicates the antioxidant ability of 70% ethanol fraction is stronger than the other extract. The radical scavenging activity of the extracts could be related to the nature of phenolics, thus contributing to their electron transfer/hydrogen donating ability. This plant contains active compounds namely tartrate acid, stenohidric, citric acid, stearic acid, palmitic acid, flavonoid and quercetin [5]. *C.cujete* leaves contain flavonoid and phenol which give antioxidant activity in vitro [8]. Flavonoid found in *C.cujete* can act as antioxidants and protect the cells of the body from radical damage [9]. Alkaloid in *C. cujete* may explain why it is being used as anti-inflammatory agents [5,10]. Other studies have shown that C. cujete contains Flavonoid-quercetin [11], tannins, phenols, saponins, anthraquinones and cardenolides [5]. tartaric acid, cyanohydric, citric acid, cresentia acid, beta-sitosterol, stigmastrol, alpa and beta amyrine, esteric acid, palmitic acid, apigenin, naphthaquinone, iridoid glycosides, 3-hydroxyoktanol glycosides [11]. Flavonoid-quercetin also found in *C.cujete* has activity as an antioxidant that protects the body cells from free radical damage that contribute to cell damage and various health-related problems [5]. *C. cujete* has significant wound healing activity [12].

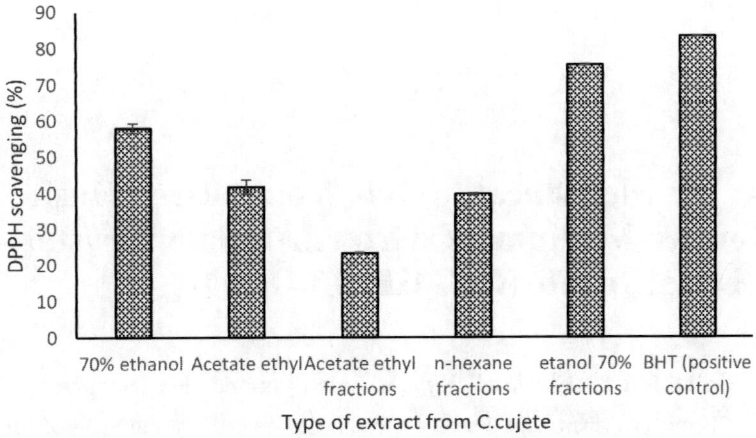

Figure 2. Percentage of DPPH Free radical scavenging from extract and fractions of *C.cujete.*

4. Conclusions

Crescentia cujete Leaves of the ethanol extract, etyl acetat extract and fractions exhibited antioxidant activities, however, ethanol fractions of leaves showing the highest antioxidant activity based on the results of DPPH tests.

References

[1] Larson R A 1988 *Phytochemistry* **4** 969-978.
[2] Cao G *et al* 1997 *Free Radicals Biol. Med.* **22** 749-760.
[3] Amarachukwo U A *et al* 2017 *Asian Pac. J. Health Sci* **4** 27- 35.
[4] Zengin G *et al* 2011 *Rec. Nat. Prod* **5** 123-132.
[5] Ejelonu B C *et al* 2011 *African Journal Biotechnology* **10** 84.
[6] Miliauskas G *et al* 2004 *Food Chemistry* **85** 231-237.
[7] Kusuma, Susanti A M 2013 *Laporan Penelitian, Fakultas Farmasi Universitas Muhammadiah Purwokerto.*
[8] Narhaedah 2018 Skripsi, Universitas Negeri Makassar.
[9] Arthur M 1992 *Human Nutr .* **55** 321-325.
[10] Michael 2004 Publishers GMBH, MNHN pp.191.
[11] Marc N O 2008 *Journal of Food Technology* **6** 267-270.
[12] Hartati, *et al* 2018 *AIP Conference Proceedings* 2030.

IOP Conf. Series: Materials Science and Engineering **551** (2019) 012112 doi:10.1088/1757-899X/551/1/012112

Synthesis of Metasilicate Gel from Rice Husk and Its Application as Medium of Growth Single Crystal Calcium Tartrate Tetrahydrate (CaC4H4O6.4H2O)

Suriati Eka Putri[1], Hasri[1], Sumiati Side[1], Yunanda FR[1], Subaer[2]

[1]Chemistry Department, Faculty of Mathematics and Natural Science, Universitas Negeri Makassar, Indonesia
[2]Physics Department, Faculty of Mathematics and Natural Science, Universitas Negeri Makassar, Indonesia

E-mail: ekaputri_chem@unm.ac.id

Abstract. Has been done a research about synthesis metasilicate gel from rice husk and its application as medium of growth single crystal. synthesis This research was carried out in several stage namely sample preparation, metasilicate gel synthesis, crystal CaTT synthesis, and metasilicate gel characterization. Rice husk ash that used content SiO_2 of 98.45 %, it was potential to be a sodium silicate filtrate. Filtrate of sodium silicate was reacted with tartrat acid ($C_4H_6O_6$) and produce metasilicate gel with pH 5,00; 5,25; 5,50; 5,75 and 6,00. The supernatant of calcium chloride ($CaCl_2$) diffuse into the gel and formed the clear white crystal CaTT as much as 0,2659 g at the optimum pH 5,25. Analysis of functional groups of metasilicate gel by FTIR provides the specific absorption of –OH group at 3400.05 cm^{-1} and 920.05 cm^{-1}, C=O group at 1622.13 cm^{-1} and C-O group at 1346.31 cm^{-1}, and the Si-O-Si group at 1064.71 cm-1. Characterization of gel metasilicate by XRD indicated that metasilicate gel produced formed on $Na_2C_4H_4O_6.2H_2O$, compound of SiO_2 and C-Grafit. Based on the analysis of functional groups by FTIR and characterization by XRD, it was concluded that the rice husk was potentially to be a sources of metasilicate gel as medium to grow single crystal of CaTT.

1. Introduction

Rice husk as a waste from the rice milling process has the potential to be exploited by the abundance of rice production in Indonesia, especially in South Sulawesi. The rice milling process produces rice husks of about 20-30 % of the grain weight [1]. Rice husk as waste, allows it to be utilized since the chemical composition contained in it can still be processed further. Rice husk ash has a high silica content of around 94.4% [2]. High silica content in rice husk ash can be used as raw material for making the material of silica based, such as for making sodium silicate [3]. Sodium silicate has been synthesized from natural ingredients by various methods. One method that can be done to synthesize sodium silicate using alkaline dissolution, which is reacted with NaOH.

Several research [3] [4], have synthesized sodium silicate from rice husk ash which is the main ingredient in making metasilicate gel and can be used as a crystal growth medium. It shows that the silica content of rice husk ash can be used as an alternative material for making metasilicate gel.

Several methods of crystal synthesis, such as hydrothermal methods, steam diffusion, coprecipitation, microwave, sol gel, and metasilicate gel. Metasilicate gel is a relatively simple method in manufacturing and production, it has a relatively stable structure [5].

IOP Conf. Series: Materials Science and Engineering **551** (2019) 012112 doi:10.1088/1757-899X/551/1/012112

Synthesized single crystal calcium tartrate tetrahydrate (CaTT) using metasilicate gel as a crystal growth medium. The synthesis of single crystals in gels is influenced by gel acidity, supernatant concentration, temperature, growth time and growth techniques [6] [7]. The effect of pH is a factor that greatly affects crystal growth.

2. Experimental

2.1. Material
The material used for sample preparation are distilled water and hydrochloric acid 0,1 M. For synthesis of metasilicate gel from rice husk, sodium hydroxide solid, tartaric acid various concentration, and distilled water.

2.2. Sample Preparation
Rice husk was heated at 700 °C for 2 hours. Rice husk ash is crushed and sieved at 200 Mesh. A total of 25 g of rice husk ash was washed with 150 mL HCl 0.1 M and stirred for 1 hour, left to stand overnight. Then filtered and washed until neutral pH. The ash is dried at 110 °C for 2 hours, and weighed. The composition of rice husk ash was analyzed using XRF.

2.3. Synthesis of Metasilicate Gel From Rice Husk
A total of 6 g of rice husk ash added 200 mL NaOH 1 M while stirring, heated for 1 hour, then left for 18 hours, filtered and sodium silicate filtrate is obtained. 5 mL of sodium silicate filtrate are added, then 0.5 M tartaric acid solution was added and stirred until pH 5.00 was obtained. The test tube is covered with aluminum foil and stored for 4 x 24 hours until a gel was formed. Then metasilicate gel was made with a pH of 5.25; 5.50; 5.75; and 6.00.

2.4. Synthesis of Single Crystal Calcium Tartrate Tetrahydrate (CaTT)
Dripped 2.5 mL of supernatant solution CaCl2 0.45 M into the gel through the tube wall and closed again with aluminum foil, then let stand for 7 x 24 hours. The crystals formed, separated from the gel, then dried and weighed. The metasilicate gel was characterized using XRD.

3. Results and Discussions

3.1. Synthesis of Metasilicate Gel From Rice Husk
The results of analysis the composition of rice husk ash ash were dominated by SiO_2 compounds of 98.45%, shown in table 1.

Table 1. Composition of rice husk ash

Compound	Percentage (%)
SiO_2	98.45
K_2O	0.913
CaO	0.218
P_2O_5	0.186
Fe_2O_3	0.099
MnO	0.0921
SrO	0.0103
ZnO	0.0088
Nb_2O_5	0.0067

Rice husk ash is reacted with NaOH to dissolve SiO_2 and produce sodium silicate filtrate based on the reaction:

$$SiO_{2\,(s)} + 2\,NaOH_{\,(l)} \rightarrow Na_2SiO_{3\,(aq)} + H_2O \quad (1)$$

IOP Conf. Series: Materials Science and Engineering **551** (2019) 012112 doi:10.1088/1757-899X/551/1/012112

Metasilicate gel consists of rice husk sodium silicate and tartrate acid as difference pH. The appearance of metasilicate gel shown in figure 1.

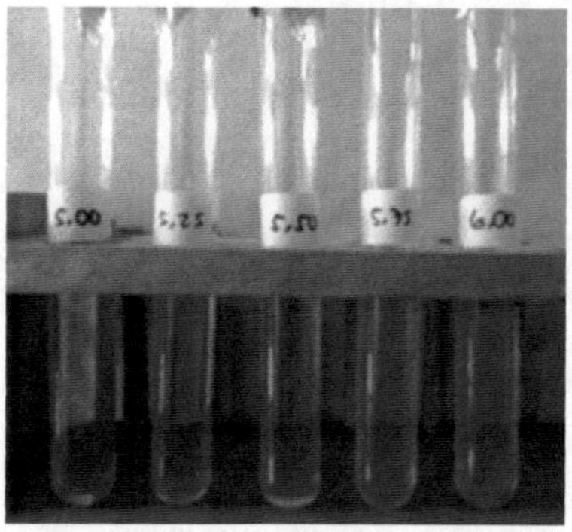

Figure 1. Rice Husk Metasilicate Gel.

Sodium silicate in water forms several monosilicate acids (H_4SiO_4) which form three-dimensional system polymers Si-O-Si and water chains based on reactions as shown in Figure 2.

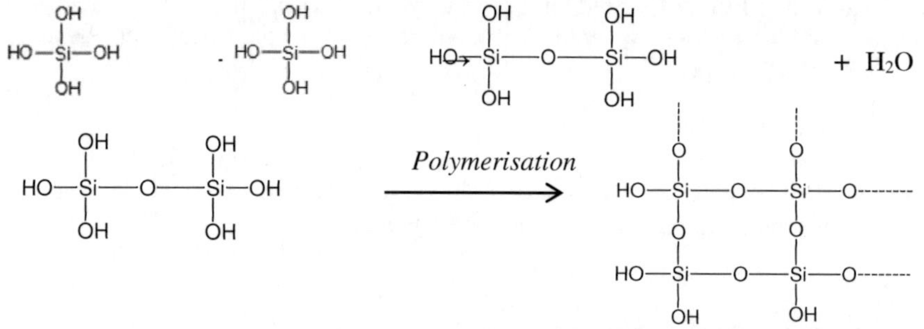

Figure 2. Hydrogen Bonding Between ≡Si-OH

The structure of the gel is strongly influenced by the pH of the gel. At pH > 5.25, the gel becomes harder then at pH <5.25. At pH>5.25 less water is produced during the formation of silanol group, as shown in Equation (2).

$$\equiv Si\text{-}O^- + H_3O^+ \rightarrow \equiv Si\text{-}OH + H_2O \qquad (2)$$

Conversely, at pH < 5.25, the addition of tartrate acid excessively all protonated silanol (≡Si-OH) groups form siliconium ions and two water molecules formed as by-products such as the reaction in Equation (3).

$$\equiv Si\text{-}OH + H_3O^+ \rightarrow \equiv Si^+ + 2H_2O \qquad (3)$$

The number of water molecules formed will increase by the number of $[H^+]$ ions. The amount of water molecules in the gel can cause the gel cavity to be more tenuous by the presence of hydrogen bonds between water molecules and two molecules ≡Si-OH as shown in Figure 3.

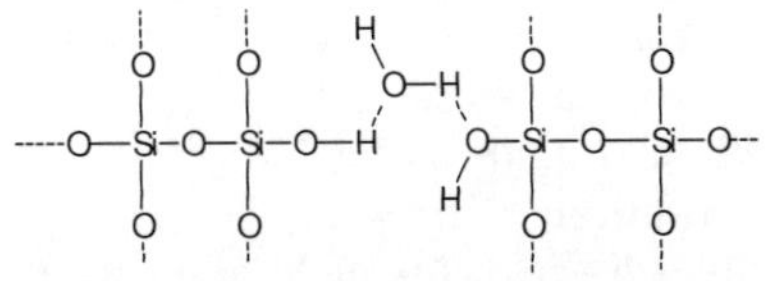

Figure 3. Hydrogen Bonding Between ≡Si-OH

3.2. Synthesis of Single Crystal CaTT by Metasilicate Gel Method

In the metasilicate gel, tartrate acid will react to form sodium tartrate based on the Equation (4):

$$2NaOH + H_2C_4H_4O_6 \rightarrow Na_2C_4H_4O_6 + 2\,H_2O \quad (4)$$

CaTT crystals are synthesized by dripping the supernatant $CaCl_2$ solution and reacting with the tartrate in the gel based on the reaction in Equation (5).

$$CaCl_2 + Na_2C_4H_4O_6 + 4H_2O \rightarrow CaC_4H_4O_6.4H_2O + 2NaCl \quad (5)$$

The optimum pH of CaTT crystal synthesis using metasilicate rice husk gel was obtained at pH 5.25 with a mass of 0.2649 g. The relationship between pH gel metasilicate rice husk and the mass of crystals produced is shown in Figure 4.

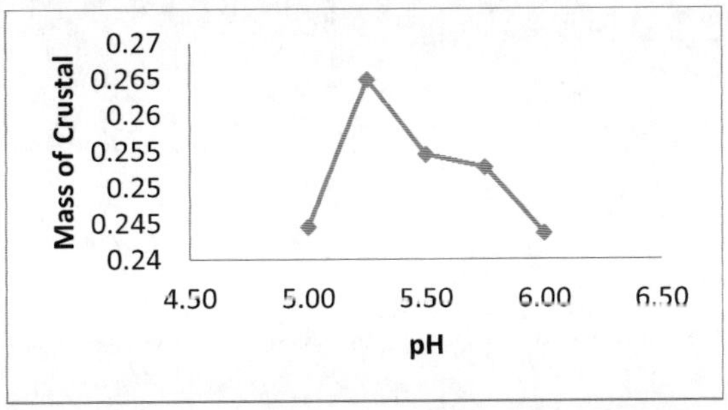

Figure. 4. The relationship between pH and mass of crystal

Crystal with the highest mass is 0,2649 g. At the pH of the gel <5.25 which is soft, the supernatant solution does not diffuse properly, while the pH of the gel >5.25 is relatively hard so that the supernatant solution is difficult to diffuse. Crystal synthesis of CaTT with supernatant $CaCl_2$ using metasilicate rice husk gel produced white crystals that were insoluble with distilled water with a size 1.5 mm x 7 mm as shown in Figure 5.

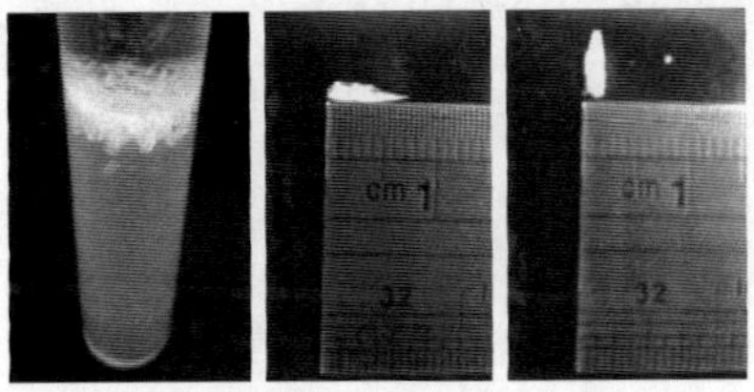

Figure. 5. CaTT crystal

3.3. The XRD Analysis

The results of X-Ray Diffraction (XRD) analysis as a qualitative analysis to identify the content of the metasilicate gel compound of rice husk and pure metasilicate gel are shown in Figure 6, it was a similar diffractogram. It has a peak with a small intensity indicating that both have a crystalline phase. The crystalline phase contained in the metasilicate gel is a crystalline sodium tartrate dihydrate ($Na_2C_4H_4O_6.2H_2O$) at 2θ = 31.922° and 32.059°, which is formed tartrate acid ($C_4H_6O_6$) with the remaining NaOH synthesized sodium silicate (Na_2SiO_3).

Three-dimensional cavity structures by crosslinking silanol and siloxane groups were identified by diffractogram in the presence of SiO_2 compounds at 2 θ = 25.517° and 25.61°. The metasilicate rice husk gel showed a peak at 2 θ = 26.27° which is a C-graphite compound, while pure metasilicate gel showed a peak at 2θ = 43.98° C-diamond compound. This is due to the presence of other oxide compounds contained in rice husks which are used as samples, so that the carbon compounds identified in the metasilicate rice husk gel and pure metasilicate gel are not similiar.

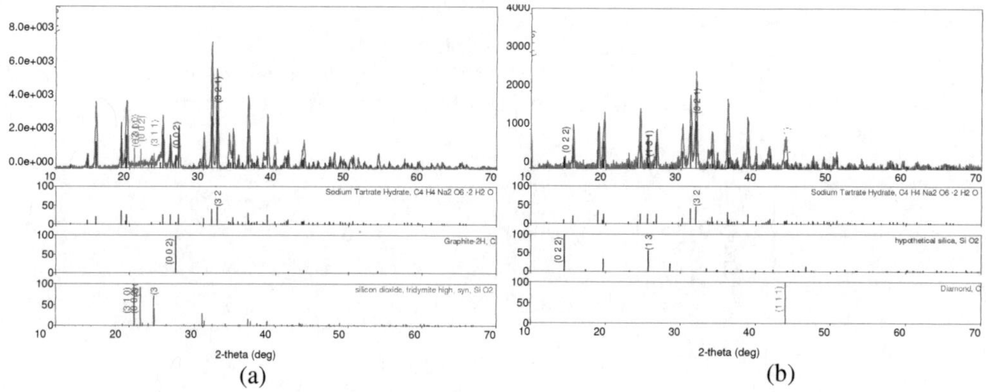

Figure 6. Difractogram of (a) rice husk metasilicate gel (b) pure metasilicate gel.

4. Conclusions

Based on the results obtained from the research conducted, it can be concluded that the optimum pH of metasilicate rice husk gel formation as a single crystal CaTT growth medium is pH 5.25, produce clear white CaTT crystals with a mass of 0.2649 g which is insoluble with distilled water. Based on XRD analysis the rice husk gel metasilicate is formed from Na2C4H4O6.2H2O salt, SiO2 and C-graphite compounds.

References

[1] Patil R *et al* 2014 IOSR-JAC 26-29.
[2] Folleto *et al* 2006 Journal of Material Research **9** 335-338.
[3] Trivana *et al* 2015 Journal of Science and Technology **7** 66-75.
[4] Kalapathy U *et al* 2000 Bioresource Technology **85** 285-289.
[5] Patel A R *et al* 1982 Material Sci. **72.**
[6] Prananto Y P *et al* 2012 Journal Trop.Life Science **2** 8-9.
[7] Ginanjar R R *et al* 2014 Seminar Nasional Hasil Penelitian 3-8.

Fabrication of Zeolite-Geopolymer Hybrid Bulk Materials Based on Laterite Soils Deposit Gowa Regency, South Sulawesi, Indonesia

Anita Dewi Permatasari[1], Nur Akifah[1], Nur Syamsih Amalia[1], Muhammad Rizal Fahlepy[1], Susilo Sudarman Desa[1] and Subaer[1]

[1]Center of Excellence of Green Materials and Technology (CEoGMTech), Physics Department, FMIPA, Universitas Negeri Makassar, 90224, Makassar, Indonesia

E-mail: subaer@unm.ac.id

Abstract. In this study, zeolite-geopolymer hybrid bulk materials were fabricated by using laterite soils deposit Gowa regency, South Sulawesi, Indonesia as a precursor for possible heavy metal absorbent. Laterite soils was calcined at 750°C for 6 hours to transform it into the amorphous aluminosilicate phase. Zeolite-geopolymer materials were produced through alkali activation method with curing temperature around 70°C for 2 hours. There were two different oxide molar ratios of SiO_2/Al_2O_3 used to produce zeolite-geopolymer hybrid bulk materials and the specimens were designated as K1 and K2. The presence of zeolitic phase within geopolymer network was examined by using of XRD, FTIR, and SEM-EDS technique. The mechanical properties of the resulting material were studied by means of compressive and tensile strength. The results showed that Gowa laterite soil was succesfully fabricated as zeolite-geopolymer hybrid bulk materials, and the type of zeolite coined with geopolymers were zeolite A in sample K1 and un-named zeolite in sample K2. The mechanical strengths of zeolite-geopolymers produced in this study were below to those of pure geopolymers.

1. Introduction

Laterite soil is commony formed in tropical area through weathering processes which suitable for the formation of iron, aluminum and titanium oxides. These processes break down silicate minerals into clay minerals such as kaolinite and illite having a variety of physical properties [1]. An important feature of these soils is their capability of hardening upon exposure to wetting and drying. The hardening process is due to crystallization of continuous aggregates of iron oxide minerals followed by dehydration [2]. Despite the fact that laterites have been recognized as a potential source of aluminosilicates for inorganic polymers (geopolymers) synthesis, very little has been reported on the development of geopolymers from laterites [1]. In Indonesia, laterite soil deposits can be easily found, one of them is in Gowa regency, South Sulawesi which is considered as aluminosilicate minerals [3].

Geopolymers are inorganic polymeric materials with three dimensional Si-O-Al networks synthesized from aluminosilicates in high alkaline environment. Apart from having excellent early and long-term mechanical strength, geopolymers also have low shrinkage value, low density, good immobilization of heavy metals and excellent fire and chemical resistance [4]. As a porous material, the durability of geopolymers depends on the ability to withstand the intrusion of aggressive agents. The transport properties of a porous materials are influenced by the pore structure characteristics [5].

It has been recognized that geopolymer is a family of zeolite which is known to have good thermal and chemical stability. Zeolite can be used as membranes or catalytic materials which offer a wide range of operating conditions [6]. Inorganic membranes such as ceramics, metals, and glass show promising properties different from the organic ones. They can be backwashed frequently without damaging the separation layer. Inorganic membranes are highly resistant to cleaning chemicals, they can be sterilized and autoclaved repeatedly at 130–180 °C and can withstand temperature up to at least 500 °C [7]. Zeolite-geopolymer hybrid bulk materials can have a wide pore size distribution, since zeolites have high microporosity and geopolymers have mesopores [8].

Zeolites have attracted intensive research due to their potential applications such as separation membrane, catalytic membrane reactor, chemical sensor, electrode, opto-electronic device, low dielectric constant material for electrical insulator, protection or insulation layer or as host for the organization of functional guest molecules [9]. In this study, zeolite-geopolymer hybrid materials were fabricated using laterite deposit Gowa regency for possible application as heavy metal membrane. The amounts and species of zeolite, pore distribution of the resultant zeolite-geopolymer hybrid bulk were investigated.

2. Experimental

The soils from Gowa regency were clean, immersed in water for 24 hours, dried, grounded and sieved 200 mesh. The soils were dehydroxylated at 750µ for 6 hours and grounded again by using ball milling at 200 rpm for 5 hours. Samples were synthesized by varying the molar ratio of SiO2/Al_2O_3, Na_2O/SiO_2, H_2O/Na_2O. The geopolymers paste was molded in accordance with the measurements requirement. All samples were cured at 70µ for 2 hours and the resulting materials were stored in open air for 7 days before conducting any measurements.

The functional groups of the resulting sample were examined by using Fourier Transform Infra-Red (FTIR). Precise Fourier Transform Infrared Shimadzu was used in Laboratorium Instrumen Jurusan Kimia FMIPA ITS Surabaya. The sample was prepared by dissolving it in KBr with ratio 1: 600, then analyzed by FTIR.

X-ray diffraction (XRD) measurements were conducted to study the structure and phase of the resulting sample [10] by using XRD Rigaku MiniflexII with an acceleration voltage of 30 kV and a current of 15 mA. Phase analysis and mineralization of geopolymer were examined by using PDXL2. In this study, the measurements were performed in range of 10°-70° 2θ.

SEM characterization was used to study the structure and morphology of the laterite sample. Analysis of geopolymer morphology by using SEM Tescan Vega3SB coupled with Bruker Energy Dispersive Spectroscopy (EDS).

The mechanical properties of the resulting materials was charaterized by means of compressive and tensile strength. The samples for compressive strength were cylindrical in shape with a diameter of 2.5 cm and height of 5 cm, whereas the samples for tensile strength measurements were rectangular in shape with a dimension of 12.0 x 2.5 x 1.5 cm in accordance with the requirement for three point bending tensile measurement.

3. Results and Discussions

The raw material for zeolite-geopolymer synthesis was in the form of aluminosilcate mineral obtained from the processing and dehydroxilation of laterites at 750 °C for 4 hours. Figure 1 shows the laterite soil from Gowa regency before and after dehydroxylation with grain size passed the 200 mesh sieve.

Figure 1. Laterite soils (A) before and (B) after dehyroxylation at 750°C about 6 hours.

The loss of water in the form of hydroxyl chain from the network of aluminosilicate changed the color of laterite before and after dihydroxylation as shown in figure 1. As a result, the structure of laterite changed from crystalline to amorphous phase. The red-brown color of laterite is also an indications of the high content of hematite (Fe_2O_3) phase. The chemical composition of laterite was examined by using SEM-EDS technique and the result is shown in table 1.

Table 1. Chemical compositions of laterite Gowa before and after dehyroxylation.

Compound	Bofore		After	
	Norm.Comp [wt.%]	C Error (3 sigma) [wt.%]	Norm.Comp [wt.%]	C Error (3 sigma) [wt.%]
Na_2O	0.69	0.17	0.89	0.18
MgO	2.16	0.24	2.35	0.23
Al_2O_3	33.85	1.79	35.72	1.67
SiO_2	32.34	1.37	33.11	1.24
SO_3	0.13	0.17	-	-
TiO_2	2.03	1.31	1.98	1.06
FeO	28.80	9.70	28.94	8.85

Tabel 1 shows that the molar oxide ratio of Al_2O_3 and SiO_2 of laterite Gowa is an exellent aluminosilicate mineral for Na-poly (ferro-sialate) type geopolymer production [3, 11]. Figure 2 showed the XRD structure and phase compositions of Gowa laterite before and after dehydroxylation at 750 ^oC for 6 hours.

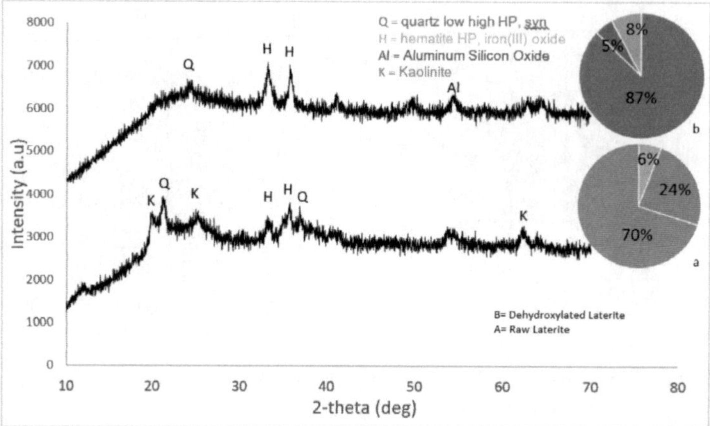

Figure 2. The diffractogram of Gowa laterite before and after dehydroxylation with phase compositions.

Figure 3 shows the FTIR spectrum of laterite before and after dehydroxylation in the range of 400 – 4000 cm^{-1}. The spectrum clearly showed the transformation of crystalline laterite into amorphous state. Figure 3(a) shows that the bands at 3622 and 3697 cm^{-1} were attibuted to the OH stretching vibrations of inner-surface hydroxyl groups. The bands at 1629 and 3448 cm^{-1} were assigned as bending and stretching vibration of H–O–H, respectively.

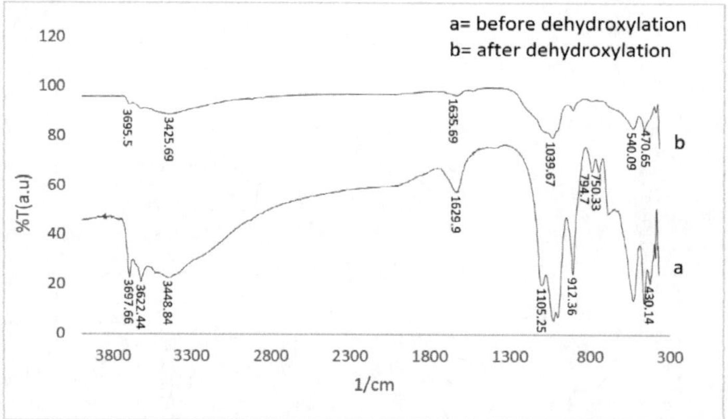

Figure 3. FTIR spectrum of Gowa laterite (before and after dehydroxylation)

The stretching band of Si–O–Si was observed at 1039 cm^{-1}. The band at 912 cm^{-1} indicated the presence of hematite, whereas quartz was observed at 794 cm^{-1}. The band of Si–O was observed at 430, 470, 540, and 698 cm^{-1} confirmed the presence of kaolinite [12]. In this study, zeolite-geopolymer materials were prepared by using two different compositions, designated as K1 and K2. The oxide molar ratios of K1 and K2 are shown in table 2.

Table 2. The oxide molar ratios of samples K1 and K2

Sample	SiO_2/Al_2O_3	Na_2O/SiO_2	H_2O/Na_2O
K1	3	0.3	10
K2	4	0.3	10

Figure 4 shows the zeolitic phase of the resulting geopolymers K1 and K2 based on XRD examinations. It can be seen that both samples contained high weight percentage of zeolite. Sample K1 was characterized by 70 wt. % of zeolite A (Na12 Al12 Si12 O48 (H2O)27) and sample K2 contained 90 wt.% of un-named zeolite (Na3 Al3 Si3 O12 (H2O)2). Other phase such as hematite and quartz were origanated from raw laterite. The presence of analcime phase (Na (Al Si2 O6) (H2O) was formed through alkali activation of kaolinite phase of laterite and become parts of geopolymer paste. These results suggested that the type of zeolite produced trough geopolymerisation process of aluminosilicate minerals such as laterite soils depends on oxide molar ratios of SiO_2/Al_2O_3 keeping other molar ratios remains constant. This results are in line with those reported by [13] who synthesized zeolite X dan Y through geopolymerisation of metakaolin. The functional groups of zeolite-geopolymers based on FTIR measurements are shown in figure 5.

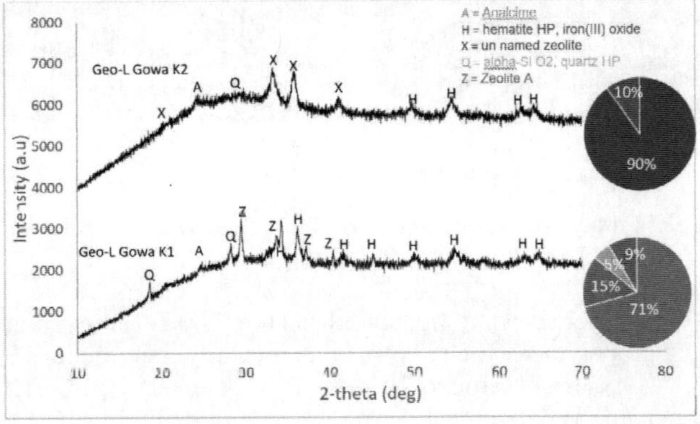

Figure 4. Diffractogram of K1 and K2 showing zeolitic phases

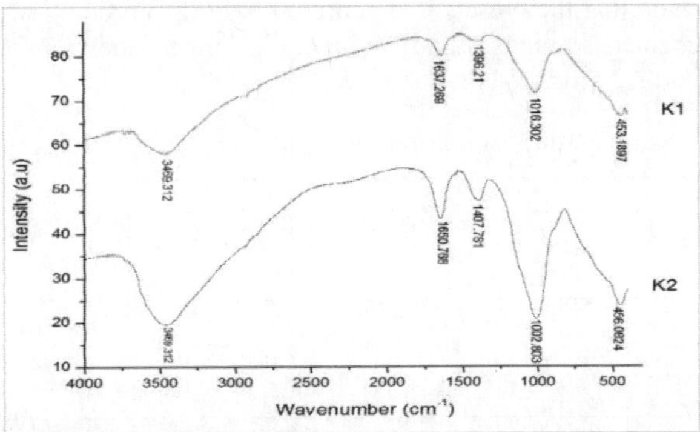

Figure 5. FTIR Spectrum of zeolite-Geopolymer for samples K1 and K2

The broad bands around 1650 and 3469 cm^{-1} were attributed as bending and stetching of water, respectively. The spectrum indicated that sample K2 contain more water than K1. The broad band at 456 and 1002 cm^{-1} were attributed to the zeolitic formation within the geopolymer network. The presence of band at 1407 cm^{-1} was due to unreacted parts of laterite during geopolymer synthesis [11].

The surface morphology of zeolite-geopolymer of K1 and K2 were examined by using by using SEM as shown in figure 6(a) and 6(b), respectively. The quality of SEM images of K1 and K2 are not the same due to the difference of SEM used during this study as the first machine for sample K1 was out of order.

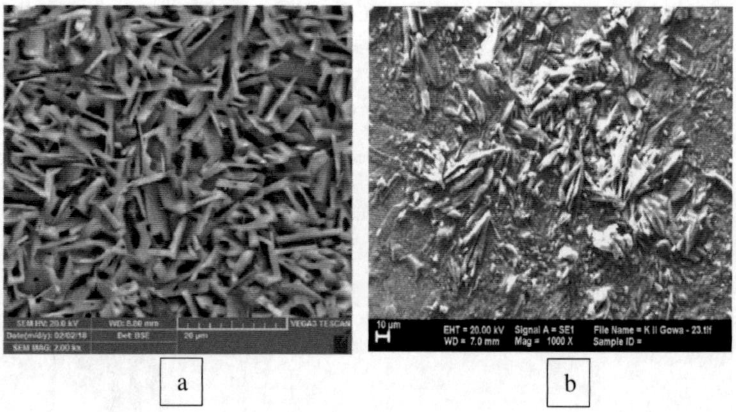

Figure 6. SEM image of Gowa laterite zeolite-geopolymer (a) K1 and (b) K2.

The morphology of K1 and K2 shown in figure 6(a) and 6(b) clearly indicate the crystalline phase of zeolite on the surface of geopolymers. It can be seen, particularly for sample K1 the highly uniform shape of zeolite crystal with a length around 10 μm and wide about 1 μm. The pore structure and distribution have not been able to determined during the time of this study. The formation of zeolitic phase through geopolymerisation route has been subject to intensive research [13–16]. This type of material offer a potential as a new mebrane reactor for methanol or alcohol production [17,18], or for molecular separation and adsorption [19].

It well known, however, that the presence of zeolitic phase within the network of geopolymer will reduced the overall mechanical strength of geopolymer [13]. Table 3 shows the compressive and tensile strength of sample K1 and K2 produced in this study.

Table 3. The magnitude of compressive and tensile strength measurements

Sample	Compressive Strength (MPa)	Tensile Strength (MPa)
K1	5.09	3.30
K2	13.77	4.93

Table 3 showed that the magnitude of compressive and tensile strength of K1 and K2 were far below to those of most geopolymer produced from metakaolin or fly ash. It is clear that these materials are not suitable for structural applications which require high mechanical strength.

4. Conclusions

The results of this study showed that laterite soils deposit Gowa Regency, South Sulawesi, which contain high wt.% of hematite has been succesfully synthesized as zeolite-geoplymer hybrid bulk material. The mineralogy of laterite soil before and after dehydroxylation was examined by means of SEM-EDS, XRD and FTIR suggesting that the material was suitable as geopolymer precursor. The variation of oxide molar ratio of SiO_2/Al_2O_3 produced zeolitic phase within geopolymer network which is suitable for various catalytic applications. The mechanical strength of zeolite-geopolymer hybrid bulk materials is low and not suitable for structural applications.

References

[1] Lemougna P.N *et al* 2017 *Appl. Clay Sci.* **148** 109-117.
[2] Gualteri. M. L *et al* 2015 *Cem. and Conc.* **67** 259-270.
[3] Subaer *et al* 2017 *MATEC Web Conf.* **97**
[4] Liew. Y. M *et al* 2017 *Constr. and Build Mat.* **156** 9-18.
[5] Nguyen. Q. H. *et al* 2018 *Constr. and Build Mat.* **191** 853-865.
[6] Dogan. H *et al* 2010 *Vacuum* **84** 1123-1132.
[7] Caro. J *et al* 2010 *Adv. in Nanoporous Mat.* **1** 1-96.
[8] Takeda. H *et al* 2013 Characterization of Zeolite in Zeolite-Geopolymer Hybrid Bulk Materials Derived from Kaolinitic Clays **6**.
[9] Caro. J *et al* 2008 *Micro. and Meso. Mat.* **115** 215-233.
[10] Subaer, "Pengantar Fisika Geopolimer" 2015.
[11] Obonyo E. *et al* 2014 *Sustainability* **6** 5535-5553.
[12] Eisazadeh. A *et al* 2012 *App. Clay Sci.* **67-68** 5-10.
[13] Subaer *et al* 2018 *Sol. Stat. Phenom.* **273** 167-174
[14] Lee. N. K *et al* 2016 *Micro. and Meso. Mat.* **229** 22-30.
[15] Arbel Haddad, M *et al* 2017 Journal of Nuclear Materials **493** 168-179.
[16] Wang. H *et al* 2018 *Engineering* **4** 406-415.
[17] Gorbe. J *et al* 2018 Separation and Purification Technology **200** 164-168.
[18] Achiou. B *et al* 2018 Journal of Environmental Chemical Engineering **6** 4429-4437.
[19] Abdelrasoul. A *et al* 2017 *Micro. and Meso. Mat.* **242** 294-348.

IOP Conf. Series: Materials Science and Engineering **551** (2019) 012114 doi:10.1088/1757-899X/551/1/012114

Influence of Manganese Substitution on Photocatalytic Activity of $Fe_{3-x}Mn_xO_4$ Catalyst by Heterogeneous Photo Fenton-like Reaction

Rasyidah Alrozi[1,2], Nurul Adzwin Noordin[1], Elsabba Edwin[1], Nor Aida Zubir[1,2*], Norhaslinda Nasuha[1,2], Hawaiah Imam Maarof[2,3], and Mohamad Anuar Kamaruddin[4,5]

[1]Faculty of Chemical Engineering, Universiti Teknologi MARA, Cawangan Pulau Pinang, 13500 Pulau Pinang, Malaysia
[2]Hybrid Nanomaterials, Interfaces & Simulation (HYMFAST), Faculty of Chemical Engineering, Universiti Teknologi MARA, Cawangan Pulau Pinang, 13500 Pulau Pinang, Malaysia
[3]Faculty of Chemical Engineering, Universiti Teknologi MARA, Shah Alam, Selangor, Malaysia
[4]Environmental Technology Division, School of Industrial Technology, Universiti Sains Malaysia, Penang, Malaysia
[5]Center of Excellence Geopolymer and Green Technology (CEGeoGTech), School of Materials Engineering, Universiti Malaysia Perlis (UniMAP), Perlis, Malaysia

E-mail: noraida709@.uitm.edu.my

Abstract. In this study, the influence of manganese substitution on the photocatalytic activity of heterogeneous $Fe_{3-x}Mn_xO_4$ catalyst were investigated. The $Fe_{3-x}Mn_xO_4$ catalysts were synthesized through co-precipitation method at varies manganese (Mn) substitution ratio (x) of 0, 0.1, 0.2, 0.3, 0.4 and 0.5, respectively. The photocatalytic performance of the resultant catalysts were tested in photo Fenton-like degradation of AOII dye in the presence of 22Mm H_2O_2 at initial pH of 3. Interestingly, discoloration of 95.9% AOII was achieved at optimized x value of 0.3. Enhancement of nearly 98% was observed compared to pristine catalyst (x=0) which was able to degrade up to 48.4% only. In addition, it also demonstrated that the AOII degradation kinetic using $Fe_{3-x}Mn_xO_4$ catalyst at x=0.3 by photo Fenton-like reaction was fitted well with the first order reaction kinetic (k = 0.018 min-1 and R^2 = 0.9862). Thus, it can be suggested that the substitution of Mn into the matrix of iron oxide provides a positive influence in enhancing the overall photocatalytic peformance of heterogeneous $Fe_{3-x}Mn_xO_4$ catalyst during the photo Fenton-like reaction.

1. Introduction

Recently, advanced oxidation processes (AOPs) have been widely used as an efficient treatment of industrial wastewater containing pollutant that can be harmful to the environment. Hydroxyl radical act as an oxidizing agent was generated during the process to remove pollutants and also improve the biodegradability of water. Last decades, Fenton-based reactions were extensively being studied in

AOP as their ability to degrade the huge amount of contaminations and show a successful result [1]. The Fenton-like reagent is a mixture of Fe^{3+} with H_2O_2 and it is an active systems for the oxidation of organic compounds. The reaction between Fe^{3+} with H_2O_2 will produce the hydroperoxyl (·OOH) radicals as the oxidizing agent [2]. In order to enhance the Fenton processes, ultraviolet light (UV) is introduced to the Fenton reaction as it can decomposed the photoactive $Fe(OH)^{2+}$ species while promoting the generation of radical in the reaction as oxidizing agent [3].

Heterogeneous photo-Fenton reactions had gained much attention in the degradation of organic pollutants due to their unique characteristics of complete mineralization of organic pollutant, simple separation of catalyst from wastewater and will inhibit metal ion contamination [4]. Magnetite (Fe^3O^4) can be used as heterogeneous catalyst as it containing both Fe^{2+} and Fe^{3+}. These iron ions have structural characteristic that peculiar than others which is the inverse spinel structure with Fe^{2+} in the octahedral site while Fe^{3+} occupied both octahedral and tetrahedral sites [5]. The introduction of transition metal such as nickel (Ni), manganese (Mn), cobalt (Co) and also chromium (Cr) remarkably enhanced the catalytic activity of Fe_3O_4. The presence of transition metals in spinel lattice of magnetite can modify the redox properties and also influence the stability of the catalyst [6].

A few researchers have reported the degradation of organic compounds by AOPs, including Fenton reagent, electro-Fenton, and also ultrasonic irradiation [7–9]. However, detailed research and investigation on the effect of Mn substitution on the catalytic activity of $Fe_{3-x}Mn_xO_4$ by heterogeneous photo-Fenton reaction has not yet been addressed. Therefore, in this study, the influence of Mn substitution on photocatalytic activity of $Fe_{3-x}Mn_xO_4$ catalyst by heterogeneous photo Fenton-like reaction as well as the kinetic study of the resultant catalyst will be investigated.

2. Methodology

2.1. *Synthesis of $Fe_{3-x}Mn_xO_4$ catalyst*

$Fe_{3-x}Mn_xO_4$ was prepared by mixing of 40 mL of 0.05M $FeCl_3.6H_2O$ and 18mL of 0.05M $MnCl_2$. The mixture was heated into a beaker at 90°C using the double boiling technique while stirred continuously for 10 minutes. Then, 1M NaOH was added dropwise to the mixture until it reach pH 10. The mixture was left for aging process for 30 minutes. Then, the precipitate formed was separated by using centrifuge at 400 rpm for 10 minutes and washed by using distilled water and ethanol several times. After washing, the precipitation was dried in an oven at temperature 90°C for 24 hours. The same procedures were repeated to produce different ratio of manganese loading at x=0.1, 0.2, 0.3, 0.4 and 0.5 respectively by using different volume of 0.05M $MnCl_2$ solution.

2.2. *Phtocatalytic performance*

The photocatalytic degradation of Acid Orange II was tested using 0.1g/L dosage of $Fe_{3-x}Mn_xO_4$ catalyst at various Mn loading (x = 0, 0.1, 0.2, 0.3, 0.4, and 0.5) with 100mL of Acid Orange II solution with concentration 35 ppm at pH 3. The reaction was carried out using a photo-reactor. Four 9W UVC lamp were inserted in the center of the photo reactor as the UV light source. Then, 22Mm of H_2O_2 was added to the mixture as an oxidizing agent and the four 9 UVC lamps were turned on to initiate the degradation reaction. The concentration of Acid Orange II was analyzed at wavelength of 484nm using DR 2800 Portable Spectrophotometer. In order to compare the performance of the catalysts under UVC light irradiation, the control reaction of dark heterogeneous Fenton-like reaction was carried out in the same photo reactor under same reaction conditions but without the presence of light source.

3. Results and Discussion

3.1. *Photocatalytic activity of $Fe_{3-x}Mn_xO_4$*

IOP Conf. Series: Materials Science and Engineering **551** (2019) 012114 doi:10.1088/1757-899X/551/1/012114

Figure 1 shows the degradation efficiency of AOII using $Fe_{3-x}Mn_xO_4$ at x value of 0, 0.1, 0.2, 0.3, 0.4 and 0.5 in both photo Fenton-like reaction and Fenton-like reaction, respectively. The performance of $Fe_{3-x}Mn_xO_4$ catalysts in both reactions were compared to determine the effect of Mn substitution in the Fenton-like reaction with and without the aid of UV irradiation. Interestingly, discoloration of 95.9% AOII was achieved at optimized x value of 0.3 in photo Fenton-like reaction. Enhancement of nearly 98% was observed compared to pristine catalyst (x=0) which was able to degrade up to 48.4% only. The degradation process was greatly enhanced in the presence of UV irradiation due to additional source of highly oxidative •OH radicals generated in the system and its ability to influence the formation of the radicals [10]. In fact, at the same x value (x=0.3), degradation of AOII in dark Fenton-like reaction (without UV irradiation) was also found higher (62%) compared to others manganese loading. It was suggested that the substituted manganese at x=0.3 enable to accelerate the redox properties of active sites between the iron and manganese species by increasing generation of interfacial •OH radicals during catalysis [11].

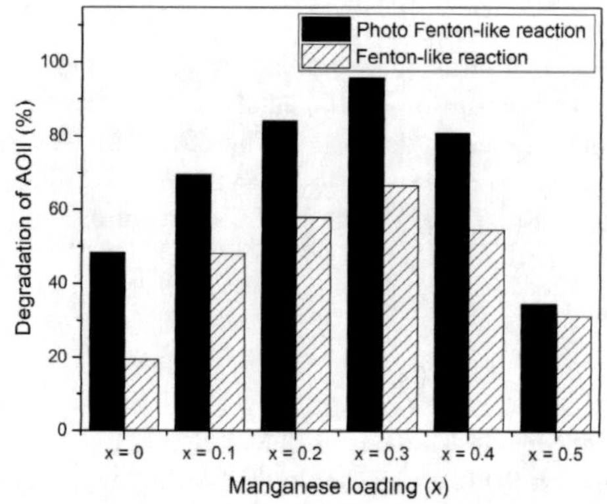

Figure 1. AOII degradation by Photo Fenton-like reaction and Fenton-like reaction (dark) using $Fe_{3-x}Mn_xO_4$ catalyst at various manganese loading (x).

In addition, the presence of Mn ions in magnetite, strongly favored the H_2O_2 decomposition as as there is changes in the structure of Mn-magnetite surface [8]. The enhancement of the percent degradation was mainly caused by the increases of active site available for photo Fenton-like reaction and also due to the unique redox loop ($Mn^{2+}/Mn^{3+)}$ that involve in single electron transfer. Hence, the Mn demonstrate a great potential of transition metal that can be substituted in Fe_3O_4 catalysts. However, at x=0.4 and 0.5, the percentage of degradation obtained were reduced to 81.0% and 34.6%, respectively. According to Wang et al. [12], high content of Mn loading has increased the turbidity of the solution, hence there were some fraction of incident rays of UV light scatter and consequently reduce the penetration of the light into the solution. With decreasing penetration depth, the photons reached the surface of catalyst decreased and the generation of •OH radicals also decreased.

Figure 2 shows the time-dependent degradations of AOII under different reaction conditions. Without using H_2O_2 as oxidant but under UV irradiation, the degradation of AOII was only took placed at the earlier 10 minutes but then no further degradation was shown for the next 150 min. A similar pattern of AOII degradation profile was obtained when H_2O_2 was added with the presence of UV irradiation but with slightly higher of AOII degradation at the end of 180 min. It can be observed that H_2O_2 contributed to the degradation of AOII dye. Interestingly, the degradation of AOII reached 95.9% (magenta line) when $Fe_{3-0.3}Mn_{0.3}O_4$, H_2O_2 and UV irradiation presence in the reaction; while the degradation efficiency of AOII was just 62% (blue line) when only $Fe_{3-0.3}Mn_{0.3}O_4$ and H_2O_2 existed in the solution. This indicates that UV irradiation also played an important role in accelerating the generation of •OH radicals in the system. The •OH radicals has stronger oxidizing ability to mineralize the organic dye into small molecule [10]. The synergistic effect between the incorporation of Mn as an active sites and the iron species under UV irradiation might also be the possible reason to this phenomenon [13].

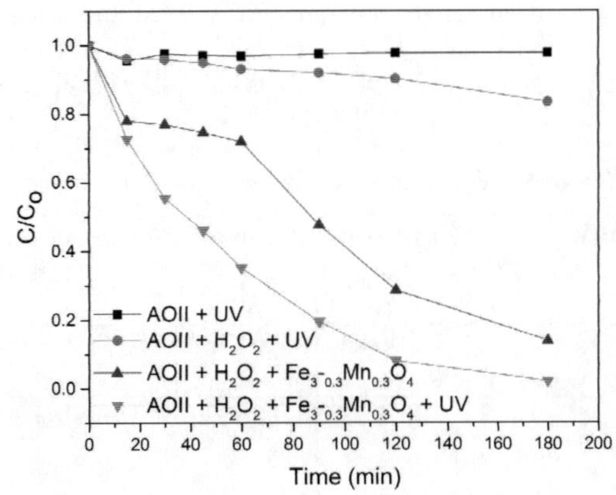

Figure. 2. Degradation profile of AOII under different reaction conditions. Experimental conditions: AOII 35 mg/L, H_2O_2 20mM, catalyst 0.1g/L and pH 3.

3.2. Kinetic study of $Fe_{3-x}Mn_xO_4$ catalysts

Kinetic models were employed to analyse the AOII degradation behaviour generated by $Fe_{3-0.3}Mn_{0.3}O_4$ catalyst in photo Fenton-like reaction and Fenton-like reaction, respectively. The kinetic analysis was conducted using the data within 30 min of reaction. In this work, two kinetic models have been used to study the degradation kinetics of AOII dye: first-order and second-order. Both model equations can be derived from n^{th} order kinetics equations [14](equations (1)-(4)).

First-order reaction:

$$\frac{dC}{dt} = -kC^n \tag{1}$$

$$\ln\left(\frac{C_o}{C_t}\right) = k_1 t \tag{2}$$

Second-order reaction:

$$\frac{dC}{dt} = -k_2 C^2 \tag{3}$$

$$\frac{1}{C_t} - \frac{1}{C_o} = k_2 t \tag{4}$$

where C_t is the AOII concentration at time t (min), C_o is the initial concentration of AOII, and k_1 and k_2 denote the first-order and second-order reaction constants, respectively.

Both kinetic model constants and linear regression coefficients are summarised in Table 1. As seen in Table 1, $Fe_{3-0.3}Mn_{0.3}O_4$ catalyst in photo Fenton-like reaction showed better linear regression coefficients as compared to the reaction in dark Fenton. It also demonstrated that the AOII degradation kinetic using $Fe_{3-0.3}Mn_{0.3}O_4$ catalyst in photo Fenton-like reaction followed the first-order kinetic model which suggests that the reaction depends on the AOII concentration only and known as unimolecular reaction. The reaction kinetic constants for both models are increased for $Fe_{3-0.3}Mn_{0.3}O_4$ in photo Fenton-like reaction compared to the reaction in dark Fenton which indicates that the reaction rate is increased as the reaction been carried out under UV irradiation. The used of external source energy such as UV has accelerated higher •OH yield compared to dark Fenton process [15]. As a result, it will enhance the photocatalytic activity of AOII dye using $Fe_{3-0.3}Mn_{0.3}O_4$ catalyst.

Table 1. First-order and second-order kinetic model constants and linear regression coefficients

Catalysts	**First-order kinetic**		**Second-order kinetic**	
	k_1	R^2	k_1	R^2
$Fe_{3-0.3}Mn_{0.3}O_4$ (Fenton-like reaction)	0.0043	0.6333	0.0002	0.5563
$Fe_{3-0.3}Mn_{0.3}O_4$ (Photo Fenton-like reaction)	0.0180	0.9862	0.0010	0.8903

4. Conclusions

As a conclusion, substitution of manganese into the Fe_3O_4 structure has an outstanding effect on the photocatalytic activity of AOII degradation. $Fe_{3-x}Mn_xO_4$ catalyst with x value of 0.3 has shown great photocatalytic activity with 95.9% of AOII degradation. The presence of manganese, H_2O_2 and UV irradiation led to the enhancement of photocatalytic activity due to the high formation of •OH reactive radicals during the heterogeneous photo Fenton-like reaction. The AOII degradation kinetic using $Fe_{3-0.3}Mn_{0.3}O_4$ catalyst in photo Fenton-like reaction followed the first-order kinetic model which suggests that the reaction depends on the AOII concentration only and known as unimolecular reaction. The reaction rate is increased as the reaction been carried out under UV irradiation. Hence, such findings suggested that the substitution of manganese provide positive influence in enhancing the overall photocatalytic performance of resultant catalyst during photo Fenton-like reaction as well as the degradation kinetics of organic pollutants.

Acknowledgements
The authors gratefully acknowledged the Ministry of Higher Education Malaysia (MOHE) and Universiti Teknologi MARA (UiTM) for financial support under Fundamental Research Grant Scheme (600-IRMI/FRGS 5/3 (110/2016).

References

[1] Su C. et al. 2016 *Bioresour. Technol.* **200** 1065
[2] Hassan H *et al.* 2011 *Int. Conf. Environ. Sci. Eng. Proc.* **8** 232
[3] Kakavandi B *et al* 2016 *RSC Adv.* **6** 84999
[4] Li H. *et al.* 2015 *J. Hazard. Mater.* **287** 32
[5] J. Feng *et al.* 2019 *Sep. Purif. Technol.* **67** 213
[6] G. Wei et al. 2015 *J. Mol. Catal. A Chem.* **398** 86
[7] Oliveira L.C.A *et al.* 2004 *Appl. Catal. A Gen.* **259** 253
[8] Kiwi J *et al.* 1994 *Appl. Catal. B, Environ.* **3** 335
[9] Wu K. *et al.* 1999 Y. Xie, J. Zhao, H. Hidaka, *J. Mol. Catal. A Chem.* **144** 77
[10] Shi J, *et al.* 2014 *Water Res.* **59** 145
[11] Alrozi R *et al.* 2018 *Mater. Today Proc.* **5** 21874
[12] Wang S. *et al.* 2008 Dye. Pigment. **76** 714

IOP Conf. Series: Materials Science and Engineering **551** (2019) 012114 doi:10.1088/1757-899X/551/1/012114

[13] Liang X et al. 2014 *J. Colloid Interface Sci.* **426** 181
[14] Shi H. C. *et al.* 2005 *Tetrahedron.* **61** 297
[15] He Y. *et al.* 2018 *J. Hazard. Mater.* **344** 230

IOP Conf. Series: Materials Science and Engineering **551** (2019) 012115 doi:10.1088/1757-899X/551/1/012115

Development of Hydro-Debris 2D Model

Puji Harsanto[1], Jazaul Ikhsan[1], Nursetiawan[1], Mohd Remy Rozainy[2,3], and Khairy Abdul Wahab[2]

[1]Department of Civil Engineering, Faculty of Engineering, Universitas Muhammadiyah Yogyakarta,55183 Yogyakarta, Indonesia.
[2]School of Civil Engineering, Universiti Sains Malaysia, Engineering Campus, 14300 Nibong Tebal, Penang, Malaysia.
[3]Center of Excellence Geopolymer & Green Technology, Universiti Malaysia Perlis 01000, Perlis, Malaysia.

Corresponding author: ceremy@usm.my

Abstract. The principal aim of a vertical two-dimensional numerical model development is for estimating the particle tracing and mechanism of 10mm and 2.5mm particles. The model development in this study is based on a model developed by Yamashiki et al. (1997). The particle tracing movement can be visually analyzed by using HSVC. A numerical model was developed using the Marker and Cell (MAC) method, which involves a SGS (Subgrid-Scale) model and the PSI-Cell (Particle Source in Cell) Method. The transportation processes of debris and air bubble were simulated in Lagrangian form by introducing air bubble and debris markers.

1. Principle Background of Hd2dm

Air bubble movement characteristics were simulated by this numerical model. The staggered Marker and Cell (MAC) grid was first publish by Harlow and Welch, 1965 which has become one of the most popular ways to simulate fluid flow in computer graphics. The MAC grid method discritizes space into cubical cells with width h. Each cell has pressure p, defined at its center. It also has a velocity, u =(ux, uy, uz), but the component of the velocity are placed at the centers of three of the cell faces, ux, on the x-min face, uy, on the y-min face and uz on the z-min face as shown in Figure 1. Staggered position of the velocity in this manner tends to produce more stable simulations than storing the velocity at the cell centers. In addition to spatial grid cells, set of marker particles (point in space) is used to represent the fluid volume. As the simulation progresses, the marker particles are moved through the velocity field and then used to determine which cells contain fluid.

The Marker and Cell method was modified, combining SGS model and PSI-CELL method (Gotoh, 1992 and Crowe et al., 1977). A lagrangian solution for the air bubble transport was determined by surrounding water pressure with assumption of initially uniform air bubbles diameters. The mechanism of picking up and transport process of bed material in macro turbulence zone was simulated using the lagrangian model based on Tchen's equation (Tchen, 1947 and Gotoh, 1992), assuming that bed materials is uniform. The computer code was written in C language, using reference "103 Y0/VIFMAC" of the computer library of Kyoto University, created by Takemoto (Takemoto et al., 1983) and developed by Tetsuro Sakai et al. (1987). The systems of governing equations are the grid-filtered time-dependent three-dimensional compressible (with low Mach number) mixed flow Navier-Stokes, liquid phase continuity equations. SGS (Sub Grid Scale) model for only liquid phase is introduced. The effect of

Lagrangian sediment particle into liquid phase is being considered using PSI-CELL (Particle Source in Cell) method (Gotoh, 1992 and Crowe et al., 1977).

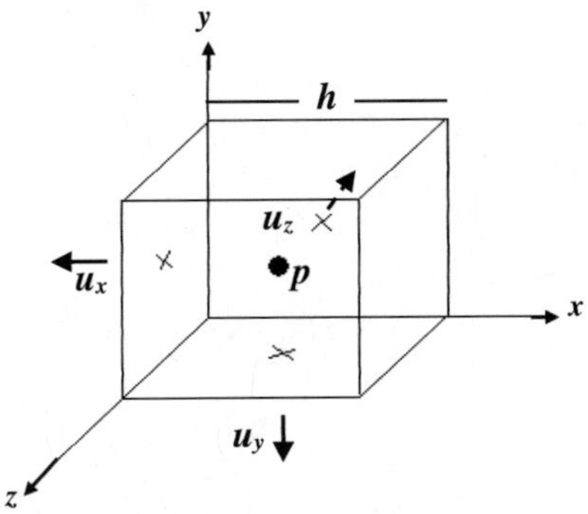

Figure 1: A MAC grid cell. Velocity components u_x, u_y, and u_z are stored on the minimal faces of the cell. Pressure, p is stored at the cell center.

1.1. The Lagrangian model

Models of sediment transport in turbulent flow are determined with another aspect. The model of van Rijn can be classified as a deterministic model and Eulerian. There is another aspect of the modeling of sediment transport, the model probabilistic and, sometimes, Lagrangian. The fundamentals of this model were made by using the Einstein (1950) parameters pick up rate and step length. More sophisticated model was developed by Nakagawa and Tsujimoto (1980) that explains about the mechanism step length using a probabilistic function. This model was known as a Eulerian model. The model of Nakagawa and Tsujimoto is very sensitive especially in term of changing the input and favorable in explaining the phenomenon. The mechanisms of the sediment pulling up and transport of saltation and rolling had been searched. Gotoh clarified the mechanism of transport based on the model of Nakagawa and Tsujimoto. Lagrangian model was used in this study with following reasons 1) Allows the simulation of motion of each particle of the sediment and 2) Unifies the phenomenon of sediment transport.

The first reason was used because the existences of coarse sediment near the downstream area. The Lagrangian model has its advantage when the diameter of the material is large. The second reason was it easily explained about transition of the sediment transport phenomenon. When a particle of sediment is pulled up and picked up by runoff, it will be transported by the macro turbulence. The equation of motion of the particle in the flow, according to Newton's second law can be written as:

$$M\overrightarrow{a_s} = \overrightarrow{F_D} + \overrightarrow{F_L} + \overrightarrow{F_g} \tag{1}$$

where

M : virtual mass of sediment particle

$\overrightarrow{a_s}$: sediment particle acceleration

$\overrightarrow{F_D}$: drag force hydrodynamic

$\overrightarrow{F_L}$: ascending hydrodynamic forces

$\overrightarrow{F_g}$: force of gravity

Assumed that the particle is a sphere, the quantities above are:

$$M = \rho(\frac{\sigma}{\rho} + C_M)A_3 d^3 \tag{2}$$

$$\vec{F}_g = \rho(\frac{\sigma}{\rho} + C_M)\vec{g}A_3 d^3 \tag{3}$$

$$F_D = -\frac{1}{2}C_D \rho A_2 d^2 |\vec{u}_x| \vec{u}_x + \rho(1 + C_M)A_3 d^3 \frac{D\vec{u}_f}{Dt} \tag{4}$$

where

$\vec{u}_r = \vec{u}_p - \vec{u}_f$:relative velocity of the particle

According to the work of Van Rijn (1984), the hydrodynamic forces in turbulent flow and upward:

$$\vec{F}_L = -\alpha_L \rho v^{1/2} d^2 \left(\frac{d\vec{u}}{dn} \right) \tag{5}$$

where

n: vector perpendicular to the vector $\vec{u}_r$

α_L : ancestry coefficient, varying according to the number

1.2. Particle equation

In the present study was performed using particle tracking equations which is lagrangian equation. Basic equations are as follows:

$$\rho\left(\frac{\sigma}{\rho} + C_M\right)A_3 d^3 \frac{d\vec{u}_p}{dt} = \varepsilon\left(-\frac{1}{2}C_D \rho |\vec{u}_r| \vec{u}_r A_2 d^2 + \rho(1 + C_M)A_3 d^3 \frac{d\vec{u}_f}{dt}\right) + \rho\left(\frac{\sigma}{\rho} - 1\right)A_3 d^3 (g_i - \mu_f g_j) \tag{6}$$

$$C_D = 0.4 + \frac{24}{R_e}; R_e = \frac{|u - u_p| d}{v} \tag{7}$$

In which C_D: drag coefficient, $\vec{u}_p$: velocity vector of sediment particle (for each diameter), $\vec{u}_f$: water phase velocity vector $\vec{u}_r = \vec{u}_p - \vec{u}_f$ $A_2 = \pi/4$, $A_3 = \pi/6$, d : diameter of the sediment C_M: virtual mass coefficient (=0.2, 0.35 and 0.5), s :density of sediment particle, ρ water density. ε: shading coefficient determined only when sediment is in the bed and shaded by other sediment, μ_f friction coefficient only works when sediment is at the bottom. This term was determined according to Nakagawa et al. (1978) for the simulation of successive saltation movement as followed Gotoh method. The lagrangian air bubble transport equation is introduced similarly as:

$$\rho\beta A_3 a^3 \frac{d\vec{u}_a}{dt} = -\frac{1}{2}C_{Da}\rho|\vec{u}_{ra}|\vec{u}_{ra}A_2 a^2 + \rho\beta A_3 a^3 \frac{D\vec{u}_f}{Dt} - \rho g A_3 a^3 \tag{8}$$

where β is virtual mass, given as 0.5, ρ is water density, a is a air bubble diameter, C_{Da}: friction coefficient of air bubble, as given as 2.6. The Adam-Bashforth method was used for time integration. The air entrainment process was treated as follows: The entrainment from the water surface was treated by distributing air bubble marker particles in the sub-cells of the surface cells with no water marker cell. These distributed air bubble marker particles are transported according to the surface air drag coefficient which is assumed to 1.0. The air entrainment by the air bubble capitation is simulated by distributing the air bubble marker particles in the inner air cells. The air bubble pairing and dividing cannot be simulated in this study.

IOP Conf. Series: Materials Science and Engineering **551** (2019) 012115 doi:10.1088/1757-899X/551/1/012115

1.3. Preliminary numerical simulation results by HD2DM

Figure 2 shows the simulation results by HD2DM. The flume length is 5m with 20cm height and width respectively. The composed of mixed material (grain size 2.5mm and 10mm) is shaped near the upper part of the channel. For the purposed of numerical simulation investigation, only 25° slope angle is considered. In this case, those mixed materials were set as follows (a) The ratio between small particles (2.5mm) to big particles (10mm), 9:1 and (b) The ratio between small particles (2.5mm) to big particles (10mm), 4:1

During preliminary study of particle segregation, the HSVC was placed near the downstream of the flume channel. After a good understanding of particle routing characteristics at this part, the HSVC was moved near the upstream (erodible bed) of the flume channel. The 25° slope angle is fixed and the water flow (discharge) from upstream was set to be 150cm/s for all simulations case. The preliminary results can be divided into two groups which are (a) and (b). Result of (a) shows the gradual collapse process of mixed grains material by the stream water. The simulation results of composed material movement characteristics were shown in (b). In order to reproduce the experiment condition, numerical simulations were introduced with a stopper which was set at the tip of the mixed material. Simulations shows the collapse which did not start from the tip of the material and the stopper did not move during the simulation. The water stream which flows around the stopper and the bubble entrainment is reproduced clearly. For preliminary results, we can declare that simulation results and high-speed camera pictures are in fairly good agreement. The ratio between small particles (2.5mm) to big particles (10mm) is 9:1. From the results, the ratio of the particles shows a good routing mechanism. For that reason, the ratio of small particle is high; therefore, the numbers of small particles are many as comparison to big particles.

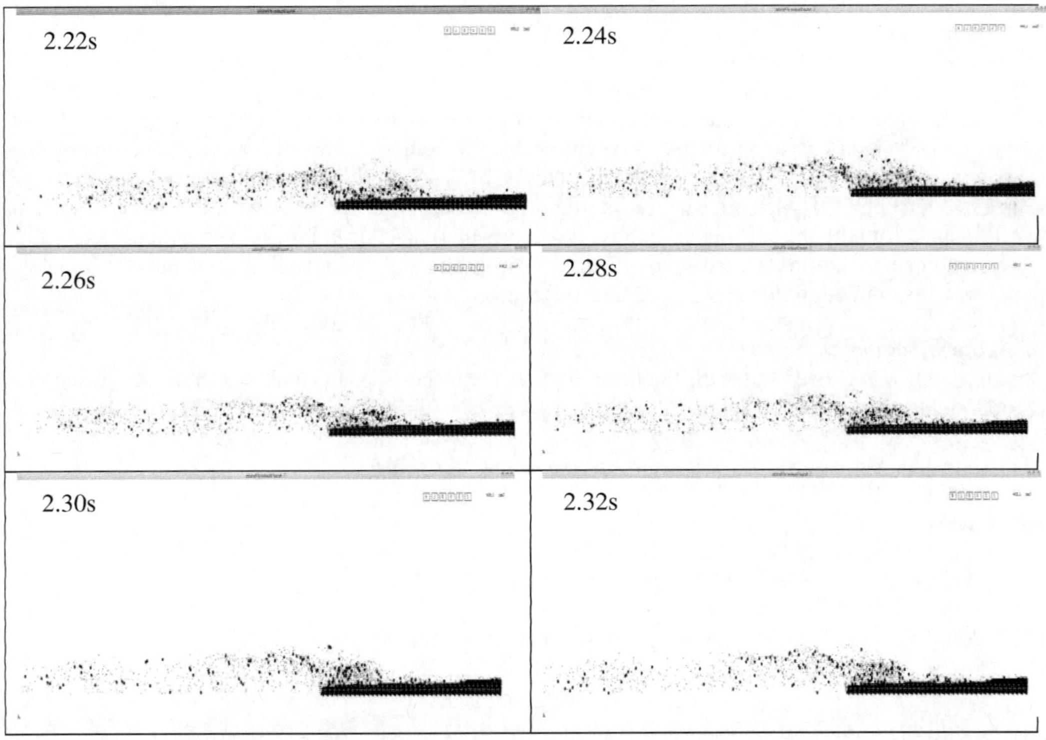

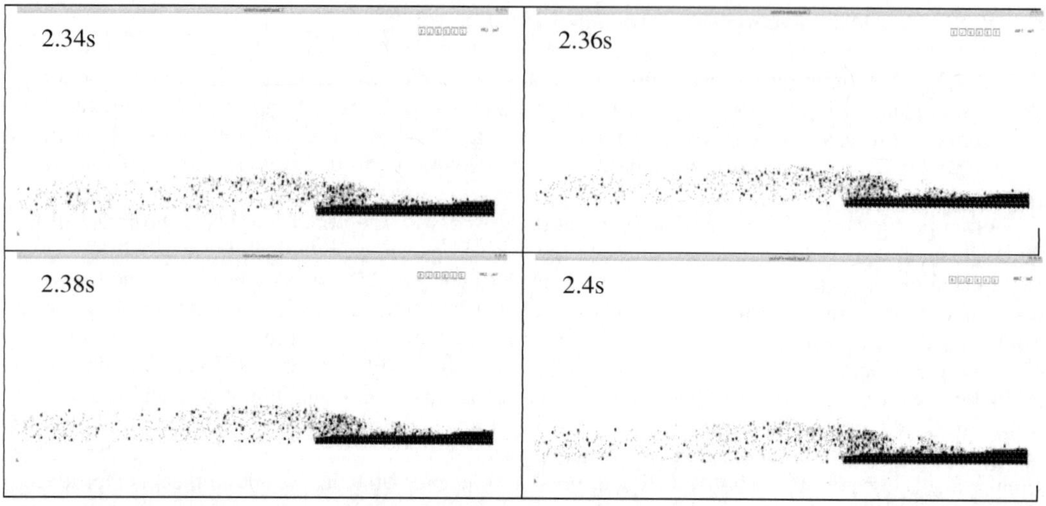

Figure 2. Numerical simulation results of the flow mechanism from 2.22s to 2.40s, image captured at every 0.02s.

2. Conclusions

A two-dimensional numerical model was developed for computing the characteristics of the particle routing of debris flow to simulate the classification effects by applying the particle tracking method which is the most suitable for qualitative simulation of the two-phase flow, Lagrangian suspended load momentum equations was considered to be applicable. A numerical model was developed using the Marker and Cell (MAC) Method, which involves a SGS (Subgrid-Scale) model and the PSI-Cell (Particle Source in Cell) Method. The transportation processes of debris and air bubble were simulated in Lagrangian form by introducing air bubble and debris markers. Air bubble movement characteristics were simulated by this numerical model. The flow condition of the low fluid content rate and low porosities was omitted for the simulation based on reasons above.

3. Acknowledgments

This research was carried out with financial support from the Fundamental Research Grant Scheme (FRGS) under Ministry of Higher Education of Malaysia (Grant No. 7611800357). A very special thanks goes to the School of Civil Engineering, Universiti Sains Malaysia and Universitas Muhammadiah Yogyajarta for the support on conducting this research.

References

[1] Injection Molding: An Introduction, 1995.
[2] Annicchiarico D *et al* 2014 *Mater. Manuf. Process* **29** 662.
[3] Lam Y C et al 2004 *Int. J. Prod. Res.* **42** 2047.
[4] John Wiley & Sons, Design and Analysis of Experiments 2005.

IOP Conf. Series: Materials Science and Engineering **551** (2019) 012116 doi:10.1088/1757-899X/551/1/012116

Acoustic Doppler Current Profiler Measurement of River Discharge at Sungai Perak, Bota Kiri

Wan Ruslan Ismail[1,a)], Muhammad Shazril Idris Ibrahim[2,b)], Ismail Abustan[3,c)], Mohd Remy Rozainy[4,6,d)], Sumayyah Aimi Mohd Najib[5,e)] and Muhammad Faiz Syazwan[6,f)]

[1]School of Humanities, Universiti Sains Malaysia, 11800 Minden, Penang, Malaysia.
[2]School of Civil Engineering, Universiti Sains Malaysia, Kampus Kejuruteraan, 14300 Nibong Tebal, Penang, Malaysia.
[4]Department of Civil Engineering, Faculty of Engineering, University of Malaya, 50603 Kuala Lumpur, Malaysia
[5]Department of Geography and Environment, Faculty of Human Science, Universiti Pendidikan Sultan Idris, 35900 Tanjung Malim, Perak, Malaysia
[6]Center of Excellence Geopolymer & Green Technology, Universiti Malaysia Perlis 01000, Perlis, Malaysia.

E-mail: faizsyazwan05@gmail.com

Abstract. Nowadays, the mining of sand resources from rivers areas especially in Perak state is a common practice and may lead to destruction of public assets as well as impacts or increase stress on commercial and noncommercial living resources that utilize these areas. This study was carried out to determine possible river sand removal capacity and their consequences towards Sungai Perak. Acoustic Doppler Current Profiler (ADCP) is utilized to make river profiler in term of velocity meshes and river bed depth. It also could measure the total discharge of a particular cross-section. In this study, 8 locations of the river were chosen to analyze the river discharge profile either symmetrical or non-symmetrical. From the result and analysis, only one out of eight location showed symmetrical profile. This result showed that the flow of water in the river was not in a good condition and river sand mining in a controlled condition need to be conducted to improve the quality of the river. Other than that, the average flow of Sg. Perak at Pendiat (7770 km^2) is 220 m^3/s which is same magnitude that was reported by Department of Irrigation and Drainage. Based on this result, it is possible to mine sand island and this activity will reduce local scouring.

1. Introduction

River sand and gravel have long been used as aggregate for construction of roads and building. Today, the demand for these sources has increased in a gretaer amount aligned with the rapid growth of construction and building. The main activity to extract these sources is by the activity of sand mining [1]. Sand mining is the removal of sand from their natural configuration. Rapid activity of sand mining could give impact to the environmental and cause bad impact to the ecosuystem. Environmental problems occur when rate of extraction of sand, gravel and other material exceeds the rate at which natural processes generate these materials. Excessive removal of sand may significantly distort the natural equilibrium of a stream channel The magnitude of the impact basically depends on the magnitudes of the extraction relative to bed load sediment supply and transport through the reach.

Many Perak streams, rivers and their floodplains have abundant quantities of sand and gravel that is mined conveniently and economically for a variety of uses. In recent years, rapid development has led to an increased demand for river sand as a source of construction material. This has resulted in a mushrooming of river sand mining activities which have given rise to various problems that require urgent action by the authorities [2]. However, unregulated and unmonitored sand mining has taken place without a clear regulatory framework and this has aggravated environmental problems. In addition, sand prices have increased several times over thanks to a ban imposed on traditional sources of supply. Elevated sand prices have driven the exploitation of sand from fragile and remote places, causing more damage to the road infrastructure due to heavy loads carried on weak rural roads. This situation has been further complicated by politics. No tangible benefit can be derived from current institutional activities unless there is a mechanism that meets both demand for sand and the need for conservation goals. In addition, the needs of the rural poor, who benefit from river sand mining, need to be considered [3].

Therefore, there is an urgent need to identify appropriate policy guidelines that guarantee environmental protection with minimum regulatory costs and high levels of public cooperation [4]. This paper discuss and explain on the appropriate method to handle the sand mining activity to recover and improve the quality of Sungai Perak, Bota Kiri and highlighted the usage of ADCP as one of the important parameters and instruments to determine the allowable amount of river sand that can be extracted based on the data collection.

2. Acoustic Doppler Current Profiler

An Acoustic Doppler Current Profiler, or Acoustic Doppler Profiler, is often referred to with the acronym ADCP. The instrument is used to measure how fast water is moving across an entire water column. An ADCP anchored to the seafloor can measure current speed not just at the bottom, but also at equal intervals all the way up to the surface. The instrument can also be mounted horizontally on seawalls or bridge pilings in rivers and canals to measure the current profile from shore to shore, and to the bottoms of ships to take constant current measurements as the boats move. In very deep areas, they can be lowered on a cable from the surface. ADCP discharge measurements are made from moving boats; therefore, the boat velocities must be subtracted from the ADCP measured water velocities. Figure 1 shows the evaluation of acoustic doppler current profiler measurements of river discharge ADCP's can compute the boat speed and direction using "bottom tracking" [5].

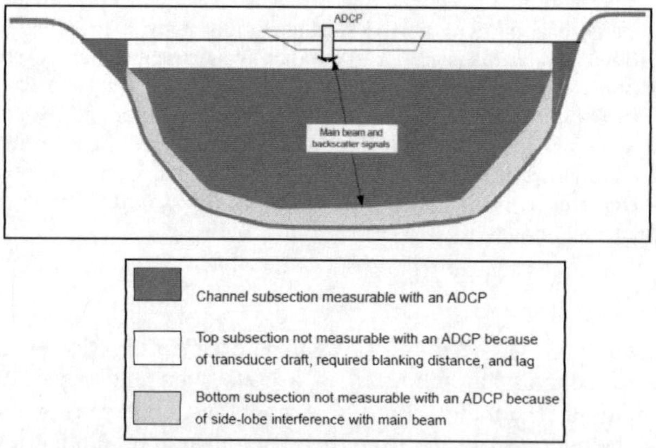

Figure 1. Evaluation of acoustic doppler [5]

3. Design of Experiment Setup

IOP Conf. Series: Materials Science and Engineering **551** (2019) 012116 doi:10.1088/1757-899X/551/1/012116

In this study, eight different locations were chosen to be analysed using the ADCP instruments. The eight locations were shown in Figure 2 below. These locations were chosen randomly to measure the average flowrate and also to determine the quality of flow of the river weather symmetrical which indicate good flow or non-symmetrical which indicate bad flow due to improper management of sand mining. All data were collected and analysed for all locations.

Figure 2. Eight Location of ADCP lines

4. Results and Discussions

4.1. Results

Figure 3 shows several ADCP data taken along the Sungai Perak, Bota Kiri. These data were selected based on the consistency of the data and their influences toward the result of this research. Line R1LB, R4RB and R5LB are the most crucial data for this study. These locations illustrate the real behaviour of the river and sediments underneath. R1LB located at the upstream of the site area. Meanwhile line R4RB and R5LB were located between a river island. Based on the data, the average flow of the river was 220 m^3/s.However for line R4RB and R5LB, the flow recorded for each line was half of the average flow of the river. This was due to the existence of the river island which divided the total flow in half. The average depth of the river is 1.50 m. The maximum flow of the river had affected the rate of the sediment transport proportionally.

The shape of river cross section is not symmetrical. Furthermore, most of the data collected shows that the maximum flow of the river didn't occur at the centre of the river cross section. This is caused by a poor river management and uncontrolled developments and mining by the sand operators and local authorities. Therefore, appropriate river sand mining can be done in a specific area which is needed to be diminished (supported by the result of this research) in order to improve the quality of the river

IOP Conf. Series: Materials Science and Engineering **551** (2019) 012116 doi:10.1088/1757-899X/551/1/012116

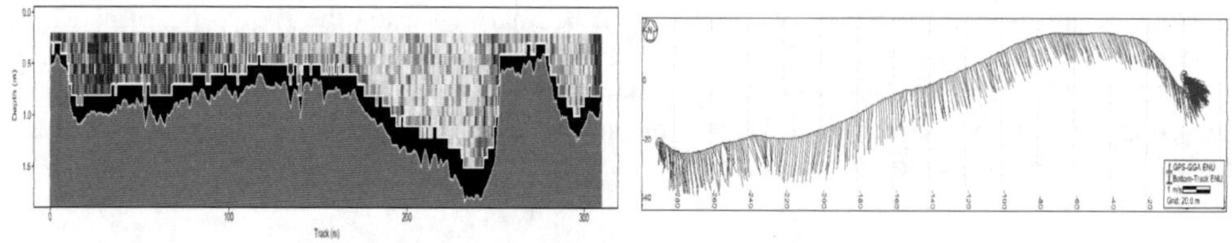

R1LB River Profile and Alignment

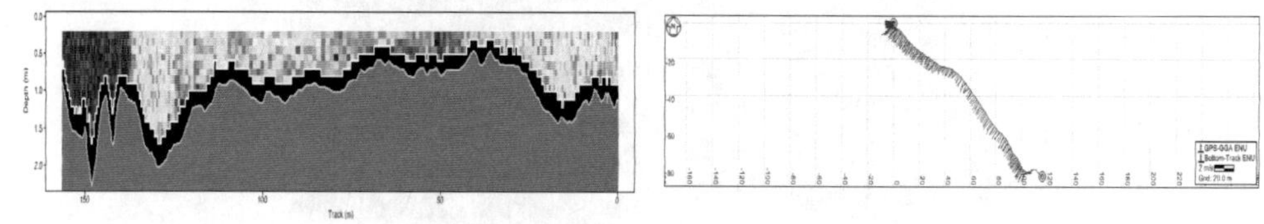

R4RB River Profile and Alignment

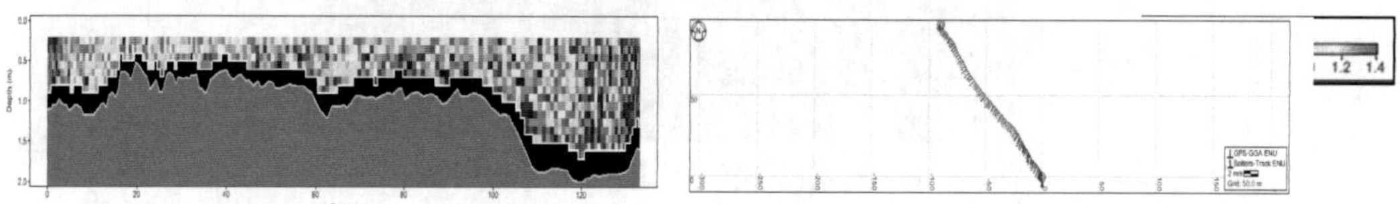

R5LB River Profile and Alignment

Figure 3. ADCP result for three different lines

5. Conclusions

From this study, it is concluded that the discharge flow for this river is non-symmetrical which considered as low quality river management. Most of the data collections showed the maximum flow of the river didn't occur at the centre of the river cross section. The average flow of the river was 220 m^3/s. However, line R4RB and R5LB, the flow recorded for each line was half of the average flow of the river. This was due to the existence of the river island which divided the total flow in half. All of these issues occur due to poor river management and uncontrolled developments and mining by the sand operators and local authorities. Therefore, appropriate river sand mining should be conducted in controlled condition to improve the river quality.

References

[1] Arbogast B. F *et al* "The human factor in mining reclamation: U.S. Geological Survey Circular," 2010

[2] Ashraf M. A *et al* 2010 *Sci. Res. Essays.* **6(6)** 1216-1231

[3] Harrison D. J *et al* 2005 *Geol Soc, London, Spec Publ.* **250(1)** 35-45

[4] Gunaratne L. H. "Cost-Benefit Studies of Natural Resource Management in Southeast Asia" 2010

[5] Callede J *et al* 2000 *Hydrological Sci J* **45 (6)** 911-924

IOP Conf. Series: Materials Science and Engineering **551** (2019) 012117 doi:10.1088/1757-899X/551/1/012117

An Experience Of Hydraulic Physical Model Study In Malaysia.

Zainordin Firdaus Zulkefly[1], Ismail Abustan[2], Mohd Remy Rozainy[3] and Khairy Abdul Wahab[4]

[1,2,3,4]School of Civil Engineering, Universiti Sains Malaysia, Engineering Campus, 14300 Nibong Tebal, Penang, Malaysia.
[3]Center of Excellence Geopolymer & Green Technology, Universiti Malaysia Perlis 01000, Perlis, Malaysia.

E-mail: mjumain0686@ppinang.uitm.edu.my

Abstract. Water Resources Section at School of Civil Engineering, Universiti Sains Malaysia have involved more than 20 projects in pump sump physical modeling, 4 projects in hydraulic spillway study and more than 20 projects related to hydraulic physical model. The total cost of overall projects involved is more than 1 million USD. Hydraulic physical modeling has proven to be important and efficient tool in conceptual as well as detailed hydraulic design. Pump sump model provide effective solution to complex hydraulic problems with unmatched reliability. Meanwhile spillway study is often necessary to determine the possible occurrences of hydraulic jumps, splashing over the site walls and undesired flow vortices. This research can be done accurately because the equipment is complete and sophisticated. Problems related to hydraulic physical model can be solved wisely and accurately with aid of accurate design, data collection and data analyses.

1. Introduction

Resources Section of School Civil Engineering in University Science Malaysia had involve many project expertise in physical hydraulic modelling since 1999 until 2015 such as pump sump modelling, and spillway modelling, officially there has over 24 project was awarded. All these physical hydraulic model was developed in Integrated Research Space (Ruang Penyelidikan Bersepadu). Figure 1 show signage of Integrated Research Space (Ruang Penyelidikan Bersepadu). Meanwhile Figure 2 show building of Integrated Research Space (Ruang Penyelidikan Bersepadu).The physical hydraulic modelling activities need a large space for fabrication and specialized apparatus and it compatible with RPS.

Figure1. Signage of Integrated Research Space (Ruang Penyelidikan Bersepadu) Equipment facilities

IOP Conf. Series: Materials Science and Engineering **551** (2019) 012117 doi:10.1088/1757-899X/551/1/012117

Figure 2. Building of Integrated Research Space (Ruang Penyelidikan Bersepadu)

2. Equipment facilities

Valeport Velocity Meter (VVM) show in Figure 3. Used of this tools to measured velocities. For this study VVM used to get mean velocity only. VVM probe measure the velocity by the center of each node. The point of measurement near water intake are very detail compared to entrance part.

Fig.3. Show velocity meter (Valeport Velocity Meter)

The flow measurement show in Figure 4, is used to flow measure in the tank. Flow was pull through pump model into sectional side circulating pumps send to two calibrated ultrasonic flow meter and control valve to the intake structure modelled. The flow reading before and after was noted and be monitor to avoid fluctuation during sampling process.

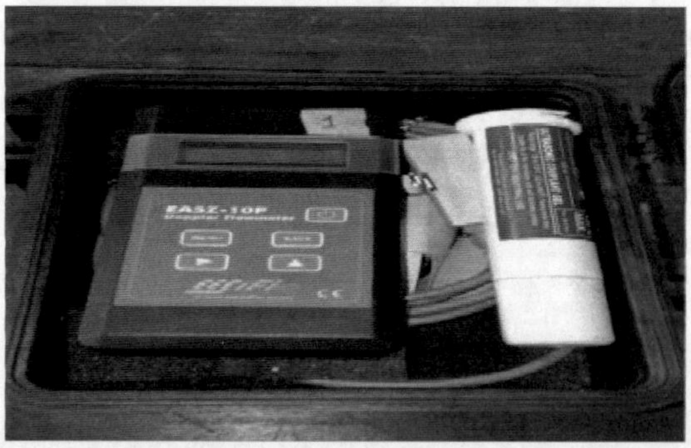

Figure 4. Flow measurement equipment

IOP Conf. Series: Materials Science and Engineering **551** (2019) 012117 doi:10.1088/1757-899X/551/1/012117

The swirl meter or vortimeter show in Figure 5, is tools to calculate the swirl angle or angle of flow with due to the pump axis. While a swirl angle is a flow parameter to determine either the amount of swirl existing in the intake is acceptable for operation. The swirl angle measured in the pump sump model will be compared to the standard which is ± 5° [1].

Figure 5.Swirl meter.

The dye is in Figure 6 shows dosed into water surface, intake floor, intake wall intake entrance and under the bell mouth so that visualization of the flow distribution can be made. Vortices attached to fixed boundaries below the surface are difficult to detect and may require a careful search with the dye tracer technique [2].

Figure 6. Dye tracer technique for observation approach.

Miniature Current Meter shows in Figure 7. These system is highly sensitive to LOW velocities5.0 cm/s up to 150 cm/s. The probe is accurate reading (± 2.5% comprised of an acoustic sensor with a high reading above 0.5 m/s, ± precision instrument that can be relied upon to give accurate reading ± 2% of true velocity.

IOP Conf. Series: Materials Science and Engineering **551** (2019) 012117 doi:10.1088/1757-899X/551/1/012117

Figure 7. Miniature current meter.

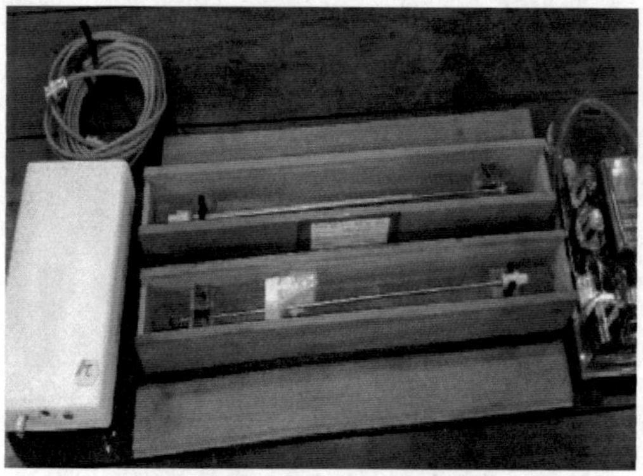

Figure 8.Flow meter with set.

A flow meter show in Figure 8 is a tools used to measure linear and nonlinear, mass or volumetric flow rate of a liquid.

3. Physical Hydraulic Model

A physical model is to predict physical phenomenon behavior. The model can be trusted as a reliable if the design are contain correct principle, although sophisticated measurement-methods only to increase the accuracy of wrong interpretation [3].The physical hydraulic model is an indispensable because has ability to do something again in exactly the same way to many complicated flow situation and very useful when deal with expensive and complex hydraulic works [4].The main purpose of physical hydraulic is to optimize the design, maximize the operation with safety consideration aspect. Visualization from hydraulic physical model give true information for researcher to make right and better decision in designing [5]. The physical hydraulic represented in a scale and still reliable method to evaluate the designs of hydraulics structures where the hydraulic flow cannot immediately be computed mathematically in turbulent [6]. A scale model enable modification and remedial task to be quickly and urgently investigated. Combination between stimulation model and scaled down physical models with certain ratio are used to validate hydraulic structures design.

3.1. Pump Sum Model

Generally water pumps is very important in operational irrigation drainage, agricultural, storage plant and industry plant. However the pump always breakdown thus the rational to design of pump-sump was developed. The main problem for pump are surface and subsurface vortex. There was no initial method to predict these problem especially pre-rotation, uncontinuous distribution of flow and cavitation. Free surface vortices may pull air from surface into the pump and cause impeller unbalance loading and repeated vibration. All this disturbing will decrease pump capacity and capability [7]. Meanwhile subsurface vortex formation occur by little in submergence, but more in sump configuration, allocation pump in the sump, upstream situation, and channel with velocity, Subsurface vortices can be monitor using model test. Subsurface vortices which may produce coming from floor, site or back walls get into the pump and cause vibration and cavitation. The computational fluid dynamics (CFD) can be apply to predict the flow characteristics in physical model of pump sump [8].The pump sump modelling for flood mitigation before design improvement show in Figure 9a, while Figure 9b shows improvement had been taken by redesign and problem had been solve.

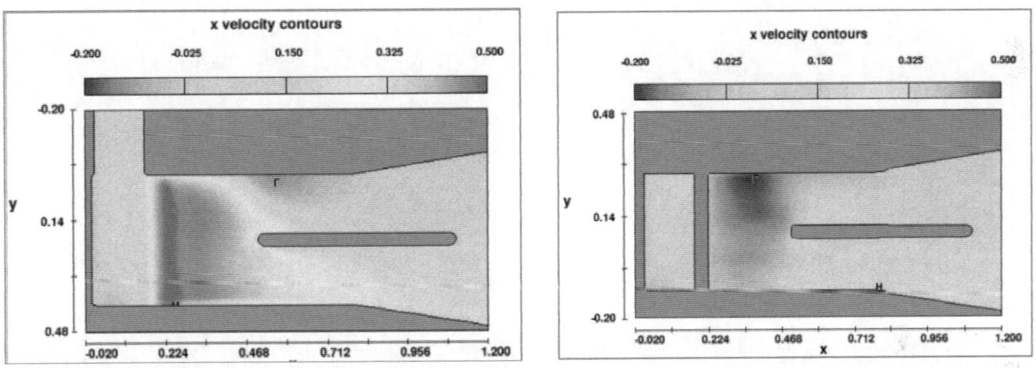

Figure. 9. (a)Pump sump modelling for flood mitigation purpose before modification and (b) Pump sump modelling for flood mitigation purpose after modification and redesign

3.2. Spillway Dam

Dam is important public structure with multifunction such as water storage, flood control and power electricity supply [9]. Dam was construct with many part, among the other spillway is the most important structure in term of safety. Spillway is a structure control release flow from dam or bypass flow during high water level in dam. Design guideline use for spillway dam is refer to USBR [10]. The gravitational forces are dominant (most important) in free water surface work so the Froude Number must be made equal to the prototype [11] . Model similitude for this study 1:20, length 1:20, velocity 1:4.47 and flow 1:1788. The construction of physical model of spillway done base on down scale 1:20 from large scale. Perspex and pvc as material to construct because has ability to scale proportionally. Physical Model of spillway consist of inlet portion from the reservoir, the labyrinth weir, the chute, the spillway approach channel the stilling basin and a portion of downstream channel.

4. Conclusion

Both stimulation model and physical modelling are interrelated on each other advantages, thus, are recommended to be used together compare to separately. Combination stimulation model and physical hydraulic model, is the best solution of hydraulic designing. Experience being involve in to physical hydraulic project will increase the understanding precision of hydraulic phenomena and many problem can be solve realistically and economically.

IOP Conf. Series: Materials Science and Engineering **551** (2019) 012117 doi:10.1088/1757-899X/551/1/012117

References

[1] Matahel A *et al* 2001 *J. Hydr. Eng.* **127** 10
[2] Padmanabhan M *et al* 1984 *J.Hydr. Eng.* **110** 1540
[3] Yildiz A. *et al.* 2018 S 1989 *Rec. Adv. Hydr. Phy. Mode.* **10** 105
[4] Ismail A *et al* 2016 *Nat. Sem. on Phys. Mod. Appl. Tech.* **3** 58
[5] Hubert C 1999. "The Hydraulics of Open Channel Flow" Published in by Arnold, 338 Euston Road, London NW1 3BH, UK.
[6] Ansar, 1997 US Army Corps of Engineers, 1988.
[7] Dicmas J L 1987. "Vertical Turbine, Mixed Flow, and Propeller Pumps." McGraw-Hill, New York
[8] Ismail A *et al* 2012 *Sem. Phys. Model. Appl. Tech.* **21** 504
[9] United States Department of the Interior Berau of Reclaimation Design of Small Dams (USBR) .
[10] Mohd R. R *et al* 2016 Sem. Phys. Mod. Appl. Tech **2** 75
[11] Mohd R. R *et al* 2016. Int. Sust. Tech. Energ. & Civil. Conf. **14** 65

IOP Conf. Series: Materials Science and Engineering **551** (2019) 012118 doi:10.1088/1757-899X/551/1/012118

Sustainable Extraction of Micronutrient of Food Waste by Thermal Treatment under Elevating Temperature

M.A. Kamaruddin[1,*], Z.N. Azman[2], N.F Ahmad[3], and M.H. Hanif[4]

[1]Environmental Technology Division, School of Industrial Technology, University Sains Malaysia, 11800 Pulau Pinang, Malaysia

E-mail: zuhayransrn@gmail.com

Abstract. Food waste management by biological treatment is widely used due to the eco-friendly and more attractive than physical and chemical treatment. However, the use of biological treatment takes longer time for the food waste to degrade. To overcome this problem the pretreatment of food waste is proposed. The aim of this study is to determine the characteristic of food waste when subject to pretreatment, to determine the effect of thermal pretreatment on chemical compositions of food waste, and to determine the relationship between the effect of temperature and time on food waste carbon to nitrogen ratio, pH and micronutrient by using statistical analysis. The water bath method was carried out in this experiment. The parameter measured were Carbon to Nitrogen ratio (C;N), pH and micronutrient with the effect of temperature and time. The result of the food waste before treatment and after pretreatment was compare. The laboratory results showed that the chemical composition of the food waste was affected by the thermal pretreatment of water bath method for C:N and pH which in the range 11.18 to 25.94 and 5.32 to 5.72 respectively. The statistical analysis indicates that the interaction effect between retention time and temperature was not significant for pH, Fe, Cu and Mn except for C:N ($F=4.394$, $P= 0.003$). Overall, the thermal pretreatment applied on the food waste revealed that temperature below 100 °C did not suitable to use as a pretreatment before the biological treatment, in order to expedite the degradation process. Instead, high temperature which above 100°C and long heating duration are beneficial for the release and reduction of organic compounds.

1. Introduction

Food waste is a discarded or uneaten food which is part of municipal solid waste. Food waste can cause environmental problem due to the high organic content that can lead to odour problem and the growth of the pathogenic organism. To overcome this problem its need an appropriate management in order to reduce the environmental pollution and the risks to human health.

There are several treatment techniques on the food waste such as composting, landfilling and anaerobic digestion. Pretreatment is the most effective treatment in speeding up the degradation process. According to the previous study, in order to enhance the productivity of biological treatment, pretreatment is needed to speed up the hydrolysis process [1]. Pretreatment is a chemical break down of compound due to the reaction with heat and water. The end product of the food waste pretreatment can be used as a fertilizer and a solid compostable residue, which can be used as a compost and soil conditioner.

There are four different types of thermal pretreatment technique available which is hot air oven, microwave, autoclave and hot water bath [2]. In previous study, thermal pretreatment was used to facilitate the solubilisation of food waste and improve the biodegradability of the substrates [3]. Thus, thermal pretratment is important in order to expedite the degradation process.

Thermal pretreatment is used to enhance the decomposition of the food waste. The product of this treatment also will be used before composting which is to speed up the time for a compost to become stable and mature.

2 Methodology

2.1. Food Waste Collection

Food waste that used in this study was collected from Baktisiswa Cafeteria at Universiti Sains Malaysia. The organic and inorganic waste was separated from the waste and the method used as recommended by [4]. Only the organic waste was used in this treatment

2.2. Preparation Of Food Waste

First, the food waste that have been collected was dried for 24 hours at 130°C in an oven dried [5,6]. In this study the dry food waste were blended in the heavy duty blender (WARING, Heavy-duty 1.5 HP motor) as shows in Plate 3.1 to homogenize the composition of the food waste. The samples were then stored in air-tight plastic bags for further use

2.3. Thermal pretreatment

The dried food waste sample was placed in a 250ml of conical flask with deionised water at solid to liquid ratio 1:10 [7,8].The conical flasks were sealed with aluminium foil before being placed in Memmert Hot Water bath (WNB 7) [2]. The temperature of pre-treatment was varied from 60oC to 100oC with 10oC increments for 30min to 60min with 30min increments [9].

2.4. Chemical Characteristics

The chemical characteristics performed for the food waste before and after treatment in this study were pH, C:N and micronutrient (Fe, Cu, Mn).

2.4.1. pH determination

The pH of the food waste before and after treatment was determined by using HACH Sension 3 PH meter as shows in Plate 3.3. First, the pH meter was calibrated by using three pH buffers which 4, 7 and 10. Then, the dried food waste sample were mixed with deionizes water in 1:10 ratio to make a solution. Then, the solutions were stir uniformly for five (5) minute until the solvent dilute. The solutions were measure after one (1) hour by using pH meter (HACH Sension 3). The experimental data was repeated three times in order to ensure the accuracy and reproducibility of data.

2.4.2 Carbon to Nitrogen ratio (C:N)

Determination of carbon to nitrogen ratio was carried out based on the literature study as recommended by Iqbal et al. (2014). The carbon and nitrogen contents of the sample were determined by using Total organic carbon method and Total Khejadal Nitrogen.

2.4.3 Micronutrient content (Fe, Cu, Mn)

Micronutrient content which all Fe, Cu and Mn of the food waste were determined by using Flame Atomic Absorption Spectroscopy (FAAS).

IOP Conf. Series: Materials Science and Engineering **551** (2019) 012118 doi:10.1088/1757-899X/551/1/012118

2.5. Data analysis
The collected data from laboratory was calculated and assessed graphically by the use of Microsoft Excel and IBM SPSS. Data were analysed by two-way ANOVA to determine interactions between temperature and time. The strength of relationship between independent variable and dependent variable was analysed by using Correlation analysis. The significance of differences in the average of C:N, pH, and micronutrient element concentration were determined.

3. Results and Discussion

3.1. Effect of thermal pretreatment on chemical characteristics of food waste

3.1.1. Carbon to Nitrogen Ratio
Carbon to nitrogen ratio (C:N) is the ratio between the percentage of carbon to the percentage of nitrogen. Based on the literature the value of C:N was found in the range 15.72 to 24.4. The C:N of the treated food waste in this study was compared to the range as recommended by Malaysia standard of Commercial organic fertilizer. In general, the value of C:N will increase by increasing the temperature and residence time. However, in this study the C:N values shows different pattern.

As shown in Figure 1, the value of C:N was decreases by temperature from untreated (20.56) to 60oC for both 30 and 60 minutes of thermal pretreatment at 11.18 and 12.81. At 70oC of thermal pretreatment for 30 minutes of retention time has lower C:N (11.8) than 60 minutes of retention time (15.92). In the next level of temperature which was 80Co the values of C:N for 30 minutes of treatment was higher (25.94) than C:N for 60 minutes (23.97). For temperature at 90oC, 30 minute and 60 minutes of retention time have the same level of the C:N value which is 10.84 and 10.94 respectively. Then, 60 minutes of retention time at 100oC of thermal pretreatment has higher C:N (7.67) value than 30 minutes (15.23). It was apparent that the carbon and nitrogen content varied greatly when increasing the heating temperature and duration time. The higher the temperature, the higher of C:N became. According to a previous study, temperature below 100oC was described as low temperature thermal pretreatment [10]. Based on the C:N result presented it could be concluded that at the lower temperarature under thermal pretreatment, the organic matter in the food waste may not effective in transforming the food waste into small molecules as result a decrease in C:N value except for 80oC of thermal pretreatment for both 30 and 60 minutes of retention time. Decrease in C:N value for both control time and temperature, it due to the incomplete mineralization of organic matter [10], on the other hand, according to [11] reduction in the C:N was due to the degradation of organic matter. However, all of the C:N of the food waste were within the range that suggested by the Malaysia standard of Organic Fertilizer (3.8-42.7) which is in the range 7.67 to 25.94.

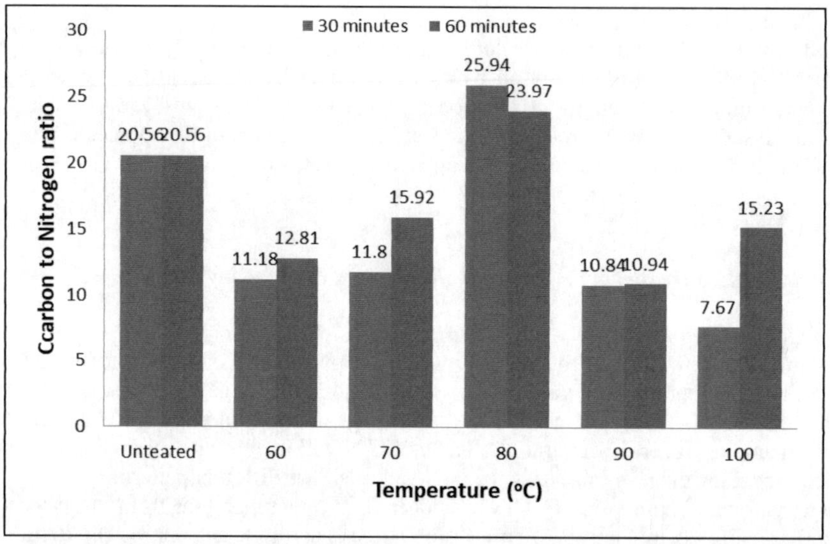

Figure 1. Carbon to Nitrogen ratio (C:N) for untreated and treated food waste

3.1.2. pH value

pH value was conducted out to measure the acidity or alkalinity of the food waste, in order to determine whether the treated food waste is suitable for direct application to the soil or compost. Based on the literature study the pH value of the treated food was in the range 5.98 to 8.74 [12]. In general, the pH value of the food waste will decrease by increasing the temperature and residence time. However, in this study the pH values shows different pattern.

As shown in Figure 2, pH value was increase by temperature, from untreated to treated food waste of thermal pretreatment. At 60oC to 90oC of thermal pretreatment for 30 minutes of retention time has higher pH values than 60 minutes of retention time which is in the range of 5.36 to 5.72 and 5.32 to 5.65 respectively. In the next level of of temperature which was 100oC the value of pH for 30 minutes of treatment was lower (5.34) than pH value for 60 minutes of retention time (5.4). The entire sample for untreated and treated was mild acidic in nature. The longer duration of pretreatment was formed to increase the pH except for 100oC of thermal pretreatment for both 30 and 60 minutes of retention time. The decrease in pH was limited compared to pH at higher temperatures which was at 100oC. Higher temperatures (100oC) increased the pH reduction rate because organic acids were continuously released into the liquid phase from the solid phase in the food waste during thermal pretreatment [13, 14].When the treatment temperature increased to 100oC, the decrease in pH became more apparent. These results also suggested that both as the rate and amount of organic acid release increased, in a lower pH value was expected. However, all of the pH values of the food waste was within the range (5.21-5.72) that suggested by the Malaysia standard of Commercial Organic Fertilizer (4.5- 9.8).

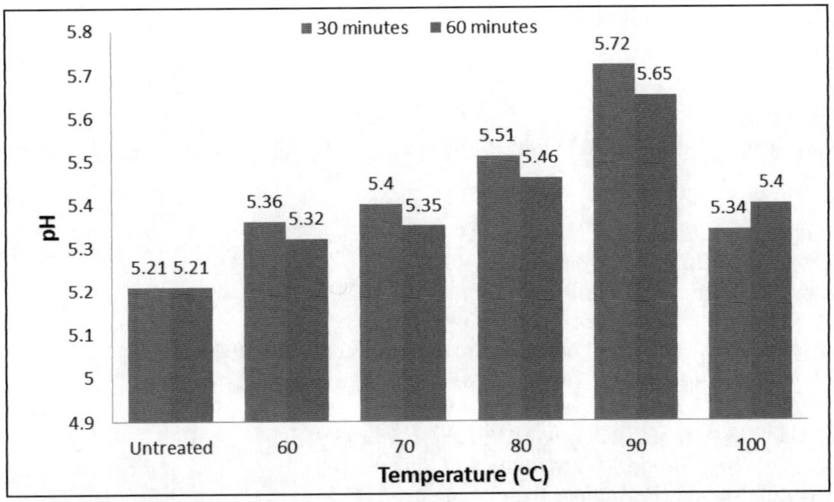

Figure 2. Carbon to Nitrogen ratio (C:N) for untreated and treated food waste

3.1.3. Micronutrient

Micronutrient is an essential for plant growth and soil in order to balanced crop nutrition. They include copper (Cu), iron (Fe) and manganese (Mn). The result indicates that the amount of micronutrient content in the food waste was really limited due to the composition of the food waste. The thermal pretreatment also does not affect the micronutrient content in the food waste which is may due to the unhomogenized of the food waste during the food waste preparation. Thus, the thermal pretreatment by water bath method give a flattuated result to the micronutrient content in the food waste.

3.2 Interaction between duration time and temperature by using statistical analyses

Table-1 Interaction between duration time and temperature

Interaction	**Parameter**				
	C/N ratio	**pH**	**Fe**	**Cu**	**Mn**
Time * Temperature	P = 0.003	P = 0.287	P= 0.345	P= 0.120	P=0.801
Time	P=0.004	P=0.155	P= 0.970	P= 0.109	P= 0.514
Temperature	p=0.00	P=0.000	P= 0.845	P= 0.801	P= 0.532

4. Conclusions

From the research, we can conclude several conclusions which are;

i. pH of untreated and pretreated food waste was more or less similar which in the range 5.21 to 5.72. The higher C/N ratio (25.94) of the food waste was at 80°C of thermal pretratment for 30 minutes of retention time. For micronutrient, it shows that thermal pretreatment does not give much effect on Fe, Cu and Mn.

ii. Interaction effect between retention time and temperature was not significant for pH, Fe, Cu and Mn except for C/N ratio.

Acknowledgements
This work was financially supported by School of Industrial Technology USM, Pulau Pinang.

References

[1] Ma, Y *et al* 2017 *Bioresour. Technol.* **241** 448.
[2] Barua, V.B *et al* 2017 *Bioresour. Technol.***227** 147
[3] Taherzadeh, M.J. *et al* 2008 *Int. J. of Mol. Sci.* **9** 1621
[4] Chang, J.I. *et al* 2008. *Bioresour. Technol.***99** 8068
[5] Kamaruddin, M.A *et al* 2018, AIP Conference Proceedings **2030** 20199
[6] Ding, L *et al* 2017. *Bioresour. Technol.***241** 491
[7] Agu, R., *et al* 1997 *Waste management* **17** 91
[8] Kamaruddin, M.A *et al* 2012 *Applied Water Science*, **3** 359
[9] He, M., *et al* 2012 *Procedia Environmental Sciences* **16** 85
[10] Kamaruddin, M.A 2016 *Applied Water Science.* **5** 113
[11] Liu, X., *et al* 2012 *Waste Management* **32** 249
[12] Sundberg, C *et al* 2013. *Waste Management* **33** 204
[13] Chu, C.-F *et al* 2008 *International Journal of Hydrogen Energy* **33** 4739
[14] Yan, Z. *et al* 2015 *Bioresour. Technol.***177** 266

Recovery of iron from electric arc furnace slag: effect of heating temperature and time

Norhaslinda Nasuha[1,2*], Nor Shaziela Nobri@Mohd Shukri[1], Nurizani Edora Rohizat[1]. Rasyidah Alrozi[1,2], Nor Aida Zubir[1,2], Hawaiah Imam Maarof[2,3]

[1]Faculty of Chemical Engineering, Universiti Teknologi MARA, Cawangan Pulau Pinang 13500 Pulau Pinang, Malaysia
[2]Hybrid Nanomaterial, Interface & Simulation (HYMFAST), Faculty of Chemical Engineering, Universiti Teknologi MARA, Cawangan Pulau Pinang 13500 Pulau Pinang, Malaysia
[3]Faculty of Chemical Engineering, Universiti Teknologi MARA Shah Alam, 40450 Shah Alam, Selangor, Malaysia

E-mail:.norhaslinda.nasuha@ppinang.uitm.edu.my

Abstract. This work presents the iron oxide recovery from electric arc furnace slag with different temperature and time using acid leaching method. The results showed that formation of iron oxide from electric arc furnace (IO-EAFS) was influenced by the temperature and time of heating. This is evidenced by constant acid concentration and liquid to solid ratio, 5 hours of heating time was exhibited the highest amount of iron ion in extracted solution. The maximum of iron extract in heating temperature was at 100 °C. Throughout the characterization using Scanning Electron Microscopy (SEM), the iron oxide illustrated the microstructures were agglomerates predominantly with irregular shapes and XRD results further colobrated the presence of hematite combined with other metal oxide exists in the IO-EAFS. From the TGA analysis, it was identified that the recovered iron oxide was totally decomposed at 525°C. This results indicate that electric arc furnace slag is potential material of iron oxide for the future appliction.

1. Introduction

Malaysia is a growing developed country with population of 31.19 million in 2016 where demand of development is high [1]. Therefore, the demand of the raw materials for building and architecture constructions such as ceramic, polymers and steel are also on the rise [2]. Hence, it will increase the waste generation. It was assessed around 2 to 4 tons of wastes are being created per ton of steel delivered [3]. The assortment of wastes such as slags and sludges that are generated from steel plants are known as blast furnace flue dust (BFFD) and sludge (BFFS), blast furnace slag (BFS) and electric arc furnace slag (EAFS). Ca-silicates, Ca-Al-ferrites, and molten oxides of calcium, magnesium, manganese and iron are some profitable metals that contain in the sludges [1]. The increment of production of steel every year leads to inflation in slag generation and usually they are disposed via landfilling [4].

Landfill is necessary for Malaysia and waste production is predicted to increase from 292 kg/capita to 511 kg/capita within year 2000 to 2025 [5]. Furthermore, most landfills in Malaysia are operating with lacking legitimate defensive measures, for example, lining frameworks, gas venting and leachate treatment [6]. In recent time, landfilling turns out to be less positive as it is will involve a considerable measure of terrains and subsequently required a huge expense [3]. Nonstop opening another land for landfill likewise will ultimately cause long-lasting corruption to the natural flora and fauna in encompassing region [2] [7].

Hence, reusing of the wastes for different profitable item or product is regards as an attractive and environmental-friendly solution for legitimately deal with the yearly enormous measure of the wastes. Recovering these wastes to be reused as auxiliary crude materials or other application in industry are vital for managing metals and mineral assets as well as for ensuring the earth sustainability. To recover the EAFS, some physical or chemical mineral processing techniques may be applied such as crushing, grinding, classification, hydro cyclone, magnetic separation, flotation, roasting or leaching [3].

In this research, leaching process are used to recover iron oxide from EAFS. Iron oxide is chosen as the metal that are going to recover because the EAFS are an iron oxide rich product and iron oxide can be used as secondary raw material in steel production [7] [8]. By utilizing the iron oxide in EAFS, the disposal and production cost for steel production will be reduced others than reducing the waste.

2. Methodology

2.1. Experimental procedure

The crude EAFS in the form of rock were grinded by using jaw crusher, cone crusher and ring mill and then was sieved to obtain a specific range of 40 to 60 μm. The sieved EAFS was weighted for 100 g before soaked in 300 mL of 50wt% sulfuric acid at a liquid to solid ratio of 3:1. The mixture was then heated at 80 °C for 4 hours. After heating process, the mixture was filtered and the filtrate was collected for the precipitation process. A few drops of hydrogen peroxide were added slowly into the filtrate in order to quench the reaction until it is completely oxidised and turn from greenish to brownish solution. To adjust the pH at 3, ammonia solution was added into the solution. The precipitated of iron oxide will form and filtered before drying it at 80°C for 24 hours. The dried iron oxide was then calcined by using muffle furnace at temperature of 400°C for 2 hours before further analysis. The process was then repeated by varying the temperature during acid leaching process for 90°C, 100°C and 110°C.

2.2 Characterization

The recovered iron oxide from EAFS was characterized to obtain its physical properties by using TGA, SEM and XRD.

3. Results and Discussion

3.1. Effect of heating temperature

A variation of temperature in the range of 80°C to 110°C with 10°C temperature interval during sulfuric acid leaching were conducted. The weight of iron oxide recovered at each temperature were weighted and tabulated in Table 1. The optimum reaction temperature for the leaching process is found to be 100°C which 0.56 g of iron oxide was leached. According to F. Kukurugya et al., 95°C is the optimum temperature in sulfuric acid leaching in recovery of iron oxide [9]. This claim can also be supported by similar research done by Biao et al.[10] and Lim et al. [11]. However, these claims can be contended by T. Havlik et al. who indicated that the optimum temperature is at 100°C [12].

Table 1. Effect of acid leaching temperature

Leaching temperature (°C)	Weight of iron oxide after dry (g)
110	0.28
100	0.56
90	0.41
80	0.11

3.2. Effect of heating time

Six samples were carried out with different heating time during leaching process such as 2, 3, 4, 5, 6, and 7 hours respectively. The iron content in the leaching solution was analysed using ICP-AAS accordingly. Figure 1 illustrated the iron recovered from the leaching process. Li et al stated as the time increases, more iron being leached out from the slag in sulfuric acid solution [13]. It can be seen the highest optimum condition for the iron recovery is at 5 hours sample with constant 90°C. After 5 hours of leaching at fixed temperature, an abrupt decrease of iron concentration in the solution can be observed followed with 7 hours accordingly. According to studies from Jinping et al, the efficiencies of Fe recovery will increased from 2 to 24 hours at 80°C with the ratio of 1:5 [13].

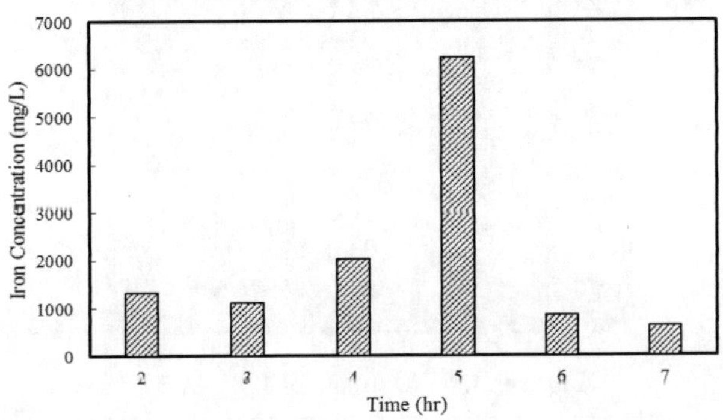

Figure 1. Recovery of iron from EAFS at Different Time

3.3. Iron Oxide Characterization

The thermal behaviour of recovered iron oxide from EAFS was shown in Figure 2. The declining TGA thermal curve indicates the weight loss of recovered iron oxide occurred after being operated at temperature range of 25°C to 600°C. Referring to the TGA thermal curve, it shows that the decomposition step of recovered iron oxide occurred within the temperature range of 25°C to 525°C. The major weight loss as much as 40% takes place at temperature region of 300°C to 400°C and it is complete decomposed at 525°C. According to Divine et al, the major weight loss occurs at temperature 225°C and iron oxide decompose completely at temperature 500°C [14]. This finding had identified that the recovered iron oxide was totally decomposed at 525°C. Figure 3 illustrated the agglomerates predominantly with irregular shapes with low sphericity and sharp edges. The particles examined where extremely rough surface texture with platy, crystalline structure. Moreover, Yao et al studied the microstructure of the particle is in irregular shapes after leaching process. The pores that are found within agglomerates indicate the particle of EAFS have been leached by H_2SO_4 during leaching process [15]. Figure 4 represents the XRD patterns of phases in iron oxide. Besides, peaks appear overlapping in XRD pattern that represents the crystalline phases in the sample. It was stated that the more peaks appear that showed its crystalline structure and presents mascagnite ($(NH_4)_2SO_4$),

IOP Conf. Series: Materials Science and Engineering **551** (2019) 012119 doi:10.1088/1757-899X/551/1/012119

manganese iron phosphide ($Mn_{0.1}Fe_{0.9}P$), hematite (Fe_2O_3), and iron aluminium silicate ($Fe_3(Al_{0.2}Fe_{1.8})(SiO_4)_3$).

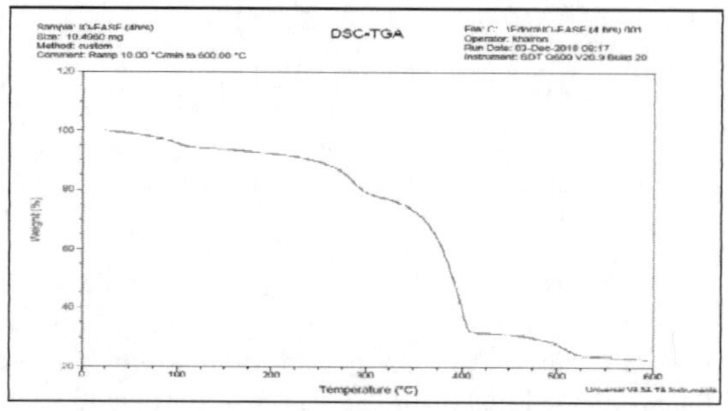

Figure 2. TGA Thermal Curve of IO-EAFS

Figure 3. SEM image of IO-EAFS

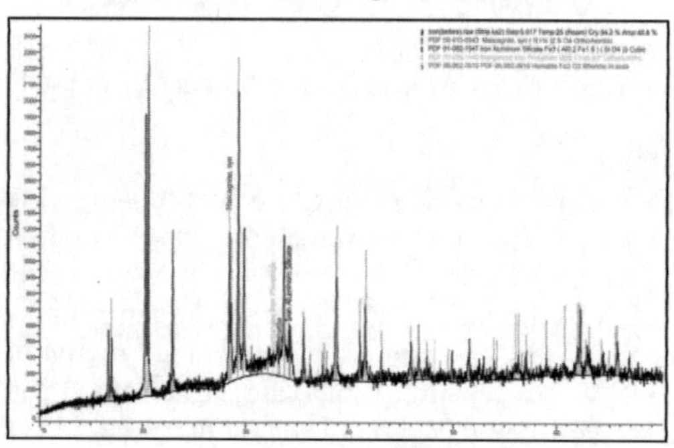

Figure 4. XRD patern of IO-EAFS

4. Conclusion

In this study, the effect of heating temperature and time were carried out. The best results in iron recovery were obtained when the condition for acid leaching at temperature of 100 °C for 5 hours. Regarding to the characterization it is observed that iron oxide existed in form of hematite and other metals oxide. Thus, it can be concluded that iron oxide from EAFS via method of acid leaching can be maximixed with appropriate of conditions.

Acknowledgements
This work was supported by Faculty of Chemical Engineering UiTM Cawangan Pulau Pinang.

References

[1] WorldBank, Population, Total Data, World Bank. (2016). Retrieved from https://data.worldbank.org/indicator/SP.POP.TOTL (accessed May 1, 2018).
[2] Ter T.P *et al* 2014 *Waste Manag.* **34** 2697
[3] Das B *et al* 2007 *Resour. Conserv. Recycl.* **50** 40
[4] Koros P.J. *et al* 2003 *Mater. Process. Sci.* **34** 769
[5] Lau V.L. *et al* 2004 *Forum Geookol.* **15** 7
[6] Manaf L.A. *et al* 2009 *Waste Manag.* **29** 2902
[7] Ter T.P. *et al* 2016 *Procedia Chem.* **19** 842
[8] Nasuha N. *et al* 2017 *J. Environ. Manage.* **196** 323
[9] Kukurugya F. *et al* 2015 *Hydrometallurgy.* **154** 20
[10] BiaoDeng W.S. *et al* 2017 *Metals* **8,** 8
[11] Wei L.J *et al* 2018 *E3S Web of Conf.* **34** 1
[12] Havlik T. *et al* 2005 *Hydrometallurgy.* **77** 41
[13] Li J. *et al* 2007 *Wuhan Univ. J. Nat. Sci.* **12**, 541
[14] Yufanyi D.M. *et al* 2005 *Am. J. Chem.* **5**, 1
[15] Affandi A.F *et al* 2017 *Int. Conf. Environ. Res. Technol Proc.* **1**, 240

Synthesis Of Epoxidized Oleic Acid- Based Palm Oil by Peracid Mechanism

Mohd Jumain Jalil[1*], Aliff Farhan Mohd Yamin[1] Mohd Saufi Md Zaini[2], Mohd Azahar Mohd Ariff[1], Siu Hua Chang[1], Norhashimah Morad[3], and Abdul Hadi[1]

[1]Faculty of Chemical Engineering, Universiti Teknologi MARA (UiTM) Pulau Pinang, 13 500 Permatang Pauh, Pulau Pinang, Malaysia
[2]Faculty of Chemical Engineering, Universiti Teknologi MARA Cawangan Terengganu, Kampus Bukit Besi, 23200 Dungun, Terengganu, Malaysia
[3]School of Industrial Technology, Universiti Sains Malaysia

E-mail: mjumain0686@uitm.edu.my

Abstract. Commercial epoxides available in the market are mainly petroleum and animal based which are non-environmental friendly and often non-halal. Oleic acid (C18:1) derived from palm kernel oil contains the unsaturation double bond (C=C) in its long alkyl chain makes it an attractive fatty acid for the production of eco-friendly epoxide. The epoxidation of oleic acid was carried out by using in situ generated performic acid (HCOOOH) to produce epoxidized oleic acid. Performic acid was formed by mixing formic acid (as oxygen carrier) and hydrogen peroxide (as oxygen donor). The results indicated that optimum conditions occurred at moderate agitation speed 300 rpm and by using 30% hydrogren peroxide. Sulfuric acid was found to be the most efficient catalyst. It was found that a maximum relative conversion to oxirane (RCO) achieved was 87.61 % at optimal condition.

1. Introduction

Palm oil is one of the vegetable oil that exist in the industry and it is widely used in the development of polymeric materials [1]. The growth of vegetable oil in the demand has increased compared to petroleum-based polymer. This is because vegetable oil is eco-friendly and low cost as well as the petroleum has given an environmental pollution concern and increasing in cost [2]. So, the vegetable oil becomes an alternative in the production of epoxide. In the epoxidation of vegetable oil reaction, oxirane ring-opening or known as epoxide ring degradation will occurs [3]. The oxirane ring-opening will reduce the yield and in order to gain high yield and high peroxide values of epoxidized vegetable, the opening need to be minimized [4]. Moreover, palm oil containing low in unsaturated double bond contents and this will limit synthesis and oxirane cleavage reaction of EPO[2]. The commercial production of palm oil is quite high compared to other vegetable oil and thus make the study of synthesis and oxirane cleavage of palm oil more practical and economical [5].

Oleic acid or also known as cis-9-octadecenoic acid has good characteristics such as have low melting point which is 13.4 ºC, boiling point is 360 ºC, colourless to pale yellow, crystalline form and long chain of carboxylic acid [6]. While, in term of composition, oleic acid from palm oil and palm kernel oil is 39.2% and 15.4% respectively. This value shows that oleic acid contained in palm oil is higher than palm kernel oil which is the ratios between these two is about 18:1. So, the potential to obtain oleic acid as a material that will be used in epoxidation is not complicated [7]. Normally the usage of palm oil is often for edible purposes while palm kernel oil is used for non-edible purposes such as making soaps and detergents [8].

IOP Conf. Series: Materials Science and Engineering **551** (2019) 012120 doi:10.1088/1757-899X/551/1/012120

Oleic acid was chosen in this study because it is one of the organic chemical raw material and can produce epoxidized of oleic acid after epoxidation. There are many applications of oleic acid such as plastic plasticizer, as production of azelaic acid by oxidation, raw material for manufacture carbon paper, manufacture of soap and as an analysis reagents and solvents [9] [10]. The epoxidized of rapeseed oil is highly commercial importance as a plasticizer and stabilizer for plastics, lubricants, and an intermediate for the synthesis of surfactants. Previous studies were focused on epoxidation of vegetable oils (complete triglyceride) [11],[12],[13]. The present study however focuses on the epoxidation of fatty acid, a smaller molecule as compared to a triglyceride molecule. The objective of the present study is to investigate the optimum reaction conditions with respect to the yield of epoxidized oleic acid.

2. Methodology

Crude oleic acid, formic acid and sulfuric acid catalyst were added simultaneously into the reactor. The mixture was stirred continuously at fixed speed and the mixture was gradually heated to the desired temperature. The reaction time for each experiment was 50 minutes and it was recorded at the beginning of the introduction of hydrogen peroxide. Hydrogen peroxide was added drop by drop. Sampling for analysis was done by withdrawing approximately 5ml of reaction mixture for every 5 minutes. The samples were quenched and cooled for 2 minutes before being analyzed using oxirane oxygen content (OOC) analytical method. The OOCexp value was determined by direct method using hydrobromic acid (HBr) solution in glacial acetic acid (AOCS Cd 9-57). The percentage of conversion can be calculated based on the theoretical OOC and the iodine value was determined according to Wijs method.

3. Results and Discussion

3.1 Effect of Aqueous Hydrogen Peroxide Concentration on the Rate of Reaction

In this study, hydrogen peroxide serves as the oxygen donor, which contributes active oxygen to the formic acid in order to form performic acid. According to Oftring et al., [14] any hydrogen peroxide aqueous solutions that contains 10 - 55% of hydrogen peroxide by weight can be used to form peracid solution. Therefore, experiments to study the effect of hydrogen peroxide concentration were carried out using hydrogen peroxide at a concentration of 30 and 35% in aqueous solution. The effect of aqueous hydrogen peroxide concentration on the percentage relative conversion to oxirane (RCO) for oleic acid epoxidation with performic acid at 75 °C reaction temperature and 1:1 formic acid to hydrogen peroxide mole ratio is presented in Figure 1. It was found that hydrogen peroxide concentration of 30% had a positive effect on the rate of epoxidation, achieving higher RCO of around 87.61%. A concentration of 35% resulted in a lower rate of epoxidation and lower RCO which renders it unsuitable for use in this study which focussed on improving the reaction rate and epoxide yield. Hence, the hydrogen peroxide with a concentration of 30% in aqueous solution was chosen for all subsequent reactions.

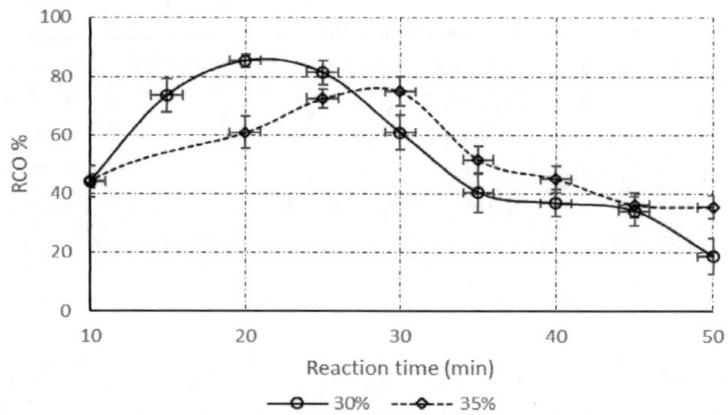

Figure 1. Effect of Hydrogen Peroxide Concentration on the Percentage Relative Conversion to Oxirane.

3.2. Effect of Oxygen Carrier on the Rate of Reaction

Two types of oxygen carrier (peracid) investigated were in situ peracetic and performic acids. Their influences on the epoxidation of oleic acid are shown in Figure 2. It shall be noted that epoxidation can be carried out using either formic acid or acetic acid [15]. It can be seen from Figure 2 that the maximum yield of epoxide was achieved when formic acid was used as the oxygen carrier compared to acetic acid. It is likely that the high epoxide yield obtained was due to the immediate reaction between the performic acid formed with the C=C double bonds in the oleic acid chain as well as minimization of oxirane ring degradation. According to Jalil et al [16] formic acid has the ability to minimize oxirane ring-opening by preventing the deprotonation of performic acid.

Furthermore, it can be observed that the time taken to produce maximum yield of epoxides was reduced when formic acid was used as the oxygen carrier. Nevertheless, should the in situ performic acid concentration becomes too high, it will contribute to a strong acidic condition which causes severe loss of epoxide by ring opening route. In this case, a use of inert solvents such as heptanes and toluene was suggested [17] and the addition of acid-neutralizing compounds was recommended during the epoxidation process in order to maintain a pH value of 3 – 6 [18]. Selection of oxygen carrier does not only affect the rate of conversion to oxirane, but also the stability of oxirane produced

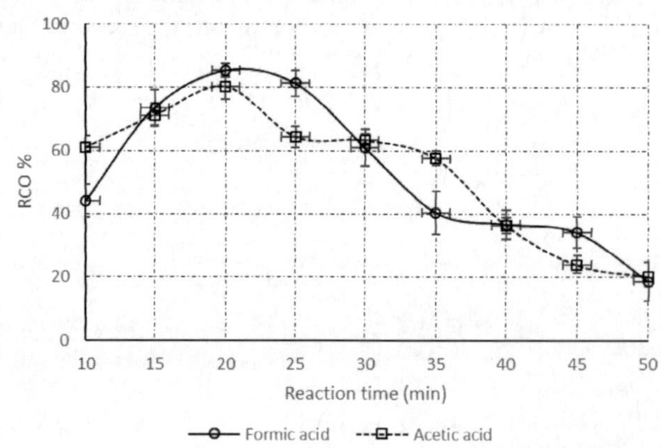

Figure 2. Effect of Oxygen Carrier on the Percentage of Relative Conversion to Oxirane.

3.3. Effect of Catalyst Loading on the Rate of Reaction

The effect of catalyst loading (sulphuric acid concentration) on the percentage of relative conversion (RCO) to oxirane for oleic acid epoxidation was also investigated in this study. The catalyst function is to lower the activation energy and increases the rate of reaction, which indirectly decreases operational cost resulting from prolonged reaction. In this study, the sulphuric acid concentration was varied from 0.2 to 1.0 g and the results are illustrated in Figure 3. Clearly, the rate of epoxidation reaction increases with an increase in sulphuric acid loading. It is also apparent that high RCO (~88%) was achieved at a shorter reaction time using 0.2g sulphuric acid. Increasing the catalyst loading to 0.6 and 1.0g resulted in higher epoxidation rate but lower RCO at around 70%. This result is noteworthy when compared to the findings of Campanella et al., [19] who investigated the epoxidation of fatty acid methyl esters by in situ performic acid without the addition of catalyst. The results of Campanella et al., [20] showed a conversion to oxirane of 83.5% at 40 °C was only achieved after a reaction time of 600 minutes.

In contrast, the maximum conversion to oxirane was attained within 25 minutes at 55 °C in this study. It is believed that the extra hydrogen ions from the external acid catalyst contributed to the increase in an oxirane ring formation. The high RCO obtained from the addition of 0.2 g of sulphuric acid may be attributed to the hygroscopic nature of the sulphuric acid [21]. Since the formation of performic acid is a reversible process, the performic acid acts as the substrate which is responsible for subsequent epoxidation, along with oleic acid. The hygroscopic nature of the sulphuric acid results in the absorption of water causes equilibrium constant (Keq) was maintained. As a result, higher amount of performic acid was produced which in turn leads to a higher RCO, in accordance to the Le Chatelier principle [22]. However, the performic acid begins to decompose once the equilibrium balance favouring performic acid formation is not met, thus decreases the performic acid concentration and concurrently increases the water content in the reacting liquid system [8]. The rapid increment of water minimizes the hygroscopic effect of the sulphuric acid on the water molecules. A sulphuric acid concentration of 0.6 and 1.0 g results in a high acidic condition which causes premature rupture of the oxirane rings and thus decreases the RCO. Under both conditions, the hygroscopic effect of the sulphuric acid is minimum compared to the degradation of oxirane rings caused by the high acidity of the reaction medium.

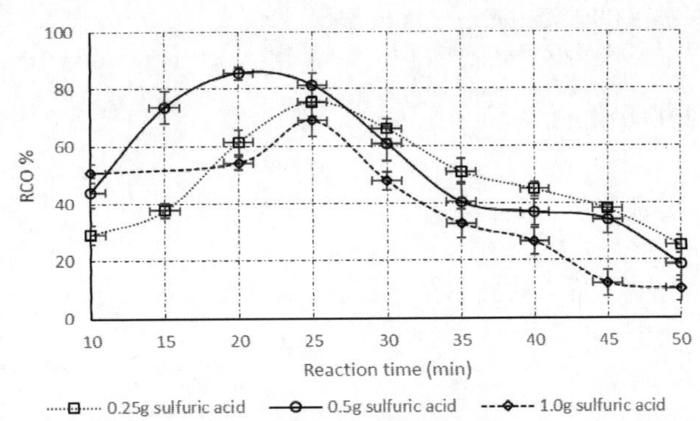

Figure 3. Effect of Catalyst Loading (Sulphuric Acid Concentration) on the Relative Conversion to Oxirane.

4. Conclusions

In this study, crude oleic acid derived from palm oil has been epoxidized using in situ formed performic acid (HCOOOH) in order to produce epoxidized oleic acid. Epoxidized vegetable oils have gained great importance in recent years due to the fact that they are derived from renewable and sustainable resources and therefore perceived as environmental friendly. In this study, crude oleic acid derived from palm oil has been epoxidized using in situ formed performic acid (HCOOOH) in order to produce epoxidized oleic acid. A good conversion to oxirane was observed at 300 rpm stirring speed as well as 30%(wt/wt) aqueous hydrogen peroxide concentration, 0.2g of sulphuric acid catalyst and formic acid as an oxygen carrier were used.

Acknowledgements

This work was financially supported Universiti Teknologi MARA, Cawangan Pulau Pinang, Malaysia

References

[1] Jumain M.J. *et al* 2014 *Adv. Mater. Res.* **906** 125
[2] Jalil M.J *et al* 2018 *AIP Conf. Proc.* **2030** 020189
[3] Santacesaria E e*t al* 2011 *Chem. Eng. J.* **173**, 198
[4] Farhan A. *et al* 2018 *Inter. J. Eng. Technol.* **7** 123
[5] Jalil M.J. *et al* 2017 *Recent Innov. Chem. Eng.* **10** 1
[6] Awang R. *et al* 2001 *J. Am. Oil Chem. Soc.* **7** 1249
[7] Zheng J.L. *et al* 2015 *Int. J. Eng. Technol* **62** 1
[8] Jalil M.J *et al* 2017 *Industrial Cmbustion,* **20** 1606
[9] Jalil M.J. *et al* 2018 *Int. J. Eng. Technol.***7** 152
[10] Scala J.L. *et al* 2002 *J. Am. Oil Chem. Soc.* **79** 373
[11] Darfizzi D. *et al* 2016 , *Sains Malaysiana* **45** 1149
[12] Zhang C *et al* 2014 *Macromol. Rapid Comm.*, **35** 1068
[13] Vianello C *et al* 2017 *Chem. Eng Trans.* **57** 1
[14] Milchert E *et al* 2008 *Czas. Tech.* **1** 284
[15] Chou T *et al* 1986 *Chem. Eng. Comm.* **41** 253
[16] Jalil M.J. *et al* 2018 *Chem. Chem. Technol.* **12** 296
[17] Cai X, *et al* 2018 *Inter. J. Chem. Kinet.* **50** 1
[18] Ren Q *et al* 2019 *Materials* **8** 6623
[19] Campanella A. *et al* 2008 *Chem. Eng. J.* **144** 466
[20] Campanella A. *et al* 2005 *Lat. Am. Appl. Res.* **216** 211
[21] Goud V. *et al* 2007 *Chem. Eng. Scien.* **62** 4065
[22] Rafizan A. *et al* 2014 *J. Appl. Sci. Agric.* **9** 86

IOP Conf. Series: Materials Science and Engineering **551** (2019) 012121 doi:10.1088/1757-899X/551/1/012121

A preliminary study of thermal comfort and lighting in IHL library building

Norasikin Hussin[1,*], Azli Abd Razak[2] , Faizal Baharum[3] and Y. Yaakob[1]

[1]Faculty of Mechanical Engineering, Universiti Teknologi MARA Cawangan Pulau Pinang, Kampus Permatang Pauh, Pulau Pinang, Malaysia.
[2]Faculty of Mechanical Engineering, Universiti Teknologi MARA Shah Alam, 40450 Shah Alam, Selangor, Malaysia.
[3]School of Building, Housing and Planning, Universiti Sains Malaysia, 11800 Pulau Pinang, Malaysia.

E-mail: mjumain0686@ppinang.uitm.edu.my

Abstract. IHL (Institute of Higher Learning) library building is greatly utilized to access information from IT connection and materials. This building as a place to study, reading and writing for students and academics in the university. However, the indoor air quality and lighting system in the library building is not comfort and not satisfactory. In this paper, the investigation of thermal comfort and lighting in IHL library buildings. The library building of LBA, LBB and LBC has been selected. This study has been done with distribution of 350 set of questionnaires. The result shows that 63% of respondents claimed that uncomfortable with the environment because is very cold, which is indoor temperature in average of 20ºC. Meanwhile, 54.3% of respondents are unsatisfied with lighting level because it is too dimmed and most of them prefer to study near the window because of daylight. By lux meter reading, the illumination in average of 225 lux. From the reading, recommendations were made in order to enhance the thermal comfort and potential energy saving.

1. Introduction

Library design is one of the major factors to ensure the comfortable environment. The architects, internal designers, engineers and who is responsible to library design are adhere to the guidelines by ASHRAE and Malaysia Standard. One of the most efficient design is a sustainability building design which is built in an environmentally to ensuring the health and well-being of occupants. Activities such as study, writing, and reading usually done in a closed environment in order to protect people from weather and uncomfortable conditions. The poor quality of indoor temperature and relative humidity could affect the occupants, books and building material itself. Mould and pest activity can growth in high relative humidity, while the low relative humidity leads to desiccation. This mould can be harmful to humans, so that the occupants need to be protected with good indoor air quality.

One of the initial factors in offering acceptable environment for occupants is thermal comfort. There are four environmental parameters such as air temperature, air velocity, relative humidity and mean radiant temperature that need to be accessible in adequate scopes for occupants to be comfortable in their living area. Besides that, air temperature also often taken as the key design

IOP Conf. Series: Materials Science and Engineering **551** (2019) 012121 doi:10.1088/1757-899X/551/1/012121

parameter for thermal comfort[1]. Based on the definition by the US Environmental Protection Agency, good indoor air quality is symbolized by concentrations of pollutants and thermal (temperature and relative humidity) circumstances that do not negatively affect the health, comfort, and performance of occupants. Indoor environment quality has mediating effects and helps to increase the energy efficiency and overall occupant's satisfaction.

Good lighting is required for good visibility of the environment and should provide a luminous environment to perform task perfectly, increase in productivity and prevention of health effects. The quality of lighting can affect the motivation, emotion, and physical health and avoidance of glare. Traditional library more to reading activities. According to literature [2], modern libraries similar to most office buildings because the purpose is not for reading only but also incorporated widely with IT based workplaces. Lighting design should be considered more like in offices rather than in traditional libraries. Proper light makes the occupants makes all the tasks with easier and accurately. The human receive about 85% of their information through their sense of sight [2]. Inappropriate light encourages to human health emotion such as eye strain, headache. According to previous researcher [3], luminance level and work tasks can form a link to each other so as to determine that luminance level is main factor for working efficiency.

The main objective of this study is to determine the thermal comfort and lighting in IHL library building. In order to achieve the objective, this preliminary study involves measuring the temperature, relative humidity and lux. The scope of this study is to identify the thermal comfort such as indoor air temperature and relative humidity, and lighting level of library building in location A, B and C. This study focuses only on the library users as a respondents. As a preliminary study, the respondents had been provided with question surveys and walk-through assessment was carried out.

2. Methodology

In this case study, the library building in location A has been selected in order to identify the thermal comfort. The library building has been investigated which is a standalone building and located at the Merbok campus (5.7138° N, 100.4505° E). Fig. 1.(a) presents a view of the library in location A. The library building consists of 3-storey and total area of library building is 2143 square meters. Meanwhile, the second building is located at the location B which is Bertam campus (latitude 5.5314°, longitude 100.4755°) experiencing hot and humid weather trough out the year. The library building consists of 3-storey and the total area of library building is 6244.32 square meter [4]. A library building is shown in Fig. 1.(b). The third building is located at the location C, Permatang Pauh campus, Fig. 1.(c). This building consists three stories and construction area 5707.02 square meters, the total volume is 25681060 cubic meters and the ceiling height of each level is 4.5 meter [5].

A set of questionnaire was designed in order to identify the thermal comfort in the library building, which consisted of three sections. The first section is to present respondent's background such as type of programs, gender, and frequency of incoming to library building, while in second section is about thermal comfort and third section for lighting in library building.

(a) Library Building A (LBA)

(b) Library Building B (LBB)

(c) Library Building C (LBC)

Figure 1. View of the library building.

3. Result and Discussion

The questionnaire had been distributed to 350 respondents. 276 respondents in LBA are participated in this survey and complete the questionnaire while 74 did not, thus the overall completion rate was 78.9%. More than 80% are participated in order to complete the question survey in LBB and LBC, and less than 20% did not complete it. Based on the analysis, the frequency of students incoming to the library building in twice a weeks is 22% and more than 3 to 6 times a weeks is 52%. Mostly, the students like coming to the library building because they like using the library internet connection especially in LBA. The library internet connection is very good and faster for the students and their studies. However, the 44% respondents claims that incoming once a month in the LBC. The reasons of their comings because to complete their assignments.

The second section in question survey is about thermal comfort in library building. The purpose of this section is to know the satisfaction of the lighting level and indoor temperature. This satisfaction analysis is needed in order to improve the thermal comfort and utilization of library building. Table 1 shows the percentage of occupants satisfaction of indoor temperature and lighting level.

The indoor temperature and humidity are needed to measure in order to identified the occupants satisfaction. For the LBC, 63% of respondents are unsatisfied with indoor temperature. Most of them complained about their feeling uncomfortable because it is very cold. The setting temperature in LBC is 17ºC [5] and indoor temperature in average of 20ºC. Because of that, the students cannot proceed their activities such as reading and writing not more than 2 hours. The temperature should not below than 22.5C or the occupants will feel discomfort with the environment [6]. But, in average 50% respondents are unsatisfied with indoor temperature in LBA and LBB. In LBC, the indoor temperature is in range 22ºC to 23ºC, the relative humidity is 68% to 71%. Table 2 is shows the setting temperature for each library building.

From Table 1, more than 70% of respondents are satisfied with lighting level in LBB and LBC. Most of them agreed that lighting level is good for them in doing activities such as reading and writing. However, the 54.3% of respondents in LBA are claimed that feeling unsatisfied with lighting level in LBA. Most of them complained that the environment is too dimmed for them to doing activities. According to literature [7], the lux, lumen/m^2 that suitable for study in the library is 500 lux, so that the occupants has feeling uncomfortable with lighting. In other words, the LBA need to improve the lux by increasing the numbers of lamps. The lack of lighting will lead to the depression in condition such as seasonal adjustment disorder [8]. In this case, 53 zones has identified in order to measure the light level in LBA. Based on reading by lux meter (Testo), the result shows that average of illumination is 224 lux at ground floor, 226 lux at first floor and second floor with area 228 meter square is 153 lux. But, a few zones with windows have more than 300 lux and 500 lux. Based on the question survey, more than 51.4% of respondents like to choose the reading area nearest to window glass which is limited. The LBA more to 70 to 80% of concrete and the rest is window glass. From literature [9], the office workers prefer the daylight to source lighting and most of them like to work near windows because it enhances their productivities compared to the ones sitting far from window . Because of that, the LBA installing the variuos types of lamps for lighting system such as fluorescent lamps and downlight. The purpose is to give more bright and comfort for occupants. There is no

IOP Conf. Series: Materials Science and Engineering **551** (2019) 012121 doi:10.1088/1757-899X/551/1/012121

complaint about lighting is too dimmed in LBB and LBC. Observation by lux meter shows that the illumination in LBB and LBC is 500lux and 900lux, respectively.

Table 1. Percentage of satisfaction in library building

Building	Indoor temperature		Lighting level	
	Satisfied	Unsatisfied	Satisfied	Unsatisfied
LBA	45.3	54.7	45.7	54.3
LBB	46.2	53.8	70.5	29.5
LBC	37.0	63.0	88.3	11.7

For the setting temperature and lux level in library building as shown in Table 2, the human body temperature needs to be maintained at a constant 37±5ºC regardless of the prevailing ambient condition. According to to MS1525, the guideline for a standard indoor environment design for Malaysian building is in the range of 24ºC to 26ºC [5]. Previous researcher [10] shows that the raising set point from 22ºC to 26ºC encourages to energy saving. The light level for library building is more in the range 500 to 1000 lux but it is depending on the type of activities.

Table 2. Setting temperature and lux level in library building

Building	Setting temperature	Lux level
LBA	21ºC	225
LBB	22ºC	500
LBC	17ºC	900

4. Conclusion

This paper use a set of question survey in order to identify the occupants satisfaction regarding on thermal comfort in library building. This case study analysed three library building such as LBA, LBB and LBC. From the question surveys, the library building need to improve their thermal comfort especially indoor temperature and lighting. Around 63% of respondents had expressed their uncomfort feeling with the environment because is was too cold. In order to provide the comfort, the best solution is to re-setting the temperature. Most of the previous research works are studied on the resetting temperature for thermal comfort and potential energy saving. In LBA, 54.3% of respondents are unsatisfied with lighting level and most of them complained that the environment is too dimmed. Illumination in LBA in average 225 lux and need to improve with referring to ASHRAE90.1-2013 which is 500 lux. Installation of LED fluorescent lamps or green lighting is recommended to improve the situation. In addition, it contributes to minimize energy consumption.

Acknowledgements

The authors would like to express and appreciation to respective personnel in the involved libraries including the technical staff. A special thanks to my FYP students for their assistance in distributing the questionnaire and collecting the data. We would also like to thank all the repondents who provided their time to complete the questionnaires for this study.this research was also supported and funded by the Universiti Teknologi MARA Cawangan Pulau Pinang.

References

[1] Ibrahim S. H *et al* 2013 *Int. J. Energy Environ.* **4** 1003
[2] Altan H *et al* 2010 *Chapter 3: Low Cost Sustainable Building Solutions report*
[3] Weston,H.C. Industrial Healath Research Board, Report No. 87, London (1945).
[4] Hussin N *et al* 2018 *AIP Conf. Proc.* **2030** 020044
[5] Hussin N *et al* 2018. *Eng. Sci. Technol. Colloquium* **1** 63
[6] ANSI/ASHRAE Standard 55-2010 Thermal Environmental Conditions for Human Occupancy. vol. 2010, 2011.
[7] Preto S. *et al* 2018 *AHFE 2018: Adv. Design Inclusion* **776** 180
[8] *http://www.flanderstoday.eu/education/poor-lighting-can-affect-pupils-health-and-happiness. Retrieved on 8 Nov 2018.*
[9] Lottrup L *et al 2015 J. Landscape Res.* **40** 57
[10] Chua S. C *et al* 2010 *Sustain. Energy Rev*, **14** 2916

IOP Conf. Series: Materials Science and Engineering **551** (2019) 012122 doi:10.1088/1757-899X/551/1/012122

Removal of Nickel, Zinc and Copper from Plating Process Industrial Raw Effluent Via Hydroxide Precipitation Versus Sulphide Precipitation

Nur Ain Zainuddin[1], Tengku Azwan Raja Mamat[1], Hawaiah Imam Maarof [1], Siti Wahidah Puasa[1] and Siti Rohana Mohd Yatim[2]

[1]Faculty of Chemical Engineering, Universiti Teknologi MARA (UiTM), 40450 Shah Alam, Selangor, Malaysia.
[2]Department of Environmental Health, Faculty of Health Science, Universiti Teknologi MARA (UiTM), 40450 Shah Alam, Selangor, Malaysia.

E-mail: nurain1465@uitm.edu.my

Abstract. Electroless plating wastewater contains mixed metals element. Although precipitation is the most common method applied to remove metals from the wastewater, it fails to completely precipitate all the metals at a single pH. The present work evaluates the effectiveness of the sulphide and the hydroxide precipitation for removal of nickel, zinc and copper from industrial wastewater, namely acid rinse and nickel rinse samples. The optimum pH and coagulant dosage were determined via Jar Test for both types of samples. Sulphide precipitation removed heavy metals more efficient than hydroxide precipitation. For the acid rinse sample, 95.32% of nickel was removed using sulphide precipitation while 76.66% removal using hydroxide precipitation. For nickel rinse sample, 93.75% of zinc was removed using sulphide while 68.8% removal using hydroxide. 65.75% of nickel was removed using both sulphide and hydroxide. Total removal of copper was achieved for both hydroxide and sulphide precipitation, either from acid rinse or nickel rinse samples.

1. Introduction

Hydroxide treatment is the most common and widely used precipitation method for removal of heavy metal from industrial wastewater [1]. However, this method has its limitation in treating mixed metal effluents since all metal hydroxides do not completely precipitate at a single pH. Each type of metal hydroxide is favourable to precipitate at a certain pH range. According to the solubility-pH relationship, as shown by Fig. 1, the best-compromised pH for hydroxide precipitation process is pH 10-10.5 for nickel, pH 9.0-9.5 for zinc and pH 8.5-9.5 for copper. It shows that these heavy metals cannot be completely removed at single pH range. Hydroxide precipitates tend to resolubilize if the solution pH is not in the compromised range [2]. Besides, hydroxide precipitation produces a large amount of sludge which required further treatment before disposal. For sulphide precipitation, the optimum pH for copper is pH 8 while both nickel and zinc are at a range of pH 10-10.5. Therefore, nickel and zinc can be removed at a single pH using sulphide precipitation.

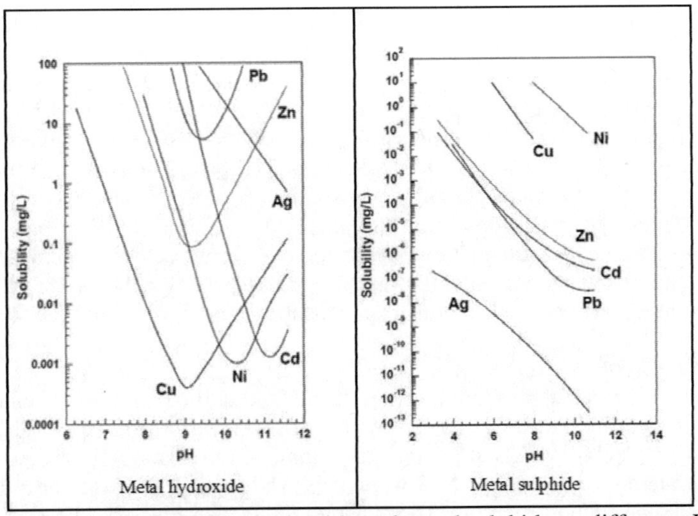

Figure 1. The solubility of metal hydroxide and metal sulphide on different pH [3].

Sulphide precipitation has advantages over hydroxide precipitation [4]. Sulphide can remove a high degree of metals compare to hydroxide due to the lower solubility of metal sulphide over a broad pH range [5]. Its reaction rate is faster than hydroxide precipitation due to the high reactivity between sulphide and heavy metal ions. The sludge generated is less compare to hydroxide precipitation. However, sulphide precipitation gains little attention because of the sulphide dosing and process control difficulty due to the sensitivity of the process [6]. It is found in limited application because of the potential of releasing toxic gaseous, hydrogen sulfide (H2S) emissions and operational difficulties [7].

The concentration of heavy metal in the electroless plating industry wastewater commonly not complies with the acceptable discharge concentration. For instance, aging wastewater at a nickel plating plant contained 2900 ppm of nickel ions [8]. According to the Environmental Quality (Industrial Effluent) Regulations 2009 by the Department of Environment (DOE), Malaysia [9], the acceptable discharged concentration for nickel and copper is 1.0 mg/L, while zinc is 2.0 mg/L. The purpose of this research is to evaluate the performance of hydroxide precipitation and sulphide precipitation for removal of heavy metals, namely nickel, zinc and copper from electroless plating wastewater. The optimum pH and coagulant dosage were also determined.

2. Methodology

2.1. *Wastewater Description and Analysis*

The wastewater was obtained from electroless copper plating Company A and Company B. Two different samples were obtained which were acid rinse from Company A and nickel rinse from Company B. Approximately 10 litres of each sample was collected. The heavy metal concentration in the wastewater was determined using multimeter portable colorimeter (HACH DR900), and HACH Method 8009, 8506 and 8105 was employed for zinc, copper and nickel, respectively.

2.2. *Jar Test*

Jar Test was carried out to evaluate the optimal pH and coagulant dosage of the precipitation process. The optimum pH and coagulant dosage for precipitation are required to enhance the coagulation process and to avoid overdose of the chemicals. The coagulant used in these experiments was ferric

IOP Conf. Series: Materials Science and Engineering **551** (2019) 012122 doi:10.1088/1757-899X/551/1/012122

chloride (FeCl3). Firstly, the wastewater sample (300 mL) was poured into a series of ten beakers. The pH of the sample in each beaker was adjusted to ten different values (3, 4, 5, 6, 7, 8, 9, 10, 11 and 12) by adding H2SO4 or NaOH. Then, 0.5 mL of FeCl3 was added to each beaker/jar. A rapid mix was applied (150 rpm) for 3 minutes, and then it was reduced to 60 rpm for 15 minutes to promote larger floc formation. The mixer was turned off between 30 to 45 minutes to allow the flocs to settle down. Then, the turbidity of the sample was measured. The pH value that gave the lowest turbidity value was chosen as the optimum pH of the precipitation process. The Jar Test was repeated to determine the optimum coagulant dosage at the obtained optimum pH. Five beakers of the sample were prepared and different coagulant dosage was added in the respective beaker (0.2, 0.4, 0.6, 0.8 and 1.0 mL). At the end of the experiments, the mass of the suspended solid produced by each test was measured by the gravimetric method and the result was identified as the weight of sludge.

2.3. Hydroxide and Sulphide Precipitation

The removal of heavy metal by precipitation was carried out at the optimum pH and optimum dosage of coagulant. It was carried out by a stirring process at 150 rpm for 3 minutes. Then, the precipitant (either hydroxide or sulphide) was added into the sample and the mixing process was continued at similar speed for 5 minutes. NaOH and N2S were employed as the hydroxide and sulphide precipitant, respectively. Next, the anionic polymer (flocculant) was added and the speed of the mixing was reduced to 30 rpm and maintained for 15 minutes. For sulphide precipitation, the process was conducted in alkaline condition (pH 8 to 12) to prevent H2S gas formation [1, 7]. Therefore, the pH of the sample was adjusted to pH 8 to 12 using NaOH prior to adding the Na2S in the sample. The removal efficiency of heavy metals by hydroxide and sulphide precipitation was calculated.*Research Procedure*

3. Results and Discussion

3.1. Acid Rinse Sample

Figure 2 shows the turbidity and the weight of sludge at various pH using hydroxide and sulphide precipitation and (b) for acid rinse sample. Using hydroxide precipitation, the turbidity value at pH 3 was very high and was excluded in Figure 2(a). It is also found that pH 8 and pH 4 have the lowest turbidity which is about 0.10 mg/L. However, a higher weight of sludge was produced at pH 4 compared to pH 8. Furthermore, pH 4 is not complied with Standard B discharge requirement by the local authority. The pH of discharged wastewater needs to be controlled between pH 5.5 to pH 9 [5]. Therefore, the optimum pH for acid rinse sample using hydroxide precipitation is at pH 8 and was similar reported by Kurniawan, et al. [10].

Sodium sulphide (Na2S) was employed as the precipitant in sulphide precipitation process. However, directly mixing the Na2S into the sample will produce harmful hydrogen sulphide gas [7, 11]. The sample will turn black with a high turbidity value. This is associated with the acidic nature of the sample. To prevent this from happen, the sample was neutralized using NaOH prior to adding the Na2S in the sample. Figure 2(b) shows that the lowest turbidity value was achieved at pH 8, which was similar to the result obtained using hydroxide precipitation. However, sulphide precipitation produced 50% less sludge compares to hydroxide precipitation. At pH 8, hydroxide and sulphide precipitation produce 0.02 and 0.01 g of sludge, respectively. Blais, et al. [7] also reported less amount of sludge was produced by sulphide than hydroxide treatment.

Figure 3 shows the turbidity and weight of sludge obtained at various coagulant dosage. It was observed in Figure 3(a) that the lowest turbidity value was achieved using 0.4 mL coagulant by hydroxide precipitation while the lowest sludge production was obtained using 0.2 mL coagulant. The dosage of 0.4 mL was chosen as the optimum condition as 0.2 mL coagulant gave a slightly higher value of turbidity compared to 0.4 mL. Besides, choosing less than 0.4 mL dosage may reduce the efficiency of heavy metal removal. Underdosing of coagulant contributes to colloidal particle cannot be destabilized so it will disrupt the flocculation process, hence more amount of sludge will be produced. The amount of sludge produced at the optimum coagulant dosage was found to be 0.02 g. Increasing the coagulant dosage more than 0.4 mL increase the turbidity value. This is caused by attributed to charge reversal and destabilization of colloidal particles due to overdosing [12].

On the other hands, Figure 3(b) shows that the optimum coagulant dosage for sulphide precipitation was using 0.6 mL of coagulant. It gave the lowest turbidity value and sludge produced. The amount of optimum coagulant needed in sulphide precipitation is higher compared to hydroxide precipitation. More amount of coagulant is required due to low solubilities of most metal sulphide salt [13]. However, the volume of sludge production remains low compared to hydroxide precipitation.

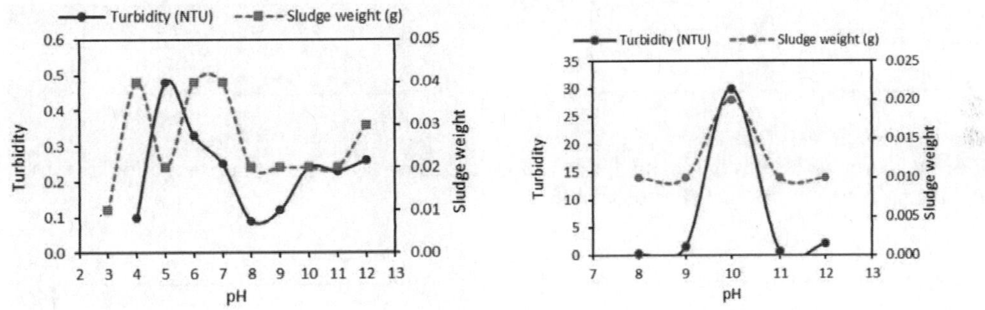

Figure 2. Turbidity and weight of sludge at various pH using (a) hydroxide precipitation and (b) sulphide precipitation for acid rinse sample.

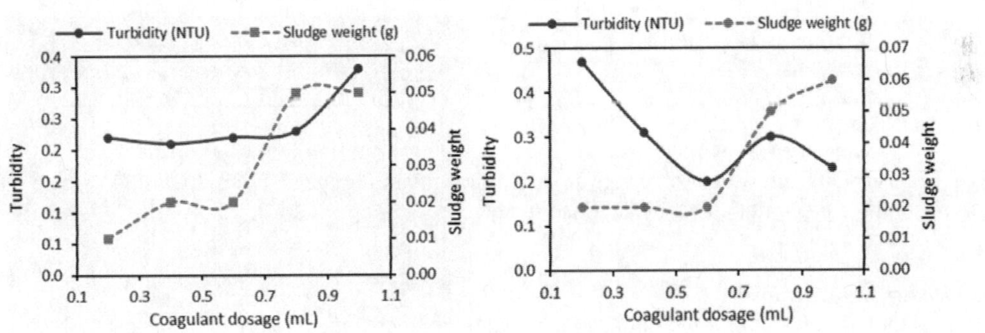

Figure 3. Turbidity and weight of sludge at various coagulant dosage using (a) hydroxide precipitation and (b) sulphide precipitation for acid rinse sample.

3.2. Nickel Rinse Sample

Figure 4 shows the obtained turbidity and sludge produced by the hydroxide and sulphide precipitation of nickel rinse sample. The turbidity values at more acidic conditions are excluded in the plot as their values are too high and exceeding the upper limit detection of the turbidity meter. For both types of precipitations, it is observed that the lowest turbidity was obtained at pH 10. However, hydroxide precipitation produced more sludge than sulphide precipitation at this pH.

For the effect coagulant dosage, the optimum amount of coagulant for nickel rinse sample by hydroxide precipitation was 0.6 mL, as shown in Figure. 5. Using this amount of coagulant, the obtained turbidity value and the sludge produce was the lowest (0.01 g). On the other hands, the

IOP Conf. Series: Materials Science and Engineering **551** (2019) 012122 doi:10.1088/1757-899X/551/1/012122

amount of sludge gradually increases with the increase of coagulant dosage in sulphide precipitation process, as illustrated by Fig. 5(b). However, higher coagulant dosage (more than 0.8 mL) promotes a high value of turbidity. Although 0.2 mL produced the lowest amount of sludge, the turbidity was too high (127 NTU) and was excluded from the plot. Therefore, the optimum coagulant dosage was 0.4 mL for which it gave the lowest turbidity value with a reasonable amount of sludge produced.

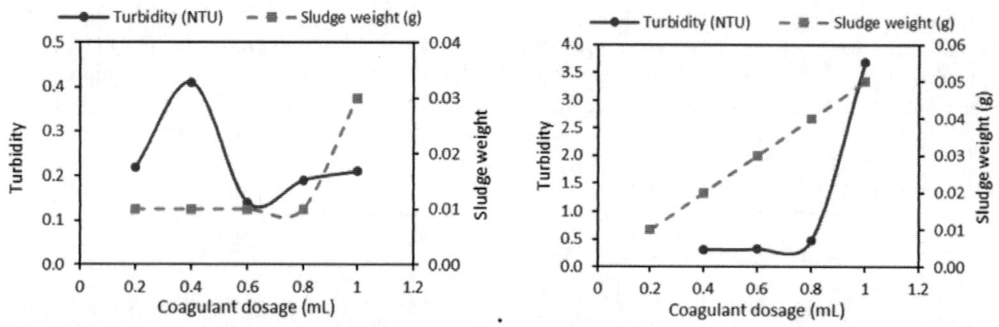

Figure 4. Turbidity and weight of sludge at various pH using (a) hydroxide precipitation and (b) sulphide precipitation for nickel rinse sample

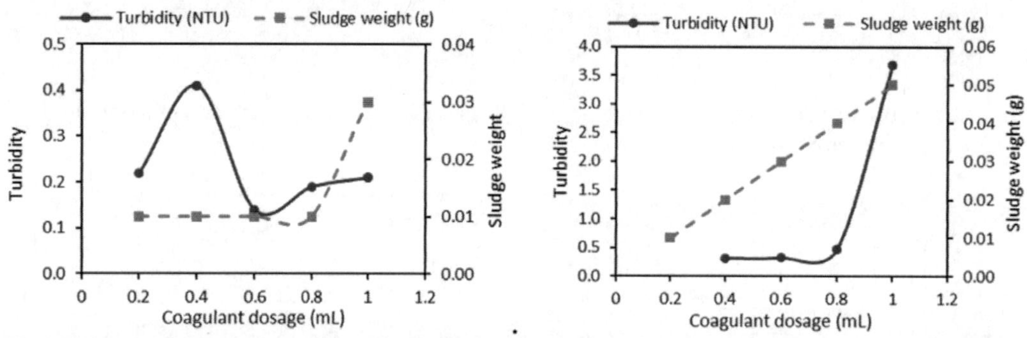

Figure 5. Turbidity and weight of sludge at various coagulant dosage using (a) hydroxide precipitation and (b) sulphide precipitation for nickel rinse sample.

3.3. Heavy Metal Removal

Table 1 tabulates the data for the initial concentration and percentage removal of metals by hydroxide and sulphide precipitation at their respective optimum pH and coagulant dosage. Acid rinse sample does not contain zinc, but it contains a high concentration of nickel and copper which is 12.32 and 15.52 mg/L, respectively. Both nickel and copper concentration do not comply with local authority standard limit which is 1.0 mg/L. The removal efficiency of nickel was 76.66% after the hydroxide precipitation process. Higher removal was obtained using sulphide precipitation (95.32%). A similar observation was found for removal efficiency of Zn in nickel rinse sample. According to Chen, et al. [5] and Fu and Wang [6], sulfide precipitation can achieve a high degree of metal removal over a broad pH. Additionally, total removal of copper was obtained for both type of samples using either hydroxide or sulphide precipitation. Removal of copper removal is very efficient in both method because at pH 8 copper is least soluble compared to nickel. It can be concluded that pH 8-9 is the optimum pH for copper removal while nickel is at pH 10-10.5 (as illustrated by Figure 1).

Table 1. Heavy metals removal from acid and nickel rinse samples by hydroxide and sulphide precipitation.

	Acid Rinse			Nickel Rinse		
	Ni	Zn	Cu	Ni	Zn	Cu
Initial concentration (mg/L)	12.32	ND	15.52	1.06	6.40	0.20
Percentage removal by hydroxide precipitation (%)	76.66	-	100	65.79	68.75	100
Percentage removal by sulphide precipitation (%)	95.32	-	100	65.79	93.75	100

ND - Non-detectable

4. Conclusions
In this research work, the optimum pH and coagulant dosage for hydroxide and sulphide precipitation of heavy metals were determined. At these optimum conditions, the performance of hydroxide and sulphide precipitation was compared. Sulphide precipitation was found to remove heavy metals (zinc, nickel and copper) more efficient than hydroxide precipitation. For acid rinse sample, 95.32% of nickel was removed using sulphide precipitation while 76.66% removal was obtained using hydroxide precipitation. For nickel rinse sample, 93.75% of zinc was removed using sulphide while 68.75% removal using hydroxide. A similar result was obtained for removal of nickel by both precipitation method. The removal efficiency of nickel was 65.79%. Additionally, total removal of copper was achieved for both sulphide and hydroxide precipitation in acid rinse and nickel rinse samples. As a conclusion, the performance of sulphide precipitation treatment was better than hydroxide precipitation as it removes a high amount of nickel, zinc and copper.

Acknowledgements
We acknowledge the facilities provided by the Universiti Teknologi MARA, Institute of Quality and Knowledge Advancement (InQKA) and Integrated Separation Technology Research Group (i-STRonG). The authors are grateful for financial support provided by the Faculty of Chemical Engineering, UiTM Shah Alam.

References
[1] Gunatilake S.K. 2015 J. Multidiscip. Eng. Sci. Stud. 1 12 .
[2] Brbooti M.M. et al. 2011 J Eng. Technol 29, 595
[3] Control and Treatment Technology for the Metal Finishing Industry. Industrial Environmental Research Laboratory. EPA 625/8-80-003, 1980.
[4] Lewis A et al. 2006 81, 197.
[5] Chen et al 2018 J. Water Process Eng. 26, 289
[6] Fu F et al. 2011 , J. Environ. Manage. 92, 407
[7] Blais J. F. et al. 2008 Practice Periodical of Hazardous, Toxic, and Radioactive Waste Management 12, 135
[8] Shih Y.J. et al. 2013 Sep. Purif. Technol. 104, 100
[9] Environmental Quality (Industrial Effluent) Regulations 2009, P.U.(A) 434/2009.
[10] Kurniawan T.A et al. 2006 Chem. Eng. J. 118, 83
[11] Veeken A.H.M et al. 2003 Water Res. 37, 3709
[12] Baghvand A.D et al. 2010 J. Environ. Sci 6, 442
[13] Mokone T.P. et al. 2012 Water Res. 46, 2088

IOP Conf. Series: Materials Science and Engineering **551** (2019) 012123 doi:10.1088/1757-899X/551/1/012123

Assessment of Ambient Air Pollution in Langkawi Island, Malaysia

Nur Elis Sharmila Zulazmi[1], Ahmad Zia Ul-Saufie[2,3], Wan Nur Shaziayani Wan Mohd Rosly[2], Norazian Mohamed Noor[4], Nor Aida Zubir[1,3]

[1]Faculty of Chemical Engineering, Universiti Teknologi MARA, 13500 Pulau Pinang, Malaysia
[2]Faculty of Computer and Mathematical Sciences, Universiti Teknologi MARA, 13500 Pulau Pinang, Malaysia
[3]Hybrid Nanomaterials, Interfaces & Simulation (HYMFAST), Faculty of Chemical Engineering, Universiti Teknologi MARA, Cawangan Pulau Pinang, 13500 Pulau Pinang, Malaysia
[4]School of Environmental Engineering, Universiti Malaysia Perlis (UniMAP), Perlis, Malaysia

E-mail: ziaulsaufie@gmail.com

Abstract. For many years, Southeast Asia region encountered several ongoing air pollution episodes. These episodes had indirectly significant impact on the air quality of neighboring countries. Hence, this study aims to assess the characteristics and trend of air pollution in Malaysia particularly at Langkawi Island. Air pollution hourly monitoring records from 2003 to 2017 were used in analyzing the statistical data analysis. Eight parameters were selected in this study which known as PM_{10}, CO, SO_2, NO_2, O_3, relative humidity (RH), temperature (T) and wind speed (WS). The results showed that the hourly trends of PM_{10}, CO, SO_2, NO_2 and O_3 in Langkawi Island were below the MAAQG standard. The results also indicated that dispersion of PM_{10} in Malaysia were significantly affected by temperature, wind speed and relative humidity.

1. Introduction

For many years, Southeast Asia region particularly to Malaysia has encountered ongoing air pollution episodes [1]. The worst air pollution episode was reported on 2015 as very unhealthy air pollution levels were detected in Malaysia due to the transboundary haze pollution originated from uncontrolled agriculture burning land and activities in Indonesia [2]. Such episodes posed on high potential of adverse impacts to human health and environment.

Air pollutions can be generated by numerous sources such as power plant, factories, vehicles, windblown dust, haze and open burning of forests or any garbage by human activities as well [3]. In fact, it been reported that the rapid growth of power plant in Malaysia in the past few years [4,5] has contributed to plausible decrease of air quality in Malaysia recently. This is due to the facts that the high amount of waste gas stream emissions from power plants have contributed to continuous release of dust, chemicals, organic and inorganic pollutants into atmosphere during operational hours [6].

The air quality can be represented by the air pollution index (API) as summarized in Table 1. The types of air pollutants being monitored in API are PM_{10}, sulphur dioxide (SO_2), nitrogen dioxide

IOP Conf. Series: Materials Science and Engineering **551** (2019) 012123 doi:10.1088/1757-899X/551/1/012123

(NO_2), carbon monoxide (CO) and ozone (O_3) which are harmful to human health and environment [7]. When API value is below 50 and 100, it is considered as good and moderate which means that the air surrounding is healthy and safe for the people. If the API exceeds 100, it considered as unhealthy condition to human and environment. Meanwhile once the API reached to >300 the area is claimed to be hazardous that posed dangerous effects to human health.

Hence, the aims of this works are aims to assess the characteristics and trend of air pollution in Malaysia particularly at Langkawi Island. The air pollution databases from from 2003 to 2017 were used in the statistical analysis and compared with the Malaysian Ambient Air Quality Guidelines (MAAQG) as reference.

Table 1. Malaysian Ambient Air Quality Guidelines (MAAQG) [8]

Malaysia's API	
Air Pollution Index (API)	**Air Quality Category**
0-50	Good
51-100	Moderate
101-200	Unhealthy
201-300	Very unhealthy
301+	Hazardous

2. Methodology

2.1. Data Acquisition

Data of air quality monitoring in Langkawi Island was acquired from Department of Environment (DOE), Malaysian. The duration of data collection were from 2003 to 2017. The air pollutant parameters used in this work were PM_{10}, CO, SO_2, NO_2, O_3 and several of meteorological factors which known as wind speed (WS), relative humidity (RH) and temperature (T).

2.2. Site location

Langkawi Island is located on the north of peninsula Malaysia at latitude of 6.3500° N and longitude 99.8000° E with coverage area of 478.5 km^2. Langkawi Island has been chosen since the island is known as one of the most attractive tourism destination in Malaysia. For instance, it's been reported that the number of tourists in Langkawi kept increasing from 2.4 million in 2010 to 3.63 million in 2016 [9]. In facts, the projection of these numbers is expected to increase until 5 million in 2020 [10]. Figure 1 shows the map of Malaysia highlighting Langkawi Island.

Figure 1. Map of Langkawi Island [11].

IOP Conf. Series: Materials Science and Engineering **551** (2019) 012123 doi:10.1088/1757-899X/551/1/012123

2.3. Statistical Analysis

Several statistical analysis were conducted in order to compare the concentration of the air pollutants and to determine the distribution of air pollutants concentrations. The Pearson correlation analysis was conducted to determine the relationship between air pollutants concentration and the meteorological factors. These data were analyzed using IBM SPSS statistical software and the hourly trends for each air pollutant parameters from June 2003 to 2017 were compared to MAAQG.

3. Results and Discussion

3.1. Long Term Record of Air Quality Data

The compilation of air quality data at Langkawi Island are summarized in Table 2. Interestingly, PM_{10} was found to be the most dominant air pollutant compared to others. The averaged concentration of PM_{10} recorded was 39.26 μg/m^3 and it was below the MAAQG standard for the 24 hours concentration. CO was found to be the predominant air pollutant after PM_{10} and followed by O_3, NO_2 and SO_2 which had their average concentration at 0.525 ppm, 0.02 ppm, 0.005 ppm and 0.001 ppm. The averaged concentration of these air pollutants were also below the MAAQG standard limit as well.

Table 2. Overall data on air quality at Langkawi Island.

Parameters	Mean	Standard deviation	Min	Max	Skewness
PM_{10} (μg/m^3)	39.259	16.016	7.52	386.94	5.506
CO (ppm)	0.525	0.207	0	3.2050	1.238
NO_2 (ppm)	0.005	0.004	0	0.0677	2.269
SO_2 (ppm)	0.001	0.001	0	0.0118	2.020
O_3 (ppm)	0.02	0.014	0	0.147	0.949

Figure 2 shows that the long-term monthly records of PM_{10} data. The emission of PM_{10} was found higher in February to March and June to July. Such observations were plausible to the severe haze episodes originated from uncontrolled forest fires [12] in Indonesia particularly to 2005. The concentration of CO, SO_2 and NO_2 was due to the emissions of motor vehicles used by local as well as tourists. Meanwhile, the O_3 is formed from photochemical reactions of NO_2 that involved sunlight and heat [13].

The profile of relative humidity, CO and SO_2 possessed with similar pattern profiling throughout the years. For example, the increased in relative humidity would decreased the temperature of Langkawi Island from September to December because of the monsoon season.

IOP Conf. Series: Materials Science and Engineering **551** (2019) 012123 doi:10.1088/1757-899X/551/1/012123

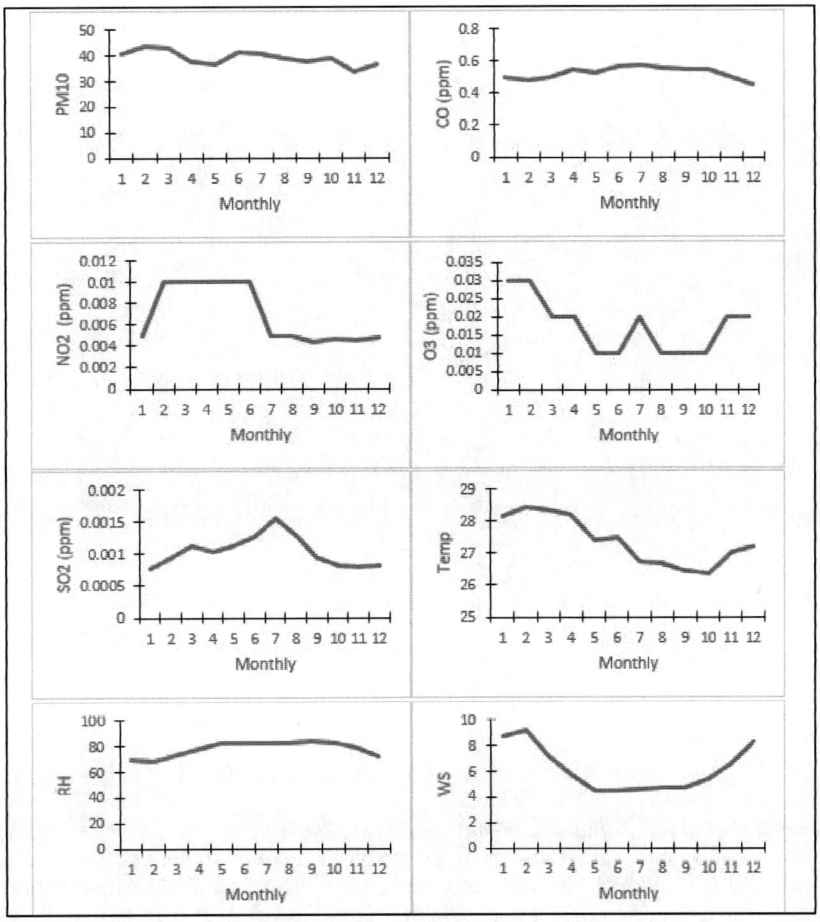

Figure 2. Monthly trends of air quality in Langkawi Island.

3.2. Daily Trends

Figure 3 shows the hourly air data quality of each pollutants in Langkawi Island. There was few peaks on each air pollutants' concentration within 24 hours trend. It is clearly exhibited that the concentration of PM_{10}, NO_2 and SO_2 were increased from 7 am to 9 am. This observation can be attributed by traffic congestion as people were going out to work and children going to school during these hours. Similar pattern was observed in the evening session as well. There was small increased in concentration of PM_{10} and NO_2 from 5 pm to 6 pm as people were going back from work and school. The concentration of O_3 were highest from 12 pm to 5 pm due to the hot weather and the existence of sunlight during the daylight. The concentration of NO_2 and SO_2 has the same pattern where it has an increased and decreased phase at specific time.

Uniquely, the meteorological factors such as temperature and wind speed possessed the same pattern of profiling where it increased slowly from 9 am and decreased slowly to 6 pm because of the sun's warming effect that directly influencing the increased in wind speed. On contrary, the relative humidity decreased with the increased in the temperature and wind speed.

IOP Conf. Series: Materials Science and Engineering **551** (2019) 012123 doi:10.1088/1757-899X/551/1/012123

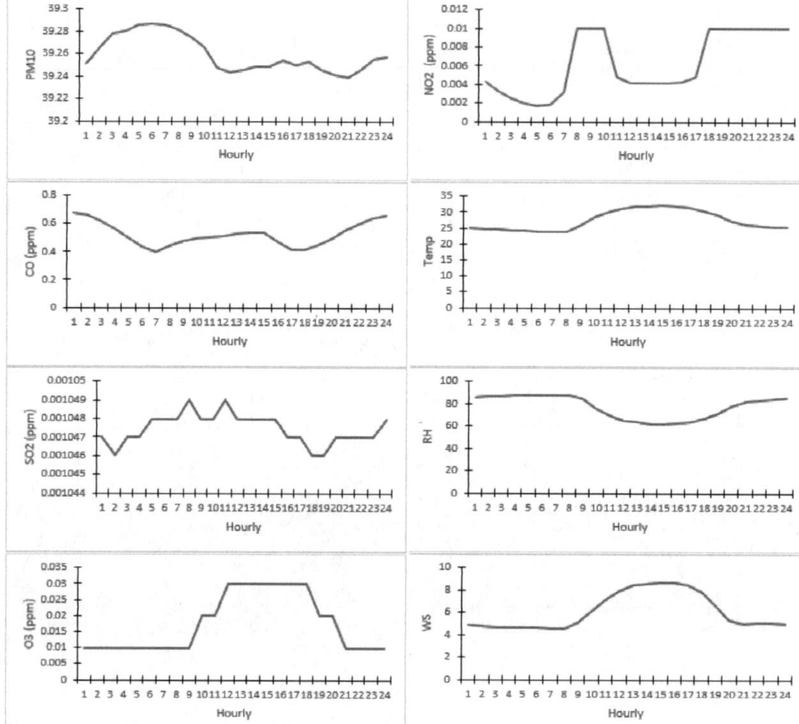

Figure 3: Hourly trends for air quality in the Langkawi Island.

3.3. Correlation between air pollutants and meteorological factors

Table 3 shows detailed interaction and correlation between the air pollutants. Interestingly, all air pollutants have a positive correlation with each other. PM_{10} has a positive low correlation with CO, NO_2, SO_2 and O_3 which were ($r = 0.283$, $p < 0.01$), ($r = 0.480$, $p < 0.01$), ($r = 0.070$, $p < 0.01$) and ($r = 0.215$, $p < 0.01$), respectively. The correlation of PM_{10} with O_3 became positive was due to the hot weather conditions that induced the formation of O_3 [14]. Although CO has a positive low correlation with NO_2 and SO_2 at ($r = 0.166$, $p < 0.01$) and ($r = 0.144$, $p < 0.01$), it also has a negative low correlation with O_3 ($r = -0.120$, $p < 0.01$). A negative relationship suggested the likely hydrocarbon oxidation which consumes O_3 and subsequently producing CO [15].

Table 3: Correlation between parameters.

Parameter	PM_{10}	CO	NO_2	SO_2	O_3
PM_{10}	1				
CO	0.283**	1			
NO_2	0.480**	0.166**	1		
SO_2	0.070**	0.044**	0.008*	1	
O_3	0.215**	-0.120**	-0.081**	-0.046**	1

**. Correlation is significant at the 0.01 level (2-tailed).
*. Correlation is significant at the 0.05 level (2-tailed).

Meteorological factors have a significant correlation with all the pollutants as presented in Table 4. PM_{10} has a positive low correlation with temperature and wind speed with ($r = 0.076$, $p < 0.01$) and ($r = 0.069$, $p < 0.01$) while posed a negative low correlation with relative humidity ($r = -0.119$, $p < 0.01$).

High relative humidity caused the wash-out processes in the atmosphere according with the low numbers of rain occasions [16,17]. Therefore, there is a negative correlation between PM_{10} and relative humidity. In addition, temperature also has a negative correlation with relative humidity ($r = -0.748$, $p < 0.01$) but has a positive linear correlation with wind speed ($r = 0.411$, $p < 0.01$). The wind speed correlation with PM_{10} and temperature possessed positive low correlation at ($r = 0.069$, $p < 0.01$) and ($r = 0.411$, $p < 0.01$), respectively.

Table 4: Correlation between meteorological factors.

Parameter	PM_{10}	T	RH	WS
PM_{10}	1			
T	0.076**	1		
RH	-0.119**	-0.748**	1	
WS	0.069**	0.411**	-0.049**	1

**. Correlation is significant at the 0.01 level (2-tailed).

4. Conclusions

In conclusions, the average concentrations of air pollutants in Langkawi Island were still not exceed the standard limit of MAAQG. All the meteorological factors (temperature, relative humidity and wind speed) were significantly correlated with the concentrations of air pollutants. PM_{10} turned out to be the dominant type of air pollutants in Langkawi Island which followed by CO, O_3, NO_2 and SO_2 where the hot weather conditions contributed to such emission of PM_{10} concentration. In addition, the wind speed and temperature influenced the distribution of PM_{10} concentration within the vicinity due to south-west monsoon season. Such assessment's findings are imperative in protecting the human health as well as environment.

Acknowledgements

The author would like to thank to Universiti Teknologi MARA Pulau Pinang and the Department of Environment (DOE) for providing the air quality monitoring data in this work.

References

[1] https://news.mongabay.com/2005/08/indonesian-forest-fires-again-cause-haze-in-malaysia/. Retrieved on 16-Oct-2018.
[2] Wen Y S *et al* 2016 *Nova J. Eng. App. Sci.* **5** 1.
[3] https://www.malaysia-traveller.com/air-pollution.html. Retrieved on 16-Oct-2018.
[4] http://sci219haze.blogspot.com/2013/08/how-is-haze-monitored.html?m=0. Malaysia Environemtal Quality Report (2017). Retrieved on 16-Oct-2018.
[5] Norazian M N et al 2011 Physics Procedia 22 318.
[6] Noor N M et al 2011 Australian Journal of Basic and Applied Sciences 5 2796.
[7] Noor N M et al 2015 Revista de Chimie 66 1443.
[8] http://www.mplbp.gov.my/en/citizens/services/tourism/page/0/1. Retrieved on 22-Oct-2018.
[9] http://www.astroawani.com/berita-malaysia/lima-juta-pelancong-dijangka-kunjungi-langkawi-pada-2020-185830. Retrieved on: 22-Oct-2018.
[10] Ul-Saufie A Z *et al* 2013 *Atmos. Environ.* **77** 621.
[11] Awang M. B. *et al* 2000 *Respirol.* **5**, 183.
[12] Atkinson R W 2000 *Atmos. Environ.* **34** 2063.
[13] Atkinson R W *et al* 2012 *Environ. Health Perspect.* **120** 1411.
[14] Real E E 2008 *Atmos. Chem. Phys.*, **8** 7737.
[15] Azmi S Z *et al* 2009 *Air Qual. Atmos.* Health **3** 52.
[16] Rahman S R A *et al* 2015 *World Environ.* **5** 1.

Bioynthesis of silver nanoparticles using *Graptophyllum pictum* leaves extracts

Mohamed Syazwan Osman[1,2], Muhammad Ismail[1], Khairunnisa Khairuddin[1], Najwa Azis[1], Nur Syifa Salim[1], Nur Aqilah Zaini[1], and Norfezah Md Noor[3]

[1]Faculty of Chemical Engineering, Universiti Teknologi MARA, 13500 Permatang Pauh, Penang, Malaysia.
[2]Molecular System Design & Optimization Research Group (MSORG), Faculty of Chemical Engineering, Universiti Teknologi MARA, 13500 Permatang Pauh, Penang, Malaysia.
[3]Faculty of Hotel Management & Tourism, Universiti Teknologi MARA, 13500 Permatang Pauh, Penang, Malaysia.

E-mail: syazwan.osman@uitm.edu.my

Abstract. Silver nanoparticles (AgNPs) have found significant application in industry especially in medical and catalysis field. Considerable attention has been made to synthesis AgNPSs via various routes. However, in recent years, focus have been shifted to synthesis AgNPs using more environmentally friendly route. Attempts to synthesis AgNPs by using plant extracts as potential complex reducing agent have found significant interest to date. Hence, this work will explore the feasibility to synthesis AgNPs by using locally available *Graptophyllum pictum* leaves (*puding hitam*). The extracts were evaluated on the their total phenolic content to shed potential insights on the mechanism prior AgNPs synthesis. The feasibility of the AgNPs were evaluated by using UV-Vis and Transmission electron microscopy (TEM). In conclusion, *Graptophyllum pictum* leaves (*puding hitam*) could be a viable alternative for the environmentally friendly synthesis of AgNPs.

1. Introduction

The mainstream preparation of functional nanomaterials employs physical, chemical and mechanical techniques. However, the biotechnological approach for the synthesis of metallic NPs is becoming more important and it will hopefully overcome the abovementioned disadvantages of non-biotic methods. There is considerable recent demand for change in the way biosynthesized NPs are applied. The current literature not only suggests new protocols for biosynthesis of NPs which use various types of biomass but it also recommends immediate testing of developed nanostructures in numerous applications – most frequently in catalysis or as antimicrobial agents [1-3]

The biosynthesis of NPs is gaining greater attention; especially because the synthesis protocols are user-friendly and there is no special equipment, financially demanding devices or harmful chemicals required. The main advantages of NPs biosynthesis are; only a metal precursor is needed in most cases; reduction or chemical transformation and eventual NPs embedding occurs in one step; is mediated only by the (bio)chemicals contained in the biomass.Therefore, using plant extracts as a reducing agent in the synthesis of nanoparticles has received increasing interest. Being readily scalable and biocompatible, along with the availability of simple protocols, makes this approach a favorite procedure for the synthesis of nanoparticles. Moreover, compared with the plant in vivo synthesis of nanoparticles, the bio-reduction of metal ions by plant extracts provides a more flexible control over the size and shape of the nanoparticles [4-5] Generally, biochemical reduction is initiated by addition of the metal salts solution to the plant extracts, and is followed by storing the mixture at room temperature with or without agitation until the reaction is complete (within minutes) depending on the nature of the plant extract. Color changes in the reaction mixture indicate the formation of the nanoparticles. For example, AgNPs solutions appear dark (yellow-brown) in color because of the surface plasmon resonance phenomenon. Plus, the formation of the nanoparticles can be confirmed by UV–visible spectrophotometer.

Graptophyllum pictum is a traditional medicine plants from Papua, Indonesia [7] containing alkaloids [7]. Its secondary metabolite contents act as an alternative base source and capping agent in MNPs synthesis as hazardous chemicals substitute. In previous study, *Graptophyllum pictum (GP)* leaves extract was reported contains alkaloids and flavonoids [7]. Alkaloids as a weak base are needed in MNPs preparation, and flavonoids have hydroxyl group to prevent MNPs agglomeration [7]. Both of functional materials are contained in GP as base source and capping agents.

2. Methodology

2.1. Materials
Silver nitrate ($AgNO_3$) was purchased from Fisher Scientific and used as received without further treatment. *Graptophyllum pictum* (GP) leaves were purchased from a local supplier. The leaves were washed thoroughly with tap water to remove dirt and washed again with distilled water before being dried in oven at 40 °C for 48 hours. The dried MCV leaves were grinded into powder by using vertical pulverising machine (Hong Chun model RT-34) and stored at room temperature for further use. All the aqueous solutions were prepared by MilliQ water (18.2 Ωcm^{-1}) [8].

2.2. Preparation & characterization of MCV leaves extract and biosynthesis of silver nanoparticles
Dried GP leaves (0.5 g) was boiled in 20 mL deionized water at 100 °C for 30 min and then cooled to room temperature (30 °C). The final obtained extract was filtered through filter paper and used as reducing/stabilizing agents for the following experiments. In a scintillating vials, 20 mL of the GP leaves extract was reacted with 10 mM of $AgNO_3$ solution at room temperature under stirring conditions of 250 rpm.The colour change of the reaction was observed and the time taken for the changes was noted. The solution colour changes immediately from dark brownish to purple red colour indicating the formation of silver nanoparticles (Ag-NP). The Ag-NPs solution obtained was kept safely at 4 °C. Concentration of the $AgNO_3$ as precursor were varied from 10-50 mM to understand the influence of it from colloidal stability perspective. The Ag-NPs were then characterized using TEM for its morphology and UV-Vis to monitor the AgNP formation kinetics. Dynamic light scattering (DLS) method by using Malvern ZS were employed to study its colloidal stability [8].

3. Results and Discussion

3.1. Ag-NPs visual observation and structural morphology
Gross examination of the potential Ag-NPs formation were performed. After 6 hours of reaction, the solution color changes from dark brownish to purple red indicating potential reaction occur (Figure 1a). This observation were in agreement with findings from Shakeel and co-workers[4]. The sample from 50 mM $AgNO_3$ precursor concentration were then subjected to structural investigation with TEM (Figure 1b). It was observed the formation of AgNPs with satisfactorily shperical shape with approximate diameter around ~68 nm (based on N=30 AgNPs). This finding were consistently with other finding from Tippayawat and co-worker using aloe vera as reducing agent [1].

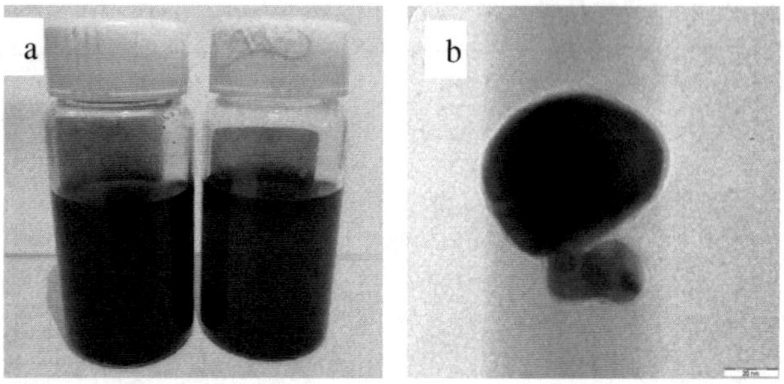

Figure 1. (a) left vial- GP leaves extract before reaction; right vial – AgNPs at the end of reaction (~6hrs) (b) TEM micrograph of AuNPs derived from GP leaves extract.

3.2. UV-Vis spectrophotometry

Parallel changes were observed when $AgNO_3$ concentration were subjected to different concentration range from 10 mM- 50 mM respectively. The experiment were conducted at room temperature. The interaction at room temperature (25 °C) has been reported to generate stable AgNPs [2]. It was observed that there is slight peaks consistently at around 450 nm correspond to AgNPs surface excitation band (Figure 2). Upon incremental of the $AgNO_3$ concentration, the intensity increase correspondingly. UV–Vis absorption spectrum of the reaction solution approved the formation of silver nanoparticles from silver ions with the obtained peak at 450 nm [6]. The nature and characteristics of nanoparticles can be detected from SPR peak. Spherical and uniform nanoparticles in the solution have a single SPR band in the absorption spectra. In other words the number of peaks develops as anisotropy increases [5]. These results consistent with the finding from TEM morphology above. So, it is concluded, the biosynthesized Ag NPs are satisfactorily spherical in shape.

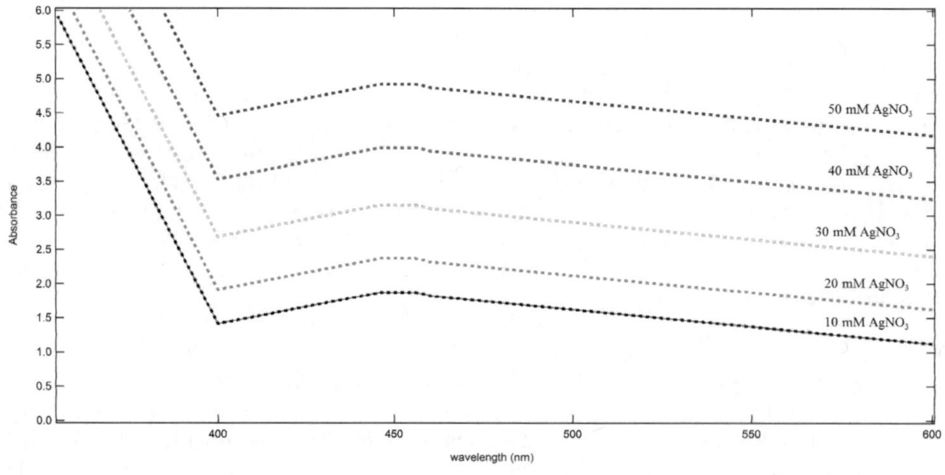

Figure 2. Spectrophotometric observation of AgNPs samples with different initial concentration of $AgNO_3$.

3.3. DLS measurement

The aim of dynamic light scattering measurements was carried out to determine the size distribution and polydispersity of produced Ag-NPs with different pH. This was clearly elucidated by the intensity obtained during the DLS measurements. From the Table 1 shown below, there sre significant differences when different concentration of $AgNO_3$ been employed. Hydrodynamic size to indicate colloidal size of Ag-NPs whereas zeta potential suggested the stability of the colloidal Ag-NPs respectively. It was suggested that higher concentration of $AgNO_3$ as precursor will result higher stability of the nanoparticles. For an instance, 50 mM $AgNO_3$ results in hydrodynamic diameter size around ~120 nm and zeta potential reading of -40 MV. The indicator whether the colloids are stable or not is based on zeta potential value above +/- 30 mV [8]. The Ag-NPs samples were then be subjected to same measurement after 7 days to gauge their long term stability and the same trends were observed. Even though the hydrodynamic size significantly increased across all samples potentially due to particles agglomeration, is still within threshold of colloidal stability for sample 50 mM. Further optimization are required for other process factors such as initial pH, reaction temperature and time in order to understand the behavior of the synthesis better [1].

Table 1. DLS measurement of hydrodynamic size and zeta potential of AgNPs derived from GP leaves extracts

Sample ID	**Hydrodynamic size (nm)**		**Zeta Potential (mV)**	
	24 hours	**7 days**	**24 hours**	**7 days**
10 mM	311.5	678.5	-18.5	-2.5
20 mM	256.4	564.6	-21.6	-11.7
30 mM	180.8	496.2	-27.8	-17.6
40 mM	145.7	386.5	-35.1	-25.6
50 mM	119.5	245.8	-40.6	-33.5

4. Conclusions

From the research, we can conclude several conclusions which are;

i. Ag-NPs derived from GP leaves extracts were synthesized successfully by facile biological method.
ii. The morphology and surface plasmon resonance behavior of the as-made Ag-NPs were characterized.
iii. Colloidal stability of the as made Ag-NPs for different precursor initial concentration were investigated.

Acknowledgements

This work was financially and technically supported by Faculty of Chemical Engineering, Universiti Teknologi MARA cawangan Pulau Pinang

References

[1] Kratosovaa G *et al* 2019 *Biotechnol. Advan.* **37** 154
[2] Ajitha B. *et al* 2016 *J. Mol. Liq.* **219** 474
[3] Nadhirah M. *et al* 2014 *Adv. Mater. Res.* **832** 350
[4] Ahmed S. *et al* 2016 *J. Radiation Res. App. Sci.* **9** 1
[5] Fahimirada S *et al* 2019 *Ecotoxicol. Environ. Safety* **168** 260
[6] Pirtarighata S. *et al* 2019 *Mater. Sci. Eng. C* **98** 250

[7] Sari I.P. *et al* 2017 *Mater. Sci. Eng.* **191** 1
[8] Osman M.S. *et al* 2018 Material Today Proc. **5** 22050

IOP Conf. Series: Materials Science and Engineering **551** (2019) 012125 doi:10.1088/1757-899X/551/1/012125

Synthesis of dimercaptosuccinic acid (DMSA) coated magnetic nanoparticles: feasibility on dye decolorization and heavy metal adsorption

Mohamed Syazwan Osman[1,2], Noorfarah Asyiqin Noorossaidy[1], Najwa Azis[1], Nur Syifa Salim[1], Nur Aqilah Zaini[1], Nur Alwani Ali Bashah[1]

[1]Faculty of Chemical Engineering, Universiti Teknologi MARA, 13500 Permatang Pauh, Penang, Malaysia.
[2] Molecular System Design & Optimization Research Group (MSORG), Faculty of Chemical Engineering, Universiti Teknologi MARA, 13500 Permatang Pauh, Penang, Malaysia.

E-mail: syazwan.osman@uitm.edu.my

Abstract. As highlighted in UN sustainable development goals 2030 report, one of the important challenge in future is to provide clean water. Thus, enormous efforts have been made to improve water quality from recalcitrant pollutants such as organic dyes, heavy metal and others. Magnetic nanoparticles is one of viable candidate to degrade the above pollutants due to its unique properties such bifunctional (catalytic and adsorption) capability & ease for magnetic separation. However, one of the main challenge is to maintain its colloidal stability to exploit its advantages for water remediation. Thus, this work explored its colloidal stability by coating it with heavy metal chelating agent namely Dimercaptosuccinic Acid (DMSA). The resultant clustered magnetic nanoparticles were evaluated its feasibility by using Rhodamine B dye and lead (Pb) heavy metal as model system. It is revealed that DMSA-coated magnetic nanoparticles is a viable candidate for this model system.

1. Introduction

Due to emerging concern in protecting the public health sustainability, water security is crucial towards achieving this goal. Enormous efforts have been made to ensure water cleanliness via various technologies. In the effort to mitigate water contamination issues, there is a growing interest on using magnetic nanoparticles (MNPs) as an adsorbent and/or catalyst for pollutant removal [1]. Magnetite (Fe_3O_4), maghemite (γ-Fe_2O_3), hematite (α-Fe_2O_3), and goethite (α-FeOOH) [2-3] are some of the most commonly used MNPs. On top to the exceptionally high surface to volume ratio that provides large area of loading sites for pollutant removal, their intrinsic magnetic property allows further separation of this nanoagent from the treated water.

The most significant challenge related to the formulation of MNP suspension is the occurrence of particle agglomeration. Agglomeration is a process in which the individually dispersed particles are clumped into a single cluster [1]. Agglomeration starts with the formation of doublets, which later on grow into triplets, quartet, and so forth. Therefore, the number concentration of the primary particles decreased accompanying with the increase in the number concentration of larger agglomerates, as the time progresses [6] One of the strategies to enhance their colloidal stability is by functionalized with polymer so that the MNPs is been clustered to ensure their collective forces intact.

Among the functionalized IONP, those coated with meso-2,3- dimercaptosuccinic acid (DMSA, Figure 1A) present great potential in both target-drug delivery and remediation of heavy metal pollution [4]. DMSA, a dithiol, is a derivative and an analogue of dimercaprol that contains two sulfhydryl groups (–SH) [7]. Oxidation of DMSA molecules around IONP generates a cage of disulfide-cross-linked DMSA molecules around the IONP core, which establishes a negative surface charge of the particles due to an excess of carboxylate groups17 (Figure 1B). The ability of sulfhydryl-containing compounds to chelate metals, e.g. lead, mercury, arsenic, and cadmium, has been well studied. DMSA is not toxic to humans and has been approved for clinical use for chelation therapy [4].

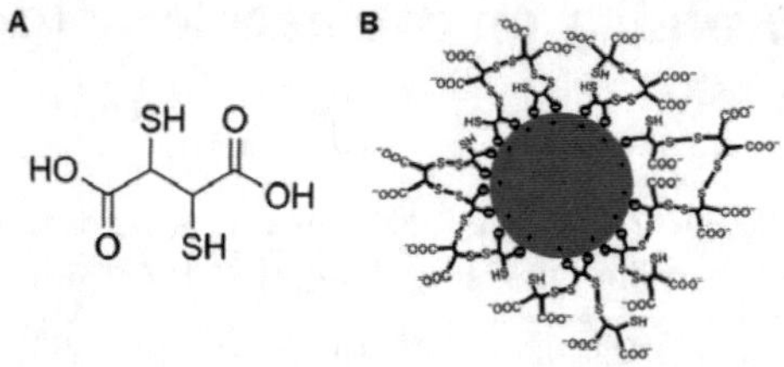

Figure 1. A. Structural formula of meso-2,3-dimercaptosuccinic acid (DMSA); B. DMSA-MNPs: DMSA molecules bind to MNPs surface via their carboxylate groups and form a cage by disulfide bridges.[4]

2. Methodology

2.1. Materials

For the preparation of bare iron oxide nanoparticles, chemicals used were Iron (II) Chloride Tetrahydrate (Merck), Iron (III) Chloride 6-Hydrate (HmbG Chemicals). In order to bind this magnetic nanoparticles, meso-2,3 Dimercaptosuccinic Acid (Sigma Aldrich) used. For heavy metal adsorption, lead nitrate Pb $(NO_3)_2$ (Sigma Aldrich) was used while for dye degradation, Rhodamine B Base (Sigma Aldrich), Sodium Hydroxide (Sigma Aldrich) and Hydrogen Peroxide (Systerm) were used as it is.

2.2. Synthetic procedures of MNPs

MNPs were synthesized by using co-precipitation method to obtain the magnetic iron oxide nanoparticles [4]. The preparation steps of iron oxide nanoparticles is by mixing the ferric chloride and ferrous chloride all together with the molar ratio of 2:1. Then the distilled water added to the solutions containing both Fe^{2+} and Fe^{3+} and heat up to 50°C for 10 min. After heating, the solutions precipitated by ammonia. In this moment, black color particles being precipitated which denoted iron oxide nanoparticles. These particles separated by using a strong magnet and then wash with distilled water for a few times. The particles is left in the water in order to make it in the state of colloidal.

2.3. Synthesis of MNPs-DMSA

The MNPs were coated with DMSA. For this 0.13 g (0.7 mM) of DMSA were dissolved in 150 mL double-distilled water by stirring at 50 °C and then added to 100 mL 40 mM iron in form of MNPs during vigorous stirring. After 30 min at room temperature, the particulate content was separated by centrifugation at 800g for 5 min. After the removal of the supernatant the particles were resuspended in 80 mL double-distilled water. The pH of the dispersion was adjusted to 10 with NaOH and then to 7.4 with HCl. The resulting dispersion was finally filtered through a 0.2 μm filter [4].

2.4. Nanoparticles characterization and colloidal stability test

Nanoparticles structural investigation was carried out using a transmission electron microscopy (EFTEM, Libra 120 Carl Zeiss) operated at 100 kV. Micrographic grids were prepared by placing a drop of diluted sample dispersion in ethanol onto a carbon-coated copper grid and dried at room temperature. On the other hand, Dynamic light scattering (DLS) method by using Malvern ZS were used to study the hydrodynamic diameter of the as-made nanoparticles.

2.5. Feasibility for the heavy metal adsorption and dye decolorization
Solutions containing Pb (II) were prepared at concentration 50 ppm. Then, 1 g/L of the MNPs cated DMSA was added to 50 mL lead aqueous solution under vigorous stirring of 250 rpm. The experiments were performed in triplicate. After a specified time, the MNPs coated DMSA were magnetic separated and the initial and residual concentrations of arsenic were measured by inductively coupled plasma-optical emission spectroscopy (ICP-OES) (Perkin Elmer Optima DV 7000).The RHB degradation experiments were carried out by which the 1 g/L MNPs coated DMSA were added into 50 mL of a 100 mg/L RHB solution in the presence of H_2O_2 as electron donor, and the suspension was shaken in a thermostated shaker at stirring rate of 250 rpm under room temperature. The experiments were performed in triplicate. After the reaction, the microcapsules were separated using an external neodymium boron ferrite (NdBFe) cylindrical magnet. The concentration of RHB was monitored spectrophometrically recorded using a Perkin Elmer Lambda 25 UV–vis spectrophotometer, with a maximum absorbance wavelength for RHB at 554 nm

3. Results and Discussion

3.1. Nanoparticles charactetization & colloidal stability
Nanoparticles morhplogy were characterized by using TEM as shown in Figure 2 revealed that the MNPs coated DMSA were clustered into smaller groups. By rough dimension, the size of nanocluster MNPs-coated is approximately ~60nm. It was observed, nanocluster MNPs-coated were encapsulated by a layer of polymer indicating potential successful functionalization and stabilization of the nanoparticles. The potential mechanism for the colloidal stability is the steric repulsion in which DMSA inhibit attraction between the nanoparticles [1]. Further investigation using DLS measurement, further confirmed the hydrodynamic diameter of MNPs coated DMSA ~113.6 nm much smaller to as-made MNPs by 3-fold. This might be due to MNPs tendency to agglomerate due to high magnetic dipole moment between particles [1]. In addition, zeta potential value of -37.8 mV after 7 days indicated the stability of MNPs coated DMSA samples (Table 1). The agglomeration of as-made MNPs were confirmed by the incremental value of zeta potential after 7 days of synthesis.

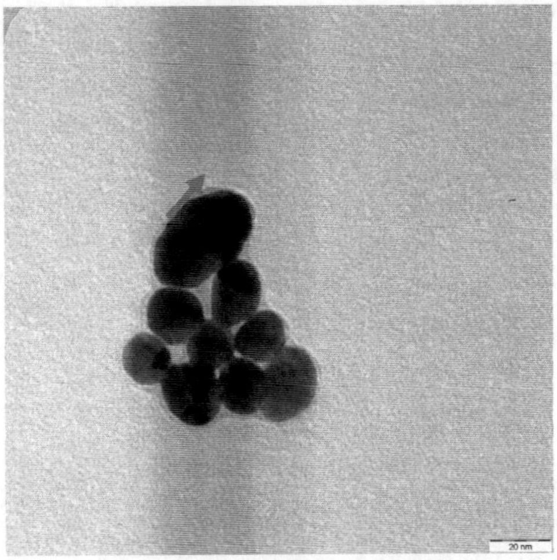

Figure 2. TEM micrograph of MNPs coated DMSA. DMSA successfully clustered the MNPs

IOP Conf. Series: Materials Science and Engineering **551** (2019) 012125 doi:10.1088/1757-899X/551/1/012125

Table 1. DLS measurement of hydrodynamic size and zeta potential

Sample	Hydrodynamic size (nm)		Zeta Potential (mV)	
	24 hours	**7 days**	**24 hours**	**7 days**
Bare MNPs	327.5	548.8	-25.8	-18.4
MNPs coated DMSA	113.6	208.1	-41.7	-37.6

3.2. Heavy metal adsorption feasibility

Heavy metal chelating capacity for the MNPs coated DMSA were investigated using lead (II) as model system. The adsorption process were performed for 6 hours experiment as presented in Figure 3. It was clearly observed the adsorption capacities of MNPs coated DMSA were at maximum around 130 mg/g after 3.5 hours adsorption time in comparison just 80 mg/g for as-made MNPs respectively. The sorption binding capacity of MNPs coated DMSA is far more superior might be due to additional availability of ligand shells from the DMSA [5]. The availability of active sites and its pH value are crucial in optimizing the process of the sorption. Furthermore, thiol group attached to DMSA is potentially have strong binding afffnity with heavy metals. DMSA also is non-toxic for human being which make it an attractive candidate for MNPs surface functionalization and stabilization [5].

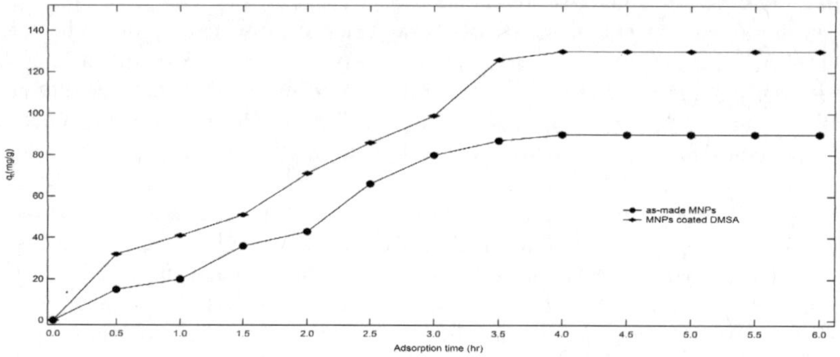

Figure 3. Comparison of adsorption capacities (mg/g) between as-made MNPs and MNPs coated DMSA

3.3. Dye degradation feasibility

One of the advantages of MNPs is their ability to perform bi-functional wich is adsorptive and catalytic properties. With the introduction of hydrogen peroxide as oxidizing agent, MNPs will act as catalyst to degrade the organic contaminants such as dyes. From the experiment using RHB dye as model system as observed in Figure 4, clearly depicted the superiority of the MNPs coated DMSA compared to as-made MNPs to catalytically degrade the dye. This might be due to colloidal stability behavior MNPs-DMSA which render to availability of active sites for the oxiding agent and pollutant to react onto surface of catalyst [1]. Interesting to point out, the imrpovement is not more than 0.1 of C/C_0 which potentially indicated inhibitory behavior of DMSA for the RHB dye to penetrate onto MNPs surfaces which explain less dramatic improvement of the degradation. However, this hypothesis need to be confirmed further.

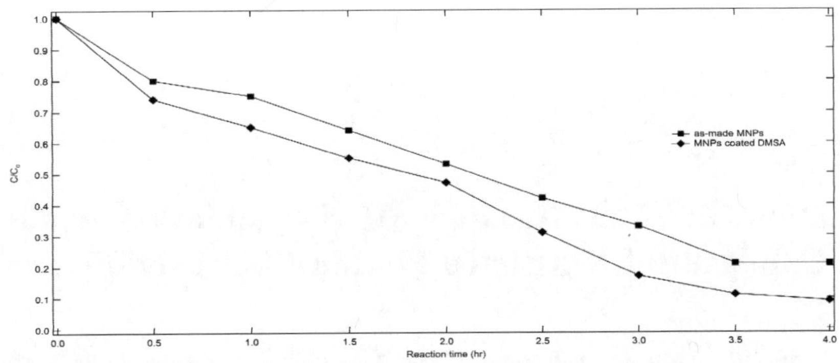

Figure 4. Comparison of catalytic performance between as-made MNPs and MNPs coated DMSA

4.Conclusions

From the research, we can conclude several conclusions which are;

i. Magnetic nanoparticles coated with DMSA were synthesized successfully by the electrostatic interaction technique
ii. The stability of clustered MNPs-coated DMSA were better compared to as-made MNPs in term of hydrodynamic diameter size after 7 days of synthesis (~208nm)
iii. As the result, lead (II) adsorption and RHB dye decolorization proof that statistically increment of performance were observed by using MNPs-coated DMSA model system.

Acknowledgements

This work was financially supported by IRAGS research grant (600-RMI/DANA5/3/IRAGS(47/2015) from Universiti Teknologi MARA, Selangor and technical support from Faculty of Chemical Engineering, Universiti Teknology MARA cawangan Pulau Pinang

References

[1] Yeap S.P. *et al* 2017 *J Nanopart Res,* **19** 367
[2] Andreu J.S. *et al* 2011 Soft Matter, **7**, 2336–2339
[3] Aredes S.*et al* 2012 *Clean Prod.* **29** 208
[4] Zhang Y.Q. *et al* 2016 *Environ. Sci.: Nano* **3**, 754
[5] Yantasee W. *et al* 2007 *Environ. Sci. Technol.* **41** 5114
[6] Henry C *et al* 2013 *Langmuir* **29** 13694
[7] Valois C.R.A *et al* 2010 Biomaterials **31** 366

IOP Conf. Series: Materials Science and Engineering **551** (2019) 012126 doi:10.1088/1757-899X/551/1/012126

Optimization of Acid Orange II Degradation using $Fe_{3-x}Mn_xO_4$ Catalyst in UV assisted Fenton-like reaction

Elsabba Edwin[1], Nor Aida Zubir[1,2*], Mohamad Khairul Azam Selamat[1], Rasyidah Alrozi[1,2], Norhaslinda Nasuha[1,2], Hawaiah Imam Maarof [2,3]

[1]Faculty of Chemical Engineering, Universiti Teknologi MARA, Cawangan Pulau Pinang, 13500 Pulau Pinang, Malaysia
[2]Hybrid Nanomaterials, Interfaces & Simulation (HYMFAST), Faculty of Chemical Engineering, Universiti Teknologi MARA, Cawangan Pulau Pinang, 13500 Pulau Pinang, Malaysia
[3]Faculty of Chemical Engineering, Universiti Teknologi MARA, Shah Alam, Selangor, Malaysia

E-mail: noraida709@uitm.edu.my

Abstract. In this work, detailed investigation on the optimization of acid orange II (AOII) degradation using $Fe_{3-x}Mn_xO_4$ catalyst in UV assisted Fenton-like reaction was performed. The Central Composite Design (CCD) based on Response Surface Methodology (RSM) with four independent operational parameters (catalyst dosage, oxidant concentration, pH and UV irradiation) and one response (percentage of AOII removal) was applied for statistical and optimization analysis. The maximum AOII removal (~100%) was found at optimal conditions of 0.4 g/L catalyst dosage, 35 mM oxidant concentration, pH 2.6 and 3 x 9W UVC. Interestingly, validation of these optimal conditions was confirmed by the experimental findings of having less than 2% absolute deviation. Such findings confer that the derived empirical quadratic model was in good agreement with the experimental findings.

1. Introduction

Water pollution due to untreated dye-contaminated wastewater from textile dyeing industries posed a serious environmental problem. It has been estimated that 17~20% of industrial water pollution originated from textile [1] industries and approximately 15% of synthetic dyes are lost during the dyeing process [2] which subsequently turns to be dye-contaminated wastewater.

Treatment for wastewater that containing dye is considered to be one of the challenging areas in the environment fraternity. Physico-chemical treatments are effective in the removal of dye but high operational cost and production of secondary sludge limits their application [3,4]. Moreover, biological wastewater treatment is no longer suitable for the treatment of dye-contaminated wastewater due to the recalcitrant and inhibitory nature of these dyes. Hence, advanced oxidation process (AOPs) such as UV assisted Fenton-like reaction have end up as an increasing area of researchers interest for eliminating recalcitrant organic pollutants in wastewater because it is more efficient, non-selective and environmental friendly [5,6].

In UV assisted Fenton-like reaction, strong oxidizing agent of hydroxyl radicals (•OH) [5,7] is produced in the reaction mixture of during catalysis. The strong oxidizing and reactivity of •OH radicals may turn a complex recalcitrant organic pollutant into harmless compounds such as carbon dioxide (CO_2) and water (H_2O) [11]. For instance, the use of manganese substituted magnetite [8,9] as heterogeneous catalyst in UV assisted Fenton-like reaction have been highlighted recently due to its notable enhancement in both of catalytic performance and actitivity during catalysis.

In enhancing the overall catalytic performance of dye degradation in UV assisted Fenton-like reaction; multiple operational parameters are required to be optimized at the optimal values. In past, many researchers used one-factor-at-a-time (OFAT) approach [8-12] for such pursposes. However, this approach possessed of several limitations such as i) unable to provide detailed interaction between operational parameters towards the responses, ii) requires numerous experimental runs and iii) time-consuming and costly. In circumventing these limitations, an interactive approach known as response surface methodology (RSM) has been introduced to facilitate the detailed global interactions between operational parameters in process optimization. Therefore, the present work focuses on the determination of optimal operational conditions particularly on acid orange II degradation using manganese substituted magnetite ($Fe_{3-x}Mn_xO_4$) catalyst in UV assisted Fenton-like reaction.

2. Methodology

2.1. Synthesis of manganese substituted mangnetite ($Fe_{3-x}Mn_xO_4$) catalysts

The $Fe_{3-x}Mn_xO_4$ catalysts at x value of 0.3 were synthesized through precipitation-oxidation method. First, 40 mL of 0.05 M $FeCl_3.6H_2O$ was mixed with 17.4 mL of 0.05 M $MnCl_2$ solution. The mixture was heated at 90°C using double boiling technique whilst continuously stirred for 10 min. Then, 1 M NaOH was added dropwise into the mixture until pH 10. The mixture was aged for another 30 min under constant stirring at 90°C. Then, the mixture was let to be cooled to room temperature. The resulting precipitates were centrifuged and washed several times with deionized water and dried at 90°C for 24h.

2.2. Optimization of UV assisted Fenton-like reaction using CCD-RSM

The photocatalytic activity of manganese substituted mangnetite ($Fe_{3-x}Mn_xO_4$, x=0.3) catalysts were tested in heterogeneous UV assisted Fenton-like reaction using acid orange II (AOII) as the model pollutant. Central composite design (CCD) in RSM was applied as experimental design to determine the optimized operational parameters of UV assisted Fenton-like reaction. This experimental design was mathematically modelled and optimized by Design Expert 11.0.5.0 software (State-Ease, Inc.). A five-level-four-factor CCD was chosen with rotatable α value fixed at 2 [13].

Four operational parameters which also known as independent variables have been identified as the pH of solution, oxidant (H_2O_2) concentration, irradiation intensity and catalyst dosage. Detailed experimental range and level or each independent variable are presented in Table 1. The response variable of this experimental design was the percentage of AOII removal. A total number of 30 experimental runs were performed comprising 16 factorial points, 8 axial points and 6 replications at the center point. Experimental results were analyzed using analysis of variances (ANOVA) in interpreting the relationships between four independent variables and response variable of the experimental design. Data having p-value lower than 0.05 was considered statistically significant.

Table 1. Independent variables and coded levels for CCD-RSM.

Parameter	Unit	Code	- α	-1	0	+1	+α
pH		A	2.0	2.5	3.0	3.5	4.0
Oxidant concentration	mM	B	5	15	25	35	45
Catalyst dosage	g/L	C	0.2	0.4	0.6	0.8	1.0
UV Irradiation intensity	Watts	D	0	1 x 9	2 x 9	3 x 9	4 x 9

The heterogeneous UV assisted Fenton-like reaction was performed using a custom made photoreactor set-up equipped with 4 x 9W UV-C lamp (Sankyo-GPX9, λ_{max} of 257.3 nm). The reaction was initiated by switching on UV lamps after the addition of specific H_2O_2 concentration into 100 mL AOII solution consisting of specific catalyst dosage and pH accordance to the CCD-RSM experimental design (Table 1). The reaction was prolonged for 180 min at stirring rate of 200 rpm. Sampling solution was filtered through 0.2 μm syringe filters and the filtrates concentration were analyzed using portable spectrophotometer at λ_{max} of 484 nm. The percentage of AOII/color removal can be calculated using Equation 1:

$$\text{Percentage of removal, } Y_1\ (\%) = \frac{C_o - C_t}{C_o} \times 100 \qquad (1)$$

where C_o is the initial concentration of AOII and C_t is the final AO7 concentration at 180 min of reaction.

3. Results and Discussion

3.1. ANOVA analysis

Table 2 summarizes the detailed set of CCD-RSM's experimental design together with its response variable (percentage of AOII/color removal). In this experimental design, the highest and lowest removal of AOII were found to be at 99.83% and 22.99%, respectively. The statistical analysis of variance (ANOVA) was perfomed based on the acquired response variable and presented in Table 3. The ANOVA analysis exhibited that the derived quadratic model was statistically significant with F-value and p-value of 30.40 and 0.0001, consecutively. The lack of fit p-values of 0.3914 implied that the lack of fit of this model was not significant relative to the pure error. The non-significant term indicated that the lack of fit is good and acceptable. Moreover, the model exhibits a decent regression coefficient (R^2) of 0.966 which suggested that the acquired experimental data were well-fitted to the derived quadratic model that could provide a good estimation of predicted response (percentage of AOII removal) as depicted in Fig. 1

Interestingly, small p-value of less than 0.05 were found in the operational parameter of pH (A), catalyst dosage (C) and UV irradiation (D) that signify these pameters to be significant in the regression of empirical quadratic model (equation 2). The coefficient with one coded factor represent the effect of the specific operational parameter, whilst the coefficients with two coded factors as well the second-order terms represent the direct interaction between operational parameters and its quadratic effect, respectively [14].

$$Y_1\ (\%) = 47.69 - 21.06A - 0.19B - 8.72C + 2.98D + 0.05AB + 12.96AC - 0.06AD + 0.95BC + 1.43BD - 0.49CD + 2.95A^2 - 1.68B^2 + 7.49C^2 + 0.85D^2 \qquad (2)$$

Table-2. Experimental design of CCD-RSM with experimental values of AOII removal.

Run	Factor 1 A: pH	Factor 2 B: Oxidant concentration (mM)	Factor 3 C: Catalyst Dosage (g/L)	Factor 3 D: UV Radiation (W)	Response AOII/ Color Removal (%)
1	3.5	15	0.4	1	32.68
2	3.0	25	0.6	0	35.67
3	2.0	25	0.6	2	98.39
4	3.0	45	0.6	2	37.02
5	3.0	25	0.2	2	99.77
6	3.5	15	0.8	1	40.97

7	2.5	15	0.4	1	99.43
8	2.5	15	0.4	3	99.77
9	3.0	25	0.6	2	53.01
10	2.5	15	0.8	1	59.48
11	2.5	35	0.4	3	99.65
12	2.5	35	0.8	1	55.26
13	3.0	25	0.6	4	63.65
14	3.5	35	0.8	3	42.93
15	2.5	35	0.8	3	64.56
16	3.5	35	0.4	1	22.99
17	4.0	25	0.6	2	17.75
18	3.0	25	0.6	2	47.28
19	3.0	25	0.6	2	51.09
20	3.5	35	0.8	1	45.76
21	3.5	35	0.4	3	35.9
22	3.0	25	1.0	2	52.64
23	3.0	25	0.6	2	36.93
24	3.0	25	0.6	2	47.11
25	3.0	25	0.6	2	50.72
26	2.5	15	0.8	3	58.27
27	3.5	15	0.4	3	31.34
28	2.5	35	0.4	1	99.83
29	3.0	5	0.6	2	42.06
30	3.5	15	0.8	3	39.45

Table-3. ANOVA analysis for percentage of AOII removal (Y_1).

Source	Sum of Squares	df	Mean Square	F-value	p-value	
Model: Quadratic model	17312.00	14	1236.57	30.40	< 0.0001	**Significant**
A-pH	10647.52	1	10647.52	261.78	< 0.0001	
B-Oxidant concentration	0.8778	1	0.8778	0.0216	0.8852	
C-Catalyst Dosage	1823.00	1	1823.00	44.82	< 0.0001	
D-UV Irradiation Intensity	212.59	1	212.59	5.23	0.0372	
AB	0.0390	1	0.0390	0.0010	0.9757	
AC	2686.09	1	2686.09	66.04	< 0.0001	
AD	0.0663	1	0.0663	0.0016	0.9683	
BC	14.42	1	14.42	0.3546	0.5604	
BD	32.86	1	32.86	0.8079	0.3829	
CD	3.99	1	3.99	0.0981	0.7584	
A^2	239.19	1	239.19	5.88	0.0284	
B^2	77.37	1	77.37	1.90	0.1881	
C^2	1537.42	1	1537.42	37.80	< 0.0001	
D^2	19.84	1	19.84	0.4878	0.4956	
Lack of Fit	444.78	10	44.48	1.35	0.3914	**Not significant**

Source	Sum of Squares	df	Mean Square	F-value	p-value
Std. Dev.	6.38				
Regression Coefficient, R^2	0.9660				
Adjusted R^2	0.9342		Predicted R^2		0.8438

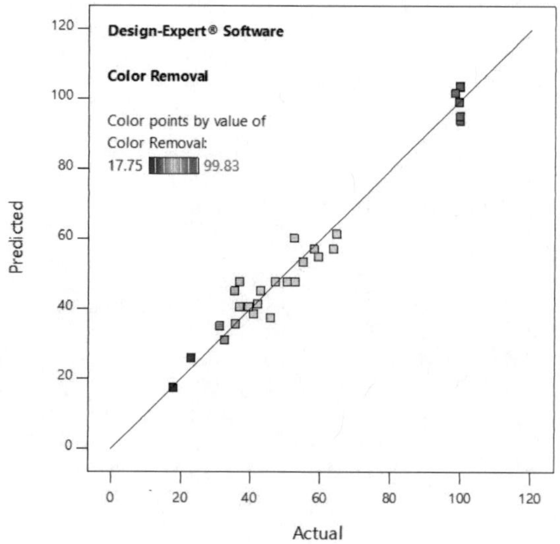

Figure 1. Plot of predicted vs. experimental values of AOII removal.

3.2. Interactions effect between operational parameters and optimization

Detailed interactions between these operational parameters was presented by the 3D response surface plots as shown in Figure 2. These plots reveal direct interaction effect between the operational parameter towards the percentage of AOII removal. Interestingly, significant interaction effect those operational parameters can be well observed between pH (A), catalyst dosage (C) and UV irradiation (D) rather than oxidant concentration (B) at the optimum conditions. For instance, high AOII removal can be attained at low pH range (Figure 2a), low catalyst dosage (Figure 2b) and high UV irradiation intensity (Figure 2 c,d) disregard to oxidant concentration (Figure 2a, Table 3 (p-value of 0.8852)). Similar observations were also reported in the previous work [11,15,16] related to degradation of dyes.

IOP Conf. Series: Materials Science and Engineering **551** (2019) 012126 doi:10.1088/1757-899X/551/1/012126

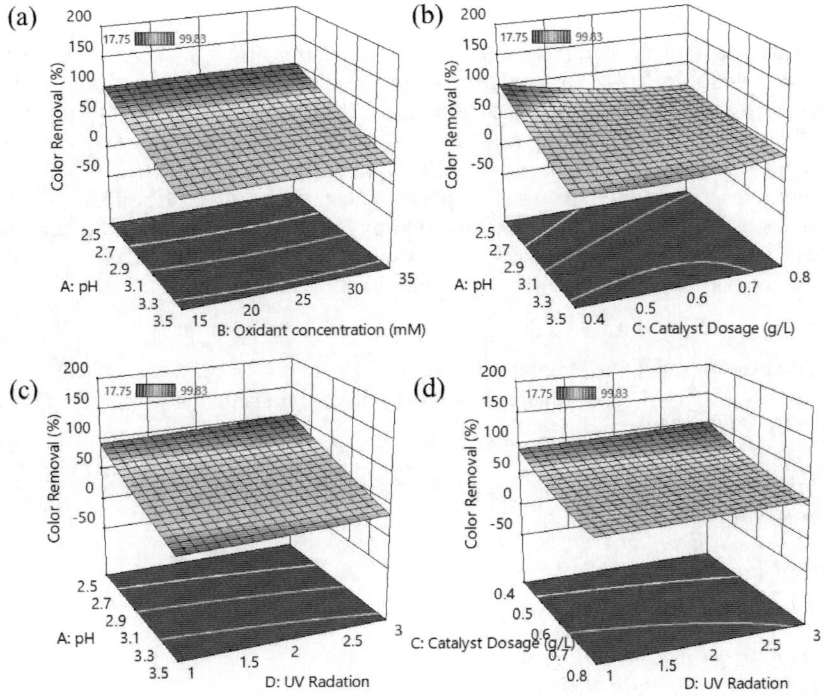

Figure 2. 3D response surface plots of (a) AB, (b) AC, (c) AD and (d) CD interaction effects towards AOII/ color removal (Y_1).

Such observations are in accordance with the proposed optimal operational conditions based on derived empirical quadratic model at desirability of 1. The maximum AOII (~100%) was anticipated at pH (A) solution of 2.6, 35 mM oxidant concentration (B), 0.4 g/L catalyst dosage (C) and 3 x 9W irradiation intensity (D). Validation of model reliability at these conditions were performed in duplicate runs. Interestingly, the percentage of AOII removal in both experimental runs were found to be at 99.78% and 97.45%. Small absolute deviation of less than 2% between the experimental and predicted values of AOII removal. Such findings confer that the derived empirical quadratic model was in good agreement with the experimental findings.

4. Conclusions

In conclusions, the optimal operational parameters for AOII degradation using $Fe_{3-x}Mn_xO_4$ catalyst were determined to be of solution pH (A) 2.6, oxidant concentration (B) 35mM, catalyst dosage (C) 0.4 g/L and UV irradiation (D) of 3 x 9W. At these conditions, the maximum predicted AOII removal was close to 100%. ANOVA analysis confers that interaction between pH, catalyst dosage and UV irradiation possessed significant influence towards the AOII removal during the reaction. In addition, the high regression coefficient value of 0.966 was in good agreement between the predicted and actual/experimental values of AOII removal. Hence, such findings present new insights into detailed global interactions between operational parameters for optimization of AOII degradation in UV assisted Fenton-like reaction particularly with utilization of $Fe_{3-x}Co_xO_4$ catalyst.

Acknowledgements

The authors gratefully acknowledged the Ministry of Higher Education Malaysia (MOHE) and Universiti Teknologi MARA (UiTM) for financial support under Fundamental Research Grant Scheme (600-IRMI/FRGS 5/3 (110/2016).

References

[1] Kant R. 2012 *Natural Sci.* **4** 22
[2] M.P. Prasad *et al* 2013 *J. Ecol. Environ. Sci.* **4** 4
[3] Cao X. et al. 2018 *Chemosphere* **208** 219
[4] Ramirez J.H *et al* 2007 *Appl. Catal B* **75**, 312
[5] Ahmed B *et al* 2011 *Ind. Eng. Chem. Res.* **50** 6673
[6] Elboughdiri N *et al* 2015 *Adv. Chem. Eng. Sci. B,* **5** 111
[7] Ahmadi M *et al* 2015, *Adv. Environ. Technol.* **2** 59
[8] Zhong Y. *et al* 2014 *Appl. Catal B,* **150** 612
[9] Liang X. *et al* 2014 *J. Colloid Inter. Sci.,* **426** 181
[10] Zhou L. *et al* 2018 *Appl. Catal. B,* **237** 1160
[11] Palas B *et al* 2016 *J. Photochem. Photobiol. A Chem.* **324** 165
[12] Ramirez J.H. *et al* 2005 *Catal Today,* **107** 68
[13] Bezerra M.A *et al* 2008 *Talanta.* **76** 965
[14] Abdullah N.H. *et al* 2018 *Mater. Today Proceed,* **5** 21956
[15] Wei G.T *et al* 2012 *Catal. Commun.* **17** 184
[16] Azami M *et al* 2012 *J. Serbian Chem. Soc.* **77** 235

Screening of factors influencing Pectin Extraction of Pomelo peels using 2-Level Factorial Design

Halifah Pagarra[1], Roshanida A. Rahman[2], Abd. Muis[1] dan Arifah Novia Arifin[1]
[1]Department of Biology, Faculty of Mathematics and Natural Sciences Universitas Negeri Makassar Indonesia
[2]Department of Bioprocess, Faculty of Chemistry and Energy Universiti Teknologi Malaysia, Skudai Johor Bahru Malaysia

E-mail: halifah.pagarra@unm.ac.id

Abstract. A study was conducted to vary the parameters affecting the extraction of pectin from pomelo peels (Citrus maxima) using a 2-level factorial design (2LFD). Experimental design was carried out to screen for significant environmental factors for extraction results. The factors involved consisted of pH, temperature and extraction time. The results of the variance analysis (ANOVA) found three main factors that had a statistically significant effect on the results of pectin extraction, namely pH, temperature and extraction time. The following important factors are the interaction factors between pH and temperature, the interaction between temperature and extraction time, the interaction between pH and extraction time, the interaction between temperature and extraction time. Statistical analysis shows that the linear model is significant with R2 value of 0.9715. It was found that pH, temperature and extraction time are the most important parameters that affect the results of pomelo peels pectin extraction.

1. Introduction

Pomelos's scientific name is Citrus Maxima in the Citrus group and Rutaceace. about 30% of the total fruit weight is pomelo peels (like white spongy flowers), is a potential source of pectin [1]. Pectin is a complex family Galacturonic acid polysaccharide present in the primary and middle cell walls lamela plant tissue. Pectin is widely used as a gelling agent and stabilizer in various food products, pharmaceuticals, cosmetics [2]. Pectin is usually extracted from some resources, like orange peel, pomace apple, lemon and sugar beet by inorganic acids such as HCl [3] or HNO3 [4], and organic acids such as citric acid [5], in the pectin industry. It is well-known that pectin has a role as a polysaccharide structure in the main cell walls and middle lamellartall plants which contributes to the structural strength of a plant and also to various cell functions including [6].

One of the main components of the primary cell wall and is generally considered to be around one third of all primary cell wall macromolecules is Pectin. Pectin is a group of polysaccharides consisting mostly of D-galacturonat acid and some carboxylic groups of galacturonic acid molecules in the pectin chain are methyl esterified and the percentage of the group esterified is expressed as DE (degree of esterification). Pectin is divided into two majors depending on the level of esterification; pectin ester is higher with ED more than 50%, and low pectin ester with DE lower than 50% [7]. The results and quality of pectin depend largely on the source, as well as the method used for pectin extraction. The results of pectin extracted from sugar beet by acid (HCl or HNO3) at varying pH (1-3), temperature

IOP Conf. Series: Materials Science and Engineering **551** (2019) 012127 doi:10.1088/1757-899X/551/1/012127

(75-90oC) and time (30-90 minutes) show galacturonic acid content and extraction time varies from 295 to 528 mg / g (dry weight) and 34% to 94%, respectively [8]. This work aims to extract pectin from pomelo peels (Citrus maxima)

2. Experimental

2.1. Material

The selected pomelo peels (Citrus maxima) from Pangkep district, South Sulawesi. Furthermore, the peels part used as the experimental material in this study. Pomelo peel is cut into small pieces and then washed and dried in an oven at 60°C for 5 days until the dry weight of the leaves is stable. This dry sample of pomelo peel is ground and filtered to get a small particle size using a 1 mm filter.

2.2. Pectin Extraction

Extraction methods based modification from [8] and [9]. A total of 10 g pomelo peels flour was added to aquadest with ratio of 1:20. Determination of pH 1.5 and 2.5 with 0.5 N HCl solvent, then incubated of temperature at 60°C and 100°C, extraction time of 60 minutes and 120 minutes. Stirring is carried out during incubation. After incubation, filtering is done to get the filtrate using filter paper. Furthermore, the filtrate was precipitated with 95% ethanol concentration for 24 hours. The precipitate is washed twice also with 95% ethanol. The obtained pectin is dried in an oven at 50°C until a constant weight is reached. The dried is calculated as the weight of dry pectin (g) per 100 g of dried pomelo peels.

2.3. Experimental Design and Statiscal Analysis

The two-level factorial design made by software DESIGN EXPERT (State-Ease Inc., Statistics simplified, Minneapolis, MN, USA, Version 6.0.4) was used to filter out the factors that influence the extraction of pomelo peels pectin. There are 3 variables used, so this design contains a total of 24 experimental experiments. Each independent variable is investigated for duplicates at high (+1) and low (-1). The variables used were pH (A), temperature (B), extraction time variables coded -1 (low code) and +1 (high coded) (Table 1). Variables that had a large effect on the results of pomelo peels pectin extract were identified based on a confidence level above 95% ($P < 0.05$). Important factors and interactions identified from half-normal plot analysis were chosen to produce a first-order model for texture responses after effects and interactions were evaluated. The significance of the linear effect of the three variables was evaluated by analysis of variance (ANOVA). The coefficient of determination R^2 (pronounced R-Square) and the adjusted R^2 coefficient are used to evaluate the suitability of the model. The statistical significance of the second-order model equation is determined by the value of F which is significant and not significant-of-fit F-value [10].

Table 1. Range of variable and their coded levels independent

Factor	Name	Low Coded	High Coded
		(-1)	(+1)
A	pH	1.5	2.5
B	Temperature (°C)	60	100
C	Extraction Time (min)	60	120

3. Results and Discussions

3.1. Screening using the Two Level Factorial Design method

Screening to determine which of the three variables that affect the results of pomelo peels extract using the Two Level Factorial Design method. Variables or factors have the most significant influence on the results of pomelo peel extract. To find out the main ingredients that significantly affected the results of pomelo peels extraction, the relative significance of the three factors was investigated by a 2-level factorial design. The design consists of 24 experiments. Filtering design is used to detect factors or independent variables that have a higher impact on the response variable resulting from pomelo peels extraction. Table 2 shows the predicted level of pomelo peels extraction from experimental data. Half the normal plots can be used to determine the significant factors that affect the response. In a study of the extraction results of pomelo peels extract, the experimental design began with a screening experiment involving 3 factors which reviewed the literature having a high impact on pectin production. In this review, there are 6 factors that have been studied using 2LFD experiments as shown in Table 2 and the value of the results of pomelo peels extraction is predicted by mathematical models and experimental data.

Interpretation of research analyzed statistically using analysis of variance (ANOVA) and complete results is shown in Table 2. The probability values (P-values) of each term and interaction factors are listed in Table 3. P-values less than 0.005 mean that these factors are significant to the response under study. This is based on the confidence level set at 95%. This model is significant with a probability of <0,0001. This means that the regression model is produced to explain the correlation of the results of pomelo peels extraction with the factors tested correctly in the statistics. Table 2 Analysis of variance (ANOVA) for the results of pomelo peels extraction using a two-level factorial design. pH is one of the most important and very significant variables obtained from the results of statistical analysis through this study. Temperature and extraction time are also the next important factor identified in this statistical analysis. The importance of temperature in the development of biological processes is very clear because temperature can determine the effects of the extraction of pomelo peels.

Table 2. Analysis of the variance (ANOVA) for yield of pectin from pomelo peels using 2LFD

Source	Sum of squares	Degrees of freedom	Min square	Value > F	P>F
Model	191.14	6	31.86	96.53	<0.0001*
A	15.20	1	15.20	46.06	<0.0001
B	5.90	1	5.90	17.88	<0.0006
C	51.33	1	51.33	155.55	<0.0001
AB	54.90	1	54.90	166.36	<0.0001
AC	54.30	1	54.30	164.53	<0.0001
BC	9.50	1	9.50	28.79	<0.0001
Pure Error	5.60	16	0.35		
Cor Total	196.75	23			
$R^2 = 0.9715$					
Adj R-Squared = 0.9614					
Std. Dev.= 0.57					

According to [11], showed that the results of extracting cocoa peels pectin with HCl solvent was thehighest pectin results at pH 2.5. Likewise, the extraction temperature is very influential, where extraction temperatures of 50°C - 95°C using citric acid at pH 2.5 or 4.0 and hydrochloric acid at pH 2.5 or 4.0 increase significantly (p <0 , 05) from the results of pectin. Furthermore [12], showed that

IOP Conf. Series: Materials Science and Engineering **551** (2019) 012127 doi:10.1088/1757-899X/551/1/012127

the optimum conditions for extraction of pectin was found at a temperature of 100°C by 60 min on the base of pectin yield from lemon pomace.

Two-level factorial design studies have been used to filter out significant factors affecting the results of pectin extraction of pomelo peels. The selected and filtered factors include pH, temperature, and time of extraction. The performance of pectin extraction was studied in terms of pectin yield. The results of this experiment suggest that the proposed mathematical model can explain the capabilities of the factors studied with the determination coefficient at 0.9715 (R2 = 97.15%). Two variables tested were significant factors influencing pectin yield obtained 18.2% with P value less than 0.05 ($P < 0.05$). A half-normal plot is used to determine the significant factors affecting the response in this study. This plot determines the absolute range of factor impression and shows the symbols of factors far below the linear line is a significant factor in the response [13]. In Figure 1, a half-normal plot indicates that the factor symbols far from the linear line are a significant factor in the yields of pomelo peels pectin extract. The effect of A (pH), B (temperature), and C (extraction time) is clearly located far from the line and thus shows a strong signal. This decision is related to P-value <0.0001 (Table 2). The next significant factor was the interaction factor between pH and temperature (AB factor), interaction between pH and extraction time (AC factor), and interaction between temperature and extraction time

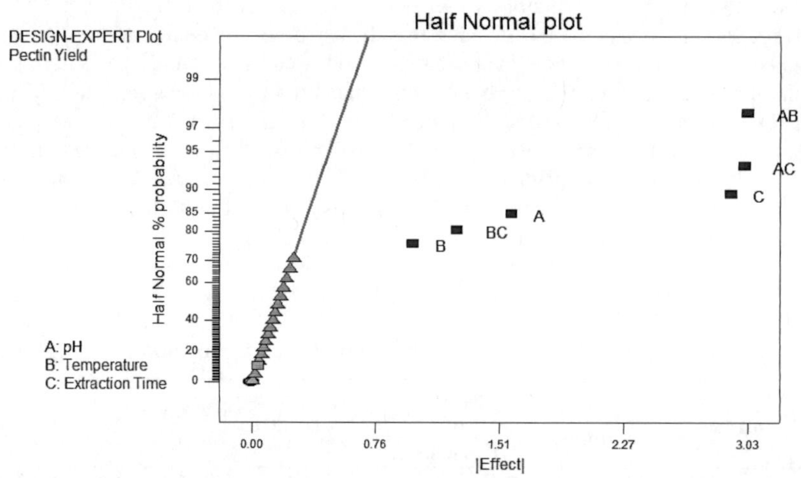

Figure 1. The plot is half normal for the effect of pH (A), temperature (B), extraction time (C) of pectin yield from pomelo peels.

3.2. Effect of Main Factors and Interaction Factors

Effect of Main Factors and Interaction Factors on the results of pomelo peels pectin extract. Responses to each of the main factors and also the interaction between factors were also examined statistically and also patented the response that these factors were also known to be definite in response. Fig.2, from the results of this study shows that there are 3 main factors that have a very statistically significant effect on the results of Pomelo peels extract, A (pH) and B (temperature). and factor C (extraction time) was found to be significant, that is, statistically giving an impression on the extraction process rather than the results of pomelo peels pectin extract in this study.

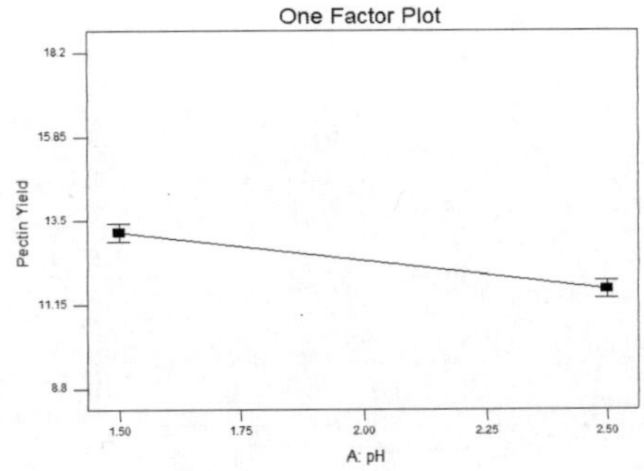

Fig. 2 a. Plot the main effect A (pH) on the results of pomelo peels pectin extract

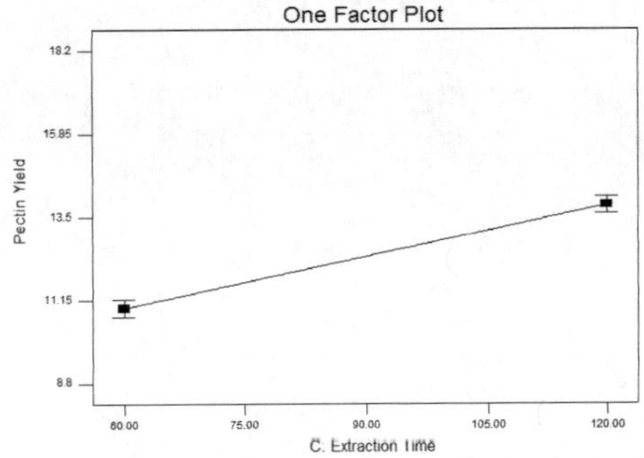

Fig. 2b. Plot the main effect B (Temperature °C) on the results of pomelo peels pectin extract

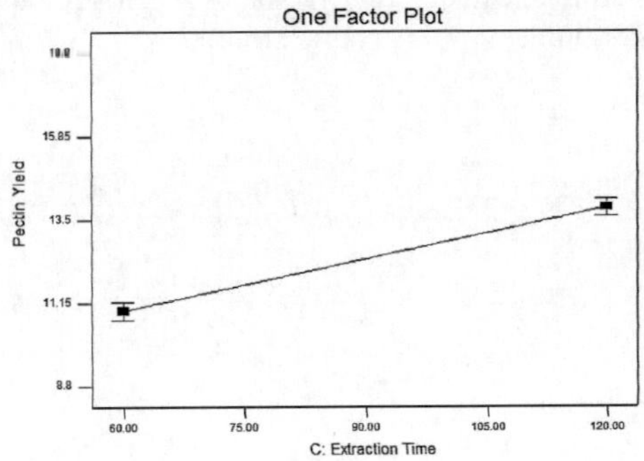

Fig.2c. Plot the main effect C (Extraction time) on the results of pomelo peels pectin extract

Figure 2 (a-c) also shows that this main factor is important in the extraction of pomelo pectin peels. Instead of Figure 2 (a-c), it can be seen that the increase in extraction activity factors in the ranking (-1 and +1) studied has had a significant effect on the results of pectin extract. Increasing the value for pH, temperature and extraction in the position studied has led to an increase in the results of pectin extract or is said to have given a significant impression of the response studied.

IOP Conf. Series: Materials Science and Engineering **551** (2019) 012127 doi:10.1088/1757-899X/551/1/012127

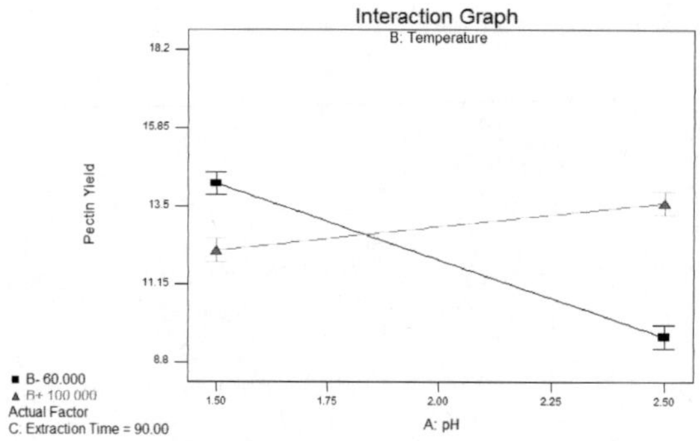

Fig. 3a. Plot significant interaction factors between pH and Temperature on the results of pomelo peels pectin extract (%)

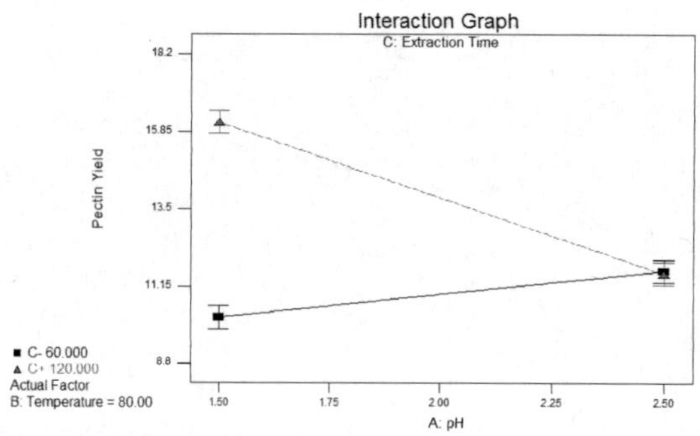

Fig. 3b. Plot significant interaction factor between pH and extraction time on yields of pomelo peels pectin extract (%)

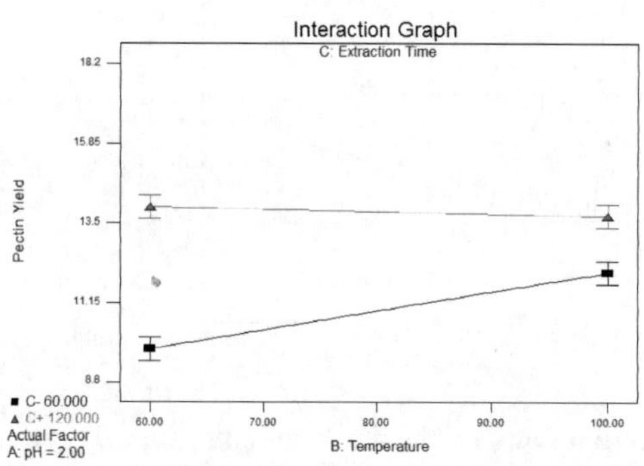

Fig. 3c. Plot significant interaction factor between temperature and extraction time on yields of pomelo peels pectin extract (%)

In addition to the impact of the main factors, the interaction between factors also shows different effects as a result of the statistical analysis that has been carried out. Figure 3a shows the interaction effect between 2 significant factors on pectin grapefruit skin extraction. The interaction of AB (pH with temperature), interaction of AC (pH with extraction time) and interaction of BC (temperature with extraction time) had a significant effect on the responses studied with P-values <0,0001 smaller than 0.05.

Figure 3a shows the last interaction between two factors, pH (A) and temperature (B). The slope for high concentration B - (temperature of 60°C) and low concentration of B+ (temperature of 100°C) to both steep. This shows that the increase of the pH with the temperature (B -) showed the yield of pectin extract was higher.

Figure 3b shows the last interaction between two factors, pH (A) and extraction time (C). The slope for high concentration C+ (extraction time of 120 minutes) and low concentration of C- (extraction time of 60 minutes) to both steep. This shows that the increase of the temperature with the extraction time (C+) showed the yield of pectin extract was higher.

Figure 3a shows the first interaction between B (temperature) and C (extraction time). The interaction effect of BC has the interaction, where effect B is greater when B+ is at a temperature of 100°C. This plot shows that when the data is bound to B+, the effect is more flat than B -. However, in this study, the target is the result of pectin extract. Therefore, B + is the best choice based on the target. This interaction may occur because both factors contribute

4. Conclusion

The results of pectin from pomelo peels has been evaluated using a full two-level factorial Design. The effect of the operating parameters from experimental studies, the pH factor (A) and temperature (B) and extraction time (C) are identified as the most significant factor of pomelo peels pectin yield. Three interactions between AB, AC, and BC affect one factor to another in the pectin extraction process. The combination effect between A and C shows the highest percentage pectin yield contribution. Therefore, it can be concluded that the formation of pectin depends on this two factors. Therefore, pH, temperature and extraction time are those significant factors and are used in subsequent optimization experiments using a centralized composite design (CCD). The screening of this factor was aimed at identifying significant factors and subsequent research using RSM was to use these significant factors to obtain optimum pectin yields of pomelo peels in the studied stages.

References

[1] Pawadee M et al 2014 *F. Hydr.* **35** 383.
[2] Le P T Q et al 2014 *Ban. J. Biotech.* **9** 67.
[3] Koubala B B et al 2008 *F. Hydr.* **22** 1345.
[4] Cristina L V *et al* 2011 *Carb. Poly.* 1236 **84** 1230.
[5] Jiangga Y *et al* 2012 *Carb. Poly.* **87** 1663
[6] Pagarra H *et al* 2018 *Americ. Inst. Phy.* **1** 1
[7] Pagarra H *et al* 2014 *Appl. Mech. Mat.* **625** 920
[8] Levigne S *et al* 2002 *Carb. Poly.* **49** 145
[9] Rehman Z U 2004 J. *Chem. Soc. Pakistan.* **26** 73
[10] Myers R H *et al* 2009 *Montgomery.* **3** 1
[11] Chan S Y *et al* 2013 *F. Chem.* **141** 3752
[12] Azad A K M *et al* 2014 *J.Food Nutrit. Sc.* **2** 30
[13] Abdul W S *et al* 2007 *J. Appl. Therm. Eng.* **27** 413

IOP Conf. Series: Materials Science and Engineering **551** (2019) 012128 doi:10.1088/1757-899X/551/1/012128

The Effect of N-Hexane Extract of Sambiloto Leaves (*Andrographis paniculata*) on Spermatozoa Quality of Male ICR Mice (*Mus musculus*)

Andi Asmawati Azis[1], Adnan[1], Sitti Nurmaidah[1] and Andi Citra Pratiwi[1]

[1]Department of Biology, Faculty of Mathematics and Natural Sciences, Universitas Negeri Makassar, Indonesia

E-mail: asma.azis@gmail.com

Abstract. The study aimed to determine the effect of N-Hexane extract of *Andrographis paniculata* leaves on testis weight of male mice (*Mus musculus*), along with the amount, motility, viability, and abnormality of its sperm. The tested animals used in this study were male *Mus musculus* aged 10-12 weeks with the average body weight (bw) of 21 g. The study was an experimental study with a Completely Randomized Design (CRD), which consists of one control group (A0, without extract) and three experimental groups (A1, A2, A3, with the extract concentration of 125 mg/kg bw, 250 mg/kg bw, and 500 mg/kg bw, respectively). The volume of extract given to each experimental mice was 1 cc per day, for a total of 18 days. The parameter of observation were the weight of testis, along with the amount, motility, viability, and the total of abnormal sperm. The data collected were analyzed by using F Test (α 0,05) and BNT Test (α 0,05). The result of data analysis showed that the N-Hexane extract of *A. paniculata* which was given for 18 days consecutively has affected the spermatozoa production of male mice. Each concentration of N-Hexane extract of *A. paniculata* (125, 250, and 500 mg/kg bw) had caused the decreasing of testis weight; the decreasing of spermatozoa production; the decreasing of motility and viability of spermatozoa; and the increasing of abnormal sperm. Hence, the study suggested that the N-hexane extract of *A. paniculata* has the potential to be used as antifertility treatment as it decreased the reproductive fuction of male *Mus musculus*. The extract concentration which cause the biggest effect is 500 mg/kg bw.

1. Introduction

Indonesia is a country rich in natural resources. The natural resources are mostly efficacious as medicine. Since the 19th century, this type of medicinal plant had begun to give a scientific touch, both in terms of planting and in terms of research on its content and properties. Indonesia has abundant natural resources. Many plants contain bioactive compounds which can be used as medicines. Generally, these compounds originate from steroids, isoflavonoids, alkaloids, and xanthons, all of which can affect reproductive function in mice [1].

Phytosterol is a sterol which is naturally obtained from plants. Chemically, phytosterol is similar to cholesterol obtained from animals. Sterol consists of three cyclohesan ring combinations with various types of sterols (more than 40 phytosterol). In plants, there are more than 40 compounds dominated by the phytosterol group [2]. Phytosterol (also known as plant sterol) is a cholesterol-like molecule found in plants. Phytosterol is a group of steroid alcohols, phytochemicals that occur naturally in plants. Plants contain a number of phytosterols which act as structural components of cell membranes. In mammalian cells, this role is replaced by cholesterol. Three types of compounds commonly referred to as "phytosterol" are sitosterol (better known as beta-sitosterol), stigmasterol and campestero [3].

Plants have a great potential to be used as a source of fertility regulating compunds. Various bioactive compounds in plants, especially groups of steroid compounds, alkaloids, isoflafonoid, triterpenoids and xanthons have properties as a regulator of fertility. Preclinical tests conducted by researchers showed that bioactive compounds in plants affect the fertility of test animals, such as disturbing spermatogenesis, reducing the power of conception, increasing the percentage of gestation loss, preventing/ inhibiting implantation and reducing the number of progeny. Bioactive ingredients from plants have activities that can be reversed, and if their use is stopped, fertility returns to recovery [4].

Today, many researchers in the world are beginning to highlight medicinal plants to be used as alternatives to overcome reproductive problems, since the use of medicinal plants (herbs) is considered as natural and not harmful. One example is Sambiloto (*Andrographis paniculata)*, a bitter plant which is suspected as a plant that have the potential to overcome reproductive problems. *A. paniculata* is one of the plants that can be found easily in Indonesia, as it is high in abundance and has been widely known as a traditional medicine. The use of the plant is pervasive, since people in Indonesia already familiar with its use [5].

The testing of various types of drugs is usually conducted prior to its application for livestock or human. Thus, tested animals are needed for investigating the effect of a certain drugs. For this purpose, mice are commonly used as tested animals. A study on oral administration of *A. paniculata* have shown an antifertility effect as it decreases the quality of the ejaculated sperm on male mice [6]. A study conducted by Akbarsha and Murugaian [7] also found that the treatment of *A. paniculata* leaf to male mouse has caused the seminiferous epithelium was thoroughly disrupted and the fully differentiated spermatozoa were far too limited. Thus, it supports the possible prospective use of *A. paniculata* leaf as as antifertility compound.

The potential benefits of *A. paniculata* to be used as antifertility, along with the scientific objectives to strengthen previous researches, has put an urge to conduct further research on the effect of *A. Paniculata* leaf extract on the quality of spermatozoa, particularly on mice (*Mus musculus*).

2. Research Method

2.1. Time and Place of Research

The research was conducted from May to June 2013. The implementation of this study took place at the Biology Microtechnics Laboratory, Faculty of Mathematics and Natural Sciences, UNM.

2.2. Type of Research

The research is an experimental research which consists of two stages, namely (i) the preparation stage, which included the maintenance of test animals, the copulation of test animals, the making of extracts; and (ii) the implementation stage, which included the treatments, surgery, and the preparation of wet mount slide to be observed under the microscope.

2.3. Research Design

The research design used was a Completely Randomized Design (CRD) consisting of four treatment groups with five replications to obtain twenty combinations. Group I (was given the symbol A0) is a control group that was not given the n-hexane extract of *A. paniculata* leaves. Group II, III, and IV (were given the symbols A1, A2, and A3, respectively) are the groups which were given the n-hexane extract of *A. paniculata* leaves based on the prescribed dose. Group II was given a dose of 125 mg/ kg bw. Meanwhile, group III and IV was given a dose of 250 mg/ kg bw and 500 mg/ kg bw, respectively.

2.4. Research Procedure

2.4.1. Test Animals
The test animals used in this study were male ICR (*Mus musculus*) which were obtained from the Maros Regency Animal Research Agency (BPH). The age of the mice were 10-12 weeks with a body weight of 21 g. The food given for the test animals were AD2 which was obtained from Sinar Harapan store in Veteran Selatan and the drinking water used was tap water.

The handling of the test animals was carried out in the microtechnical laboratory of Biology Department, Faculty of Mathematics and Natural Sciences, UNM with 12 hours of lighting and 12 hours of dark conditions. The mice were placed in a rectangular box that has been given husk to prevent the base from the dirt and also a cover made from a wire mesh. The husk is replaced twice a week.

2.4.2. Extraction
The material to be extracted was the *A. paniculata* leaves obtained from the Green House experimental garden, Faculty of Mathematics and Natural Sciences, UNM. The extraction process was carried out at the Microbiology Laboratory of Biology Department. In order to prepare the leaf extract of *A. paniculata* leaf extract, the leaves of *A. paniculata* was washed with a wet weight of 2000 g, then dried (aerated) until it reached a dry weight of 1700 g. Dried leaves were then mashed and soaked with n-hexan for 24 hours, then could be filtered with filter paper. The extract was then evaporated for 6-8 hours to obtain a pure extract. The result of extraction was 1 gram of a semi-liquid extract with a deep green colour and a bitter stinging smell

2.5 Implementation of Research.

2.5.1. Preparation
Before being given to the tested animals, the extract was weighed using an analytical balance based on the required dose, and then diluted with 0.5% Carboxil Methyl Cellulosa (CMC). Prior to the dilution, 0.5 g of CMC was weighed with an analytical balance and dissolved in 100 ml of distilled water. The mixture of CMC and extract were homogenized by using Hot Plate Magnetic stirrer and stirring rod so that the solution becomes homogeneous and ready to be used. The dosage of solution for each treatment was made once for three days. The volume of n-Hexan extract from *A. paniculata* leaves given to each mouse was 1 cc.

2.5.2. Treatment
Each experimental group (A1, A2, and A3) was treated by giving n-Hexan extract of *A. paniculata* leaves once a day. Each experimental group was given different doses, i.e. A1 = 125 mg / kg bw, A2 = 250 mg / kg bw, and A3 = 500 mg / kg bw. All male mice were weighed using Ohaus balance from the first day of handling until the 18th day of extract treatment. The extract was given every day at 08.00 – 10.00, by using Feeding Tube No.4 and 1 cc size spoit.

2.5.3. Surgery
On the 18th day, the injection of n-hexane extract of *A. paniculata* was stopped. The tested animals were then subjected to euthanasia through cervical dislocation. In order to produce the dislocation, the mice were restrained in a normal standing position on a flat surface. The right hand then quickly pushed the mice head forward and down, while the left hand pulling the tail backward. The surgery was performed by stretching the mice body on a surgical board, holding the leg by using a pin, then cutting it through the abdomen slowly.

2.5.4. Observation
The observational parameters of this study, including: the weight of testis, along with the amount, motility, viability, and the total of abnormal sperm.

IOP Conf. Series: Materials Science and Engineering **551** (2019) 012128 doi:10.1088/1757-899X/551/1/012128

a). The Weight of Testis
The male mice were euthanized through cervical dislocation, then were subjected to surgery by using surgical instruments, such as surgical scissors, surgical tweezers, and pins. The reproductive organs were cut then stored in a petri dish containing 2 ml of NaCL 0.9%. The testis was then cleaned from fat, dried with tissue, and then weighed using an electronic balance scale. The reproductive organ that have been taken must be stored in 0.9% physiological NaCl solution as soon as possible, so that the spermatozoa inside the testicles do not die and the consistency of the organ could be maintained.

b) Total Sperm
Spermatozoa were obtained from the cauda epididymis. Both Cauda epididimymis squeezed into a watch glass containing 1 ml of 0.9% NaCl. It was cut into smaller parts by using scissors until it reached a smooth consistency, and then stirred with a stirring glass. This solution is called a spermatozoa suspension. The number of spermatozoa was calculated by using an improved Neubauer count chamber (haemocytometer). The spermatozoa suspension that has been diluted with 1 mL of physiological saline (0.9% NaCl) was taken by using the haemacytometer pipette until it reached 0.5 scale and then NaCl was taken using the same pipette until it reached the scale 11, and shaken until homogenous then put into counting chamber (heemocytometer), after which be closed with a cover glass. While closing with a cover glass, the air bubble should not be formed. The haemocytometer which contained a suspension of spermatozoa was then observed under a light microscope with a magnification of 400x. The result of calculation was then entered into the formula of sperm count = N/ 2 spermatozoa/ ml suspension. In which, N = number of sperm counted in boxes 1, 2, 3, 4 and 5 (middle part).

c) Sperm Motility
According to WHO, sperm motility is classified into four levels, namely: Class A, sperm that moves forward quickly in a straight line; Class B, sperm that moves forward but in curved or wavy lines or in a straight line but slowly; Class C: sperm that moves its tail but does not move; and Class D, sperm that doesn't move at all. In accordance with WHO motility assessment standards, spermatozoa are said to be normal if the sperm percentage of criteria A and B is greater or equal to 50%. For this reason, the data analysis in this study only used the percentage of sperm criteria A + B. The total number of Class A and Class B sperm were then analysed for its average along with its percentage.

d). Sperm Viability
The spermatozoa suspension obtained can be used to analyse the viability of the sperm. Sperm viability analysis was carried out with supravital staining, through which 1 drop of sperm was placed on the glass object, added with 1 drop of eosin solution, then stirred. After that, a smear preparation was made and dried in air. The preparation was observed under a microscope with magnification of 400 x. Living spermatozoa are colorless, while dead ones are red. A total of 100 spermatozoa were calculated, and the results are presented in percent.

e). Sperm Abnormalities
Observations were carried out by looking at the forms of abnormalities that exist. Observation on the abnormal characteristics of sperm were conducted by calculating the number of abnormal spermatozoa from 100 observed spermatozoa. The characteristics of abnormal spermatozoa including double heads, large heads, small, broken body, broken heads, curly tail, and so on.

2.6 Data Analysis Techniques

Data obtained from the observation were analysed by using inferential analysis of F test with the significance of α 0.05, followed with LSD test with the significance of α 0.05.

3. Result and Discussion

3.1. Research Result

The results of observation on the average body weight (g), testicular weight (g), sperm count (%), sperm motility (%), sperm viability (%), and the average number of abnormal sperm (%) in control mice and in mice treated with n-hexane extract of *A. paniculata* for 18 days at a dose of 125 mg/ kg bw, 250 mg/ kg bw, and 500 mg/ kg bw respectively are presented in Table 1 to Table 6.

Table 1. The average body weight (g) on various treatment with n-hexane extract of *A. paniculata* for 18 days.

No.	Dosage (mg/bw)	Average ± SD	LSD α 0,05
1	0	24,546 ± 0,3242c	
2	125	22,622 ± 1,2726b	1,527
3	250	21,083 ± 1,1403b	
4	500	22,483 ± 0,9538a	

Note: the numbers followed by the same letter are not statistically different on α 0,05 of LSD test.

Table 2. The average weight of mice testis (g) on various treatment with n-hexane extract of *A. paniculata for* 18 days

No.	Dosage (mg/bw)	Average (g) ± SD	LSD α 0,05
1	0	0,609 ± 0,075b	
2	125	0,567 ± 0,009ab	0,108
3	250	0,515 ± 0,396a	
4	500	0,500 ± 0,062a	

Table 3. The average number of mice spermatozoa (105 million/ml) on various treatment with n-hexane extract of *A. paniculata* for 18 days

No.	Dosage (mg/bw)	Average ± SD	LSD α 0,05
1	0	59,863 ± 2,9744d	
2	125	41,162 ± 1,6996c	9,354
3	250	26,275 ± 2,8712b	
4	500	20,790 ± 3,1819a	

Note: the numbers followed by the same letter are not statistically different on α 0,05 of LSD test.

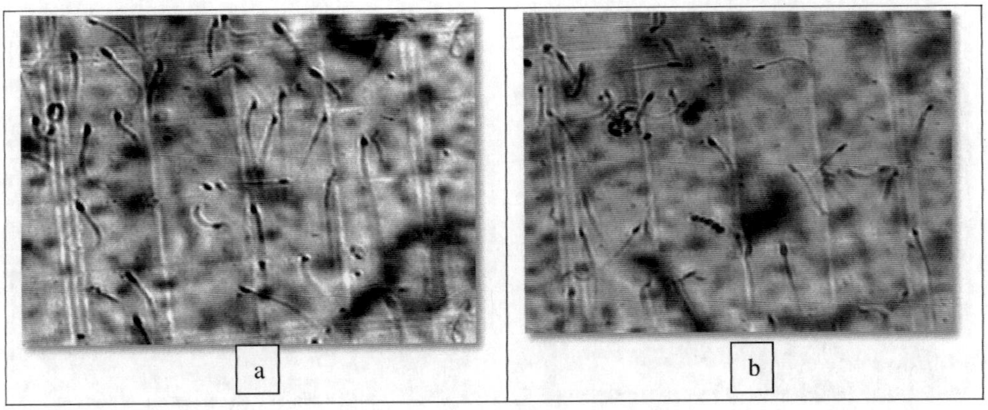

Figure 1. The counting of total number of spermatozoa on haemocytometer with the magnification of 400 x: a. control group, b. experimental group.

Table 4. The average motility of spermatozoa (%) on various treatment with n-hexane extract of *A. paniculata* for 18 days

No.	Dosage (mg/bw)	Average (%) ± SD	LSD α 0,05
1	0	34,662 ± 3,4119d	
2	125	24,490 ± 1,0464c	2,29
3	250	16,832 ± 0,3350b	
4	500	12,997 ± 0,2735a	

Note: the numbers followed by the same letter are not statistically different on α 0,05 of LSD test.

Table 5. The average viability of spermatozoa (%) on various treatment with n-hexane extract of *A. paniculata* for 18 days

No.	Dosage (mg/bw)	Average (%) ± SD	LSD α 0,05
1	0	18,330 ± 1,1885a	
2	125	36.162 ± 3,5559b	
3	250	49,495 ± 3,4179c	4,484
4	500	46,283 ± 1,1940c	

Note: the numbers followed by the same letter are not statistically different on α 0,05 of LSD test.

Table 6. The average abnormal spermatozoa (%) on various treatment with n-hexane extract of *A. paniculata* for 18 days

No.	Dosage (mg/bw)	Average (%) ± SD	LSD α 0,05
1	0	16,332 ± 1,4150a	
2	125	30,662 ± 1,6557b	3,872
3	250	36,165 ± 3,7170c	
4	500	44,580 ± 2,6017d	

Note: the numbers followed by the same letter are not statistically different on α 0,05 of LSD test.

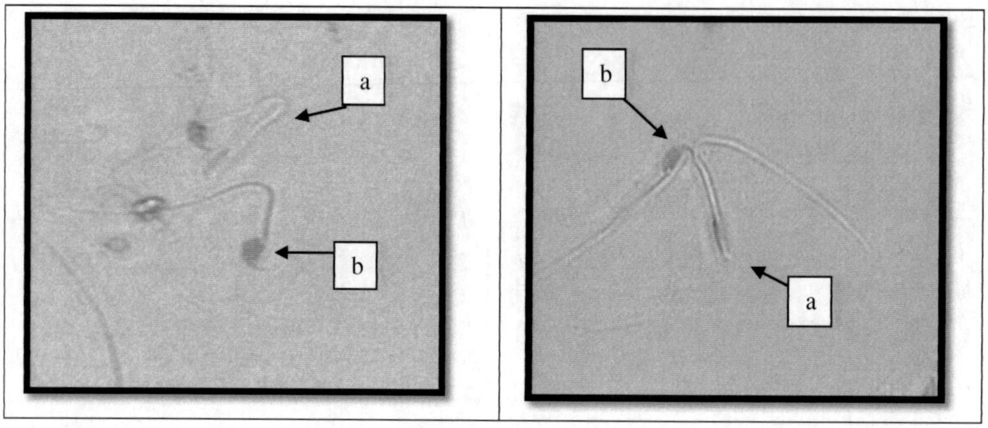

Figure 2. Observation of spermatozoa viability, a. alive spermatozoa = do not absorb dye, b. dead spermatozoa = absorb dye.

3.2. Discussion

The present study indicates that the treatment of N-Hexane *A. paniculata* extract results in the decreasing of spermatozoa quality in the treated *Mus musculus*, as it caused the decreasing of testical weight, spermatozoa counts, motility, and viability of spermatozoa, while increasing the total number of abnormal sperm. The extract concentration of *A. paniculata* which cause the biggest effect is 500 mg/kg bw.

The decreasing weight of testis along with the decreasing quality of spermatozoa after the treatment with N-Hexane *A. paniculata* extract for 18 consecutive days could occur due to several factors, including (i) the androgenic and cytotoxic characteristics of *A. paniculata* extract, (ii) *A. paniculata* extract works directly on germinal cells, and (iii) *A. paniculata* extract works to interfere with steroidogenesis. The decreasing of mice body weight along with the shrinkage of its testis may occur due to the presence of androgenic and cytotoxic compounds in *A. paniculata* extract. The compound is β-sitosterol which belong to the group of phytosterol and cytotoxic saponins. Sterols are derivatives of steroid compounds. The excessive use of steroids can cause side effects, such as the increasing risk of infertility and the shrinkage of testis.

A decreased in testicular size occurs due to the loss of germinal epithellial cells which cannot undergo regeneration. It can be seen from testicular size and weight of test treated and control mice which is influenced by the level of estrogen hormone in male mice. It is argued that cytotoxic effects of a substance can cause the death of spermatogenic cells in seminiferus tubule, thus causing a decrease in testicular weight.

The decrease in counts, motility, and viability of sperm is caused by the presence of saponin compounds, β-sitosterol, in *A. paniculata* extract which can cause the increase of testosterone levels along with the inhibition of energy formation in the body of tested animals. The increase of testosterone levels has a negative feedback effect on hypothalamus and anterior pituitary. The negative feedback effect on hypothalamus will stop the secretion of GnRH (Gonadothropins Releasing Hormone), thus inhibiting gonadotropin secretion, i.e. Luteinizing Hormone (LH) and Follicle Stimulating Hormone (FSH), by the anterior pituitary. LH functions to stimulate Leydig cells to produce testosterone, while FSH functions to stimulate spermatogenesis [8].

The decrease in the number of motile sperm may occur due to the presence of compounds in *A. paniculata* n-hexane extract which can inhibit the formation of energy, namely β-sitosterol, which has a chemical structure similar to the hormone testosterone as a hydrocarbon compound. A substance can work as a hormone because it contains substances whose molecular arrangement is similar to hormones. Thus, it is suspected that β-sitosterol also works like testosterone.

Testosterone is an essential hormone in spermatogenesis. However, it has a nature of negative feed back when it reaches a high level of concentration in the blood. It is suspected that β-sitosterol compounds contained in the extract of *A. paniculata* leaf also cause testosterone levels in the body of tested animals to increase.

The decreasing count of motile sperm is thought to be due to the presence of compounds in *A. paniculata* which can inhibit the formation of energy. Namely β-sitosterol, which has a chemical structure similar to the hormone testosterone which is a hydrocarbon compound. A substance can work as a hormone because it contains substances whose molecular arrangement is similar to hormones. Thus, it is suspected that β-sitosterol is also like testosterone. Testosterone is an essential hormone in spermatogenesis, but in high levels in the body it has the nature of negative feedback. It is suspected that β-sitosterol compounds contained in the extract of bitter leaf also cause testosterone levels in the body of test animals to increase [9]. The increasing level of testosterone in the blood, will cause a negative feedback to occur on the hypothalamus and the anterior pituitary. Thus, the secretion of gonadotropin will be inhibited. If FSH production is stopped or reduced due to the effect of negative feedback, then spermatogenesis will also stop [10].

Testosterone is the primary reproductive hormone that has a role in the process of spermatogenesis (sperm formation). During the process of spermatogenesis, testosterone in collaboration with FSH and LH will affect the formation of spermatozoa. However, if there is a significant increase of testosterone level in blood plasma, it will inhibit the release of FSH and LH. Accordingly, it is suggested that there must be a balance between the three groups of hormone. As a hormone that plays a role in the process of spermiogenesis, testosteron will show a different pattern if its concentration is low. In this case, the low level of testosterone will cause the process of spermiogenesis to be disrupted. Moreover, it may also lead to primary abnormalities, such as the size of spermatozoa's head is too big or too small, the tail is double or has a different appearance than the general structure of spermatozoa. Testosterone biosynthesis involves various substances, enzymes, and other steroid hormones, including hormones in the group of the estrogen and androgen. The presence of flavonoid inhibited the activation of kinase protein, thus inhibiting the testosteron production and the process of spermiogenesis [8, 11, 12].

The treatment of male mice with n-hexane extract of *A. paniculata* has affected the count of normal sperm. The higher the dosage given, the lower the count of normal sperm. The decrease in the number of normal sperm may occur due to the interference during spermatogenesis. As stated earlier, that the chemical content of *A. paniculata* leaf extract is antiproliferative (saponin and B-sitosterol), thus interferes with the spermiogenogenesis process which causes the spermiogenesis process to not proceed properly. Consequently, the counts of normal spermatozoa decreased.

The results of microscopic observation showed that there were several types of abnormal spermatozoa found on treated mice, including spermatozoa with branched head; branched tail; twisted tail; doubled head; enlarged body; folded body; broken tails; curly tails; and rolled tails. The additional parts of spermatozoa may inhibit its effectiveness to move toward the ovum. Thus, fertilization is not likely to occur. A good morphology of spermatozoa including a 'comma' shaped head with a normal magnitude and a long tail that is not circular or double. The tail is used by spermatozoa to do movements. The main part of the tail contains most of the motility power of the spermatozoa and has a vital role for motility [11].

4. Conclusions

Administration of n-hexane extract of *A. paniculata* leaf at a dose of 125 mg / kg bw, 250 mg / kg bw, and 500 mg / kg bw for 18 consecutive days has affected the spermatozoa quality of male mice (Mus musculus), as it causes the decrease of testis weight; the decrease of count, motility, and viability of spermatozoa; while increasing the number of abnormal spermatozoa.

References

[1] Sasidarn S *et al* 2011 *Afr. J. Tradit. Complemet. Altern. Med.* **8** 1.
[2] Mustafa G *et al* 2017 *Mat. Sc. Pharm,* **1** 17.
[3] Ogbe R J *et al* 2015 *As. J. of Plnt and Res.* **5** 10.
[4] Olajedi O 2016 *Nat Prod. Ind. J.* **12** 8.
[5] Widyawati T 2007 *Maj. Ked. Nus.* **40** 216.
[6] Sumarmin R *et al* 2018 *IOP conf. series: Earth and Env. Sci.* **130** 1.
[7] Akbarsha M A *et al* 2000 *Phytother. Res.* **14** 432.
[8] Hansson V *et al* 1978 *Steroids.* **21** 457.
[9] Halim V S *et al* 2004 *Maj Farm Ind* **15** 136.
[10] Tilbrook A J 2001 *Bio Rep* **64** 735.
[11] Claire L *et al* 2010 *Hum Rep Upd* **16** 205.
[12] Oyedemi S *et al.* **2010** *Exp. Toxicol.* **29** 377.

Study Photodegradation of Acid Orange 7 Using ZnO Waste Batteries

Hasri[1], Sugiarti[1], Sadriadi[1]

[1]Chemistry Department, Universitas Negeri Makassar, Makassar, Indonesia.

E-mail: hasriu@unm.ac.id; sadriadiabdullah@yahoo.com

Abstract. This research aims to determine the ability of ZnO waste batteries to degrade Acid Orange 7. The first, synthesized zinc acetate precursors from Zn waste batteries. Next, synthesized ZnO from the precursor of zinc acetate and impregnated with zeolite to form a ZnO-Zeolite composite used the sol-gel method. Photodegradation test for Acid Orange 7 dye used a time 30, 60, 120 , 240, 360 and 480 minute, concentration of 30, 50,75, 100, 150ppm with variations in pH 3, 5, 7, 9 and 11. Measurement of degradation resultsconcentration of Acid Orange 7 dye used a UV-Vis Spectrophotometer at a wavelength of 484.45 nm. The results showed that ZnO-Zeolite composites were able to degrade Acid Orange 7 dyes at pH 7, 480 minute, concentration 50 ppm with a degradation percentage of 94.15%.

1. Introduction

The textile industry has become one of the industries that have experienced rapid development in Indonesia and has a positive impact on the welfare of the community, including reducing the number of unemployed, increasing income, and the output of the industry. However, many textile industries dispose of staining waste into the environment without prior processing, which results in inhibiting the penetration of sunlight into the water, thus disrupting the photosynthetic activity of microalgae and threatening the sustainability of aquatic ecosystems. Based on the consequences, it is necessary to conduct an effort to overcome the problem of textile dye waste before being disposed into the environment. Some methods that can be done include the method of ultraviolet radiation, biodegradation and ozonolysis, [3]. This method requires relatively high operational costs. Other methods that can be used, such as coagulation, electrocoagulation and chlorination [5]. However, the method also has a disadvantage, such as the emergence of problems due to the production of new phases containing more concentrated pollutants. One method that is easy and relatively inexpensive to apply is the photodegradation method, this method can decompose dyes into components that are simpler and safer for the environment. The color of photodegradation can be conducted using photocatalyst materials and ultraviolet (UV) radiation. Photocatalyst materials such as ZnO are able to absorb the solar spectrum and quantum light more than TiO_2. The researcher tried to make a photocatalyst with a better size distribution by dispersing semiconductor particles on solid support materials such as zeolite. Zeolite can be as a supporting of solid material which is able to disperse semiconductor materials, thus preventing the sintering process in semiconductor materials, because zeolites have a porous structure, a large surface area, so that more catalysts can be developed [8]. pH has an important role during the photodegradation process to produce hydroxyl radicals. The more hydroxyl radicals formed, the greater the ability of photocatalysts to oxidize organic compounds [1]. Indigo Carmines dye degradation results in 44.24% using TiO_2-Zeolite catalysts occurring at pH 5 [2].

IOP Conf. Series: Materials Science and Engineering **551** (2019) 012129 doi:10.1088/1757-899X/551/1/012129

ZnO-Zeolite catalyst was able to degrade azo Acid Blue92 at pH 6 [6]. The results of the degradation of blueberry methylene using ZnO-activated carbon occur at pH 11 [9].

The use of ZnO catalyst impregnated with zeolite and activated carbon as mentioned above, generally uses ZnO which is synthesized from pure zinc acetate precursors. For this reason, the researcher was interested to analyze the effect of pH on the photodegradation of Acid Orange 7 azo dyes using ZnO, which was synthesized from battery waste and impregnated with zeolite.

2. Experimental

2.1. Material

The materials used are anode of zinc-carbon primary battery waste, natural zeolite, 96% ethanol, sodium carbonate solids (Na2CO3), glacial acetic acid (CH3COOH), concentrated hydrochloric acid (HCl), sodium hydroxide (NaOH) 2 M, zinc chloride (ZnCl2) 25%, silver nitrate solids (AgNO3), distilled water, aqua bides, Acid Orange 7 solids, filter paper, aluminum foil, and Whatman filter paper.

2.2. Preparation of Zeolite

Natural zeolite is crushed and sifted to a size of 150 mesh. Then washed, the washing results were filtered and dried in an oven at a temperature of 110^0 C for 2 hours. The result of washing zeolite was added 75 mL $ZnCl_2$ 25%, then shaker for 4 hours. Filtered zeolite, washed distilled water. Then oven at 110^0 C.

2.3. Preparation of Zinc Acetate

A total of 10 grams of Zinc-Carbon battery anode is dissolved in 10 ml of concentrated HCl. The solution is left until most of the anodes dissolve. The solution is filtered, the filtrate is added with a solution of 0.75 M sodium carbonate as much as 200 mL. The suspension formed is then filtered and dried at room temperature. The suspension formed is then added with 20 mL of glacial acetic acid until a clear solution is formed. The solution is then heated to form a gel and solidifies at room temperature. The solid is then mashed and stored in an airtight container.

2.4. ZnO-Zeolite Composite Synthesis

ZnO-Zeolite composites are made with a mass ratio of ZnO: Zeolite 5: 1. Zn (CH3COO)2 weighed 8.989 grams respectively and added zeolite as much as 0.664 grams, then dissolved into 16 mL ethanol. The mixture is reflux while stirring for 2 hours at 760C. Then 45 mL of 2 M NaOH was added and stirred for 1 hour with a magnetic stirrer. The mixture was left overnight and then filtered [9]. The precipitate in the oven at 600C and continued with calcination at 4000C. ZnO-Zeolite composites were characterized using Scanning Electron Microscopy (SEM-EDX).

2.5. Determination of optimum pH degradation of Acid Orange 7 dye

Acid Orange 7 dye solution was made with a concentration of 50 ppm, then 20 mL was taken and then put in 5 beakers. Acid Orange 7 solution is regulated by pH to pH 3, pH 5, pH 7, pH 9 and pH 11. Next, 50 mg of ZnO-Zeolite is added to the beaker. Then, stir it with a magnetic stirrer and irradiate it with UV light for 480 minutes. After being irradiated with UV light, then centrifuged at a speed of 7000 rpm for 15 minutes, then measured the absorbance using a UV-Vis spectrophotometer.

2.6. Determination optimum concentration of acid orange 7 dye degradation

A total of 20 ml of Acid Orange solution using the optimum pH with variations in concentrations of 30, 50, 75, 100, 150. Then, 50 mg of ZnO-Zeolite was added to the beaker. Then, stirred and irradiated with UV light for 480 minutes. Then centrifuged for 15 minutes, and measured the absorbance using a UV-Vis spectrophotometer.

IOP Conf. Series: Materials Science and Engineering **551** (2019) 012129 doi:10.1088/1757-899X/551/1/012129

2.7. Determination of optimum time interaction of acid orange 7 dye substance
A total of 20 mL of Acid Orange solution uses optimum pH and concentration with a variation of time 30, 60, 120, 240, 360, and 480 minutes. Then, 50 mg of ZnO-Zeolite is added to the beaker. After that, stirred and irradiated with UV light the sample was centrifuged for 15 minutes, and measured the absorbance using a UV-Vi spectrophotometer.

3. Results and Discussions
ZnO-Zeolite photocatalysts are synthesized using zeolite and zinc acetate precursors derived from battery waste. The synthesis of ZnO-Zeolite photocatalyst was carried out using the sol gel method, which is the process of phase change from sol to gel based on the hydrolysis process and precursor condensation. The growth of ZnO particles from the precursor of zinc acetate generally experiences 4 stages, namely solvation, hydrolysis, polymerization and transformation into ZnO [7]. The reaction mechanism is:

Solvation step:

$$Zn(CH_3COO)_2 \xrightarrow{\text{ethanol}} Zn^{2+} + 2CH_3COO^-$$

Hydrolysis step

$$Zn^{2+} + 2CH_3COO^- + 2NaOH \xrightarrow{\text{etanol + air}} Zn(OH)_2 + 2\,CH_3COONa$$

Polymerization step

$$Zn(OH)_2 + 2H_2O \rightarrow Zn(OH)_4^{2+} + 2H^+$$

Transformation step

$$Zeolit + Zn(OH)_4^{2+} \rightarrow ZnO\text{-}Zeolit + H_2O + 2OH$$

Figure 1. Results of ZnO-Zeolite Composite Synthesis

The characterization results show a structure that clumps with the clump surface of the material are small-sized scattered randomly. The clump is thought to be a zeolite material while the small material scattered randomly on the surface of the clot is thought to be a Zinc Oxide (ZnO) material.

Figure 2. SEM analysis of the Results of ZnO – Zeolite magnification 10.000x

ZnO - Zeolite EDX analysis showed the mass percentage of O element = 22.67%, Al = 1.77%, Si = 6.19%, and Zn = 64.36%. The data prove that the ZnO-Zeolite composite was successfully synthesized.

3.1. Determination of optimum pH degradation of Acid Orange 7 dye

PH parameters are very influential in the process of degradation of dyes. Each dye has an optimum pH, depending on the type of dye and the conditions at that time. pH 7 is the optimum pH with a degradation percentage of 94.15%. (Figure 3). At pH 7 OH radicals formed in the photodegradation process better decomposite the Acid Orange 7 dyes than in acidic or basic conditions. The same thing was found in the Pertiwi study (2018) which obtained the results of dyestuff degradation at an optimum pH 7.

Under acidic conditions, the exclusive force between the photocatalyst and the dyestuff will cause a decrease in adsorption and a decrease in the dyestuff photodegradation. In addition, ZnO has a tendency to dissolve by decreasing the pH of the solution [4].

$$ZnO + 2H^+ \rightarrow Zn^{2+} + H_2O$$

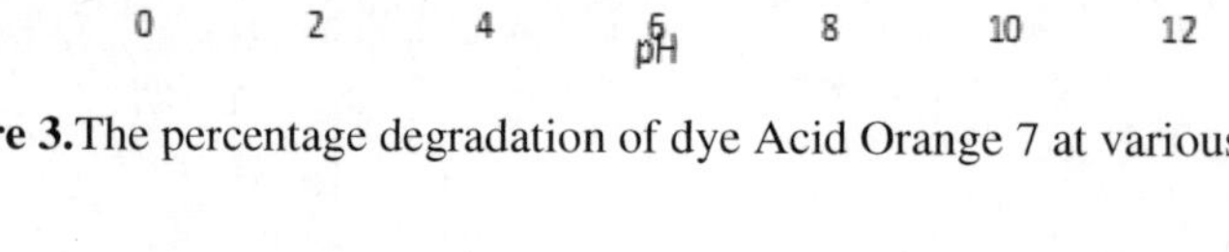

Figure 3. The percentage degradation of dye Acid Orange 7 at various pH.

IOP Conf. Series: Materials Science and Engineering **551** (2019) 012129 doi:10.1088/1757-899X/551/1/012129

Under acidic conditions, the surface of the catalyst is covered by the Acid Orange 7 dye molecule which causes absorption of UV radiation on the surface of the catalyst to decrease. As in alkaline conditions, the degradation of Acid Orange 7 less effective. This is because under alkaline conditions the dye will cause many OH- bound to the surface of the catalyst. The amount of OH- which is bound to the surface of the catalyst, will prevent the absorption of dyes to the surface of the catalyst. So, the process of photodegradation decreases. Increasing pH will reduce the presence of OH radicals obtained from the results of dye degradation by photocatalysts [2].

The best time to degrade Acid Orange 7 is for 380 minutes with a degradation percentage of 75.33%. The structure of Acid orange 7 is quite stable and has one azo group (N = N). So that the time needed by hydroxyl radicals is longer to break the unsaturated bonds found in Acid Orange 7.

Determination of optimum concentration to know the best concentration on the photodegradation of Acid Orange 7. At concentrations of 150 ppm and 300 ppm there was a decrease in the percentage of degradation of 73.56% and 34.95% respectively. This occurs because the higher concentrations of dyes, the number of molecules is more.Therefore, it requires a higher number of hydroxide radicals. The percentage of degradation decreases with the increasing of concentration. This is because the catalyst has reached the maximum limit in degrading dyes. the concentration needed to degrade Acid Orange 7 dyes is 50 ppm with a degradation percentage of 94.15%.

4. Conclusions

It was concluded that pH affected the degradation activity of Acid Orange 7 using ZnO impregnated zeolite (ZnO-Zeolite) with a degradation percentage 94.15% at optimum pH 7.

References

[1] Afrozi *et al* 2014 *IX- Batan, ISSN : 1978.*
[2] Ansori *et al* 2016 **19**(2).
[3] Bhernama *et al* 2017 *Journal Of Islamic Science and Technology* **3**(1).
[4] Delshade *et al* 2011 *Monatsh Chem* **142** 119.
[5] Modirshahla *et al* 2011 *Journal Elsevier Desalination* 271.
[6] Mohaghegh *et al* 2014 *Material Science in Semiconductor Processing* 167.
[7] Rani S *et al* 2008 *Solar Energy Material & Solar Cells* **92** 1639.
[8] Rianto *et al* 2012 *Alchemy* **2**(1).
[9] Wismayanti *et al* 2015 *Chemistry Journal* **9**(*1*)

IOP Conf. Series: Materials Science and Engineering **551** (2019) 012130 doi:10.1088/1757-899X/551/1/012130

Research Design and Identification of the Project Methodology Solutions Using the Finite Element Method

Alina Bianca Pop[1], Aurel Mihail Țîțu [2,3], Gheorghe Ioan Pop[4], Costel Ceocea[5] and Ștefan Țîțu[6]

[1]SC TEHNOCAD SA, 72, Vasile Alecsandri Street, Baia Mare, Romania
[2]Lucian Blaga" University of Sibiu, 10, Victoriei Street, Sibiu, Romania
[3]The Academy of Romanian Scientists, 54, Splaiul Independenței, Sector 5, Bucharest, Romania
[4]SC UNIVERSAL ALLOY CORPORATION EUROPE, 244A, Principală Street, Dumbrăvița, România
[5]Vasile Alecsandri" University of Bacău, 157, Calea Mărășești, Bacău, România
[6]The Oncology Institute "Prof. dr. Ion Chiricuță", 34-36, Republicii Street, Cluj-Napoca, România

Abstract. The anticipation of the process variables (temperature, cutting forces, and stress distribution) plays an important role in the tool geometry design and cutting conditions optimization. The researchers found that the anticipation of these variables experimentally is costly both from a time requirement and financial point of view. Having this information, the finite element model becomes the most favourable alternative and most useful in all aspects. Existing Empirical models are very useful in demonstrating these matters, but if tests have not been carried out to create these models types, then it brings benefits in the improvement specifically to the process itself. This paper is focus to conduct the research and discover a methodical design solution based on the finite element method. This entails defining the input data used within the research framework, including: application type, finite element analysis, material model, and, of course, choosing the process parameters. After determining the input data, following the completion of the theoretical models of the experiment, respecting the finite element model, the output will take into account the final results of the selected research. In order to obtain optimal solutions and to increase the precision and accuracy of the obtained results, it is necessary to validate this using the experiment.

1. Introduction

For this research, a wide bibliographic material, which includes over 200 national and international references, was studied. Throughout this paper, reference will be made to the most important among them. Since the 2000s, the FEA experimental method has been advanced [1]. Regarding to the modelling and experimental technique chosen to be used, the researchers is very important to understand, not only their capability, but also the existing theory. Due to the cutting process complexity, more and more frequently, numerical approaches have been developed and adopted to replace direct experimental approaches; which are costly and time-consuming. Among the numerical techniques, the most frequently encountered during the past two decades is the Finite Element Method [2]. Using the Finite Element applications, the cutting process has focused particularly on: setting of the material properties related to the deformation, strain rate and temperature [3]; modelling tool - chip interface depending to the friction [4]; the functions and variables of the analysis, such as the cutting forces, the chip geometry, stresses, deformation, temperature distribution in the workpiece, etc. [5]. Therefore, in order to achieve a research focused on the experiment with the finite element, certain conditions must be fulfilled and reached, such as: selection of the FEA application; selection of the simulation method; introducing the material properties and the cutting tool properties in the FEA application; selection of the material model; selection of the cutting process, the cutting regime and the geometric parameters of the tool which will constitute the influence factors of the cutting process; selection of geometric, state and actions parameters that are analysed in the experiment.

2. Research design using the finite element methods

2.1. Modelling type selection

Currently, there is a wide range of finite element applications. These applications are based on two modelling types: the implicit modelling [6] and the explicit modelling [7]. The finite element applications which assume an implicit modelling, offers a variety of factors such as: bars, plates, membranes, axial symmetrical, 3D general solid. When choosing certain items types, the imposing automatic mathematical models on which they are based should be taken into account. The explicit modelling involves the selection of the mathematical model of the physical problem. In this situation, there are two possibilities: either searching for a finite element analysis program which can be used to solve the mathematical model, or creating such a program. Therefore, the explicit modelling involves much more experience in programming and resources than the implicit modelling. Generally a combination between implicit and explicit modelling is used. Physical problem to be modelled is divided into sub problems. Therefore, the conventional sub problems for which exist corresponding resolution pro-grams may be resolved by implicit modelling, while those requiring special processing can be resolved only by explicit modelling.

2.2. Selection of the FEA application

Among the applications that can be used to the simulation achievement, DEFORM is one of the most used. In the Figure 1 it is indicated as a percentage grade used by FEA applications during the past two decades.

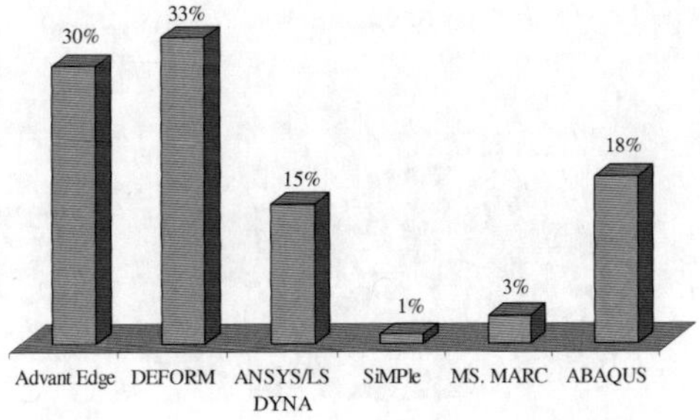

Figure 1. A percentage grade used by FEA applications during the past two decades (1998 – 2018).

2.3. Selection of the modelling technique

In the finite element method application the calculation of the main structures, there are mainly two methods of describing the material motions associated with the mesh network of the structure, which are the Euler and Lagrange models [8]. The strong point of each method are actually the weak point of the other, and for this reason the two approaches actually complement each other. Arbitrary Lagrange - Euler (ALE) has been developed primarily to benefit from the advantages of the two traditional techniques and to avoid their disadvantages, using each technique when it may be advantageous to do so. From another perspective of both described methods, both Lagrange and Euler are based on the definition of a meshing network [9]. To re-mesh the bodies which recorded large deformation, it is recommended to use the SPH technique (Smooth Particle Hydrodynamics) in which, unlike the previous models, there no longer appear issues related to severe distortion of the elements which would lead to numeric instability, [9].

In Figure 2, an analysis of the methods justifying the FEA experiments during the past two decades has been shown. As is clear from the graph shown in Figure 2, Lagrange was the most used technique, due to the fact that the procedure has advantages in terms of meeting the desired objectives pursued in carrying out experiments with the cutting processes simulation.

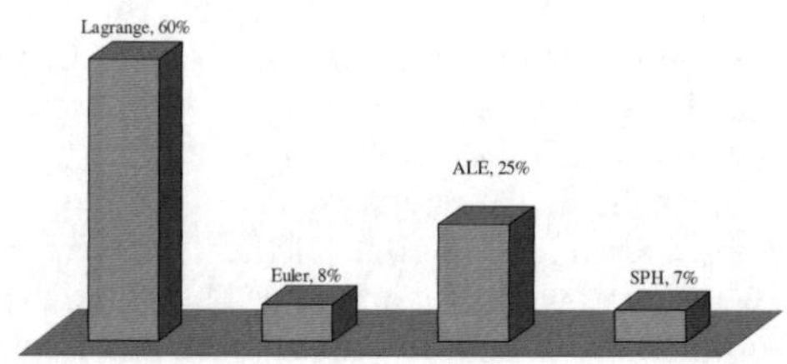

Figure 2. Percent comparison of FEA experiments during the past two decades.

2.4. Material type selection

Once the conditions determining the application of the finite element analysis and the experimental methods necessary to conduct the research, another step consists in choosing the material type. The selection and use of a particular material model is one of the most important components of the simulation process using the finite element method, since the material selection fundamentally influences the analysis results. In the cutting process, the more the chosen material model approaches the behaviour of the workpiece, the more accurate the results are [12]. The evolution of the use of different types of material models from 1998 to the present is highlighted in Figure 3.

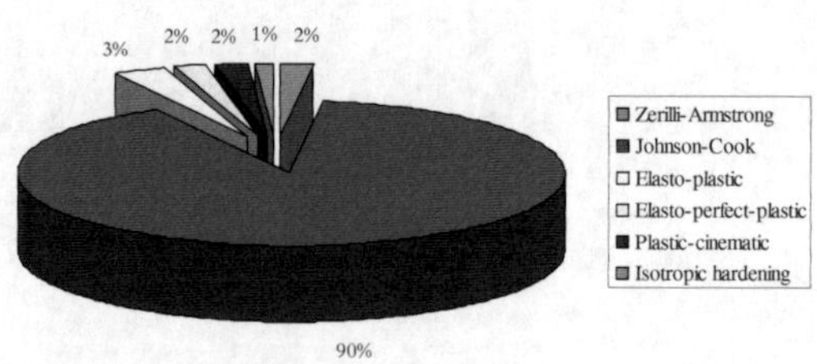

Figure 3. Comparison of the use of the various types of material models.

2.5. Selection of the workpiece material

The next necessary condition to perform the finite element simulation of the cutting process and experiment is constituted by the selection of the workpiece material [13]. Figure 4 shows the percentage comparison of the frequency of the work-piece materials used, from which it can be seen that steels were most often used, accounting for more than 60% of the total.

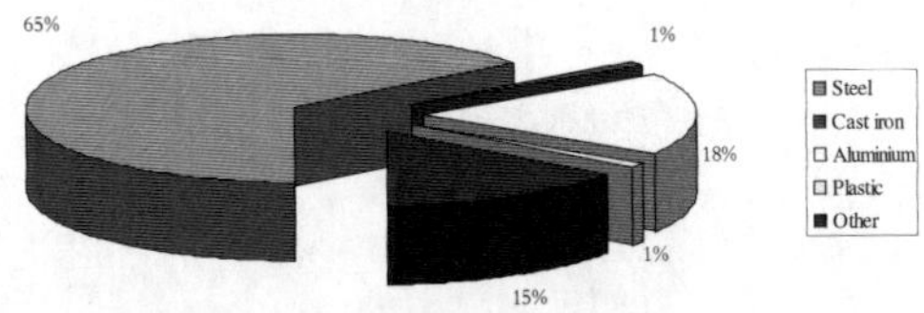

Figure 4. Percentage comparison on the frequency of the workpiece materials.

2.6. *Selection of the cutting operation and process parameters*

Another necessary step to be accomplished in carrying out the researches pursued in this paper is the selection of the cutting operation. According to the bibliographic study it was found that turning is the most frequent operation approached in the cutting processes study. In figure 5 is highlighted the percentage comparison on the frequency of the cutting operations used, clearly the superiority of the turning process, consisting of 50% of the total studies identified between 1998 and 2018 [14, 15].

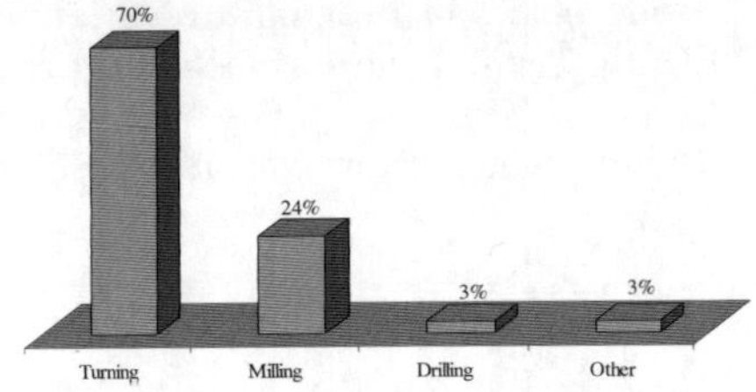

Figure 5. Percentage comparison on the frequency of the cutting operations.

2.7. *Selection of the influence factors*

Another aspect to be defined is to determine the tool geometry parameters died in the simulation. Figure 6 shows that the most commonly used element was the clearance angle of the tool, which was identified in 54% of all the bibliographic material studies. At the same time, it is necessary to establish the cutting regime parameters. It was found that the most used parameter in this respect is the cutting speed, which accounts for almost half of the total (Figure 7) [15, 16].

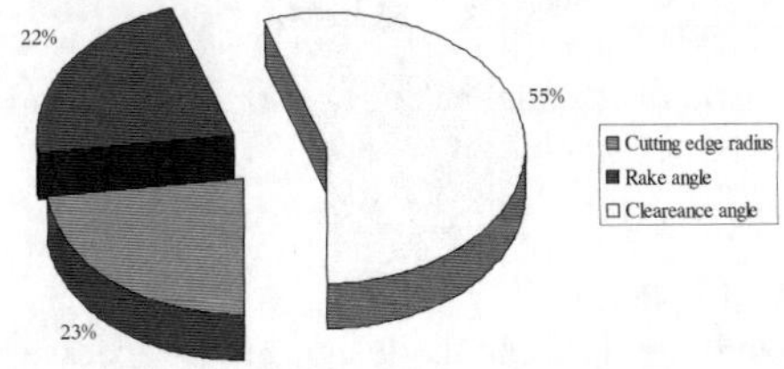

Figure 6. Percentage comparison of the geometrical parameters of the tool studied in the simulation.

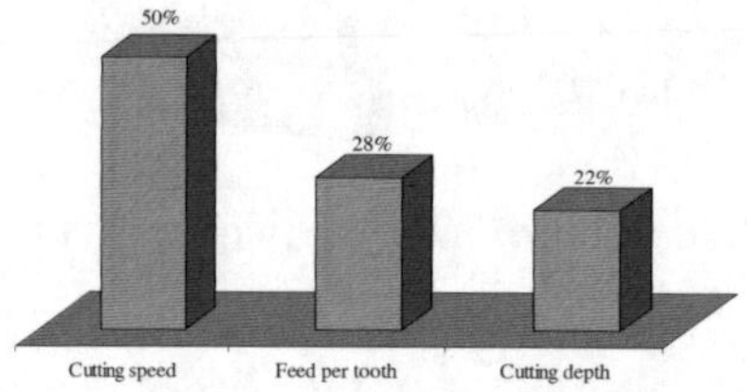

Figure 7. Percentage comparison of the studied cutting regime parameters.

2.8. Response selection

Once the influence factors have been established, it is necessary to set the research response. Figure 8 shows the most common responses. It was found that the most studied geometric factor is the chip formation, which appears as a study object in over 60% of the research papers identified in the bibliographic material [17, 18, 19, 20]. The influence of the cutting regime parameters is also felt on the state and action factors. As Figure 9 shows, following the frequency comparison of these parameters, it can be seen that the most studied factors were tensions, forces, and temperature [21].

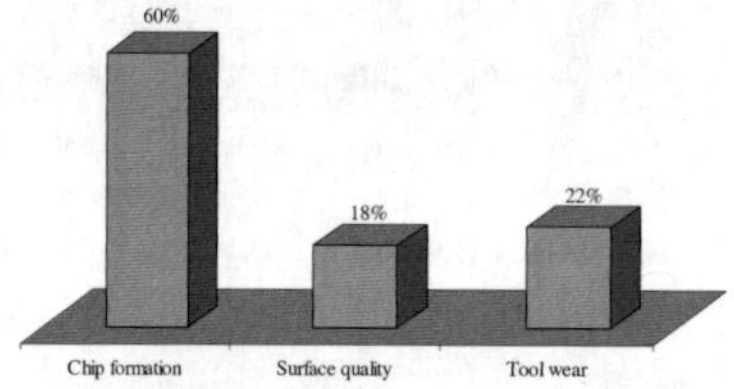

Figure 8. Percentage comparison of the studied geometrical factors.

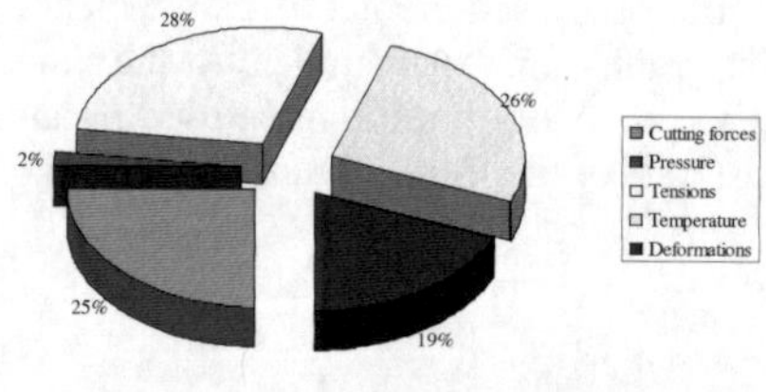

Figure 9. Percentage comparison of the analysis of the state and action factors in the cutting process.

3. Designing the experimental research

In order to achieve the experimental research design, it is necessary to go through some stages consisting of: theoretical model achievement; experimental method identification; the workpiece material selection; the tool type and cutting regime selection; measuring devices selection; design of fixing devices; performing the experiment (preparation of the workpiece, tools, devices and CNC machine; conduct an experiment simulation; necessary adjustments; start the experiment in a predefined order of the types and number of measurements; completing the experiment and making the final measurements; recording the obtained values);data collection and information centralization.

4. Conclusions

This paper presents the needed steps to design a research and identify methodical design solutions using the finite element method. This approach has been chosen, given that the variables anticipation of the experimental cutting process is costly both in terms of time as well as financially. Therefore, the finite element simulation becomes the most favourable and most convenient alternative in all aspects. In order to achieve the proposed objective, it was necessary to establish the input data used in the research, which includes: the application type used, the finite element analysis method, the material model, and, of course, the process parameters selection. After determining the input data following the theoretical models of the experiment, respectively the finite element simulation, the output data will take into account the determination of the studied research direction. In order to obtain the optimal solutions and at the same time to increase the accuracy of the obtained results, at the simulations completion, it is necessary to perform the results validation with the experiment.

References

[1] Markopoulos A P (2012) *Finite Element Method in Machining, Springer Science & Business Media.*
[2] Pop A B *et al* 2015 *AJME* **13** 1.
[3] Arulkirubakaran D *et al* 2016 *Int. J. Refrac. Met. And. Hard. Mater.* **54** 165.
[4] Thepsonthi T *et. al* 2015 *J Mater. Proc. Technol.* **221** 128.
[5] Kohir V *et al* 2014 J. Eng. Fundamentals **1** 30.
[6] Harewood F J *et al* 2007 *Comp. Mater. Sci*, **39**(2), 481.
[7] Klocke F 2010 *Manufac. Proc. 1 Cutting*. Springer Verlag.
[8] Coman G *et al* 2007*An. Adv. Course*. Ed. Presa Universitara Clujeana.
[9] Houcque D 2008 Eng. and App Sci **12** 1.
[10] Wang F *et al* 2015 J. Alloys. and Comp., **633** 220.
[11] LSTC, LS-DYNA 2014 Keyword User's Manual Volume II R7.1 (Vol. II).
[12] Zhang Y *et al* 2015 Procedia CIRP, **31**, 112
[13] Pop A B 2018 Procedia-Social and Behavioural Sciences **238**, 355.
[14] Kalyan C *et al* 2015 J Mat. Proc. Tech. **216**.
[15] Pop A B *et. al* 2017 MATEC Web of Conferences (EDP Sciences 2017)
[16] Țîțu A M 2017 MATEC Web Conf., **112**, 01009
[17] Mabrouki T *et al* 2016 Comptes Rendus Mécanique, **344** 335.
[18] Filippi E *et al* 2015 Procedia CIRP **31** 124.
[19] Camargo J C *et al* 2014 Int. J. Ref. Met. Hard Mat. **47**, 61.
[20] Chiappini E *et al* 2014 IJMTM, **77**, 16
[21] Sui S C *et al* 2015 IJAMT, **2** 9.

IOP Conf. Series: Materials Science and Engineering **551** (2019) 012131 doi:10.1088/1757-899X/551/1/012131

Reduction of PM10 Emission from A Retrofitted Multicyclone

M. Dewika[1,*], M. Rashid[2], M.P. Khairuninisa[3], J. Ruwaida[4] and Y.Y. Sara[5].

[1]Centre of American Education, Sunway University, Bandar Sunway, 47500 Selangor, Malaysia.
[2,3,4]Air Resources Research Laboratory, Malaysia Japan International Institute of Technology, 54100 UTM Kuala Lumpur, Malaysia.
[5] School of Environmental Engineering, University Malaysia Perlis, 02600 Arau, Perlis, Malaysia.

E-mail: dewikan@sunway.edu.my

Abstract. Multi-cyclone is one of the particulate control systems commonly used in the industry due to its low capital, operating, and maintenance cost as well as its reliability under a wide range of operational conditions. Unfortunately, it has low performance in collecting fine particulate especially of less than 10 μm in diameter or PM10. Thus, this limitation motivated a cost-effective retrofit of the existing conventional multi-cyclone. The retrofit was performed by creating a more negative pressure at the dust hopper of a conventional multi-cyclone, by drawing out a small portion of the air using an Induced Draft Fan. The particulate emission concentration and overall collection efficiency of the unit at various effective inlet velocities ranging from 6.2 to 19.8 m/s with particulate loading of 30 mg/m^3 of 100 μm sample size diameter of Palm Oil Mill Boiler Fly Ash were investigated in this study. Interestingly, the introduction of suction increased the overall performance of the multi-cyclone by 10 to 20%. The finding suggests that a simple retrofit managed to increase the overall performance of the conventional multi-cyclone even for PM10.

1 Introduction

Air pollution is defined as an introduction of particulate matter (hereinafter PM), chemicals or biological materials that cause harm or discomfort to humans and other living organisms. These discomforts are damaging to the natural and built environment as the PM is released into the atmosphere [1]. PM is a by-product of industrial processes such as combustion, grinding, crushing and many more, which needs to be controlled and removed from being emitted into the atmosphere as it creates adverse effects to human.

There are many types of particulate control systems available in the market such as cyclone, electrostatic precipitator, bag house filters and scrubber. Generally, cyclone is the most common among all, and it is used for removing particulates from contaminated flue gas due to its low capital, operating, and maintenance cost and as well as its reliability under a wide range of operational conditions [2].

In industry, conventional multi-cyclone (hereinafter CMC) which is comprised of several small-diameter cyclones arranged in parallel with a common gas inlet and outlet is widely used commercially.

It is preferred due to its capability of handling massive volumetric flow of contaminated flue gas which cannot be handled by a single cyclone [3], [4].

Unfortunately, multi-cyclone is more efficient in collecting coarse PM rather than fine particulate. Thus, in many cases, multi-cyclone is used merely as a pre-cleaner unit upstream of a more efficient air pollution control system as it is not effective in meeting the stringent particulate emission requirement [5]. Therefore, it is the primary reason for being utilised as a pre-cleaner at the upstream of other devices to reduce dust loading and remove larger particulates from the stream [6].

Thus, the performance of a retrofitted multi-cyclone (from now on refer as, RMC) on the collection of fine particulate size fraction of PM10 was investigated in this study. The experimental findings obtained from a pilot plant unit of the CMC and RMC were compared and presented in this study.

2 Methodology

2.1 Particulate Sample

Palm Oil Mill Boiler Fly Ash (POMBFA) generated from the combustion of palm fiber and shell in the palm oil mill boiler was used as a dust sample in the experiments. The sample collected from a mill was mixed thoroughly to ensure homogeneity and dried for 24 hours at 110 °C to remove moisture. The dried POMBFA was sieved and categorized into size fraction below of below 100 μm via sieving and used for experimental runs. The particulate size distribution of the sample was further determined by using a micro-sieve (Sieve vibrator Model Octagon Endecotts).

2.2 The Pilot Plant

The CMC pilot plant consists of a dust feeder, pressure meter, induced fan (IDC), and stack. It is fitted with four identical miniature cyclones in a single compartment and equipped with a control panel [7]. The speed of the IDC fan, which controls the volumetric airflow rates of MC is regulated by the control panel. Figure 1 presents the schematic diagram of the CMC and its retrofitted part shown in blue, which includes an additional suction inlet pipeline with a suction valve, an ID fan (IDS) connected to a stack with a sampling port. A separate control panel is attached to manipulate the IDS fan as a mean to control the volumetric air flow rate of the suction.

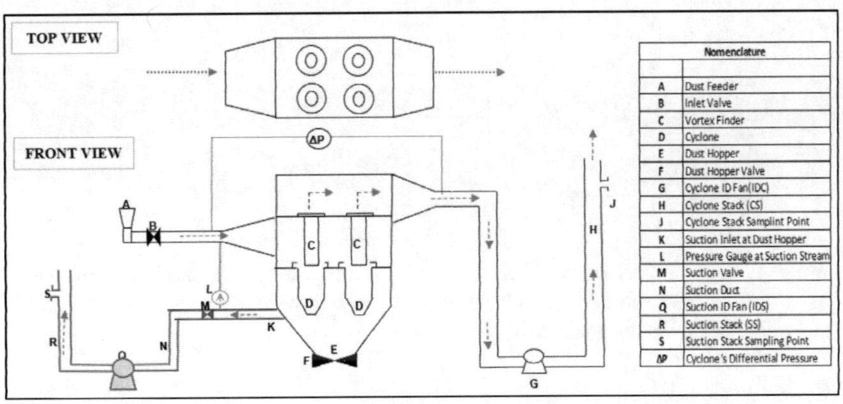

Figure 1. The schematic diagram of the CMC with its retrofitted component in blue.

2.3 Performance Test

The performance of the CMC and RMC was determined through the total particulate emission concentration measured iso-kinetically at the cyclone stack following the USEPA Method 17: Determination of Particulate Emissions from Stationary Sources In Stack Filtration Method. In addition, the particulate emission and its size distribution at the cyclone stack was determine using GRIMM

Portable Laser Aerosol Spectrometer dust monitor (Model 1.108/1.109). The experiments were conducted at varies volumetric air flow rate ranging between 0.11 to 0.35 m^3/s corresponding to effective inlet velocity of 6.2 to 15 m/s with constant mass loading of 30 mg/m^3 using POMBFA as the test sample. Similar experimental procedures were repeated on the RMC applying suction by withdrawing a portion of volumetric airflow at the dust hopper of the unit. The percentage suction was from 0 to 20%, representing the ratio of airflow rate extracted from the dust hopper to the total amount of airflow rate passing through the cyclone.

3 Results and Discussion

3.1 Particulate Distributions

Figure 2 presents the mass cumulative percentage of POMBFA sample fed and emitted from the CMC and RMC, which showed a similar pattern between them but shifted apart. The mass cumulative distribution of the sample fed into the cyclone was determined using sieving method while the mass cumulative distribution of the cyclone was based on GRIMM's readings. It is observed that the particulate emitted from the stack represents smaller size fraction with 50% mass cumulative size 2.7 μm compared to 45 μm of the raw sample. Interestingly, Hasyimah et al., (2017) also found that the particulate 50% cut size of the CMC was 2.5 μm, which is similar to this study. This construes that the existing CMC is efficient in removing fine particulate size fraction. However, introduction of suction has further improved the efficiency of CMC by shifting the cut size to 2.1 μm as shown in Figure 2.

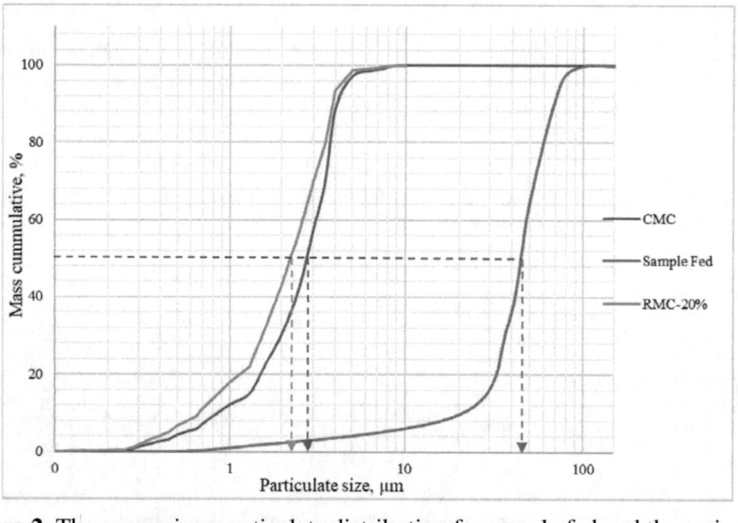

Figure 2. The comparison particulate distribution for sample fed and the emission concentration from the CMC and RMC (i.e. with 20% suction) at cyclone flow rate 0.11 m^3/s with constant mass loading of 30 mg/m^3

3.2 Experimental run on CMC

The overall efficiency is equal to the ratio of the difference between the inlet and emitted particulate concentration of the cyclone separator to the inlet particulate concentration [8]. Figure 3 presents the collection efficiency of the CMC with respect to the effective inlet velocity at mass loading of 30 mg/m^3, which showed that the collection efficiency increases with effective inlet velocity. The effective inlet velocity is the ratio of volumetric air flow rate by the entry area of the axial cyclone successively means an increase in effective inlet velocity also indirectly increase the volumetric air flow rate.

As in Fig 3 and based on the best-fit curve, the there is an increase in collection efficiency with effective inlet velocity till 14 m/s, thenceforth the collection efficiency started to decline. Hence, the maximum collection efficiency at 14 m/s was 88% for 30 mg/m^3. As the inlet velocity increases, the centrifugal force of the particulate increased contributing to greater efficiency [9]. However, too high inlet velocity decreases the collection efficiency because of increased in turbulence and re-entrainment of particulate.

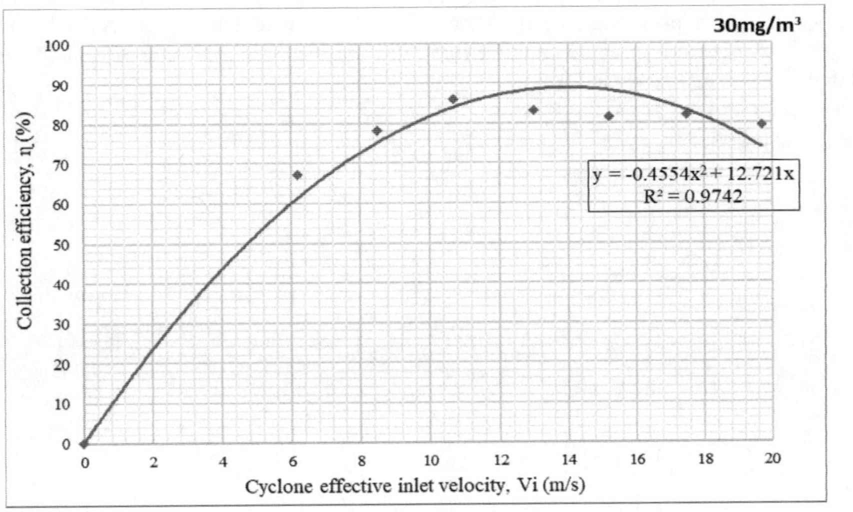

Figure 3. Overall collection efficiency of CMC at mass loading 30 mg/m^3 at different inlet velocities.

3.3 Experimental run on the RMC

Figure 4 depicts the collection efficiency of the retrofit cyclone at varied cyclone flow rates with different percentage of suctions where similar trend lines are observed in the experiment. As in the figure, the RMC with suction percentage of 10, 15 and 20% showed a positive upward trend representing an increase in collection efficiency with the increase in percentage of suction. The intercept values of 67, 78 and 86 in Figure 4 represent the collection efficiency of the CMC namely at 0% suction, where collection efficiency of the CMC increases with the flow rate of the cyclone. The collection efficiency of the RMC at 20% suction was 83, 93 and 95% for the flow rate of 0.11, 0.15 and 0.19 m^3/s, respectively. This affirms that the cyclone's total overall performance for the fine particulate control has increased with the presence of the retrofit by 9 to 16% compared to the CMC alone. It is observed that the increment is higher at lower cyclone flow rate compared to higher cyclone flow rate, which is represented by the high slope value in Figure 4. Therefore, it is anticipated that suction pull is more effective at lower cyclone flow rate because the suction force needs to overcome lower swirling force of the cyclone at lower flow rate. Huang et al., (2018) stated that the particulate concentration distributions can be roughly classified into two regions in the cyclone where at the cylindrical part of the cyclone, the particulate concentration is lower and follows the gas spiraling downward flow patterns. While at the conical part of the cyclone, the reducing of the cone radius caused the acceleration of the gas tangential velocity. The enhanced centrifugal force by the increasing of the gas tangential velocity moved the particulate toward the conical wall. With the decreased of the conical volume and the increased particulate numbers around the wall region, the particulate concentration was higher at the conical part of the cyclone. Since the retrofit creates a pull of air at the dust hopper by having suction, these particulates along with air will travel further into the dust hopper, creating more collision among particulate, increasing collection at dust hopper which explains the increase in cyclone performance with presence of suction.

As in Figure 4, the suction created at the dust hopper creates a pull of air toward the suction stream line leaving a reduction of mass and enhanced the negativity of pressure in the hopper and the suction duct. Naturally, a negative pressure area would be balanced by flow of air from a higher pressure to a lower to balance the pressure at that point. Similarly, to replace the mass of air that is being pulled, it is anticipated that the air flow in the conical tip would flow downwards into the hopper carrying the vortex further down into the hopper. Higher pull of air due to high flow rate at suction stream which is created by higher suction percentage would create a very high-pressure difference. This phenomenon explains the higher collection efficiency at 20% suction compared to the lower ones.

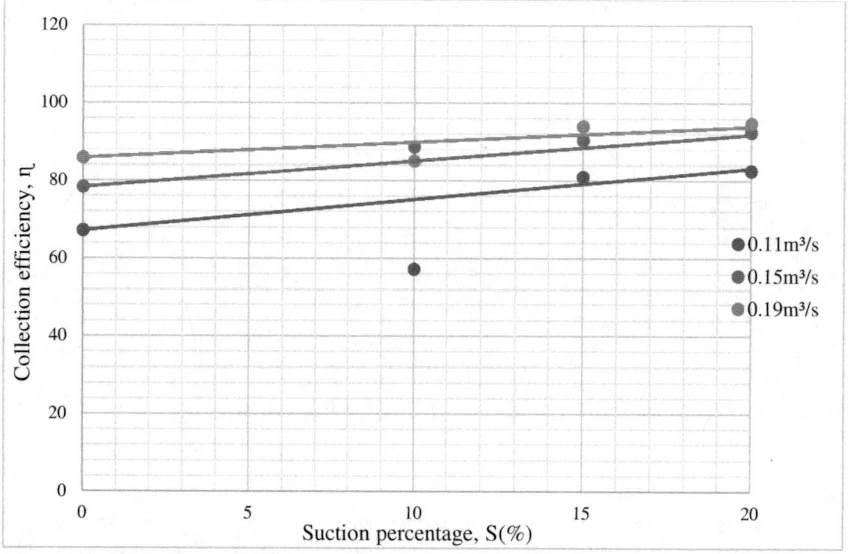

Figure 4. The collection efficiency of retrofitted multi-cyclone at different cyclone flow rate with various suction percentage.

Figure 5 depicts the emission concentration of PM10 sampled at the cyclone stack under influence of varied suction percentage at constant cyclone flow rate 0.11 m^3/s with fixed inlet mass loading of 30 mg/m^3 which showed the reduction in emissions with increase in suction percentage. This figure clearly depicts that the CMC (i.e. 0%) has higher PM10 emissions compared to RMC for PM10 size fraction. This downward trend affirms that PM10 has reduced by 50% with the presence of 20% suction at the RMC.

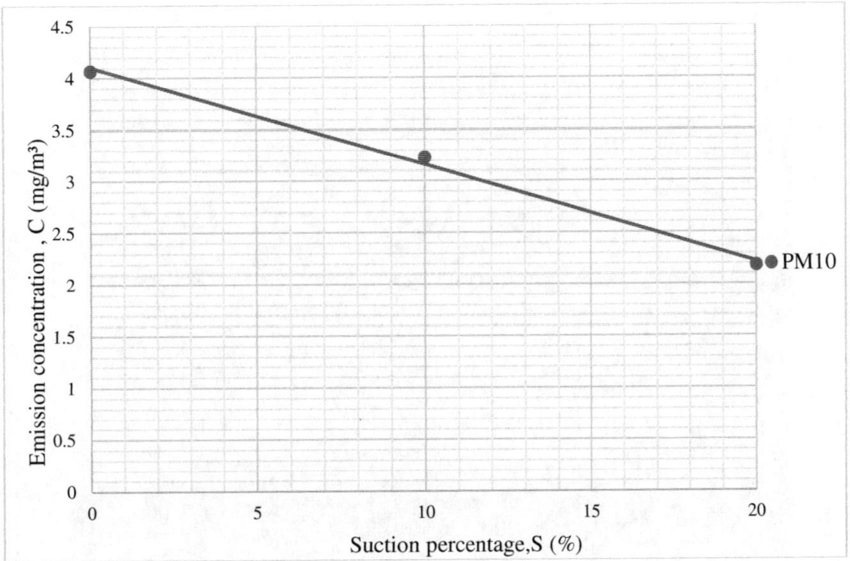

Figure 5. PM10 emission concentration at the cyclone stack under various suction percentage at cyclone flow rate $0.11 m^3/s$.

3.4 Morphology of Particulate Collected by CMC and RMC

Figure 6 depicts the SEM images of the POMBFA particulates collected on the filter media, scanned for their diameters randomly to determine the particulate cut-size of the cyclone at the respective operating conditions. Moreover, the SEM was used to support and verify the results from GRIMM. Figure 6 (a) displays the image captured on a blank filter while, Figure 6 (b, c, and d) showed particulates on CMC and RMC sampled filter under 10 and 20% suction. These images clearly showed that the particulate size emitted is reduced with RMC compared to CMC. An increased in suction flow rate decreases the amount of fine particulate and its particulate cut-diameter. Smallest particulate diameter 3 μm is measured at volumetric air flow rate of 0.11 m /s at suction 20% while the largest is 9 μm at volumetric air flow rate of 0.11 m /s at zero suction.

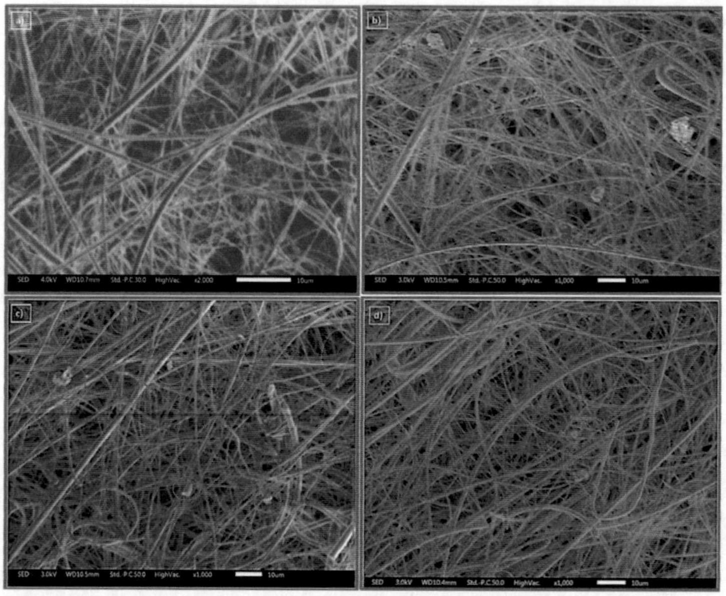

Figure 6. Filter paper sampled at cyclone stack a) blank filter; b) CMC (0% suction); c) RMC -10% suction and d) RMC - 20% suction.

4 Conclusions

Several conclusions can be drawn from the study and these are;

i. The CMC demonstrated increase in collection efficiency with the effective inlet velocity. However, beyond 14 m/s, the collection efficiency declined due to particulate re-entrainment.
ii. The introduction suction at the RMC has further increased the collection efficiency of the existing CMC by decreasing the cyclone's cut size from 2.7 to 2.1 μm.
iii. In addition, RMC also demonstrated reduction in PM10 emissions at the cyclone stack by 50% for constant cyclone flow rate of 0.11 m^3/s at fixed inlet mass loading of 30 mg/m^3.

Acknowledgements

This work was financially supported by FRGS from Ministry of Education (MOHE), Malaysia and MJIIT UTMKL, Malaysia. The authors would like to express appreciation for the support of the GUP Tier 2 Grantt of Q.K130000.2643.15J50

References

[1] P. O. Boamah, *et al.* 2012 *Glob. J. Biosci. Biotechnol.* **1(2)** 124.
[2] T. C. Hsiao *et al.* 2015 *J. Aerosol Sci.* **86** 1.
[3] P. Baltrėnas and A. Chlebnikovas 2015 *Sep. Sci. Technol.* **50(8)** 1257.
[4] K. S. Lim *et al.* 2004 *J. Aerosol Sci.*, **35(6)** 743.
[5] P. S. Sankar and R. K. Prasad 2015 *Part. Sci. Technol.***33(4)** 385.
[6] M. Sangkhamanee *et al.* 2011 *Part. Sci. Technol.* **29(4)** 345.
[7] N. Hasyimah 2017 *Modelling and Optimization of Mr-Deduster as a Fine Dust Emission Control.*
[8] Z. Ji, *et al.* 2009 *Powder Technol.* **191(3)** 254.
[9] A. Avci and I. Karagoz 2003 *J. Aerosol Sci.* **34** 937.
[10] A. N. Huang 2018 *J. Taiwan Inst. Chem. Eng.* **0**, 1.

Characteristics of Wild Orchids in Mallawa Resort at Bantimurung Bulusaraung National Park, South Sulawesi, Indonesia

Siti Fatmah Hiola[1], Gufran Darma Dirawan[2], Muhammad Wiharto[1], Syamsiah[1]

[1]Biology Department, Universitas Negeri Makassar, South Sulawesi, Indonesia
[2]Environmental Education Studies, Postgraduate Program, Universitas Negeri Makassar, South Sulawesi, Indonesia.

E-mail: fatmah.hiola@unm.ac.id

Abstract. This study aims to determine the characteristics of wild orchids in the area of Resort Mallawa, Bantimurung Bulusaraung National Park. This type of research is descriptive with the exploration method, which is tracking the study area which is known to have a fairly high distribution of wild orchids. Exploration of wild orchids is carried out at 85 distribution points, with an altitude of 432 meters above sea level to 849 m above sea level. The results showed that there were 40 types of orchids, 33 species identified, and 7 species not identified. Based on the growth properties of wild orchids found were dominated by 35 species of epiphytic orchids, while for the type of growth was dominated by 24 species of sympodial orchids. The most found locations of wild orchids are in Mallenreng Hamlet I. The most common species is *Trichoglottis geminata*. Based on the height of the place, the species that grows at the highest position is *Coelogyne* sp., while in the lowest position is *Dendrobium* sp. 1. Based on the visual intensity of light, orchids were found in the shade as many as 5 species, half sheltered as many as 21 species, while in the open as many as 4 species. There are 5 orchid species that can grow in the shade and half shade, 3 species that grow in half shade and open places, and 2 species that can grow in three places. By knowing the presence and diversity of wild orchids in the study area shows that the ecosystem in the area is still quite good, considering that wild orchids are plants that are quite susceptible to interference.

1. Introduction

The diversity of Indonesian orchids that are spread in various regions of the archipelago is a potential that has not been fully explored. The study of wild orchids is essential to increase knowledge about the possibility of flora in Indonesia.

Experts say that in the world there are an estimated 50,000 species of wild orchids collected in 1,200 genera [1]. Orchids are classified as members of the Orchidaceae genera. In Indonesia, there are around 5,000 orchid species which are spread in almost all parts of Indonesia. The islands in Indonesia which have known the number of species of orchids, among others in Java as many as 731 species, Sumatra as many as 1118 species, Borneo as many as 2000 species, Sulawesi and Maluku as many as 820 species, with 548 species of which are in Sulawesi [2]–[6].

Recorded species of orchids that are endemic in Sulawesi are estimated to be around 253 species. It shows that 80% of the total types of orchids found on Sulawesi Island are endemic [7]. The diversity of types of wild orchids in Sulawesi, especially in Bantimurung Bulusaraung National Park, Maros

Regency, South Sulawesi, recorded 90 species of wild orchids with 43 genera, 5 species have not been identified. The types of orchids that dominate the area are epiphytic orchids of 70 species [8].

One of the conservation areas in South Sulawesi that is often found in the presence of wild orchids is in the Bantimurung Bulusaraung National Park area, commonly abbreviated as Babul National Park (Babul NP). It occupies an area of 43,750 hectares between 119°34'17"-119°55'13" East and 4°42'49"-5°06'42" South. The area in Babul NP has eight resorts including, Mallawa Resort, Pattunuang-Karaenta Resort, Bantimurung Resort, Camba Resort, Balocci Resort, Minasate'ne Resort, Tondong Tallasa Resort, and Butterfly Park Resort. Based on the results of interviews from the officer of Babul NP, stated that Mallawa Resort has a better diversity of wild orchids than other resorts including the area in Babul NP. Therefore, it is necessary to collect data or record the diversity of orchids to become a database for the development of wild orchids in Bantimurung Bulusaraung National Park. [8].

2. Research Method

This type of research is descriptive with exploratory methods, where observations of objects to be studied are carried out by tracing the study area, to collect images of objects (photo documentation), then identify each wild orchid found through orchid characterization.

This research was carried out in the Bantimurung Bulusaraung National Park area in the Mallawa Resort area. This research was conducted for four months (February-June 2017). Data retrieval is done by exploring and observing directly in the research area. This exploration area consists of 4 locations, namely: Barugae Village; Bentenge Village, Mallenreng I Hamlet, and Mallenreng II Hamlet, Samaenre Village.

The tools used in this study were digital cameras (optical zoom 63x, 20.1MP), magnifying glass, GPS Garmin Montana 655, binoculars, maps, literature/orchid identification reference books and writing instruments. The material used is orchid specimens, tally sheet, label, and plastic bags. Before conducting research, first, determine the wild orchid exploration route based on information from local people and field officers.

Characterize each specimen with identification, determine the nature and type of growth, and visual needs of light. The identification stage of orchid species is based on direct observation and matching documentation images with literature. Identification is carried out based on plant organs including roots, stems, leaves, flowers and fruit (if any).

The literature / book regarding the reference consists of Flora [9], The Orchid Book: A Guide to the Identification of Cultivated Orchid Species [10], Orchid of Sulawesi [11], Native Orchid of Indonesia [6], Koleksi Anggrek Kebun Raya Bogor [12], Orchids of Java [13], An Alphabetical List of Indonesian Orchid Cultivated in Bogor Botanic Garden [14], Orchids of Indonesia [3], and Tropical Orchids of Southeast Asia [2]. Besides that, it also uses several website sources, such as http://plants.usda.gov; http://flnativeorchids. com; www.flora.dempstercountry.org; www.theplantlist.org/;

3. Results and Discussions

Based on the results of the research, it is known that the types of wild orchids at Resort Mallawa, Bantimurung Bulusaraung National Park Region have quite diverse types. The types of wild orchids found in the area are presented in Table 1 below.

Table 1. The types of wild orchids found at Resort Mallawa in Bantimurung Bulusaraung National Park.

No.	Latin Name	Indonesian/Local Name	Habitus	Growing Type
1	*Abdominea minimiflora* (Hook. f.) J. J. Sm.	Anggrek peru	Epiphytic	monopodial
2	*Aerides inflexa* Teijsm. & Binm.	Anggrek lilin	Epiphytic	monopodial

No.	Latin Name	Indonesian/Local Name	Habitus	Growing Type
3	*Aporum uncatum* (Lindl.) Brieger	-	Epiphytic	Sympodial
4	*Arundina* sp.	Anggrek bambu	Litophytes	Sympodial
5	*Bulbophyllum* sp.	-	Epiphytic	Sympodial
6	*Calanthe triplicata* (Rumph.) Ames	Ahan abal Anggrek popocongan Bunga tida lapis Lemba utan	Terrestics	monopodial
7	*Cleistoma* sp.	-	Epiphytic	monopodial
8	*Coelogyne* sp.	Anggrek hitam	Epiphytic	Sympodial
9	*Cymbidium finlaysonianum* Wall. Ex Lindl.	Anggrek lidah ular Anggrek pandan Anggrek perahu	Epiphytic	Sympodial
10	*Cymbidium* sp.1	-	Epiphytic	Sympodial
11	*Cymbidium* sp.2	-	Epiphytic	Sympodial
12	*Dendrobiumphalaenopsis* Fitzg.	Anggrek larat	Epiphytic	Sympodial
13	*Dendrobium* sp.1	-	Epiphytic	Sympodial
14	*Dendrobium* sp.2	Anggrek merpati	Epiphytic	Sympodial
15	*Dendrobium* sp.3	-	Epiphytic	Sympodial
16	*Dendrobium* sp.4	-	Epiphytic	Sympodial
17	*Flickingeria* sp.	-	Epiphytic	Sympodial
18	*Gastrochillus* sp.	-	Epiphytic	monopodial
19	*Hebenaria radiata* (Thunb.) Spreng.	Anggrek bangau putih	Terrestics	monopodial
20	*Lipariscondybulbon* Rchb. f.	-	Epiphytic	Sympodial
21	*Lipariselegans* Lindl.	-	Epiphytic	Sympodial
22	*Liparis* sp.	Anggrek kutilang	Epiphytic	Sympodial
23	*Luisia* sp.	-	Epiphytic	monopodial
24	*Oberonia costeriana* J.J. Sm.	-	Epiphytic	monopodial
25	*Phaius* sp.	-	Epiphytic	Sympodial
26	*Phalaenopsis amabilis* (L.) Blume	Anggrek bulan	Epiphytic	monopodial
27	*Phalaenopsis* sp.1	-	Epiphytic	monopodial
28	*Phalaenopsis* sp.2	-	Epiphytic	monopodial
29	*Phalaenopsis* sp.3	-	Epiphytic	monopodial
30	*Phreatia* sp.	Anggrek upil	Epiphytic	Sympodial
31	*Trichoglottis geminata* (Teijsm. & Binn.) J. J. Sm.	Anggrek madu Anggrek kobra	Epiphytic	Sympodial
32	*Trichoglottis* sp.	-	Epiphytic	Sympodial
33	*Vandopsis lissochiloides* (Gaudich) Pfitzer	Anggrek vanda merah	Litophytes	monopodial
34	Spesies 1	-	Epiphytic	Sympodial
35	Spesies 2	-	Epiphytic	monopodial
36	Spesies 3	-	Epiphytic	Sympodial
37	Spesies 4	-	Epiphytic	Sympodial
38	Spesies 5	-	Epiphytic	Sympodial
39	Spesies 6	-	Epiphytic	monopodial

No.	Latin Name	Indonesian/Local Name	Habitus	Growing Type
40	Spesies 7	-	Epiphytic	monopodial

Table 1 shows the number of species of wild orchids found in Resort Mallawa as many as 40 types. There are 13 types of wild orchids identified to species level, 20 species identified only to genera level, and 7 species that have not been identified. This is because the type of orchid found has not had a generative organ making it difficult to identify. According to Hiola, et al [4], to distinguish between one type of orchid with another orchid, one of which is seen from the flower.

Based on the type of growth, there are three types of orchid habitus, namely epiphytes, terrestics, and litophytes. The habitus is dominated by epiphytes as many as 35 species, while based on the type of growth are mostly sympodial orchids as many as 24 species.

These wild orchids are found in different conditions. There are orchids were found in the shade as many as 5 species, half sheltered as many as 21 species, while in the open as many as 4 species. There are 5 orchid species that can grow in the shade and half shade, 3 species that grow in half shade and open places, and 2 species that can grow in three places (Table 2).

Table 2. Types of Wild Orchid Based on Visual Light Intensity at Resort Mallawa at Bantimurung Bulusaraung National Park Area.

No.	Orchids Name	Visual Light Intensity		
		Shade	Half Shade	Open
1	*Abdominea minimiflora* (Hook. f.) J. J. Sm.	+	-	-
2	*Aerides inflexa* Teijsm. & Binm.	+	+	-
3	*Aporum uncatum* (Lindl.) Brieger,	-	+	+
4	*Arundina* sp.	+	+	+
5	*Bulbophyllum* sp.	-	-	+
6	*Calanthe triplicata*(Rumph.) Ames	-	+	+
7	*Cleistoma* sp.	-	+	-
8	*Coelogyne* sp.	+	+	+
9	*Cymbidium finlaysonianum* Wall. Ex Lindl.	+	+	-
10	*Cymbidium*sp.1	+	-	-
11	*Cymbidium*sp.2	-	+	-
12	*Dendrobiumphalaenopsis* Fitzg.	-	+	-
13	*Dendrobium* sp.1	-	+	-
14	*Dendrobium* sp.2	-	+	-
15	*Dendrobium* sp.3	-	+	-
16	*Dendrobium* sp.4	-	+	-
17	*Flickingeria* sp.	-	+	+
18	*Gastrochillus* sp.	-	+	-
19	*Hebenaria radiata* (Thunb.) Spreng.	-	-	+
20	*Lipariscondybulbon* Rchb. f.	+	+	-
21	*Lipariselegans* Lindl.	-	+	-
22	*Liparis* sp.	-	+	-
23	*Luisia* sp.	-	+	-
24	*Oberonia costeriana* J.J. Sm.	-	-	+
25	*Phaius* sp.	-	+	-
26	*Phalaenopsis amabilis* (L.) Blume	-	+	-
27	*Phalaenopsis* sp.1	-	+	-

No.	Orchids Name	Visual Light Intensity		
		Shade	Half Shade	Open
28	*Phalaenopsis* sp.2	-	+	-
29	*Phalaenopsis* sp.3	-	+	-
30	*Phreatia* sp.	+	-	-
31	*Trichoglottis geminata* (Teijsm. & Binn.) J. J. Sm.	+	+	-
32	*Trichoglottis*sp.	+	-	-
33	*Vandopsis lissochiloides* (Gaudich) Pfitzer	-	-	+
34	Spesies 1	+	+	-
35	Spesies 2	-	+	-
36	Spesies 3	-	+	-
37	Spesies 4	-	-	+
38	Spesies 5	+	-	-
39	Spesies 6	-	+	-
40	Spesies7	-	+	-

Note: + = exist

Based on the results of research, there were 40 types of wild orchids at Resort Mallawa. This was almost equivalent to the previous findings by TN. Babul in 2011 found 39 species. Thomas and Schuiteman (2002) in Sulistriani (2008) stated that there were about 548 types of Sulawesi orchids. When compared with the total number of orchid species in Sulawesi, the species found at Resort Mallawa only reached 7.3%.

In Table 1, it shows several types of wild orchids that do not yet have an Indonesian name or area name, this is because not all types of wild orchids have names that are popular or are still very rarely cultivated by the community. Each region has different terms and languages, so that one type of orchid can have more than one regional name.

The results of exploration of wild orchids were found in 85 distribution points. The characteristics and types of orchid growth that dominate are epiphytes and sympodial types. *Trichoglottis geminata* is the most growing species among the other species, namely as many as 10 points. Based on height, *Coelogyne* sp. at an altitude of 849 m above sea level which is the highest growth position compared to the other species found in Bentenge Hamlet, while the lowest level is *Dendrobium* sp.1 with an altitude of 432 m above sea level in Mallenreng I. Hamlet Based on the distribution location, Mallenreng I Hamlet is the location of the most abundant orchid plants, followed by Mallenreng II, Bentenge and Ballanglohe. According to Hiola et al. [15], the type of forest and the presence of vegetation can be a limiting factor for the distribution of these types of orchids. Each type of orchid has a different distribution level, so that each forest area contains different varieties of orchids. orchids as a place of refuge, taking nutrients, developing and regenerating. Therefore, the structure and diversity of vegetation types of tree stands at the study site will also affect the variety of types of orchids that exist.

Generally, wild orchids are found to grow in half-shade conditions. However, there are some that can grow in 2 or 3 types of conditions. Types of *Aerides inflexa, Cymbidium finlaysonianum, Liparis condybulbon, Trichoglottis geminata* and species 1 are types of orchids that can grow in shaded and half-sheltered conditions. Types of *Aporum uncatum, Calanthe triplicata* and *Flickingeria* sp. is a type that can grow in half-shade and open space. Type of *Arundina* sp. and *Coelogyne* sp. is a type that can grow in all three conditions.

According to Suwila [16] states that one of the differences in the way of life of epiphytic and terrestrial plants is in their light needs, so species of orchids that like bright light will grow as epiphytic plants, while those who like shade will grow on the forest floor. According to Wiharto et al. [17], tree vegetation that is not too tight causes the intensity of sunlight to reach the surface of the ground. Furthermore Hiola, et al. [4] stated that physiologically the light energy has a large influence

IOP Conf. Series: Materials Science and Engineering **551** (2019) 012132 doi:10.1088/1757-899X/551/1/012132

on orchids, either directly or indirectly. Direct influence is in the process of photosynthesis, while the indirect effect is on its growth, germination and flowering.

4. Conclusions

Based on the results obtained, it can be concluded as follows.

- The number of wild orchids found at Resort Mallawa is 40 types with 13 species-level identified, 20 genera-level identified, and 7 species unidentified.
- The habitus of wild orchid is dominated by epiphytes as many as 36 species.
- The type of orchid growth is mostly sympodial as many as 24 types.
- Found 5 types of wild orchids in shade, 21 types in half shade, 4 types in open places, 5 types in shade and half shade, 3 types in half shade and open, and 2 types in all three places.

References

[1] Parnata A S 2005 *Agromedia Pustaka, Jakarta* 23.
[2] Banks D P 1999 *Singapore Periplus Ed. Ltd 64p.*
[3] Handoyo F *et al* 2010, *Orchids of Indonesia*. Indonesian Orchid Society.
[4] Hiola S F *et a*l 2015 *J. Trop. Crop Sci* **2** 28.
[5] Thomas S *et al* 2002 *A Preliminary Catalogue*. American Orchid Society.
[6] Handoyo F *et al* 2006 *Native Orchids of Indonesia*. Indonesian Orchid Society of Jakarta.
[7] Byrne P O 1999 *Malayan Orchid Rev*. **33** 43
[8] BTNBB, "Keanekaragaman Jenis Anggrek Alam di Taman Nasional Bantimurung Bulusaraung," Makassar, 2012.
[9] C. G. G. Van Steenis, *J,. 2005*. 2005.
[10] Cullen J 1992 *The orchid book: a guide to the identification of cultivated orchid species*. Cambridge University Press.
[11] Handoyo F *et. al* 2012 *Orchids of Sulawesi*. Jakarta: Perhimpunan Anggrek Indonesia (PAI).
[12] Puspitaningtyas D M *et al 1999*, *Koleksi Anggrek Kebun Raya Bogor*. UPT Balai Pengembangan Kebun Raya Lembaga Ilmu Pengetahuan Indonesia
[13] Comber J B 1990 *Kew Bentham-Moxon Trust. R. Bot. Gard. Kew 407p.-col. illus.. ISBN*, vol. 947643214.
[14] Supardi D *et al* 1999 Botanic Gardens of Indonesia, Indonesian Institute of Sciences.
[15] Hiola S F *et al* 2017 *Int. J. Sci. Res.* **6** 328.
[16] Suwila M T 2015 *Florea J. Biol. dan Pembelajarannya* **2** 1.
[17] Wiharto M *et al* 2018 *2nd International Conference on Statistics, Mathematics, Teaching, and Research.*

AUTHOR INDEX

AUTHOR INDEX

AUTHOR INDEX

AUTHOR INDEX

Institute of Physics
Dirac House, Temple Back
Bristol BS1 6BE UK

ISSN: 1757-8981
ISBN 978-1-5108-9385-6

CURRAN ASSOCIATES INC.
proceedings.com

American College of Veterinary Internal Medicine (ACVIM) Forum 2016

Denver, Colorado, USA
8 - 11 June 2016

Volume 1 of 2

ACVIM
American College of Veterinary Internal Medicine